TECTONIC ARCHAEOLOGY

About Access Archaeology

Access Archaeology offers a different publishing model for specialist academic material that might traditionally prove commercially unviable, perhaps due to its sheer extent or volume of colour content, or simply due to its relatively niche field of interest. This could apply, for example, to a PhD dissertation or a catalogue of archaeological data.

All *Access Archaeology* publications are available as a free-to-download pdf eBook and in print format. The free pdf download model supports dissemination in areas of the world where budgets are more severely limited, and also allows individual academics from all over the world the opportunity to access the material privately, rather than relying solely on their university or public library. Print copies, nevertheless, remain available to individuals and institutions who need or prefer them.

The material is refereed and/or peer reviewed. Copy-editing takes place prior to submission of the work for publication and is the responsibility of the author. Academics who are able to supply print-ready material are not charged any fee to publish (including making the material available as a free-to-download pdf). In some instances the material is type-set in-house and in these cases a small charge is passed on for layout work.

Our principal effort goes into promoting the material, both the free-to-download pdf and print edition, where *Access Archaeology* books get the same level of attention as all of our publications which are marketed through e-alerts, print catalogues, displays at academic conferences, and are supported by professional distribution worldwide.

The free pdf download allows for greater dissemination of academic work than traditional print models could ever hope to support. It is common for a free-to-download pdf to be downloaded hundreds or sometimes thousands of times when it first appears on our website. Print sales of such specialist material would take years to match this figure, if indeed they ever would.

This model may well evolve over time, but its ambition will always remain to publish archaeological material that would prove commercially unviable in traditional publishing models, without passing the expense on to the academic (author or reader).

Archaeopress Access Archaeology

Preface

This volume is a compilation of my work in the fields of Geoarchaeology and Geology developed through my research in Japanese archaeology. The result has been the formulation of an approach that is rooted in Plate Tectonics, as Japan is located in one of the world's many subduction zones (active margins) where an oceanic plate is being subducted below (drawn under) a continent. Japanese archaeologists themselves have developed several sub-disciplines that derive from subduction zone processes: volcanism, earthquakes, and tsunami. These can be bundled into a higher order unit that I term 'Tectonic Archaeology', which has a different focus on the archaeological remains than the sister sub-discipline, 'Disaster Archaeology'. By using Plate Tectonics to frame life and death in the Japanese Islands, one comes to a greater understanding of the deep Earth processes affecting the archipelago and the people who have lived through natural disasters (or not) and continue to deal with their occurrences and useful products. Many of the processes are applicable to other areas of the world, either at plate edges or intraplate locations. Therefore, whether one is specifically interested in Japan or not, this book is a first attempt at approaching archaeology through Plate Tectonics, using Japan as a case study. Neither a texbook nor a reference book, it should be taken as an exploration to see what new insights tectonics can offer archaeology. As stated by Cavazza *et al.* "A good knowledge of the interactions between deep processes (geodynamics) and surface processes (climate, erosion transport and deposition of sediments, as well as biosphere) is *a prerequisite* for risk assessment in such highly populated area [sic] as the Mediterranean shores and their hinterlands" (2004: 28, italics added). Although Japan's geology is complicated, it is nothing like the past and present compilation of the Mediterranean region, but the former can serve as a model on which to build understanding.

Readership

The chapters are bundled into an initial exploratory Chapter 1, three substantive Parts, and extended Appendices and Glossary/Indexes. The Parts are relevant to different readerships. **Geoarchaeologists** may be more interested in Chapter 1 – a survey of how tectonic processes are dealt with in the current geoarchaeological literature – followed by Part II which discusses the Japanese sub-disciplines of Tephroarchaeology, Earthquake Archaeology, and Tsunami Archaeology as well as containing chapters on hazard risk mitigation and jade formation. However, much of the terminology used in these chapters is defined and discussed in Chapter 2, accompanied by an extended Appendix and a detailed Glossary of geological terms. Part I, presents subduction zone geology through Japan as a case study and tracks the formation of the Japanese landmass through time from the Jurassic to present-day problems of volcanic soils in Japan. For both **beginning geologists and archaeologists**, Japan can serve as an introduction to subduction zones and deep Earth geodynamics and can illustrate how expertise in the tectonic archaeology subdisciplines widens understanding of the archaeological record. **Japanologists** may find Part I challenging but interesting vis-à-vis their background work in various parts of Japan. Parts I and II aim to give **Japanese archaeologists** a short course in Plate Tectonic processes and present the work of their colleagues in the tectonic sub-disciplines. **Gemologists** may be particularly interested in Chapter 12, which deals with the formation of the true jades (nephrite and jadeitite) in relation to subduction zone processes. It was questions about jade that first got me interested in geology and led me to Plate Tectonics. **Landscape archaeologists** may appreciate my struggle to understand Nara Basin geology and geoarchaeology (Part III) from the time of my dissertation research to the present, while the historical overview in Chapter 1 may give pause for reflection. Two chapters are devoted to correcting misconceptions about volcanic soils: Chapter 7 reveals that not all volcanic soils are beneficial, while Chapter 15 rectifies views on why Japan has a dearth of human and faunal remains.

Most of the chapters herein have been previously published as journal articles or chapters in edited books; they appear here with original publisher permissions. Much of the original texts have been reproduced in this volume with their original bibliographic citations. Where new information supersedes or expands these citations, additional bibliographical references have been supplied. The chapter texts have thus been rearranged, updated, and edited to appear here as an integrated presentation. Their derivations are discussed below, but please see the original publications for their individual acknowledgments.

Chapter Outline

In Chapter 1, I argue that Tectonic Archaeology, as I define it, is a foundation as well as an umbrella for Geoarchaeology. This chapter was originally published under the title "**Tectonic archaeology as a foundation for geoarchaeology**" in *Land* **2021**, 10, 453, 20pp. [https://doi.org/ 10.3390/land10050453]. It is reproduced here virtually unchanged. Rather than an introduction to the volume, this is an exploration of how 'tectonics' and associated processes have been represented in past geo-archaeological literature. It was discovered that there is a lack of discussion about what 'tectonics' actually means and how the term is applied to archaeology. This realization led to Chapter 2, which lays out the basics of subduction zone processes.

PART I A Geological Introduction to the Japanese Islands

Part I is designed to give a brief introduction to most aspects of Japanese geology. Although it can be used independently of the volume, many of the terms are defined in the volume's Glossary of geological terms – useful for non-geologists reading about Japanese geology for the first time – and the contents of the geological belts in Japan are given in Appendix 6. Many of the other appendices are designed to give more concrete information and context to the geological sequences described in the text.

The chapters in Part I draw firstly on three articles originally published in *Japan Review*, with permission from the International Research Center for Japanese Studies in Kyoto (Nichibunken): "**Origins of the Japanese Islands: the new 'Big Picture'**", *Japan Review* 15: 3-48 (2003); "**The making of the Japan Sea and the Japanese mountains: understanding Japan's volcanism in structural context**", *Japan Review* 20: 3-52 (2008); and "**Origins of Japan—the 'Big Picture' Revisited: A Review of New Plate Tectonics Research**", *Japan Review* 25: 169-184 (2013). These have been chopped up, edited, up-dated, and redistributed; the 2003 article forms the basis of Chapters 2 and 3 and Part I Reflections. The 2008 article forms the basis of Chapters 4 and 5. Much of the revised concepts in the 2013 article have been represented in these three chapters and Part I Reflections.

Essentially, the storyline of Chapter 2 documents Japanese involvement in Plate Tectonics research and applies Plate Tectonics principles to understanding tectonic activities in Japan. Chapter 3 discusses the geotectonic construction of the current Japanese Islands and traces the creation of the Japanese landmass at the edges of the China cratons from ca. 510 million years ago to the Miocene. Chapter 4 is about the physical rifting, together with rifting volcanics, of that landmass from the continental edge 15 million years ago to form the present-day Japanese archipelago, then how the main island of Honshu is being impacted by the collision of the Izu Arc. Chapter 5 looks at the evolution of the landscape to form the mountains, basins, and plains that are inhabited today.

Chapter 6 is a newly compiled chapter with material on igneous activity, partly drawn from ***TephroArchaeology of the North Pacific***, edited by GL Barnes & T SODA [Oxford: Archaeopress (2019) with permission of Editor David Davison, but much new material has been gleaned from the general literature]. This chapter describes the igneous rock composition of the Japanese Islands that complements its basement structure of sedimentary Accretionary Complexes as presented in Chapter

3. And it takes as its jumping off point the collision of the Izu Arc with Honshu from 8 million years ago to examine the different ways the two oceanic plates, Pacific and Philippine, are subducting under northeast and southwest Japan respectively, providing very different volcanic and seismic structures to these areas. The more amazing issue is how the two plates overlap under central Honshu which, together with the Izu Arc collision, give this region an unusual tectonic character. The structures in each region from Hokkaido in the north to Okinawa in the south are examined for effects of plate movement and volcanism and how these may have changed over time. The latter is particularly of interest to archaeologists for the distribution of obsidian resources.

Chapter 7 is reproduced here in entirety from its original location as "**Tephra-derived soils of Japan in comparative context**", pp. 202-233 in *TephroArchaeology in the North Pacific*, ed. by GL Barnes & T SODA [Oxford: Archaeopress (2019) with permission]. It provides a counterexample to the generalized notion that volcanic soils are fertile and always productive for farming. Japanese volcanic soils are classed as andosols which occur there in two forms: one incorporating the amorphous clay called allophane and other colloids or the other not having these. Either way, the soils are not fertile, with nutrients captured in the colloids and made unavailable to plant growth or aluminium toxicity when the colloids are absent. There is tremendous archaeological interest in how the andosols were formed, which reflects on the habits of prehistoric peoples intentionally firing stands of *Miscanthus* reeds to increase hunting productivity.

PART II The Tectonic Archaeologies of Japan

Part II consists of three chapters reviewing the work of Japanese archaeologists on evidence of the occurrence of volcanic eruptions, earthquakes, and tsunami. These are also important to modern-day occupants, so that a chapter has been included on hazard determination and response. The final chapter on the formation processes of the true jades, nephrite and jadeitite, illustrates the close relationship between archaeological materials and Plate Tectonics.

Chapters 8 to 10 are reproduced here fairly faithfully except where reformatted, updated, and expanded where necessary. Some of the statements on general geology that appeared in the originals have been moved to other relevant chapters or appendices.

Chapter 8 is based on material in ***TephroArchaeology of the North Pacific***, edited by GL Barnes & T SODA Oxford: Archaeopress (2019) and my article on "**Vulnerable Japan: the volcanic setting of life in the archipelago**", pp. 21-42 in *Environment and Society in the Japanese Islands*, ed. by Philip Brown & Bruce Batten [Corvallis, OR: Oregon State University Press (2015), with permission of Tom Booth, Director of OSU Press]. It explains why volcanic archaeology in Japan is dominated by tephra and how tephra cover changes the landscape as well as impacts on habitation. Several case studies are drawn from the 'homelands' of tephroarchaeology in Tohoku, Kanto and southern Kyushu in the above publications.

Chapter 9 originally appeared as "**Earthquake archaeology in Japan: an overview**", pp. 81-96 in *Ancient Earthquakes*, ed. by M Sintubin, IS Stewart, TM Niemi & E Altunel. Geological Society of America Special Paper 471 [Boulder, CO: GSA (2010); it appears here with permission from Jeanette Hammann, GSA Director of Publications.] It presents the development of archaeoseismology in Japan in parallel with that in the Mediterranean and demonstrates why and how earthquake evidence and damages are so different in the two areas. Recognizing sediment deformation at archaeological sites is an acquired skill for archaeologists. Moreover, this Chapter introduces the distinction between Active Fault earthquakes and subduction earthquakes – a distinction applicable around the world.

Chapter 10 reproduces my article "**The search for tsunami evidence in the geological and archaeological records, with a focus on Japan.**" *Asian Perspectives* 56.2: 132-165 (2017) [with permission of Pamela Wilson of Asian Perspectives and the University of Hawaii Press]. The sub-discipline of Tsunami Archaeology is very new and has mainly been conducted in New Zealand, but the 2011 Tohoku-oki Earthquake and subsequent tsunami has jump-started both archaeological and geological work in identyfying tsunami evidence in the archaeological record.

Chapter 11 draws on my article "**Vulnerable Japan: the volcanic setting of life in the archipelago**" noted above but expands to include hazard risk mitigation for earthquakes and tsunami as well as volcanic eruptions. It also deals with the interrelationships of these three tectonically based hazards in addition to examining typhoons and the fear of landslides in Japan. The overlap of Tectonic Archaeology with Disaster Archaeology is acknowledged and discussed because often tectonic and non-tectonic risks combine together to make hazardous living for populations past and present.

Chapter 12 takes us in a completely different direction but ties the work back directly to Plate Tectonics. It reviews the nature of false jades and nephrite as originally presented in my article "**Understanding Chinese jade in a world context**." *Journal of British Academy* 6: 1-63 (2018) [with reproduction permission by James Rivington, Head of Publications and Editor of the JBA]. But it has been expanded to include the problems with jadeite, using Itoigawa jade from Japan as an exemplar, and Chinese *feicui*, a relatively new entrant to the gemological world.

PART III Nara Basin Studies

Part III reviews the geology of the Nara Basin, where I did most of my state formation research, with up-to-date resources. Two geoarchaeological projects (coring and excavation) in the Nara Basin carried out by myself and colleagues tested my dissertation hypotheses, and the summary of that work is reproduced here. The last chapter goes beyond the Nara Basin to provide an explanation why bones are not regularly recovered in Japanese soils. This brings together various strands of tephroarchaeology, soil science, and climate data, and it corrects a critical misunderstanding about the nature of Japanese soils.

Chapter 13 is an updated version of an appendix to my dissertation, "**Nara Basin Geomorphology**," Appendix I to *Yayoi-Kofun Settlement Archaeology in the Nara Basin, Japan*. PhD dissertation, Department of Anthropology, University of Michigan. Ann Arbor: University Microfilms International (UMI, now ProQuest) (1983). This chapter has been vastly rewritten and updated to include reference to current resources available online for Japanese geology, which provide a much richer and detailed view of Nara Basin geology than accessible in the 1970s. Nevertheless, the original appendix underwrote two projects on landscape transformation in Nara that I summarized in 2005 "**Nara Basin Geoarchaeology**", with NISHIDA Shiro and OKITA Masa'aki, *Geoarchaeology: An International Journal* 20.8: 837-860, [DOI:10.1002/gea.20085, reproduced with permission]. This summary appears only slightly updated as Chapter 14.

Finally, Chapter 15 consists of a reflection on why human and faunal remains are relatively rare in Japan by rebutting a common misconception about the nature of Japanese volcanic soils in Japan. Originally presented as "**Acid Soils and Acid Rocks: Implications for Bone Preservation**," at the 2004 Society for American Archaeology (SAA) conference in Montreal, Canada, it was published as a working paper in the panel proceedings (JAA 2004).

Acknowledgments

The acknowledgments for each published article incorporated herein are not reproduced here, but I thank all my colleagues who were involved in commenting on my work and offered constructive criticism as well as permissions to use their illustrations. For the early chapters in Part I, I wish to personally thank TAIRA Asahiko and ISOZAKI Yukio for their time and effort in discussing my original articles of 2003 and 2008.

In writing this new version, I have had the good fortune of communications with John Firth, a participant in the Ocean Drilling Project (ODP) investigating offshore geology of Japan. He has been a tireless reader of Part I and I have benefitted from his many comments, suggestions, and long reading lists. Other geologists and archaeologists, both in Japan and abroad, have kindly answered my questions and provided me with illustrative material. For this final draft, I am particularly indebted to Geoff Bailey for constructive criticisms and encouragement.

Many of the illustrations in PART I were redrawn by Linda Bosveld of the Durham Archaeological Services. I have updated and/or modified these from their original publications listed above as needed. For the remaining illustrations, I have modified figures found in the literature and offer them here for wider understanding. Colleagues who have donated illustrations are ISOZAKI Yukio, OKAMURA Yukinobu, SANGAWA Akira, MATSUDA Jun-ichirō, IIZUKA Yoshiyuki, Harald Furnes, Inna Safonova, Steve Smith, Denice Cabanban, and staff of Kagawa Prefectural Product Promotion Organization, Itoigawa Jade Workshop, and Kamitsukeno-sato Museum; the illustrations were gratefully received, and I also thank those who gave permission to reproduce their figures. All illustrative material is acknowledged in the Figure & Table Sources found at the end of each chapter.

A Personal Journey

My personal journey into geology is detailed here (optional reading) in the spirit of providing context to my research career in archaeology. It demonstrates to **early career researchers** how serendipity and curiosity can lead to unexpected changes in research directions.

My interest in Geoarchaeology began in the mid-1970s, at a time when that field was not yet fully developed and indeed the term was not yet well-known. Geology, on the other hand, has been my early retirement project, culminating in a BSc in Geosciences (Geology) in 2012 at age 65. In between these times, my major focus has been on state formation in East Asia (Barnes 1988, 1993, 2001, 2007, 2015). This volume is thus a record of a personal journey away from and beyond my original thematic study of state formation in Japan, which itself began with a Freshman Year Abroad at International Christian University in Tokyo when I studied Japanese art and archaeology under J. Edward Kidder and became enamoured with Kofun-period elite material culture (*haniwa* sculptures, crowns, gilt-bronze horse-trappings, jade curved beads, and the like). Having begun my Japanese language studies at ICU, I continued at the University of Colorado, taking a BA in Japanese Language & Literature with a double major in East Asian Civilizations. This equipped me with the background to do research in Japanese archaeology at the graduate level, using the Japanese language, once I caught up with archaeological training in the University of Michigan's MA programme.

Landscape archaeology & excavation

From the beginning of my graduate research in 1972 at the University of Michigan, I had two interests: one in the theory of state formation, and the other in the landscape within which the early state developed in Nara Prefecture, Japan (cf. Apx 2: Fig. A). The latter interest was stimulated by my auditing a course in "Archaeological Geology" in 1977, given by Bill Farrand in UofM's Geology Department. This covered dating methods, Plate Tectonics, rocks & minerals, weathering and soils, site sedimentology, geomorphology, paleoclimates, and Quaternary stratigraphy, laying the groundwork for my later geological interests. I dedicate this work to him and was lucky enough to tell him so before he unexpectedly passed away in 2011.

My dissertation fieldwork in the late 1970s in Nara (Barnes 1983, 1988) entailed landscape reconstruction carried out by aerial photograph analysis under the supervision of Prof. Y. Takehisa at Nara Women's University; then, fieldwork in the mid to late 1980s, while teaching East Asian archaeology at Cambridge University (1981–1995), provided opportunities to test that reconstruction (Barnes 1992, 2005) with a geological coring project (JRG 1985, 1986) and an archaeological excavation (Barnes & Okita 1993). The appendix to my dissertation, updated and rewritten, and review of the two test projects are incorporated into Part III herein.

Jade & Plate Tectonics

While next teaching at Durham University (1996–2006), I became involved in a British Museum conference where I presented on jade (Barnes 1996), stimulating my interest in jade as a material – made into important prestige goods in the pre- and proto-historic societies of China, Korea, and Japan. I began by reading about jade; but having no background in Earth Sciences at the time, I did not understand the mineralogy. I thus resorted to reading books on Japanese geology in English, but these were even more impenetrable for reasons described in Chapter 2.1. Out of frustration, I pestered my colleagues in the Earth Sciences Department at Durham University to allow me to sit in on their classes, and when I exhausted those, I enrolled for courses in geology with the Open University. Japan did not form any part of these courses, but everything I learned was applicable to it.

The first thing that I became aware of was that the main books on Japanese geology available when I first started reading about it were still written within the former paradigm of geosyncline theory (Tomari 2005; see Chapter 2.1). As I learned the details about Plate Tectonics at the OU, I rewrote Japanese geology in the new paradigm for myself, just so that I could understand it. I published these study papers on the formation of the Japanese Islands in *Japan Review* (Barnes 2003, 2008, 2013), and they are incorporated here in PART I. These papers do not even qualify as Archaeological Geology since, as my husband complained, "but, there weren't even any human beings at that time". They are pure Geology. However, an understanding of Plate Tectonics – and the geology and Earth Sciences that it encompasses – underwrites the "wide variey of research methods and an eclectic approach to data" that is allegedly employed in Geoarchaeology and Archaeological Geology (Rapp 1982: 45); moreover, it makes the collection of data integrative rather than 'eclectic'.

In the late noughties, I became involved in the Seismological Society of America, via contacts with Iain Stewart and Manuel Sintubin, and gave a paper at the Santa Fe SSA conference in 2008 on earthquake archaeology in Japan, subsequently published in 2010 (Barnes 2010; Chapter 9 herein). My views on volcanic archaeology were published in 2015 (Barnes 2015; incorporated into Chapter 8), while tsunami archaeology followed in 2017 (Barnes 2017; Chapter 10 herein). The Japanese discipline of 'volcanic ash archaeology' has received my most recent attention. In 2016, I convened a panel together with my colleague SODA Tsutomu on tephroarchaeology (his translation of the Japanese term *kazanbai kōkogaku*, 'volcanic ash archaeology') at the World Archaeological Congress (WAC8) in Kyoto 2016. The papers from this conference together with other invited additions form our edited book, *TephroArchaeology in the North Pacific* (Barnes & Soda 2019, e-book available for free from the publisher's website). Material that I wrote for this volume has been integrated here in Chapters 6 and 8. It was only upon returning to the problem of jade in order to give a lecture, in 2017 for the Elsley Zeitlyn Lecture Series on Chinese Archaeology and Culture at the British Academy (Barnes 2018), that I realized Plate Tectonics was not a diversion from the study of jades – my initial stimulus for studying geology. True jades have *everything* to do with Plate Tectonics, and so the last Chapter of PART II (12) presents these findings — a fitting closure to a 20-year journey into geology that began with a question about jade.

Of course, I realized along the way that the most salient works on Japanese geology did not appear in English-language *textbooks* but in Japanese books and journals or in specialist English-language geology journals which were often buried behind paywalls of the big science publishers. Thus, it was imperative for me, in order to continue this line of studies, to have a continuing academic affiliation. Having retired from Durham University in 2006, with my Department of East Asian Studies being closed down by the University in 2007, I was set adrift — but rescued by the late Jon Davidson, whose untimely death in 2016 was a great loss to us all and to the field of volcanology. In 2006, Maurice Tucker and Jon Davidson, both of the Durham Earth Sciences Department, agreed to nominate me as a Fellow of the Geological Society of London (FGS) on the basis of my previous geoarchaeological work, and then in 2012, I became a more legitimate Fellow upon completion of my BSc. Jon also supported my work through his Department, and his colleagues John Gluyas, Colin Macpherson, and Andrew (Andy) Aplin, Yaoling NIU, and Mark Allen, plus the able Department Manager Jill Hoult in the Earth Sciences Department have continued that support so that I can access the Durham University Library holdings remotely. I am eternally grateful for the opportunity to have continued my research under their umbrella.

2022 marks 50 years since I arrived at the University of Michigan to study archaeology and met my future husband, David Hughes, who taught me Japanese historical linguistics. Throughout this past half-century, David has been my constant and generous supporter in my career, often acting as my production editor and always as my proofreader for my many publications. I owe him my life as it has been, which I now dedicate to him for the rest of our years together.

Gina L Barnes BA, BSc, MA, PhD, FGS
Emeritus Professor, Durham University
Project Associate in Earth Sciences, Durham University
Professorial Research Associate, SOAS University of London
Fellow of the Geological Society of London

Further Reading

For those interested in pursuing the original research and updates for this book, I include comments here on the most critical resources. Two printed sources were especially useful in the composition of Part I. The first consists of thematic sections of the journal *The Island Arc* entitled "Geology and Orogeny of the Japanese Islands" (vol. 5.3, September 1996) and "Orogeny of the Japanese Islands" (vol. 6.1, March 1997); these represented the first holistic representation of Plate Tectonics and palaeogeography following the final abandonment of geosyncline theory by Japanese geologists. The *Island Arc* publications, in English, are notable for the colour plate reconstruction sequences of the continental clusterings (Maruyama 1997*b*), which could not be reproduced here but were especially enlightening. The second source was the multi-volume series on *Nihon no Chikei* (Japanese Landforms) published in Japanese by the University of Tokyo Press from 2000; the first volume *Sosetsu* (General Introduction) (Yonekura 2001) is especially useful for reviewing neotectonic processes. The descriptions of individual regions in different volumes of this series are informative interpretations of the lay of the land.

Within the last fifteen years since my initial two publications, in addition to innumerable scientific papers in published print and online journals, two sources have been particularly useful in revision. The most recent is a volume on *The Geology of Japan*, edited by Moreno *et al.* (2016), from the Geological Society of London; it is a specialist publication that informs the more accessible presentation here. The initial chapter in that volume (Taira *et al.* 2016) is an update and expansion of "Tectonic evolution of the Japanese island arc system" (Taira 2001) which I used in my original publication of 2003. The other source was a series of articles, in Japanese with English titles and abstracts, on 'New Paradigms' in Japanese Plate Tectonics published in the journal *Chigaku Zasshi* [Journal of Geography]. A review article in Engish accompanied them (Kasahara *et al.* 2010). Drawing on these *Chigaku Zasshi* publications, I wrote an update for my 2003 article in 2013, both published in *Japan Review* as listed above. Many of those comments therein are integrated here.

The New Paradigms series consists of thirty-five articles published in three parts under the title "Nihon Rettō Keiseishi to Jisedai Paradaimu" [Geotectonic Evolution of the Japanese Islands under New Paradigms of the Next Generation] in three issues of *Chigaku Zasshi*: 119.2 (2010), 119.6 (2010), and 120:1 (2011). All articles but one are in Japanese with English titles and abstracts, and most illustrations have captions in both Japanese and English for further perusal. Free PDFs of these articles are available via the *Chigaku Zasshi* website, or directly from J-Stage (Japan Science and Technology Agency [www.jstage.jst.go.jp]). The English abstract for each article gives a good indication of content, while every issue has its own preface in Japanese, in addition to the one overview in English (Kasahara *et al.* 2010).

References (omitting those detailed above)

Barnes, Gina (**1982**) "Prehistoric landscape reconstruction and spatial analysis of artifact discovery locations." *Kokogaku to Shizen Kagaku* 15: 113-131 (in Japanese with English abstract).

---- (**1983**) *Yayoi-Kofun settlement archaeology in the Nara Basin, Japan.* Ann Arbor, MI.: University Microfilms (ProQuest). Published as Barnes 1988.

---- (**1988**) *Protohistoric Yamato: archaeology of the first Japanese state.* Anthropological Papers 78 and Modern Papers in Japanese Studies 17. Ann Arbor: Museum of Anthropology and the Center for Japanese Studies.

---- (**1993**) *China, Korea and Japan: the rise of civilization in East Asia.* London & New York: Thames & Hudson.

---- (**1996**) "China: questions in jade." *The Times Higher Education Supplement* 6(Dec): ii-iii.

---- (**2001**) *State formation in Korea: historical and anthropological perspectives.* Richmond: Curzon.

---- (**2007**) *State formation in Japan: emergence of a 4th-century ruling elite.* London: Routledge.

---- (**2015**) *Archaeology of East Asia: the rise of civilization in China, Korea and Japan.* Oxford: Oxbow Books.

BARNES, Gina L & OKITA, Masaaki (eds) (**1993**) *The Miwa Project: survey, coring and excavation at the Miwa site, Nara, Japan.* BAR International Series 582. Oxford: Tempvs Reparatvm.

BARNES, Gina L & Tsutomu SODA (eds) (**2019**) *TephroArchaeology in the North Pacific.* Oxford: Archaeopress.

CAVAZZA, William; François ROURE & Peter A ZIEGLER (**2004**) "The Mediterranean area and the surrounding regions: active processes, remnants of former Tethyan Oceans and related thrust belts", pp. 1-30 in *The TRANSMED Atlas: the Mediterranean region from crust to mantle*, ed. by W CAVAZZA. Berlin: Springer.

ISOZAKI, Yukio; Shigenori MARUYAMA & Kazumasa AOKI *et al.* (**2010**) "Geotectonic subdivision of the Japanese Islands revisited: categorization and definition of elements and boundaries of Pacific-type (Miyashiro-type) orogen." *Chigaku Zasshi* 119.6: 999-1053 (in Japanese with English title and abstract).

JAA (Japanese Archaeological Association) (**2004**) *Recent Palaeolithic studies in Japan: proceedings for tainted evidence and restoration of confidence in the Pleistocene archaeology of the Japanese archipelago.* Tokyo: Japanese Archaeological Association.

JRG [Joint Research Group on the Geomorphological Recognition and Land Utilization of Pre- and Protohistoric Japanese Peoples] (co-author) (**1986**) "Natural environments in the Nara Basin through the pre- and protohistoric ages I: geology and geomorphology." *Kobunkazai Kyoiku Kenkyu Hokoku* 16: 1-30 (in Japanese).

---- (1987) "Natural environments in the Nara Basin through the pre- and protohistoric ages II: descriptions of core samples and analysis on biogenic materials." *Kobunkazai Kyoiku Kenkyu Hokoku* 16(March): 23-74 (in Japanese).

KASAHARA, Junzo; Osamu SANO & Nobuo GESHI *et al.* (**2010**) "Overview of a Special Issue on 'Geotectonic Evolution of the Japanese Islands under New Paradigms of the Next Generation (Part I-IIII)'." *Chigaku Zasshi* 119.6: 947-958 (in English).

MARUYAMA, Shigenori; Yukio ISOZAKI & Gaku KIMURA *et al.* (**1997**) "Paleogeographic maps of the Japanese Islands: plate tectonic synthesis from 750 Ma to the present." *The Island Arc* 6.1: 121-142 (in English).

MORENO, Teresa; Simon WALLIS & Tomoko KOJIMA *et al.* (eds) (**2016**) *The geology of Japan.* London: Geological Society of London.

RAPP, George Jr & John A GIFFORD (**1982**) "Archaeological geology." *American Scientist* 70.Jan-Feb: 45-53.

TAIRA, Asahiko (2001) "Tectonic evolution of the Japanese island arc system." *Annual Review of Earth and Planetary Sciences* 29: 109-134.

TAIRA, Asahito; Y OHARA & SR WALLIS *et al.* (**2016**) "Geological evolution of Japan: an overview", pp. 1-24 in *The geology of Japan*, ed. by T MORENO *et al.* London: Geological Society of London.

TOMARI, Jiro (**2005**) "The concept of geosynclines and plate tectonics in Japan." *Kagakushi Kenkyū* 44.233: 23-32.

YONEKURA, Nobuyuki; Sohei KAIZUKA & Michio NOGAMI *et al.* (eds) (**2001**) *Regional geomorphology of the Japanese Islands, Vol. 1: Introduction to Japanese geomorphology [Nihon no Chikei].* University of Tokyo Press.

Style Notes

• Abbreviations and acronyms can be found in Appendix 1.
• Special effort has been made to explain geological terms in the text; they appear in sans serif font on first use in each Chapter and are defined and indexed in the Glossary.
• Japanese administrative units (prefectures and districts) are presented in Apx 1: Fig. A, while Japanese archaeological periods are given in Apx 1: Tables A, B.
• Geological periodization is given in Apx 3.
• East Asian names are usually given surname first; the surname is in SMALL CAPS when accompanied by the personal name.
• British spelling is adopted but using 'z' as in *The Times* newspaper. Punctuation is generally American-style except for abbreviations (British abbreviations do not take a full stop [period] if the abbreviation includes the last letter of the word: e.g., ed. [editor] but eds [editors]).
• Dates are given coming forward for all BP, BC, and AD dates: e.g. 10,000–6000 BP = from 10,000 to 6,000 years ago; dates in the thousands BC are given without commas, as in dates AD. Millions of years *ago* are abbreviated as 'Ma' (*mega annum*); thousands of years *ago* are 'ka' (*kilo annum*); and billions of years *ago* Ga (*giga annum*); mya may also be used for million years *ago*.
• Date ranges (*duration*) are given as 'years', e.g., ky = a thousand years, my = a million years
• Japanese words and placenames are given in modified Hepburn (e.g. Tanba, rather than Tamba; Sanbagawa, rather than Sambagawa), though personal choices of name spellings are maintained if different from Hepburn romanization (e.g., Wadati, rather than Wadachi).
• Macrons for long vowels in Japanese are generally not shown except for italicized terms in the text or in bibliographical references.
• Certain terms are given capital letters for emphasis, as with Plate Tectonics, in that they reflect important concepts that have specific definitions in Japanese geology: e.g., Accretionary Complexes, Active Faults, Active Volcanoes.

Referencing System

• Acronyms are given in Apendix 1.
• Items in the glossary of Geological Terms appear in the text in sans serif font.
• The Glossary of East Asian Words is keyed to italicized terms in the text; kanji and meanings are given together with Chapter occurrence.
• Cross-references to Chapters and Figures in this volume are capitalized (Chapter, Figure, Table); those in other works are lower case (ch., fig., table); 'cf.' is used to mean 'see'.
• Figure and Table titles are by chapter, e.g., Figure 1.1 (first Figure in Chapter 1); their sources are given at the end of each chapter text before the bibliography.
• Figures and Tables in the Appendices are given letter referents, e.g., Apx 2: Figure A., specific to that appendix.
• Bibliographies for each chapter are included after that Chapter or Appendix; there is no overall book bibliography.
• Items in the bibliography that have no date (n.d.) are, if possible, given the date of the latest cited reference therein, e.g. ≥2012 (published in or later than 2012).
• DOI (digital object identifier) numbers: these can be inserted into the 'find' field at [doi.org] to find the document online.
• Beware of URLs (uniform resource locators) and DOI references split over lines; they may contain a space that needs removing when copying and pasting.

Table of Contents

BOXES (including illustrations)

FIGURES

APPENDIX FIGURES

TABLES

APPENDIX TABLES

CHAPTER 1

Tectonic Archaeology vs Geoarchaeology

In a nutshell,
Geoarchaeology, Archaeological Geology, and Landscape Archaeology
deal with the Earth's *surface*,
whereas Tectonic Archaeology begins with *deep* Earth processes
which contribute to the creation of that surface.

This volume is an experiment in interdisciplinary study. It is aimed at geoarchaeologists, either students becoming one or professional practioners. It offers a perspective from a region that has little exposure in the general archaeological literature but which offers general geological lessons to all parts of the world. By examiming the case study of Japan, a subduction zone in geological terms, many geological processes are brought together – faulting and folding, magma production, earthquake generation, and resource potentialities – that also are useful in other regions, especially in 'fossil subduction zones' that run through otherwise stable continental areas. By understanding the deep Earth processes and where they occur(ed) past and present, expectations about the archaeological record can be fine-tuned, and heretofore unrecognized problems can be anticipated.

In addition to this initial chapter, which argues why knowledge of deep Earth processes can underwrite geoarchaeological understanding, the text is divided into three parts. Part I tells the story of the formation of the Japanese landmass – as it formed on the continental edge and after it was separated from the continent a mere 15 million years ago. These chapters may be of wider interest to geology students for a broad overview of events from the Jurassic onwards that occurred in East Asia, as wholistic coverage of Japan is also not well represented in the geological literature. Chapters in Part II deal with processes associated with tectonic plate subduction that affect the archaeological record. Part III is a retrospective on the Nara Basin, assessing previous work on a specific geographical area in terms of the processes identified in Part I.

A Comparison, For Starters

Plate Tectonics and Geoarchaeology both emerged as new fields of study in the early 1970s after many decades of incubation. By the mid-1980s, references were often made to tectonics[1] in geo-archaeological works, particularly in the assessment of coastal and shoreline change as affecting human habitation. However - as indicated in the opening to this Chapter - the focus of Geoarchaeology developed to deal explicitly with the Earth's *surface*, entailing areal geomorphology, local sediments, site formation processes, the living environment, and accessible raw materials. In contrast, Tectonic Archaeology, as I propose, deals first with the *deep* Earth - particularly mantle convection in creating tectonic plates, in governing their movements, and in generating land parcels –

[1] Terms in the Glossary are given in sans serif font on first mention in each chapter.

An earlier version of this Chapter was published as GL Barnes (2021) "Tectonic archaeology as a foundation for geoarchaeology" in *Land* 10: #453, 20pp.

and then with the nature of archaeological investigation that is constrained by tectonic processes affecting the local area. The creation of this term was stimulated by the development in Japan of three subdisciplines that deal with the effects of Plate Tectonics on archaeology: Earthquake Archaeology, TephroArchaeology, and Tsunami Archaeology. A holistic treatment of Tectonic Archaeology in Japan is offered in this volume.

How different are Geoarchaeology and Tectonic Archaeology? It is proposed here that the fields of Geoarchaeology and its mirror discipline Archaeological Geology are *centripetal*: they begin with the archaeology and bring any disciplines and techniques into them that are helpful for analyzing the anthropogenic record; we might also call it *inductive* for this reason because it begins with the archaeology, specifically the archaeological sediments (as noted also by King & Bailey 2006: 274), and works outwards. Tectonic Archaeology, on the other hand, is *centrifugal*: it begins with the geology and sets up expectations for what we might encounter in the material record emanating from it according to the location of the archaeological remains; in this sense it is *deductive*. Tourloukis (2010), for example, offers a model based in tectonically active regimes to predict the locations of Palaeolithic sites. King & Bailey (2006: 274) rightly note that tectonic phenomena characteristically operate over broad areas that reach far beyond an archaeological site and its catchment in what is called "dynamic topography" (Hager *et al.* 1985) or "dynamic landscapes" (King & Bailey 2010; Bailey & King 2011). Tectonics also generate what are known as "shallow geohazards" (Banks 2021), which affect human habitation and routines. These concepts are substantially different from the more traditional acknowledgment of geomorphologically dynamic landscapes and subsurface archaeology (Stafford 1995).

It is important to note, however, that there are many surface landforms and hazards that are 'nontectonic' in formation; these occur within 1000 m of the surface and are due to the forces of gravity out-powering tectonic stress (Nagata 2018). Examples include glacial and peri-glacial movements and deposits, and various types of landslides, slumping, and land creep, not to mention erosional processes and their related depositions. Tectonics can explain a great many surface phenomena but must be recognized as not all encompassing.

The naming of a field is critically important, as it ostensibly defines the boundaries as well as the contents of work done under its name. One of the earliest uses of the term 'geo-archaeology' occurred in Beale (1973: 136), where archaeological stone and mineral sourcing was referred to as "a kind of geo-archaeology". As a named sub-discipline of archaeology, Geoarchaeology was formalized with the publication of an edited volume in 1976 entitled *Geoarchaeology: earth science and the past* by Davidson & Shackley. The term 'Archaeological Geology' was consolidated in 1978 with the establishment of the eponymous division with the Geological Society of America, followed by Rap & Gifford's compilations in 1982 and 1985. While Davidson & Shackley's volume focussed on sediments in archaeology, Rap & Gifford (1982) named several areas coming under the remit of Archaeological Geology: landscape and coastline reconstruction (palaeogeography), provenance studies, volcanism and tephrochronology, etc.

Regarding the two sub-disciplines based on the interface between geology and archaeology, Karl Butzer (1982) described Geoarchaeology as "archaeology pursued with the help of geological methodology" versus Archaeological Geology as "geology pursued with an archaeological bias or application" (as reported in Rapp & Gifford 1985: 15). It is often difficult to distinguish between these, but Rapp & Gifford (1982: 52) proposed that they "do not characterize two ends of a spectrum of techniques but rather two contrasting and equally legitimate research goals" – while assignment to one or the other sub-discipline often depends "on the investigator and the project", not the content!

Goldberg & Macphail (2006: 2) go even further and dismiss this debate on definitions as "irrelevant", proposing that "any issue or subject that straddles the interface between archaeology and the earth sciences" falls under the remit of geoarchaeology. In a sense, I agree with their conflation of Geoarchaeology and Archaeological Geology, as I will employ the term Geoarchaeology to include both Archaeological Geology *and* even Landscape Archaeology to act as the main discipline dealing with the Earth's surface.

Meanwhile, Tectonic Archaeology has a broader remit to include the Earth Sciences. It is a term designed to begin archaeological analysis from the deep end in order to describe and explain the resources, forms, and compositions that landmasses acquire, and the geological processes they are subject to over time. Depending on where archaeological investigation is conducted, Plate Tectonics may be more or less relevant but *never irrelevant* – because all continental and insular landmasses ride on tectonic plates. McKenzie now distinguishes between oceanic tectonics and continental tectonics (2018); he states that "The important difference between oceanic and continental tectonics is that deformation of continental lithosphere is distributed, rather than occurring on narrow plate boundaries" (2018: 6). Nevertheless, tectonic activity operates on both types of plates, whether along the margins or in the interior (intraplate). As examples of continental tectonics, even perceived 'stable' continental areas such as the Rhenish Shield in Germany can be subject to incipient rifting, generating much volcanic activity (e.g., Aber 2017; Riede 2017). In North America, there are other tectonic zones that have generated seismic and volcanic activity: the Midcontinent Rift, a horseshoe-shaped failed rift that runs from Oklahoma up around Lake Superior down to Alabama (Stein *et al.* 2018), accounting for volcanic rocks around Lake Superior; the currently active Rio Grande Rift running from Leadville, Colorado to El Paso, Texas (Murray *et al.* 2019); and the New Madrid Seismic Zone, affecting seven states in the central south (Eldridge & Wolf 2019). Thus, continents, just because they are large, are not necessarily stable or uniform. Both North America and China, for example, are composite blocks of many ancient cratons (the first of the Earth's crust to have formed) – as shown in the USGS-based map by Ciaurlec (2019) – and their accreted island arcs. Knowing the rocks of an area (volcanic, metamorphic, or sedimentary) will give a clue to what kind of processes have acted on that area but *not* the extent of the structures they belong to: that is tectonics.

The most active tectonic regions of the world are modern subduction zones. These usually consist of oceanic plates being drawn under continental plates, setting off a series of associated seismic and volcanic processes. Current subduction zones occur primarily around the Pacific Ring of Fire[2] and in the Mediterranean. But fossil subduction zones are inherent in collision zones, such as caused that causing the Himalayas, where the oceanic crust of the Tethys Ocean was subducted under Eursasia until the Indian continent collided with it. Such collision zones, also referred to as suture zones or mobile belts, thus represent former active margins i.e., fossil subduction zones. Such belts or zones may be sandwiched within present-day continental regimes, such as the Qinling Orogenic Belt in China or the Central Asian Orogenic Belt (CAOB). These zones/belts are often loaded with ores and other raw materials that have been important to previous inhabitants and often form important economic zones today. Pinpointing where both past and present subduction zones occur on the map, and understanding what they consist of, will assist the geoarchaeologist in assessing the potential resources and expected landform activity within their area of investigation – or nearby, whence

[2] This is now a long outdated term (Chester 1993: 40), but the 'Pacific Ring of Fire' is still used in popular accounts; the zone is preferably termed the 'orogenic volcanic series'.

materials could be mined or traded, or where natural hazards from subduction processes affected human occupation.

Perhaps the most outstanding works in Tectonic Archaeology without being named as such is research by Bailey *et al.* (1993), King & Bailey (2006), Force (2015), McFadgen (2011), and Force & McFadgen (2012). Working in the former Tethys subduction zone of the Mediterranean, these researchers begin their presentations with an ode to Plate Tectonics and continue to evaluate how tectonic activity has affected the landscape and, by extension, human habitation in their regions of interest. As King & Bailey have stated (2006: 271):

> Few attempts have been made to incorporate tectonics into palaeogeographical reconstructions of early hominin sites and their associated landscapes. Tectonic processes, if they are not ignored completely at this local scale, are usually treated as background events, as occasional disruption of sedimentary processes, or as sources of volcanic raw material for stone tools.

If this has been the case for studies in hominin evolution, how much truer has it been for general Quaternary archaeology? 'Background' mention only works if the reader fully understands that background, and few archaeology courses include wholistic introductions to the Earth Sciences.

Why Japan?

It is not surprising that the idea for Tectonic Archaeology, as an umbrella subdiscipline to incorporate archaeological phenomena affected by Plate Tectonics, came from a Japanese context. In the current volume, Japan is our major focus,[3] but we will also examine eastern Eurasia, where past and present life is still being influenced in a variety of ways by tectonic collision in the Tethys suture zone and the Nipponides subduction zone (inclusive of Japan).

The Japanese Islands occupy a modern subduction zone and its accompanying supra-subduction zone (the rifted Japan Sea back-arc basin); thus, they comprise an excellent case study of the importance of Plate Tectonics for Geoarchaeology. The archipelago is far less complicated geographically than the Mediterranean subduction zone, but its internal geology is quite complex: the geohistory of the Japanese landmass, presented in Chapter 3, is much more surprising than the way the islands present themselves today.

Japan is generally known as a volcanic island arc, but this obscures more than 500 million years of its formation which contribute to the Japanese landmass being 80% sedimentary in origin (Taira *et al.* 1997: 468; Isozaki *et al.* 2010). These sedimentary accumulations are called Accretionary Complexes (AC), that date to when the Japanese landmass was part of the continental edge prior to 16 Ma.[4] AC form primarily from trench fill that is 'bulldozed' into the continental shelf during subduction of an oceanic plate. AC also incorporate sporadic slices of oceanic floor (ophiolites), seamounts, and limestone reef fragments which were not subducted but obducted into the continental mass, and metamorphosed mantle rocks known as serpentinite melangés. Several AC have also been subjected to metamorphism at deep levels then exhumed to the surface, adding to the complicated structure. Few AC are forming in the world today because subduction generally causes tectonic erosion of the continental edge instead of accretion. This is why the southwestern Japanese coast is of great

[3] See Appendix 2 for district and prefectural names.

[4] Scientific notations are Ga = 'billion years ago', Ma = 'million years ago', ka = 'thousand years ago'.

scientific interest because accretion is still ongoing, forming the offshore Nankai prism. Chapter 3 recounts the palaeogeographic compilation of the Japanese Islands, beginning from the break-up of the second-to-last supercontinent, Rodinia, through the assembly and break-up of Pangea, the last supercontinent, to today's continental arrangement. A closer look at the formation of the Japan Sea Basin in Chapter 4 and Japan's mountain and basin systems in Chapter 5 are presented before investigating igneous activity as it has acted on the Japanese landmass before and after formation of the archipelago.

Japan has inherited a granite basement (batholith) formed on the continental edge before the rifting of the archipelago. These are the magma chambers of very ancient volcanic eruptions whose surface manifestations have all eroded away, and the unroofed batholiths now form the granite backbone mountain ranges of southwestern Inner Zone Japan. Present-day volcanic landforms date from several eruption phases beginning ca. 20 Ma during rifting. The footprint of Active Volcanoes is relatively small among Japan's mountainous character which is primarily due to compression folding. Nevertheless, every square inch of Japan's surface has been subjected to tephra fallout at one time or another, or indeed multiple times. Tephra, however, is not the only volcanic product that might affect inhabitants: lava and gas emissions are also of concern. These have prompted researchers to propose a Volcanic Archaeology (Riede *et al.* 2020) instead of TephroArchaeology as constituted by Japanese researchers (Arai 1993; Soda 2019; Barnes & Soda 2019), and the name Archaeological Volcanology is also under consideration. Among subduction processes, volcanics have greatly affected life in the archipelago, thus Chapter 6 describes the important aspects of volcanic distribution in Japan.

Together with volcanoes, earthquakes, and tsunami are also subduction-zone hazards that must be endured. Landslides are also important consequences of earthquakes, but like tsunami, not all landslides can be assigned to earthquake activity, and no specific archaeological approach to landslides has yet been developed. Thus, the three sub-disciplines of archaeology have developed indigenously in Japan to monitor tectonic hazards in the archaeological record are *jishin kōkogaku* ('earthquake archaeology'), *tsunami kōkogaku* ('tsunami archaeology'), and *kazanbai kōkogaku* ('volcanic ash archaeology'). These can be bundled into a higher-order perspective, to wit, Tectonic Archaeology; how they are practiced in Japan will be presented in Part II.

The 'Geo' in Geoarchaeology

In distinguishing Geoarchaeology from Tectonic Archaeology, it is good to review how the Earth is treated in Geoarchaeology. 'Geo' has several meanings as a short form of 'geology' which itself is derived from Greek: the *gê* (Greek *gê*, Doric *gâ*) meaning 'earth, land, country', with the connecting vowel *-o-* and *-logia* meaning the 'study of' (Merriam-Webster n.d.; Smith n.d.). All the translations of *gê* given above are appropriate to our discussion here, but let us select 'earth' as the most inclusive as it entails an increasingly large scope of research. The use of 'earth' in geoarchaeological writing can be considered on a scale of five levels from the particular to the general.

Level 1 is 'earth' as dirt (Renfrew 1976: 4-5) – the sediments of sites and their stratigraphic layering as recovered in archaeological excavation (e.g., Holliday 2004). Sediments formed the topic of the first conference in Geoarchaeology in 1973 (Davidson & Shackley 1976). The focus on sediments by Schiffer (1983, 1987) aimed at understanding site formation as a product of human behaviour (although he has been criticized for a law-like approach by Goldberg 1989); investigating stratigraphy was thus a tool for an 'anthropological archaeology' approach, aiming to reconstruct people's place in their environment (Level 4, see below). Goldberg & Macphail (2006) set out their approach to soils and sediments, but the attention by other geoarchaeologists given to anthropogenic layers – to the exclusion of the natural stratigraphy of archaeological sites – stimulated Karkanas & Goldberg (2019) to propose a reorientation to the 'sedimentary matrix': treating artefacts, features, and anthropogenic sediments not as the primary focus but as *components equal to natural sediments* in an archaeological deposit. Since the late 1970s, the analysis of soils has been revolutionized by the development of soil micromorphology (e.g., Courty *et al.* 1989; Macphail & Goldberg 2017).

Level 2 is 'earth' as landforms — this is a geomorphological approach which outlines the nature and form of rocks and sediments as they occur in different environments: wet, dry, glacial, desert, fluvial, colluvial, etc. The volumes by Stein & Farrand (1985, 2001) address these different forms of sedimentary sources and processes, but they mention tephra and tephrochronology only in passing.

Level 3 is 'earth' as resources — raw materials that can be turned into artefacts. Identifying sources of raw materials is a geological exercise, while matching artefacts to their sources is generally the province of archaeometry involving geochemistry and mineralogy. A good example of this focus was the session on Geoarchaeology at the Geosciences '98 Conference at Keele University; the contents of the conference was determined by the panel volume's editor (Pollard 1999*b*), who was then affiliated with the Department of Archaeological Science at the University of Bradford.

Level 4 is 'earth' as terrane — the geology of a region which produces materials as both sediments and raw materials. Terrane is different from 'terrain', the latter essentially topographic. The geological meaning of terrane is, according to ITA (2019: unpg.):

> A rock formation or assemblage of rock formations that share a common geologic history. A geologic terrane is distinguished from neighboring terranes by its different history, either in its formation or in its subsequent deformation and/or metamorphism.... An exotic terrane is one that has been transported into its present setting from some distance.

Although Japanese geologists consider the geological belts making up Japanese landmass are terranes bounded by faults, recent research elsewhere suggests that terranes need not be fault-bounded and that the boundary between terranes evolves over time (Colpron *et al.* 2007). The origin and formation of terranes, both autochthonous (here, including cratons) and allochthonous, is one aspect of what Tectonic Archaeology seeks to provide for understanding the locale of archaeological research.

Level 5 is 'Earth' as a sphere — covered by mutually moving and self-reorganizing tectonic plates which entail billions of years of Earth's history. The changing tectonic context of any particular plate or fragment thereof is what provides the geological variety created over time. That variety is the product of specific geological processes, particularly at the edges of the plates and their fragments in active subduction zones, as plates subduct one under the other, collide to form mountain ranges, or accrete intra-oceanic terranes. The various processes that occur within subduction zones (Chapter 2) include some of the natural hazards that affect society as mentioned above: volcanic eruptions, earthquakes, and tsunami – obvious targets of a Tectonic Archaeology. Once activity ceases in a former subduction zone, the geological products of those processes are frozen into the body of the Earth's crust in suture zones or fossil subduction zones – or they are eroded to provide trench fill for future AC or metamorphic belts. These zones may currently occupy inland positions, so that Tectonic Archaeology is not limited to currently active subduction zones.

Level 6 is 'earth' as environment — the living and non-living stage for habitation on the continents and oceans carried by tectonic plates. This is the realm both of Environmental Archaeology (à la Butzer 1982) and the Earth Sciences themselves, as the oceans, atmosphere, even other heavenly bodies are taken into account. Geologists no longer find themselves between a rock and a hard place but in the company of vapours, liquids, thunder and lightning...and a lot of biomatter.

The above scalar effects within the concept of 'earth' are seldom addressed in Geoarchaeology (but see Stein 1993); discussions of methods and techniques are typically applied within one of the above Levels. Moreover, Geoarchaeology did not evolve in this ordering of the Levels. Environmental Archaeology (Level 6) was an early concern as proposed by Butzer (1982), preceding Waters emphasis on Level 1; Waters (1992: 6) stated that the job of the geoarchaeologist is to reconstruct "the geological factors of the human ecosystem from the sediments and soils" (the latter a distinction not always made clearly, cf. Holliday 2004: introduction). Level 2 was the focus of Stein & Farrand (1985, 2001) in their categorizing landscapes and sediments according to their geomorphological formation, in order to assess the cultural geography in different time periods. Level 3 forms the arena of Archaeological Science, particularly in identifying the raw materials and their sources as made into artefacts. Levels 4 and 5 are the domain of Tectonic Archaeology, which has heretofore been excluded from wholistic characterization *for archaeology* and which is our current concern.

Tectonics in Geoarchaeology

The history, development, and definition of the field of Geoarchaeology have been extensively covered by many authors (e.g., Hassan 1979; Rapp & Gifford 1982, 1985; Rapp 1987*a,b*; Thorson & Holliday 1990; Stein 1993; Mandel 2000; Rapp & Hill 2009; Hill 2017; Pollard 1999*a*); these need no repetition here. The field was introduced to Japan by Matsuda (2007), who thoughtfully reviewed several of these works in relation to Japanese archaeology. He concluded that, in particular, the way that the term 'stratum' (*sō*) was used in designating a 'cultural stratum' (*bunkasō*) - what I have translated in my work as an ABS (artefact-bearing stratum) - was at odds with how it was used in several science disciplines: lithostratigraphy, pedostratigraphy, biostratigraphy, chronostratigraphy, and ethnostratigraphy. By not focussing on the microfacies of sites (à la Corty 2001), the Palaeolithic hoax of 2000 was facilitated (see Chapter 6).

Having neither long experience in the field nor resources to investigate the voluminous geoarchaeological literature, my review in the following sections takes a different tack and examines an arbitrary selection of the major geoarchaeological books and the main scientific archaeology journals. Though these works are familiar to all geoarchaeologists, I review them here *especially* to assess previous appearances (or not) and uses (or not) of tectonics in its general form as well as in the specific subduction-zone processes of volcanism, earthquakes, and tsunami. *Geoarchaeology* (Rapp & Hill 2009), the major textbook of the field, gets an in-depth look below.

We can take a definition of tectonics from Keller & Pinter as referring to "the processes, structures, and land-forms associated with deformation of the Earth's crust" (2002: 1). The authors calculate that subduction zones account for 15% of the Earth's surface, but they point out first, that deformation is distributed over a much broader zone up to several hundred kilometers wide (as in the Andes), and second, that unstable areas can occur within continents, as we noted above. Intraplate tectonic activity is often due to the incipient development of oceans: rifting, as in the Red Sea or the developing Great Rift Valley in East Africa; or in the failed North American Midcontinent Rift or the active Rio Grande Rift as noted above. Another intraplate volcanic activity is the mobile hot spot locations producing the Columbia River Basalt Province and Yellowstone, and the Siberian and Deccan Traps; or new activity around ancient cratons such as the distributed volcanic fields along the southern edge of the Colorado Plateau, or the complicated geological sub-surface fault system of the New Madrid Seismic Zone affecting seven states in the south-central United States. These examples were chosen to counter the 'faulty' idea that large continental interiors have nothing to do with Plate Tectonics. Additionally, of course, any place that has mountains is an orogenic (Greek *òros* 'mountains' + *gènesis*) zone related to an ancient or operating subduction zone.

The term 'tectonics' is used in this volume as a cover term for Plate Tectonics and all its associated processes. Geologists may object this expanded meaning and to the application of the adjective 'tectonic' to archaeology. However, this word simply refers to 'structure' or 'construction', deriving from Greek (*tectōn* 'carpenter, builder'). In the archaeological context, it can be taken to indicate that societies are often structured or constrained by tectonic processes, and archaeological sites will contain structural evidence of tectonic activities – often to the detriment of their contemporaneous occupants. Tectonic Archaeology is thus surely a more relevant term than, for example, the "tectonics of transcultural transactions" (Inaga 2014: 123). Even the use of 'tectonic plates' has moved beyond its original context when historians and journalists speak of shifts in the tectonic plates of political relations and state systems, etc. Back when the term was first being developed (McKenzie & Parker 1967; Morgan 1968), geologists objected to the use of the term 'Plate Tectonics' outside of the strict meaning of the kinetics of moving plates on a sphere. An archaeological application, therefore, seems acceptable *sensu lato*. As mentioned above, the value of an overall Plate Tectonics approach to assessing archaeological sites and remnant human behaviour can be seen in earlier work of researchers (e.g., Bailey *et al.* 1993), who nevertheless did not use the term 'Tectonic Archaeology' themselves, and in current work (e.g., Force 2015 or Dickinson & Burley 2007).

In their textbook *Geoarchaeology*, Rapp & Hill state that (2009: 188): "One of the major integrating concepts in the earth sciences is that of tectonics. Many of the geologic and biologic features associated with the archaeological record can be more fully understood within the context of plate-tectonic theory." Despite this, their textbook, discussed below, had very little explanation of Plate Tectonics itself. In promoting the study of Plate Tectonics for archaeologists, we should heed the instructive comments, on how difficult it is to grasp another field, by George 'Rip' Rapp, a prominent archaeological geologist; he stated, "I was fairly narrowly trained in mineralogy and geochemistry. It has been an uphill struggle for me to learn the necessary other earth sciences and the relevant archaeology" (Jing 2007: 11). We archaeologists need to do the opposite: we need a grounding in Plate Tectonics, and to do so, it is impossible to avoid having to "master the geoscientific jargon and literature" (Stein & Farrand 2001: xii). We have to work hard to understand sister disciplines, but the effort is well worth it. In that vein, the following Chapters do not hesitate to use geological 'jargon', for this is the language of the field that we need to understand. A full Glossary of these terms is therefore appended.

Geology and archaeology are sometimes each treated as wholistic entities; this is far from reality in that both disciplines have developed many sub-disciplines and specialisms over the decades, with much specialist jargon. Moreover, the field of geology itself has been encompassed within 'Earth Sciences' or 'Geosciences' (see an interesting distinction in NSF n.d.). Earth Sciences include geophysics, geochemistry, oceanography, seismology, petrology, sedimentology, marine geology, palaeontology, and many more. Dan McKenzie, author of plate kinetics, laments that though he studied geology (stratigraphy, sedimentology, and palaeontology), what he really needed was "fluid dynamics, earthquake seismology, petrology and geochemistry"; and more than just self-study from textbooks, he says he needed to learn to think like a scientist in those fields (2018: 15-16), a skill also promoted by Buck (2004). If it was difficult for *geologists* and *physicists*, like Rip Rapp and Dan McKenzie, to come to grips with all of Earth Sciences, then how easy is it for *archaeologists* to gain a foothold knowledge? Kearey *et al.* (2009: ix) propose that Plate Tectonics is *the key* to understanding much of the Earth Sciences:

> The initial impact of the plate tectonic concept, in the fields of marine geology and geophysics and seismology, was quickly followed by the realization of its relevance to igneous and metamorphic petrology, paleontology, sedimentary and economic geology, and all branches of geoscience. More recently its potential relevance to the Earth system as a whole has been recognized. In the past, processes associated with plate tectonics may have produced changes in seawater and atmospheric chemistry, in sea level and ocean currents, and in the Earth's climate.... This extension of the relevance of plate tectonics to the atmosphere and oceans, to the evolution of life, and possibly even the origin of life on Earth is particularly gratifying in that it emphasizes the way in which the biosphere, atmosphere, hydrosphere, and solid Earth are interrelated in a single, dynamic Earth system.

Geoarchaeology has heretofore grown out of archaeological investigation where problems are addressed or solved by techniques brought in from Earth Science disciplines, ad hoc *as needed*. Instead, we need to turn this on its head. Following McKenzie's advice, we geoarchaeologists really need to learn to think more like Earth scientists in taking whole Earth processes into account in our research planning, since several aspects of Plate Tectonics play a role in geoarchaeological assessments. Butzer's (1975, 1978) evaluations of how ecological issues were (or not) integrated into archaeology can serve as a yardstick for assessing progress in integrating Plate Tectonics with Geoarchaeology.

A survey, as offered in the next section, of how tectonics has been thus far integrated into geoarchaeological writings illustrates the extent to which it is considered an adjunct rather than a foundation, as offered in the next section. Since the concept of Plate Tectonics was not formally accepted until 1965, it is unrealistic to expect references to it in the geoarchaeological literature before the 1970s. This will be the threshold for examining the occurrence of the four aspects of Tectonic Archaeology (late Tectonics, earthquakes, tsunami, and volcanism) in the literature; certainly each has been addressed for its own sake, but to what extent? The following sections will look at the way these aspects are addressed in select edited books, textbooks, and the major journals. These publications and more have myriads of case studies in their references, which have not been cited here.

Tectonics in Major Scientific Archaeology Journals

The archives for *Archaeometry* and *Geoarchaeology* are maintained by the Wiley Online Library (see references below), while those of the *Journal of Archaeological Science* (JAS) are maintained by Elsiever SienceDirect (reference below). The numerical searches were augmented by a counting exercise for JAS, indicating some trends in the popularity of topics (Table 1.1).

1958, "Archaeometry"

Published by Wiley on behalf of the Research Laboratory for Archaeology and the History of Art at the University of Oxford, and currently edited by Mark Pollard, Edward Hall Professor of Archaeological Science, Oxford, the journal *Archaeometry* covers international research in 'applied science': applying physical and biological science techniques to archaeology, anthropology, and art history. Topics covered include dating methods, artefact studies, mathematical modelling and analysis, remote sensing technology, conservation science, environmental reconstructions, demographics, and archaeological theory. Searching all of *Archaeometry*'s volumes in Wiley Online returned 87 for tectonic(s), including 29 for Plate Tectonics – but no mention of tectonics occurred in the article titles; 39 for earthquake(s); 21 each for tephra and seismic (mainly relating to discovery surveys) but 0 for archaeoseismology; and 4 for tsunami. The surprise was that despite emphasis on dating methods, tephrochronology only garnered one mention; and volcano-related topics were by far the most numerous at 355 mentions.

Pollard states that "There is a continuing need to further develop the theoretical underpinning of materials study in archaeology, including methods for the better integration of scientific provenance studies with existing archaeological understanding" (Pollard n.d.: unpg.). Additionally, he is a member of the RESET consortium, which is charged with research on (RESET n.d.):

> human palaeontology, archaeology, oceanography, volcanology and past climate change in order to investigate how our ancestors coped with rapid changes in climate during the last 80,000 years.... RESET aims to construct a new improved chronological framework for Europe using volcanic ash layers (tephra horizons) which represent time-parallel signatures allowing archaeological and geological records to be linked.

In 2013, a new *Open Journal of Archaeometry* began publication; it is a peer-reviewed, publication-fee-charging, Open Access, international scientific journal published by PAGEPress in Pavia, Italy. It currently has no Editorial Board, and published articles have declined to one to two per issue. Do not confuse this with the time-honoured *Archaeometry* from Oxford.

1972, "Journal of Human Evolution"

This journal began print publication in 1972 by Academic Press in London and is now available online through Elsevier. Topics are varied but the journal focus is on "palaeoanthropological work, primarily human and primate fossils", according to the Elsevier website. 'Plate tectonics' was introduced early on in an article by Pomerol (1975), while 'tectonic(s)' revealed no mentions until 1986, then 85 articles between 1986 and 2010, increasing to 123 between 2011 and 2021; significantly, Bailey *et al.* (2011) was offered as the prime example. 'Earthquake' garnered one or two mentions every few years (17 between 1976 and 2017); 'volcan'... produced 45 results between 1978 and 2021; and 'tsunami' produced 9 results between 2008 and 2020. 'Tephra', either as a dating method or sedimentary locus of fossils, numbered 122 articles starting from 1987; 'seismic' occurred in 18 articles between 1975 and 2015, while 'seismology' appeared only in Bailey *et al.* (2011).

1974, "Journal of Archaeological Science"

A year after the 'first' geoarchaeological symposium in Southampton in 1973 (Davidson & Shackley 1976), a new *Journal of Archaeological Science* was initiated by a primarily British editorial board but published by Academic Press in New York. The leading article was by Butzer, a member of the board, on the analysis of Acheulian calc-pan sites and in which he used the term 'geo-archaeology' (Butzer

1974). Thus, it might be understood that Geoarchaeology is a sub-division of Archaeological Science. However, in the twelve years from 1974 to 1986, the first article *title* in the journal incorporating the term 'geoarchaeology' did not appear until March 1983, and the term *within* titles occurred only three times before 1986.

These articles were accompanied by many other Earth science topics: dating methods, stone-tool petrography and sourcing, sediment analysis and soil micromorphology, rock-mineral-metal analyses and technologies, archaeomagnetism, erosion processes, ceramic petrography, trace element and isotope analysis, and the beginnings of landscape archaeology and palaeogeography. Nevertheless, as Edwards (Edwards 1983) wrote in his 'anatomy' of the first nine years of the *Journal of Archaeological Science*, the majority of articles (52.3%) published there were biological; moreover, he discerned that most of the articles during those years were written from an inductive standpoint, i.e., without overarching problem-orientation, not even overtly of environmental reconstruction.

Edwards made a division between 'materials' and soils/sediments, that is, he distinguished between artefacts analyzed by various Earth science techniques *versus* geological treatment of the major product of archaeological excavation – Renfrew's 'dirt'. Materials analyses had been published in the journal *Archaeometry* beginning from 1958. In contrast, the fact that bioarchaeological articles in the *Journal of Archaeological Science* were in the majority meant that the latter journal was clearly catering to a new outlet in the *biological* sciences for publishing on the results of flotation and fine sieving and faunal analysis by bioarchaeologists. How has this orientation fared over the last 35 years or so?

Articles and book reviews in *JAS* are indexed online from 1986 (vol. 13); the journal is now published by Elsevier, with Marcos Martinón-Torres (a materials specialist) and Robin Torrence (an archaeological volcanologist) as current editors.

Table 1.1 Results of select journal online searches

Search term:	***Archaeometry*** 1972–2020	***JAS*** 1974–2020	**Trends** in *Journal of Archaeological Science (JAS)*	***Geoarchaeology*** 1986–2020
Tephra	21	130	increasing from 2009	97
Tephrochronology	1	22	increasing from 2011	22
Tectonic(s)	87	239	doubling from 2007	333
Volcano	355	203	common in 2009–2014	353
Volcanic		664	prolific between 2008–2015	
Earthquake	39	125	highest in 2011	137
Seismic	21	80	most in 2010–2013	312
Archaeoseismology	0	3	mentioned in 2006, 2009	12
Tsunami	4	28	most between 2012–2013	59

Stats on keyword searches in selected journals are given in Table 1.1. 'Tephra' returned 130 entries, dramatically increasing in number after 2008; 22 entries for 'tephrochronology', increasing in number from 2011; 239 entries for 'tectonics, doubling from 2007; 203 for 'volcano', most common between 2009 and 2014, and 664 for 'volcanic,' most prolific between 2008 and 2015; 125 for 'earthquakes', highest numbers in 2011; 80 for 'seismic', most common between 2010 and 2013; whereas 'archaeoseismology' only gets three mentions, in 2006 and 2009; 'tsunami' garnered 28 results, highest in 2012–2013. These stats show an interesting trend of growing interest in certain topics between 2006 and 2015 then tailing off, while volcanic issues are clearly the most prominent overall.

Elsevier has developed a new sister publication, *Journal of Archaeological Science: Reports*, which reportedly focusses on the "application of scientific techniques and methodologies to all areas of archaeology"; however, in their advice to authors, they state: "If your article describes a *new* technique or has *global significance* please consider submitting to the *Journal of Archaeological Science*" (quoted from Elsevier's website cited below, emphasis added). It can only be deduced that the Reports series is for fieldwork reports not earth-shaking research results.

1986, "Geoarchaeology, an International Journal"

The journal *Geoarchaeology* was initiated in 1986 under the inspiration of Rhodes Fairbridge (Hill 2017), then Professor of Geology at Columbia University. Fairbridge had a year earlier co-founded the *Journal of Coastal Research* (Finkl 2007); his interests in coastal geomorphology dovetailed with those of early geoarchaeologists researching sea-level changes and coastal habitation.

A similar Wiley Online keyword search for *Geoarchaeology* as carried out above for *Archaeometry* returned the following as listed in Table 1.1: 'tephra' produced 97 entries, though 'tephrochronology' garnered only 22; 'volcano'/'volcanic' beat this with 353 entries; fewer entries contained 'earthquake(s)' at 137 entries, but 'tsunami' had a respectable 59. There were only 12 entries for 'archaeoseismology' itself, but 'seismic' garnered 312.

Journal summary

Examination of Table 1.1 clearly shows that volcano-related issues have been the most common topic in these journals - most likely because volcanoes (and their magma chambers below) produce resources such as granite, obsidian, and tuff that were and are still highly sought for the production of artefacts and building material. Other tectonic processes are more likely to wreak cultural destruction; but with the rise of Disaster Archaeology (Gould 2007; Grattan & Torrence 1007; Sheets 1980; Kuwahata 2019*a,b*; Maruyama 2019; Okamura *et al.* 2013; Shimoyama 1997, 2002*a,b*; Sugiyama 2019; Torrence & Grattan 2002), these fields are set to increase as future topics (see Chapter 11). Archaeological interest is possibly stimulated by the occurrence of modern disasters such as the Tohoku-oki Earthquake and Tsunami in Japan in 2011, corresponding to the increase of 'earthquake', 'seismic', and 'tsunami' entries between 2011 and 2013 in JAS.

Books Inciting Great Expectations

Four books are reviewed here for their dealings with tectonic processes. The rise in journal articles over the decades, as demonstrated above, clearly indicates an interest and use of tectonic aspects by researchers on the ground. However, how are these conveyed to readers of such books below that aspire to be training or reference volumes?

1985, "Archaeological Geology"

Rapp & Gifford introduced the new field of Archaeological Geology first in an article (1982) and then in an edited volume (1985).[5] They obviously treated their book as the coming of age of this new subdiscipline, echoed in Farrand's review of the book entitled "Birth of a discipline" (Wright 2011: 6).

[5] My comments, amended here, on Rapp & Gifford (1985) were originally published as Barnes (1986).

In contrast to earlier works on Geoarchaeology or Environmental/Ecological Archaeology – which had mainly been conceived by and aimed at archaeologists seeking to understand the broader, natural science framework of their own discipline – the stimulus for *Archaeological Geology* came primarily from professional geologists (80% of the contributors) who saw archaeology-related problems as a new frontier for their specialist applications.

Rapp & Gifford adopted Butzer's distinction between Geoarchaeology ("archaeology pursued with the help of geological methodology") and Archaeological Geology ("geology pursued with an archaeological bias or application") (p. 15). Though choosing the latter for their title, Rapp & Gifford's book contained articles belonging to both of these categories without specifying which. Perhaps the clearest examples of *doing geology* with an archaeological bias are given in the chapters on geomorphology, coastal change, sedimentary environments, and the analysis of anthrosols. These geological processes were chosen for description *because* they affected human activities. Conversely, the chapters on *doing archaeology* using geophysical survey techniques, archaeomagnetism, tephrochronology, radiocarbon dating, and petrological analysis seem to fit precisely Butzer's definition of Geoarchaeology as applied science. Two chapters, however, on paleoenvironments and palynological applications to archaeology argued for an even broader title than geology itself, following the earlier line of enquiry into environmental reconstruction.

The volume, *Archaeological Geology*, began with a review of the historical connections between geology and archaeology, forming the introduction to the above topics. More detail was devoted to 19c collaboration, especially in the context of the emergence of archaeology as a discipline, than to recent developments. In fact, the editors postulated that there was a considerable decrease in archaeological collaboration with geologists after 1910, a state of affairs that was only rectified after 1970. Curiously, the editors omitted any serious discussion of Quaternary Studies as these relate to the investigation of all aspects of Pleistocene humans, a general problem still remarked upon 25 years later (Bailey *et al.* 2011).[6] From this omission and the exclusion of physical anthropology references in the volume's bibliography, it might be inferred that this new subdiscipline as conceived of by Rapp & Gifford was meant to deal primarily with the environmental contexts of Holocene archaeology. This is in contrast to most early work of geologists on Palaeolithic sites. In Rapp's later publication with Hill, Rapp seems to see the extension of geology into the Holocene as a positive sign, as "the initially narrow scope of geoarchaeology (the antiquity of humans in the Old and New Worlds) will continue to broaden" (Rapp & Hill 2009: 273).

2001, "Earth Sciences and Archaeology"

This is an edited volume (Goldberg *et al.* 2001) that seeks to demonstrate *how* Earth Sciences are used in archaeology, giving practical information on why certain techniques were used and what kind of problems they would solve (p. viii). It is not claimed to be comprehensive, with the editors noting what *would have been* good additions to the volume. Each chapter focusses on a specific subject or process, with Archaeoseismology being the only one that specifically deals with a tectonic process. Nevertheless, various aspects of tectonics are mentioned throughout the book. Apologies to the various chapter authors for only citing page numbers below.

Tectonics

"Tectonism" is mentioned in relation to Quaternary climate cycles but generally dismissed as not short-term enough to account for them (p. 16). Among the processes mentioned are the movement of

[6] Geological time period divisions are given in Appendix 3.

continental plates to northern latitudes; the opening of the Atlantic Ocean and the Bering Strait, connecting to the Arctic Ocean, but the closure of the Isthmus of Panama, affecting ocean circulation; and the uplift of Tibetan Plateau and other mountain ranges (p. 16, 22). Volcanic eruptions and the dispersal of aerosols can affect climate (p. 16, 22), but also "rapid climatic changes can force volcanism" (stated on p. 17 without supporting evidence).

Cyprus is discussed as "a piece of Cretaceous oceanic crust that has been wedged upward...slowing of uplift might be due to a transition from a subduction to a strike-slip tectonic regime" (pp. 128-129); unfortunately, this piece of crust is not identified as the famous Troodos Ophiolite, where you can "touch the MOHO!" (Mohorovičić discontinuity).

Tectonic activity ("uplift, subsidence, and lateral crustal movements") are noted as influencing sedimentation and erosion patterns and fluvial system adjustments, again over long timespans (p. 67). River entrenchment and terrace formation may result from tectonic movements, while subsidence may result in no riverine downcutting (p. 68). Tectonism is cited as one factor in determining regional variations in stream form (p. 83). Furthermore, it is advanced that "Tectonically active regions provide unique niches for human occupation and food resources...Thus is it not surprising that we find occupation sites, especially for nonagrarian societies, along fault scarps and other tectonic landforms" (p. 144). Nevertheless, tectonic depressions provide favourable conditions for soil formation (p. 144).

Earthquakes

Few authors mention earthquakes, though both landslide deposits and soft sediment deformation are used as evidence of seismic activity (p. 67). In Peru, earthquakes are said to have traditionally been favoured agents of landscape change without consideration of other factors (p. 116). The chapter on Archaeoseismology (pp. 143-172) is written in the vein of Archaeological Geology: using archaeology to illuminate incidences of palaeoseismology (p. 144) – answering such questions as When did the earthquake occur? What did the earthquake do? When is the next earthquake? Similar data on palaeoseismology in Japan will be reviewed in Chapter 9, herein.

2006, "Practical and Theoretical Geoarchaeology"

This is a textbook written primarily for undergraduates by Paul Goldberg (now retired from Boston University) and Richard Macphail (Institute of Archaeology, UCL) (Goldberg & Macphail 2006). The book is divided into three parts, the first of which is an introduction to geology with links to archaeological relevance and examples. The second section is entitled somewhat mysteriously as 'nontraditional' geological approaches; included here are discussions of cultural deposits, experimental archaeology, forensic Geoarchaeology, etc. The third section is a how-to introduction to field methods, lab techniques, and reporting and publishing.

Tectonics

Tectonics are referred to six times in the text: as potentially causing mass movement of earth (landslides?) (p. 84); as affecting groundwater and sea levels and erosion through tectonic movement and erosion (pp. 99, 104, 114, 155); and as controlling folding and faulting (p. 309).

Tsunami

Tsunami are mentioned twice in the Coasts chapter, as depositing sediments and boulders on beaches.

Tephra

In the Sediments chapter, tephra is defined as 'volcaniclastic debris' – including volcanic ash, lapilli, blocks, bombs, and pyroclastic flow debris – with the caveat that these are "relatively uncommon in geoarchaeological contexts, as they are restricted to volcanic areas" (p. 12), thus ignoring the widespread distribution of volcanic ash over hundreds even thousands of miles and the occurrence of distal cryptotephra in sediments.

• Tephra deposition is not considered in the Aeolian Settings chapter except as a component of loess (p. 141), despite the fact that a significant amount of fine-grained volcanic ash is distributed through the air during volcanic eruptions. One map (fig. 6.18), however, illustrates distribution distances of aeolian deposits, distinguishing maximum dry distribution from even further maximum wet distribution of 'dust'. It is now widely acknowledged that tephra can be distributed thousands of miles, and cryptotephra can now be identified in sediment samples.[7] The distinction (dry vs wet), however, has not been emphasized in tephra studies per se.

• Tephrochronology is alluded to but not named as such in relation to stratigraphic dating in Iceland (p. 60), and the term does not appear in the index.

• A particular soil type (Andic horizon) is specified as based on "the organic matter accumulation and weathering of volcanic ash (e.g., Tephra) to produce allophanes" (p. 49); this is slightly incorrect as we will see in Chapter 7 herein. Moreover, the authors cite Tan (1984) in stating that "soils formed in volcanic ashes (andosols) are easily worked and naturally fertile, as their high organic matter content encourages high biological activity" (p. 63). The Japanese experience with andosols does not comply with this characterization, as andosols are highly unproductive in Japan precisely because of the presence of allophanes, mentioned above as diagnostic of Andic horizons (see Chapter 7 herein).

• The preservative effects via rapid burial by tephra (p. 224) are mentioned in terms of Pompeii first of all but also of Edo-period (1603–1969) field systems in Japan (p. 64) – not mentioning the copius data on Kofun-period field systems excavated in the Kanto region (see Chapter 8 herein).[8] Tephra layers can be extremely thick, but ground-penetrating radar can help locate subsurface features depending on their depth and composition of orverlying strata (cf. Annan 2020).

Earthquakes

This term does not appear in the index, nor does 'archaeoseismology'. Nevertheless:

• Among sedimentary structures, convolution and load structures are mentioned but not defined (p. 24). Soft sediment deformation is pictured for lacustrine deposits and credited to earthquakes (fig. 1.10).

• Flooding is an inherent hazard with volcanic eruptions, as lava, pyroclastic flows, and lahars can clog and disrupt watercourses; a good place to have mentioned this (but was not) would have been in the listing of "major flow events, such as spring flooding or during exceptional events such as hurricanes and other storms" (p. 89).

• Landslides are mentioned as products of gravitational deposition (p. 11), but they are not specifically discussed under "Slopes and Slope Deposits" (ch. 4); however, mass movement of earth materials are

[7] For an interesting utilization of cryptotephra presence, see Lowe *et al.* (2012).

[8] For Japanese districts and archaeological periodization, see Appendix 2.

acknowledged to occur in "well-known tectonically unstable areas" (p. 84). But these areas seem to be confined to land, while the mass movements of continental slope materials to form turbidites in the deep sea are not alluded to; these deposits are important as they make up the bulk of trench fill in a subduction zone that then can be accreted to the continental edge as in the the present-day Nankai prism or the Cretaceous Shimanto Belt of Japan (see Chapters 2 & 3 herein).

2009, "Geoarchaeology", the textbook

Of all the books dealt with here, *Geoarchaeology: the earth-science approach to archaeological interpretation* (Rapp & Hill 2009) excites our greatest expectations, since it is designed as a textbook for teaching Geoarchaeology. It is a masterpiece in presentation of a vast and complicated set of geological conditions that archaeologists should be aware of. Since its original publication in 1998, it has served as the bible for geoarchaeological training.

Let us return to the authors' opinion that "One of the major integrating concepts in the earth sciences is that of tectonics. Many of the geologic and biologic features associated with the archaeological record can be more fully understood within the context of plate-tectonic theory" (p. 188). How, then, are the processes and products of Plate Tectonics dealt with by Rap & Hill? Their book is a treasure trove of geoarchaeological knowledge, but how much is devoted to tectonics?

The excerpts below, from the 2009 edition, indicate a piecemeal approach to the subjects, only as relevant to specific archaeological data or questions. That approach assumes considerable background knowledge to make sense of what is offered. Without that background, questions about relationships abound: Why is andesite named after the Andes? What governs the occurrence of metamorphic rocks? What are the relationships between mafic and ultramafic rocks, or between granite and diorite and why are they found where they are? How can oceanic rocks such as serpentinite and limestone be found on land? Why are there earthquakes in New Madrid, Missouri? Is a sand blow accompanying an earthquake merely a "patch of coarse sandy sediment", and is it different from a "soil denuded of vegetation by wind action" (p. 289)? These questions and more are addressed in the ensuing Chapters of the present volume but not in Rap & Hill.

Tectonics

The textbook only gives a brief explanation of the theory of Plate Tectonics (pp. 188-189) under the section on Paleoenvironment Reconstructions (pp. 188-189). Among the processes mentioned are faulting, rifting, plate collisions, mountain building, earthquakes, and volcanism (the latter two are further treated under "Geologic Catastrophes", pp. 257-262). Tectonic movements such as subsidence and uplift are said to sometimes be recovered from coring data (p. 260), while tectonic plate boundaries are subject to earthquakes (p. 261).

Volcanism

Rapp & Hill (2009)[9] identify volcanic eruptions as major destructive forces, causing tsunami and volcanic ash fallout leading to climate change (pp. 59, 163-165, 167) and the destruction of civilization (p. 190); but eruption products also can provide habitation sites such as in lava tubes or caves cut into tuff (pp. 85-86). They quote D. Griffiths who emphasizes the "attractions as well as the hazards of life

[9] 'Volcanic' in the Rapp & Hill index contains 36 page numbers; the paragraphs here can be viewed as breaking this mass down into topics.

in an actively volcanic zone" (p. 244) such as using volcanic flagstones in road building (p. 248), but they note that volcanic ash is a poor retainer of DNA (p. 125).

Tephra layers, deposited on a minute-to-day scale (p. 131) across the landscape (pp. 27, 77, 150), are often termed "chronostrata" (fig. 2.2); they function as key layers or marker tephra for correlating archaeological deposits across space (fig. 2.6, pp. 27, 263), and they provide for "good preservation of sites" (p. 264). One major use of tephra in archaeology is tephrochronology (pp. 136, 142-143). Tephra in its various forms (lithified into tuff or weathered into clay) is as important as igneous rock for radiometric dating by K/Ar, Ar/Ar, and Pb/U isotope decay (figs. 3.8, 3.14, p. 146). Volcanic vents can release 'dead' carbon that can skew ^{14}C dates in organic matter (p. 151).

Igneous rocks are described individually (pp. 208-210) but without noting their relationships or origins; such rocks and feldspar minerals in volcanic rocks can be dated with thermoluminescence (p. 157), while hydration of broken edges of obsidian, a volcanic glass, can be measured to determine when the surface was fresh (pp. 160-161). A full sub-section is devoted to obsidian sourcing (pp. 225-226, 229), while other igneous products (table 7.1) are defined and discussed (pp. 209-210; 226-227); many are listed individually in their index.

In their section on Soil Types, tephrogenic soils (andosols/andisols) are not represented despite their abundance in New Zealand and Japan (see Chapter 7 herein), but volcanic ash is noted as inhibiting the development of inceptisols (p. 41). Interestingly, aeolian deposits of volcanic ash (or even loess) are not included in their discussion of sediment transportation by particle size (p. 50), nor are cryptotephra mentioned in their short section on soil Micromorphology, despite their listing of "vitric (presence of glass shards or pumice)" as a modifier to describe paleosols (p. 45). Volcanic ash can be a prominent component of sedimentary layers (p. 83) and tephra, even as ground-up tuff, can be used to make cement (p. 214).

Tephra and igneous rocks decompose to form clays (pp. 212-213: kaolinite, montmorillonite, smectite; bentonite is included here as a 'rock' (pp. 145, 215). Halloysite is not mentioned, while amorphous silica is mentioned but not identified as allophane (p. 86); Volcanic ash and "sands" can be used as ceramic temper (pp. 229-230), and electron-microprobe analysis has revealed ceramic pastes from distinct volcanic sources (p. 238), while basalt in ceramic bodies can be characterized by SEM-EDS (p. 238). The geologic environments are given for native copper formation in mafic and ultramafic rocks (p. 231). Volcanic materials are stated as prohibiting magnetic surveying because they already have "strong magnetic properties" (p. 114), but granite itself can be assessed with magnetic susceptibility (p. 238). Volcanic and geothermal regions can be detected with thermal infrared imaging (p. 120).

The case studies that Rapp & Hill mention in relation to volcanics range from Tonga, Fiji, and Hawaii, across Alaska, American northwest, northern Rocky Mountains amd Great Plains, Lower Yellowstone, California, Belize, Guatemala, Mexico, Peru, Bolivia, Chile, to Iceland, Great Britain, France, Italy, Greece, Egypt, Crete, Turkey, Syria, Tanzania, Kenya, East Africa, and Indonesia (only Krakatoa) – *but not* the Philippines, Japan, nor New Zealand despite a map of volcanic eruptions (fig. 9.6) that indicates much activity in the western Pacific. Their listing of volcanic arcs (p. 225) also omit Japan and New Zealand, unless they mistakenly include them in "Southeast Asia" (Japan is in East Asia, New Zealand is in Australasia; to those that work in these areas, these distinctions are important).

Seismicity

Perhaps because of Rapp's early involvement in this subject, his textbook with Hill devotes a long subsection to seismicity, albeit under geologic "catastrophes" (pp. 257-262). Nevertheless, they propose that "Except over very small regions, devastating earthquakes historically have not caused cultural change" (p. 190). Eric Force (2015) would definitely challenge that. But they do make the valid distinction that earthquake intensities (modified Mercalli earthquake intensity scale, cf. Table 2.1 herein) are more important to geoarchaeology than earthquake magnitude (M) measurements. Some examples of archaeological destruction are given to compare with the modern intensity effects. However, they are careful to advise that earthquake damage in the archaeological record has to be proved, not assumed, as there may be multiple causes such as downslope collapse due to heavy rainfall.

Seismic [hazard] maps are available to the geoarchaeologist (pp. 103, 268); however, geological maps are stated not to have sufficient information on unconsolidated ground to assess seismic risk (p. 260). Japan is mentioned along with the US in having advanced methods to evaluate seismic risk (p. 268). Seismic profiles (pp. 117-118) are described as useful in reconstructing offshore coastal paleogeomorphology, harbours, and shorelines relevant to prior habitation; it can also locate shipwrecks. "Geologic structures" in surface sediments resulting from earthquakes are sand dikes, sand pipes, sand blows, slumps, faults, and joints, fissures, landslides, and subsidence; seismic-reflection analyses can also identify slumped sub-surface sediments due to previous earthquakes (p. 260).

Liquefaction features and convoluted bedding (soft sediment deformation) appearing in sediments are termed 'secondary structures' (fig. 2.13); liquefaction, explained on p. 54, is acknowledged as a product of earthquake activity (pp. 260-261), while non-earthquake activity resulting in "contorted sediments can also be associated with spring deposits" (p. 54). Faults (presumably small) are listed above as a *product* of earthquakes, created by stress (p. 54); movement on faults is merely treated as another factor *in addition to earthquakes* that are "geologic problems for site preservation" [45] (p. 265) but apparently not as affecting previous settlement. Nowhere is it explained that movement on faults is the *cause* of earthquakes – it is the fracturing and movement of rock through stress that sends off seismic vibrations. Without providing clear distinctions, it seems that two aspects of earthquakes are being contrasted here: seismic shaking that may cause sediment deformation, and fault activities (strike-slip, normal, reverse) resulting in landmass displacement. The most interesting example of fault (or fissure) activity relates to the escape of hydrocarbon gases through cross-faulted limestone at Mt Parnassas leading to the intoxication of the oracle of Delphi (p. 273). As above, using the archaeological record to illuminate seismic processes is described as Archaeological Geology (p. 261).

Tsunami

"Tsunami debris accumulation" is mentioned as an example of the minute-to-day scale of geological activity (pp. 131, 262), and tsunami are offered as examples of "violent cataclysms [that] normally do not prompt dramatic changes in human societies" (p. 190). This statement is arguable in light of the Boxing Day Tsunami of 2004 and the Japanese 2011 Tohoku-oki Tsunami – or maybe these would be viewed as 'abnormal' situations. The authors link tsunami to risky coastal areas (p. 257), such as accompanied the eruption of Krakatoa (p. 263), but they fail to mention the tsunami damage associated with the 1755 Lisbon Tsunami (p. 261) which was "accountable for *most* of the about 70,000 deaths in Portugal, Spain and Morocco" that accompanied the earthquake (Anon. 2018: n.p., emphasis added). Tsunami can also occur on lakes (p. 261), but the cause of such inland tsunami is not dealt with (see Chapter 10 herein). The authors list tsunami as one of the "*Geologic* problems in site preservation" (p. 265, their emphasis).

2017, "Encyclopedia of Geoarchaeology"

A hefty, authoritative, if not exhaustive contribution to the field is the *Encyclopedia of Geoarchaeology* published by Springer (Gilbert 2017). Major aspects of tectonics are covered in individual entries in this encyclopedia: Volcanoes & people, Archaeoseismology, and Tsunami. The field is moving forward! All that is missing is an explanation of Plate Tectonics to underwrite them, which will be rectified in the coming revised edition.

Reflections on books

The above books are extremely rich and valuable compilations in the field of Geoarchaeology, but it is clear that tectonic processes and products receive relatively short-shrift in the overall presentations despite the fact that they have been more inclusive through time. Aspects of Tectonic Archaeology tend to be introduced piecemeal within certain geoarchaeological contexts, but they are not tied together into a coherent whole. Trying to cover all of geoarchaeological concerns in a single volume is a monumental task, and the efforts of these authors are valuable and appreciated. But the results unequivocally illustrate that researchers in (currently or previously) tectonically active regions need specialist knowledge to anticipate and deal with their archaeological cases in regional context. There must be a wealth of geoarchaeological data out there that has not yet been recognized, collected, and analyzed precisely because tectonic processes and their products are relatively unknown. This has been demonstrated in Japan by a single geomorphologist's efforts, beginning in 1985, to teach archaeologists how to recognize earthquake damage in archaeological sites by SANGAWA Akira (1986). Within eighteen months of the 1995 Kobe Earthquake, a network of 'disaster concerned archaeologists' was developed, and 378 sites nationwide were identified as having previous earthquake damage (DCAN 1996) (see Chapters 9, 11 herein).

Conclusions & Prospects

This introductory Chapter has laid out potential differences between the proposed Tectonic Archaeology and the standard discipline of Geoarchaeology. The former is characterized as centrifugal and deductive, allowing the building of expectations about both the land and its habitation. The latter is described as centripital and inductive, bringing in Earth Science technology and analysis beginning from a focus on archaeological sites.

It is proposed that by beginning with an understanding of deep Earth processes within the construct of Plate Tectonics, we are better able to understand our fieldwork areas and equipped to ask different kinds of questions about their human habitation. Japan, as my area of research, sits in a subduction zone, which conditions many aspects of the landforms, resources, and hazards found there. As an instance of subduction zone geology, the lay of the Japanese landmass will be examined in Part I, which can stand as an example for other subduction zones, both fossil and currently active.

The literature review above has revealed a general awareness of the products of Plate Tectonics but little explanation of the processes. I argue that a little knowledge of such processes goes a long way. If the tectonic development of the fieldwork area in question is known from the beginning of a project, rather than appealed to in the analysis stage, Tectonic Archaeology *undergirds but does not undermine* geoarchaeological investigation. It is a valuable first step in understanding the regional geology, resources, and transformative processes that might have been (and maybe still are) responsible for landscape composition and change, and resource availability. Regardless of where geoarchaeologists work, they are dealing with a landscape that is grounded in Plate Tectonics. Instead of being "surface scientists", geoarchaeologists have a role to play in "subsurface science", which according to Bullough (2020: 27) "is only going to get more important in the context of a growing population, supporting sustainable development and preventing dangerous climage change."

It is reasonable, therefore, to say that if Tectonic Archaeology derives from Plate Tectonics, then all the processes and products of Plate Tectonics that occur on, contextualize, or have effects on archaeological sites are obviously subjects of Tectonic Archaeology. This includes landform changes and characteristics: as when pyroclastic flows form new ignimbrite terraces to be colonized; as when faulting displaces paths, roads, and rivers that may reroute communications across the landscape; as when tephra or tsunami sand deposition is so deep that it influences the land use and kinds of crops that can be grown on it; as when erosion due to uplift exposes resources like obsidian flows and jade veins to the surface; as when earthquakes cause landslides that add new soil to lowlands for exploitation – examples are innumerable. One of the points of Tectonic Archaeology is rather than looking beyond a landscape/site/feature/object to what natural causes beyond 'culture' and 'environment' might have influenced its location/accessibility/use, instead beginning with tectonics and assess beforehand how the tectonic history of a region might have conditioned the that portion of the Earth and biological environment in ways that cannot easily be assessed from a landscape/site/feature/object alone or in ways that would not be investigated otherwise.

Finally, it is necessary to point out the dichotomous nature of tectonic processes that may either act to *destroy* past evidence or *preserve* it, possibly within the same event. Sudden destruction is most often addressed within Disaster Archaeology, while the latter is the bread and butter of archaeological excavation. Once past disasters are acknowledged for a region, modern populations are behoved to monitor and mitigate against potential future disasters. These concerns include disasters that are not immediately tectonically based, such as climate change and flooding; but when Earth systems are considered as a whole, Plate Tectonics often lie at the base of or are inter-linked with other changes such as volcanic degassing/eruptions contributing greenhouse gases to the atmosphere, or the uplift of the Himalayas changing climate patterns in East Asia. Disaster Archaeology focusses on negative impacts (cf. Chapter 11), but the positive influences of Plate Tectonics should not be overlooked. For example, a paper on landslides of green tuff, deposited by Miocene submarine eruptions but now forming mountains in northwest Japan, identifies the green tuff as good for agriculture (Tazaki 2006) – in opposition to the generally poor quality of volcanic soils in Japan (Chapter 7).

It is always said of a Critical Graduate Student (which I once was but no longer) that it is not enough to criticize and demolish previous work but necessary to build a replacement. In response, the following Chapters offer a general introduction to Plate Tectonics for archaeologists working in subduction zones using Japan as an example. With Part I as a geological primer, I attempt to explain subduction-zone processes that relate to Japan's geographical positioning (Chapter 2), then examine the palaeogeographic compilation of the Japanese landmass from the time subduction began at 520 million years ago (Chapter 3) through the rifting of the archipelago at 16 million years ago (Chapter 4), and finally outline the establishment of neotectonics (Chapter 5) and the Quaternary volcanic system (Chapter 6). Part I ends with a consideration of volcanic soils and their *infertility* (Chapter 7). Subsequently, in Part II, introductions of the three main subdisciplines of Japanese archaeology dealing with tectonic processes are provided – TephroArchaeology (Chapter 8), Earthquake Archaeology (Chapter 9), and Tsunami Archaeology (Chapter 10) – with consideration of how they are related to Disaster Archaeology (Chapter 11). Chapter 12 takes a different tack on the relevance of

Plate Tectonics to archaeology, by examining the formation processess of jade. Part III is offered as an example of regional Geoarchaeology in a tectonic framework. It is indeed a challenge to eke out the archaeological significances of Plate Tectonics for a specific fieldwork area, and in this case, it is only retrospective. This is actually a daunting task due to the fact that tectonic processes either have a long-term effect that is not noticeable within a human lifespan or a short-term effect that often only occurs sporadically. Archaeological research recovers time-slices of life at different intervals that may not match the geological events.

The approach outlined here, from the general (Plate Tectonics) to the specific (archaeological sites), differs from normal geoarchaeological offerings which usually begin with a site and try to fit it into a larger context. Yes, human behaviour and activities are our target, but by understanding the tectonics of a region, we might understand more what to look for on both more subtle and widespread levels – for example, not removing tephra layers before understanding their timing and generation so that footprints and activity between tephra fallouts can be documented (e.g., Sugiyama 2019; Sakaguchi 2019); understanding that the deposit of chemicals across fields in a tsunami might be as devastating as sand cover in affecting crop growing (Saino 2017); or developing the ability to recognize and distinguish earthquake liquefaction traces from human-made features (Matsuda 2000); or understanding the differential distribution of precious nephrite and jadeitite resources across the globe as related to orogenic zones (Harlow *et al.* 2014, 2015). All this requires a greater engagement with Earth processes than the normal geoarchaeologist encounters in their training. By knowing the geological *history* as well as *contents* of the fieldwork region, questions might be asked that otherwise might not arise.

Further Reading

There are many resources that provide a general introduction to aspects of Plate Tectonics and its use in associated disciplines.

Earth Sciences

For a two-hour forty-minute introduction to Earth Sciences, see Lake (2020).

Plate Tectonics

For more detailed explanations of tectonic processes, the first edition of Kearey & Vine on *Global Tectonics* (Kearey & Vine 1996) is a generally accessible textbook, while the second edition is highly technical (Kearey *et al.* 2009). *Plate Tectonics* (Frisch *et al.* 2011) and *Tectonics* (Moores & Twiss 1995) give great detail and prolific citations. Chester (2008) not only presents Plate Tectonics but covers the history of the idea of continental drift leading up to the current paradigm. The volume on *Active Tectonics* by Keller & Pinter (2002) is thematically organized and very readable but focussed primarily on earthquakes and geomorphology.

Tectonic Geomorphology

Goldberg was cited above as saying that in America, geoarchaeology is geomorphology or it is nothing. From the 1970s, the field of Tectonic Geomorphology has concentrated on not just landscape description but landscape evolution – rates of uplift and landform displacement through time (Kaizuka 1981; Bull 1984; Yonekura *et al.* 1990; Suzuki 2013). Thus, much recent writing on geomorphology is based in the processes of tectonics (Keller & Pinter 2002: ch. 2; Bull 2008, 2009; Burbank & Anderson 2011; Suzuki 2013), though assessing such is not without problems (Wobus *et al.*

2006). This trend has been carried into archaeology (Bailey *et al.* 2011; Kübler *et al.* 2018). However, Oguchi (2020) explains that tectonics alone do not provide explanations – they must be considered together with climate; in his article, he reviews the dichotomous development of geomorphology in Japan (historical/descriptive vs processual).

Geoarchaeology

In addition to the works mentioned above that recount the history of geoarchaeology, the Preface of Goldberg *et al.* (2001) describes several earlier volumes on geoarchaeology that are worthy of note, some of which have also been dealt with here (Davidson & Shackley 1976; Herz & Garrison 1998; Rapp & Hill 2009; and Waters 1992/1997). Among the myriad recent offerings, the most useful are Stein & Farrand (2001); Goldberg *et al.* (2001); Goldberg & Macphail (2006); and Rapp & Hill (2009).

Geological maps

See Appendix 4 on how to read a geological map, with various national resources supplied.

References

Website citations not in article format:
Elsevier n.d. [www.journals.elsevier.com/journal-of-archaeological-science-reports]
Elsevier Science Direct [www.sciencedirect.com/search?qs=&pub=Journal%20of%20Archaeological%20 Science&cid=272428]
Elsevier Journals [www.journals.elsevier.com/journal-of-human-evolution]
RESET [http://c14.arch.ox.ac.uk/reset/index.html]
Wiley Online [https://onlinelibrary.wiley.com/loi/15206548] for *Geoarchaeology* searches
Wiley Online [https://onlinelibrary.wiley.com/loi/14754754] for *Archaeometry* searches

ABER, James S (**2012**) "Volcanism of the Eifel, Germany region." *GO 326/ES 767*, dead link: [http://academic.emporia.edu/aberjame/tectonic/eifel/eifel.htm].
ANNAN, Peter (**2020**) "Subsurface reflections." *Sensors & Software* [Sensoft.ca/gpr/depth-deep].
Anon. (**2018**) "1755 the Great Lisbon earthquake and tsunami, Portugal." *SMS Tsunami Warning* [www.sms-tsunami-warning.com/pages/tsunami-portugal-1755#.XgXy8BewlSw].
ARAI, Fusao (ed.) (**1993**) *Volcanic ash archaeology*. Tokyo: Kokoin Shoin (in Japanese).
BAILEY, Geoff; Geoff KING & Derek STURDY (**1993**) "Active tectonics and land-use strategies: a Palaeolithic example from northwest Greece." *Antiquity* 67.255: 292-312.
BAILEY, GN & GCP KING (**2011**) "Dynamic landscapes and human dispersal patterns: tectonics, coastlines, and the reconstruction of human habitats." *Quaternary Science Reviews* 30.11: 1533-1553.
BAILEY, GN; SC REYNOLDS & GCP KING (**2011**) "Landscapes of human evolution: models and methods of tectonic geomorphology and the reconstruction of hominin landscapes." *Journal of Human Evolution* 60: 257-280 [DOI: 10.1016/jjhevol.2010.01.004].
BANKS, V (**2021**) Shallow geohazards and environmental change. *GA monthly lecture*, 7 May 2021. London: Geologists Association.
BARNES, GL (**1986**) "Book review of Rapp & Gifford, *Archaeological Geology*." *Geological Magazine* 122.1: 89-90.
BARNES, Gina L & Tsutomu SODA (eds) (**2019**) *TephroArchaeology in the North Pacific*. Oxford: Archaeopress.
BEALE, Thomas W (**1973**) "Early trade in highland Iran: a view from a source area." *World Archaeology* 5.2: 133-148.
BUCK, Brenda J (**2004**) "Review of *Soils in Archaeological Research* by Vance T. Holliday." *The Journal of Geology* 113.4: 501.
BULLOUGH, Florence (**2020**) "Geoscience and the future – a timer for a reboot?" *Geoscientist* 2020 (Sept): 26-27.
BULL, William B (**1984**) "Tectonic geomorphology." *Journal of Geological Education* 32.5: 310-324.
---- (**2008**) *Tectonic geomorphology of mountains: a new approach to palaeoseismology*. Chichester, UK: John Wiley & Sons.

---- (**2009**) *Tectonically active landscapes.* Chichester, UK: John Wiley & Sons.

Burbank, Douglas W & Robert S Anderson (**2011**) *Tectonic geomorphology.* Chichester, UK: John Wiley & Sons.

Butzer, Karl W (**1974**) "Geo-archaeological interpretation of Acheulian calc-pan sites at Doornlaagte and Rooidam." *Journal of Archaeological Science* 1: 1-25.

---- (**1975**) "The ecological approach to archaeology: are we really trying?" *American Antiquity* 40.1: 106-111.

---- (**1978**) "Toward an integrated, contextual approach in archaeology: a personal view." *Journal of Archaeological Science* 5: 191-193.

---- (**1982**) *Archaeology as human ecology.* Cambridge: Cambridge University Press.

Chester, David (**1993**) *Volcanoes and society.* London: Edward Arnold.

Chester, Roy (**2008**) *Furnace of creation, cradle of destruction: a journey to the birthplace of earthquakes, volcanoes, and tsunamis.* New York: Amacom.

Ciaurlec (**2019**) "North America basement rocks." Public Domain [https://commons.wikimedia.org/wiki/File:North_america_basement_rocks.png].

Colpron, Maurice; JoAnne L Nelson & Donald C Murphy (**2007**) "Northern Cordilleran terranes and their interactions through time." *GSA Today* 17.4/5 [DOI: 10.1130/GSAT01704-5A.1].

Courty, Marie-Agnès (**2001**) "Microfacies analysis assisting archaeological stratigraphy", pp. 205-241 in *Earth sciences and archaeology*, ed. by P. Goldberg *et al.* New York: Kluwer.

Courty, Marie-Agnès; Paul Goldberg & Richard I. Macphail (**1989**) *Soils and micromorphology in archaeology.* Cambridge University Press.

Davidson, Donald A & Myra L Shackley (eds) (**1976**) *Geoarchaeology: earth science and the past.* London: Duckworth.

DCAN (Disaster Concerned Archaeologists' Network) & Maizō Bunkazai Kenkyūkai (**1996**) *Excavated Evidence of Earthquakes.* Osaka: Maibun Kankei Kyūen Renraku Kaigi (in Japanese).

Dickinson, William R & DV Burley (**2007**) "Geoarchaeology of Tonga: geotectonic and geomorphic controls." *Geoarchaeology* 22: 231-261.

Edwards, Kevin J (**1983**) "Anatomy of a publication. *The Journal of Archaeological Science*, the first nine years." *Journal of Archaeological Science* 10: 413-421.

Eldridge, Caleb M & Lorraine W Wolf (**2019**) "The tectonic framework of the New Madrid Seismic Zone from lidar, gravity, and magnetic modeling." *Symposium on the Application of Geophysics to Engineering and Environmental Problems* [DOI: 10.4133/sageep.32-052].

Finkl, Charlie (**2007**) "In Memoriam: Rhodes W. Fairbridge." *Journal of Coastal Research* 23.2: iii.

Force, Eric R (**2015**) *Impact of tectonic activity on ancient civilizations: recurrent shakeups, tenacity, resilience, and change.* Lanham, MD: Lexington Books.

Force, Eric R & Bruce G McFadgen (**2012**) "Influences of active tectonism on human development: a review and neolithic example", pp. 1195-1202 in *Climates, Landscapes, and Civilizations*, ed. by L Giosan *et al.* Washington, DC & Hoboken, NJ: Americal Geophysical Union & Wiley-Blackwell Geopress.

Frisch, Wolfgang; Martin Meschede & Ronald C Blakey (**2011**) *Plate tectonics: continental drift and mountain building.* Berlin: Springer.

Gilbert, Allan S (ed.) (**2017**) *Encyclopedia of geoarchaeology.* Dordrecht: Springer.

Goldberg, Paul (**1989**) "Review of *Formation Processes of the Archaeology Record* by Michael B. Schiffer." *Geoarchaeology* 4.3: 277-289.

Goldberg, Paul; Vance T Holliday & C Reid Ferring (eds) (**2001**) *Earth sciences and archaeology.* New York: Kluwer.

Goldberg, Paul & Richard I Macphail (**2006**) *Practical and theoretical geoarchaeology.* Malden, MA: Blackwell.

Gould, R (**2007**) *Disaster archaeology.* Salt Lake City, UT: University of Utah Press.

Grattan, J & R Torrence (eds) (**2007**) *Living under the shadow: cultural impacts of volcanic eruptions.* Walnut Creek, CA: Left Coast.

Harlow, GE; S Sorensen & VB Sisson *et al.* (**2014**) "The geology of jade deposits", pp. 305-374 in *The geology of gem deposits* (2nd edition), ed. by LA Groat. Québec: Mineralogical Association of Canada.

Harlow, GE; T Tsujimori & SS Sorensen (**2015**) "Jadeitites and plate tectonics." *Annual Review of Earth and Planetary Sciences* 43: 105-138 [DOI: 10.1146/annurev-earth-060614-105215].

Hassan, FA (**1979**) "Geoarchaeology: the geologist and archaeology." *American Antiquity* 44.2: 267-270.

Herz, N & EG Garrison (**1998**) *Geological methods for archaeology.* New York: Oxford University Press.

HILL, Christopher L (**2017**) "Geoarchaeology, history", pp. 292-303 in *Encyclopedia of Geoarchaeology*, ed. by AS GILBERT. Dordrecht: Springer [DOI 10.1007/978-1-4020-4409-0].

HOLLIDAY, Vance T (**2004**) *Soils in Archaeological Research*. Oxford Scholarship Online 2020 [DOI: 10.1093/oso/9780195149654.001.0001].

INAGA, Shigemi (**2014**) "Addressing trade from the historical perspective of pirates", pp. 123-152 in *Aporia in pre-modern East Asia*, ed. by X XU. Japanese Studies Series 8. Taipei: National Taiwan University Publishing Center (in Japanese).

ISOZAKI, Yukio; Kazumasa AOKI & Takaaki NAKAMA *et al.* (**2010**) "New insight into a subduction-related orogen: a reappraisal of the geotectonic framework and evolution of the Japanese Islands." *Gondwana Research* 18: 82-105.

ITA (Information Technology Associates) (**2019**) *Dictionary of Geology* [https://theodora.com/geology/index.html].

JING, Zhichun (**2007**) "Integration comes of age: a conversation with Rip Rapp." *Geoarchaeology* 22.1: 1-14.

KAIZUKA, Sohei (**1981**) "Tectonic landforms", pp. 39-72 in *Landforms of Japan*, ed. by T YOSHIKAWA *et al.* Tokyo: University of Tokyo Press.

KARKANAS, Panagiotis (Takis) & Paul GOLDBERG (**2019**) *Reconstructing archaeological sites: understanding the geoarchaeological materix*. Oxford: John Wiley & Sons.

KEAREY, Philip; Keith A KLEPEIS, & Frederick J VINE (**2009**) *Global tectonics*. Oxford: Wiley-Blackwell.

KEAREY, Philip & Frederick J VINE (**1996**) *Global tectonics*. Oxford: Blackwell Science.

KELLER, Edward A & Nicholas PINTER (**2002**) *Active tectonics: earthquakes, uplift, and landscape* (2nd edition). Upper Saddle River, NJ: Prentice Hall.

KING, GCP & GN BAILEY (**2006**) "Tectonics and human evolution." *Antiquity* 80.308: 265-286.

KING, GCP & GN BAILEY (**2010**) *Dynamic landscapes and human evolution*. GSA Special Paper 471. Boulder, CO: Geological Society of America.

KÜBLER, Simon; Geoffrey CP KING & Maud DEVÈS *et al.* (**2019**) "Tectonic geomorphology and soil edaphics as controls on animal migrations and human dispersal patterns", pp. 653-673 in *Geological setting, palaeoenvironment and archaeology of the Red Sea*, ed. by NMA RASUL & ICF STEWART. Cham, Switzerland: Springer Nature.

KUWAHATA, M (**2019*a***) "Volcanic disaster archaeology: comments on methodological prospects and issues", pp. 41-46 in BARNES & SODA (eds) (2019).

---- (**2019*b***) "Restoration of agricultural assets after volcanic disasters in southwest Japan", pp. 92-101 in BARNES & SODA (eds) (2019).

LAKE, DJ (**2020**) "Geology: Earth Science for everyone." *Ûdemy* [https://www.udemy.com/ course/geology-earth-science-for-everyone].

LOWE, John *et al.* (**2012**) "Volcanic ash layers illuminate the resilience of Neanderthals and early modern humans to natural hazards." *PNAS* 109.34: 13532-13537 [DOI: 10.1073/pnas.1204579109].

MANDEL, Rolfe D (ed.) (**2000**) *Geoarchaeology in the Great Plains*. Norman, OK: University of Oklahoma Press.

MARUYAMA, K (**2019**) "Volcanic disaster research using archaeological methods: 10th-century eruption and population movements in northern Tohoku, Japan", pp. 140-157 in BARNES & SODA (eds) (2019).

MATSUDA, J-i (**2000**) "Seismic deformation structures of the post-2300 BP muddy sediments in Kawachi lowland plain, Osaka, Japan." *Sedimentary Geology* 135: 99-116.

MCFADGEN, Bruce (**2011**) "Archaeoseismology – a New Zealand perspective", pp. 85-100 in *A salute to the captain: celebrating the 100th birthday of Emeritus Professor J.B. Mackie 3 September, 2010*, compiled by his friends, former students and colleagues from the University of Otago School of Mines and the Department of Surveying. Otago, NZ: self-published.

MCKENZIE, Dan (**2018**) "A geologist reflects on a long career." *Annual Review of Earth and Planetary Sciences* 46: 1-20 [DOI: 10.1146/annurev-earth-082517-010111].

MCKENZIE, DP & RL PARKER (**1967**) "The North Pacific: an example of tectonics on a sphere." *Nature* 216: 1276-1289 [DOI: 10.1038/2161276a0].

MACPHAIL, Richard I & Paul GOLDBERG (**2017**) *Applied soils and micromorphology in archaeology*. Cambridge University Press.

Merriam-Webster (**n.d.**) "Geo-combining form", in *The Merriam-Webster.com Dictionary* [www.merriam-webster.com/dictionary/geo-].

Moores, Eldridge M & Robert J Twiss (**1995**) *Tectonics*. Long Grove, IL: Waveland Press, Inc. (reprinted in 2014).

Morgan, WJ (**1968**) "Rises, trenches, great faults, and crustal blocks." *Journal of Geophysical Research* 73: 1959-1982 [DOI: 10.1029/JB073i006p01959].

Murray, KD; MH Murray & AF Sheehan (**2019**) "Active deformation near the Rio Grande Rift and Colorado Plateau as inferred from continuous global positioning system measurements." *Journal of Geophysical Research-Solid Earth* 124.2: 2166-2183.

Nagata, Hidehisa (**2018**) "Studies on nontectonic geologic structure in Japan: recent progress and future perspective." *Chishitsu Zasshi* 124.11 [DOI: 10.5575/geosoc.2018.0040] (in Japanese with English title and abstract).

NSF (National Science Foundation) (**n.d.**) "About Earth Sciences." *National Science Foundation Where Discoveries Begin* [www.nsf.gov/geo/ear/about.jsp].

Oguchi, Takashi (**2020**) "Geomorphological debates in Japan related to surface processes, tectonics, climate, research principles, and international geomorphology." *Geomorphology* 366.#106805: 14pp. [DOI: 10.1016/j.geomorph.2019.06.019].

Okamura, K; A Fujisawa & Y Kondo *et al.* (**2013**) "The Great East Japan Earthquake and cultural heritage: towards an archaeology of disaster." *Antiquity* 87.335: 258-269.

Pollard, AM (**1999*a***) "Geoarchaeology: an introduction", pp. 7-14 in *Geoarchaeology: exploration, environments, resources.* Special Publication No. 165, ed. by AM Pollard. London: The Geological Society.

---- (ed.) (**1999*b***) *Geoarchaeology: exploration, environments, resources.* Special Publication No. 165. London: The Geological Society.

---- (**n.d.**) staff page at Oxford [https://www.arch.ox.ac.uk/people/pollard-mark].

Pomerol, C (**1975**) "Plate tectonics and continental drift during the Cenozoic era." *Journal of Human Evolution* 4.3: 185-191.

Rapp, George Jr (**1987*a***) "Archaeological geology", pp. 688-698 in *Encyclopedia of Physical Science and Technology 1,* ed. by RA Meyers. New York: Academic Press.

---- (**1987*b***) "Geoarchaeology." *Annual Review of Earth and Planetary Sciences* 15: 97-113.

Rapp, George Jr & John A Gifford (eds) (1982) "Archaeological geology." *American Scientist* 70 (Jan-Feb): 45-53.

---- (**1985**) *Archaeological geology*. New Haven, CT: Yale University Press.

Rapp, George (Rip) Jr & Hill, Christopher L (**2009**) *Geoarchaeology: the earth-science approach to archaeological interpretation.* New Haven, CT: Yale University Press; 1st ed. 1998.

Renfrew, Colin (**1976**) "Archaeology and the earth sciences", pp. 1-5 in *Geoarchaeology: earth science and the past,* ed. by DA Davidson & ML Shackley. London: Duckworth.

RESET (**n.d.**) *RESET* [https://c14.arch.ox.ac.uk/reset/].

Riede, Felix (**2017**) *Splendid isolation: the eruption of the Laacher See volcano and southern Scandinavian Late Glacial hunter-gatherers.* Aarhus University Press.

Riede, Felix; Gina L Barnes & Mark D Elson *et al.* (**2020**) "Prospects and pitfalls in integrating volcanology and archaeology: a review." *Journal of Volcanology and Geothermal and Research* 401.#106977 [DOI: 10.1016/j.jvolgeores.2020.106977].

Saino, H (**2017**) *Investigation methods of tsunami disaster traces: cooperation with concerned many fields.* Tokyo: Doseisha (in Japanese).

Sakaguchi, H (**2019**) "Archaeological investigation of the seasonality and duration of the 6th-century eruptions from Mt Haruna", pp. 184-191 in Barnes & Soda (eds) (2019).

Sangawa, Akira (**1986**) "Earthquake fault displacement on Kondayama tomb." *Jishin* 39.2: 15-24 (in Japanese).

Schiffer, Michael B. (**1983**) "Toward the identification of formation processes." *American Antiquity* 48.4: 675-706.

---- (**1987**) *Formation Process of the Archaeological Record.* University of New Mexico Press.

Sheets, Payson (**1980**) *Archaeological studies of disaster: their range and value.* Boulder, CO: Institute of Behavioral Science, University of Colorado.

Shimoyama, S (**1997**) "On the range of the Disaster Archaeology—for taking measure of effects from disasters." *Hominids* 1 (March): 83-103 (in Japanese with English title and summary).

---- (**2002*a***) "Volcanic disasters and archaeological sites in southern Kyushu, Japan", pp. 326-341 in TORRENCE & GRATTAN (eds) (2002).
---- (**2002*b***) "Basic characteristics of disasters", pp. 19-27 in TORRENCE & GRATTAN (eds) (2002).
SMITH, Peter (Estate) (**n.d.**) *Compound words in Greek*. Vancouver, CA: Pressbooks [https://pressbooks.bccampus.ca/greeklatinroots2/chapter/§110-A7110-some-common-greek-combining-forms].
SODA, Tsutomu (**2019**) "Tephroarchaeology and its history in Japan", pp. 24-40 in BARNES & SODA (eds) (2019).
STAFFORD, C Russell (**1995**) "Geoarchaeological perspectives on paleolandscapes and regional subsurface archaeology." *Journal of Archaeological Method and Theory* 2.1: 69-104.
STEIN, JK (**1993**) "Effects of scale on archaeological and geological perspectives", pp. 1-10 in *Effects of scale on archaeological and geoscientific perspectives*, ed. by JK STEIN & AR LINSE. GSA Special Paper 283. Boulder, CO: The Geological Society of America.
STEIN, Julie K & William R FARRAND (eds) (**1985**) *Archaeological sediments in context*. Orono, ME: Study of Early Man, Institute for Quaternary Studies, University of Maine at Orono.
---- (eds) (**2001**) *Sediments in archaeological context*. University of Utah Press.
STEIN, Seth; Carol A STEIN & Reece ELLING *et al.* (**2018**) "Insights from North America's failed Midcontinent Rift into the evolution of continental rifts and passive continental margins." *Tectonophysics* 744: 403-421.
SUGIYAMA, H (**2019**) "Investigations into the Kofun period disasters caused by Mt Haruna eruptions", pp. 168-182 in BARNES & SODA (eds) (2019).
SUZUKI, Yasuhiro (**2013**) "Tectonic geomorphological Active Fault studies in Japan after 1980." *Geographical Review of Japan, Series B* 86.1: 6-13.
TAN, KH (ed.) (**1984**) *Andosols*. New York: Van Nostrand Reinhold.
TAZAKI, K (**2006**) "Green-tuff landslide areas are beneficial for rice nutrition in Japan." *Anais da Academia Brasileira de Ciências* 78: 749–763.
THORSON, RM & VT HOLLIDAY (**1990**) "Just what is geoarchaeology?" *Geotimes* 1990 (July): 19-20.
TORRENCE, R & J GRATTAN (eds) (**2002**) *Natural disasters and cultural change*. London: Routledge.
TOURLOUKIS, Vangelis (**2010**) *The Early and Middle Pleistocene archaeology record of Greece: current status and future prospects*. Amsterdam University Press.
WATERS, MR (**1992**) *Principles of geoarchaeology: a North America perspective*. University of Arizona Press; paperback edition in 1997.
WOBUS, C; KX WHIPPLE & E KIRBY *et al.* (**2006**) "Tectonics from topography: procedures, promise, and pitfalls", pp. 55-74 in *Tectonics, climate, and landscape evolution*, ed. by SD WILLET *et al.* GSA Special Paper 398. Boulder, CO: Geological Society of America.
WRIGHT, Henry T (**2011**) "In Memoriam: William R. Farrand (1931–2011)." *Paleorient* 37: 5-8.
YONEKURA, N; A OKADA & A MORIYAMA (eds) (**1990**) *Tectonic geomorphology and tectonics*. Tokyo: Kokon Shoin (in Japanese).

Part I

A Geological Introduction to Japan

In the spirit of contextualizing a regional archaeology within the local geology, this section is offered to clarify what the Japanese Archipelago consist of and what archaeologists should be aware of in terms of geological timespans, locational continuities and discontinuities, Earth materials, and ongoing processes of transformation. For reference, the main Japanese placenames and archaeological periods are given in Appendix 2, and geological period divisions are given in Appendix 3.

An archaeologist cannot be expected to assess a local geological setting without further intensive training or without recourse to specialist advice and analyses. However, by gaining a broad knowledge of the locale in which one is working from such sources, the archaeologist might be better placed to have expectations and ask questions about archaeological site settings and contents and artefact compositions. One challenge in delving into the geological literature is to understand specialist geological terminology; many terms and their definitions are explained in Chapter 2, but judicious use of the Glossary will also be helpful in reading the successive chapters. Terms occurring in the glossary are given in sans serif font in each chapter.

Most geology textbooks begin by teaching the three Big Ideas (based on Anon. n.d.):

1) Geological Time: geological time is deep and divided into many periods whose boundaries shift with new discoveries and understandings. Most pertinent to archaeology is the re-definition of the end of the Ice Age, from 10,000 years ago to 11.7 ka (11,700 years ago). The three official abbreviations for 'years ago' will be used in this volume: Ga = billion years ago, Ma = million years ago, ka = thousand years ago. The latest version (February 2022) of the International Chronostratigraphic Chart is provided in Appendix 3. The Earth is about 4.7 billion years old, but Japan's geology proper does not begin until the Jurassic period (ca. 201–145 Ma), though the oldest Earth materials (detrital zircons) found in Japan, eroded from earlier continental locations, date to 3.8 Ga.

2) Uniformitarianism: this was a revolutionary concept promoted by James Hutton, Scottish geologist, in 1875. Essentially he proposed that the geological processes we see occurring today also occurred in the past, thereby allowing Charles Lyell, a late Scottish geologist, to paraphrase it in 1847 as 'the present is the key to the past'. More recently, however, it has been realized/discovered that many of the processes on the early Earth have *not* continued until today; thus we cannot assume that 'what you see *today* is what you get in the *past*', but we must investigate more carefully for different causes. This particularly applies to the Earth before tectonic plate subduction began or before oxygen became common in the Earth's atmosphere.

3) Plate Tectonics: this theory, formalized in 1965, provided an overarching concept that explained many disparate things about the Earth's construction and operation. It was a powerful integrator and is the dominant paradigm for thinking about Earth processes. However, as with Uniformitarianism, there are still problems to be worked out and new processes to be defined.

Another useful concept in geology is the "Rock Cycle" (Gray n.d.), which tracks the creation from magma of igneous rocks that are then broken down by weathering into sediments, which are compacted to form sedimentary rocks. These sedimentary rocks, together with igneous rocks, can then be transformed by pressure, temperature or both into metamorphic rocks, which can also be weathered into sediments. These three (igneous, sedimentary, metamorphic) rock types make up the Earth's crust, which can be melted by subduction processes, creating magma to start the cycle again.

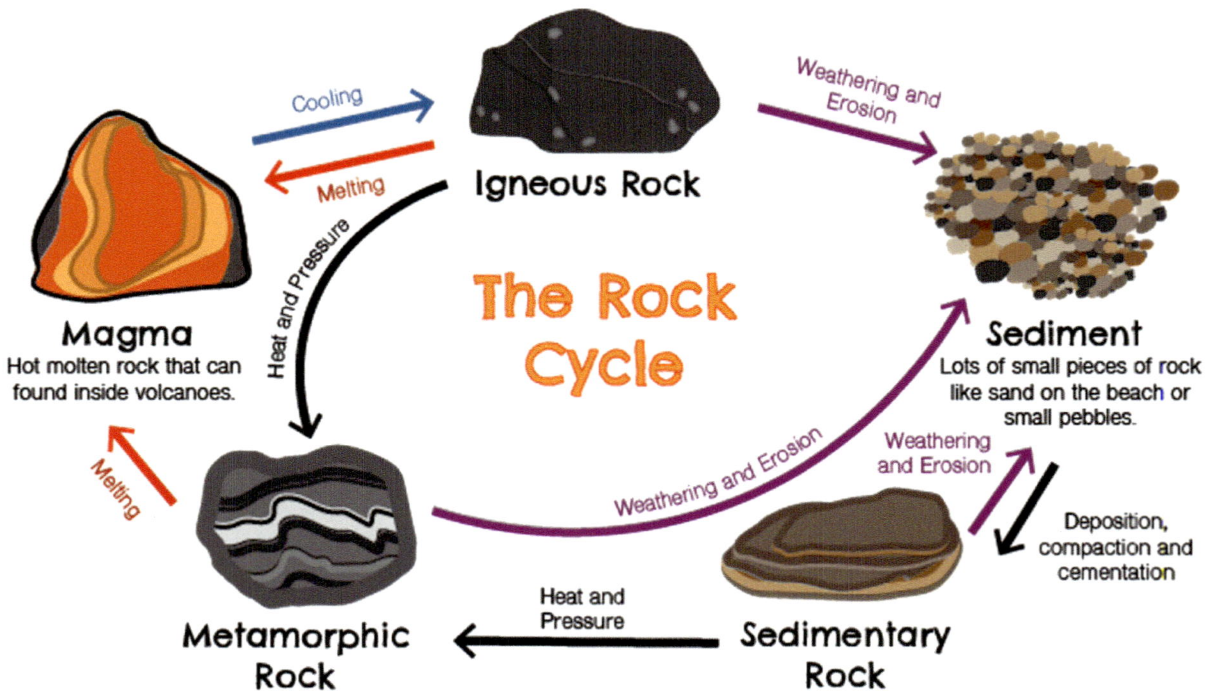

Rock cycle illustration Copyright Amy Edgington, GeoBus, Earth Sciences Department, UCL, with permission.

This volume is based on Plate Tectonics, its associated processes, and its products. There are two types of tectonic plate edges: a mid-oceanic ridge where opposing plates are expanded through time with more material added to their edges, and subduction zones where opposing plates meet and one is drawn underneath the other (collision) or down into the mantle (subduction). Although subduction is generally viewed as a destructive process, it can contribute to landmass creation through volcanism and accretion of oceanic materials. Japan, as the case study offered here, is a present-day subduction zone that exhibits all the characteristics of a land created through the various processes related to subduction (Chapter 2). These processes can be generalized to other areas of both current and fossil subduction zones around the world. In lieu of specialist courses on Japanese geology or even introductory textbooks, Chapters 3 through 7 are offered to give a whirlwind introduction to how the Japanese Islands were formed, how their history is different from what we commonly see today, and how their geology, both hard rock and soft rock, is different from some other areas of the world but represent many other subduction zones, current and fossilized. Geological terms are extensively indexed to form a study resource in themselves.

Anon. (**n.d.**) "Three big ideas: geological time, uniformitarianism, and plate tectonics", chapter 1.5 in *Physical Geology First University of Saskatchewan Edition* adapted from *Physical geology* written by Steven EARLE for the BCcampus Open Textbook Project. Saksatoon, SK: University of Saskatchewan [https://openpress.usask.ca/physicalgeology/chapter/1-5-three-big-ideas-geological-time-uniformitarianism-and-plate-tectonics/].

GRAY, Bill (**n.d.**) "The rock cycle." *Geological Society of Glasgow* [https://geologyglasgow.org.uk/local-rocks/the-rock-cycle/].

Chapter 2

A Primer in Plate Tectonics, With Specific Reference to Japan

This Chapter is aimed at archaeologists working in areas of active tectonics (producing earthquakes and volcanoes); it is written for non-geologists but from the point of view of Japanese geology. It does not purport to be a comprehensive discussion of Plate Tectonics – a job for standard publications on the subject. Geologist readers will no doubt find the presentation unusual but the citations on Japan useful. Hopefully the material provided here can be applied to archaeology in other subduction zone situations around the globe such as New Zealand or the Mediterranean. More than anything, however, it is hoped that the following gives a more holistic treatment of tectonics and its associated processes in the Japanese Islands compared to piecemeal references to Japan in major geology textbooks. Nevertheless, the following treatment is necessarily brief, and such textbooks and specialist papers should be consulted for detail. Geological terms (in **sans serif** font on first mention below and in further Chapters) are explained in a Glossary. Cross-references to numbered sections within this Chapter are referred to as e.g., Sec. 2.2.3, and in other chapters, Ch2.2.3 See Appendix 2 for Japanese placenames and archaeological period names used in Japan, and Appendix 3 for geological period divisions. For any rock types mentioned, see Strekeisen (2020).

2.1 From Geosynclines to Tectonic Plates

> All parts of the surface of the Earth are moving and changing....
> think of the view of the Earth that we acquire during our short lifetime
> as like a single frame from a continuously running sequence.
> (Butler & Bell 1988: 5)

Before Plate Tectonics, there were two competing theories of Earth's structure: **geosyncline theory**, which postulated that all Earth components have always existed where they are now, only moving vertically though there might have been some sliding sideways; and **continental drift**, which postulated the movement of continents across the face of the globe.

Geosyncline theory was stimulated by James Hall in 1857 but named by JD Dana in 1873 (Knopf 1948, 1960), and thereafter was adopted in several versions depending on the area of the world in which researchers were developing their ideas (see Dietz 1972; Mitchell & Reading 1986). Kobayashi Teiichi was the foremost proponent of this theory in Japan, his seminal work in 1941 giving rise to the Tokyo (Imperial) University school of geosynclinists (Kobayashi 1941; Takai *et al.* 1963). These researchers held that the major movement of a landmass was essentially vertical, in which mountain chains were formed by undulations or foldings of the surface. This often caused sideways slippage of materials, so that horizontal movement could not totally be ruled out; however, movement of materials on the scale proposed by Plate Tectonics – thousands of miles across the Pacific Basin to finally become incorporated into the Japanese landmass – could not be envisioned within the framework of geosyncline theory. This theory was especially powerful in Japan and lingered in geological writings long after its demise in other parts of the world (Tomari 2005, 2008). The fact that Japan was conceived by geosynclinists to have always been an island chain gave a false sense of topographic continuity for discussing and understanding the early time periods of stratigraphic formation. Syntheses of Japanese geological development under this paradigm can be found in Matsuda *et al.* (1967) and references therein.

Geosynclinal theory was challenged in Europe in the early 1900s by ideas on contintal drift, particularly by Alfred Wegener (Frankel 2012) and Arthur Holmes (Holmes 1931), and finally succumbed to data on mantle convection currents as providing the vehicle for plate movement and on sea-floor magnetic reversals (Hess 1962; Vine & Matthews 1963). The new paradigm of Plate Tectonics was compiled in papers by John Tuzo Wilson in 1963 and 1965 (Wilson, JT 1963, 1965), and the new theory was formally accepted at a Royal Institute of London conference on the topic in 1965 (Blackett *et al.* 1965). McKenzie & Parker then provided the physics of plates rotating on a sphere (1967). This new paradigm was slow to be accepted by Japanese geologists; however, as noted by Isozaki, Maruyama & Yanai (2010), several Japanese researchers were especially instrumental in the development and refinement of Plate Tectonics during the 20c (BOX 1).

2.2 Plate Tectonics in Japan

Plate Tectonics as applied to Japan began to be introduced in the early 1970s by such researchers as UYEDA (UEDA) Seiya in Japan (Uyeda 1971, 1978), and MIYASHIRO Akiho in Albany, New York (Uyeda & Miyashiro 1974), whose writings have fostered the current 'geokids' (Isozaki *et al.* 1997). It must be said that for two decades, such plate tectonicists were battling an entrenched school of interpretation in Japanese geology, and their work was often disregarded in favor of the established view of geosynclinists (Isozaki 2019). ISOZAKI Yukio, one of the 'Geokids' and now a professor at the University of Tokyo himself, has written a brief historical review of the four stages in the conversion from geosyncline theory to Plate Tectonics, describing it as "a long winding road with many arguments" between the groups of scholars (Isozaki 2019, 1996: 319). The greatest strides, he writes, came in stage four after the introduction of Plate Tectonics: in the 1980s, the refinements in biostratigraphy based on microfossils, and an acceptance of the concept of linked subduction and accretion; and in the 1990s, refinements in dating weakly metamorphosed strata. See his appendix entitled "Historical review on studies of orogeny and geotectonic subdivisions of the Japanese Islands" in Isozaki's 1996 article.

It has been traditional in Japanese geology to speak of named orogenies, i.e., 'mountain-building phases': the Akiyoshi (in the Permian-Triassic periods), Sakawa (from Jurassic to Palaeogene periods), and Oyashima (in the Miocene) (Takai *et al.* 1963). But with the move from geosyncline theory to Plate Tectonics in the early 1970s, these phases have been discarded in favour of tracking the development of the Japanese landmass by the movements of oceanic plates and particularly the subduction of mid-oceanic ridges leading to the formation of Accretionary Complexes (AC), extensively discussed below.

It was not until the mid-1980s that Plate Tectonics became "generally accepted in the Japanese geological community" (Tomari 2005: abstract, 2008). However, books retaining the geosyncline vein of thought were being published into the 1990s in Japan. For example, despite the late publication date of 1991, The *Geology of Japan* (Kimura, T *et al.* 1991) was still based on geosyncline theory; this was because it was a successor volume to The *Geology of Japan* (Takai *et al.* 1963), which had been "compiled on the occasion of the sixtieth birthday of Professor Teiichi Kobayashi", the foremost geosynclinist, as stated on the title page. A third volume, Geology of Japan, is a half-way house between geosyncline and Plate Tectonics theories; the English version, published in 1991 (Hashimoto 1991), is a translation of a Japanese text published in 1980. His chapter 10, on the "Geotectonic history of the Japanese Islands", incorporates Plate Tectonics, but most of the other chapters are cast in geosyncline terminology. Although his chapter 10 acknowledges the opening of the Japan Sea Basin late in geological history, it did not yet acknowledge the intrusion of the Izu Arc into Honshu as the cause of the reverse syntaxis structure in central Japan (see Chapter 4), and it cites Kobayashi's 1941 thesis of northeast Japan rotating eastwards to bend the geological belts (Hashimoto 1991: 236). Even when Plate Tectonics and back-arc basins were acknowledged, the palaeogeography of Japan could be written without reference to continental rifting (Kaizuka 1980).

Data on palaeomagnetism became available in the 1980s which clearly showed that the *present* Japanese archipelago rifted outwards creating the Japan Sea Basin behind it (Otofuji & Matsuda 1983; Otofuji *et al.* 1985; Otofuji 1996). This finding undermined all discussion and writing of Japanese landmass origins cast in a geosynclinal framework, but it took the Japanese government another decade to begin to replace interpretations in school texts based on this outdated theory with those purely in a Plate Tectonics framework. Only from 1993, when the dating method for low-pressure metamorphic rocks was refined (Nakajima 1997), has the 'big picture' of the compositional process of the Japanese landmass come to be rewritten.

Isozaki, Maruyama & Yanai (2010) argue that Japanese geology has gone through three previous stages of development: first non-science (1860–1900), then colonial science (1900–1980), followed by independent science (1980–2000). They now propose that it is in its latest stage of development: that of 'exporting science' (2000~), explained as the ability to provide new indigenous research technologies to other countries. Many Japanese scholars are deemed important contributors to the development of the theory of Plate Tectonics (BOX 1).

BOX 1 Early Contributions of Japanese Researchers to Plate Tectonic Theory

The Wadati-Benioff zone: Wadati [Wadachi] Kiyoo graduated in Physics from Tokyo (Imperial) University in 1922, where the first professorship in seismology in the world was occupied by one of John Milne's students in the late 1800s (Clancey 2006). Wadati took up a post in the Central Meteorological Observatory (now the JMA)[1] in 1925, publishing four papers between 1928 and 1935 that documented the existence of deep earthquakes despite the prevailing paradigm discounting earthquake occurrence below 120 km (Suzuki 2001, 2004). During this time Wadati submitted his doctoral dissertation on *Shallow and Deep Earthquakes* in 1932. The earthquake epicentres he documented followed a curved line from the Earth's surface into its depths, but plate structure was still unknown at that time. Comparable research was carried out by Hugo Benioff at Caltech in the post-war period (Benioff 1954), without reference to the three Wadati papers that had previously appeared in English. When subduction was finally understood under the new Plate Tectonics paradigm, the work of both these scholars was honoured in the naming of the Wadati-Benioff zone along the upper surface of the subducting plate.

Paired metamorphic belts: Miyashiro Akiho also took his doctorate at Tokyo (Imperial) University in 1953, his thesis addressing *Non-calcareous Garnet in Relation to Metamorphism*. After completing an Assistant Professorship at the same university, he moved to New York, first to Columbia University then to New York State University, Albany. Miyashiro was a rebel who chaffed under the scholarly paradigms prevalent in Japan at the time, and his continuing research in the United States on metamorphism produced the first full understanding of paired metamorphic belts, of weak and strong metamorphism, that form during [cont'd]

[1] The Japan Meteorological Agency (JMA) is in charge of the weather and other natural phenomena such as earthquakes and volcanic eruptions. See [http://www.jma.go.jp/jma/indexe.html] in English and Japanese.

subduction (Miyashiro 1961, 1973*a,b*; see also Zhang *et al.* 2019). Despite accolades from worldwide geological societies, Miyashiro was rebuffed by the traditionalists in Japan, only gaining recognition there in 2002. Such is the social context of scientific research. Miyashiro was a critical mentor to young Japanese geologists who came to Albany to study Plate Tectonics under him. They called themselves the 'Geokids' (Isozaki, pers. comm. 2002, him being one of them).

Identification of the Volcanic Front: SUGIMURA Arata, seismologist, received his doctorate from Tokyo University and, while initially teaching in the university's Institute of Geology, published a paper elucidating the relationship of earthquake depths with the existence of volcanic chains (Sugimura 1965). He coined the Japanese phrase translated as 'Volcanic Front' to describe the distance from the subduction trench that volcanoes would form, often in arc formation. His "isobaths of deep and intermediate earthquake foci" (Sugimura 1965: fig. 3) eerily predicted the location of Pacific Plate depths as it dips under northeast Japan, though subducting plates were not yet recognized.

Conceptualization of Accretionary Complexes: KANMERA Kametoshi studied the correlation of geosynclinal sediments off the coast of southwestern Japan compared with sediments in the Japanese landmass (Kanmera 1976). His work contributed to the concept of the Accretionary Complex, several of which are now recognized as the main components of the archipelagic basement, with igneous rocks as secondary intrusions.

A new type of orogeny, the "Pacific": The word orogeny means 'mountain-building' according to its Greek roots (*òros+gènesis*). It was traditionally used to explain the creation of the Alps and the Himalayas as the results of collision tectonics (Africa colliding with Europe, India colliding with Eurasia). The type of orogeny proposed in works by Japanese geologists directly contradicted this prevailing notion. In their view, 'mountain-building' could also be ascribed to volcanism, producing both island arcs and continental arcs of volcanic mountains in convergent margins and resulting in orogens. In 1959, MIYASHIRO Akiho proposed the concept of 'Japan-type orogeny' for this non-collisional type of mountain-building (Sugimura & Uyeda 1973; Miyashiro 1959); his ideas were revised by Matsuda & Ueda (1971) to 'Pacific-type orogeny', and then MARUYAMA Shigenori (1997) re-proposed a 'Miyashiro-type orogeny' to honour Miyashiro's inclusion of paired metamorphic belts. This concept has since been extended the concept to the entire Pacific Basin in different thermal environments (Brown, M 2010). Six stages of continental crust growth are now recognized in Pacific-type orogeny, with different parts of Japan used to represent stages 3, 5 and 6 (Maruyama *et al.* 2011).

Second continents: The idea of continental-size masses underlying our upper visible continents was proposed by KAWAI Kenji and colleagues (Kawai *et al.* 2010; Kawai *et al.* 2013). These result from the tectonic erosion of granite from volcanic arcs on the Earth's surface, which comes to rest in the lower mantle transition zone (MTZ, between 270 to 660 km deep). They estimate that about six times as much granite exists in the MTZ as on the Earth's surface and propose that the "formation and dynamics of those second continents would have controlled the Earth's thermal history over geologic time" (Kawai *et al.* 2010: 1197). Such second continents are not to be confused with 'slab graveyards', subducted oceanic floor slabs that have come to rest at the core-mantle boundary.

2.3 Tectonic Plate Construction

2.3.1 Boundaries and distributions

The world is divided up into continental plates and oceanic plates (Figure 2.1), which ride on the Earth's viscous mantle. They follow different rules of operation depending on whether they are separating, colliding, or moving past each other, and whether the edges in contact are similar (e.g., continental to continental) or different (oceanic to continental). The plates are composed of an upper solid crust (the lithosphere), riding on a viscous upper layer (the asthenosphere) of the earth's mantle. The crust and

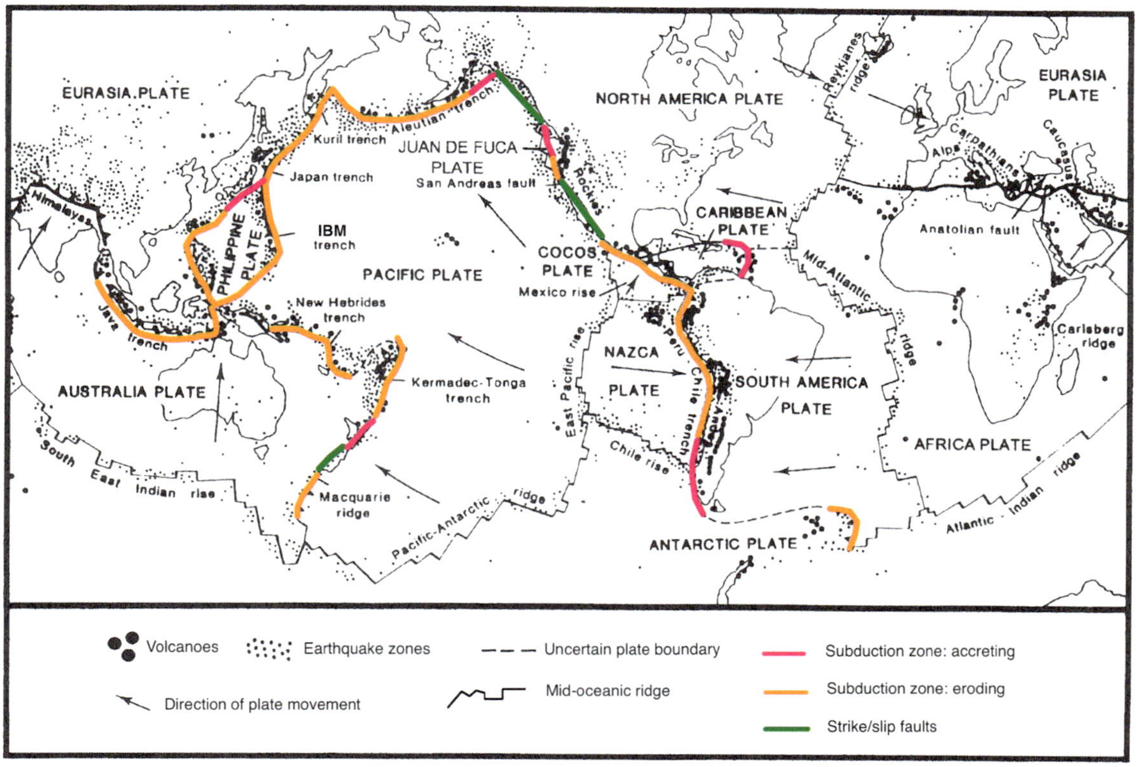

Figure 2.1 Major plate divisions of the Earth's crust.
See finer plate boundaries around the Japanese Islands in Figure 2.3

the mantle are separated by a transitional zone called the Moho, short for the Mohorovičić discontinuity (cf. Figure 2.5), the boundary recognized by changes in seismic wave velocity. There are generally two kinds of crust: continental and oceanic, described below, but neither is exclusive to the kind of plate they comprise.

Plate margins are described as either 'active' or 'passive'. Active margins are subduction zones while passive margins are not – but they can still involve a lot of activity (McClay & Hammerstein 2020). In active margins, the distinction between oceanic crust and continental crust is definitive because oceanic crust is denser and usually drawn down (subducted) under continental crust, as is generally the case around the Pacific Rim. However, in passive margins such as surround the Atlantic Basin, the oceanic crust being generated at the mid-Atlantic ridge gradually gives way to the continental crusts on either side of the Atlantic Ocean. Thus, it can be seen that the two types of crust do not occur exclusively on the different types of plates: continental plates may include today's continents as well as substantial areas of thinner ocean floor, while oceanic plates often support thick rooted volcanoes formed by magma hotspots. Thus, there are many different relationships between crustal and plate types, but Japan's case is relatively clear-cut, as we shall see.

Then there are microplates and exotic terranes. These are very often parts of continental crust that have chipped off via faulting and rifting, or the latter may be ocean island arcs – such as the Izu Island chain of Japan – that have formed via subduction processes. When these encounter other subduction zones, they may be difficult to subduct and are often accreted to the continental landmass instead, as with the Izu volcanics (see Chapter 4). Other examples include the exotic terranes around the North

Pacific (Nokleberg *et al.* 1999); the island arc Avalonia comprising the England-Wales-southern Irish basement (Nance & Thompson 1996); the island arc accreted to the Wyoming craton forming the basement rocks of Colorado and beyond (Williams & Chronic 2002); and the many accreted island arcs in western China (e.g., Xiao *et al.* 2004; Neubauer *et al.* 2010) and the CAOB (Central Asian Orogenic Belt) (e.g., Jahn 2004; Safonova & Santosh 2014).

Three forces are probably involved in moving plates across the Earth's surface: mantle convection currents caused by radioactive heat; ridge push from newly created crust at mid-oceanic ridges; and slab pull from the down-going (subducting) ocean crust (slab). The creation of new ocean floor at mid-oceanic ridges (Figure 2.2: ①), divergent margins where mantle upwells and adds material to the existing plate, increases the surface area of the ocean floor; old ocean floor is consumed through subduction at oceanic plate margins, where the crust is dragged down back into the mantle via a deep subduction trench (Figure 2.2: ②). Since the earth is finite in size, the production of new crust is always associated with the consumption of older crust somewhere else on the planet. A crude analogy would

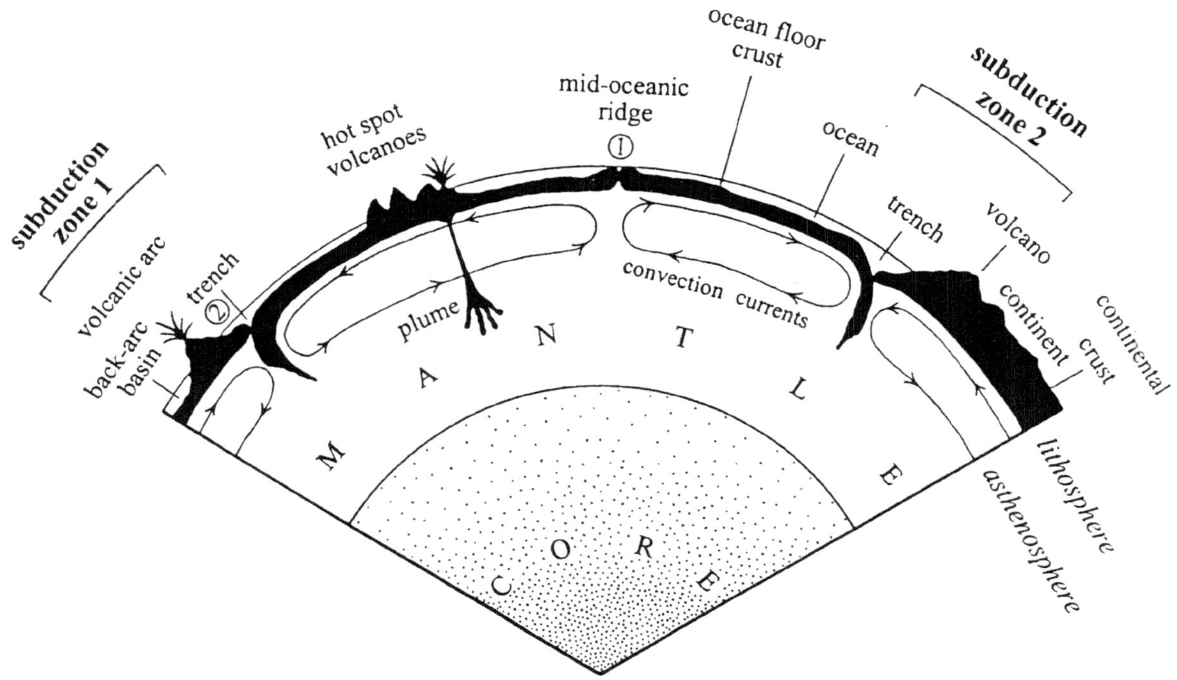

Figure 2.2 Cross-section of the Earth

Shown are the layered composition of core (solid), mantle (solid), mantle asthenosphere (viscous), and lithosphere (solid but mobile), with convection currents in asthenosphere highly exaggerated. See expanded version of subduction zone 1 in Figure 2.5.

be to think of the Pacific Plate as the surface of a conveyor belt, continuously generated upwards from one spot (an active mid-oceanic ridge), moving horizontally across a long distance (the Pacific Ocean), and disappearing downwards at the subduction zone – a trough or trench (e.g., the Japan–Izu Trenches).

The current subduction zones in the northwest Pacific (e.g., Figure 2.2: zone 1) are marked by deep sea trenches off the eastern coast of the Japanese, Kurile archipelagos, and the southern coast of the Aelutian archipelago – all constituting volcanic island arcs. The Andes subduction zone, however, has generated a continental volcanic arc without back-arc basin rifting (yet!); this was the initial state of the Japanese archipelago also before it was rifted from the Eurasian continent (see Chapter 4).

The Japanese archipelago sits at the junctures of four plates, two continental (the Eurasian and Okhotsk Plates) and two oceanic (the Pacific and Philippine Plates) (Figure 2.3). Scholarly opinion is divided on the extent and boundaries of more possible plates, for example, an Okhotsk subdivision of the North American Plate and perhaps another sub-division of a North Japan microplate (below the dashed red line), or a division of the Eurasian Plate to form an Amurian Plate (Taira 2001: fig. 1-c).

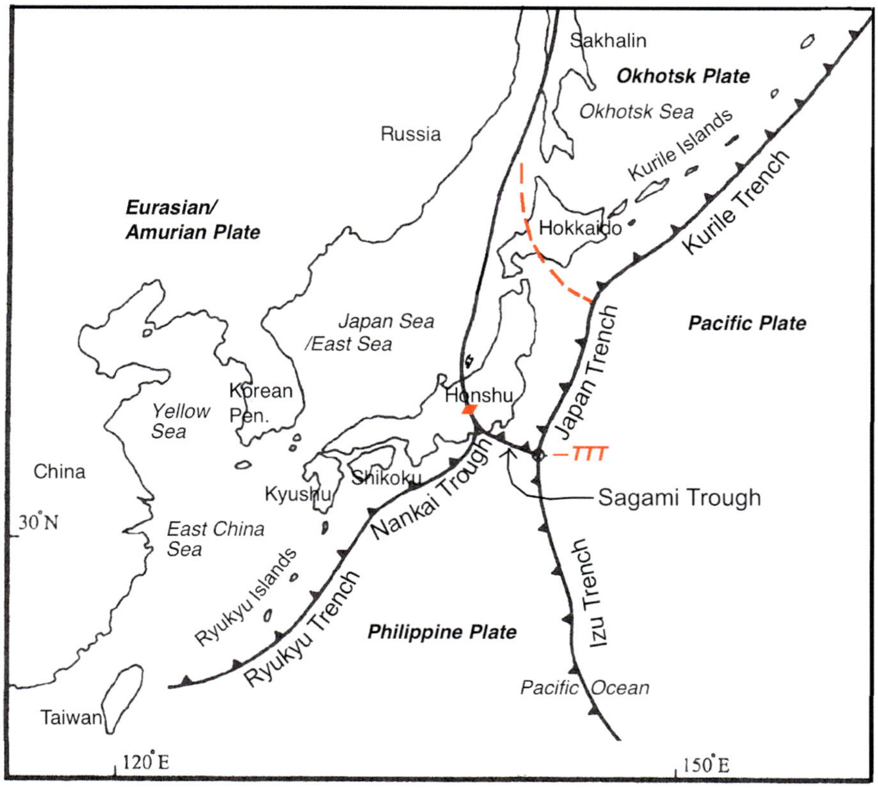

Figure 2.3 Japan at the juncture of four plates

TTT = triple trench junction
toothed line = showing direction of oceanic plate subduction
double toothed line = showing area of compression along the Itoigawa-Shizuoka Tectonic Line across Honshu, dividing Honshu into SW and NE Japan

Along the eastern edge of the archipelago, the Pacific Plate is sub-ducting into the Japan Trench in the northeast, while in southwestern Japan, the Philippine Plate is subducting both to the north into the Sagami Trough at mid-Honshu and to the west into the Nankai Trough and Ryukyu Trench. There has been some controversy through time about the number and extent of these plates (Seno 1995: 148-149), and since Seno's publication, scholarship has evolved over time, as the following points illustrate.

• The 1970s' view of the four-plate construction did not recognize the existence of the Okhotsk Plate or the Northeast Japan microplate; the boundary of the North American Plate at that time was drawn through southern Hokkaido (along the red dashed line of Figure 2.3); most of Japan was thought to be located on the Eurasian Plate.

• In the 1980s' view, earthquakes along the eastern edge of the Japan Sea resulted in the North American Plate boundary being extended through the Japan Sea on the west and passing through central Honshu via a major fault, a tectonic line named the Itoigawa-Shizuoka Tectonic Line (I-STL or ITL).

• In the 1990s, research on slip-vectors of the northern plate determined the independence of the Okhotsk Plate from the North American Plate.

• The southern boundary of the Okhotsk Plate through the I-STL was set by (Bird 2003); however, reinstating a plate boundary (red dashed line) passing through Hokkaido during the Neogene has been given seismic and geological support (Ito *et al.* 2019; Ueda 2016; Taira *et al.* 2016: fig. 1.2). This would mean that northeastern Japan and western Hokkaido belong to the Eurasian Plate, not the Okhotsk Plate in the Neogene, and the current boundary (set by Bird) has only been active since the early Quaternary (Ueda 2016)

• An Amurian Plate division within the Eurasian Plate has been identified by monitoring movement by GPS; the western boundary is unknown but the plate is defined on the south by the Qinling suture zone in central China and by the Baikal Rift and Stanovoy Mountains on the north (Heki *et al.* 1999; Apel *et al.* 2006: fig. 1), essentially taking in North China to the top of Sakhalin Island.

On the other side of the world, the Atlantic Ocean floor is expanding more rapidly than the Pacific Ocean floor, but there are no subduction zones at the Atlantic edges of either Europe/Africa or the Americas; instead, Atlantic ocean crust transitions into continental crust at passive margins, in contrast to the active margins (subduction zones) of the Pacific. Because of differential rates of crust generation in the Atlantic and the Pacific, however, the former may eventually expand at the expense of the Pacific, making it shrink in size. Oceans generally have limited lifespans, their expanditures and closures constituting the Wilson cycle (Wilson, JT 1966; Wilson, RW *et al.* 2019). Kearey & Vine once speculated (1996: 80) that the Pacific Ocean will disappear in another 300 million years if the present processes continue undeflected! These dates have since been modified, with the Pacific Ocean lifespan in the Wilson cycle estimated from 700 Ma to 200 million years in the future (Isozaki 2019); other speculations are given at the end of Chapter 3.

2.3.2 Types of Earth's crust

In general the Earth's crust exists today primarily in two forms: oceanic crust generated by igneous processes at mid-oceanic ridges, and continental crust characterizing the major landforms – greatly transformed by igneous processes and metamorphism. Oceanic and continental crusts differ in composition and chemistry, physical properties, and destinies as will be discussed below. Both, however, can be affected by hot spot volcanism such as is producing the Hawaiian Islands on an oceanic plate, and the Columbia River Basalt LIP or Yellowstone on a continental plate. Within continents there are ancient cratons and fossil subduction zones, while oceanic crust is constantly renewed at mid-oceanic ridges or spreading centres and destroyed in subduction zones. The oldest existing Pacific Plate crust at the Bering Strait subduction zone formed about 192 Ma at the mid-oceanic ridge (the East Pacific Rise), but the oceanic crust being currently subducted under northwest Japan is 120 million years old (Honda 2017).

2.3.2.1 Cratons & mobile belts

Most continents have at their hearts ancient cratons surrounded by shields, such as the Canadian Shield or Siberian Shield. The cratons date to the Archaean between >3.7 and 3.5 billion years ago (Azuma *et al.* 2017; Wiemer *et al.* 2018), and are associated with mobile belts of granite-greenstone. Shield crust formed later, in the Palaeozoic. Since there is no consensus when the Wilson Cycle and current plate subduction processes began (Azuma *et al.* 2017), Archaean mobile belts were probably not caused by subduction processes, but these were clearly in operation by the Cambrian Period ca. 540 Ma (Piper 2013) when the Japanese landmass began to accrete soon thereafter (see Chapter 3).

The term 'craton' is often used interchangeably but sometimes hierarchically (in both directions) with 'block', and several such cratons/blocks can be sutured by an orogenic zone. For example, the South China Craton (SCC) is composed of two smaller blocks: the Yangtze Block (often called the Yangtze Craton) and the Cathaysia Block joined together by the Jiangnan Orogenic Zone which marks a fossil subduction zone and an ocean closure (Charvet *et al.* 1996; Yan *et al.* 2019). The North China Craton (NCC), also called the Sino-Korean Craton/Block, also consists of two blocks (Western or Ordos, and Eastern) joined by a Central Orogenic Zone (Kusky *et al.* 2007; Santosh *et al.* 2010). In Chapter 3, because of the now known heterogeneity of the NCC and SCC, they will be referred to instead as 'blocks'.

Orogenic zones are comprised of orogens, representing fossil subduction zones, such as the Altaids (Xiao *et al.* 2003) forming a suture between the Siberian and North China cratons, or the Qinling-Dabie Orogenic Belt suturing the NCC and SCC. The latter belt generally trends from NW to SE following the Qinling-Dabie Mountains through central China, but a right-angle bend continues the suture zone to the NE, occupied by the Sulu Orogen. There is much debate how this suture zone affects the structure of the Korean Peninsula, but it is acknowledged continuing into the Japanese Islands (e.g., Oh & Kusky 2007; Omori & Isozaki 2011). Orogenic zones/suture zones may contain both continental and oceanic crust that became accreted to the cratons; some authors include all crust outside cratons in the category of mobile belts (Yang *et al.* 2015: fig. 1). The Japanese landmass formed off the North/South China blocks and is an integral part of the Nipponides Orogenic Belt (Sengör & Natal'in 1996) stretching from the northern tip of Sakhalin Island to at least Taiwan and perhaps beyond (Isozaki 2019: fig. 7).

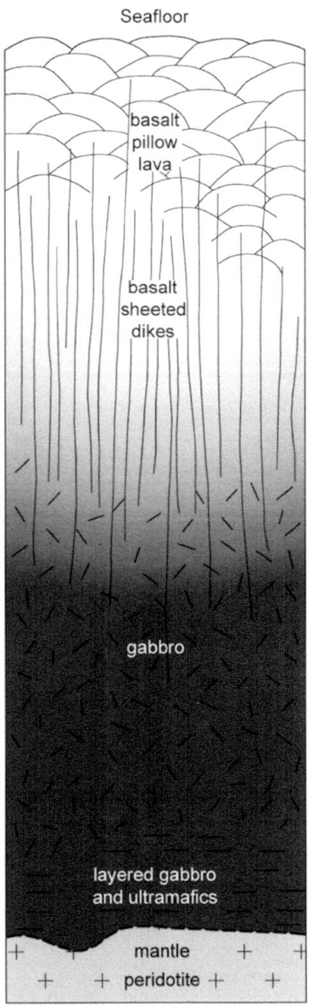

Figure 2.4 Idealized ocean plate stratigraphy (OPS)

2.3.2.2 Oceanic crust

Ocean crust, created at active mid-oceanic ridges, averages 7–10 km thick and has a high density and a specific structure (Figure 2.4), though with considerable modifications of time and place (Tucholke 1998). The ideal stratigraphic sequence of ocean plates is referred to as OPS (Ocean Plate Stratigraphy) or the 'Ophiolite Suite' (see Sec. 2.7). The mantle rock peridotite and its derivatives (lherzolite and dunite) underlie gabbroic magma chambers or mush zones along the ridges; these chambers and mush zones give rise to basaltic dykes, and when the basalt reaches the ocean floor and cold seawater, the basalt is extruded as 'pillow lava' (some look more like sausages or bolsters). The lava cools and is displaced from the ridge by a new rising lava dyke, with each dyke bearing a different magnetic signal by which seafloor spreading was originally recognized (Hess 1962). The chemical signature of Mid-Ocean Ridge Basalt (MORB) is distinct from Ocean Island Basalt (OIB) produced by magma plumes (hot spots), the latter punching through the Earth's oceanic crust to form volcanic islands, as in Hawaii.

Oceanic crust has two chronological dimensions:

a) it is stratified vertically from oldest below (first the igneous rocks, then bedded chert) with the deep-sea (pelagic) and shallow-sea (hemipelagic) sediments accumulated on top during the plate's journey across the ocean basin;

b) concerning MORB, the further away the lower strata are positioned laterally from their point of origin at the mid-oceanic ridge, the older they are going to be and the younger the overlying sediments will be.

Oceanic crust can also be created in rift zones, where continental crust is pulled apart and becomes so thin that mantle magma is able to rise to the surface, forming a oceanic floor spreading centre. The Japan Sea Basin, a typical back-arc basin, has two areas of oceanic crust formed after rifting of the continental edge to form the Japanese archipelago; the rest of the basin floor is thinned continental crust. Back-arc Basin Basalt (BABB) is transitional between MORB and IOB. Much BAB basalt is often found inland, emplaced there by the process of obduction (Sec. 2.7).

The locations of such oceanic crust together with shallow reef limestones on land has been correlated with the distribution of jades across the world (see Chapter 12).

2.3.2.3 Continental crust

Continental crust is less dense than oceanic crust and much thicker. It is primarily composed of light, bouyant igneous rocks as well as metamorphic and sedimentary rocks; its average thickness is 35–40 km. The majority of the crust is formed from medium- to high-silica igneous rocks such as rhyolite and andesite (extruded by volcanic activity) and granite and diorite (plutonic remains in magma chambers).

Appendix 5 provides several charts that illustrate the different elements, minerals, and rocks that go into making up continental and oceanic crust – particularly the different kinds of magma that result in certain rock types.

2.4 Subduction zone processes

Processes in subduction zones can be roughly described via compressional tectonics and extensional tectonics (see Stern 2002: fig. 20). Compression occurs as pressure is exerted from a subducting oceanic slab or a moving continental mass (i.e., the Amurian Plate moving eastwards) or colliding crustal masses (i.e., the Kurile Arc impacting on Hokkaido); the result is often folding and up-faulting of the landmass. Extension occurs when the landmass is thinned, either by rifted stretching and/or magma upwelling. The result is the formation of basins through down-faulting, including the Japan Sea back-arc basin (see Chapter 4).

The 'subduction factory', as it is often called, is the Earth's system for creating, reprocessing, and recycling crustal materials through the combination of "interaction between sinking slab, rising fluids, and moving mantle" (Stern 2002: 3, 18). The following only presents those aspects and processes, in a brief and condensed form, that are important for understanding Japan's subduction zone geology. For a more technical reviews of subduction processes, see Stern (2002) and Byrne *et al.* (2018).

2.4.1 Overview

In the last 450 million years, approximately 15,000 km of oceanic floor comprising several generations of oceanic plates are hypothesized to have passed under the landmass that was to become Japan, most being fully subducted (Kearey & Vine 1996: 160-161; Maruyama 1997: 114). Furthermore, it is estimated

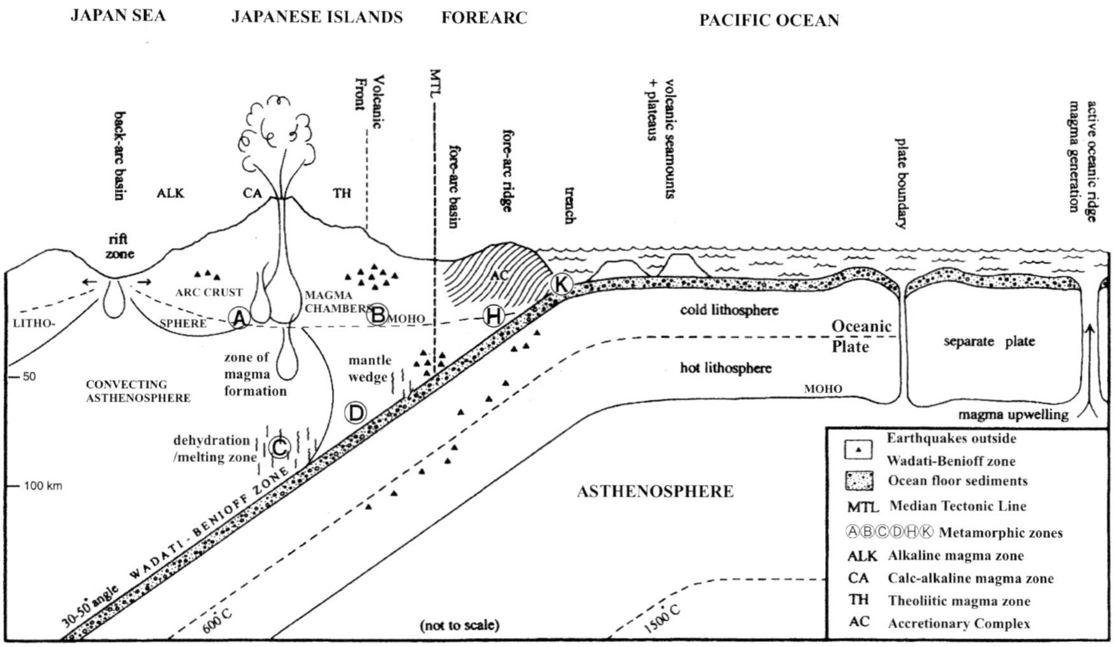

Figure 2.5 The ideal life of an oceanic plate

Shown as it nominally approaches the northern Japanese landmass today, with brief and simplified explanations below. Note the vertical exaggeration of ocean plate thickness.

that there have been a total of five plates and four ridges that have passed under the east Eurasian continental edge throughout geological history (Isozaki 1996).

• At the far right in Figure 2.5, magma upwelling (arrow) at a mid-oceanic ridge forms new basaltic dykes that adhere to both sides of the rift.

• Another hypothetical inactive plate boundary is shown to the left indicating that locations of magma upwelling can and do shift positions through time. There is no break in the Pacific Plate being subducted under the Japanese archipelago today, but such an inactive ridge exists on the Philippine Plate and is subducting under Kyushu Island (cf. Figures 5.1, 5.11). The subduction of an oceanic ridge can trigger lasting geological effects – as we shall see in the Cretaceous-period subduction of the Izanagi (Kula) ridge in Chapter 3.

• Seamounts are volcanoes produced by hotspots or near mid-oceanic ridges and thus can be characterized as OIB and MORB, respectively; large ones can cause difficulties in subduction (Cloos 1993). Two on the Pacific Plate are currently at the edge of the Japan Trench (cf. Figure 3.1), and several on the Philippine Plate are approaching the Ryukyu Trench (cf. Figure 6.11 inset: orange). Such volcanoes become defunct as the supporting oceanic plate moves away from the magma source – be it a hot spot or a mid-oceanic ridge; they then cool, shrink in size, and subside underwater where they are referred to as seamounts. A particularly impressive example is the Emperor Seamount chain in the North Pacific, existing as the tail of the hot spot volcanoes that link to the Hawaiian Island volcanoes. Most seamounts are broken up and subducted when they reach a trench, but some that are crowned with atolls have the limestone reefs shoved (obducted) into the continental edge (Taira 1990). A folded atoll exists on land in western Japan, forming the largest limestone plateau in the country, the Akiyoshidai (see BOX 3). The evolution of reef formation from fringing reef to barrier reef to atoll as the seamount sinks can be seen in Etienne (2012: fig. 2); uplift of atolls can result in limestone cliffs 100 to 150 m high.

• Volanic edifices of non-basaltic, higher silica composition may be so buoyant that they plough into the continental crust instead of being subducted (Seno 1995), as is happening with the Izu Islands in central Honshu (see Chapter 4); however, recent research has documented considerable subduction of buoyant continental crust that was formerly thought unlikely (Isozaki, Aoki & Nakama 2010).

• The subduction trench (Japan has three main ones: the Japan Trench, the Nankai Trough, and Ryukyu Trench, cf. Figure 2.3) fills with eroded continental materials, often flushed from the continental shelf through earthquake action. Among the granitic materials may exist detrital zircons, the toughest, longest lasting of minerals. These become important dating material when found in later geological contexts, e.g., incorporated into an Accretionary Complex (Sec. 2.4.4). New research has also proposed that earthquake-deposited trench fill can be dated and can help understand palaeo-earthquake sequences (Bao *et al.* 2018).

• Subduction does not always entail interaction between continental and oceanic plates but can occur between two oceanic plates. Currently, the Pacific Plate, while drawn into the Japan Trench along the continental Okhotsk (North American) Plate in the north, is also being subducted under the oceanic Philippine Plate in the west in the Izu Trench; simultaneously the north edge of the Philippine Plate is being subducted under the Amurian (Eurasian) Plate to its northwest but on top of the Pacific Plate (see Figure 6.8). These make for very complicated volcanic and seismic conditions in mid-Honshu.

• As the oceanic slab enters the trench, ocean floor sediments (including carbonates, cherts, and silt/clay) as well as terrestrial erosion products in the trench fill are bulldozed up onto the continental edge in the fore-arc region to form accretionary prisms which become part of Accretionary Complexes; slices of ocean crust are also vulnerable to obduction at this point, especially at triple junctions as exist off the coast of Tokyo (cf. Figure 2.3); the latter is unusual in that it involves the juncture of three trenches (TTT), which is inherently unstable.

• The angle of subduction of the plate (cf. Table 6.1) is important: shallowing of the angle may bring up metamorphosed material to the surface (Sec. 2.5.3), and shallow subduction itself is *less likely* to cause magma formation, as in western Honshu (see Chapter 6); steepening of the angle may cause slab rollback and rifting (Sec. 2.8) on the other side of the trench.

• ⒶⒷⒸⒹⒽⓀ The circled letters in Figure 2.5 indicate particular locations where rocks are metamorphosed into characteristic rock/mineral types (see Sec. 2.5 below). In addition, the mantle wedge is metamorphosed via fluids during the initial dehydration of the subducted slab; this soft hydrated mantle rock may then receive much oceanic plate and trench detritus pushed into it to form a serpentinite melangé.

• As the plate descends, earthquakes are generated within the oceanic slab and in the Wadati-Benioff zone. The fore-arc area between the trench and the Volcanic Front lacks volcanoes, but it is subject to heavy earthquake activity from the friction of one plate being dragged down underneath another. The Median Tectonic Line (BOX 2), a major constructional fault (Sec. 2.6.3.2), cleaves the western Japanese archipelago into Inner (continental-side) and Outer (seaward-side) zones.

• The Volcanic Front (VF) forms at a uniform horizontal distance from the deep sea trench doing the subducting (in Japan, usually between 110–200 km) (Isozaki 1997*a*: 17). Its location is determined by the depth of the down-going plate; the current Quaternary arcs in Japan (see Chapter 6) developed over plates at 100–120 km depth (Kimura, J *et al.* 2015).

• Behind the Volcanic Front, a back-arc basin may open up, in Japan's case stimulated by rifting at the continental edge (Sec. 2.8; Chapter 4) and accelerated by trench roll-back (the retreat of the oceanic slab hinge seaward as the slab steepens in subduction). This area of back-arc basin formation behind the volcanic arc is referred to as the 'supra-subduction zone' since it is horizontally far from the trench but still affected by activity in the subduction zone at depth.

2.4.2 Seismic processes

2.4.2.1 Subduction vs Active Fault earthquakes

Subduction zones are only one of four 'active margins' – plate boundaries that can cause earthquakes and do so *within* the subducting oceanic plate and as well as along the Wadati-Benioff zone (cf. Figure 2.5); the other locations are mid-oceanic ridges, transform faults (like the San Andreas fault), and collision zones (like the Himalayas). Earthquakes can also be caused my magma movement leading to volcanic eruptions, while some of the most devastating earthquakes occur in intraplate areas on Active Faults. In general, earthquakes are classified into shallow-zone earthquakes (up to 60 km depth), intermediate (60–300 km), and deep (300–700 km). Below 700 km, the mantle is too viscous for earthquakes, which require rock fracture.

Subduction zones give rise to considerable earthquakes through 1) the cracking of the oceanic crust and bending of the underlying asthenosphere as the oceanic plate is forced under the continental plate; 2) compression of the continental crust as the oceanic plate is thrust against it; and 3) deformation of the oceanic crust as it penetrates deep into the mantle (Kearey *et al.* 2009: 141). A fourth locus is under volcanoes, generated mainly by the propagation and/or inflation of dikes as magma and gases move through rock prior to and during eruptions, and then during relaxation thereafter (Roman & Cashman 2006; Roman *et al.* n.d.) These are called 'volcano-tectonic earthquakes' and usually occur in swarms. The reverse can also occur: earthquakes can activate magma through 'sloshing' (Namiki *et al.* 2016). A fifth process has recently been documented as well: earthquakes result from fluids rising through the mantle wedge, derived from the dehydration of the downgoing slab (White *et al.* 2019).

Earthquakes accompanying the subduction of an oceanic slab are generally generated in a well-defined focal plane in the upper surface of the descending slab (Figure 2.5: Wadati-Benioff zone) but also occur within the bending oceanic slab. Earthquake hypocentres generated in this focal plane may occur from the surface up to ca. 700 km in depth as the slab is subducted; beyond that depth, the mantle is viscous so movement does not produce earthquake shock waves (Condie 1997).

(Palaeo-)seismologists make a clear distinction between earthquakes caused by slab subduction (herein called 'subduction earthquakes', also called 'megathrust earthquakes') and Active Fault earthquakes, which occur in brittle, continental crust – on land or continental shelf areas. Subduction earthquakes occur in *inter*plate-type contexts, while Active Fault earthquakes cause one type of *intra*plate earthquake related to plate boundaries (Scholz 1990: 305; Satake & Atwater 2007). Crustal fault-zone intraplate earthquakes in Japan are caused by various stresses from the interaction of the four or five plates and microplates on which modern Japan rests (cf. Figure 2.3). Intraplate earthquakes may also occur in an oceanic slab as it bends into the subduction zone (Matsu'ura 2017: fig. 5).

Table 2.1 Comparison between subduction and Active Fault earthquakes

	Subduction earthquakes	**Active Fault earthquakes**
location	Wadati-Benioff zone	brittle continental crust
classification	inter-plate	intra-plate
depth	deep: ~700 km	shallow: 10~30 km
strike	>300 km	< 50 km
cyclicity	M8 100~ years	M7 every 1000 years
minimum	< M1	M6.5
maximum	M9.5 historically recorded	?
aftershocks	immediate~years	immediate, but reactived

In contrast to the 'deep' earthquakes generated in the Wadati-Benioff zone, Active Fault earthquakes are 'shallow', with their hypocentres usually located only 10–15 km deep, or at most 30 km at the base of the brittle area of the continental crust, and have a strike less that 50 km long (Scholz 1990: 125, 132; Condie 1997: 53; Altis 1999: 262). Such shallow earthquakes often result in the propagation, through brittle rock failure, of secondary cracks or 'faults' in the earth's surface along which blocks of rock move against each other. Once such faults are formed, they provide a focus for further movement. A significant difference in recurrence time characterizes these two types: interplate earthquakes recur within 100-year cycles, while plate boundary-related inland, intraplate, earthquakes recur in cycles with periodicities generally between 100 and 1000 years (Scholz 1990: 305; ERC n.d.). Active Fault earthquakes may be less frequent – on any one fault – but tend to be more destructive than subduction zone faults when they occur, as the energy released in upper brittle crust is higher (Choy & Kirby 2004). The stress loading is much higher than for subduction earthquakes, so that Active Fault earthquakes are both more infrequent and much stronger. Active Faults are said to produce a minimum earthquake magnitude (M) of 6.5, notated here as *M*=6.5 or M6.5 (Scholz 1990), and they are classified based on their mean slip rates with the most active being Class A (slip rate 1-10m/1000 years) (ERC n.d.). See more of their characteristics below in Section 2.6.3.2.

2.4.2.2 Earthquake magnitude and intensity

Calculating **earthquake magnitude** (size) has been historically variable, instrument- and distance/depth-dependent, and mathematically complex; correlations of various measurements generated from different systems around the globe are complicated (Bormann & Saul 2009). Bormann & Saul report that in Japan, a magnitude (M) measurement has been developed and used by the Japan Meteorological Agency (JMA) that is comparable to the measure of '**seismic moment**' (M_w), therefore $M_{JMA} \approx M_w$. The seismic moment represents the "total non-elastic...deformation in the seismic source volume", i.e., at the rupture hypocentre (Bormann & Saul 2009: 2474). An averaged and corrected moment measure, M_{wp}, is now used routinely in Japan, at the Alaska and Pacific Tsunami Warning Centres (ATWC and PTWC), and at the US National Earthquake Information Center (NEIC) of the USGS. These contrast with other common magnitude measures of **seismic waves**: e.g., M_b (for body waves), M_s (for surface waves), M_m

(for seismic waves in the mantle), and M_L for the old Richter scale only used within 600 km of the recording seismograph, L for 'local' (PNSN n.d.).

Another measure of earthquake magnitude is by released energy (M_e). A study of two earthquakes in Chile in 1997 (Choy & Kirby 2004) found that a subduction earthquake (M_w = 6.9, M_e = 6.1) caused little earthquake damage, contrasting to a crustal fault-based earthquake M_w = 7.1, M_e = 7.6 causing wide-spread damage. Since shaking damage is "mainly controlled by the relative amount of energy released" (Bormann & Saul 2009: 2483), and waves decay over distance and vary by regional geology. To measure regional shaking, more than 4000 Seismic Intensity Meters have been deployed across Japan. Information at each station is automatically transmitted to JMA Headquarters within a minute, and the intensity levels are mapped for the geographical regions affected. The intensity measures vary over distance from the epicentre and geological conditions of the region. These intensities are all mapped immediately after an earthquake together with the time, exact location (Lat/Long), depth of hypocentre, magnitude, and tsunami risk on JMA's real-time reporting on the JMA.go.jp website.

In addition to earthquake magnitude measures, a calculation of intensity of earthquake damage to the environment has been developed. The Environmental Seismic Intensity Scale 2007 (aka ESI 2007) proposes that the primary effects of earthquakes are surface faulting and tectonic uplift or subsidence; secondary effects include landslides, ground cracks, liquefaction, displaced boulders, tsunami, and hydrological anomalies (Guerrieri & Vittori 2007: 7). Detailed descriptions of earthquake effects on the natural environment are keyed to traditional scales such as the Modified Mercalli (MM) or JMA (Japan Meteorological Agency) scales that measure the effects on humans and the built environment.

JMA Seismic Intensity is classified into 10 categories in a Shaking Index that can be correlated with Modified Mercalli Intensity levels (Table 2.2) The JMA website does not identify which type of earthquake is being felt; however, the epicentre location and hypocentre depth may give a clue. Every

Table 2.2 JMA shaking index (cf. Apx 7) correlated with JMA intensity meter readings, Modified Mercalli Intensity

JMA Shaking Index	Intensity metre reading	MMI	Perceived shaking	Damage
0	0–0.4	I	not felt	none
1	0.5–1.4	(I)–II/III	weak	none
2	1.5–2.4	IV	light	none
3	2.5–3.4	V	moderate	very light
4	3.5–4.4	VI	strong	light
5-weak	4.5–4.9	VII	very strong	moderate
5-strong	6.0–5.4	VIII	severe	moderately heavy
6-weak	5.5–5.9	IX	violent	heavy
6-strong	6.0–6.4	X	extreme	very heavy
7	6.5<	XI~		

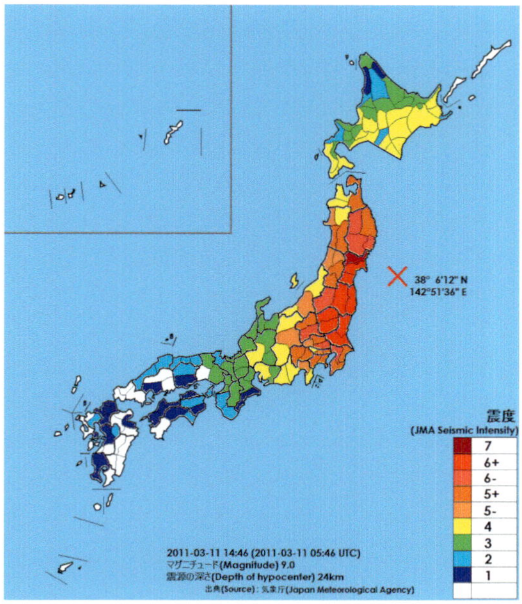

Figure 2.6 Intensity of shaking during the 2011 Tohoku-oki earthquake. X = earthquake epicentre. See Appendix 7 for illustrations of shaking intensity.

month there are more than 10,000 earthquakes in or around Japan at levels of ≥M1.5 (Zhao 2015). Figure 2.6 illustrates the intensity of shaking throughout the Japanese Islands during the Tohoku-oki subduction earthquake of 2011.

2.4.3 Igneous processes

During subduction, due to greater heat in the deep Earth, the cold oceanic slab dehydrates as it is subducted, giving off water vapour and carbon dioxide into the overlying mantle. These lower the melting temperature of the rock-forming minerals in the mantle and causes the rock to melt into magma (Middlemost 1985: 12; Kawamoto *et al.* 2013). As the magma rises to the surface through the continental crust, it is enhanced with silicates derived from the country rock, acquiring a radically different chemical signature than MORB or OIB. The magma is extruded through volcanic ducts as lava and various forms of tephra (including volcanic ash), while molten material remaining in the magma chamber crystallizes as plutonic rock.

When many magma chambers containing granite become welded into a single granitic basement, the unit is called a batholith (Brown, GC 1979). In Japan, the very early batholiths have all been eroded away except for current small exposures (Nakama *et al.* 2010): 1161-1172), but some serpentinite mélanges contain granite fragments dating between 450 Ma and 250 Ma (Suzuki *et al.* 2010: fig. 5). The Cretaceous batholiths, dating to 110 to 90 Ma and 80 to 60 Ma still exist in western Honshu (see Isozaki, Aoki & Nakama 2010: fig. 6 cont.). Uplifted, batholiths can form substantial granite mountain ranges; the Rocky Mountains are home to several batholiths named after their locations, while the Ryoke Belt of western Japan is a batholith created at the edge of the Eurasian continent before rifting from the archipelago (cf. Figure 3.7: inset); it forms the backbone geology of western Honshu, affecting the types of high-silica sediments and low-pressure/high-temperature metamorphism (Sec. 2.5.1) characterizing the region. A list of basic granite exposures in Japan can be found in Appendix 10.

In addition to magmas that develop within chambers, there are also magma mushes that underlie mid-oceanic spreading ridges, and hot-spot magmas that form plumes emanating directly from the mantle. Archaeologists may not always distinguish between subduction and hot spot volcanoes (cf. Cronin *et al.* 2008), but their effects can be radically different. Magma hot spots are responsible for many intraplate slow lava eruptions creating shield volcanoes (the Hawaiian and Reunion Islands), volcanic activity such as found in Yellowstone, and LIPs that can occur on the ocean floor (Ontong Java, Izanagi Plateau) or in continental settings (Siberian and Deccan Traps). They are also thought responsible for the break-up of some supercontinents and perhaps for the opening of several basins in the East Asian area (cf. Chapter 4).

Hot spots have been found to be associated with magma plumes generated from ancient subducted oceanic slabs that lie in the 'slab graveyard' at the core/mantle boundary. There are two primary graveyards visible in global tomography, named after prominent geologists, 'Tuzo' (for J. Tuzo Wilson) under Africa and 'Jason' (for Jason Morgan) under the Pacific! Hot spots are generally located towards the edges of these large slab graveyards (Torsvick & Cocks 2016: fig. 2.2). See Domeier *et al.* (2016); Voosen (2016); Niu (2018), and references therein.

There are more kinds of igneous rock than any other – sedimentary or metamorphic – and strangely, one kind of rock with an igneous origin can be classified as a sedimentary rock: volcanic ash (tephra) which lithifies as tuff (or weathers into clay). Tuff is only produced from volcanoes with high-silica magma; such volcanoes are extremely explosive in nature and distribute their products effusively over large areas. Beyond this sedimentary anomaly, there are many ways to classify igneous rocks: by texture, by silica content, by alkali/alkaline element content, etc. (Le Maitre 2002). These three are the main measures and will be given a brief description below.

In the northern Kanto region, the Oya Tuff (BOX 4) is a renowned building material, while in the Noto Peninsula, prehistoric villagers were known producers of ornaments and objects carved of green tuff (cf. Figure 4.5).

The silica scale, measured by SiO_2 content (see Apx 5: Table B), is the most common classifier of igneous rocks: from very high silica content (felsic) to intermediate (andesitic) to low (mafic) and to very low (ultramafic).[2] All of these can occur as either plutonic rocks or volcanic rocks as classified by mineral size (coarse/fine), depending on their circumstances of exposure and cooling. Crystals grow in a certain order called the Bowen series (Apx 5: Fig. B). Plutonic rocks consist of magma that cools inside the Earth, resulting in large, coarse crystal growth. Volcanic rocks are fine-grained rocks that are extruded into the Earth's atmosphere or oceans, either as lava or tephra – the latter conditional on the magma containing a lot of volatiles (gases) that make it explosive under decompression. Rocks with the same chemical composition on the silica scale can thus occur in textural pairs (Apx 5: Table B and Fig. F): felsic rhyolite (volcanic) and granite (plutonic), intermediate andesite (volcanic) and diorite (plutonic), mafic basalt (volcanic) and gabbro (plutonic), and ultramafic peridotite (volcanic) and komatiite (plutonic).

Volcanic rocks can be further divided based on silica vs alkali (Na_2O+K_2O) contents (Apx 5: Fig. G). Alkaline products are characteristic of the rifting volcanics discussed in Chapter 4, and occur in zone 4 of the Volcanic Fronts (cf. Figure 6.2). They are distinguished by containing alkali metals (K, Na) that are usually absent in other volcanic rocks. Calc-alkaline rocks – a subdivision of Sub-Alkaline rocks – are characteristic of subduction volcanism, discussed in Chapter 6, and form most of the volcanic rocks in Japan today.

This scheme shown in Appendix 5: Fig. G is the most simplistic for igneous rocks, whereas in reality there are hundreds if not thousands of igneous rocks – particularly granites and volcanic rocks – that can be identified by differences in chemical composition, etc.

2.4.4 Accretionary orogens

The term 'accretion' simply means 'to add to', and it has been used in a variety of ways in the geological literature; but an 'accretionary orogen' was defined as distinct from the typical Alps type of 'collisional orogen' in 1992 (Windley 1992), in that crustal growth was achieved through accretion of oceanic materials to continental edges, not collision between continental masses. Such accretionary orogens are usually found in 'mobile belts' (Yuan *et al.* 2009); however, a distinction has been made between extensional accretion, as in Japan, and compressional accretion as in the Andes (Cawood *et al.* 2009). The process of the accretionary prism (cf. Figure 2.7)[3] developing into an Accretionary Complex (AC), as the

[2] Felsic rocks used to be called 'acid' due to a mistaken understanding "based on the idea that high-silica magmas were acidic [low pH] and that low-silica magmas were basic [high pH]" (Camp n.d.: unpg., brackets added). Both 'acid' and 'basic' are now obsolete terms, replaced by felsic and mafic. The SDSU website is very useful visually for various aspects of volcanic formations and eruptions.

[3] Though some refer to accretionary prisms as 'wedges', the use of 'wedge' here is confined to 'mantle wedge' to avoid confusion (cf. Figure 1.5).

term is used in Japan, is most important for understanding Japanese geology. The basement of the archipelago is almost wholly made up of AC dating from different periods of continental edge accretionary processes and then shot through by igneous intrusions (see Chapter 3).

The importance of accretionary tectonics cannot be overstated, for these views totally overturn previous theories of the formation of the Japanese Islands (Taira, Kiyokawa & Aoike 1997). Although alien blocks of land (smaller than the usual 'exotic terranes') such as ocean floor volcanic seamounts and plateaus have been incorporated into the Japanese landmass, they are by far a minor component of the geological setting. However, a distinction must be made between accretionary tectonics as the means of forming the dominant geological belts that comprise the Japanese Islands and the chain of causation that allowed accretionary processes to occur. For accretion to take place, the subducting trench must have a sufficient amount of eroded sediments from the landmass itself to be bulldozed up again into the accretionary prism. Uplift of the continental mass from subduction processes is not enough in and of itself to cause sufficient erosion – instead, it is postulated that subducting a mid-oceanic ridge provides sufficient uplift. It is also argued that uplift stimulates magmatic movement, which creates more continental crust through granitic and volcanic emplacement; these new igneous materials, primarily granite, are then eroded and fed to the trench (Maruyama 1997). Once formed, AC are commonly weakly or strongly metamorphosed, completing the accretionary cycle.

AC primarily incorporate trench sediments and other material from the subducting oceanic plate (seamounts, ocean floor slices) as well as serpentintite mélanges. Worldwide, the survival of Accretionary Complexes is rare (cf. Figure 2.1; see also Stern 2002: fig. 16). The more common process is the loss of continental and oceanic crust through tectonic erosion during subduction (Maruyama *et al.* 2011; Tsujimori 2010; Ueda 2010; Yamamoto 2010; Isozaki 2019). ACs are estimated to have formed during one third of the time since subduction began, while erosion occurs during the other two thirds (Suzuki *et al.* 2010); today, 75% of subduction zones around the world are eroding – including the Pacific coast of Northeast Japan – while 25% are accreting, particularly in Southwest Japan (Yamamoto 2010). SWJapan is therefore a valuable resource for studying ACs as phenomena of crustal growth, especially in a world where continental crust is now decreasing in mass (Yamamoto 2010). A modelled visualization of the creation of an accretionary prism can be seen at (ATGL 2013).

Detrital zircon datings reveal seven periods of surface erosion of the NCC /NCB and SCC/SCB whose materials were eventually recycled into the proto-Japan area: 2.5–1.0 Ga (Palaeo- to Meso-proterozoic); 1.0 Ga to 800 Ma (Neoproterozoic); 520 to 400 Ma (Cambrian–Silurian); 280 to 210 Ma (Permian–Triassic); || 190 to 160 Ma (Jurassic); 110 to 90 Ma (mid-Cretaceous); || and 80 to 60 Ma (Late Cretaceous–Paleogene) (Isozaki, Aoki & Nakama 2010). The double lines || here indicate when there was a major change in the source of the erosional materials: the first at 200 Ma when the North and South China cratons collided, and then at ca. 80 Ma after the formation of the Cretaceous granite batholith-volcanic belt, which kept earlier continental materials from flowing into the proto-Japan area offshore.

ACs which do not survive are subject to tectonic erosion from the over-riding plate by the subducting plate – along with granitic continental crust – to form 'second continents'.[4] These materials are taken down into the mantle where they rest at the base of the Mantle Transition Zone (MTZ) between 520 and

[4] These granitic 'second continents' are of different source and composition (felsic) than the 'slab graveyards' (mafic) at the core-mantle boundary; the former sit at the ca. 660 km deep, while the latter are near 1000 km deep. Fukao & Obayashi (2013), however, have identified four stages of subducted slabs at different depths.

660 km deep (Senshu *et al.* 2010; Kawai *et al.* 2013; Zhao *et al.* 2017). It is estimated that five to seven times the amount of continental crust that exists on the Earth's surface is contained in these submerged second continents (Kawai *et al.* 2010). Isozaki, Maruyama & Nakama propose that accretion and erosion along the continental edge occurred alternately and repeatedly (2010: 95-100).

Tectonic erosion refers to the grinding away of the continental edge by a subducting oceanic plate, destroying existing landmass and not allowing accretionary processes to form new accretionary prisms (cf. Figure 2.1: gold). Moreover, all the tectonically eroded components from the landmass, including granitic intrusions, are themselves then subject to second continent formation or emplacement into a serpentinite mélange. Although the present-day Japanese Islands have only existed as islands since ca. 15 million years ago, it now appears that at least five series of volcanic arcs, defined by granite basement batholiths or their surviving fragments have been created at the Eurasian continental edge in the approximate position that the Japanese landmass has occupied (Suzuki *et al.* 2010). Some of these 'proto-Japan' arcs, now known as the Maizuru and Kurosegawa Belts (see Chapter 3), have existed as continental arcs and others as archipelagos similar to modern Japan (Isozaki, Maruyama & Nakama 2010): 102, fig. 12-c). Three previous arcs have been subducted and lost to second continent formation, except for blocks of granite and metamorphic rocks incorporated into – and therefore surviving in – existing serpentinite mélanges that have been exhumed (Isozaki, Maruyama & Aoki 2010; Isozaki 2019).

It has become apparent that the major crustal component of the Japanese landmass has been formed by accretion tectonics acting in an episodic fashion (Isozaki 1996: 289). Great volumes of eroded sediments from the Eurasian continent were periodically washed into the off-shore subduction trenches, but instead of the sediments being subducted, they were and still are bulldozed back onto the continental shelf together with debris off-scraped from the ocean floor as it is dragged under. This directly refutes the earlier proposition that "major volumes of continental erosion are not involved in subduction and recycling" (Brown, GC 1979: 111).

Accretion tectonics incorporate three processes as described by Maruyama (1997: 103): 1) shallow offscraping of trench sediments into an accretionary prism, 2) medium-depth accretion of sediments and oceanic floor, and 3) underplating of the mantle wedge with sediments and oceanic floor at considerable depth. Taira (1990: 193) has likened the off-scraping process to 'bulldozing' the trench sediments into the accretionary prism to form a series of imbricated tectonic units separated by fault lines (Figure 2.7; see this figure's geographical location in Figure 3.1b). The tectonic units of an accretionary prism seems to form a finite size so that successive units form underneath each other; this results in the counter-intuitive situation where the earlier units occur farther inland and stratigraphically on top of the younger units added later; this ordering is termed 'downward-younging polarity' (Isozaki 1996: 307).

The formation of an accretionary prism is only the first step in the formation of an AC, which contains far more material than just trench sediments (see photographs in Taira, Byrne & Ashi 1992). AC "are built up by a series of collisional and accretionary orogenies, including arc-arc and arc-continent collision, and the accretion of forearc and backarc subduction complexes, tectonic collages of oceanic plateaus/islands/seamounts, and exotic blocks (terranes)" (Yuan *et al.* 2009: 31). Accretion is not continuous but episodic; thus AC are individually distinct and separable because of long time-gaps between them.

Yuan *et al.* also state that "accretionary orogenesis is the fundamental process of continental evolution throughout Earth's history" (Yuan *et al.* 2009: 31). Maruyama estimates that accretionary processes have contributed to the growth of the Eurasian continental edge by ca. 400 km oceanwards in the last 450 million years (Maruyama 1997: 114). The last great Accretionary Complex in Japan, the Shimanto Belt, was formed in the Cretaceous–Early Miocene (Taira *et al.* 1982), but new accretion began in the Pliocene (Kimura, G. *et al.* 2018) and is forming the Nankai Prism (Taira, Hill & Firth 1992; Moore 2011).

Mid-oceanic ridge subduction is one of the most extreme events possible. Cold ridge subduction happens periodically when a ridge (marking the boundary between oceanic plates) becomes becomes separated from the magmatic fissure of new crust formation and is pushed along with its old plate as a new plate forms behind it from a new mid-oceanic ridge, as seen hypothetically in Figure 2.5. A hot ridge still or recently producing magma can cause much granite formation in the overlying crust. If the ridge is being subducted at an angle (obliquely) to the coastal edge, the magma produces laterally continuous batholiths. Conspicuous granite production within certain dates is thus a clue that a mid-oceanic ridge may have been subducted in the past (Tsuchiya *et al.* 2014). In addition, high P/T metamorphic belts may undergo exhumation during ridge subduction (Isozaki, Maruyama & Aoki 2010) (Sec. 2.5.3).

Increased rates of accretion during ridge subduction are not due to the incorporation of ocean floor so much as to the sudden availability of eroded sediments from the landmass itself. It is theorized that the

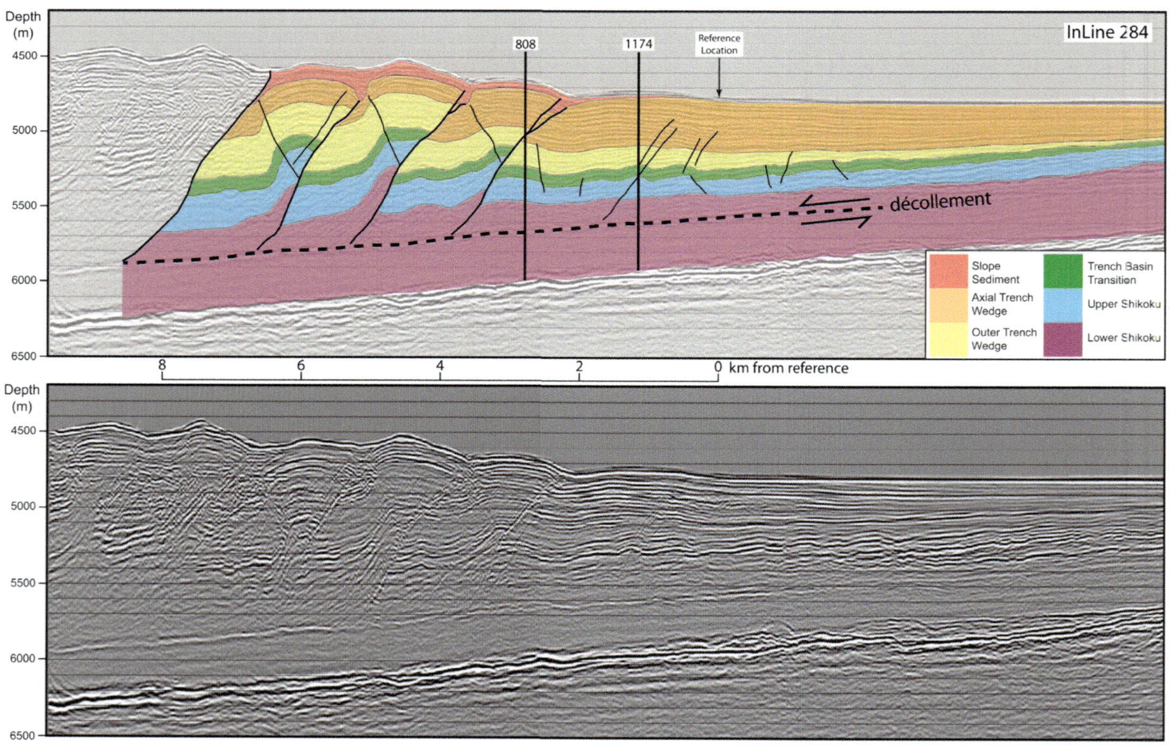

Figure 2.7 Seismic profile (lower) and interpretation (upper) of the Nankai Prism in southwest Japan. The trench sediments (gold & yellow & green) are bulldozed under the continental shelf (orange). The Shikoku Basin layers of the Philippine Plate (blue & purple) above the décollement fault running through of the ocean floor are being accreted, while those below (purple) are being subducted. Tectonic units are bound by 'backthrust' faults, interpreted here via black lines running from upper right to lower left.

subduction of oceanic ridges tends to cause uplift of the continental landmass and volcanic activity as the ridge passes into the trench and under the continental edge, as has and is happening currently with the Kyushu-Palau Ridge subducting under Kyushu (cf. Figures 5.1, 5.11). The increased height of the landmass is due not only to the raised height of the interplate ridge but to the buoyancy of hot, young oceanic crust; the corresponding uplift stimulates magma production and the formation of granitic intrusions (Maruyama 1997: 91). Such uplift also increases erosion processes, by which much terrestrial (terrigenous) material of granitic origin is washed into the ocean. The subduction trench tends to serve as a sump for these sediments eroded off the landmass, which are then bulldozed back up onto the continental shelf through accretionary processes dubbed "semi-autocannibalistic-material recycling" (Isozaki 1997*a*: 19). The time-gaps between AC thus result from the episodic decrease in accretion rate and volume caused especially during tectonic quiet times after mid-oceanic ridges are subducted into trenches.

The recognition of the integral relationships among ridge subduction, continental edge uplift, subsequent erosion, and increased accretion has resulted in the understanding that in times of uplift, the continental edge will grow proportionally to the amount of terrestrial sub-aerial erosion in an accretionary setting, as opposed to the erosion of the continental edge itself in an tectonic erosion setting. This recycling of terrestrial materials in an accretionary setting accounts for a large proportion of the Accretionary Complexes, and by implication, it makes for an extremely complicated geology. In Japan, the oceanic components (reef limestone, serpentinite/gabbroic/basaltic greenstones, bedded chert) make up only an estimated 30% of an Accretionary Complex (Isozaki 1997*a*: 16), with the majority comprised of eroded sediments from terrestrial sources bulldozed back up on land from the trench and intruded by later granites. Moreover, it is known that once an AC is in place, it can be deformed and reformed so that up to 70% of the materials in such a complex may be chaotically organized as a 'mélange', as known in the Jurassic AC belts, rather than representing the primary accreted stratigraphy (Isozaki 1997*b*: 43) (see Chapter 3).

Two advances in the research of AC are noted by Isozaki (Isozaki, Maruyama & Yanai 2010) as new exportable techniques from Japan. The first involves new dating of detrital zircon crystals in Accretionary Complexes to reconstruct the subduction record (Isozaki, Maruyama & Aoki 2010; Nakama *et al.* 2010; Isozaki 2019), using U-Pb (radioactive uranium decaying into stable lead) and LA-ICP-MS (Laser Ablation Inductively Coupled Plasma Mass Spectrometry). These methods are applied to detrital zircons residing in the AC protolith (that is, AC sediments or sedimentary rocks before being metamorphosed) and in igneous rock fragments incorporated into the AC; the detrital zircons derive from eroded granite bodies and sandstones on the continent that have been recycled into these AC. By investigating where they came from by matching their dates with known dates of rock bodies on the continent, the nature of the inner continent behind the subduction zone where the AC was forming can be investigated. These methods can also be applied to continental fragments, as well as AC, in the Japanese landmass (see Chapter 3).

A second export involves using seismic tomography to reconstruct both second continent regimes and the three-dimensional nature of Accretionary Complexes (AC) and Metamorphic Belts (MB) (Isozaki, Maruyama & Yanai 2010; Kawai *et al.* 2013). Germane to this is the use of seismic tomography to establish the importance of faults according to their structural roles, with geotectonic unit boundaries taking priority over other faults which are perhaps more recent, more visible, and account for current re-activities rather than fundamental geotectonic arrangements.

2.4.5 Collision tectonics

Subduction at land's edge is often followed by collision when the oceanic plate being subducted is pulling along behind it another continental landmass. The ocean between the two continental masses is then in the processes of closing, part of the Wilson Cycle of ocean opening and closing on a 500 myr cycle (Wilson, JT 1966; Wilson, RW *et al.* 2019). When the oceanic plate between two continental landmasses has been consumed through subduction and the continents are brought into crushing contact with each other, large-scale mountain folding and uplift can occur – witness the Himalayas as the Indian subcontinent smashes into Eurasia. Although most geological processes in Japan are due to subduction tectonics, there are several small-scale examples of collision tectonics within the landmass. Nevertheless, Japan as a whole is not the product of large-scale collision tectonics, as suggested in the general Japanology literature (e.g., Totman 2000: 13) or as asserted by Alpine geologists such as Charvet (2013). The Japanese landmass was primarily built by accretion of oceanic materials then riddled with magmatic instrusions.

If the oceanic plate is carrying an island arc on it that is positioned parallel to the trench, when the arc reaches the subduction zone, it may be accreted to the continent rather than subducted. This results in an elongate belt of new material added to the continental edge that is not an Accretionary Complex. These have traditionally been treated as 'exotic terranes' (Howell 1995), such as occur in profusion along the northwestern coast of North America (Colpron *et al.* 2007), the margins of China (e.g., Xiao *et al.* 2004; Zhou *et al.* 2006; Ma *et al.* 2018), and even appear in Japan as the South Kitakami Terrane (e.g., Aitchison *et al.* 1991; Takagi & Arai 2003). Such exotic terranes must be distinguished from other indigenous geotectonic terranes that have been developed in place but are now bounded by faults; both types are present in Japan and are discussed in Chapter 3.

2.4.6 Paired metamorphic belts

Miyashiro's work (1973*a*, 1973*b*) on paired metamorphic belts (MB) in Japan led him to propose that such pairs were diagnostic of all subduction zones. The paired metamorphic belts in Japan are adjacent and differentially metamorphosed: one resulting from low-pressure metamorphism around granitic intrusions (Ryoke) and one resulting from high-pressure metamorphism in the subduction channel (Sanbagawa) (see Chapter 3). Miyashiro used the term 'regional metamorphism' to describe this phenomenon, pointing to the subduction zone as the 'region', which was accurate in the sense that both weak and strong metamorphism were products of subduction zone processes. However, he was mistaken in other assumptions as detailed below. Since Miyashiro's work, regional high-pressure MB have been discovered also in collision zones but not paired to a weak low-pressure MB as in subduction zones (Maruyama *et al.* 2010).

The Ryoke MB is now considered a case of 'contact metamorphism', while the Sanbagawa MB is a product of 'regional metamorphism', trajectories illustrated in Figure 2.8 and defined below. The first develops at low-pressure levels from contact of intrusive plutonic bodies with country rock, activated by high-temperature magma formation from the subducting slab, while the latter results from high-pressures when sediments and rock are drawn down in the subduction channel. However, why these two belts are positioned so close together in SW Japan, separated only by the palaeo-MTL (cf. Figure 3.1b) is a conundrum. In other subduction zone settings – in northwest North America and the Andes – the low-P and high-P MB are spaced further apart. Wallis & Okudaira (2016) review several models that might account for their juxtaposition in Japan, though nothing definitive is yet known. The possibilities are reviewed in more detail in Chapter 3: Cretaceous.

2.5 Metamorphism

2.5.1 Metamorphic facies & facies series

The strengths of metamorphism in terms of pressure (P) and temperature (T) are representated in terms of metamorphic facies (Figure 2.8). When rocks are newly subjected to these various P/T conditions, their minerals experience solid-state recrystallization under dehydration to produce minerals stable at those new P/T conditions. The mineral suite and relict textures belonging to the unmetamorphosed parent rock (protolith) can reveal whether that parent rock was sedimentary or igneous; both whole minerals and individual elements can be inherited from the parent rock. Moreover, rocks may be subject to other processes (oxidation, graphitization, metasomatism) that change their mineral assemblages and mineral compositions. In all types of metamorphism, successions of new minerals form competition with each other, depending on the composition of the protolith, the sequential changes in pressure, temperature, and fluid composition – with available elements being exchanged or taken up as needed.

The trajectories (metamorphic gradients) idealized in Figure 2.8 are often referred to by their main P/T characterization and major facies (e.g., low P/high T = hornfels), but specific discussions involve critical mineral transformations. The trajectories imply a succession of facies in which minerals form along those lines at increasing temperature and pressure, progressive metamorphism resulting in the zonation of minerals in the facies series (Table 2.3). However, the trajectories shown are simplifications, and the rocks may take more winding paths in reality, including divergent return paths upon exhumation. The types of minerals formed in the different facies vary with the composition of the rock being metamorphosed. Two well-known sequences are the Barrovian for sedimentary protoliths and the Buchan for basaltic protoliths, but each type of rock (e.g., sedimentary, calcareous, mafic, felsic) entails its own sequence of minerals in each of the facies, resulting in a proliferation of mineral successions (see facies names in Wikipedia for simple descriptions and mineral contents or Strekeisen 2020 for details).

Rocks that have been metamorphosed are sometimes prefaced with 'meta-', e.g., meta-sedimentary, which reveals the protolith but not the grade or facies of metamorphism achieved. It should also be noted that the types of rocks being metamorphosed may vary both laterally and vertically within an MB; moreover, not all of an MB exists within one facies at any point in time and therefore the whole will retain different facies minerals (and their transformations) upon exhumation. A complicating factor is that during exhumation of the metamorphic belt, the minerals again undergo transformation according to temperature, depth, and increasing re-hydration (retrograde metamorphism), therefore the surface manifestation of the belt does not represent the deepest burial metamorphism which survives only if exhumation is very fast. This is one reason why jadeitite sources are so rare, as they are formed in deep burial but rarely exhumed quickly. Mineral zonation often records and preserves successive P/T changes; Maruyama *et al.* (2010) cite garnet as an ideal mineral for investigating such zoned records – while noting that the Barrovian sequence is essentially an exhumed stage retrograde metamorphic sequence, not one of progressive metamorphism. They also emphasize the role of re-hydration during exhumation, with fluid influx as a major contributor to mineral recrystallization.

A further classification distinguishes low- intermediate- and high-grade metamorphism whose boundaries cross-cut the facies (Figure 2.8). Note that the contents of each grade are different according to which trajectory a particular rock is on, and the starting and ending P/T conditions for each grade are different on each trajectory.

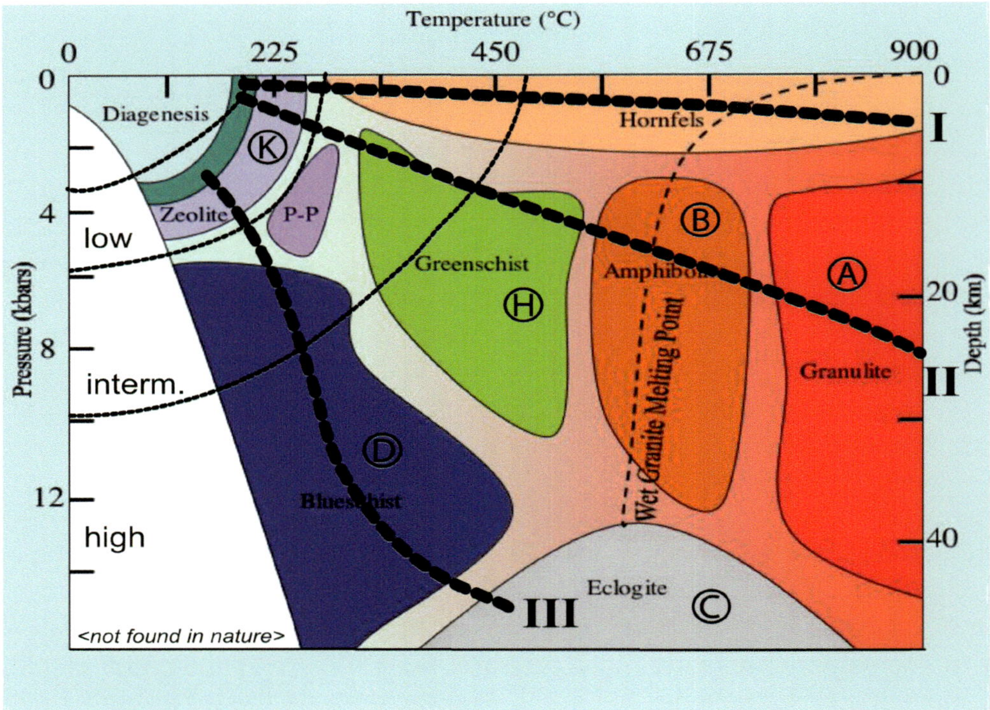

Figure 2.8 Metamorphic facies, facies series, and metamorphic grades

The top left corner of the graph represents the Earth's surface, at nominal P/T = 0/0. Three trajectories along different P/T gradients are indicated by I, II, and III, occurring in specific tectonic contexts: I in shallow contact metamorphism and rift zones, II in regional burial, and III in subduction channels. These trajectories pass through named metamorphic 'facies', (e.g., greenschist facies, P-P = prehnite-pumpellyite facies), forming serial transformations. These P/T conditions cause different minerals characteristic of each facies to form during metamorphism depending on the composition of the rock being metamorphosed. A different categorization indicates the degree of metamorphism beginning during diagenesis: low-grade, intermediate-grade, and high-grade metamorphism. The facies with circled letters can be coordinated to their locus of formation via Figure 2.5.

Table 2.3 Trajectories of metamorphic facies

Metamorphic gradient	**Major facies**	**Pressure/ Temperature**	**Named mineral succession zones**
I	hornfels	**low** P/high T	Buchan (mafic): biotite>cordierite>andalusite>sillimanite
II	greenschist	**intermediate** P/T	Barrovian (sedimentary): chlorite>biotite>garnet>staurolite>kyanite>sillimanite
III	blueschist	**high** P/low T, aka high P/T	Franciscan (incredible numbers of mineral successions in each facies as formed from different kinds of protoliths)

2.5.2 Types of metamorphism

Best lists, in general, five types of metamorphic conditions and six types of metamorphic terranes (Best, 2003: secs. 14.2.2, 14.2.3). The following descriptions focus on those combinations that commonly affect subduction zones such as Japan. It is estimated that metamorphic rocks comprise approximately 60%

of continental regions (Philpotts & Ague 2009: 414); however, the figure for Japan is much lower at less than 5% (Tanaka & Nozawa 1977: fig. 2.1). Metamorphic rocks in Japan currently occur in four general contexts: as geological belts, as metamorphic zones around granite intrusions, as metasomatized oceanic floor, and as serpentite mélanges.

Several types of metamorphism concern us here, described below. The major difference is between ***solid-state metamorphism***, involving mineral recrystallization in rock affected by pressure and temperature without melting except at very high temperatures, and ***metasomatism***, involving chemical change primarily via reaction with fluids (dissolution and precipitation). Although designated as a separate class, metasomatism is now recognised to occur in "virtually all rocks" (Newton 2014: 155) and can accompany other types of metamorphism to greater or lesser extents, as noted above for retrograde metamorphism. For this reason, metasomatism is described first below.

Metasomatism is a class of metamorphism that has been sadly neglected for the first half of the 20c but is "roaring back into the vocabulary of petrology", since 1958 to be exact (Nelson 2011: unpg.). Metasomatism warrants only four pages in *Igneous and Metamorphic Petrology* (Best 2003), scattered references in *Principles of igneous and metamorphic petrology* (Philpotts & Ague 2009), and one page in *Earth Materials* (Klein & Philpotts 2013), for example; but an 806-page book, *Metasomatism and the Chemical Transformation of Rock*, is now available on that topic alone (Harlov & Austrheim 2013). Metasomatism works via fluid transport and chemical replacement, i.e., the chemical reaction and exchanges of elements between the fluids and minerals existing in rocks. Fluids can be of various origins: e.g., circulating seawater, dehydration processes, volcanic–hydrothermal venting, or fluids generated in fault zones. Two important types of metasomatism to be considered below are contact metasomatism and serpentization.

Contact metamorphism (Trajectory A in Figure 2.8) occurs when an igneous body (magma) intrudes into the upper crust and causes chemical changes in country rock, primarily through heating. Heat from the igneous intrusion bakes the nearby rocks, causing mineral recrystallization to rocks with a hornfels texture and resulting in a local aureole of a graded series of metamorphic mineral types (albite/epidote > hornblende > pyroxene > sanidinite) around the intrusion (see Philpotts & Ague 2009: fig. 16.1). The Ryoke Belt in Japan exhibits typical contact metamorphism of a Jurassic Accretionary Complex (the Mino-Tanba) from Cretaceous igneous batholith emplacement underneath.

Contact metasomatism occurs when an igneous intrusion not only bakes the country rock but fluids are also released into the country rock which will change its mineralization. Classic skarn, for example, is a metasomatic rock accompanying igneous intrusions into carbonate rock; it contains many mineral progressions and element deposits (such as gold, copper, iron, zinc, nickel).

This form of metasomatism is credited with producing most of China's nephrite that is derived from granitic intrusions into dolomitic marble as the country rock (see Chapter 12).

Cataclastic (dynamic) metamorphism occurs in fault zones where rocks grind against each other; the friction causes mineral recrystallization, and fluids are often generated, causing some metasomatic changes.

In China, both Luanchuan and Dushan false jades are products of cataclastic metamorphism involving fluids.

Regional metamorphism

Trajectory B in Figure 2.8 occurs on a large regional scale where relatively warm rocks and sediments have been drawn down into the subduction channel and subsequently unearthed (exhumed). Two types of metamorphic belts have been identified (Maruyama *et al.* 2010: see figs. 3a,b, 25):

1) in oceanic-plate subducting channels, where the materials being metamorphosed are trench and oceanic plate rocks and sediments; these are subjected to temperatures up to 700°C and pressures to ca. 23 kbar. The Sanbagawa Belt in Japan is the type example of this kind of intermediate metamorphism, paired with the Ryoke Belt to form the paired metamorphic belts characteristic of subduction zones.

2) in collision zones where oceanic plate subduction has finished and one continent is being drawn under the other (e.g., in the Eurasian–Indian plate collision); in this case, the materials are sediments and rocks on the subducting continent, subjected to temperatures up to 900°C and 32 kbar. Such metamorphic belts in collision zone occur singly, not paired.

In both these types, the metamorphic belt is like a thin skin that is sandwiched between weakly metamorphosed layers, with a thickness and width ratio of 1:100 but which can extend for hundreds of miles in an subduction zone all along the subduction plane (Maruyama *et al.* 2010: fig. 12). How these belts are unearthed is controversial, but all acknowledge that the path taken downwards through the metmorphic facies is not the same path that the rocks follow on the way to the surface. Whether that path is taken clockwise or counter-clockwise is one point of debate; another point is to what degree fluids influence the minerals formed and then reformed (Maruyama *et al.* 2010).

Trajectory C in Figure 2.8 occurs when a cold oceanic plate is drawn down into the subduction zone to deep levels (>15 km); blueschist and eclogite facies mineral formation is dependent on high pressures (>7 kbar) and low temperatures (<550°C).

It is within the blueschist and eclogite facies that jadeite forms (cf. Chapter 12). Jadeitite resources throughout the world are associated with blueschist facies metmorphic belts (see Harlow et al. 2014: fig. 10).

Serpentinization: a form of metasomatism where water (sea water or meteoritic water) intrudes into ultramafic and some mafic rocks – usually peridotite, the main rock type of the Earth's upper mantle, and gabbro and basalt, major components of the oceanic crust.

Mantle peridotites are serpentinized with the addition of water, which reacts with the anhydrous silicates to produce new minerals, mainly in the serpentine series, and which expands the rock's volume. This can occur via fractures in the oceanic crust or as water is drawn into the subduction channel. The minerals that form during serpentinization are serpentine and its polymorphs antigorite, lizardite, and/or chrysotile (the last widely known as the major asbestos mineral)[5] (Malpas 1992: 8), with the general chemical formula $(Mg,Fe)_3(Si)_2O_5(OH)_4$ (Friedman 1997-2020). The serpentines are relatively soft minerals, ranging from 2.5 to 3.5 on the 'Mohs scale'. Serpentinite itself and the serpentine minerals are highly variable in colour from green to brown to black–green; the green is caused by iron, and black spots are usually caused by magnetite. Other minerals formed during serpentinization are minor magnetite and variable minor brucite, magnesite, calcite, dolomite, and talc. Minerals in serpentinite

[5] Chrysotile is considered a growth 'habit' of lizardite, not a separate mineral (George Harlow pers. comm. 2 March 2017); it has been redefined (Rd) in the IMA mineral list (IMA-CNMNC 2021).

remaining as relics from the parent rock (the rock type before it underwent metasomatism) include chromite, clinopyroxene, orthopyroxene, and olivine.

A rock formed primarily of any of the three serpentine minerals is termed serpentinite, a valued form of false jade, while jadeite (a 'true jade') can form from serpentinite via fluid interactions under conditions of subduction channel metamorphism (cf. Chapter 12). Axes made of serpentinite were common in Jomon cultures of northeastern Japan (Bausch 2003).

Serpentinite mélange can form during subduction, as the oceanic plate scrapes down the edge of the mantle wedge; the mantle rock peridotite is serpentinized there with seawater drawn down by the plate, thus becoming soft and acting as a trap for other rocks separated from the ocean plate (basalts, cherts, eroded granite blocks, etc.) and pushed into it. The mixture of different rocks in a serpentinite matrix forms a serpentinite mélange, which can later be exhumed to the surface. Such a mélange is often treated as an ophiolite (*sensu lato*) (Ishiwatari 1991) (see Sec. 2.7 below), but Harlow *et al.* (2014) caution against this practice. Several serpentinite occurrences in Japan contain granite fragments dating to 450 Ma and 250 Ma, deriving from batholith erosion and inclusion of granite debris in the subduction trench (Suzuki *et al.* 2010: fig. 5).

2.5.3 Burial & exhumation

Burial of geological strata can occur in several contexts. Continual erosional deposition and volcanic deposition can bury underlying layers ever deeper; this is the beginning of the process of rock formation called diagenesis (upper left corner of Figure 2.8) which leads imperceptibly into low-pressure metamorphism. Accreted materials can also be buried within the subduction zone by being drawn down to higher pressure metamorphic levels as seen above. Such matrials are slanted parallel to the subducting slab, with their leading edges tilted downwards into the mantle. The paths of exhumation of slabs of metamorphic rock, which are now harder and less pliable than their protoliths, can either be on the subduction plane or on a sub-parallel splay fault; exhumation may be relatively fast (a few million years) or extremely slow (taking many millions of years). A change in the angle of subduction of the oceanic plate to a shallower angle has been identified as a prime mover in exhuming MBs from the depths (Maruyama *et al.* 1997: 98). Once they reach mid-crust levels, they may be subject to doming and erosion on a more horizontal plane (see the subduction–exhumation model in fig. 1 of (Maruyama *et al.* 2010).

Exhumations both of high-pressure metamorphic slabs and of bodies of deeply-seated igneous rocks have been documented in Japan, being squeezed up during non-accretionary, erosive periods (Ueda 2010: fig. 9-B). Such high-pressure slabs of metamorphic rock commonly come to rest in an accretionary 'sandwich' where the surrounding belts have undergone very low, weak or no metamorphic change (Maruyama *et al.* 1997: 97; Maruyama *et al.* 2010). Moreover, the current crystalline contents of these metamorphosed units represent not the peak pressure/temperature of metamorphism but a retrogressive crystallization as the unit rehydrated during exhumation (Isozaki, Aoki & Nakama 2010: 89; Maruyama *et al.* 2010).

Exhumed metamorphic belts of blueschist facies are the source of the true jade, jadeitite (see Chapter 12).

2.6 Vertical Movements & Recycling

Despite the discarding of geosynclinal theory, which was based almost entirely on hypothesizing upward movements (defined by anticlines) and downward movements (defined by synclines) (Figure

2.9), vertical movements still have a role to play in Plate Tectonics. The burial and exhumation of materials in the subduction zone as described above are one example of such vertical movement, as are folding and faulting of certain types (e.g., Bunge & Glasmacher 2018). These all can be described as the results of opposite forces acting on the Earth's crust: compression and extension.

2.6.1 Extensional and compressional tectonics

Structural geology is the branch dealing with stresses on rock and the way rock reacts to those stresses with different forms of strain (Bond & Lebit 2020). Both stress and strain can be calculated mathematically, though we will leave the maths to the structural geologists and concentrate on the nature of the physical results here. In short, there are three different degrees of strain (changes in volume or shape) which may succeed each other temporally – all of which can be envisioned in terms of a rubber band: elastic (one that regains its former shape), plastic (having lost its elasticity, resulting in permanent deformation), and brittle (having lost even its plasticity and thus simply breaking). Which degree of strain will develop in rocks under pressure is partly determined by their temperature: those buried deep in the earth are hotter (closer to the Earth's core) and therefore more ductile (malleable), while those on the surface are cold and quite brittle. Another factor that determines the rock's reaction to stress is composition and strength: weak layers will react in a more ductile manner than strong layers, which will tend to break rather than bend, and folds will often develop faulted surfaces under continued compression. See the active modelling of a fold/fault zone at Virginia Tech (ATGL 2014).

Both folding and faulting can play roles in crustal 'lengthening' and 'shortening', terms which refer to the horizontal extent of a piece of the Earth's crust and which have implications for changes in thickness of that crust – thinning or thickening, respectively. Lengthening occurs as a result of extensional stress (also called 'tension' or 'tensile' stress) and results in crustal thinning through normal faulting, while shortening results from compressional stress and results in crustal thickening through reverse faulting (see Bailey & King 2011: fig. 1). There are other processes which contribute to such changes in crustal thickness, but all these processes together account for the great variability in continental crust thickness (anywhere between 10 and 75 km thick at a particular point) compared to the relatively uniform thickness of oceanic crust (between 5 and 7 km).

Extensional and compressional regimes are often contemporaneous and sometimes are even required together for complementary distribution of stress. Moreover, a particular stress environment can be confined to a very small region, so that in the Plio-Pleistocene of Japan from 2 Ma, for example, different stresses were acting on different regions, resulting in different forms of strain. Several landscape features are associated with differential faulting and folding and their associated areas of uplift and subsidence; these surface modulations can have great impact on human development and habitation (e.g., Bailey & King 2011), as we will assess for the Nara Basin in Japan (Part III herein).

2.6.2 Folds

Folding (Figure 2.9) is a plastic form of strain that occurs in hot or weak rock. Normally it occurs when rocks are buried and subject to high temperatures deep in the Earth's crust; by the time these folded rocks have risen to the surface, much time has passed and they have cooled in their passage upwards. Folded rocks exposed at the surface therefore are *usually* evidence of folding processes that occurred in the distant past at deep levels. However, in Japan, folding due to the compressional regime that developed around 2 Ma has resulted in the formation of long-wave and short-wave surface undulations exposed in the current landscape (Chapter 5).

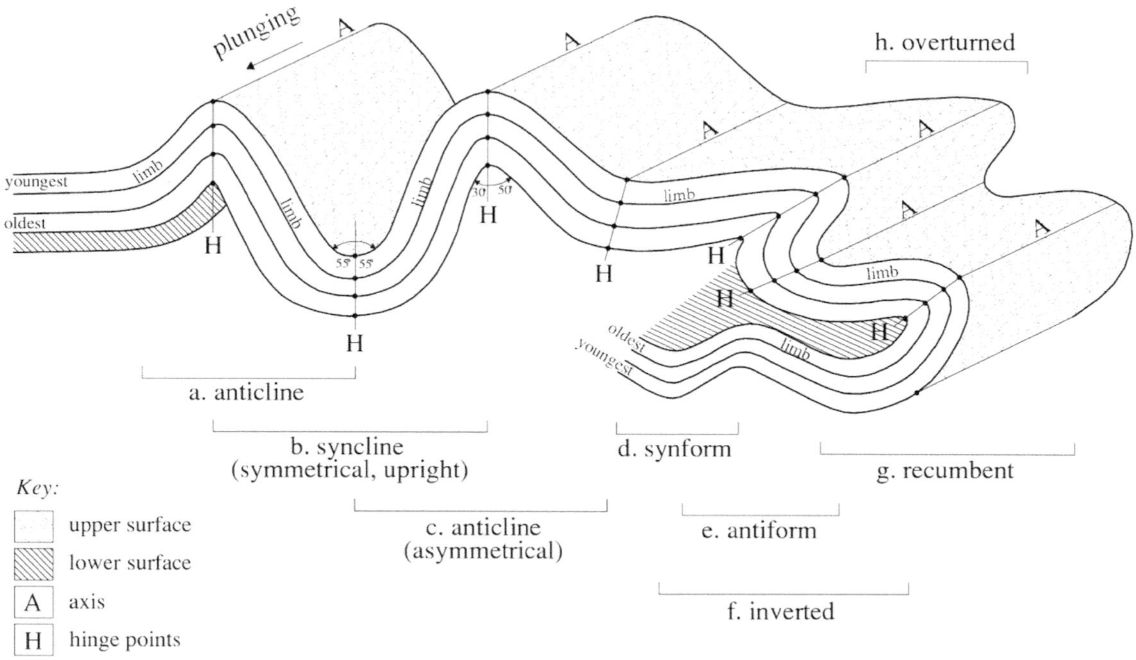

Figure 2.9 Fold geometry and terminology

(a) If three stratigraphic layers are folded as shown, a paired anticline and syncline are formed, with the youngest layer on top and the oldest layer on the bottom. However, if the stratigraphic succession is not known, then the terms antiform and synform are used. Folds may be symmetrical as in (a), with the limbs of equal length and angle, or (b) asymmetrical, with limbs describing different angles and being different lengths. The maximum point of fold is called the hinge point or zone, and the successive hinge points of the different layers form the hinge line or fold axis. If the hinge line is vertical, the fold is called an upright fold, as in (a); but it may be lying horizontal to the ground surface, in which case the fold is on its side and termed a recumbent fold (c), or it maybe a completely overturned or inverted fold, reversing the stratigraphic layering (d). On each stratum, the hinge points across its width at maximum fold define the fold axis, and often this angle is not horizontal to the surface but plunges downward at an angle (e). All these variables go to describing the type of fold seen in the geological matrix.

2.6.3 Faults

2.6.3.1 General typologies

In contrast to folding, faulting is a brittle form of strain that occurs in the upper, cold crust (Figure 2.10). We have already met a form of faulting in the tectonic lines that form the boundaries of many of the geotectonic units comprising the Japanese landmass. Four main types of faults configure the landscape:

1 strike-slip faults (Figure 2.10: far left), which allow large areas of land to slip by each other (such as the San Andreas fault); in such faults, movement occurs along its 'strike' (the directional orientation of the fault plane) (from NW to SE for San Andreas);
2 normal faults (Figure 2.10: right centre), which result from extensional stress and are an agent of crustal lengthening;
3 reverse faults (Figure 2.10: left centre), which result from compressional stress and are an agent of crustal shortening;
4 very low-angle reverse faults are called 'thrust faults' (Figure 2.10: lower right), which are common in subduction and collision zones and are the cause of much uplift and crustal thickening by

displacing the upper rock body great distances over the underlying rock. The contact between the over-riding continental plate and the subducting oceanic plate is a form of megathrust fault.

Three 19c terms from Alps geology have been incorporated into plate-tectonic discussions of thrust faults: nappe, klippe, and fenster. Nappe (from the French for 'tablecloth') describes a piece of crust that has been thrust between 2 to 5 km over an unrelated body of crust – now termed a 'thrust sheet'. Klippe (from the German for 'rocky prominence') describes a piece of nappe that is isolated from its nappe through intervening erosion or exposure – now referred to as an 'outlier'. Fenster (from the German for 'window') indicate an erosional hole in a nappe that exposes the underlying layers – now referred to with the English term 'window'.

In types 2 to 4 above, the footwall nominally belongs to the rock body that stays stable while the hanging wall belongs to the one that moves, although in some cases both may move – just one more than the other. A type 1 strike-slip fault moves laterally either right (dextral) or left (sinestral), determined by standing on one side of the fault looking across to the other and seeing which way the far side is moving: to one's right, or to one's left (it doesn't matter which side of the fault one stands on!).

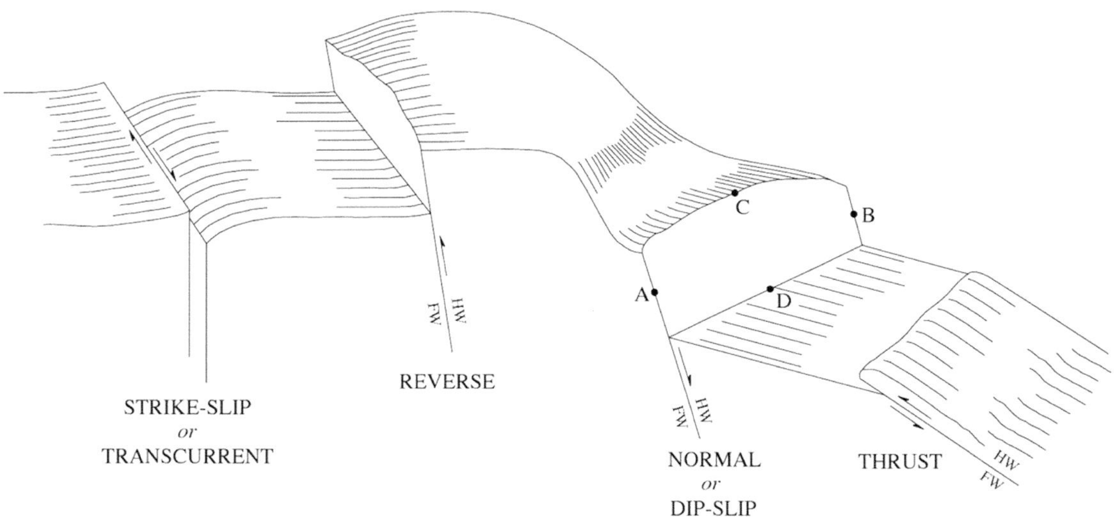

Figure 2.10 Major types of faults

The sense of movement between the footwall and hanging wall can be vertical, horizontal or diagonal at various angles, giving rise to further naming patterns. The fault may have a linear exposure (the fault line) on the ground surface as well as an embedded plane of movement (the fault plane), but many embedded faults do not have sub-aerial exposure (blind faults) and cannot be identified on the ground.

FW Footwall (the rock body underneath the fault plane)

HW Hanging wall (the rock body above the fault plane)

A—B The distance from A—B across the fault plane indicates the 'strike' direction; strike-slip faults move *along* the strike; they are described as 'left-lateral' (sinistral) or 'rightlateral' (dextral) depending on which way the rock is moving observed from either side of the fault line

C—D The distance from C—D down the fault plane indicates the 'dip' direction; normal faults move down dip, while reverse and thrust fault move up dip.

A horst is a topographically high, elongate block of crust bounded by normal faults, while a graben is an area of down-faulting, bounded by normal faults. Horst and grabens often occur alternately, though one-sided grabens (half-grabens) are also possible. A pop-up block is bounded by reverse faults.

2.6.3.2 Fault types specific to Japan

In Japan, another type of fault is recognized as a tectonic line (TL): a major constructional fault. The Itoigawa-Shizuoka TL and the Median Tectonic Line (MTL) have been mentioned above; the latter especially has undergone changes in type (from 4 to 1 above), slip-direction, and dip through time (see Chapter 3). A further category unique to Japan is 'Active Faults'.

Quaternary neotectonics (cf. Figure 5.1) have given rise to thousands of small Active Faults in Japan; these can be traced in today's landscape as scarps, depressions, bulges, and offset landforms. 'Active Faults' are defined in Japan as "the faults which have been repeatedly active during the Quaternary and probably will be active in the future" (RGAFJ 1992: 3). The Research Group for Active Faults of Japan has compiled a map of Active Faults occurring throughout the archipelago – on land and on the adjoining ocean floor (RGAFJ 1980, 1991, 1992); the map can now be seen online (GSJ 2020). However, these only represent faults which can be ascertained from surface evidence within one of the three degrees of certainty (listed below), and many more deeply buried blind faults that do not reach the present surface (e.g., Figure 5.6) may be the locus of future activity.

Three levels of certainty are used in identifying an Active Fault (RGAFJ 1992: 11, 15-17, their brackets):

I "When it is certain, based on conclusive evidence, that the fault has been active during the Quaternary."

II "When a fault is thought to be active because the sense of displacement as well as the location can be estimated, but there is no conclusive evidence."

III "When the lineaments may be associated with active faulting but the sense of displacement is unknown or [other factors caused] the lineaments." Certainty level III is designated as a 'concealed' or blind fault.

Most research has heretofore focussed on faults generated by tectonic processes, but it is important to note the growing field of 'nontectonic fault' research. Proposed in 1998 by WR Lettis and colleagues, there are four types of such faults: those deriving from widescale landslides, those formed by release of stress, those due to land subsidence and depression, and those caused by volcanic tremors (Nagata 2018: 900). Moreover, Active Faults may cause further nontectonic fault formation, as can volcanic tremors, simply from earth-shaking and land sliding.

2.6.4 Uplift

Uplift has been mentioned above as an important process in Japanese Island development, but it cannot be ascribed to a single cause: various factors operate under different conditions. A mantle plume can uplift and stretch a whole region of crust above it through thermal expansion – thought by some to be responsible for the rifting of eastern Eurasia and eventual back-arc basin formation of the Japan Sea (cf. Chapter 4). Magmatic inflation of a magma chamber can also cause local expansion of the ground surface – as seen in the spectacular 300 m rise of the Usu Volcano over 20 months in 1944–1945 *before* it erupted (Nunn 1994: fig. 3.15). If magma is being fed into a chain of volcanoes, then the whole chain will tend to rise regionally through magma infiltration and thermal expansion of the country rock. Granites are much lighter than mafic and ultra-mafic rocks, so they tend to rise up with any encouragement such as erosion of overlying strata. Granite magma chambers tend to be located between 6 to 10 km below the surface (Huber *et al.* 2019), but when they are exposed through erosion, they are considered **unroofed** – as are metamorphic belts when exhumed.

Subduction can also uplift areas: a slow-moving subducting slab jack up an overriding plate to a greater degree than a fast-moving slab (Nunn 1994: 119). Areas of collision also characteristically involve thrust faults which result in uplift through crustal thickening around the collision boundary (cf. Figure 5.8). The Oiso Hills and coastal terraces, on the eastern edge of the Izu Arc intrusion (cf. Chapter 4) are undergoing rapid uplift at 1.3–3.1 m/thousand years since 6500 years ago (Dilek *et al.* 2021). The extrusion of a metamorphic belt from the subduction channel in its final doming stage can elevate the continental edge, as is suspected happening currently with the Kii Peninsula in Japan (Maruyama *et al.* 2010). **Underplating** – the transfer of material from the subducting slab to the base of the overriding plate – can also act to uplift the leading edge of the continent, and underplating by magmatic activity will also contribute to uplift. These ocean floor and igneous rock additions to the continent from underneath add thickness to the crust. Another process that uplifts land surfaces is folding; the greater the fold amplitude, the thicker the crust. The convex parts of folds (anticlines) (cf. Figure 2.9: a,c,e; Figure 4.3) represent initially uplifted areas. An increase in altitude of the land surface through these thickening activities serves to encourage erosion, which feeds sediment accumulation and subsidence. The dating of uplift is most often accomplished by monitoring the start of deposition of coarse eroded terrestrial materials in the trench.

2.6.5 Subsidence

Subsidence is the reverse of uplift, first exemplified by the concave parts of folded areas (often termed **geosynclines**) (cf. Figure 2.9: b,d). These concave areas can serve as sumps for deposited material and subside even further through the weight of added sediments through time. Sediment loading of the crust is facilitated by the deposition of both erosional and volcanic products. Another cause of subsidence is thermal cooling, causing contraction so that extinct volcanic ranges will naturally subside as they cool. Faulted basins and grabens also represent areas of subsidence and can continue to subside through repeated faulting and sediment loading.

Terrestrial erosion is one of the equilibriating processes that serves to decrease the volume of the landmass when it exhibits **mass excess** by eroding away exposed surfaces and to increase crustal mass in areas of **mass deficiency** by depositing material in sedimentary basins. The Earth always strives for equilibrium, so uplift and subsidence can be thought of as processes that always trigger their counterparts into action, though the time frame may be very long indeed for these processes to act. Rates of uplift and erosion during the Quaternary in Japan are positively correlated; at ca. 10 cm per 1000 years, uplift operates at nearly ten times the world average (Nunn 1994: 156). A relatively high rate of precipitation is the chief erosional agent of terrestrial sediments in Japan, at 1668 mm/yr in 2014, ranking 45th of 186 countries in the world (IndexMundi 2017). In contrast, tectonic erosion, described above, operates submarinely.

2.6.6 Isostacy

The above are fairly localized examples of subsidence and uplift, but they may be discussed within the larger principles of **isostasy**. Most of us are acquainted with isostasy in terms of the rebound (uplift) of landmasses after being depressed by the great ice sheets. Ice loading will depress a land surface; so will sediment loading or even seawater on the continental shelf. When these overlying layers are removed (by melting, erosion, or lowering of sea level), the land surface underneath will rise again. Crustal thickening and/or uplift begin the erosional cycle again. The main principle is that the mass of the crust (and whatever is on top of it – water, ice, lava) should be in equilibrium with the mass of the underlying mantle, so that when the crust grows heavier, the mantle is squashed away, and when the crust grows lighter, the mantle flows in under it to replace the missing mass. As these movements of mantle material take place in solid but ductile (viscous) rock, they take enormous amounts of time, so that most landmasses show mass excesses or deficits for long periods before they attain equilibrium, if ever.

Whether a landmass is currently in equilibrium or not can be read from **gravity** measurements (see Smith & Rimer 2001). Such measurements around the Japanese Islands today (Figure 2.11: left) show mass deficits underneath the Japan Trench and the Japan Sea (not enough crust to balance the underlying mantle) and a mass excess in the archipelago (mass above that has yet to displace the mantle). The fact that these areas are not now in equilibrium with the mantle but are either in excess or deficit means that they may still be moved upwards or downwards, eventually attaining that equilibrium. '**Basin inversion**' is one aspect of such a transition, when a sedimentary basin is inverted to a topographic high through folding or faulting and thus subjecting its sediment load to erosion.

Interesting examples of rapid-acting isostacy have been recorded in Taiwan. The 2009 Morakot typhoon (cyclone, hurricane) dropped about 3 m of rain on the southern mountains, causing severe landslides

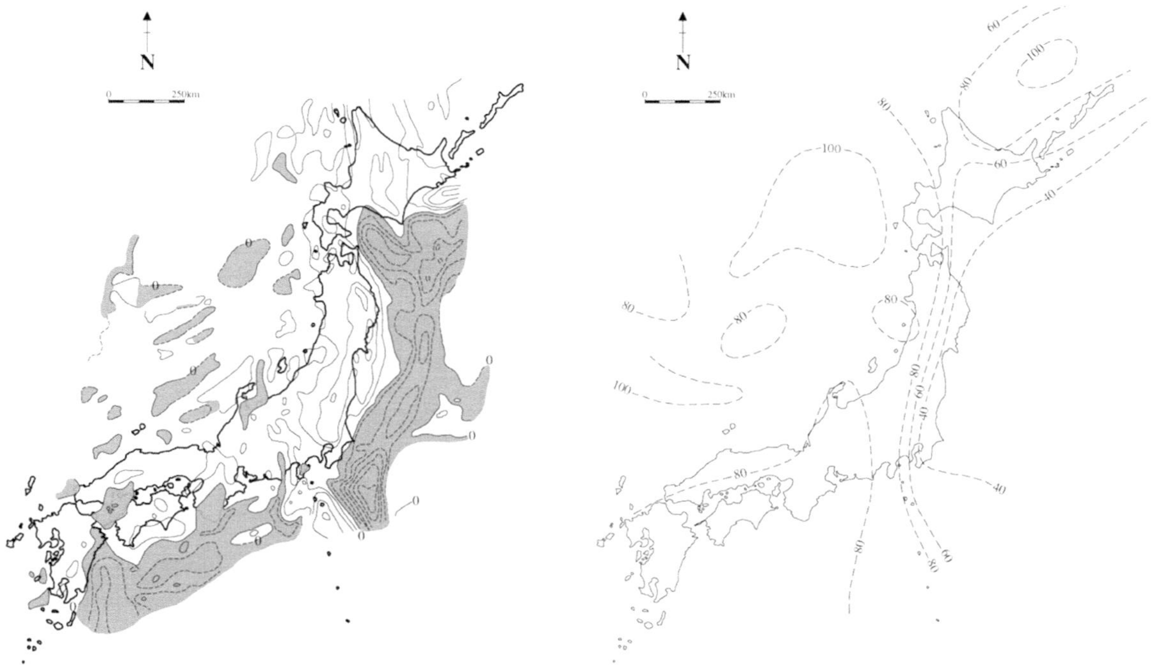

Figure 2.11 Present-day gravity and heat measurements around Japan

Left: Free-air gravity measurements at contour intervals of 50 milligals. Solid-line contours indicate positive anomalies of mass excess; dashed-line contours (shaded areas) indicate negative anomalies of mass deficiency.

Right: Heat flow distribution shown in mW m^{-2}

Note that the areas of overlapping high heat measurements correspond to hot Philippine Plate in southwestern Japan and magma presence/intrusion in the Japan Sea Basin, while mass deficiency measurements correspond to the subduction trenches along the eastern seafront and thin crust and spreading centres in the Japan Sea.

with the soil washed into the sea; for two and a half years after the typhoon, the normal earthquake regime increased in magnitude and frequency until some equilibrium had been re-established (Steer *et al.* 2020). Studies have noted that earthquake frequency can increase with tidal, ice, and snow loading and heavy rainfall, or with changes in atmospheric pressure and surface temperature (Heki 2003; Ueda & Kato 2019).

2.8 Obduction Processes & Ophiolites

Slices of ocean floor can be incorporated into landmasses as 'ophiolites'. Because of its higher density, oceanic crust is usually subducted under a continental plate, but there are occasions where the ocean crust is thrust over continental crust instead via a process called 'obduction'. Different types of ophiolite (*sensu lato*) are formed in different tectonic settings, but most commonly they occur in a 'supra-subduction' zone – i.e., in a back-arc basin setting – or on the continental side of a subduction trench in the forearc zone (Dewey 2003; Dewey & Casey 2011; Robinson & Zhou 2007; Jolivet *et al.* 2016). The most complete ophiolite (*sensu stricto*) (e.g., the Troodos ophiolite in Cyprus) is a vertical slice of oceanic crust and underlying mantle rock (peridotite) preserving the vertical stratigraphy from top to bottom, as seen in Figure 2.4; from top down: ocean floor sediments → cherts → basalt → gabbro → peridotite, but few ophiolites exhibit all of these components, and often the peridotite has been converted by seawater intrusion (metasomatism) into serpentinite. It is recommended that to be called an ophiolite, a rock body should have at least peridotite, basalt, and basalt pillow lava in clear relations – otherwise they should be called 'possible ophiolites' (Robinson & Zhou 2007), but "sliver[s] of serpentinite, gabbro, dolerite, or basalt" from oceanic sources occurring on land have all been termed ophiolites (Dewey & Casey 2011: 431). Different types of ophiolites have been recognized in various classification schemes, depending on their geochemistry, locus of formation, and methods of emplacement (Dewey 2003; Dewey & Casey 2011; Dilek & Furnes 2011, 2014; Dilek, Y. 2015). True sections of oceanic crust can be recognized by the arrangement of basaltic dykes (Figure 2.12) resulting from seafloor spreading and associated pillow lavas, etc.

Figure 2.12 Diagonally oriented basalt dykes of the ophiolite in the Isua Supercrustal Belt, Greenland (scale provided by Harald Furnes on left)

Ophiolites *sensu lato* are scattered throughout the Japanese Islands (Ishiwatari 1991). These are likely to have been formed in backarc or forearc positions around previous proto-Japans. The oldest ophiolite in Japan (Nomo Ophiolite at 580 Ma) is thought to derive from a passive margin setting of oceanic plate (similar to the current Atlantic Ocean shores) before subduction began (Isozaki, Aoki & Nakama 2010).

While ophiolites are defined by igneous rocks characteristic of oceanic crust, associated oceanic sedimentary rocks may occur in ophiolite belts on land. Sediments settling on ocean floors include skeletons and shells of sea fauna; some of these are made of silica (SiO_2), such as siliceous diatom and radiolarian shells. Radiolarian chert has been used to make artefacts in North America, at least (Gauthier *et al.* 2012); but most chert for archaeological use is secondary chert (flint) formed within limestone/chalk; see the discussion at Mindat (2010-12) and the final document at Mindat (n.d.).

Sedimentary depositions made of calcium carbonate ($CaCO_3$) include carbonate sediments, limestones, and dolostones (magnesium limestone) (see James & Jones 2016); chalk is a sedimentary rock comprised mainly of carbonate plates of coccolithophore diatoms, but chalk also contains siliceous sponge spicules which contribute to the formation of the flint nodules and beds. The problem is that the survival of both shelly and muddy carbonate sediments is only possible above the carbonate compensation depth (CCD, ca. 4000 m deep) below which calcium carbonate dissolves (e.g., Bao *et al.* 2018). Therefore, carbonate sediments do not form on the deep seafloor – the average depth of the Pacific Ocean is 4280 m – or in deep subduction trenches; because these depths are below the CCD, carbonates may not survive to be incorporated into the accretionary prism. For comparison, the Japan Trench is approximately 8000 m deep, while the Marianas Trench is about 11,000 m deep. Therefore, limestones/dolostones found on land in Japan are usually shallow-sea formations that have been obducted with ophiolite slices or sliced off subducting seamounts (see BOX 3).

Ophiolite belts become important when discussing geological sources of jade due to their dual igneous and sedimentary potentials for producing different kinds of jade, as discussed in Chapter 12.

2.9 Rifting Processes: combined extension, faulting, and volcanics

Rifting is a special case of brittle faulting of the Earth's surface. In the lay imagination, rifting might be visualized as the pulling apart of two rock bodies, leaving an empty canyon in between. But this is not what happens in reality. Rifting consists of a complex series of normal faulting and block rotation, so that successive pieces of rock at the edges of the ever-widening rift slip further down into the developing graben (Figure 2.13). Meanwhile, the continental crust is stretched thinner and thinner until it breaks completely, and melt generated by the rising mantle under the thinned layers intrudes to start building oceanic crust at a spreading centre. Such areas are marked by higher heat flow, as seen in the Japan Sea Basin (cf. Figure 2.11: right).

Theoretically, the rifting of blocks can happen symmetrically (Figure 2.13a) or asymmetrically (Figure 2.13b), forming grabens or half-grabens which often become sea basins such as the Red Sea, Japan Sea, and Okhotsk Sea. The fault planes separating the rotated blocks characteristically curve towards the centre of the valley from either side of the rift. The subsided blocks become covered with erosional sediments from the rift edge, effusive volcanics erupting from below and marine sediments if the valley becomes flooded. At the eastern edge of the Japan Sea, normal faults have been identified in the vicinity of Sado Island (cf. Figure 4.4) – presumably marking where rifted blocks have slipped down towards the continent into the opening sea basin. Interestingly, such normal faults are now highly outnumbered by reverse faults, indicating compression. These are the first indications of the formation of an incipient subduction zone, which will close the Japan Sea, following the edge of the Okhotsk Plate down the west side of northern Japan (Okamura *et al.* 1995: fig. 1; Apel *et al.* 2006: fig. 1).

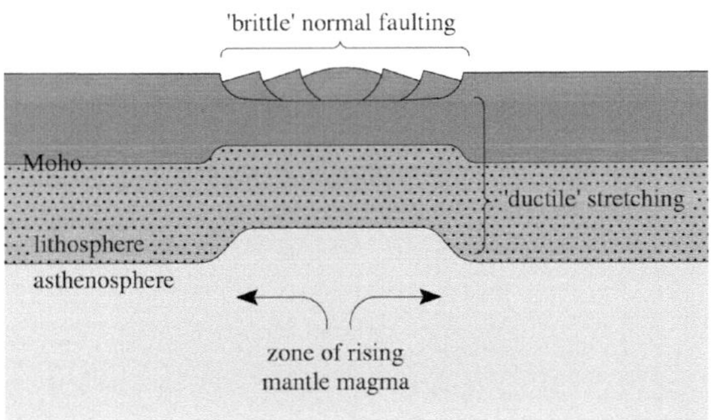

a

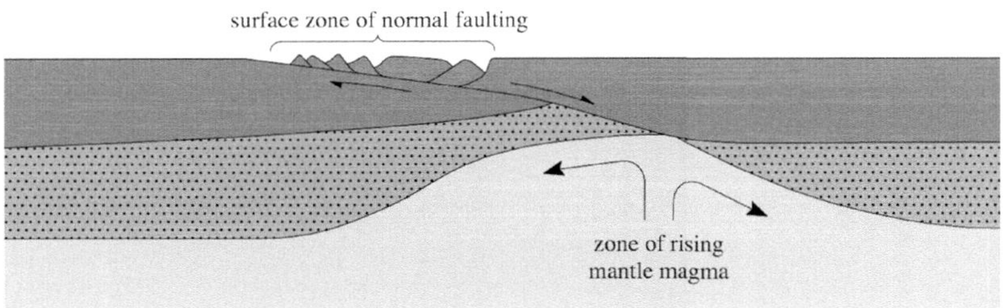

b

Rotation of rift blocks into a developing basin:

a symmetrical or 'graben' structure

b asymmetrical or 'half-graben' structure

Figure 2.13 Mechanics of rifting

The study of worldwide rifting, and the volcanics that accompanied them, was revolutionized in 1985 with the application of specialized deep seismic techniques (Rogers 2001: 52-8), producing visual images of a rift's deep structure. Such techniques have not yet been fully applied to back-arc basin rifts behind the Japanese archipelago, but ocean seismology has begun to provide views of what the Japan Sea-floor stratigraphy looks like (Figure 2.14). Layers of marine sediments drape over volcanic mounds of magma escaping through the various minor faults (see similar 'knolls' in hot spot chains in Chase 1977).

Initial rifting is often accompanied by voluminous but slow extrusions of lava on land (flood basalts) together with explosive felsic volcanic ash (tephra). The final outcome of successful rifting is considered to be the creation of upper oceanic crust in the rift valley floor via a new spreading balsaltic centre, indicating that the continental crust had completely bisected at that point. The extremes of mafic basalts (low silica content) and felsic tephra (high silica content) which are extruded in rifting contexts are characteristic 'bimodal' products of rifting – quite different from subduction products in which intermediate to felsic silica content is characteristic (cf. Apx 5: Table B).

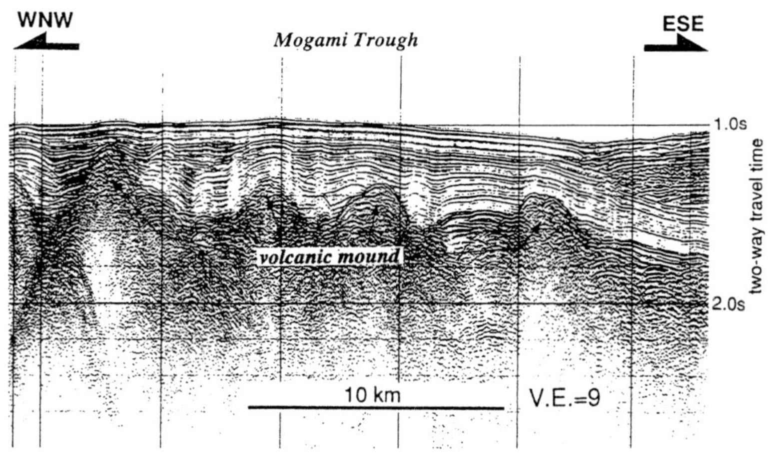

Figure 2.14 A seismic profile of the Japan Sea floor

Taken in the in the Mogami Trough located just east of the Sado Ridge (cf. Figure 4.4). Sub-vertical arrows mark the peaks of subsurface volcanic mounds, covered by sub-horizontal sedimentary layers. This is a seismic wave generated diagram with vertical scale denoting travel time of the seismic soundings, not depth of the features.

The best example of present-day rifting actually in progress is the Great Rift Valley in East Africa. The Red Sea, floored by newly created ocean crust, is a partially completed arm of that rift system. The cause of continental rifting is generally recognized as the initiation of a mantle plume, as under eastern Africa (Rogers 2001: ch. 3), supporting the interpretation of a plume under eastern Eurasia causing rifting there (but on a much smaller scale) (see Chapter 4).

2.10 Prospectus

With these general terms and processes in mind, the rest of Part I will now examine Japanese geology through time, beginning with the geological belts making up the current Japanese landmass and their palaeogeographic compilation, to the isolation of that landmass as an archipelago through rifting, and the subsequent folding, faulting, and volcanics that have occurred since that rifting event. Part I concludes with a discussion of the salient points and controversies that have arisen in the journey from the Japanese landmass to form the Japanese archipelago.

Sources of Figures & Tables:

Figure 2.1 modified by GLB after (Chester 1993: 28, fig. a; Kasahara *et al.* 2010: fig. 2)
Figure 2.2 Redrawn by DAS after (Chester 1993: fig. 2.1b; Taira 1990: fig. 1-4a), modified by GLB
Figure 2.3 compiled from (Seno 1995: figs. 5.1.1-2, 5.2.1; RGAFJ 1991: map II; Nakamura, K. 1989: 244; Okamura, Y. *et al.* 1995: 168) and drawn by DAS, modified by GLB; note that the Ryukyu Trench was previously referred to mistakenly as the Okinawan Trough in (Barnes 2003: fig. 3)
Figure 2.4 after (Tucholke 1998: 1), redrawn by DAS
Figure 2.5 compiled from (Taira, 1990: fig. 1-4b; Chester 1993: figs. 2.8-9; Kearey & Vine 1996: fig. 8.21), drawn by DAS, modified by GLB: circled letters and from Strekeisen (2020)
Figure 2.6 from JMA, license CC BY-SA 3.0 via Wikimedia Commons [https://commons.wikimedia.org/wiki/File:Shindomap_2011-03-11_Tohoku_earthquake.png]
Figure 2.7 from (Moore 2011: fig. 4), with permission of Gregory MOORE
Figure 2.8 colour illustration by David MAGRASS & Brian WELLS, Public Domain via Wikimedia Commons [https://en.m.wikipedia.org/wiki/File:Metamorphic_Facies.jpg]; grade divisions after (Nelson 2018); trajectories after (Strekeisen 2020: Blueschist fig. 2: I=1, II=2, III=5 in his diagram)
Figure 2.9 sketch by GLB redrawn by DAS
Figure 2.10 after (Lomnitz 1974: fig. 3.7), redrawn by DAS
Figure 2.11 (a) after (Uyeda 1971: fig. 4-2; (b) after (Miyashiro 1973*a*: fig. 15-4), redrawn by DAS, modified by GLB
Figure 2.12 courtesy of Harald FURNES, with permission
Figure 2.13 after (Rogers 2001: fig. 3.24), with permission from OU Publications
Figure 2.14 from (Okamura *et al.* 1995: fig. 15), Courtesy of OKAMURA Yukinobu, with permission

Table 2.1 compiled from references in text
Table 2.2 based on (Kunugi 2000: fig. 4; Goff *et al.* 2012: fig. 2)
Table 2.3 compiled from (Maruyama *et al.* 2010)

References

AITCHISON, JC; S HADA & S YOSHIKURA (**1991**) "Kurosegawa terrane: disrupted remnants of a low latitude Palaeozoic terrane accreted to SW Japan." *Journal of Southeast Asian Earth Sciences* 6: 83-92.
ALTIS, Sungat (**1999**) "Interpretations of a Middle Miocene and late Quaternary steady dextral transpression in SW Japan and the opening tectonics for the Japan Sea." *Tectonophysics* 302: 257-285.
APEL, EV; G Steblov BÜRGMANN & N VASILENKO *et al.* (**2006**) "Independent active microplate tectonics of northeast Asia from GPS velocities and block modeling." *Geophysical Research Letters (Solid Earth)* 33.11 [DOI: 10.1029/2006GL026077].

ATGL (Active Tectonics and Geomorphology Lab) (**2013**) "Growth and erosion of an orogenic wedge". Department of Geosciences, Virginia Tech [www.youtube.com/watch?v=J750DoyVRfw].
---- (**2014**) "Fold/thrust belt: internal structure, scale, and erosion," ed. by P PRINCE. Department of Geosciences, Virginia Tech [www.youtube.com/watch?v=8L-6WivVikI] (dead link).
AZUMA, S; S YAMAMOTO & H ICHIKAWA *et al.* (**2017**) "Why primordial continents were recycled to the deep: role of subduction erosion." *Geoscience Frontiers* 8.2: 337-346.
BAILEY, GN & GCP KING (**2011**) "Dynamic landscapes and human dispersal patterns: tectonics, coastlines, and the reconstruction of human habitats." *Quaternary Science Reviews* 30.11: 1533-1553.
BAILEY, GN; SC REYNOLDS & GCP KING (**2011**) "Landscapes of human evolution: models and methods of tectonic geomorphology and the reconstruction of hominin landscapes." *Journal of Human Evolution* 60: 257-280 [DOI: 10.1016/jjhevol.2010.01.004].
BAO, Rui; Michael STRASSER & Ann P MCNICHOL *et al.* (**2018**) "Tectonically-triggered sediment and carbon export to the Hadal zone." *Nature Communications* 9.121: 8pp. [DOI: 10.1038/s41467-017-02504-1].
BARNES, Gina L (**2003**) "Origins of the Japanese Islands: the new 'Big Picture'." *Japan Review* 15: 3-50.
BAUSCH, Ilona (**2003**) *Jade, amber, obsidian and serpentinite: the social context of exotic stone exchange networks in central Japan during the late Middle Jomon period.* Ph.D. dissertation, University of Durham.
BENIOFF, Hugo (**1954**) "Orogenesis and deep crustal structure – additional evidence from seismology." *Bulletin of the Geological Society of America* 65: 385-400.
BEST, Myron G (**2003**) *Igneous and metamorphic petrology*. Malden, MA: Blackwell.
BIRD, Peter (**2003**) "An updated digital model of plate boundaries." *Geochemistry Geophysics Geosystems* 4.3: #1027, 52pp. [DOI: 10.1029/2001GC000252].
BLACKETT, PMS; E BULLARD & SK RUNCORN (eds) (**1965**) *A symposium on continental drift, held in 28 October 1965.* Philosophical Transactions of the Royal Society A. 258. The Royal Society of London.
BOND, CE & HD LEBIT (eds) (**2020**) *Folding and fracturing of rocks: 50 years of research since the seminal text book of J. G. Ramsay*. GSL Special Publications 487. Geological Society of London.
BORMANN, Peter & Joachim SAUL (**2009**) "Earthquake magnitude", pp. 2473–2496 in *Encyclopedia of Complexity and Applied Systems Science, vol. 3.* New York: Springer.
BROWN, GC (**1979**) "The changing pattern of batholith emplacement during earth history", pp. 106-115 in *Origin of granite batholiths: geochemical evidence*, ed. by MP ATHERTON & J TARNEY. Orpington, Kent: Shiva.
BROWN, Michael (**2010**) "Paired metamorphic belts revisited." *Gondwana Research* 18: 46-59.
BUNGE, Hans-Peter & UA GLASMACHER (**2018**) "Models and observations of vertical motion (MoveOn) associated with rifting to passive margins: preface." *Gondwana Research* 53: 1-8 [DOI: 10.1016/ j.gr.2017.07.005].
BUTLER, BCM & JD BELL (**1988**) *Interpretation of geological maps*. Harlow, Essex: Longman Science & Technical.
BYRNE, Timothy; Michael B UNDERWOOD & Donald FISHER *et al.* (eds) (**2018**) *Geology and tectonics of subduction zones: a tribute to Gaku Kimura*. GSA Special Papers 534. Boulder, CO: Geological Society of America.
CAMP, Vic (**n.d.**) "Historical note on rock terminology (basic/acidic vs. mafic/felsic)." *How volcanoes work* [http://sci.sdsu.edu/how_volcanoes_work/rock%20term%20note.html].
CAWOOD, Peter A; Alfred KRÖNER & William J COLLINS *et al.* (**2009**) "Accretionary orogens through Earth history", pp. 1-36 in *Earth accretionary systems in space and time*, ed. by PA CAWOOD & A KRÖNER. GSL Special Publications 318. The Geological Society of London.
CHARVET, Jacques (**2013**) "Late Paleozoic-Mesozoic tectonic evolution of SW Japan: a review – reappraisal of the accretionary orogeny and revalidation of the collisional model." *Journal of Asian Earth Sciences* 72: 88-101.
CHARVET, Jacques; Liangshu SHU & Yangshen SHI *et al.* (**1996**) "The building of south China: collision of Yangzi and Cathaysia blocks, problems and tentative answers." *Journal of Southeast Asian Earth Sciences* 13.3-5: 223-235.
CHASE, RL (**1977**) "J. Tuzo Wilson knolls: Canadian hotspot." *Nature* 266.5600: 344-346.
CHESTER, David (**1993**) *Volcanoes and society*. London: Edward Arnold.
CHOY, George L & Stephen H KIRBY (**2004**) "Apparent stress, fault maturity and seismic hazard for normal-fault earthquakes at subduction zones." *Geophysical Journal International* 159: 991-1012.
CLANCEY, Gregory (**2006**) *Earthquake nation: the cultural politics of Japanese seismicity, 1868-1930.* Berkeley: University of California Press.

Cloos, Mark (**1993**) "Lithospheric buoyancy and collisional orogenesis: subduction of oceanic plateaus, continental margins, island arcs, spreading ridges, and seamounts." *Bulletin of the Geological Society of America* 105: 715-737.

Colpron, Maurice; JoAnne L Nelson & Donald C Murphy (**2007**) "Northern Cordilleran terranes and their interactions through time." *GSA Today* 17.4/5 [DOI: 10.1130/GSAT01704-5A.1].

Condie, Kent C (**1997**) *Plate tectonics and crustal evolution.* Oxford: Butterworth-Heinemann.

Cronin, Shane J; Karoly Nemeth & Vincent E Neall (**2008**) "Volcanism and archaeology", pp. 2185-2196 in *Encyclopedia of archaeology*, ed. by DM Pearsall. Amsterdam: Elsevier.

Dewey, John (**2003**) "Ophiolites and lost oceans: rifts, ridges, arcs, and/or scrapings?", pp. 153-158 in *Ophiolite concept and the evolution of geological thought*, ed. by Y Dilek & S Newcomb. Boulder, CO: Geological Society of America.

Dewey, JF & JF Casey (**2011**) "The origin of obducted large-slab ophiolite complexes", pp. 431-444 in *Arc-continent collision*, ed. by D Brown & PD Ryan. Heidelberg: Springer.

Dietz, Robert S (**1972**) "Geosynclines, mountains and continent-building", pp. 124-131 in *Continents Adrift*, ed. by JT Wilson. San Francisco: W.H. Freeman.

Dilek, Yildirim (ed.) (**2015**) *Ophiolites: mantle sources, melt evolution and emplacement mechanisms.* Episodes 4 (Special Issue).

Dilek, Yildirim & H Furnes (**2011**) "Ophiolite genesis and global tectonics: geochemical and tectonic fingerprinting of ancient oceanic lithosphere." *Bulletin of the Geological Society of America* 123: 387-411.

---- (**2014**) "Ophiolites and their origins." *Elements* 10: 91-100.

Dilek, Yildirim; Yujiro Ogawa & Yasukuni Okubo (**2021**) "Characterization of modern and historical seismic-tsunamic events and their global-societal impacts", in *Characterization of modern and historical seismic-tsunamic events, and their global-societal impacts*, SP No. 501, ed. by Y Dilek *et al.* Geological Society of London [DOI: 10.1144/SP501-2021-17].

Domeier, Mathew; Pavel V Coubrovine & Trond H Torsvik *et al.* (**2016**) "Global correlation of lower mantle structure and past subduction." *Geophysical Research Letters* 43.10: 4945-4953.

ERC (Earthquake Research Committee) (**n.d.**) "Seismic activity in Japan: types of earthquakes in the Japanese archipelago" [www.hp1039.jishin.go.jp/eqchreng/eqchrfrm.htm].

Etienne, Samuel (**2012**) "Marine inundations hazards in French Polynesia: geomorphic impacts of Tropical Cyclone Oli in February 2010", pp. 21-40 in *Natural hazards in the Asia-Pacific region: recent advances and emerging concepts*, ed. by JP Terry & J Goff. GSL Special Publications No. 361. The Geological Society of London.

Frankel, Henry R (**2012**) *The continental drift controversy, vol. 1: Wegener and the early debate.* Cambridge University Press.

Friedman, Hershel (**1997-2020**) "The mineral & gemstone kingdom" [minerals.net].

Fukao, Yoshio & Masayuki Obayashi (**2013**) "Subducted slabs stagnant above, penetrating through, and trapped below the 660 km discontinuity." *Journal of Geophysical Research: Solid Earth* 118: 5920-5938 [DOI: 10.1002/2013JB010466].

Gauthier, Gilles; Adrian L Burke & Mathieu Leclerc (**2012**) "Assessing XRF for the geochemical characterization of radiolarian chert artifacts from northeastern North America." *Journal of Archaeological Science* 39.7: 2436-2451 [DOI 10.1016/j.jas.2012.02.019].

Goff, James R; Catherine Chagué-Goff & James P Terry (**2012**) "The value of a Pacific-wide tsunami database to risk reduction: putting theory into practice", pp. 209-220 in *Natural hazards in the Asia-Pacific region: recent advances and emerging concepts*, ed. by JP Terry & J Goff. Special Publications No. 361. The Geological Society of London.

GSJ (Geological Survey of Japan) (**2020**) "Active Fault database" [https://gbank.gsj.jp/activefault/search].

Guerrieri, L & E Vittori (eds) (**2007**) *Intensity scale ESI 2007.* Memorie Descrittive Della Carta Geologica d'Italia No. 74. Italy: SystemCart.

Harlov, Daniel & Hakon Austrheim (eds) (**2013**) *Metasomatism and the chemical transformation of rock: the role of fluids in terrestrial and extraterrestrial processes.* Lecture Notes in Earth System Sciences. Heidelberg: Springer.

HARLOW, GE; T TSUJIMORI & SS SORENSEN (**2015**) "Jadeitites and plate tectonics." *Annual Review of Earth and Planetary Sciences* 43: 105-138 [DOI: 10.1146/annurev-earth-060614-105215].

HARLOW, George E; Sorena S SORENSEN & Virginia B SISSON *et al.* (**2014**) "Chapter 10: The geology of jade deposits", pp. 305-374 in *The geology of gem deposits*, ed. by LA GROAT. Québec: Mineralogical Association of Canada.

HASHIMOTO, Mitsuo (ed.) (**1991**) *Geology of Japan*. Tokyo & Dordrecht: Terra Scientific Publishing Company & Kluwer Academic Publishers (in English).

HEKI, Kosuke (**2003**) "Snow load and seasonal variation of earthquake occurrence in Japan." *Earth and Planetary Science Letters* 207: 159-164.

HEKI, K; Shin'ichi MIYAZAKI & Hiroaki TAKAHASHI *et al.* (**1999**) "The Amurian Plate motion and current plate kinematics in eastern Asia." *Journal of Geophysical Research* 104.B12: 29, 147-129, 155.

HESS, HH (**1962**) "History of ocean basins", pp. 599-620 in *Petrologic studies: a volume to honor A.F. Buddington*, ed. by AEJ ENGEL *et al.* Boulder, CO: Geological Society of America.

HOLMES, Arthur (**1931**) "Radioactivity and earth movements." *Transactions of the Geological Society of Glasgow (for 1928-29)* 18: 559-606.

HONDA, Satoru (**2017**) "Geodynamic modeling of the subduction zone around the Japanese Islands." *Monographs on Environment, Earth and Planets* 5.2: 35-62.

HOWELL, David G (**1995**) *Principles of terrane analysis: new applications for global tectonics*. [1st ed. title: *Tectonics of suspect terranes* (1989)]. London: Chapman & Hall.

HUBER, Christian; Meredith TOWNSEND & Wim DEGRUYTER *et al.* (**2019**) "Optimal depth of subvolcanic magma chamber growth controlled by volatiles and crust rheology." *Nature Geoscience* 12: 762-768.

IMA-CNMNC (International Mineralogical Association – Commission on New Minerals, Nomenclature and Classification) (**2021**) "The new IMA list of minerals – a work in progress – updated: January 2021" [http://cnmnc.main.jp/imalist.htm].

IndexMundi (**2017**) "Average precipitation in depth (mm per year): country ranking." [www.indexmundi.com/facts/indicators/AG.LND.PRCP.MM/rankings].

ISHIWATARI, Akira (**1991**) "Ophiolites in the Japanese Islands: typical segment of the circum-Pacific multiple ophilite belts." *Episodes* 14.3: 274-279.

ISOZAKI, Yukio (**1996**) "Anatomy and genesis of a subduction-related orogen: a new view of geotectonic subdivision and evolution of the Japanese Islands." *The Island Arc* 5.3: 289-320.

---- (**1997*a***) "Contrasting two types of orogen in Permo-Triassic Japan: accretionary versus collisional." *The Island Arc* 6.1: 2-24.

---- (**1997*b***) "Jurassic accretion tectonics of Japan." *The Island Arc* 6.1: 25-51.

---- (**2019**) "A visage of early Paleozoic Japan: geotectonic and paleobiogeographical significance of Greater South China." *The Island Arc* 28.e12296, 17pp.

ISOZAKI, Yukio; Kazumasa AOKI & Takaaki NAKAMA *et al.* (**2010**) "New insight into a subduction-related orogen: a reappraisal of the geotectonic framework and evolution of the Japanese Islands." *Gondwana Research* 18: 82-105.

ISOZAKI, Yukio; Shigenori MARUYAMA & Kazumasa AOKI *et al.* (**2010**) "Geotectonic subdivision of the Japanese Islands revisited: categorization and definition of elements and boundaries of Pacific-type (Miyashiro-type) orogen." *Chigaku Zasshi* 119.6: 999-1053 (in Japanese with English title and abstract).

ISOZAKI, Yukio; Shigenori MARUYAMA & Gaku KIMURA (**1997**) "Preface to thematic issue: geology and orogeny of the Japanese Islands." *The Island Arc* 6.1: 1-2.

ISOZAKI, Yukio; Shigenori MARUYAMA & Shuichi YANAI (**2010**) "At the stage of 'Exporting Science': a historical review of studies on the geotectonic subdivision and orogeny of the Japanese Islands." *Journal of Geography* 119.2: 378-391 (in Japanese with English title and abstract).

ITO, Chihiro; Hiroaki TAKAHASHI & Mako OHZONO (**2019**) "Estimation of convergence boundary location and velocity between tectonic plates in northern Hokkaido inferred by GNSS velocity data." *Earth, Planets and Space* 71.86: 8pp.

JAHN, Bor-ming (**2004**) "The Central Asian Orogenic Belt and growth of the continental crust in the Phanerozoic", pp. 73-100 in *Aspects of the tectonic evolution of China*, ed. by J MALPAS *et al.* The Geological Society of London.

JAMES, Noel P & Brian JONES (**2016**) *Origin of carbonate sedimentary rocks*. Chichester, UK: John Wiley & Sons, Ltd.

JOLIVET, Laurent; Claudio FACCENNA & Philippe AGARD *et al.* **(2016)** "Neo-Tethys geodynamics and mantle convection: from extension to compression in Africa and a conceptual model for obduction." *Canadian Journal of Earth Sciences* 53.11: 1190-1204.

KAIZUKA, S **(1980)** "Late Cenozoic palaeogeography of Japan." *GeoJournal* 4.2: 101-109.

KANMERA, Kametoshi **(1976)** "Correspondences between past and present geosynclinal sediments I,II." *Kagaku* 46: 284-291, 371-378 (in Japanese with English title).

KASAHARA, Junzo; Osamu SANO & Nobuo GESHI *et al.* **(2010)** "Overview of a special issue on "Geotectonic evolution of the Japanese Islands under new paradigms of the next generation (Part I-IIII)." *Chigaku Zasshi* 119.6: 947-958 (in English).

KAWAI, Kenji; Taku TSUCHIYA & Shigenori MARUYAMA **(2010)** "The second continent." *Chigaku Zasshi* 119.6: 1197-1214 (In Japanese with English title and abstract).

KAWAI, Kenji; Shinji YAMAMOTO & Taku TSUCHIYA *et al.* **(2013)** "The second continent: existence of granitic continental materials around the bottom of the mantle transition zone." *Geoscience Frontiers* 4: 1-6 [DOI: 10.1016/j.gsf.2012.08.003].

KAWAMOTO, Tatsuhiko; Masako YOSHIKAWA & Yoshitaka KUMAGAI *et al.* **(2013)** "Mantle wedge infiltrated with saline fluids from dehydration and decarbonation of subducting slab." *PNAS* 110.24: 9663-9668 [DOI: 10.1073/pnas.1302040110].

KEAREY, Philip & Frederick J VINE **(1996)** *Global tectonics*. Oxford: Blackwell Science.

KEAREY, Philip; Keith A KLEPEIS & Frederick J VINE **(2009)** *Global tectonics*. Oxford: Wiley-Blackwell.

KIMURA, Gaku; Hiroaki KOGE & Takeshi TSUJI **(2018)** "Punctuated growth of an accretionary prism and the onset of a seismogenic megathrust in the Nankai Trough." *Progress in Earth and Planetary Science* 5.#78, 12pp. [DOI: 10.1186/s40645-018-0234-1].

KIMURA, Jun-Ichi; Yoshitaka NAGAHASHI & Yasufumi SATOGUCHI *et al.* **(2015)** "Origins of felsic magmas in Japanese subduction zone: geochemical characterizations of tephra from caldera-forming eruptions <5 Ma." *Geochemistry, Geophysics, Geosystems* 16: 2147-2174 [DOI: 10.1002/2015GC005854].

KIMURA, Toshio; Itaru HAYAMI & Shizuo YOSHIDA **(1991)** *Geology of Japan*. Tokyo University Press (in English).

KLEIN, Cornelius & Anthony R PHILPOTTS **(2013)** *Earth materials: introduction to mineralogy and petrology*. Cambridge University Press.

KNOPF, Adolph **(1948)** "The geosynclinal theory." *Bulletin of the Geological Society of America* 59: 649-670

---- (1960) "Analysis of some recent geosynclinal theory." *American Journal of Science, Bradly Volume* 258-A: 126-136.

KOBAYASHI, Teiichi **(1941)** "The Sakawa orogenic cycle and its bearing on the origin of the Japanese Islands." *Journal of the Faculty of Science, Imperial University of Tokyo* Sec. 2.5: 219-578.

KUNUGI, Takashi **(2000)** "Relationship between Japan Meteorological Agency instrumental intensity and instrumental Modified Mercalli intensity obtained from K-NET strong-motion data." *Jishin* 53: 89-93 (in Japanese with English title).

KUSKY, TM; BF WINDLEY & M-G ZHAI **(2007)** "Tectonic evolution of the North China Block: from orogen to craton to orogen", pp. 1-34 in *Mesozoic sub-continental lithospheric thinning under Eastern Asia*, ed. by M-G ZHAI *et al.* The Geological Society of London.

LE MAITRE, RW (ed.) **(2002)** *Igneous rocks. A classification and glossary of terms. Recommendations of the International Union of Geological Sciences Subcommission on the Systematics of Igneous Rocks*. Cambridge University Press.

LOMNITZ, Cinna **(1974)** *Global tectonics and earthquake risk*. Developments in geotectonics 5. Amsterdam: Elsevier Scientific Pub. Co.

MA, Shouxian; Zongqi WANG & Yingli ZHANG *et al.* **(2018)** "Bainaimiao arc as an exotic terrane along the northern margin of the North China Craton: evidences from petrography, zircon U-Pb dating, and geochemistry of the Early Devonian deposits." *Tectonics* 38: 2606-2624 [DOI: 10.1029/ 2018TC005426].

MALPAS, J **(1992)** "Serpentine and the geology of serpentinized rocks", pp. 7-30 in *The ecology of areas with serpentinized rocks: a world view*, ed. by BA ROBERTS & J PROCTOR. Dordrecht: Kluwer Academic Publishers.

MARUYAMA, Shigenori **(1997)** "Pacific-type orogeny revisited: Miyashiro-type orogeny proposed." *The Island Arc* 6.1: 91-120.

MARUYAMA, Shigenori; Yukio ISOZAKI & Gaku KIMURA *et al.* **(1997)** "Paleogeographic maps of the Japanese Islands: plate tectonic synthesis from 750 Ma to the present." *The Island Arc* 6.1: 121-142.

MARUYAMA, Shigenori; J MASAGO & I KATAYAMA *et al.* (**2010**) "A new perspective on metamorphism and metamorphic belts." *Gondwana Research* 18: 106-137.

MARUYAMA, Shigenori; Soichi OMORI & Hiroaki SENSHU *et al.* (**2011**) "Pacific-type orogens: new concepts and variations in space and time from present to past." *Chigaku Zasshi* 120.1: 115-223 (in Japanese with English title and abstract).

MATSUDA, T; K NAKAMURA & A SUGIMURA (1967) "Late Cenozoic orogeny in Japan." *Tectonophysics* 4.4-6: 349-366.

MATSUDA, TE & S UYEDA (**1971**) "On the Pacific-type orogeny and its model: extension of the paired metamorphic belts concept and possible origin of marginal seas." *Tectonophysics* 11: 5-27.

MATSU'URA, RS (**2017**) "A short history of Japanese historical seismology: past and the present." *Geoscience Letters* 4.3: 15pp.

MCCLAY, KR & JA HAMMERSTEIN (eds) (**2020**) *Passive margins: tectonics, sedimentation and magmatism*. GSL Special Publications 476. Geological Society of London.

MCKENZIE, DP & RL PARKER (**1967**) "The North Pacific: an example of tectonics on a sphere." *Nature* 216: 1276-1289 [DOI: 10.1038/2161276a0].

MIDDLEMOST, Eric AK (**1985**) *Magmas and magmatic rocks: an introduction to igneous petrology*. Harlow, UK: Longman Scientific & Technical.

Mindat (**2010-12**) "Chert vs. flint" [https://www.mindat.org/mesg-207246.html]

---- (n.d.) "Chert" [www.mindat.org/min-994.html].

MITCHELL, AHG & HG READING (**1986**) "Sedimentation and tectonics", pp. 471-519 in *Sedimentary environments and facies*, ed. by HG READING. Oxford: Blackwell Scientific.

MIYASHIRO, Akiho (**1959**) "Abukuma, Ryoke and Sanbagawa metamorphic belts." *Journal of the Geological Society of Japan* 65.769: 624-637 (in Japanese with English abstract).

---- (**1961**) "Evolution of metamorphic rocks." *Journal of Petrology* 2: 277-311.

---- (**1973*a***) *Metamorphism and metamorphic belts*. London: George Unwin & Allen (English translation of 1965 Japanese text).

---- (**1973*b***) "Paired and unpaired metamorphic belts." *Tectonophysics* 17: 241-154.

MOORE, Gregory F (**2011**) "Structural restoration of thrusts at the toe of the Nankai Trough accretionary prism off Shikoku Island, Japan: implications for dewatering processes." *Geochemistry Geophysics Geosystems* 12.5: 15pp. [DOI: 10.1029/2010GC003453].

MORENO, Teresa; Simon WALLIS & Tomoko KOJIMA *et al.* (eds) (**2016**) *The geology of Japan*. Geological Society of London.

NAGATA, Hidehisa (**2018**) "Studies on nontectonic geologic structure in Japan: recent progress and future perspective." *Chishitsu Zasshi* 124.11 [DOI: 10.5575/geosoc.2018.0040] (in Japanese with English title and abstract).

NAKAJIMA, Takashi (**1997**) "Regional metamorphic belts of the Japanese Islands." *The Island Arc* 6.1: 69-90.

NAKAMA, Takaaki; Takafumi HIRATA & Shigeru OTOH *et al.* (**2010**) "Paleogeography of the Japanese Islands: age spectra of detrital zircon and provenance history of the orogen." *Chigaku Zasshi* 119.6: 1161-1172 (in Japanese with English title and abstract).

NAKAMURA, Kazuaki (**1989**) *Volcanoes and plate tectonics.* Tokyo University Press (in Japanese).

NAMIKI, Atsuko; Eleonora RIVALTA & Heiko WOITH *et al.* (**2016**) "Sloshing of a bubbly magma reservoir as a mechanism of triggered eruptions." *Journal of Volcanology and Geothermal Research* 320: 156-171.

NANCE, R Damian & Margaret D THOMPSON (eds) (**1996**) *Avalonia and related peri-Gondwanan terranes of the circum-North Atlantic*. GSA Special Paper 304. Boulder, CO: Geological Society of America.

NELSON, Stephen A (**2011**) "Inosilicates (pyroxenes and amphiboles)." *EENS 2120 Petrology* [www.tulane.edu/~sanelson/eens211/inosilicates.htm].

NELSON, Stephen A (**2018**) "Types of metamorphism." *EENS 2120 Petrology* [www.tulane.edu/~sanelson/eens212/typesmetamorph.htm].

NEUBAUER, Franz; Yongjiang LIU & Johann GENSER *et al.* (**2010**) "Paleozoic to Jurassic terrane accretion along the northeastern margin of Tibet plateau." *Geophysical Research Abstracts*, vol. 12. EGU General Assembly 02–07 May 2010. Vienna: European Geosciences Union.

NEWTON, Robert C (**2014**) "Book review: Metasomatism and the chemical transformation of rock." *Elements* 2014 (April): 155.

Niu, Yaoling (**2018**) "Origin of the LLSVPs at the base of the mantle is a consequence of plate tectonics: a petrological and geochemical perspective." *Geoscience Frontiers* 9: 1265-1278.

Nokleberg, Warren J; Leonid M Parfenov & James WH Monger *et al.* (**1999**) "Circum-North Pacific tectonostratigraphic terrane map." Open-File Report 94-714 pamphlet. Washington, DC: US Department of the Interior & USGS.

Nunn, Patrick D (**1994**) *Oceanic islands.* Oxford: Blackwell.

Oh, Chang Wan & Timothy Kusky (**2007**) "The Late Permian to Triassic Hongseong-Odesan Collision Belt in South Korea, and Its Tectonic Correlation with China and Japan." *International Geology Review* 49.7: 636-657 [DOI: 10.2747/0020-6814.49.7.636].

Okamura, Yukinobu; Mahito Watanabe & Rie Morigiri *et al.* (**1995**) "Rifting and basin inversion in the eastern margin of the Japan Sea." *The Island Arc* 4: 166-181.

Omori, Soichi & Yukio Isozaki (**2011**) "Paleozoic Japan and the eastern extension of the collisional suture between the North and South China cratons." *Chigaku Zasshi* 120.1: 40-51 (in Japanese with English title and abstract).

Otofuji, Yo-ichiro (**1996**) "Large tectonic movement of the Japan Arc in late Cenozoic times inferred from paleomagnetism: review and synthesis." *The Island Arc* 5.3: 229-249.

Otofuji, Y & T Matsuda (**1983**) "Paleomagnetic evidence for the clockwise rotation of southwest Japan." *Earth and Planetary Science Letters* 62: 349-359.

Otofuji, Y; T Matsuda & S Nohda (**1985**) "Opening mode of the Japan Sea inferred from the paleo-magnetism of the Japan arc." *Nature* 317: 603-604.

Philpotts, Anthony R & Jay J Ague (**2009**) *Principles of igneous and metamorphic petrology*. Cambridge University Press.

Piper, John DA (**2013**) "A planetary perspective on Earth evolution: lid tectonics before plate tectonics." *Tectonophysics* 589: 44-56.

PNSN (**n.d.**) "Volcanic earthquakes." *Pacific Northwest Seismic Network* [https://pnsn.org/outreach/earthquakesources/volcanic].

RGAFJ (Research Group for Active Faults in Japan [Katsudansō Kenkyūkai]) (ed.) (**1980**) *Active faults in Japan, sheet maps and inventories.* University of Tokyo (in Japanese).

---- (**1991**) *Active faults in Japan - distribution maps and explanatory materials.* Tokyo University Press (in Japanese).

---- (**1992**) *Maps of active faults in Japan with an explanatory text.* University of Tokyo Press [condensed version of Katsudansō Kenkyūkai 1991 in Japanese and English].

Robinson, Paul T & Mei-fu Zhou (**2007**) "The origin and tectonic setting of ophiolites in China." *Journal of Asian Earth Sciences* 32: 301-307.

Rogers, Nick (**2001**) *From rifting to drifting: mantle plumes and continental break-up*. Milton Keynes, UK: The Open University.

Roman, Diana C & Katherine V Cashman (**2006**) "The origin of volcano-tectonic earthquake swarms." *Geology* 34.6: 457-460 [DOI: 10.1130/G22269.1].

Roman, Diana; Mel Rodgers & Matt Gardine *et al.* (**n.d.**) "The origin of volcano-tectonic (VT) earthquake swarms." Powerpoint, 33 slides [vhub.org › resources › download › Roman_VT].

Safonova, I Yu & M Santosh (**2014**) "Accretionary complexes in the Asia-Pacific region: tracing archives of ocean plate stratigraphy and tracking mantle plumes." *Gondwana Research* 25: 126-158.

Santosh, M; Dapeng Zhao & Timothy Kusky (**2010**) "Mantle dynamics of the Paleoproterozoic North China craton: a perspective based on seismic tomography." *Journal of Geodynamics* 49: 39-53.

Satake, Kenji & Brian F Atwater (**2007**) "Long-term perspectives on giant earthquakes and tsunamis at subduction zones." *Annual Review, Earth and Planetary Sciences* 35: 349-374]DOI: 10.1146/annurev.earth.35.031306.140302].

Scholz, Christopher H (**1990**) *The mechanics of earthquakes and faulting.* Cambridge University Press.

Sengör, AMC & Boris A Natal'in (**1996**) "Turkic-type orogeny and its role in the making of the continental crust." *Annual Review of Earth and Planetary Sciences* 24: 263-337.

Seno, T (**1995**) *Foundation of plate tectonics.* Tokyo: Asakura Shobo (in Japanese).

Senshu, Hiroki; Shigenori Maruyama & Shuji Rino (**2010**) "Innovation of thermal evolution model of mantle." *Chigaku Zasshi* 119.6: 1215-1227 (in Japanese with English title and abstract).

SMITH, Peter J & Hazel RYMER (**2001**) *The physical and chemical properties of the earth.* Milton Keynes, UK: The Open University.

STEER, Philippe; Louise JEANDET & Nadaya CUBAS *et al.* (**2020**) "Earthquake statistics changed by typhoon-driven erosion." *Scientific Reports* 10.10899 [DOI 10.1038/s41598-41020-67865-y].

STERN, Robert J (**2002**) "Subduction zones." *Reviews of Geophysics* 40.4: 3.1~13.

STREKEISEN, Alex (**2020**) "Alex Strekeisen: I vetrini della mia fantasia" [www.alexstrekeisen.it/english/index.php].

SUGIMURA, Arata (**1965**) "Distribution of volcanoes and seismicity of the mantle in Japan." *Bulletin of the Volcanological Society of Japan* 10.(special issue): 37-58 (in Japanese with English title and abstract).

SUGIMURA, Arata & Seiya UYEDA (**1973**) *Island arcs: Japan and its environs.* Developments in Geotectonics 3. Amsterdam: Elsevier (reprinted in 2013).

SUZUKI, Kazue; Shigenori MARUYAMA & Shinji YAMAMOTO *et al.* (**2010**) "Have the Japanese Islands grown? Five 'Japan's were born, and four 'Japan's subducted into the mantle." *Chigaku Zasshi* 119.6: 1173-1196 (in Japanese with English title and abstract).

SUZUKI, Yasumoto (**2001**) "Kiyoo Wadati and the path to the discovery of the intermediate-deep earthquake zone." *Episodes* 24.2: 118-123 (in English).

---- (**2004**) "Kiyoo Wadati's contribution to seismology and study of ground subsidence." *Chikyu Kagaku* 58: 61-64 (in Japanese).

TAIRA, Asahiko (**1990**) *The Birth of the Japanese Islands.* Iwanami Shinsho 148. Tokyo: Iwanami (in Japanese).

---- (**2001**) "Tectonic evolution of the Japanese island arc system." *Annual Review of Earth and Planetary Sciences* 29: 109-134.

TAIRA, Asahiko; Timothy BYRNE & Junichiro ASHI (eds) (**1992**) *Photographic atlas of an accretionary prismn: geologic structures of the Shimanto Belt, Japan.* Tokyo: University of Tokyo Press.

TAIRA, A; I HILL & J FIRTH *et al.* (**1992**) "Sediment deformation and hydrogeology of the Nankai Trough accretionary prism: synthesis of shipboard results of ODP Leg 131." *Earth & Planetary Science Letters* 109: 431-450.

TAIRA, Asahiko; Shoichi KIYOKAWA & Kan AOIKE *et al.* (**1997**) "Accretion tectonics of the Japanese islands and evolution of continental crust." *Earth & Planetary Sciences* 325: 467-478 (in English with French title and abstract).

TAIRA, Asahito; Y OHARA & SR WALLIS *et al.* (**2016**) "Geological evolution of Japan: an overview", pp. 1-24 in T MORENO *et al.* (eds) (2016).

TAIRA, A; H OKADA & JH McD WHITAKER *et al.* (**1982**) "The Shimanto belt of Japan: Cretaceous-lower Miocene active-margin sedimentation", pp. 5-26 in *Trench-forearc geology: sedimentation and tectonics on modern and ancient active plate margins*, ed. by JK LEGGETT. Oxford: Blackwell Scientific.

TAKAGI, Hideo & Hiroyoshi ARAI (**2003**) "Restoration of exotic terranes along the Median Tectonic Line, Japanese Islands: overview." *Gondwana Research* 6.4: 657-668.

TAKAI, Fuyuji; Tatsuro MATSUMOTO & Ryuzo TORIYAMA (eds) (**1963**) *Geology of Japan.* Tokyo University Press (in English).

TANAKA, Keisaku & Tamotsu NOZAWA (**1977**) *Geology and mineral resources of Japan.* Kawasaki, Japan: Geological Survey of Japan, 3rd edition (in English).

TOMARI, Jiro (**2005**) "The concept of geosynclines and plate tectonics in Japan." *Kagakushi Kenkyu* 44.233: 23-32 (in Japanese with English title and abstract).

---- (**2008**) *Rejection and acceptance of plate tectonics - a history of Earth science in postwar Japan.* University of Tokyo Press (in Japanese).

TORSVIK, Trond H & L Robin M COCKS (**2016**) *Earth history and palaeogeography.* Cambridge University Press.

TOTMAN, Conrad (**2000**) *A history of Japan.* Malden, MA: Blackwell.

TSUCHIYA, Nobutaka; Tomoyo TAKEDA & Kenichiro TANI *et al.* (**2014**) "Zircon U–Pb age and its geological significance of late Carboniferous and Early Cretaceous adakitic granites from eastern margin of the Abukuma Mountains, Japan." *Journal of the Geological Society of Japan* 120.2: 37-51 (in English).

TSUJIMORI, Tatsuki (**2010**) "Paleozoic subduction-related metamorphism in Japan: new insights and perspectives." *Chigaku Zasshi* 119.2: 294-312 (in Japanese with English title and abstract).

TUCHOLKE, Brian E (**1998**) "Discovery of 'megamullions' reveals gateways into the ocean crust and upper mantle." *Oceanus Magazine* 41.1 [www.whoi.edu/oceanus/viewArticle.do?id=2392&archives=truel].
UEDA, Hayato (**2010**) "Structural evolution of forearcs by subduction erosion of accretionary complex." *Chigaku Zasshi* 119.2: 362-377 (in Japanese with English title and abstract).
---- (2016) "Hokkaido", pp. 201-221 in T MORENO *et al.* (eds) (2016).
UEDA, T & A KATO (**2019**) "Seasonal variations in crustal seismicity in San-in district, southwest Japan." *Geophysical Research Letters* 46: 3172-2179.
UYEDA, Seiya (**1971**) *New view of the earth.* Iwanami Shinsho 779. Tokyo: Iwanami (in Japanese).
---- (1978) *The new view of the earth: moving continents and moving oceans.* New York: Freeman.
UYEDA, Seiya & Akiho MIYASHIRO (**1974**) "Plate tectonics and the Japanese Islands: a synthesis." *Bulletin of the Geological Society of America* 85: 1159-1170.
VINE, FJ & DH MATTHEWS (**1963**) "Magnetic anomalies over oceanic ridges." *Nature* 199: 947-949.
VOOSEN, Paul (**2016**) "Graveyard of cold slabs mapped in Earth's mantle." *Science* 354.6315: 954-955.
WALLIS, Simon Richard & Takamoto OKUDAIRA (2016) "Paired metamorphic bests of SW Japan: the geology of the Sanbagawa and Ryoke metamorphic belts and the Median Tectonic Line", pp. 101-124 in MORENO *et al.* (eds) (2016).
WHITE, Lloyd T; Nicholas RAWLINSON & Gordon S LISTER *et al.* (**2019**) "Earth's deepest earthquake swarms track fluid ascent beneath nascent arc volcanoes." *Earth & Planetary Science Letters*: 25-36.
WIEMER, Daniel; Christoph E SCHRANK & David T MURPHY *et al.* (**2018**) "Earth's oldest stable crust in the Pilbara Craton formed by cyclic gravitational overturns." *Nature Geoscience* 11: 357-361.
WILLIAMS, Felicie & Halka CHRONIC (**2002**) *Roadside geology of Colorado.* Missoula, MT: Mountain Press.
WILSON, John Tuzo (**1963**) "Evidence from islands on the spreading of ocean floors." *Nature* 197.4867: 536-538.
---- (**1965**) "A new class of faults and their bearing on continental drift." *Nature* 207.4995: 343-347.
---- (**1966**) "Did the Atlantic close and then re-open." *Nature* 211: 676-681.
WILSON, Robert W; GA HOUSEMAN & KJW MCCAFFREY *et al.* (eds) (**2019**) *Fifty years of the Wilson Cycle concept in plate tectonics.* GSL Special Publication 470. Geological Society of London.
WINDLEY, B (**1992**) "Proterozoic collisional and accretionary orogens", pp. 419-446 in *Developments in Precambrian Geology, vol. 10*, ed. by KC CONDIE. Amsterdam: Elsevier.
XIAO, Wenjiao; Brian F WINDLEY & Ji HAO *et al.* (**2003**) "Accretion leading to collision and the Permian Solonker suture, Inner Mongolia, China: termination of the Central Asian orogenic belt." *Tectonics* 22.6: #1069 [DOI: 10.1029/2002TC001].
XIAO, Wen-jiao; Liang-chang ZHANG & Ke-zhang QIN *et al.* (**2004**) "Paleozoic accretionary and collisional tectonics of the eastern Tianshan (China): implications for the continental growth of Central Asia." *American Journal of Science* 304: 370-395.
YAMAMOTO, Shinji (**2010**) "Tectonic erosion: new perspectives on Pacific-type orogeny and continental growth models." *Chigaku Zasshi* 119.6: 963-998 (in Japanese with English title and abstract).
YAN, Chaolei; Linangshu SHU & Michel FAURE *et al.* (**2019**) "Time constraints on the closure of the Paleo-South China Ocean and the Neoproterozoic assembly of the Yangtze and Cathaysia blocks: insight from new detrital zircon analyses." *Gondwana Research* 73: 175-189.
YANG, Jingsui; Paul T ROBINSON & Yildrim DILEK (**2015**) "Diamond-bearing ophiolites and their geological occurrence." *Episodes* 39.4: 344-364.
YUAN, Sihua; Guitang PAN & Liquan WANG *et al.* (**2009**) "Accretionary orogenesis in the active continental margins." *Earth Science Frontiers* 16.3: 31-48.
ZHANG, L; Z ZHANG & H-P SCHERTL *et al.* (**2019**) *HP-UHP metamorphism and the tectonic evolution of orogenic belts.* GSL Special Publications SP474. Geological Society of London.
ZHAO, Dapeng (**2015**) "The 2011 Tohoku earthquake (Mw 9.0) sequence and subduction dynamics in Western Pacific and East Asia." *Journal of Asian Earth Sciences* 98: 28-49.
ZHAO, Dapeng; Yukio ISOZAKI & Shigenori MARUYAMA (**2017**) "Seismic imaging of the Asian orogens and subduction zones." *Journal of Asian Earth Sciences* 145: 349-367.
ZHOU, Mei-fu; Yuxiao MA & Dan-Ping YAN *et al.* (**2006**) "The Yanbian Terrane (Southern Sichuan Province, SW China): a Neoproterozoic arc assemblage in the western margin of the Yangtze Block." *Precambrian Research* 144.1/2: 19-39.

Chapter 3

The Palaeogeographic Compilation of the Japanese Basement

Beginning with *Nihon Shoki*'s mythical view of the creation of Japan, this Chapter introduces the geotectonic belt system of the current Japanese Islands, then goes back to the Precambrian era to track the formation of these geological belts as compiled at the edge of the Eurasian continent. The Chapter ends with the rifting of the Japan Sea Basin in the Miocene, isolating those belts in an off-shore archipelago that became the Japanese Islands.

From Myth to Plate

The *Nihon Shoki* 8c-chronicles of Japan state that (Aston 1972: 10-12): the male and female gods Izanagi and Izanami...

> stood on the floating bridge of Heaven, and held counsel together, saying:
> "Is there not a country beneath?" Thereupon they thrust down the jewel-spear
> of Heaven, and groping about therewith found the ocean. The brine which
> dripped from the point of the spear coagulated and became an island...

In such a manner were the eight major islands formed, according to the Japanese creation myths. It is significant in this context that one of the early tectonic plates recognized as having a role in the formation of the Japanese Islands has been dubbed the Izanagi Plate by Japanese researchers. This plate existed between 220 and 75 million years ago and was one of the forerunners of the Pacific Plate, now the largest player in the Ring of Fire.

However, the history of the Japanese landmass is far older than the Izanagi Plate, and the current islands retain materials from several previous eras of 'supercontinents' (Evans 2016). Supercontinents (and semi-supercontinents) are thought to be brought together by cold downwelling mantle plumes and then break up through upwelling magmatic activity underneath them (Murphy & Nance 2013; Santosh 2013). It is logical that during all this movement over such an enormous amount of time, the Japanese landmass did not always have the shape it possesses today – in contrast to the assumption in geosyncline theory that it was always an island arc off Eurasia. Nevertheless, how dramatically that landmass has changed through the successive geological eras has just recently come to light. One of the more exciting recent discoveries is that the current Japanese Islands had several archipelagic predecessors, formed and consumed during geological history (Isozaki & Maruyama *et al.* 2010).

The geotectonic units that make up Japanese basement rocks are traditionally called 'belts' (*tai*). These may be termed 'terranes' (cf. Taira *et al.* 2016), but this imported term is avoided by some Japanese geologists – who have referred to it as "terrane plague" (Kasahara *et al.* 2010: 955) – because it too often implies an exotic (allochthonous) origin and is testimony to the nature of 'colonial geology' in Japan (see Chapter 2.2) (Isozaki & Aoki *et al.* 2010). Here, 'belt' will be retained in the formal names of the geotectonic units, but they may be referred to as 'terranes' in discussion; in some cases, several belts, however, have been grouped into a more inclusively named South Kitakami Terrane and treated as an 'exotic terrane' (e.g., Tazawa 2001).

This Chapter begins by presenting the several geological belts that make up the archipelago today (listings are given in Appendix 3.1). These belts form the basement rocks of the Japanese landmass but do not include the more recent sedimentary cover in which volcanism played a large part. Below, the formation of the geotectonic belts is traced successively through time, beginning approximately 520 Ma and ending with the collision of the Indian sub-continent with Eurasia from ca. 55 Ma (cf. Apx 3). This collision is thought to have resulted in significant structural realignments of the geotectonic belts along the Eurasian coast; this theme is continued in Chapter 4 which provides details of the rifting processes and the topographic development of the islands thereafter. Volcanism is a phenomenon that comes to the fore during the rifting of the Japanese archipelago from the continent (cf. Chapter 4), and it accounts for much of the Quaternary sedimentary cover of modern-day Japan (cf. Chapter 6). The sequence of events covered in Chapters 3 and 4.

Japan's Geological Belts

Classifications

The geotectonic belts (Figure 3.1) that make up the Japanese landmass are neither homogeneous in content nor coherent in structure. They constitute Japan's basement rocks, and because they are relatively old, their original portions are most likely divided up and distributed across the landscape through processes described in Chapter 2: successive faulting, tectonic rearrangement, transformation through metamorphism, igneous intrusion, and erosion. These, in addition to the original heterogeneity of contents within Accretionary Complexes (AC), make for an extremely complicated geology: the 'devil is in the detail'. Nevertheless, Plate Tectonics allows for a processual grasp of why and how such complex structures arose, and in doing so, it provides the tools for greater understanding of Japanese geology than allowed for by the standard descriptions of lithostratigraphic Groups, Formations, Members, and rock types occurring in different areas.

The take-away for archaeologists upon knowing the dates, constituents, and transformations of the different geotectonic belts is the ability to develop some expectations of the lay of the land and available resources where they are working, to recognize similar conditions elsewhere in the country for comparison, and to acknowledge differences in resources that might affect potential trade & exchange.

Comparing Figure 3.1 with other renditions reveals many differences: belts are differently named, divided, combined, newly identified, or given different abbreviation tags, etc., as research progresses (cf. Isozaki 1996; Isozaki & Maruyama *et al.* 2010; Taira *et al.* 2016). This is as it should be but makes it rather difficult to understand their relationships: a few tectonically rearranged terranes have different names in different parts of the country but are of the same origin, while others have been divided and given different names. The 2010 map version is chosen here for representation since the colours clearly show relationships over distance, though the map is not as detailed as that in Taira *et al.* (2016: fig. 1.6); or Isozaki (2019: fig. 2). Descriptions of each geological belt are given in Appendix 6, though their complicated and transformative histories (Figure 3.2) make it difficult to describe them in brief; see Wakita *et al.* (2018) for narrative descriptions.

Very early (Palaeozoic) belts altogether occur in very tiny amounts (ca. 5% of Japan's basement) and are very shallow in depth (Isozaki 2019); they are thought to have been subjected to significant tectonic erosion. Most of the archipelagic basement dates to the Jurassic–Cretaceous Periods and is overlain by more recent (Palaeogene–Neogene Period) rocks and sediments of volcanic and sedimentary origin. Any of these original rocks might have been affected by repeated metamorphism resulting in overprinting of mineral types.

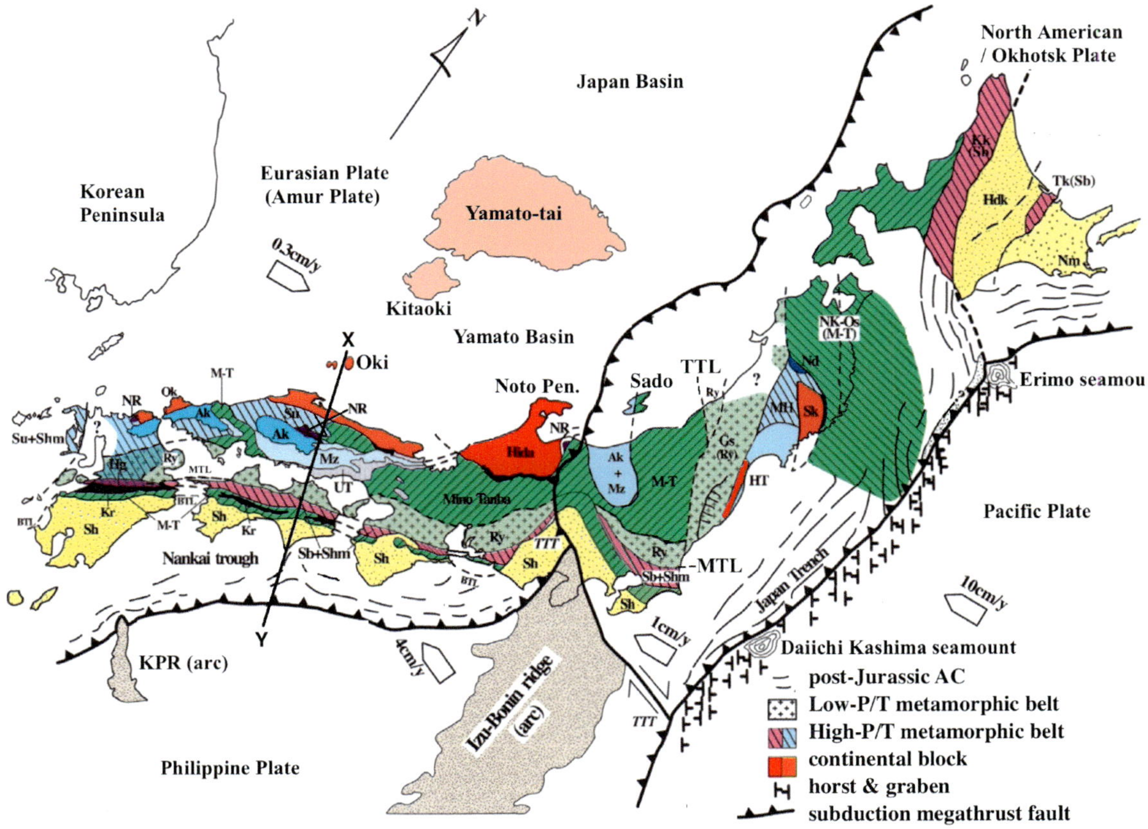

3.1a: As the belts appear in Japan today, but not in the location they were originally compiled. Belt name abbreviations and brief descriptions are given in Appendix 3.1; contents of select belts are shown in Figure 3.2; for more explanations of individual belts, see Wakita (2013) and Wakita *et al.* (2018); for very detailed regional accounts see Moreno *et al.* (2016). The X–Y transect is shown below.

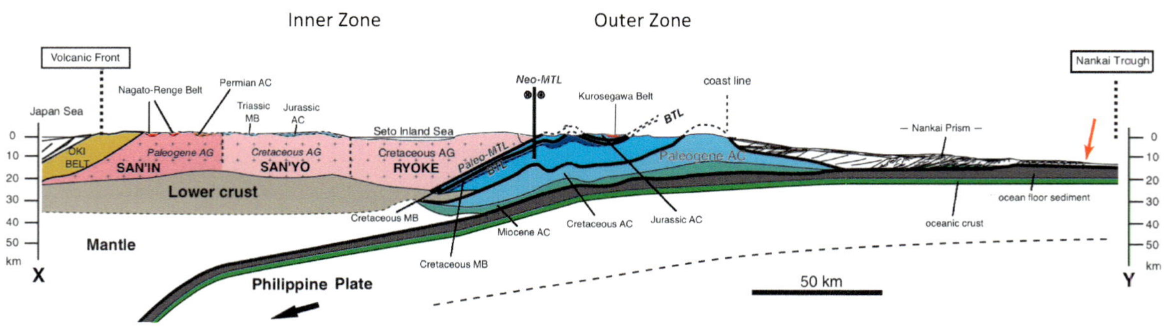

3.1b: Section through Southwest Japan, divided by the neo-MTL on the surface into the landward Inner Zone and oceanward Outer Zone. Arc granite (AG) basements (pink) are thick in the Inner Zone; note that the youngest AG (Palaeogene) are furthest west (cf. Chapter 6, Apx 10). These AG are topped by shallow Accretionary Complexes (AC) that have mostly been metamor-phosed; in the Outer Zone are the thick AC bordered by the palaeo-MTL at depth. For the structure of the currently accumulating submarine Nankai Prism (red arrow), see Figure 2.7.

Figure 3.1 The geotectonic belts of Japan

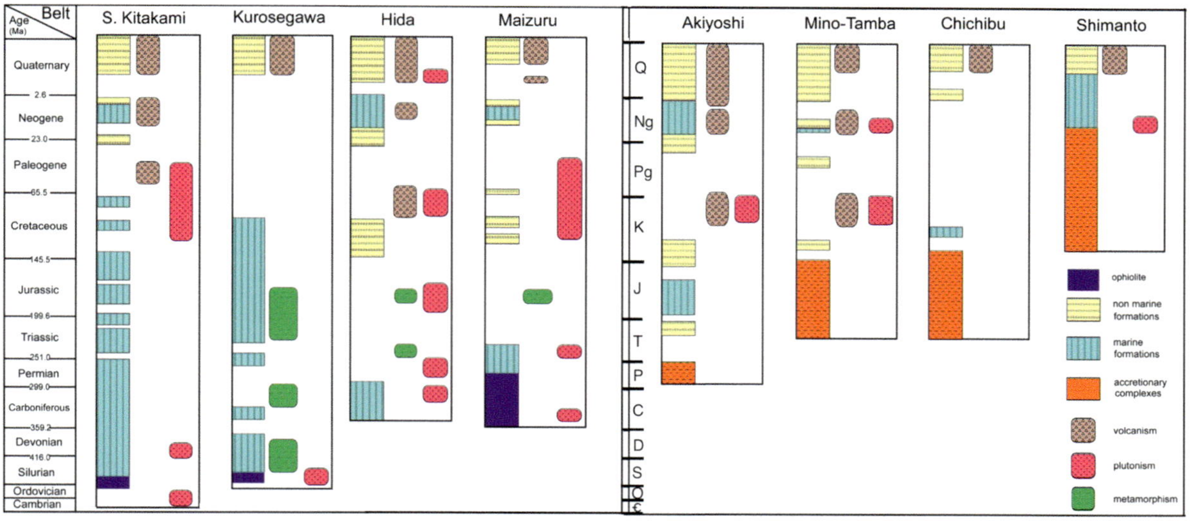

Figure 3.2 Examples of the complex contents of select geotectonic belts, as they were compiled and transformed through time

80% Accretionary Complexes

Accretionary processes (Chapter 2.4.4) have created the basic fault-bounded structure of geotectonic units in Japan, resulting in Accretionary Complexes (AC) that generally run parallel to the plate edge or continental coastline. AC account for 80% of the current archipelagic basement (Taira *et al.* 1997: 468; Isozaki & Aoki *et al.* 2010); many of these AC have been weakly metamorphosed. Geographically, the terranes are ordered by age from west to east: the oldest are in the west and higher in the stratigraphy; the youngest terrane lies to the east (oceanward) and is stratigraphically lowest. These positions are due to the nature of accretionary prism formation, with successive prisms added below the preceding ones by the plate subducting from the east; this is Isozaki's 'downward-younging polarity' (1996: 307). Accretionary activities can be divided into six stages and their remains are coordinated from the Primorye to Australia, including Japan (see Wakita *et al.* 2018).

Taira *et al.* (2016) divide rocks in AC into 1) those comprising shelf sediments (limestone, clastics, tuff, but no chert or mafic/ultramafic rocks) and 2) those comprising ocean floor contents (chert, seamount greenstone and their limestone reefs, and ocean mafic/ultramafic rocks (cf. Figure 2.4). Erosional products characteristically comprise trench fill – products eroded from the landmass that accumulate in the subduction trench. Although shelf sediments may occur as stratified units, most other rocks occur as small particles or larger blocks in chaotic (mélange) organization. However, ophiolites – which can include ultra-mafic mantle rocks of the ocean floor or fragments of large igneous provinces (LIPs) produced by hot spot volcanism (Hirsh & Ishida 2002) – can be very large additions to the AC. Their contents ceased to be added to after the end of that accretionary episode; however, they may later be intruded by igneous rocks and overlain by more recent rocks.

From these descriptions, it should be clear that AC contain a variety of rocks of different ages and cannot be briefly annotated. Moreover, to determine the age of accretion, the dates of all earlier rocks must be discounted. For example, the Sanbagawa protolith contained mid-ocean ridge basalt (MORB), ocean floor rocks (bedded chert, etc.), and trench fill of eroded continental material (Okamoto *et al.* 2000). Dates from detrital zircons contained in these materials occur in the following proportions: Archean

22.2%, Permian 8.9%, Triassic 8.9%, Jurassic 48.9%, Early Cretaceous 4.4%, Late Cretaceous 6.7% (Nagata *et al.* 2017; see also Wakita *et al.* 2018).

20% Metamorphics and granites

It is logical to think that the remaining 20% of basement consists of metamorphic belts (MB) and later granitic intrusives from different periods (cf. Apx 10). The 'protoliths' of metamorphic belts may be AC of earlier formation, with their transformed mineral assemblages dependent on the depth (pressure) and temperature where they were metamorphosed (cf. Figure 2.8). Thus, some AC may fit into both categories, once as AC in terms of original contents and MB in terms of transformation. A different but relatively common type of metamorphic rock existing in the geological belts is serpentinite, formed via metasomatism: water intrusion into ultra-mafic rocks (cf. Apx 5: Table B). Serpentinite melangés are common within AC, and serpentinized mantle rock characteristically occur in ophiolites. In any case, it is important to recognize that two situations prevail: whole belts can be metamorphic in nature (MB), and previously metamorphosed rocks can occur within AC. By volume, various kinds of gneiss and schist account for the majority of metamorphic rocks in Japan (GSJ 1992: 103-108).

Metamorphic Belts (MB)

Isozaki (Isozaki 2019: fig. 3) proposes that five instances of high P/T metamorphism in different geological terranes are potentially linked to ridge subduction. The most prominent of these is the Sanbagawa Metamorphic Belt (MB), the high P/T belt contrasted with the Ryoke low P/T in Miyashiro's original formulation of "paired metamorphic belts" (cf. BOX 1). The Ryoke MB is a result of contact metamorphism of the Mino-Tanba AC into which the Ryoke granites intruded. Thus, a distinction must be made between the Ryoke granites (igneous) and Ryoke MB (metamorphosed AC) as discussed below. The Sanbagawa MB is similarly complicated: it has been discovered to incorporate at least two different ACs, formed and metamorphosed at different times; also, the geographical extents of these separate components have now been revised (Tsutsumi & Miyashita 2009; Aoki *et al.* 2010; Otoh *et al.* 2010; Isozaki & Aoki *et al.* 2010). Although these scholars refer to the two components as the Sanbagawa Metamorphic Belt *sensu stricto* (from the Sanbosan AC) and the Shimanto Metamorphic Belt (a portion of the North Shimanto AC) (Isozaki & Aoki *et al.* 2010: 8), others recommend a different naming pattern with Jurassic, Early Cretaceous, and Late Cretaceous prefixes to Sanbagawa MB portions in order to reflect their different formation times (Knittel *et al.* 2014). Research published before, or relying on articles published before, 2010 will refer to the Sanbagawa MB in a wholistic manner (*sensu lato*) without the internal division, and it is not clear whether all authors thereafter have adopted the divisional terminology. Moreover, the formation and metamorphism of the belt(s) are undergoing radical re-dating, with a consensus yet to be built, as discussed further below. Finally, the belt(s) have different compositions in different areas of Japan over its 1000 km length; thus, one has to be familiar with the structure and contents of the local presentation to understand the local use of its resources.

The Sanbagawa schists and steatite were particularly important for artefact manufacture in the Yayoi and Kofun periods. The schists served well as reaping tools before iron was introduced, while exposures in the Kii River valley south of the Nara Basin (cf. Chapter 13) and sources in the Kanto were particularly mined for steatite to make imitation objects and disc beads.

Other belts that consist of metamorphosed Accretionary Complexes are the Chizu MB, Kamuikotan MB, Nagato-Renge MB, Ryoke MB, Shimanto MB, Suo MB, and Tokoro MB (see Apx 6). These are variously metamorphosed weakly to strongly in contexts varying from subduction–exhumation to contact metamorphism. In addition, continental basement rocks that were attached to the Japanese landmass during rifting in the Miocene are MB: "Hida and Oki blocks [belts] consist mainly of metamorphic rocks such as psammitic [sandy] and pelitic [silty/clayey] biotite gneiss" which were metamorphosed from "a wide variety of protoliths (granite, basalt, sandstone, mudstone and limestone) of Permo-Carboniferous age" (Watanabe *et al.* 2016: 431, square brackets added).

Faults and tectonic lines (TL)

All of these geotectonic terranes are fault-bounded. Considerable work has been expended in identifying the kinds of faults (cf. Figure 2.10) that separate each type of unit, how they have activated over time, and how important they have been in constraining the developmental sequence of the Japanese landmass (Isozaki & Maruyama *et al.* 2010; Yanai *et al.* 2010). Certain structural faults are designated as 'tectonic lines' (TL), though interestingly, these big, recently active faults are classified as steep-angle normal strike-slip faults of minor fourth-class significance, such as the neo-MTL, Itoigawa-Shizuoka TL, Kanto (Toné, or Kashiwazaki-Choshi) TL, and the Tanakura TL. They represent boundaries of microplates which were activated relatively recently during and after the opening of the Japan Sea. Interestingly the latter three are *right-lateral* strike-slip faults cutting Honshu from NW to SE; but in Tohoku, several parallel NW–SE *left-lateral* strike-slip faults dating to the late rifting stage have been identified and were found to have exerted considerable control of quakes/aftershocks during the 2011 Tohoku-oki Earthquake (Baba & Yoshida 2020). Structurally more important are the older major E-W Tectonic Lines that divide the Japanese landmass from oldest to youngest, north to south: the palaeo-MTL, Nagato-Hida, Osayama, Omi, Ishigaki-Kuga, Butsuzo, and Aki TL; they are essentially sub-horizontal thrust faults which represent previous subduction planes (Isozaki & Aoki *et al.* 2010: fig. 8).

• The Tanakura Tectonic Line (TTL) is a strike-slip fault, which facilitated the movement of a continental fragment (Figure 3.1: Sk) into its current position on the eastern edge of Tohoku during back-arc basin formation. The TTL is considered the boundary between two distinct geological provinces in modern Japan: the western semi-continuous terrane distribution and the northeastern area of terrane fragmentation and repositioning – clearly visible in Figure 3.1. According to the restoration by Takagi & Arai (2003), the TTL was a continuation of the MTL before tectonic rearrangement.

• The MTL (Median Tectonic Line): much has been learned about the MTL since the early publication of Sugiyama (1973); however, there is disagreement about the nature of the fault, reviewed by Sato *et al.* (2015). Many authors regard it as a previous a subduction thrust-fault plane (Sato *et al.* 2015), running from the southern China to Siberia (e.g., Itoh *et al.* 2013: fig. 2a; Oshika Museum 2018), while others consider it a low-angle "mid-crustal detachment surface" which developed during the post-Palaeocene and became the "essential contact" between the Ryoke and Sanbagawa Belts (Isozaki *et al.* 2020: fig. 2); Wallis & Okudaira seem to follow the former view but show how it may have developed into the exhumation plane for the Sanbagawa MB (2016: fig. 2c.5).

In any case, the MTL is now discussed in two chronological senses, the palaeo-MTL (low-angle old plate interface/detachment surface) and the neo-MTL (dextral strike-slip fault), though the dates for these are highly contested. Many researchers consider the palaeo-MTL to date to the Cretaceous (e.g., Taira *et al.* 2016: 12; Sato *et al.* 2015: fig. 6-c) against the more recent view that it dates post-Palaeocene (Isozaki & Aoki *et al.* 2010). The neo-MTL (BOX 2) seems to have developed during the opening of the Japan Sea Basin (ca. 20–16 Ma) (Yanai 2010) (see Chapter 4), but it was reactivated in the Late Pliocene (ca. 3–2 Ma) as a right-lateral strike-slip fault (Toda 2016). Alternatively, Sato *et al.* (2015) propose that the neo-MTL might have developed as late as 800 ka in the Pleistocene in conjunction with as shift in the direction of subduction of the Philippine Plate to the WNW.

BOX 2 The Median Tectonic Line Museum

Japan's Unique Presentation of a Geological Fault

The Median Tectonic Line (MTL) is a major constructional fault in Japan, running from Kyushu to the Kantō region. How to put it in a museum? The Ohshika Geological Museum of Median Tectonic Line, established in 1993 in Ohshika Village, Nagano Prefecture, showcases this massive fault line. The museum's location is remote: not chosen for easy access but for proximity to the fault's excellent exposure at two nearby outcrops. The museum building itself is of recent modern construction, built in 1992 and funded by Oshika Village. This is a one-man museum (see mtl-muse.com); Mr. Kazuro KAWAMOTO served as Curator, researcher, and guide at the time of my visit to his museum; and it was a delight to avail myself to his fathomless enthusiasm for Japanese geology.

The ground floor exhibits tell of the formation of the Japanese Islands and describe the several geological belts that make up the archipelago. Small rock samples of every kind are on display, while much larger rocks sit outside the museum. Upstairs there is a model of the region with buttons to push elevating the landforms along the fault. Though the rocks are labeled in both Japanese and English, most written explanations are all in Japanese; however, Mr. Kawamoto spoke English from his five years in the US for those who could not read Japanese.

The rocks in the front courtyard are arranged in groups derived from the five geological belts that conveniently converge near the museum. These belts are are described in the ensuing sections (and in 3.1):

- the Mino-Tanba belt: a sedimentary belt (Accretionary Complex, AC) formed in the Jurassic
- the Ryoke belt: a Jurassic AC sedimentary belt with Cretaceous volcanics lightly metamorphosed by high-temperature/low-pressure
- the Sanbagawa belt: a pelitic and mafic AC schist belt metamorphosed at depth by low-temperature/high-pressure
- the Chichibu belt: another JurassicAC
- the Northern Shimanto belt: a Cretaceous AC

One might note from Figure 3.1 that these belts twist up and around the Kanto region via the Kanto Syntaxis, pushed out of line by the intrusion of the Izu Arc into central Honshu (see Chapter 4). The MTL generally forms the boundary between the Ryoke and Sanbagawa belts.

Outcrops of the Median Tectonic Line in Ohshika Village, Nagano

The fault exposures are somewhat removed from the museum, necessitating car travel. Going 3 km downstream from the Museum along the Koshibu River to the village hall and then 14 km upstream along the Kashio River, one comes to the Kitagawa outcrop; here the fault zone is only 20 m wide (left). At the second Ankoh outcrop, the MTL clearly separates the reddish granite Ryoke from the greenschist Sanbagawa Belt. Generally, the MTL fault zone can be characterized as consisting of high-angle right lateral faults and low-angle reverse-slip faults, but in the Ohshika area, the fault dips almost vertically. DWH 185 cm, 6'1" for scale

The neo-MTL divides SW Japan into Inner and Outer zones (cf. Figure 3.1b), and it demarcates the northern boundary of the Nankai forearc microplate (Isozaki & Aoki *et al.* 2010: 95), which is moving westwards with the Philippine Plate (cf. Figure 5.10). The organization of terranes in the Inner and Outer zones differ, especially in western Japan: Inner Zone AC thinly overlie granite foundations, while in the Outer Zone, AC are imbricated as geotectonic slabs overlying the subducting plate.

The paleo-MTL was earlier believed to have shifted Outer Zone strata more than 1000 miles northwards from off the southern China coast to their present position in Japan (Taira 1990, 2001; Tazawa 2001). This interpretation of "the once popular but unrealistic strike–slip fault-controlled tectonics in SW Japan" has now been discarded in favour of *in situ* sub-horizontal stacking of ACs by some researchers (Isozaki & Aoki *et al.* 2010: 87; Ito & Sato 2010).

Looking Forward

In the following sections, the development of these different geotectonic belts will be presented in a time sequence from Precambrian to the Miocene; the chronostratigraphic chart available in Appendix 3 will be useful for those like myself initially, who are unfamiliar the geological period successions. These time-period names will be used at a necessary minimum, but it is quite the case that most of the following discussion deals with events prior to the migrations of humans to the Japanese Islands, now estimated to have been about 42,000 years ago. At that point, they entered a landscape that had been half a billion years in the making but had been an archipelago for no longer than 16 million years.

Tracking Japan Throughout the Ages

Setting the stage

Palaeogeography, which is based on palaeomagnetic data, has been identified by Isozaki as the current "leading concept" in research on the formation of the Japanese landmass (Isozaki 2019: fig. 1) and more important even than Plate Tectonics itself. Scores of researchers are involved in mapping the development of the continents through billions of years of Earth's history, most emphatically presented in the map animations by the Paleomap Project (Scotese 2003). The work on reconstructing the palaeographic compilation of proto-Japan began in the 1990s (e.g., Maruyama *et al.* 1997). In the following sections, this work on the gradual piecing together of the geological belts will be summarized in a generalized manner necessitating vast oversimplifications (see Moreno *et al.* 2016 for a more detailed treatment). Much of the palaeogeographic proveniencing is often confusing and controversial, with researchers offering differing dates, interpretations of locations and affiliations, and geotectonic models – and the maps produced at various times will be conflicting. These will no doubt be refined in future and datings continue to be improved; meanwhile, the data offered below give only an outline of events. Moreover, it must be kept in mind that Figure 3.1 represents the current archipelagic form of Japan, whereas the belts themselves – *except for* the current submarine Nankai Accretionary Prism (cf. Figure 2.7) – all formed *before* the rifting of the archipelago between 20 and 16 Ma.

Several supercontinents frame the following story (Table 3.1). The earliest subduction evidence in Japan dates to ca. 500 Ma, at the edge of Gondwana continent. The position of the Japanese landmass in relation to the shifting of continental positions can be seen in Wakita *et al.* (2018: fig. 13). Once the North China Block (NCB) and South China Block (SCB) broke off from Gondwana, accretion continued along their margins. These blocks are important as the Japanese landmass incorporates material from both, and it was affected by the amalgamation of the China blocks to form 'the continent': eastern China and the Korean Peninsula. From the Permian onwards (300~ Ma), accretionary tectonics occurring periodically along the continental edge of the SCB resulted in the formation of Accretionary Complexes which now comprise the majority of Japan's basement geology. Another, perhaps unanticipated, aspect of this history is that the current oceanic plates subducting under Japan (Pacific Plate, Philippine Plate) are late arrivals. They were preceded in time by several other plates (Isozaki 2019: fig. 3) which supported different 'oceans' (Table 3.1). The chronological presentation of the development of the Japanese landmass below is given up to the point of archipelagic formation at 15 Ma, for which further details are presented in Chapter 4.

Table 3.1 Significant plates, continents, and oceans through time in relation to Japan

Period	**Plate**	**Ma**	**Ocean/Sea**	**Continent**
Neo-proterozoic		1580–1400 900 900~≥750		Nuna Yangtze+Cathaysia > SCB Rodinia
Palaeozoic	Nomo Renge Farallon	520–450 450–340 340–240 300	proto- Pacific Panthalassa	Gondwana ca. 520–300 Ma Island arc accretion to NCB SCB+NCB > EChina/Korean Peninsula
Palaeoz—Mesoz		320~180/160		Pangaea
Mesozoic	Izanagi	240–120	Panthalassa	
	Kula	120–80	Pacific	Current continental arrangements
Cenozoic	Pacific	80–present	Pacific	
	Philippine	60–present	Philippine	

The main point of tracing the development of the Japanese landmass here is to emphasize that it was part of Eurasia prior to the rifting of the archipelago and that aspects of its composition are closely related to that of present-day China Mainland and the Korean Peninsula. Santosh & Senshu (2011) are correct in declaring that Japanese geology cannot be understood in insular isolation but its analysis must take in global processes. Moreover, seeing how the archipelago was compiled allows us to see how complicated the regional geology is. Prof. Yoneta ICHIKAWA, of Nara Educational University, once remarked to me how much he admired the formation of the Rocky Mountains (my home area) – so simple compared to Japan (pers. comm. 1978).

Precambrian events (before 541 Ma)

The oldest rock fragment in Japan dates to 3.8 Ga and next oldest is 2.0–1.8 Ga; Santosh & Senshu (2011) suggest these might relate to the Nuna/Columbia supercontinent, though Kenorland? may be closer to that time period (Evans 2016). There is considerable controversy whether individual belts derived from the North China Block (NCB) or the South China Block (SCB). Detrital zircons are used to determine their sources: NCB zircon dates in China range from 2.6–1.6 Ga and the dominant zircon ages from the SCB are 1.2–0.7 (Wakita et al. 2018: 6). Detrital zircons found in Japan that yield these ages *pre-date* the formation of Japan's earliest geological belts but occur *within* them (Isozaki *et al.* 2015; Aoki *et al.* 2012). Invariably they point to erosional processes on continental blocks whose pre-existing products were later incorporated into AC or MB or even in extruded lavas. Wakita and colleagues (2018: 12) propose that the South Kitakami Terrane formed as an "immature island arc" along the SCB while it was still part of Gondwana; once the SCB and NCB broke away from Gondwana, then "all tectonic units became homogeneous, and were mainly from North China in Jurassic and Cretaceous times" (2018: 7, fig. 12a). At this time Gondwana contained parts of the NCB that would become the Hida and Oki Belts.

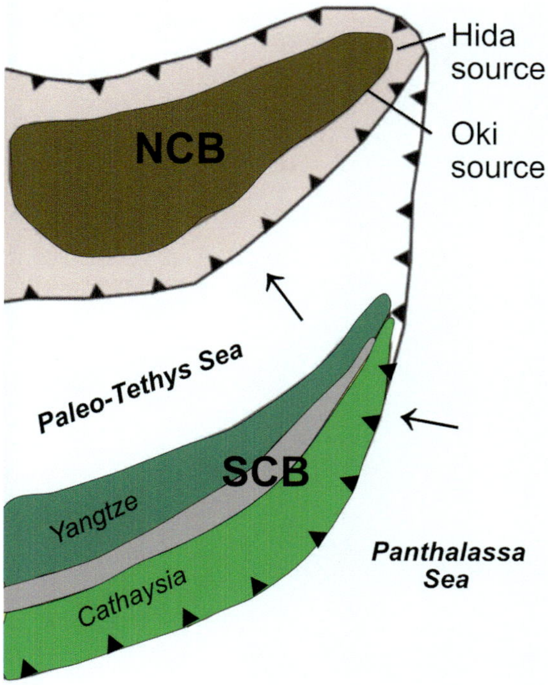

Figure 3.3 One Reconstruction of the possible positions of Hida and Oki while the North China Block and South China Block were still separated by the Palaeo-Tethys Sea but after the beginning of subduction processes.

Teeth indicate direction of subduction; arrows indicate direction of plate movement.

The oldest geological belts in Japan are the Hida and Oki which were detached from the continent during Miocene rifting; together these belts are "essentially a remnant fragment of *moderately deep-level* continental crust that was once part of the Asian continental margin" (Eihiro *et al.* 2016, italics added). They were previously understood to have had different origins: Hida deriving from the NCB and Oki deriving from the SCB (Yangtze +Cathysia), as shown in Figure 3.3. But new datings have revised this assessment (Santosh & Senshu 2011; Nakama *et al.* 2010). Dates of detrital zircons in the Oki Belt are assessed to be as old as the Hida Belt and match the date of the NCB at 2.0–1.8 Ga, leading to the conclusion that both Hida and Oki are considered to derive from NCB. In contrast, detrital zircons recovered in Japan from the SCB date only to ca. 1 Ga, between two and one billion years later than those in Hida-Oki. However, Isozaki has proposed a new interpretation (discussed below) and states that still, "no final agreement has been yet reached for years after long-lasting debate" (Isozaki 2019: 2). So, we must continue to 'watch this space'.

The South China Block is important because, in contrast to the Hida-Oki Belts, the Accretionary Complexes of Japan formed off the SCB. The SCB is composed of two components (Figure 3.3), the Yangtze and Cathaysia Blocks, which became joined together between 1.0–0.9 Ga via the accretion of the Shibao island arc (Zhang *et al.* 2015; Isozaki & Aoki *et al.* 2010). The assembly of the SCB occurred between the break-up of Nuna and the assemblage of the next supercontinent, Rodinia. The cause of the break-up of Rodinia between 875 and 720 Ma is identified as superplume activity (Li *et al.* 2008). At this time, the conjoined SCB was subjected to rifting along the same joint axis of the Shibao suture, now marked by the Nanhua Rift (Wang *et al.* 2011) (as shown in Figure 3.3).

Palaeozoic (ca. 541–300 Ma): evidence of early subduction

Around 600 Ma, the eastern continental edges of the NCB and SCB changed from passive margins to active margins with the beginning of subduction, but at 530 Ma these blocks were still widely separated. The evidence in Japan for this new subduction regime has now been pushed back from 450 Ma to 520 Ma (Kunugiza & Goto 2010; Santosh & Senshu 2011), as evidenced by fragments in today's Japanese Islands (Table 3.2) (cf. Isozaki 2019: fig. 4). The shift from passive to active margin caused slices of oceanic floor at the passive margin and in back-arc basins to be incorporated into the developing Japanese landmass. These were obducted as massive ophiolites (listed in Table 3.2) or dismembered and incorporated into serpentinite mélanges – such as fragments of the Oeyama Ophiolite in the Hida Marginal Belt (Ishiwatari *et al.* 2016; Tsukada *et al.* 2017) (Figure 3.4). This belt has also yielded the earliest date for magmatism, stimulated by subduction, at 520 Ma (Wakita *et al.* 2018: 2).

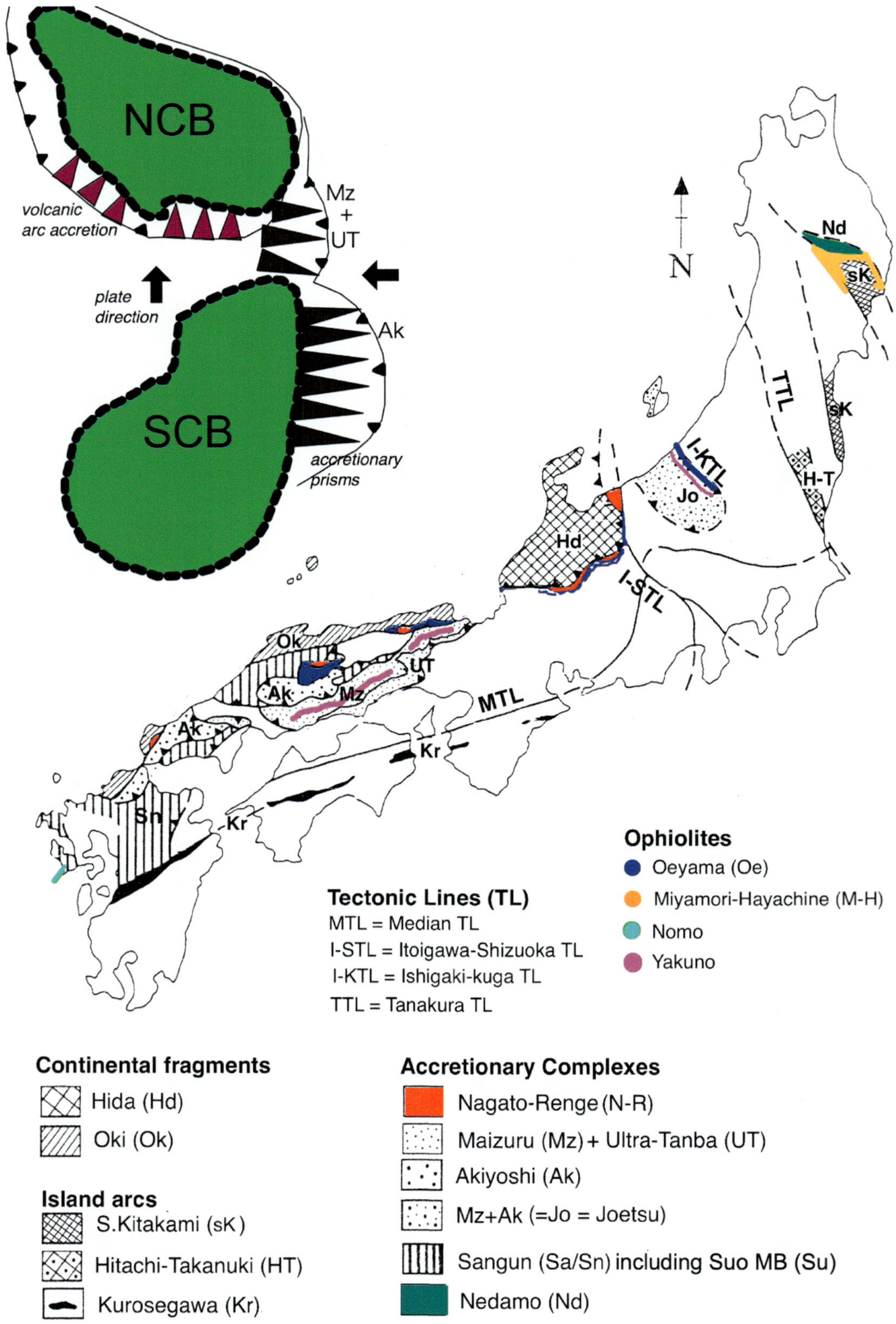

Figure 3.4 Continental fragments and earliest geotectonic belts of the Palaeozoic in Japan (ca. 540–250 Ma) Inset shows possible relations among Permian belts before the conjoining of the China Blocks; see text.

Table 3.2 Palaeozoic arc remnants, metamorphosed mantle rock (serpentinite), and ophiolites in modern Japan, indicating initial subduction processes

Unit	Date Ma	Location
island arc granite	510–500	Hitachi Belt; Isawagawa, Hikawa granites
felsic tuff	484–430	Hitoegawa
forearc sediments	472	Hida Gaien Belt
arc granites	450	Hikami, Mitaki, Yokokura, Yatsushiro
blueschist	450	Kitomyo
OPHIOLITES:		
Nomo	>500	NW Kyushu
Oeyama	530	SW Honshu Renge Belt
Miyamori-Hayachine	485–444	NE Honshu South Kitakami Belt

Jadeitite blocks occur in the Oeyama and Miyamori-Hayachine ophiolites (Taira et al. 2016), scattered through four locations along the northern edge of Honshu Island. Gemstone quality jadeite is confined to the Kotaki River drainage in Niigata Prefecture; samples can be seen in the Fossa Magna Museum in Itoigawa City, Niigata (Tsujimori & Harlow 2017). Whether this jadeitite has substantial inclusions of another mineral, omphacite, will be examined in Chapter 12. In any case, Japan has not produced the clear emerald-green jade known as 'feicui' that comes from Myanmar/Burma.

Isozaki (2019) proposes that the three Palaeozoic belts of Nagato-Renge (N-R), Kurosegawa (Kr), and South Kitakami (sK) which included the Hayachine ophiolite (Figure 3.4), all formed in the same arc-trench system that developed during subduction stage 1. These have been identified with similar compositions in the Primorye and named as a coherent 'South Kitakami Terrane', which was later tectonically rearranged (Tazawa 2001; see his fig. 3). These belts contain shelf sedimentary rocks as well as subduction-related granite intrusions and limestones of various periods (Eihiro *et al.* 2016). Around 450 Ma, a ridge subduction event occurred, resulting in the metamorphic transformation of the Kurosegawa Belt; then around 350 Ma, a second ridge was subducted, causing the metamorphism of the Renge Belt (Isozaki 2019: fig. 3); these early plates created by these mid-oceanic ridges have not been well publicized, but Isozaki names them first as the Nomo Plate and second as the Renge Plate. Thereafter, it was the Farallon Plate that was being subducted for the next 100 million years (cf. Table 3.1).

The Carboniferous (359–299 Ma) produced the oldest recognized AC in Japan (Stage 2 according to Wakita *et al.* 2018): the unmetamorphosed Nedamo Belt (Taira *et al.* 2016). It is currently sandwiched between the North and South Kitakami Belts in northeastern Japan (cf. Figure 3.1: Nd). Probably the tiniest of the geological belts, it is highly significant in that it contains small 'particles' of ophiolite and blueschist among other trench fill sediments, indicating its position in a forearc zone and confirming the existence of accretionary processes.

The multi-stage process began of joining the NCB and SCB together to form the modern China mainland began during the assembly of Pangaea (Wu & Zheng 2013). This involved the arc-continent collision of three island arcs and associated back-arc basins between 480 and 310 Ma within the Palaeo-Tethys Sea corridor, accounting for some of the ophiolites and limestone/marbles currently sandwiched between the NCB and SCB in the Qinling-Dabie-Sulu metamorphic zone (Figure 3.4: inset area of volcanic arc accretion).

Permian (ca. 299–252 Ma): Farallon Plate subduction

During the 150 million years of existence of the supercontinent Pangaea, the Permian period witnessed the active subduction of the Farallon Plate under the South China Block. Such subduction activity resulted in sequences of Accretionary Complexes along the continental margin, now identified as the Akiyoshi (Ak), Maizuru (Mz), and Ultra-Tanba (UT) Belts (Figure 3.4 and inset). These AC contain greenstones of Carboniferous date that were produced as seamounts or LIPs (large igneous provinces) further out in the ocean before being brought to the subduction trench (Sano *et al.* 2000).

• The Akiyoshi AC (Kojima *et al.* 2016) was compiled from late Permian into the Triassic, but it contains greenstones dating to the Carboniferous (ca. 360–330 Ma) and limestone accumulating on those greenstones up through the Permian. The limestones are sized from pebbles up to a ca. 3 x 5 km block, now known as Akiyoshidai (BOX 3), which originally formed a coral reef cap on a seamount. These greenstones are identified as the Akiyoshi-Sawadani palaeo-seamount chain (Isozaki 1997*a*; Sano *et al.* 2000).

• The Triassic part of the Akiyoshi AC was later metamorphosed around 240 Ma to form the Suo (Su) MB; Miyazaki *et al.* (2016) treats it as a metamorphic complex within the broader Sangun (Sn) Belt. But the name of Sangun itself has been discarded by others who split it into the Suo and Renge schists; such are the vagaries of naming practices through different research paths. The Renge MB is now treated as part of the Oeyama Belt and considered to refer only to the schists (Tsujimori 2017).

• The Maizuru Belt is named for the Maizuru igneous plateau (Isozaki 1997*a*; Wakita *et al.* 2018), the remains of which are now referred to as the Yakuno Ophiolite in the Maizuru Belt. The plateau's igneous rocks there date to the Early-Middle Permian (Ishiwatari *et al.* 2016) and were incorporated into the Maizuru Belt in Late Permian. The Maizuru Belt is referred to as an island arc system by Wakita et al. (2018: 3).

• Closely associated with the Maizuru Belt and Yakuno Ophiolite is the Ultra-Tanba Belt, comprised of imbricated layers of intercalated sandstone, mudstone, and felsic tuff, etc.; these layers have been interpreted as early Ordovician through late Devonian trench fill (Tazawa 2001) that have been bulldozed into an accretionary prism during the Permian (Kojima *et al.* 2016). That these layers remain relatively coherent is unusual in Japan's geology.

Both seamounts and plateau are thought to be volcanic features originating from a superplume in the Panthalassa Sea near the Australian continent (Isozaki 1997*a*: 20; Taira 1990: 192), much like the Emperor Seamount chain and Hawaiian Islands created from the hot spot on the northern Pacific Plate. The Farallon Plate carried these igneous seamounts and plateau to the northwest where they eventually collided with the continental margin of the China Blocks, but instead of being drawn down into the trench and subducted, their upper portions were scraped off and accreted to the continental edge. It follows that the oceanic igneous rocks and limestones are older than the AC within which they now reside, allowing time for their formation and transit.

The relationship between the four Permian belts has been conceived as an array along the eastern edge of the North and South China Blocks, with Maizuru/Ultra-Tanba situated further north than Akiyoshi/Suo (Figure 3.4: inset). However, with the collision of the two China Blocks, these belts are thought to have been juxtaposed east-west (Wakita *et al.* 2018). There is great contention how this collision played out, as discussed below under Triassic events.

BOX 3 The Akiyoshidai Karst Park of Western Honshu

Yamaguchi Prefecture, in far western Honshu, hosts a broad area 130 km^2 of exposed limestone karst, dating to the early Carboniferous to middle Permian periods. This is the Akiyoshidai (Akiyoshi Plateau), designated as a Natural Monument by the Japanese government and showcased as a geological park and site museum (https://akiyoshidai-park.com/en/).

The biggest attraction is the Shuhodo or Akiyoshi-do cavern. About 10 km of the karst cave system have been surveyed, revealing 440 caves. A distance of 1 km is open to the public, including a blue-ceiling cavern and a series of travertine stepped pools. The ground surface of the limestone plateau is pitted with conical erosion sinkholes or, more formally, dolines. These are exposed because the area of the nature reserve is fired every year (*hi-ire*) to remove tree growth. The vista over the plateau is thus eerily barren within the surrounding forested mountains.

Akiyoshidai is only one of several such limestone accretions from palaeo-seamount chains that are scattered throughout Japan. It is, however, the largest and most intact. The Akiyoshi Accretionary Complex making up this part of Honshu derives its name from this plateau. It formed in the Permian–Triassic periods, and the complex incorporates basalt, chert, shale, tuff, sandstone, and seamount greenstone as well as limestone. The AC has been weakly metamorphosed to greenschist facies.

The atoll, folded by a meteorite?

Two proposals exist in the literature concerning the nature of the limestone terrane. First, the Akiyoshi limestone was described as a ring-shaped atoll folded back on itself, like a doughnut folded in half. This interpretation was stimulated by a reversal of rock ages halfway through the stratigraphic sequences. The folding was proposed to have happened out in the ocean, before the atoll was accreted to land; the possible presence of shocked minerals suggested a meteorite strike (Sano & Kanmera 1988), but others propose that the actual folding was caused by a shock wave in the sea (Miura & Tanaka 2004). The second view is that the Akiyoshi limestone is not a coherent atoll but a collection of tectonically rendered limestone blocks from that atoll incorporated into a tectonic melangé (Kojima *et al.* 2016), "possibly fortuitously showing a consistent reversed 'stratigraphy' when viewed on the large scale" (Taira *et al.* 2016: 9). Both stratigraphically coherent units and chaotically organized units have been identified (Miyazaki *et al.* 2016).

Currently several seamounts (cf. Erimo and Daiichi-Kashima in Figure 3.1) on the Pacific Plate are approaching the Japan Trench off the coast of Ibaraki Prefecture, and a collection of seamounts and the Amami Plateau on the Philippine Plate are approaching the Ryukyu Trench (cf. Figure 6.11: inset). It is estimated that at least the latter will be smoothly subducted because their crustal thickness is less than 25 km (Yamamoto *et al.* 2009) – a threshold value governing partial subduction as in the current Izu Islands.

Triassic (ca. 250–200 Ma): uniting the China blocks

The next big event occurring in this region was the collision and final unification of the North and South China Blocks by Late Triassic times, with the SCB being subducted under the NCB (Wu & Zheng 2013). Between 240 and 225 Ma, the final Qinling suturing of the NCB and SCB took place to yield eastern China and the Korean Peninsula, referred to hereafter together as 'the continent'. The closure of the palaeo-Tethys seaway was zippered from east to west; recent geodynamic modelling (e.g., Stampfli 2013) illustrates that the history of this collision was far more complicated than represented in Figure 3.5.

The suture zone (annotated MB, metamorphic belt in Figure 3.5) is currently marked by the Qinling–Tongbai–Hong'an–Dabie–Sulu Orogenic Belt in the position of the former palaeo-Tethys Ocean (Wu & Zheng 2013). The Qinling Mountains divide present-day China into northern and southern ecological zones as well as tectonic blocks. The suture crosses into the Korean Peninsula, and four tiny parcels of mid-temperature/pressure metamorphic terranes claimed to correlate with this suture zone have recently been found at widely separated locations in Japan: in Kyushu, near the Noto Peninsula, and in eastern Tohoku (Omori & Isozaki 2011). Also, the Suo Belt (composed of Akiyoshi AC) was found to have been metamorphosed during the China block collision, and there were considerable rearrangements of geotectonic units on the continental edge (Wakita *et al.* 2018).

Several reconstructions have been proposed for the conjoined continental blocks. The traditional conception is shown (Figure 3.5: left) with the repositioning of the Hida-Oki fragments as identified in Figure 3.3. Isozaki, however, has proposed that the SCB was much larger than formerly portrayed (Figure 3.5: right) and that its collision with NCB placed SCB material along the eastern seaboard northwards into what is now the Primorye region of Russia (2019: fig. 7). He names this extension Greater South China (GSC); this scenario takes into account the geographical distribution of South Kitakami Terrane fragments, known from both Japan and Russia as discussed above.

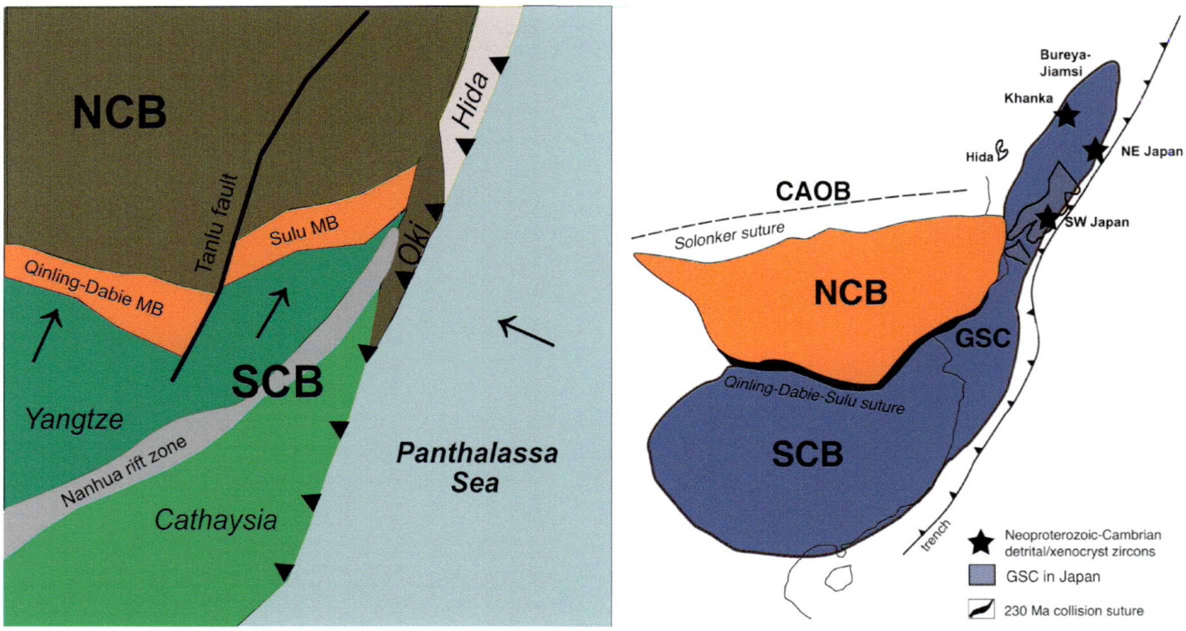

Figure 3.5 Two visions of NCB–SCB relations after collision in the Triassic, ca. 230 Ma

A ridge subduction event has been proposed for the end of the Triassic (Maruyama *et al.* 1997; Isozaki 2019: fig. 3), but Wakita *et al.* (2018) reports that dates of ocean floor basalts in AC sandwiching this time period are not young enough to provide evidence of ridge subduction. Nevertheless, the Late Triassic witnessed a change-over from the Farallon to the Izanagi Plate subducting under the eastern continental edge. Stage 4 AC formation is thought to have occurred from Middle Triassic to the Early Cretaceous, consisting mainly of ocean plate rocks (Wakita *et al.* 2018: 3).

Jurassic (ca. 200–145 Ma): meeting Izanagi

The break-up of Pangaea began by rifting at ca. 200 Ma through the centre of the landmass, in the region of today's Atlantic Ocean (Evans 2016). The now consolidated China blocks had seen the passing of the Farallon Plate and was now receiving the Izanagi Plate into its subduction trenches which ran along the continental edge for 6000 km (Isozaki 1997*b*: 42).

The Jurassic AC (Mino-Tanba, Ryoke, Chichibu Belts) (Figure 3.6) account for around 63% of AC areal coverage in Japan (Isozaki 1997*b*: 28). The incipient Japanese landmass, at the inner edge of the northwestern Panthalassa Sea trench, again experienced the accreting of bulldozed oceanic floor sediments and volcanic features. In particular, an oceanic plateau (the Izanagi Plateau) and seamount swarm (the Akasaka-Kuzuu cluster), dating from Early Carboniferous through Late Jurassic, impacted on the edge of the Japanese landmass (Hirsh & Ishida 2002; Sano *et al.* 2000).

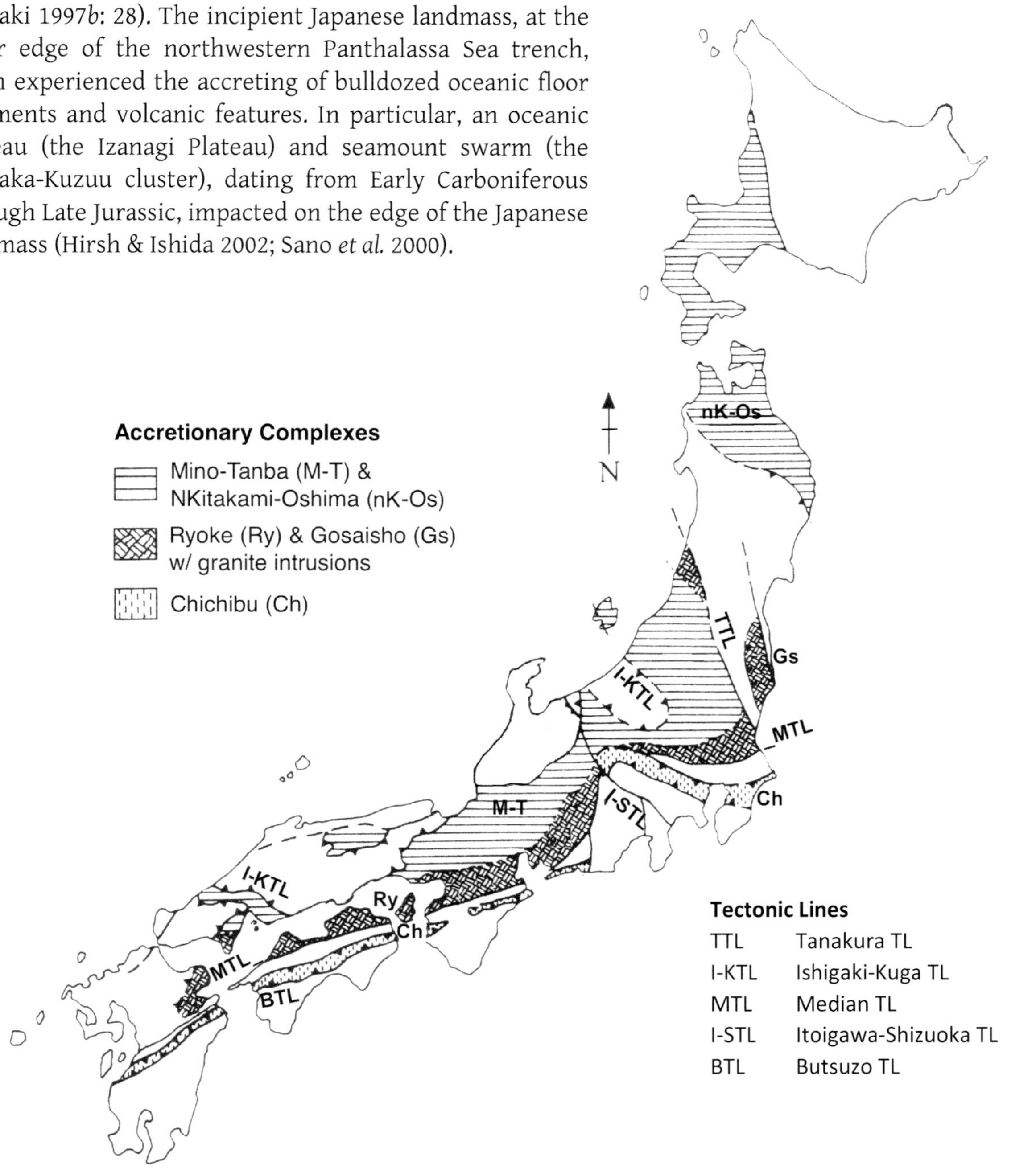

Figure 3.6 Jurassic AC Belt locations and major Tectonic Lines (TL) boundaries in modern-day Japan

Fragments were incorporated into the Mino-Tanba Belt (Maruyama *et al.* 1997) and Chichibu Belt, and they form the Mikabu greenstones associated with Chichibu and Sanbagawa Belts (Hirsh & Ishida 2002; Wallis & Okudaira 2016).

• The Mino-Tanba [M-T] AC is connected to the N. Kitakami [nK]-Oshima Belt in northern Japan. It is composed of several descrete accretionary units that were successively added to the Japanese landmass, younging to the south; these units contain basalt, limestone, chert, and sediments which occur either as stratigraphically organized or as chaotic tectonic mélanges (Kojima *et al.* 2016). The Oshima Belt, at least in Hokkaido, is composed predominantly of chert (Ueda 2016), indicating an upper ocean floor derivation.

• The accretionary units that comprise the Chichibu [Ch] Belt contain either trench fill sediments or oceanic sequences (deep-water chert or shallow-water limestone) (Kojima *et al.* 2016). Chichibu is divided into north and south segments with the Kurosegawa Belt between them. Kojima *et al.* evaluate two models to account for their separation: the surficial position of the Kurosegawa as a 'nappe' or a 'klippe' *on top of* the Chichibu Belt segments (Isozaki 1997*a*), or the sandwiching of it *between* them. They opt for the latter, arguing that the Chichibu unit formed in linear fashion but then was broken up with the two parts juxtaposed against each other through strike-slip faulting, and the Kurosegawa Belt served as the 'suture' between them (Kojima *et al.* 2016). They argue against 'nappe/klippe' emplacement of Kurosegawa because it is not a shallow Permian-Triassic unit, having basement rocks from the Cambrian–Ordovician, and they include it in the 'SKitakami terrane', which has a similar basement.

Pacific Basin plate reorganization ca. 170–65 Ma

Plate relationships in the Pacific basin between the mid-Jurassic through the Cretaceous are difficult to reconstruct and subject to widely varying interpretations (Smith 2007). At the beginning of this timeframe, the main plate affecting the developing Japanese landmass was the Izanagi Plate, which existed in the western Pacific basin together with the Farallon Plate in the eastern basin and the Phoenix Plate in the southern basin. In the mid-Jurassic, the Pacific Plate was newly born from the triple-junction of these three plates (Boschman & von Hinsbergen 2016). After 90 Ma, the Pacific Plate began pushing westwards as it grew, forcing the Izanagi and Farallon Plates into northwestern and northeastern subduction trenches, respectively; fragments of the Farallon Plate still exist as the Juan de Fuca (Gorda), Nazca, and Cocos Plates off the Americas (cf. Figure 2.1) (Neall & Trewick 2008).

Another plate, Kula, is often regarded as a fragment of the Izanagi Plate and referred to together as the Izanagi/Kula Plate (Norton 2007). However, others subscribe to a spreading ridge between the two, the latter following the former in being subducted under proto-Japan (Wallis *et al.* 2016). Isozaki also specifies that subduction of the spreading centre between Izanagi and Kula was responsible for the metamorphism of the Sanbagawa Belt in the Cretaceous (2019: fig. 3). The current Beringia Plate has been identified with Izanagi/Kula (Rea, & Dixon 1983), but other recent models consider Kula to be a sub-section of the older Farallon Plate (Smith 2007).

Cretaceous (ca. 145–66 Ma): episodic growth

In the Early Cretaceous, the last of the Izanagi Plateau and seamounts known as the Mikabu greenstones appear along the southern edge of the Sanbagawa Belt and were incorporated into the southern Chichibu Belt, while in Hokkaido the Sorachi Plateau was accreted as Sorachi greenstone (Figure 3.7). These oceanic plateaus are thought to be part of a Large Igneous Province (LIP) active in the southern Pacific Basin; they experienced partial high P/T metamorphism during subduction in the Japan Trench along the continental edge before exhumation into the Japanese landmass (Ichiyama *et al.* 2014). Also in Hokkaido, the Horokanai Ophiolite dating to the mid-Jurassic was accreted as part of the Sorachi-Yezo Belt (Ueda 2016).

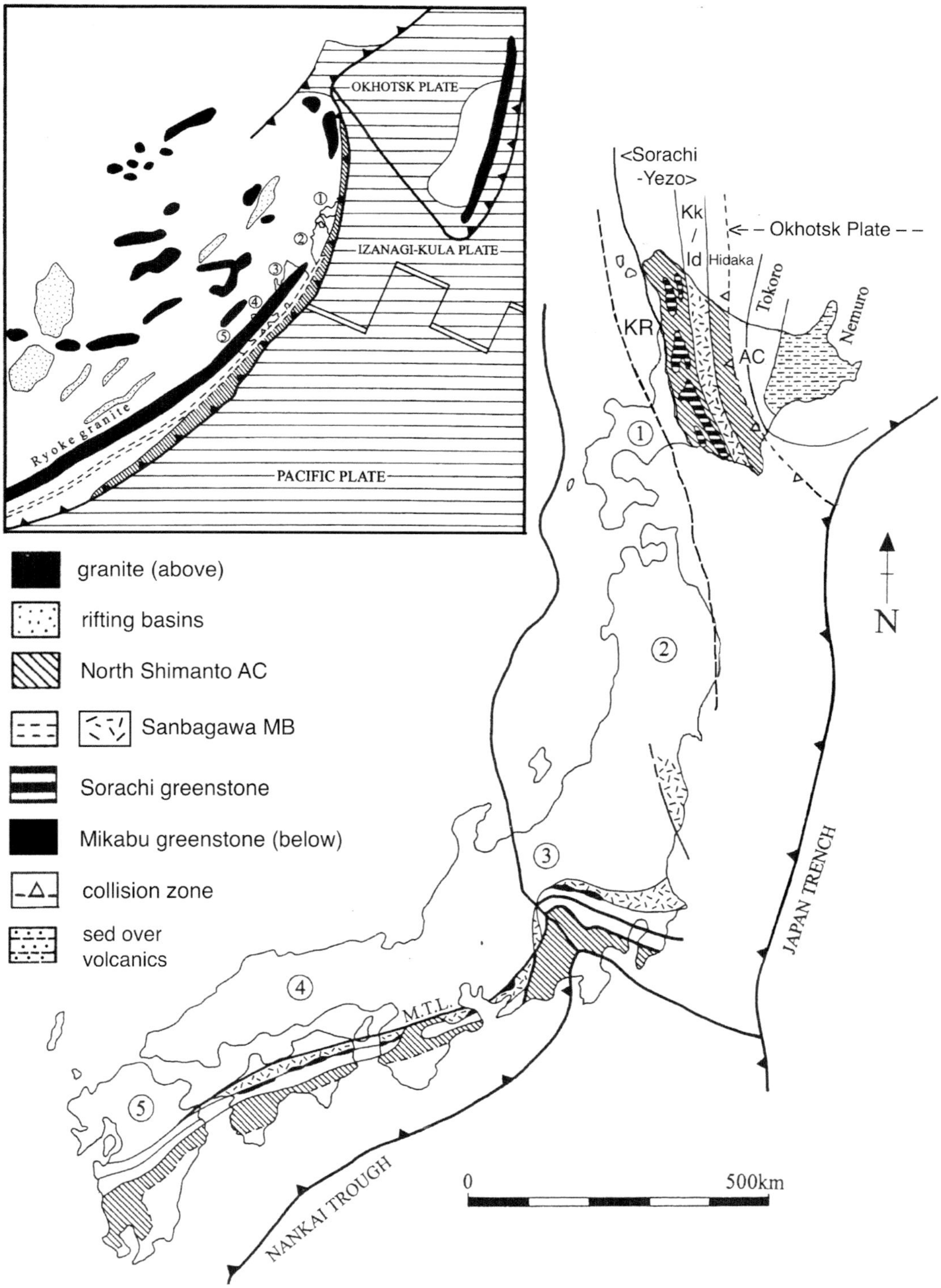

Figure 3.7 Late Cretaceous formation of the geotectonic belts of Japan. Numbers in circles indicate formation loci (inset) and present-day loci; updates in sans serif font. Subduction trenches are in their modern locations, not their contemporaneous positions.

In the early Cretaceous, the existing rocks of the South Kitakami Belt were intruded by granites from which orogenic mesothermal gold deposits formed (Ishihara & Murakami 2004). Placer gold from these deposits constituted the first gold discovered in Japan in 749 AD, and the region's gold was mined until 1300. The Great Buddha of Todaiji Temple in Nara, made of gilt bronze in 752, was facilitated by this discovery of placer gold in Tohoku (Ii 2002).

At ca. 120 Ma, the Izanagi Plate changed directions to move more northerly, slipping along the continental rather than directly into it (Maruyama *et al.* 1997). This sideways movement along the continental shelf halted the accretionary and igneous processes that accompanied subduction. Such tectonic quiescence lasted until 90 Ma when the Izanagi Plate changed directions again to the northwest. Maruyama and colleagues (1997) have proposed that the northward movement of the spreading ridge between the Izanagi-Kula and Pacific Plates not only created the igneous intrusions leading to the Ryoke batholith along the continental edge but also exhumed the deep-seated Sanbagawa MB (cf. Figure 3.7: inset). In addition, the igneous activity accelerated the cycle of uplift, erosion, and increased accretionary activity to form the Northern Shimanto Belt.

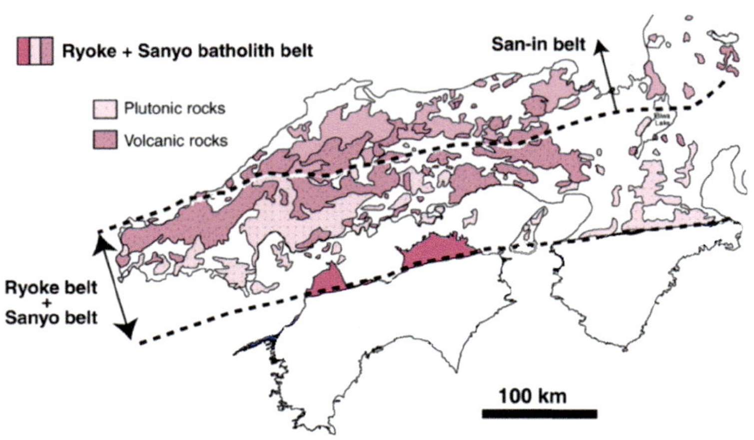

Figure 3.8 Early and Late Cretaceous granite belts in western Japan

The Ryoke batholith is composed of Older and Younger Ryoke granites; complementary granites also occur in the Kitakami and Abukuma Mountains of north-eastern Honshu (Nakajima *et al.* 2016) (cf. Apx 10). Batholiths form between 6 and 10 km below the land surface where magma chambers are maintained at pressures of 2 ± 0.5 kbar until their contents crystallize (Huber *et al.* 2019); individual chambers coalesce into larger groupings (batholiths) and can be uplifted and unroofed during tectonic movement. The Inner Zone of Western Japan houses three batholiths: Ryoke, formed prior to 65 Ma; San'in granite plutons cooling between 70 and 45 Ma, and the San'yo granite plutons cooling between 40 and 30 Ma. Between ca. 110 and 55 Ma, the Younger Ryoke granites and the San'yo granites also intruded into the Akiyoshi, Maizuru, and Ultra-Tanba belts (Nakajima *et al.* 2016) (cf. Figure 3.8). Having been uplifted during the Miocene–Pleistocene tectonic rearrangements, these granite bodies now form the backbone ranges of western Honshu (Kimura, J *et al.* 2014).

The Ryoke Belt is actually the southern part of the Mino-Tanba Belt that was intruded by the Older Ryoke granite; this was subjected to high-grade contact metamorphism (low P/high T) in the mid-Cretaceous (ca. 100 Ma) (Wallis & Okudaira 2016: 112). The granite occupies 70–80% of the Ryoke Belt, while the metamorphosed portions of the Mino-Tanba AC in the Ryoke Belt account for only 20–30%.

From the Middle Cretaceous to the early Palaeogene is designated as stage 5 in AC formation (Wakita 2018). The Sanbagawa Belt was first formed as the Sanbosan AC (Okamoto *et al.* 2000; Isozaki *et al.* 2010*a*) and then metamorphosed into an MB all within the Cretaceous period. The wholistic Sanbagawa is an exhumed high-grade MB that stretches over 1000 km and is up to 30 km wide and 2~5 km thick (Masago *et al.* 2005); it currently lies sub-horizontal to the surface and is bounded on the top by a normal fault and on the bottom by a thrust fault (cf. Figure 2.10) (Okamoto *et al.* 2000). However, as stated above, the Sanbagawa MB has been divided into two parts, Sanbagawa MB *sensu stricto* and the Shimanto MB.

The Sanbosan AC was formed between 140 and 130 Ma (Masago *et al.* 2005) and contained oceanic materials: mainly trench fill that lithified as sandstone and shale but also some oceanic basalt (including the Mikabu greenstones), chert, and carbonates (Wallis & Okudaira 2016; Watanabe *et al.* 2016). Different parts of it were metamorphosed into the Sanbagawa MB at different times and to different maximum facies (cf. Figure 2.8), as portions of the AC were differentially drawn down into the subduction trench. As recently as 2016, the date of eclogite facies metamorphism is stated to have been achieved as early as 120 Ma (Masago *et al.* 2005; Wallis & Okudaira 2016), but recent detrital zircon dating indicates between 80 and 70 Ma (Shimura *et al.* 2020; Nagata *et al.* 2017; Y. Isozaki pers. comm. 6 March 2021). Following maximum metamorphism, the materials were exhumed to greenschist facies levels (see the P/T paths in fig. 2c.3 of Wallis & Okudaira 2016). These authors identify those metamorphic 'paths' as consistent with the subduction of a young (and hot) oceanic plate, indicating a nearby spreading ridge (see their fig. 2c.5). This event is illustrated as one of several ridge subductions in the history of the Japanese landmass (Isozaki 2019: fig. 3: pink lines). Weller *et al.* (2015: 581, fig. 2) have confirmed that the whole was metamorphosed over a 40-million-year timespan during the approach and subduction of a spreading ridge, most likely of the Izanagi/Kula Plate (cf. Table 3.1), thus concurring with Isozaki (2019) and Wallis & Okudaira (2016). Dates of exhumation are somewhat indicated by greenschist facies overprinting, given in Wallis & Okudaira (2016: fig. 2c.3) as occurring at (90–80 Ma), and completed exhumation is indicated by sedimentation on eroded surfaces after 50 Ma (Isozaki & Itaya 1990).

The Sanbagawa Belt hosts the major copper and zinc resources of Japan, present throughout the Belt's traverse of Shikoku Island and present in spots in eastern Kyushu, lower Kinki, and Tokai regions; these resources are described as a "volcanogenic massive sulphide Cu deposit" of Jurassic age (Watanabe et al. 2016: fig. 10.1). Hydrothermal activity near an active oceanic ridge as proposed above caused the deposition of thick sulfides on the ocean floor between 160–155 Ma (Nozaki et al. 2010). When drawn into the subduction channel, metamorphosed then exhumed to the continental edge, the metals remained stable, but the basalts were transformed to green schists. The copper and zinc of the Besshi unit of the Sanbagawa Belt in Shikoku were exploited at the Besshi Copper Mine, founded by the Sumitomo Group and in operation between 1690 and 1973.

The Ryoke and Sanbagawa Belts formed the basis of Miyashiro's proposal for Pacific-type Orogeny (cf. BOX 1); he postulated that a high-T weakly metamorphosed belt (e.g., Ryoke) paired with a high-P strongly metamorphosed belt (Sanbagawa) were characteristic of Pacific Rim subduction zones. This has mainly proven true, but in most cases such belts are far apart in the current geography. In Japan, however, the belts are spaced close together, separated by the neo-MTL, which serves as a divider between Inner and Outer Zone Japan despite the fact that the Ryoke and Sanbagawa Belts on either side of the neo-MTL belong to the same orogenic event (i.e., ridge subduction). Wallis & Okudaira (2016) note that the main problem with Miyashiro's interpretation is the absence of 100–200 km of forearc sediments that originally separated the pair (see Yanai *et al.* 2010: fig. 7). Wallis & Okudaira review five hypotheses to account for this absence, but they temporarily conclude that strike-slip faulting removed part of the forearc region brought the Sanbagawa zone into parallel alignment with Ryoke and normal faulting brought the high P/T metamorphic rocks (Sanbagawa) to the surface. Isozaki and colleagues (Isozaki *et al.* 2010*b*), however, argue for foreshortening of the crust through thrust faulting.

The other Cretaceous AC is the Shimanto, the northern edge of which became metamorphosed as the Shimanto MB adjacent to the Sanbagawa MB. Ironically, the so-called Sanbagawa unit in the Kanto, where the MB was named after the Sanba River, has been re-identified as the Shimanto MB (Aoki *et al.* 2010). The Shimanto AC accumulated along the continental edge as the Early/North (NShimanto) and Late/South (SShimanto) Cretaceous units, including a tectonic mélange (Kimura *et al.* 2016). These Cretaceous terranes continue into Hokkaido as the Hidaka Belt but are absent in northeastern Japan due to tectonic erosion (Wakita *et al.* 2018: 5). The Aki Tectonic Line separates the Northern and Southern Shimanto Belts and represents a 6 myr gap between the two. The Southern Shimanto Belt is the last major AC exposed on land; it is intruded by a seamount of the Kinan chain, which has uplifted an off-shore bank, and is heading towards the eastern gap between Shikoku and Honshu Islands. Part of it is also exposed on the Boso Peninsula of Chiba Prefecture, uplifted by the collision of the Izu Peninsula with Honshu, and it is exposed on Okinawa Island in the far southwest (Ujiie 2002).

Peripheral developments from Cretaceous to the Palaeogene (66–23 Ma)

Hokkaido assembly

Concurrently, a new plate was forming off the northern edge of the continent through 'trench jumping' – the relocation of the subduction trench to the oceanward side of a block called the Okhotsk Plate (cf. Figure 3.7: inset). This new plate contained the landmasses that would eventually become the Kamchatka Peninsula, the northern tip of Sakhalin Island, the Kurile Arc, and the eastern tip of Hokkaido Island. The final alignment of Hokkaido took place following the opening of the Japan Sea Basin in the Neogene, but the Island's structure will be discussed as a whole here. Hokkaido has an extremely unusual structure: it consists of an arc-arc collision along the collision zone in the Hidaka Belt nominally shown in Figure 3.7. Descriptions below follow Ueda (Ueda 2016, see his fig. 2g.2):

- On the far western side of the collision zone, geological strata are related to those of northern Honshu; the Early Cretaceous Kabato-Rebun (KR) Belt is a newly recognized volcanic chain at the eastern edge of the Oshima Belt, and it cuts along the NKitakami Belt in Tohoku. To its east is the Sorachi-Yezo Belt, sediments of a forearc basin associated with the KR volcanics and is continuous with the NShimanto Belt further south. In addition to the oceanic materials mentioned above, Sorachi-Yezo Belt contains two parallel sedimentary sub-belts, the Kamuikotan (Kk) and Idonappu (Id). These are accretionary prisms dipping to the west that have been differentially metamorphosed: Kamuikotan, to the west, accumulated earlier, then was more deeply buried and therefore more highly metamorphosed; and Idonappu is of later accumulation at more shallow levels with strong deformation.
- The east side of the collision zone belongs to the Okhotsk Plate. The eastern tail of Hokkaido consists of the Nemuro Belt, including the intrusive head of the Kuril Arc which descended into position from the northeast. To its west, is the Tokoro Belt, consisting of arc and forearc sediments over oceanic basement incorporating basalts, limestone, and chert. These accretionary prisms dip to the east.
- It is the middle section, the Hidaka Belt, incorporating the collision zone that is of much interest. The Hidaka Belt is divided down the middle into west-dipping accretionary prisms on the west and east-dipping accretionary prisms on the east – all of early Palaeogene date (in contrast to the Cretaceous Sorachi-Yezo Belt). These indicate subduction zones that are oriented in opposite directions before collision. The different terranes of west and east Hokkaido meet in the centre of the Hidaka Belt, representing the closure of an ocean in the Miocene and resulting in folding and thrusting and the exhumation of a metamorphic belt in the southern Hidaka Belt.

Philippine Plate & IBM Arc creation

The Philippine Plate is calculated to have originated by 60 Ma in the southern Pacific east of Java; by 50–45 Ma, the Pacific Plate was being subducted under the Philippine Plate's northern edge (Figure 3.9), forming the beginnings of the IBM[1] volcanic arc (Hall 2002; Lallemand, 2016). From then on, the Philippine Plate began to migrate north from below 11°N while rotating clockwise between 60 and 90° (Yamazaki 2010: fig. 4; Fang *et al.* 2011: fig. 5; Mahony *et al.* 2011: 2211), with two-thirds of that amount occurring between 40 and 15 Ma (Wu *et al.* 2016: fig. 27). Due to this migration and rotation, the Philippine Plate was not in the vicinity of the Japanese landmass much prior to 15 Ma; thus, the plate that was being subducted under that landmass during the early rifting of the Japan Sea Basin was the Pacific Plate. The effect of Philippine Plate encroachment was ultimately to cause the collision of the Izu Arc with Honshu, discussed more fully in Chapter 4.

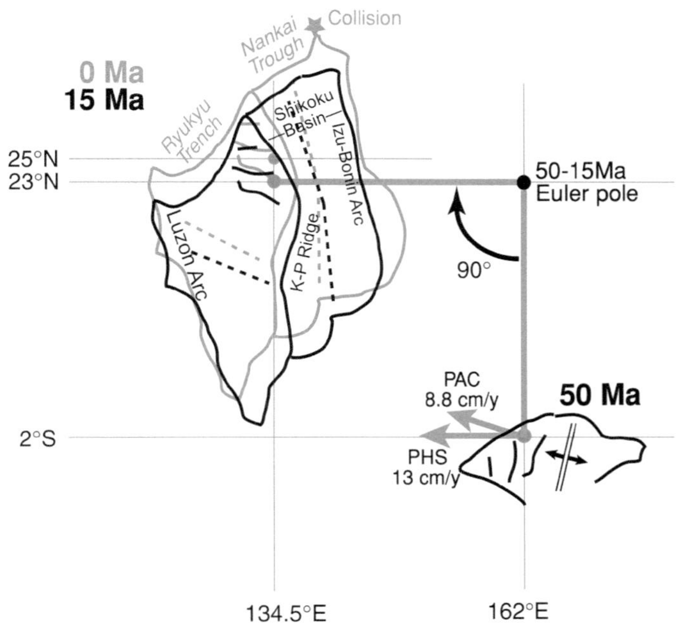

Figure 3.9 Model of Philippine Plate rotation and movement northward between 50 and 15 Ma

During the Philippine Plate migration, between 30 and 25 Ma, the Kyushu-Palau (K-P) Ridge split off the Izu Arc and the Shikoku Basin opened as a back-arc basin between them (Lallemand 2016; Pickering *et al.* 2013). The young, hot oceanic crust of the basin contrasted with the older, colder floor of the West Philippine Basin, which had formed between 60 and 40 Ma. On the western side of the West Philippine Basin, two further volcanic arcs, the Manila-Luzon volcanic arcs formed through opposite polarity subduction relations with the Sunda Plate to its west. Among these arcs, the Izu and Luzon arcs are especially important for the geology of Honshu and Taiwan, respectively, as they collided with these units later in time, while the Kyushu–Palau Ridge greatly affects tectonics in Kyushu.

The collision of the Luzon Arc and Taiwan was instrumental in providing present-day nephrite jade resources which have supplied Southeast Asian jade working throughout prehistory (e.g., Iizuka & Hung 2005).

Structural realignments

During this time period changes on both the eastern and western fronts of the Japanese landmass resulted in great transformations. Oceanward, with the passing of the Izanagi Plate, the Pacific Plate began subduction in the Japan Trench under the Eurasian continent after 56 Ma (Smith *et al.* 2007). Landward, the impact of the Indian sub-continent on Eurasia from 55 Ma set off several modifications

[1] IBM = Izu-Bonin-Mariana arc; of these, the Izu Arc is most relevant here. Various portions of the IBM are also referred to as the Izu-Ogasawara, Izu-Bonin, Izu-Mariana, Shichito-Mariana, or Tanzawa-Izu Arcs.

and rearrangements of tectonic plates in the western Pacific, giving rise to the island arc topography that characterizes this area today. These processes as they affected the Japanese landmass along the continental edge will be reviewed in Chapter 4.

Conclusions

This brief review has accounted for the formation and location of the major geological belts of Japan in their palaeogeographic settings. These belts comprise the basement rocks of the Japanese landmass, and they consist of mainly Accretionary Complexes (AC), many of which have undergone later metamorphism to form Metamorphic Belts (MB). In addition, through time these basement rocks have been intruded by both mafic and felsic magmas in processes characteristic of subduction zones, as well as being composed of igneous rocks derived from mantle plumes during Japan Sea Basin opening (Chapter 4). The Cretaceous granites discussed above (cf. Figure 3.8) are the most widespread, but the intrusion of igneous rocks has affected most other geotectonic belts (Apx 10, 11). Many of the geological terranes of Japan have been conceptually reorganized as more information about their contents and relationships is discovered. Appendix 6 provides an incomplete, but hopefully useful indication of the natures of these terranes, which in the main are correlated with Figure 3.1. Appendix 8 gives a chronological listing of the major events described above.

Word of warning: the geotectonic belts can be briefly described in terms of their formation and transformations, but their contents are extremely complicated and contain material from many geological periods, as indicated by Figure 3.2. For use in the field, archaeologists should familiarize themselves with the myriad contents, distributions, outcrops, and their potential for influencing habitation in their fieldwork area. Moreno et al. (2016) gives further details, and the issues raised above and geological belt contents have been dealt with by several other authors (e.g., Wakita et al. 2013; Isozaki & Aoki et al. 2010).

In updating the overviews on the formation of the basement rock belts, several previous research results have been discarded by Japanese scholars; these are presented below in Table 3.3 together with my assessments of new understandings and trends. Important revisions include:

- the beginning of subduction in the proto-Japan area has been pushed back to 520 Ma;
- five previous proto-Japan arc-granite batholiths have been detected;
- the Hida and Oki Belts are most likely to derived from the North China craton;
- portions of the Qinling-Dabie-Sulu suture have been identified in four locations in Japan;
- the Sanbagawa Metamorphic Belt has been divided into two terranes;
- the Median Tectonic Line acted in two phases, the paleo-MTL and neo-MTL, on vastly different scales;
- the idea that the Outer Zone moved into place 1000 km from the south via the neo-MTL has been mainly discarded in favour of in situ development (Isozaki & Maruyama *et al.* 2010), though Wallis & Okudaira (2016: 118) continue to argue for displacement of 800 km or more.

For those interested in the distant future, some researchers (Maruyama *et al.* 1997; Safonova & Maruyama 2014) argue for Japan rejoining the Eurasian continent to form 'Amasia' in 1.5 million years' time (see Isozaki 2019: fig. 3) – due to the Japan Sea already beginning the process of closing up as of 2 Ma (Kamata & Kodama 1999). Then in 50 million years the Australian continent – which is moving northwards – is expected to arrive on Japan's doorstep. However, there are differing views on the future of the Pacific Ocean. Maruyama's scenario has the North American continent closing onto the eastern edge of newly positioned Australia, leaving the Pacific Ocean trapped 250 million years into the future as a small sea in the northwest below the present-day Bering Strait (Maruyama 1997: fig. 9). In another scenario, Scotese has revised his view that the Atlantic would close; instead, he now proposes it would widen, and Australia would collide with Southeast Asia in 50 million years' time (Scotese 2003). The by-word for the future, as it has been in the past, is "all change!".

Table 3.3 Revisions in Japanese plate tectonics research: 'mistakes' as drawn from Isozaki & Maruyama *et al.* (2010) with corrections given by the author. X (*batsu*) indicates 'wrong', O (*maru*) indicates 'correct'.

X Mistaken research results X	O Corrections O
X–That all of Japan formed and developed off the North China craton	O–Instead, most of the Japanese landmass formed off the South China craton
X–That the most important faults are the Median Tectonic Line (MTL) and those bordering the Fossa Magna (see Chapter 4)	O–Instead, the sub-horizontal thrust faults separating the various ACs and MBs are the major structuring agents of the Japanese landmass
X–That the origin of the MTL can be traced to a Cretaceous strike-slip fault running along the continental edge and connected with the Tanakura Tectonic Line (TTL)	O–The Paleo- and Neo-tectonic MTL operated differently at different times, with only the latter being a shallow strike-slip fault related to the TTL
X–That the Japan Sea Basin opened (like French doors) from the middle, anchored at fixed points (Euler poles) at opposite ends	O–Instead, the Japan Sea Basin rifted along two strike-slip faults, E and W, and the Japanese landmass moved directly away from the continent
X–That the most important faults in Japan are steep angle normal faults	O–Instead, the most important are the sub-horizontal thrust faults separating ACs
X–That Japan's several strike-slip faults are over 1000 km long and some parts of Japan moved up along them from the area of Vietnam	O–Instead, the strike-slip faults are relatively shallow, recently activated, and affect the landmass in situ
X–That Japan is composed of strike-slip fault bounded terranes that do not share a common genesis	O–Instead, it is composed of sub-horizontal thrust sheets that formed successively in similar subduction environments
X–That when ACs were not forming, the oceanic plate was subducting very obliquely	O–The absence of ACs is more likely due to loss of ACs through tectonic erosion
X– That the growth of ACs occurred continuously seaward	O–ACs formed intermittently
X–That once arc crust is formed, it is not destroyed and continental growth continues seaward	O–Instead, arc crust can be destroyed through tectonic erosion, and rather than growing in size, the landmass can be reduced in size
X–That when an arc collides with an orogenic zone, continental crust is not increased	O–Granite basements of arcs can be subducted, as is currently occurring with the Izu Arc

Figure & Table Sources

Figure 3.1a after (Isozaki *et al.* 2010*b*: fig. 1), modified by GLB
Figure 3.1b after (Isozaki *et al.* 2010*b*: fig. 4B), modified by GLB
Figure 3.2 extracted from (Wakita *et al.* 2013: fig. 3), and minimally modified by GLB; CCC license #5271530142503
Figure 3.3 extracted from (Takahashi *et al.* 2017: fig. 12), and redrawn by GLB
Figure 3.4 after (Isozaki 1997*a*: figs. 1, 16) redrawn by DAS and modified by GLB, with inset redrawn by GLB based on (Wakita *et al.* 2018: fig. 11a)
Figure 3.5 left: extracted and redrawn from (Takahashi *et al.* 2017: fig. 12); right: extracted from (Isozaki 2019: fig. 7), modified by GLB
Figure 3.6 after (Isozaki, 1997*b*: fig. 2), redrawn by DAS, modified by GLB
Figure 3.7 main after (Kimura, G 1997: fig. 1); inset after (Maruyama *et al.* 1997: fig. 14), all redrawn by DAS, modified by GLB
Figure 3.8 modified by GLB from (Isozaki *et al.* 2010*a*: fig. 6)

BOX 1 photo by the author
BOX 2 photo by the author

Table 3.1 compiled from (Li *et al.* 2008; Isozaki 2019: fig. 3; Wu & Zheng 2013; Evans 2016)
Table 3.2 compiled from (Isozaki *et al.* 2010*a*: fig. 3; Isozaki *et al.* 2011; Matsumoto *et al.* 2011; Ishiwatari *et al.* 2016; Taira *et al.* 2016; Watanabe *et al.* 2016)
Table 3.3 compiled from (Isozaki & Maruyama *et al.* 2010)

References

AOKI, Kazumasa; Shigeru OTOH & Shuichi YANAI *et al.* **(2010)** "Existence of new independent regional metamorphic belt in the Sanbagawa Metamorphic Belt: orogenic evolution of Japan from Cretaceous to Tertiary." *Chigaku Zasshi* 119.2: 313-332 (in Japanese with English title and abstract).

AOKI, Kazumasa; Yukio ISOZAKI & Shinji YAMAMOTO *et al.* (2012) "Tectonic erosion in a Pacific-type orogen: detrital zircon response to Cretaceous tectonics in Japan." *Geology* 40: 1087-1090 [DOI: 10.1130/G33414.1].

ASTON, WG **(1896/1972)** *Nihongi: chronicles of Japan from the earliest times to A.D. 697*. Tokyo: Charles E. Tuttle Company, reprint 1972.

BABA, Kei and Takeyoshi YOSHIDA **(2020)** "Geological structures controlled the rupture process of the 2011 M9.0 Tohoku-Oki earthquake in the Northeast Japan Arc." *Earth, Planets and Space* 72.94: 14 pp. [DOI: 10.1186/s40623-020-01212-3].

BOSCHMAN, Lydian M & Douwe JJ VON HINSBERGEN **(2016)** "On the enigmatic birth of the Pacific Plate within the Panthalassa Ocean." *Scientific Advances* 2: e1600022.

EIHIRO, Masayuki; Tatsuki TSUJIMORI & Kazuhiro TSUKADA *et al.* **(2016)** "Palaeozoic basement and associated cover", pp. 25-60 in T MORENO *et al.* (eds) (2016).

EVANS, DAD; ZX LI & JB MURPHY **(2016)** "Four-dimensional context of Earth's supercontinents", pp. 1-14 in *Supercontinent cycles through Earth history*, ed. by ZX LI *et al.* London: The Geological Society of London.

FANG, Yinxia; Jiabiao LI & Mingbi LI *et al.* **(2011)** "The formation and tectonic evolution of Philippine Sea Plate and KPR." *Acta Oceanologica Sinica* 30.4: 75-88.

GSJ (Geological Survey of Japan) (ed.) **(1992)** *An illustrated guide to the rocks and minerals of Japan*. Tokyo: Tokai University Press (in Japanese).

HALL, Robert **(2002)** "Cenozoic geological and plate tectonic evolution of SE Asia and the SW Pacific: computer-based reconstructions, model and animations." *Journal of Asian Earth Sciences* 20: 353-431.

HIRSH, F & K Ishida **(2002)** " Izanami Plateau: Japan's Triassic low latitude carbonates." *Eclogae Geologicae Helvetiae* 95: 43-55.

HUBER, Christian; Meredith TOWNSEND & Wim DEGRUYTER *et al.* **(2019)** "Optimal depth of subvolcanic magma chamber growth controlled by volatiles and crust rheology." *Nature Geoscience* 12: 762-768.

ICHIYAMA, Y; A ISHIWATARI & JI KUIMURA *et al.* (**2014**) "Jurassic plume-origin ophiolites in Japan: accreted fragments of oceanic plateaus." *Contributions to Mineralogy and Petrology* 168.1: 1-24 (in Japanese with English title and abstract).

II, Hiroyuki (**2002**) "An outline of Japanese gold and silver production." *Wat on Earth: Canadian newsletter for the Earth Sciences* 24 Aug 2002 [https://uwaterloo.ca/wat-on-earth/news/outline-japanese-gold-and-silver-production].

IIZUKA, Yoshiyuki & Hsiao-Chun HUNG (**2005**) "Archaeomineralogy of Taiwan nephrite: sourcing study of nephritic artifacts from the Philippines." *Journal of Austronesian Studies* 1.1: 33-79.

ISHIHARA, Shunso & Hiroyasu MURAKAMI (**2004**) "Granitoid types related to Cretaceous plutonic Au-quartz vein and Cu-Fe skarn deposits, Kitakami Mountains, Japan." *Resource Geology* 54.3: 281-298.

ISHIWATARI, Akira; Kazuhito OZAWA & Shoji ARAI *et al.* (**2016**) "Ophiolites and ultramafic rocks", pp. 223-250 in T MORENO *et al.* (eds) (2016).

ISOZAKI, Yukio (**1996**) "Anatomy and genesis of a subduction-related orogen: a new view of geotectonic subdivision and evolution of the Japanese Islands." *The Island Arc* 5.3: 289-320.

---- (**1997*a***) "Contrasting two types of orogen in Permo-Triassic Japan: accretionary versus collisional." *The Island Arc* 6.1: 2-24 (in English).

---- (**1997*b***) "Jurassic accretion tectonics of Japan." *The Island Arc* 6.1: 25-51 (in English).

---- (**2019**) "A visage of early Paleozoic Japan: geotectonic and paleobiogeographical significance of Greater South China." *The Island Arc* 28.e12296: 17 pp.

ISOZAKI, Y & T ITAYA (**1990**) "Chronology of Sanbagawa metamorphism." *Journal of Metamorphic Geology* 8: 401-411.

ISOZAKI, Yukio; Kazumasa AOKI & Takaaki NAKAMA *et al.* (**2010**) "New insight into a subduction-related orogen: a reappraisal of the geotectonic framework and evolution of the Japanese Islands." *Gondwana Research* 18: 82-105.

ISOZAKI, Y; M EHIRO & H NAKAHATA *et al.* (**2015**) "Cambrian plutonism in northeast Japan and its significance for the earliest arc-trench system of proto-Japan: new U-Pb zircon ages of the oldest granitoids in the Kitakami and Ou Mountains." *Journal of Asian Earth Sciences* 108: 136-149.

ISOZAKI, Yukio; Shigenori MARUYAMA & Kazumasa AOKI *et al.* (**2010**) "Geotectonic subdivision of the Japanese Islands revisited: categorization and definition of elements and boundaries of Pacific-type (Miyashiro-type) orogen." *Chigaku Zasshi* 119.6: 999-1053 (in Japanese with English title and abstract).

ISOZAKI, Yukio; Shigenori MARUYAMA & Takaaki NAKAMA *et al.* (**2011**) "Growth and shrinkage of an active continental margin: updated geotectonic history of the Japanese Islands." *Chigaku Zasshi* 120.1: 65-99 (in Japanese with English title and abstract).

ISOZAKI, Yukio; Ryo HASEGAWA & Harue MASUDA *et al.* (**2020**) "Finding Paleogene beds in the uppermost Izumi Group in western Kii Peninsula, SW Japan." *Chishitsugaku Zasshi* 126.11: 639-644 (in Japanese with English title and abstract).

ITO, Tanio & Hiroshi SATO (**2010**) "Crustal structure of the trench—island arc—back-arc sea system from the Nankai Trough to the northern margin of the Yamato Basin, Southwest Japan." *Chigaku Zasshi* 119.2: 235-244 (in Japanese with English title and abstract).

ITOH, Yasuto; Keiji TAKEMURA & Shigekazu KUSUMOTO (**2013**) "Neotectonic intra-arc basins within southwest Japan – conspicuous basin-forming process related to differential motion of crustal blocks", pp. 191-207 in *Mechanism of sedimentary basin formation - multidisciplinary approach on active plate margins*, ed. by Y ITOH. London: IntechOpen (in English).

KAMATA, Hiroki & Kazuto KODAMA (**1999**) "Volcanic history and tectonics of the Southwest Japan Arc." *Island Arc* 8: 393-403.

KASAHARA, Junzo; Osamu SANO and Geshi, Nobuo *et al.* (**2010**) "Overview of a Special Issue on 'Geotectonic evolution of the Japanese Islands under new paradigms of the next generation (Part I-IIII)'." *Chigaku Zasshi* 119.6: 947-958 (in English).

KIMURA, G; Yoshitaka HASHIMOTO & Asuka YAMAGUCHI *et al.* (**2016**) "Cretaceous-Neogene accretionary units: Shimanto Belt", pp. 125-137 in T MORENO *et al.* (eds) (2016).

KIMURA, Jun-ichi; James B GILL & Tomoyuki KUNIKIYO *et al.* (**2014**) "Diverse magmatic effects of subducting a hot slab in SW Japan: results from forward modeling." *Geochemistry Geophysics Geosystems* 15: 691-739 [DOI: 10.1002/2013GC005132].

KNITTEL, U; S SUZUKI & N NISHIZAKA *et al.* (**2014**) "U–Pb ages of detrital zircons from the Sanbagawa Belt in western Shikoku: additional evidence for the prevalence of Late Cretaceous protoliths of the Sanbagawa Metamorphics." *Journal of Asian Earth Sciences* 96: 148-161.

KOJIMA, S; Y HAYASAKA & Y HIROI *et al.* (**2016**) "Pre-Cretaceous accretionary complexes", pp. 61-100 in T MORENO *et al.* (eds) (2016).

KUNUGIZA, Keitaro and Atsushi GOTŌ (**2010**) "Juvenile Japan: hydrothermal activity of the Hida-Gaien Belt indicating initiation of subduction of proto-Pacific Plate in ca. 520 Ma." *Chigaku Zasshi* 119.2: 279-293.

LALLEMAND, Serge (**2016**) "Philippine Sea Plate inception, evolution, and consumption with special emphasis on the early stages of Izu-Bonin-Mariana subduction." *Progress in Earth and Planetary Science* 3: #15 [DOI: 10.1186/ s40645-016-0085-6].

LI, ZX; SV BOGDANOVA & AS COLLINS *et al.* (**2008**) "Assembly, configuration, and break-up history of Rodinia: a synthesis." *Precambrian Research* 160: 179-210.

MAHONY, SH; LM WALLACE & M MIYOSHI *et al.* (**2011**) "Volcano-tectonic interactions during rapid plate-boundary evolution in the Kyushu region, SW Japan." *GSA Bulletin* 123.11/12: 2201-2223 [DOI: 10.1130/B30408.1].

MARUYAMA, Shigenori (**1997**) "Pacific-type orogeny revisited: Miyashiro-type orogeny proposed." *The Island Arc* 6.1: 91-120.

MARUYAMA, Shigenori; Yukio ISOZAKI & Gaku KIMURA *et al.* (**1997**) "Paleogeographic maps of the Japanese Islands: plate tectonic synthesis from 750 Ma to the present." *The Island Arc* 6.1: 121-142.

MASAGO, Hideki; Kazuaki OKAMOTO & Masaru TERABAYASHI (**2005**) "Exhumation tectonics of the Sanbagawa high-pressure metamorphic belt, southwest Japan—constraints from the Upper and Lower Boundary Faults." *International Geology Review* 47.11: 1194-1206 [DOI: 10.2747/0020-6814.47.11.1194].

MATSUMOTO, Ken-ichi; Kazuko SUGIMURA & Izumi TOKITA *et al.* (**2011**) "Geology and metamorphism of the Itoigawa-Omi area of the Hida Gaien Belt, central Japan: reconstruction of the oldest Pacific-type high P/T type metamorphism and hydration metamorphism during exhumation." *Chigaku Zasshi* 120.1: 4-29 (in Japanese with English title and abstract).

MIURA, Y & S TANAKA (**2004**) "Akiyoshi limestone blocks transported by the P/T boundary event to Japan Islands." *Lunar and Planetary Science* 35.#2150: 2pp.

MIYAZAKI, Kazuhiro; Masaki OZAKI and Makoto SAITO *et al.* (**2016**) "The Kyushu-Ryukyu Arc", pp. 139-174 in T MORENO *et al.* (eds) (2016).

MORENO, Teresa; Simon WALLIS & Tomoko KOJIMA *et al.* (eds) (**2016**) *The geology of Japan.* London: Geological Society of London.

MURPHY, J Brendan & R Damian NANCE (**2013**) "Speculations on the mechanisms for the formation and breakup of supercontinents." *Geoscience Frontiers* 4.6: 667-680.

NAGATA, Mitsuhiro; Yoshikazu TAKACHI & Tsuyoshi YAMAMOTO (**2017**) "New Zircon U-Pb dates for the Nagasaki metamorphic rocks (*sensu lato*) of the Nomo Peninsula." S-GL36 in *JpGU-AGU Joint Meeting 2017, 20-25 May* [https://confit.atlas.jp/guide/event/jpguagu2017/subject/SGL2036-P2010/detail?lang=ja] (dead link).

NAKAJIMA, Takashi; Masaki TAKAHASHI & Teruyoshi IMAOKA *et al.* (**2016**) "Granitic rocks", pp. 251-272 in T MORENO *et al.* (eds) (2016).

NAKAMA, Takaaki; Takafumi HIRATA & Shigeru OTOH *et al.* (**2010**) "Paleogeography of the Japanese Islands: age spectra of detrital zircon and provenance history of the orogen." *Chigaku Zasshi* 119.6: 1161-1172 (in Japanese with English title and abstract).

NEALL, Vincent E. & Steven A. TREWICK (**2008**) "The age and origin of the Pacific islands: a geological overview." *Philosophical Transactions of the Royal Society B* 363: 3293-3308.

NORTON, Ian O. (**2007**) "Speculations on Cretaceous tectonic history of the northwest Pacific and a tectonic origin for the Hawaii hotspot", pp. 451-470 in *Plates, plymes and planetary processes*, ed. by GR FOULGER & DM JURDY. Boulder, CO: Geological Society of America.

NOZAKI, Tatsuo; Yasuhiro KATO & Katsuhiko SUZUKI (**2010**) "Re–Os geochronology of the Iimori Besshi-type massive sulfide deposit in the Sanbagawa metamorphic belt, Japan." *Geochimica et Cosmochimica Acta* 74: 4322-4331.

OKAMOTO, Kazuaki; Shigenori MARUYAMA & Yukio ISOZAKI (**2000**) "Accretionary complex origin of the Sanbagawa, high P/T metamorphic rocks, central Shikoku, Japan – layer-parallel shortening structure and greenstone geochemistry." *Chishitsugaku Zasshi* 106.1: 70-86 (in English).

OMORI, Soichi & Yukio ISOZAKI **(2011)** "Paleozoic Japan and the eastern extension of the collisional suture between the North and South China cratons." *Chigaku Zasshi* 120.1: 40-51 (in Japanese with English title and abstract).

Oshika Museum **(2018)** "The MTL from its beginnings to the present-day " *Oshika Museum of Japan Median Tectonic Line* [https://mtl-muse.com/mtl/aboutmtl/mtl-history/] (in Japanese).

OTOH, Shigeru; Masanori SHIMOJO & Kazumasa AOKI *et al.* (2010) "Age distribution of detrital zircons in the psammitic schist of the Sanbagawa Belt, southwest Japan." *Chigaku Zasshi* 119.2: 333-346 (in Japanese with English title and abstract).

PICKERING, Kevin T; Michael B UNDERWOOD & Sanny SAITO *et al.* **(2013)** "Depositional architecture, provenance, and tectonic/eustatic modulation of Miocene submarine fans in the Shikoku Basin: results from Nankai Trough Seismogenic Zone Experiment." *Geochemistry Geophysics Geosystems* 14.6: 1722-1739.

REA, David K & John M DIXON **(1983)** "Late Cretaceous and Paleogene tectonic evolution of the North Pacific Ocean." *Earth & Planetary Science Letters* 65: 145-166.

SAFONOVA, Inna & Shigenori MARUYAMA **(2014)** "Asia: a frontier for a future supercontinent Amasia." *International Geology Review* 56.9: 1051-1071 [DOI: 10.1080/00206814.2014.915586].

SANO, Hiroyoshi & Kametoshi KANMERA **(1988)** "Paleogeographic reconstruction of accreted oceanic rocks, Akiyoshi, southwest Japan." *Geology* 16.7: 600-603.

SANO, Sakae; Yasutaka HAYASAKA & Koichi TAZAKI **(2000)** "Geochemical characteristics of Carboniferous greenstones in the Inner Zone of Southwest Japan." *The Island Arc* 9: 81-96.

SANTOSH, M (2013) "Evolution of continents, cratons and supercontinents: building the habitable Earth." *Current Science* 104.7: 871-879.

SANTOSH, M & Hiroaki SENSHU **(2011)** "History of supercontinents and its relation to the origin of the Japanese Islands." *Chigaku Zasshi* 120.1: 100-114 (in Japanese with English title and abstract).

SATO, Hiroshi; Naoko KATO & Susumu ABE *et al.* **(2015)** "Reactivation of an old plate interface as a strike-slip fault in a slip-partitioned system: Median Tectonic Line, SW Japan." *Tectonophysics* 644-645: 58-67.

SCOTESE, Christopher R **(2003)** "Paleomap Project." [scotese.com].

SHIMURA, Yusuke; Tetsuya TOKIWA & Makoto TAKEUCHI **(2020)** "Tectonic evolution of the Sanbagawa Metamorphic and Shimanto Accretionary Complexes: approaches from field survey and zircon U–Pb age." *Nihon Chikyū Kagakukai Nenkai Yōshishū* 67.PRO175: S1 (in Japanese with English title) [DOI: 10.14862/geochemproc. 67.0_172].

SMITH, AG **(2007)** "A plate model for Jurassic to Recent intraplate volcanism in the Pacific Ocean basin", pp. 471-495 in *Plates, plymes and planetary processes*, ed. by GR FOULGER & DM JURDY. Boulder, CO: Geological Society of America.

STAMPFLI, GM; C HOCHARD & C VÉARD *et al.* **(2013)** "The formation of Pangea." *Tectonophysics* 593: 1-19.

SUGIYAMA, Riuji (ed.) **(1973)** *Median Tectonic Line*. Tokyo: Tokai University Press (in Japanese with some English summaries).

TAIRA, Asahiko **(1990)** *The Birth of the Japanese Islands*. Iwanami Shinsho 148. Tokyo: Iwanami (in Japanese).

---- **(2001)** "Tectonic evolution of the Japanese island arc system." *Annual Review of Earth and Planetary Sciences* 29: 109-134.

TAIRA, Asahiko; Shoichi KIYOKAWA and Kan AOIKE *et al.* **(1997)** "Accretion tectonics of the Japanese islands and evolution of continental crust." *Earth & Planetary Sciences* 325: 467-478 (in English with French title and abstract).

TAIRA, Asahiko; Y OHARA & SR WALLIS *et al.* **(2016)** "Geological evolution of Japan: an overview", pp. 1-24 in T MORENO *et al.* (eds) (2016).

TAKAGI, Hideo & Hiroyoshi ARAI **(2003)** "Restoration of exotic terranes along the Median Tectonic Line, Japanese Islands: overview." *Gondwana Research* 6.4: 657-668.

TAKAHASHI, Yutaka; Deung-Lyong CHO & Jianren MAO *et al.* **(2017)** "SHRIMP U–Pb zircon ages of the Hida metamorphic and plutonic rocks, Japan: implications for late Paleozoic to Mesozoic tectonics around the Korean Peninsula." *The Island Arc* 27: e12220 [DOI: 10.111/iar.12220].

TAZAWA, Jun-ichi **(2001)** "Middle Permian brachiopod faunas of Japan and South Primorye, Far East Russia: their palaeobiogeographic and tectonic implications." *Geosciences Journal* 5.1: 19-26.

TODA, S **(2016)** "Crustal earthquakes", pp. 371-408 in T MORENO *et al.* (eds) (2016).

TSUJIMORI, Tatsuki (**2017**) "Early Paleozoic jadeitites in Japan: an overview." *Journal of Mineralogical and Petrological Sciences* 112: 217-226.

TSUJIMORI, Tatsuki and George E HARLOW (**2017**) "Jadeitite (jadeite jade) from Japan: history, characteristics, and perspectives." *Journal of Mineralogical and Petrological Sciences* November [DOI: 10.2465/jmps.170804].

TSUKADA, Kazuhiro; Koshi YAMAMOTO & Onon GANTUMUR *et al.* (**2017**) "Early Paleozoic subduction initiation volcanism of the Iwatsubodani Formation, Hida Gaien belt, Southwest Japan." *International Journal of Earth Sciences* 106: 1429-1451.

TSUTSUMI, Yukiyasu and Atsushi MIYASHITA (**2009**) "SHRIMP U-Pb dating of detrital zircons from the Sanbagawa Belt, Kanto Mountains, Japan: need to revise the framework of the belt." *Journal of Mineralogical and Petrological Sciences* 104: 12-24.

UEDA, Hayato (**2016**) "Hokkaido", pp. 201-221 in T MORENO *et al.* (eds) (2016).

UJIIE, Kohtaro (**2002**) "Evolution and kinematics of an ancient décollement zone, mélange in the Shimanto accretionary complex of Okinawa Island, Ryukyu Arc." *Journal of Structural Geology* 24: 937-952.

WAKITA, K (**2013**) "Geology and tectonics of Japanese Islands: a review – the key to understanding the geology of Asia." *Journal of Asian Earth Sciences* 72: 75-87.

WAKITA, Koji; Takanori NAKAGAWA & Masahiro SAKATA *et al.* (**2018**) "Phanerozoic accretionary history of Japan and the western Pacific margin." *Geological Magazine* [DOI: 10.1017/S0016756818000742].

WALLIS, Simon Richard & Takamoto OKUDAIRA (**2016**) "Paired metamorphic bests of SW Japan: the geology of the Sanbagawa and Ryoke metamorphic belts and the Median Tectonic Line", pp. 101-124 in T MORENO *et al.* (eds) (2016).

WANG, Xuan-Ce; Zheng-Xiang LI & Xian-Hua LI *et al.* (**2011**) "Geochemical and Hf–Nd isotope data of Nanhua rift sedimentary and volcaniclastic rocks indicate a Neoproterozoic continental flood basalt provenance." *Lithos* 127: 427-440.

WATANABE, Yasushi; Tetsuichi TAKAGI & Nobuyuki KANEKO *et al.* (**2016**) "Mineral and hydrocarbon resources", pp. 431-456 in T MORENO *et al.* (eds) (2016).

WELLER, OM; SR WALLIS & M AOYA *et al.* (**2015**) "Phase equilibria modelling of blueschist and eclogite from the Sanbagawa metamorphic belt of southwest Japan reveals along-strike consistency in tectonothermal architecture." *Journal of Metamorphic Geology* 33: 579-596.

WU, Yuan-Bao & Yong-Fei ZHENG (**2013**) "Tectonic evolution of a composite collision orogen: an overview on the Qinling–Tongbai–Hong'an–Dabie–Sulu orogenic belt in central China." *Gondwana Research* 23.4: 1402-1428 [DOI: 10.1016/j.gr.2012.09.007].

WU, Jonny; John SUPPE & Renqi LU *et al.* (**2016**) "Philippine Sea and East Asian plate tectonics since 52 Ma constrained by new subducted slab reconstruction methods." *Journal of Geophysical Research: Solid Earth* 121: 4670-4741 [DOI: 10.1002/2016JB01].

YAMAMOTO, Shinji; H SENSHU & S RINO *et al.* (**2009**) "Granite subduction: arc subduction, tectonic erosion and sediment subduction." *Gondwana Research* 15.3/4: 443-453.

YAMAZAKI, T; M TAKAHASHI & Y IRYU *et al.* (**2010**) "Philippine Sea Plate motion since the Eocene estimated from paleomagnetism of seafloor drill cores and gravity cores." *Earth Planets Space* 62.6: 495-502.

YANAI, Shuichi; Kazumasa AOKI & Yoshimitsu AKAHORI (**2010**) "Opening of Japan Sea and major tectonic lines of Japan: MTL, TTL and Fossa Magna." *Chigaku Zasshi* 119.6: 1079-1124 (in Japanese with English title and abstract).

ZHANG, Yuzhi; Yuejun WANG & Yanhua ZHANG *et al.* (**2015**) "Neoproterozoic assembly of the Yangtze and Cathaysia blocks: evidence from the Cangshuipu Group and associated rocks along the Central Jiangnan Orogen, South China." *Precambrian Research* 269: 18-30.

CHAPTER 4

Shaping the Japanese Archipelago

To refute:

The Japanese Islands have experienced no drastic change from the earliest Miocene to the present

(Sugimura & Uyeda 1973: 92)

Fifty years ago it was thought, by even the most progressive Japanese geophysicists of the day as the above quotation shows, that the Japanese archipelago and its accompanying Japan Sea Basin had existed at least from earliest Miocene times (ca. 23 Ma) and perhaps even from Palaeogene (ca. 65–23 Ma) or Cretaceous times (145–65 Ma) (Smith 1982). However, revolutionary advances in palaeomagnetic dating in the 1980s dramatically changed this picture. On the basis of those new data, it was postulated that the Japanese archipelago did not exist in its current form until 15 Ma in the Middle Miocene, when the processes of continental rifting to make the Japan Sea Basin pushed the Eurasian fringe out to form a series of offshore islands. Since then, *everything* in Japan has changed – and so has our knowledge of these transformations changed in only the last few decades. This chapter will provide an overview of those changes, including the opening of the Japan Sea Basin, the lasting effects of Miocene volcanics, the collision of the Izu Arc with Honshu, and creation of the Kanto Syntaxis and Fossa Magna.

Revolutionary Advances

The most fascinating part of Japan's recent geological history is the larger scale magmatic processes that led to the detachment of the Japanese landmass from the continent to form the current island arc (Figure 4.1). As we have seen in Chapter 3, the Japanese landmass originally formed part of the continental coastline of Eurasia throughout most of its accretionary history. Beginning from around 21 Ma, part of this coastline rifted into an arc-like formation with the development of a partially ocean-floored back-arc basin behind it (cf. Bogatikov et al. 2009). The full extension of the archipelago was achieved by 15 Ma. Only from that time can we say that current Japanese Archipelago existed as a separate geographical entity.

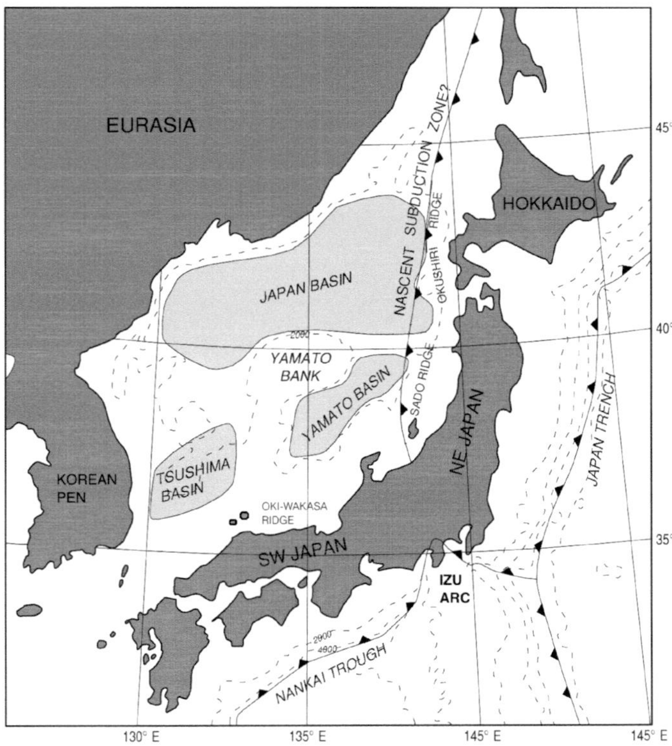

Figure 4.1 Geography of the Japan Sea Basin today

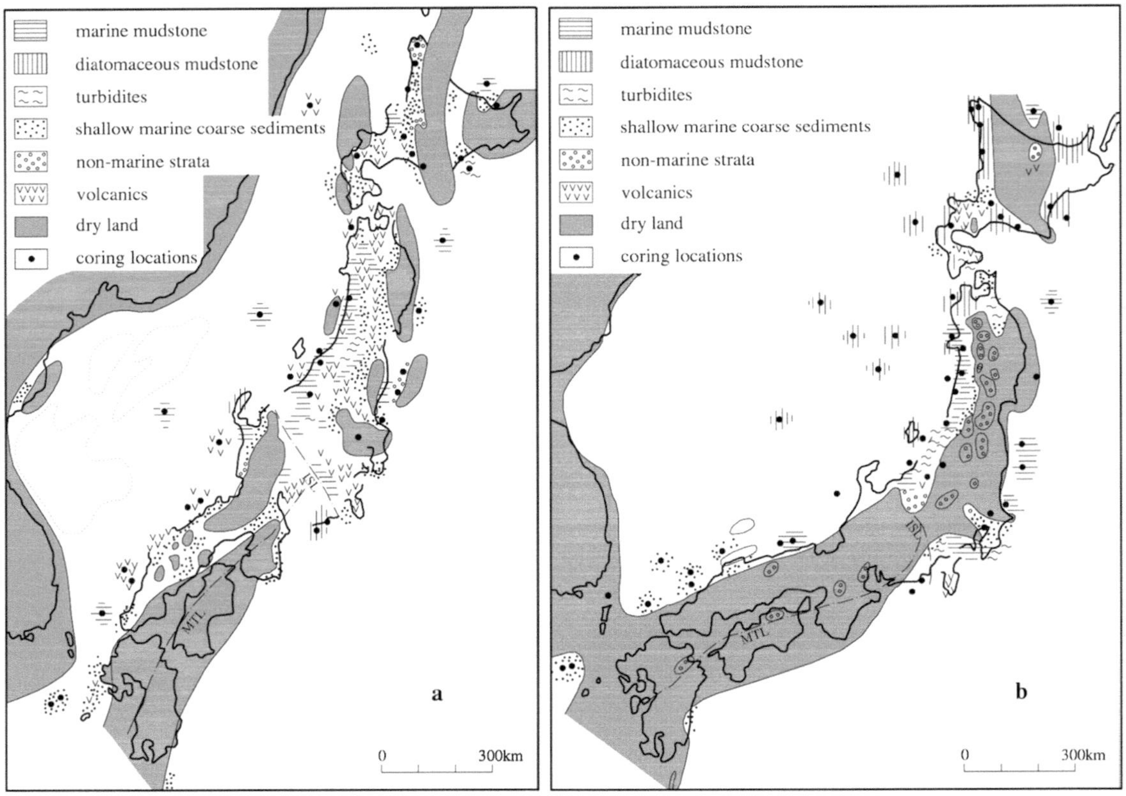

Figure 4.2 One conception of the post-rifted Japanese Islands

Left: half-submerged at 16 Ma with few emergent areas
Right: between 8 and 3.5 Ma with a landbridge connection to the continent

The rifted arc, as first detached from the Eurasian continent, and the archipelago of today bear little resemblance to each other. Half the landmass was originally underwater, and there were no land-based volcanoes (Figure 4.2). After rifting, the archipelago underwent regionally distinct compression and extension regimes to form new mountain ranges and faulted basins. The collision of the Izu Arc with central Honshu after back-arc basin formation has been a major contributor to the modern Japanese landscape. Not only did it enable the formation of the Fuji volcano, it indented the geotectonic belts of central Honshu to form the Kanto Syntaxis. The Izu Arc divides Japan into two subduction regimes, resulting in a higher level of present-day volcanic activity in northeastern than in southwestern Japan – except for southern Kyushu where most of the recent Quaternary volcanoes are located (cf. Figures 5.11, 6.9). *Nothing could be further from the truth than the idea that the Japanese landmass has not changed in the last 24 million years.*

The first section below traces the important mechanics of rifting that formed the Japanese archipelago. Rifting did not simply mean splitting rocks off the continental edge and setting them adrift as islands; it also involved much generation of magma, which rose through the cracks as the Earth's Eurasian crust was being stretched thinner. This magma, whether extruded as lava or as volcanic ash, was generally bimodal in nature, involving both mafic (basaltic) and felsic (rhyolitic) extrusions and thus differing in chemical composition to the later intermediate volcanic products associated with the subduction of oceanic plates (cf. Apx 5: Table B).

The second and third sections review the different kinds of volcanism that accompanied the final phases of rifting: the generation of 'green tuff' and the establishment of andesitic volcanoes in the Setouchi (Inland Sea) region.

The fourth section investigates the consequences for Honshu of back-basin opening via the intrusion of the Izu Arc. With the Japanese archipelago pushed eastwards away from the continent, the Izu volcanic arc carried on the edge of the Philippine Plate began to plough into Honshu, causing the Kanto Syntaxis: the bending of earlier geological belts – originally formed parallel to the NE-SW trending Eurasian coastline – around the intruding Izu Arc. The western edge of the arc intrusion is marked by the I-STL (Itoigawa-Shizuoka Tectonic Line), along which formed the Fossa Magna, the 'big ditch' that marks a cultural and linguistic divide between NE and SW Japan (Ono 1970).

Rifting, Magmatism & Japan Sea Basin Formation

Opening of the Japan Sea Basin

There is little more dramatic than the opening of a back-arc basin. An extensional back-arc basin forms through crustal thinning and rifting behind an igneous arc (cf. Figure 2.13), pushing out that section of continental coast into an offshore arc, with the new basin behind the arc filling with sea water and often having a floor built of oceanic crust. Both the Japan Sea and the Japanese Islands resulting from rifting have thinner (15–20/23 km thick) than normal continental crust (Bogatikov *et al.* 2009), with oceanic crust forming in localized regions in the Japan Sea Basin. Back-arc basins can also develop within oceans as well, such as the Shikoku Basin behind the Izu Arc on the Philippine Plate. That basin's opening governed the positioning of the Izu Arc, which had great ramifications (literally) for Honshu development. Around 4 Ma, the rate of subduction of the Philippine Plate accelerated perhaps in conjunction with the formation of the Okinawa Trough west of the Ryukyu Archipelago (Kimura *et al.* 2005: 983); thus an incipient back-arc rift zone may develop into a back-arc basin behind the Ryukyu Archipelago in future. For mainland Japan, the igneous arc in question here is the Cretaceous Ryoke Belt that marked the volcanics accompanying subduction of the Izanagi ridge in the Late Cretaceous (cf. Figure 3.7: inset). It comprises the backbone of the arc of crust that was rifted from the continent. Yanai and colleagues identify 31 separate continental blocks that were rifted at this time, and they have attempted to fit the pieces back into their original position against the continent (see Yanai *et al.* 2010: fig. 7). Kawai *et al.* (1971) attempted an early explanation of the curved shape of the archipelago; however, their reconstructions do not account for the tectonic rearrangement of the geological belts north of the Tanakura Tectonic Line (cf. Figure 3.1a).

The final configuration of the Japan Sea Basin (cf. Figure 4.1) is divided into internal basins, and its topographic highs, such as the Yamato Bank and various ridges are continental remnants stranded during rifting. The Tsushima Basin is still floored by continental crust, but in the Japan Basin, that crust has been completely rifted through, and a spreading centre has developed to form 100% basaltic oceanic crust in the basin's floor (Yanai 2010: fig. 3). This ocean floor is 8.5 km thick overlain by 2 km of sediments (Hiratas N *et al.* 1992). It has magnetic lineations similar to those paralleling mid-oceanic ridge spreading centres, and dredgings and corings from them have provided dates of their formation (Tamaki 1995: 415; Nohda 2009). In contrast, the basaltic oceanic crust in the Yamato Basin is anomalously thick (>14 km); its formation is not attributed to a spreading center but to the propagation of multiple rift zones (Hirahara *et al.* 2015). Yanai *et al.* (2010) estimate the basalt only covers 50% of the basin's floor.

Northeastern and southwestern Japan have different rifting histories, as presented by Yoshida *et al.* (2014) and Kimura *et al.* (2005). Moreover, the opening of the Japan Sea Basin was heterogeneous according to basaltic ocean floor dates, with the Japan Basin probably opening before the Yamato Basin (Nohda 2009; Nakajima 2013). The differing viewpoints are noted in Table 4.1 (see also Apx 9); Kimura and colleagues give more detailed chemical and typological descriptions through the various stages for southwest Japan. It is clear even from the simplified descriptions below that rifting volcanics consist of a wide range of rock types with equally various chemical characteristics in terms of trace elements and isotopic compositions. For archaeological purposes, the tuff is more important than the basalts and will be discussed in detail below under the "Green Tuff Movement".

As in other instances of continental rifting around the world, flood basalts of alkali composition (Apx 5: Fig. G), characteristic of the early stages of rifting (Yoshida's Period I), were first deposited on dry land – the Eurasian mainland in this case, as known from North China (Kusky *et al.* 2007: 340). Subsequently, volcanic products were deposited into freshwater lakes as rifted blocks subsided and the depressions filled with water. Here we are concerned with Kimura's Stages I–III during which rifting volcanic products are characteristically 'bimodal' – either mafic or felsic but with little intermediate composition (cf. Apx 5: Table B). The basalts (mafic) were extruded as lava or sills and form the floors of the the Japan and Yamato Basins; the tuff (felsic) was extruded submarinely as tephra and lies on top of the basalt floors (Nohda 2009). The kinds of basalt recovered from those basins' drilling cores is judged to range from theoliitic basalt to basaltic andesite (Apx 5: Fig. G) – common in zones of crustal extension (Philpotts & Ague 2009: 144), while Tamaki *et al.* noted that "a variety of parental magmas derived from different and compositionally heterogeneous mantle sources was required to produce the range of igneous units sampled at each site " (1992: 1341).

The opening of the Japan Sea basin caused some fragments of the original North China Block to be detached and carried along with the existing Accretionary Complexes. Those fragments now form the earliest dated and westernmost geological terranes of the Japanese Islands: the Hida and Oki Belts (Chapter 3; Apx 6). Also tagging along were the Oeyama ophiolite in the Hida marginal belt, and metamorphosed rocks in the Qinling suture zone resulting from the Triassic collision of the North and South China Blocks, appearing in the Higo/Hitachi-Takanuki Belts, the Hida/Marginal Belts (Chapter 3; Apx 6), and Mt Sefuri in northern Kyushu (Omori & Isozaki 2011). New 'tectonic lines' also came into being at this time: among them are those discussed in Chapter 3 (the neo-MTL, TTL, I-STL and K-CTL [or KTL]) with new ones introduced bordering the Japan Sea on the northeast and southwest.

During the tectonic volcanism accompanying rifting, hydrothermally driven argentiferous galena veins intruded through altered Miocene quartz porphyry and surrounding country rock in the Korean Strait (Mindat n.d.). Veins subsequently exposed on Tsushima Island comprised the first discovery of silver in Japan, in 674 AD. To process this silver from lead, cupellation techniques were probably acquired from Tang Dynasty China (Nakanishi & Izawa 2014).

The newly formed archipelago and sea

At 15 Ma, as the rifting process ended, the southwestern Japanese landmass was uplifted by the action of the Philippine Plate and became dry land reconnected to the continent (Yonekura *et al.* 2001: 302). The northeastern Japanese landmass was left mostly underwater (cf. Figure 4.2: left); it became directly subject to subduction of the Pacific Plate but was only uplifted to become dry land due to a shift from extension to tectonic compression around 3.5 Ma (Itaki 2016) (cf. Figure 4.2: right).

Initially, the NE and SW sections of Japan were tucked against the Eurasian continent. Takahashi (2006) has identified the Volcanic Front at that continental margin that was active between 18 and 16 Ma; he proposes that the VF was formed entirely by the subduction of the Pacific Plate, with Philippine Plate only beginning later. The erupted andesites of that VF are now found on the inner (northern) edge of SW Japan and the outer (eastern) edge of NE Japan (Takahashi 2006: fig. 4) with their adjoining ends now 230 km apart. This odd arrangement can only be explained by differential rifting of these two segments of Japan. How they took their current relative positions is highly debated (see below). During this process, the geological belts in both NE and SW Japan were tectonically rearranged, hence the modern occurrence of the Hitachi-Takanuki Belt in central Tohoku isolated from its consanguinous Higo Belt, and hence the disappearance of 150 km of forearc in the Outer Zone; Hokkaido compilation via the descent of the Okhotsk Place following the Japan Sea opening (as shown in the modern configuration in Figure 3.7).

Table 4.1 Simplified chronology of Japan Sea Basin opening and volcanics

Abbrev: alk = alkaline, calc-alk = calc-alklaline, N = North, NE = northeast, SW = southwest; Ma = million years ago

<table>
<tr><th colspan="2">Continental</th><th colspan="2">Back-arc basin</th><th colspan="2">Archipelago</th></tr>
<tr><td colspan="2">Yoshida et al. (2014) NE Japan
Period I (stages 1–4)
continental margin 66–21Ma
mafic–felsic subaerial volcanism</td><td colspan="2">Period II (stages 5–7)
back-arc basin formation
21–13.5Ma
submarine bimodal volcanism</td><td colspan="2">Period III (stages 8–13)
island arc 13.5–present
felsic caldera eruptions, then andesitic stratovolcanoes</td></tr>
<tr><td colspan="2" rowspan="2">Kimura et al. (2005) SW Japan</td><td>Stage I
initial rifting
25–17Ma</td><td>Stage II
17–12Ma
back-arc basin opening</td><td>Stage III
12–4Ma Miocene volcanic arc</td><td>Stage IV
4Ma– present
LatePliocene–Holocene arc</td></tr>
<tr><td>bimodal sub-alk on land before opening</td><td>from 16.5Ma:
bimodal alk & calc-alk off N coast & outer zone, but continuous calc-alk on both Honshu coasts</td><td>bimodal alk off N coast; but on land mostly alkaline basalt/andesite; ultramafics on N coast</td><td>bimodal; alk off N coast but mixture of alk+andesite /dacite on land</td></tr>
<tr><td>Nohda (2009) mainland volcanics</td><td>40Ma subduction volcanics; 33Ma ocean intrusion into rift basin</td><td></td><td>lower basalt:
21.2–17.7Ma
upper basalt:
17.7–14.9Ma, 11Ma;
tuff 14.9</td><td>tuff
7.6, 6.3Ma</td><td></td></tr>
<tr><td>Nakajima (2013)</td><td>core-complex mode 35–24Ma:
incipient rifting</td><td>wide-rift mode 24–15Ma: Japan Sea opening</td><td></td><td>narrow-rift mode 15–13.5Ma: late syn-rift system</td><td></td></tr>
</table>

Once rifted, the incipient archipelago was flooded by high sea level, called the Nishi-Kurosawa Transgression in Japan (Apx 9), during the Miocene climatic optimum (cf. Figure 4.2: left) (Tsuchi 1992: 237-43). Shells recovered from the marine layers are tropical, and evidence of mangrove swamps and coral reefs abounds (Tsuchi 1992: fig. 2). Diatom studies of sediments have revealed that the southern end of the basin was closed between 8 and 3.5 Ma due to intra-arc compression (Itaki 2016) (Figure 4.2: right), so that it received only cold water from the north during that time. At 3.5 Ma, the Tsushima

Strait opened to again allow entry of warm-water species into the Japan Sea Basin on a palaeo-Tsushima current (Koizumi 1992: 18; Itaki 2016).

Competing hypotheses for rifting cause

The ultimate cause of Japan Sea back-arc basin formation has been highly debated for a long time (Marsaglia 1995). Since the early 1980s, the Japan Sea opening has been acknowledged to relate to distant events and to other occurrences of back-basin opening (Tapponnier *et al.* 1982), with the Japan Sea Basin being proposed as a pull-apart basin (Lallemand & Jolivet 1985). In 1997, Maruyama and colleagues postulated that between 40 and 25 Ma, the East Asian landmass was subject to extension tectonics, which were "probably associated with a domal uplift of the crust induced by an uprising mantle plume from the deep mantle" (Maruyama *et al.* 1997: 137). The rifting that followed was cited as accounting for several basins: Bohai, Baikal, Japan Sea, and Kuril Basins. Recent research on the Bohai Basin assumes such mantle upwelling as leading to extension tectonics (Chen *et al.* 2011); but a study on the Baikal Basin found very little magmatism associated with its rifting and instead cites compression tectonics from the impact of the India Plate on Eurasia from 55 (30) Ma resulting in pull-apart basins (Petit & Déverchère 2006). Modelling, however, has demonstrated that rifting with or without magma eruptions may stem from the same geotectonic environment (Koptev *et al.* 2016).

The source of the aforementioned superplume may related to the Indian subcontinent (Tapponnier *et al.* 1982), postulating that the collision and intrusion of the Indian sub-continent into Eurasia between 55 and 40 Ma is now making itself felt on the eastern Eurasian rim. The progressive subduction of the Tethys Ocean floor between India and Eurasia resulted in the collision of the two landmasses, creating the Himalayan mountains (Harris & Whalley 2001). The down-going ocean floor slab, however, was not destroyed in this process but was fully pulled down by subduction and disappeared into the Earth's mantle; and now the Indian continent itself is being drawn under Eurasia. Seismic tomographic studies of Earth (analogous to CAT scans of one's body) suggest that once oceanic slabs descend as far as the core-mantle boundary, there may be a reflection effect sending a mantle plume up at a correspondingly opposite angle (Blake & Argles 2003: 13). Thus, the northeastward descent of the Tethys Ocean floor slab, once separating India from Eurasia, may have sent off a plume[1] to rise underneath eastern Eurasia – where extensional rifting became obvious on the Earth's surface after 40 Ma.

Another hypothesis suggested that the loss of the continental 'root' under the north China Mainland through delamination stimulated the upwelling of new magma, causing thinning of the crust and leaving it vulnerable to Pacific Plate tension (Zhai *et al.* 2007). This argument postulates that only the eastern half of the North China Block was affected, and that area is indeed marked off by a distinct GGL (Great Gradient Line) dividing the NCB into a western region with ≥150 km thick lithosphere (beneath the loess plateau) and an eastern region with only ≤80 km thick lithosphere (Niu *et al.* 2015: fig. 1). Liu *et al.* (2019) provide an 8-stage model of complicated slab subduction between 160 and 20 Ma that involves delamination processes.

A new contender in this debate is the role of second continents (of granitic composition) in generating heat to produce plumes. The low gravity measurements cited for delamination show a mass deficit continuing further north-south than just the North China Block; indeed, they match an area where a

[1] The term "plume" is used advisedly here, in respect of Foulger's challenging of the plume/hot spot concept (Foulger 2010).

second continent is hypothesized to lie under eastern China and may be detected by monitoring geoneutrino emissions (Roskovec *et al.* 2018). Multiple plumes are clearly invoked as instrumental in extension and back-arc basin formation since the Eocene (56–43 Ma) and Early to Middle Miocene (23–15 Ma) by Yanai *et al.* (2010: fig. 12). Two of those plumes resulted in alkaline volcanics, at the eastern edge of the Eurasian continent and the western edge of the Yamato Basin. Alkaline volcanics (cf. Apx 5: Fig. G) signify a mantle source for the magma, and a wet mantle plume has continued erupting even after rifting ended, with varying alkaline chemical composition of volcanics in SW Japan (Tatsumi 1983; Iwamori 1991; Kimura *et al.* 2005).

Repositioning the Japanese landmass

Rifting did not take place in one smooth movement; Yanai *et al.* (2010) emphasize that two large fault systems and many smaller faults were responsible for moving at least 31 blocks from the continental edge, the largest of which were NE and SW Japan. How and where these pieces of modern Japan were put together is controversial. Altis (1999) reviewed the two contenders, the 'double door' and 'drawer models' (Figure 4.3), but further discussion has since taken place.

• In the mid-1980s, palaeomagnetic data were used to propose that NE and SW Japan rotated down from the continent like a 'double door' opening outwards (Figure 4.3: green) (Otofuji *et al.* 1985).

• Yanai and colleagues term this the 'fan model' which should require a counterbalancing formation of a convergent margin in the Yellow Sea area – for which there is no evidence (see Yanai *et al.* 2010: fig. 13).

• The high rate of rotation for SW Japan through 56° clockwise was also rejected as impossible by (Lallemand & Jolivet 1985).

• In response, a second model was proposed to operate like an 'opening drawer' directly away from the continent along two strike-slip faults (Figure 4.3: red): the Tanakura Tectonic Line (TTL) in Tohoku and the Ululun Tectonic Line (UTL) running down the west of Kyushu (Jolivet *et al.* 1994). This movement was proposed to involve very little rotation. However, the next year Jolivet and colleagues revised their model to incorporate some rotation (Jolivet *et al.* 1995).

• Baba and colleagues reanalyzed the earlier palaeomagnetic results and determined that the rapid clockwise rotation of SW Japan did occur between 14.8 to 14.2 Ma, while the counterclockwise rotation of NE Japan occurred between 16.5 to 14.4 Ma (Baba *et al.* 2007).

• Nohda confirms that these movements are consistent with the "intensive basaltic volcanism...and tuff activity at 14.86 Ma", though early basalts for the Japan Basin had not yet been found (as of Nohda 2009: 608). Kimura also accepts the proposed rotation for SW Japan (Kimura *et al.* 2005), while Nakajima notes that the late volcanics (Table 4.1) indicate extension on the Japan Sea side of Tohoku while the Pacific Ocean side was under compression (Nakajima 2013).

It is not inconceivable that these two models (Figure 4.3) acted together to produce the current Japanese archipelago. If they did not, it would be hard to explain the transformation of the fairly *linear* layout of the reconstructed original positions of NE and SW Japan forming the continental edge (cf. Itoh *et al.* 2013: fig. 2b) now taking a significant *arc* shape. Moreover, the significance of the Fossa Magna zone, discussed below, as dividing NW and SW Japan is more difficult to explain within the 'drawer' model if Honshu moved altogether southeastwards without parts of it rotating. Baba & Yoshida (2020) describe the Fossa Magna as a 'pull-apart' basin that formed during rifting and that continues into the Japan Sea separating the Yamato Basin from the Japan Basin (see their fig. 1). The occurrence of *kuroko* (lit. 'black-ore', aka black smoker) beds in the Fossa Magna signifies that there was once open sea between the two halves of Honshu; these are now subject to compression along the Itoigawa-Shizuoka Tectonic Line bordering the Fossa Magna on the west. The volcanics that

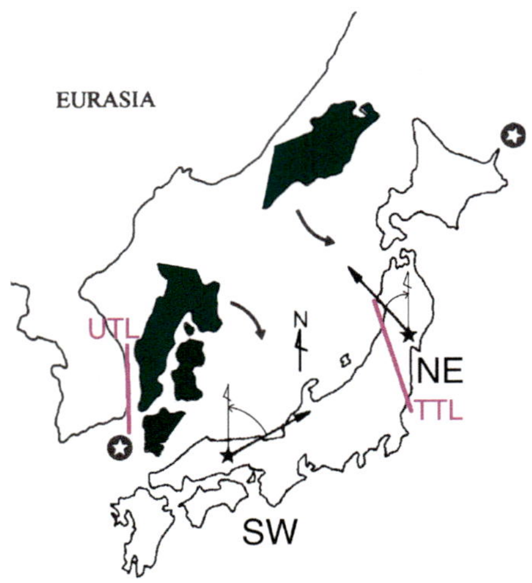

Figure 4.3 Models for the opening of the Japan Sea Basin

'Double-door model' (green): The *starred arrows* in Figure 4.3 indicate the locus for palaeomagnetic readings pointing North, while the *circled stars* are the Euler pole anchors for the rotation; the *curved arrows* indicate the direction of rotation outwards in the upper right, and the direction of rotation inwards necessary to bring the starred arrows back in line with current North. Both indicate a substantial degree of rotation of the NE and SW landmasses needed to form today's arc-shaped archipelago (Otofuji 1985).

'Drawer model' (magenta): two strike-slip fault systems, one in the west and the Tanakura Tectonic Line (TTL) in the east, are thought responsible for the movement of the Japanese landmass away from the continent. The western system has been variously identified as the Yangsan Fault (Jolivet *et al.* 1994), the Ululun Tectonic Line (UTL) (Isozaki *et al.* 2010: 95), the Yangsan-Ulsan Fault (Yanai *et al.* 2010), and the West Kyushu Tectonic Line (Yanai *et al.* 2010).

accompanied rifting were typical bimodal products, characteristic of rifting processes, extruded in a ratio of 3:2 felsic:mafic (Watanabe *et al.* 2016; Shikazono 1994). Volcanic mounds can be seen in the Japan Sea Basin floor stratigraphy (cf. Figure 2.14).

Altis (1999) proposed a quite different model for basin opening involving the collision of the Okohtsk Plate with Eurasia and setting off of NNE-trending Early to Middle Miocene rifts. He argues that after a hiatus between 15 and 5 Ma, these trends began again in SW Japan between 1.5 to 1 Ma. These thoughts will be revisited in Chapter 13 in relation to the Nara Basin formation.

The 'Green Tuff' Movement

As the Japan Sea basin opened up between 19 and 15 Ma during the Nishi-Kurosawa Transgression (Apx 9), tephra was extruded under water down the northwest coast and across middle Honshu, forming the Green Tuff Zone (Figure 4.4). The tuff has a silica range of andesitic to dacitic (cf. Apx 5: Table B & Fig. F) (Shikazono 2003). These pyroclastics took on a characteristic green colour through submarine hydrothermal reaction (metasomatism) due to the absorption of magnesium-Mg from seawater, causing green minerals such as chlorite and smectite to form in the lithifying tuff (Shikazono 1994). *Kuroko* deposits, now found throughout the east-northeastern Green Tuff Zone on land, contain significant copper, lead, and zinc (Shikazono 2003). The Green Tuff and *kuroko* deposits were emplaced during localized rifting between 16 and 13.5 Ma (Nakajima 2013) – the final extensional tectonics before the Philippine Plate began subducting under SW Japan.

The Green Tuff is an excellent example of volcanic rock classed as a sediment. The term 'Green Tuff' was originally the name of a stratigraphic unit in northeast Japan within the Nishikurosawa Formation (Watanabe *et al.* 2016), but it has been extended to account for a wide zone of Miocene alkaline volcanics (Bogatikov 2000: 98) stretching from the Kuril Islands down through western Honshu into the Ryukyu Islands and across Honshu in the Fossa Magna zone as seen in the Oya region (BOX 4) into the Izu Arc; it is also a term used abroad (e.g., in Italy, Liszewska *et al.* 2018).

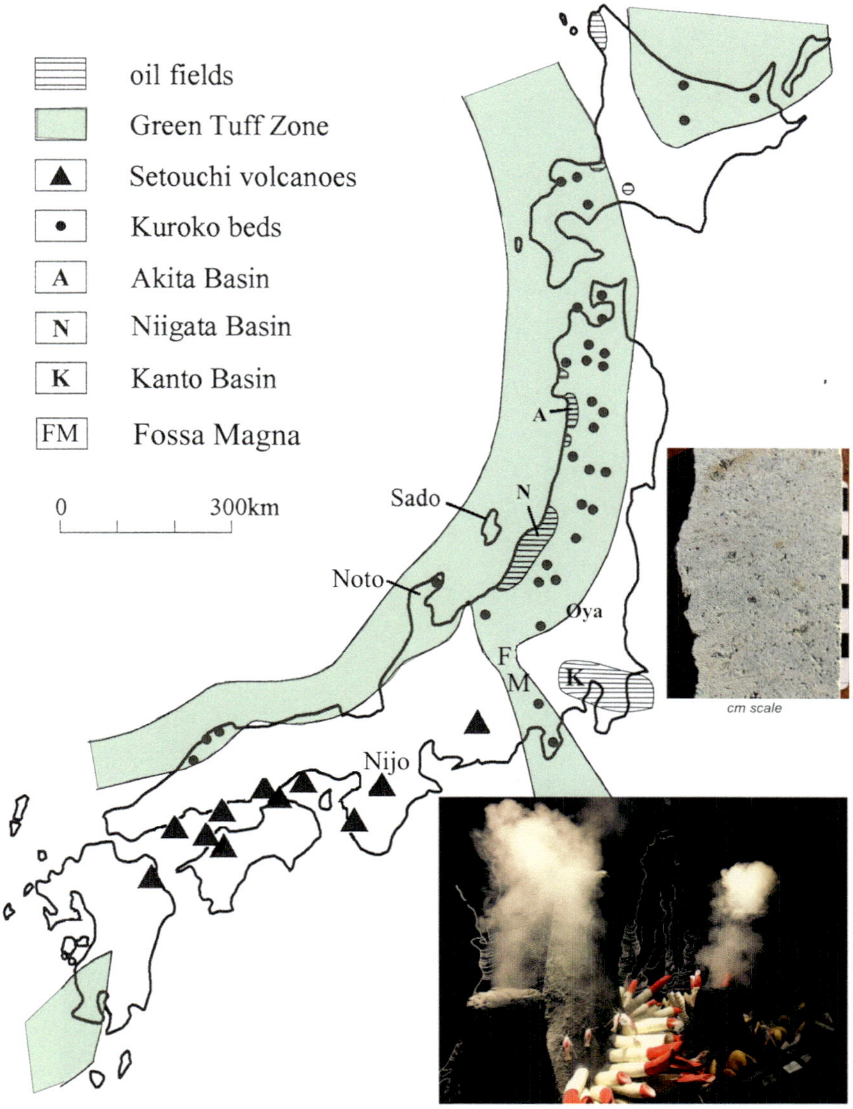

Figure 4.4 The Green Tuff Zone and associated structures: southwestern Japan's Miocene volcanoes; northwestern oil shale basins, and *kuroko* ore beds
Insets: a museum diaorama of live black smokers and their fauna; green tuff from Iwate Prefecture

The designation 'Green Tuff Zone', though, is unfortunate because: a) not all the volcanics lithified as tuff – there were lavas as well; b) not all the rocks in the zone are even igneous – many are non-igneous sedimentary rocks; c) not all green rocks in the zone are tuff, and d) not all tuff in the Green Tuff Zone are green. Nevertheless, the term lives on. The quantities of volcanic materials produced during this submarine stage are staggering: it is estimated that 170,000 km^3 of volcanic materials were extruded between 25 and 2 Ma (Shikazono 1994), and deposits between 2 and 6 *kilometers* deep are known, including within the Fossa Magna part of the Green Tuff Zone (Figure 4.4: FM) (Taira *et al.* 2016; Yanai *et al.* 2010). This indicates continuing and rapid subsidence of submarine rift zones and basins during extensional tectonics, which then kept collecting the volcanic effusions. These products are interdigitated with the marine muds, sands, and coastal gravels that were generated by erosion following tectonic uplift. The west coast of Sado Island (Figure 4.5: upper left) is a particularly good area to see these bluish-green tuff layers exposed along the seashore.

> *On the western side of Noto Peninsula, villagers of the Kofun period were known producers of ornaments and objects carved of green tuff (Figure 4.5: lower left), often conflated with jasper. These objects were important status symbols for early rulers and were deposited as grave goods in the mounded tombs of that era (cf. Barnes 2007). And unexpectedly, green tuff nutrients enhance agriculture in the Green Tuff Zone, unlike Quaternary tephrogenic soils — as we shall see in Chapter 7. Because of their green colour it is thought that many of these objects substituted for jade, a precious stone in the Kofun period that usually took the form of a curved bead. As it were, most jade in Japan is white, though it often has quantities of the green mineral omphacite giving it a jade-like appearance (see Chapter 12). The most common green tuff object is a 'bracelet', which comes in three different forms. The* kuwagataishi *('hoe-shaped' stone) below is an ususual form derived from a shell prototype: a conch shell cut vertically retaining the flange. The circular 'bracelets' may have been inspired by the Chinese* bi, *a jade perforated disc. In the case of very small holes, some* sharinseki *may have had other functions, though experimental archaeology suggests that slipped onto the arm of a child, they may have been worn as a status signifier.*

Figure 4.5 Green Tuff landscapes and artefacts
Kofun-period bracelets displayed at The National Museum of Japanese History, Chiba Prefecture, and modern building stone extraction at Marusho quarry in 2003, Iwate Prefecture

BOX 4 The Geology of Oya-ishi Tuff and its Quarry Museum

Oya stone (*ishi*) is a Miocene tuff made famous for its sculptural qualities by Frank Lloyd Wright in his Japanese architectural creations. The quarry at Oya ('big valley') near Utsunomiya, Tochigi Prefecture, is one of the best known sources of tuff building-stone in Japan. The underground mine is now disused and has been transformed into a museum.

The Oya Formation, extending over 100 km^2, lies north of the town of Oya in the Sugata River drainage; it is comparable to the Nagaoka Formation which occupies a central portion of the Utsunomiya Hills north of Utsunomiya City. The two formations are divided by the Tagawa River, which runs north to south between them. The Oya and Nagaoka Formations consist mainly of rhyolite and dacite pumice tephra between 120 to 300 m thick. The 'formation' in their names indicates they are considered sedimentary rocks; once deposited, the pumice tuff then solidified into tuff.

These pumice layers were laid down underwater in the Mid-Miocene during the Green Tuff Movement accompanying the rifting of the Japanese landmass. Much of the tephra generated during rifting was extruded underwater, bestowing a particular chemical signature – the green mineral chlorite. A marine shale layer separates the middle and lower Oya strata. The strata themselves dip approximately 10° to the east-southeast, abutting to the west on the Taki andesite with which they are thought to be associated (Ota 1949).

Oya-ishi architectural properties and uses

Ota (1949) discerned three *Oya-ishi* strata, only some of which were useful as building stone. The upper and middle layers of the middle stratum produce the best tuff: *aome* (greenish from minerals of the celedonite-glauconite series), and *shirome* (whitish). The products from these were graded and identified with unique symbols to mark their quality. Layers which include large lava blocks and bombs are useless as building stone, as are layers with a sandy texture (*suname*) comprised of quartz and feldspar crystals and fragments of rhyolite and glass (Ota & Sudo 1950). Tephra alters to clay minerals through time, and the pumice commonly forms a clay textured like fermented soybeans (*miso*). The lowest stratum is heavily weathered, having a *miso*-like texture.

Over 100 locations around Oya produced Oya stone; several other named tuffs – Tachi'iwa, Tashimo, Sakurada – come from the same area; but Oya stone has become the generic name for all. In its heyday in the 1960s, the tuff quarry at Oya was producing 200,000 tons of stone per year; quarried leavings were further ground up and used in cement and fertilizer.

Oya stone was used for centuries throughout Japan but not necessarily for constructing buildings. Tuff is an easily worked stone, cut to form blocks for walls, foundations, and pavements – or carved into stone lanterns for gardens and animal guardians for shrines and even sarcophagi. Not until the Meiji period did whole buildings begin to be made of tuff. A storehouse standing at an Oya Town bus stop (photo) is an example of a traditional *kura* storehouse built of tuff instead of plastered rubble-wattle-and-daub (Treib 1976).

The former quarry as the Oya Stone Museum

The Oya quarry is now disused and has been made into a park and geological museum: the above-ground mined out areas leave towering cliffs surrounding new parks, and the below-ground mines extend 30 m underground through a series of block-cut 'rooms' and 'corridors'. The Oya Keikan Park is dominated by a Buddhist sculpture, the Heiwa Kannon (below left), carved out of the tuff cliff after WWII to commemorate the war dead and promote peace.... [cont'd]

However, it is not the first Buddhist monument to inhabit the site. It is said that the monk Kukai established a rock-cut temple (right) in 810 in the cliffs at Oya, which can now be visited as well. The Buddhist images at the temple, one of several *magaibutsu* (carved-cliff Buddha) sites, have been designated as three Special Historic Sites by the Japanese Government.

The underground mine closed in 1986 after nearly 70 years of operations but soon reopened as a museum. The efforts of early 20th-century miners to carve out blocks at increasing depths can be seen as one walks through 20,000 m^2 of cave. It has also become a venue for arts performances and wedding ceremonies, bringing this ancient geological formation into modern use.

Frank Lloyd Wright's creations are preserved as historic buildings: the Museum Meiji-Mura in Inuyama City, Aichi Prefecture, now houses the main entrance hall and lobby of the Imperial Hotel, built in Tokyo by Wright between 1916-22; it was dismantled and moved in 1968. Typically, Wright employed Oya stone for decorative outdoor facades, and for carved pillars and beams in the hotel's interior, attesting to its light weight and sculptural properties. The YAMAMURA Tazaemon Mansion in Kobe City was also built by Wright using Oya stone.

Objects made of *Oya-ishi* and related tuffs are still being made and sold by artisans as distant as Japan Arts and Crafts of Norwood, Connecticut and Oya Stone Inc. in New York City.

The Setouchi Volcanic Zone

Kimura *et al.* (2005, 2014) have compiled a history of volcanic activity in the forearc zone of Japan that developed during the rotation of SW Japan between 17 and 15 Ma and continued until 10.5 Ma. A wide variety of lavas were emplaced: MORB, alkali basalts, felsic pyroclastic flows forming tuffs, and what are called HMA – high-magnesium andesites (see Kimura *et al.* 2005: fig. 3). These all represent the initiation of subduction of the Philippine Plate under SW Japan. What are of interest here are the andesites that comprise the first post-rifting Volcanic Front, formed in the Setouchi region from 16.5 to 12.5 Ma (Kimura *et al.* 2014) (cf. Figure 4.4: triangles). All these early volcanoes are now extinct and eroded and no longer appear as conical mountains, but they form important components of some current mountain ranges, such as Mt Nijo on the border between Nara and Osaka Prefectures.

The type of HMA produced in the Setouchi zone came to be known as sanukite (BOX 5), named by E. Weinschenk in 1891 after the pre-modern province of Sanuki in northern Shikoku Island (Tatsumi &

Ishizaka 1982: 294). It can be divided into six types depending on the identity of large crystal (phenocryst) inclusions: a) augite-olivine, b) bronzite-olivine, c) hornblende-olivine, d) olivine-augite-bronzite, e) bronzite, and f) aphyric (lacking phenocrysts) (Tatsumi & Ishizaka 1982).

BOX 5 Sanukite Past and Present

Sanukite has an unusual fracture pattern for a glassy volcanic material, making for different artefact types than obsidian. In the Palaeolithic period, sanukite was knapped in a special sequence to produce side-blow flakes (below, middle right; see also Barnes 2015: box 4.3); these were made into 'knives'. In the prehistoric periods, sanukite was used to make arrowheads: small in the Jomon period for hunting small birds and animals, but larger in the Yayoi period when turned into weapons. Also in the Yayoi period, reaping knives of either polished or flaked surfaces (below, left) were used to harvest ripe grainheads of rice. Some reaping knives were drilled with holes to pass a cord through for slipping the fingers into for a grip; others had end-notches as shown here ≫≪ to secure a cord around them (upper right). The latter example also shows the detachment surface of an earlier side-blow flake; embedded crystals can also be seen – which wear proud along the blade during use. The significant chemical variations along the extensive zone of Miocene volcanics allow for chemical sourcing to trace manufacturing and distribution of stone tools made of sanukite (e.g., Higashimura & Warashina 1975; Warashina *et al.* 1978; Higashimura & Warashina 1981; Takehiro 2013). Whereas in the prehistoric period, sanukite was a major resource for stone-tool making, today it is used to make stone chimes (lithophones) (Hasegawa *et al.* 2014). Near Mt Nijo, a quarry still mines sanukite for road works.

Chipped sanukite reaper ca. 10 cm wide

Sakukite side-blow flake with hinge fracture

A memorial sundial of sanukite on Medake ridge near Mt Nijo; it is a popular spot for hikers to rest.

Mt Nijo has also supplied great amounts of corundum (aluminum oxide Al_2O_3) in the form of the stone corundite/emery. It is an important material used in grinding and polishing since it has a MOHS hardness value of 8, greater than quartz at 7. Before the advent of synthetic corundum, the area around Mt Nijo was renown for its excellent natural corundum. Sakai City in Osaka Prefecture was a major producer, with distribution controlled by the Ha-no-mono-kai (Association for Bladed Instruments). The corundum was historically used for sharpening samurai swords and kitchen knives.

Honshu–Izu Arc-Arc Collision & Ramifications

The collision of the Izu Arc with Honshu is one of the more spectacular processes in Japanese tectonic history (Figures 4.6, 4.7). Exactly when the Izu Arc began impacting Honshu is "highly controversial and widely debated" (Wu et al. 2016: 4730). Essentially, opposing models utilize a 'fixed' triple junction (TTT, located as today off Tokyo) or a 'migrating' TTT that dragged along the southern edge of SW Japan before settling off Tokyo (Kimura *et al.* 2005, 2014). Since the 'fixed TTT' hypothesis does not account for the migration northwards of the Philippine Plate from the southwest Pacific nor the rotation of the plate as ascertained through palaeomagnetic data (cf. Figure 3.9) (cf. Lallemand 2016; Wu *et al.* 2016), the 'migrating' model (cf. Figure 3.9) is detailed here.

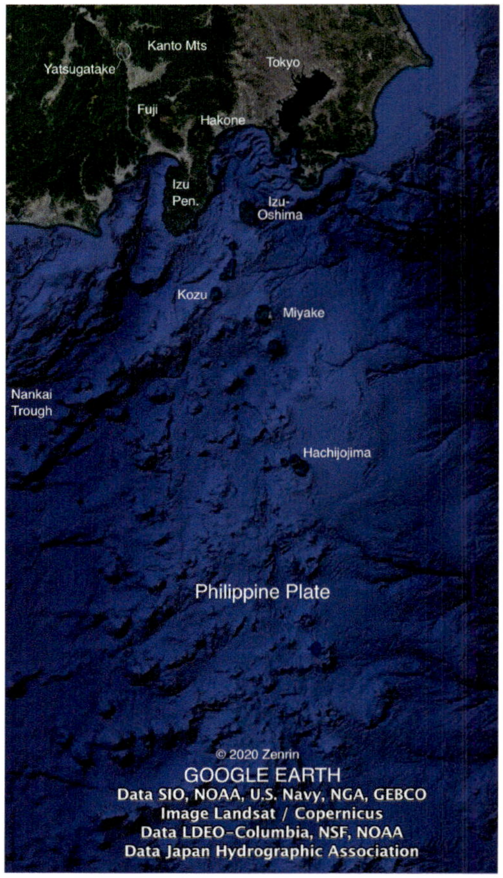

Figure 4.6 The Izu Arc on the Philippine Plate, extending from the Izu Peninsula southwards, including the islands of Izu-Oshima, Kozu, Miyake & Hachijojima

The 'migrating' model presented by Wu *et al.* (2016: fig. 30) differs in important details with respect to that presented by Kimura *et al.* (2005: fig. 9): compare positions, timing, and plate rotations. Wu *et al.* have contributed the most recent interpretation which may again change in future. At the end of its travels northwards from below 12°N, the Philippine Plate is modelled by Wu *et al.* (2016: fig. 30) as encountering the edge of the continental plate around 17–16 Ma in the vicinity of the Ryukyus where a northwest-dipping subduction zone developed, causing the rise of a mountain belt in the Ryukyu region and forearc volcanism in Kyushu (2016: fig. 30b). The subsequent spreading of the Shikoku Basin in a fan shape pushed the Izu Arc, located on the northeastern side of the plate eastwards (fig. 30c) – this is a crucial detail affecting the timing of the impact of the Izu Arc on Honshu. Further plate subduction, causing the Miocene volcanism in the Setouchi Volcanic Zone and continued subduction, finally brought the Izu Arc head into contact with central Honshu.

Other scholars favour earlier or later dates for the beginning of Izu Arc collision with Honshu. Hoshi & Sano (2013) have obtained palaeomagnetic data from the western arm of the Kanto Syntaxis, described below, indicating its counter-clockwise rotation between 17.5 and 15 Ma; they attribute this to the Izu Arc collision. Pickering *et al.* (2013: 1724) identify initial subduction effects between 17 and 15 Ma, while D. Hirata *et al.* (2010) opt for 15 Ma. The arc is thought to have actually served as a doorstop to rifting from the continent during the Japan Sea Basin opening at 15 Ma (Yonekura *et al.* 2001: 304). In contrast, Mahony *et al.* (2011) argue for a much later date for collision of between 8 and 6 Ma for the impact. This difference in perception may be due to a period of 'slow subduction' of the Philippine Plate proposed between 12 and 4 Ma and then a sudden acceleration subduction around 4 Ma (Kimura *et al.* 2005; Pickering *et al.* 2013). The collision created the Kanto Syntaxis, effected mountain-building around the Fossa Magna, and defined the two-plate subduction regime in Japan today ' (cf. Table 6.1). It is interesting that the Izu Arc chose the weakest point in SW Japan to have indented: into the Itoigawa-Shizuoka fault zone (I-STL) between NE and SW Japan filled with Green Tuff volcaniclastics.

Arc-arc collision & accretion

The collision of the Izu Arc with Honshu ostensibly occurred in intermittent accretion events. Four blocks of the volcanic Izu Arc have been incorporated into the impact zone in succession: Kushigata, Misaka, Tanzawa, and Izu. The Izu block constitutes today's Izu Peninsula, which is actually located on the intruding head of the Philippine Plate. The Kushigata block is incorporated into the Koma Mountains, while the Misaka and Tanzawa blocks now form eponymous mountain ranges. Amano (1989: table 4) records differing views on the timing of accretion of these different blocks, but Sueoka *et al.* now give collision dates of the "Kushigatayama Block at 15–13 Ma, the Misaka Block at 13–8 Ma, the Tanzawa Block at 8–5 Ma, and the Izu Block since ~1 Ma" (2017: 6788).

The Tanzawa Mountains were once part of the deeply buried (plutonic) middle crust in the Izu chain but which now reach a current altitude of around 1700 m msl, while sudden uplift is timed between 3.5 and 2 Ma (cf. Figure 5.2: upper #4). Initially, the collision took place all underwater, when northeastern Japan was still in the main submerged; the resultant intensive volcanic activity created much basaltic lava and pyroclastic deposition in the collision zone. The western edge of the collision zone is formed by the Itoigawa-Shizuoka Tectonic Line, a fault which developed in the early Miocene at the beginning of continental rifting (Yanai *et al.* 2010).

Unlike the cold, dense, basaltic seamounts and ultramafic ophiolites that had been accreted to the Japanese landmass in the distant past (e.g., in the Chichibu and Maizuru Belts, cf. Chapter 3; Apx 6), the Izu Islands are young, buoyant, and hot. Ploughing almost perpendicularly into Honshu, they proved difficult to wholly subduct into the trench. When an arc collides obliquely, it subducts smoothly, as is currently the case with the Kyushu-Palau Ridge, also on the Philippine Plate, subducting under Kyushu (Isozaki *et al.* 2010: 95) (cf. Figures 5.1, 5.11). Recent seismic studies on the Izu Arc indicate that buoyant material such as an island arc can be easily subducted if less than 25 km thick (Yamamoto *et al.* 2009). About 700 km of Philippine Plate has already been subducted, including the majority of the Izu Arc (Lallemand 2016).

The Kanto Syntaxis

Several geological belts of southwestern Japan bend around the Izu Peninsula in what is technically referred to as the Kanto Syntaxis (Takahashi & Saito 1997), an indentation caused by the intrusion of the head of the Izu Arc into central Honshu. There is disagreement when this bending actually took place, though all dates are far later than originally proposed at 100 Ma in the Cretaceous (cf. Ogawa 1982: fig. 6). Takahashi & Saito (1997) propose two stages of bending: 15–12, and 12–6; Yonekura *et al.* postulate 11 Ma (2001: 303), while Yamamoto *et al.* (2009) consider the bending did not start until 5 Ma based on palaeomagnetic data. Taira (1990: 150) reported that the two arms of the bend moved independently at different times and rates, with the eastern arm taking longer to attain its present position.

The area around the Kanto Syntaxis is being thickened through thrust faulting, from ca. 22 km to greater than 40 km thick (Kimura *et al.* 2005: 972). In addition to plutonic segments of the Izu arc, volcanic sediments have been accreted to the edge of the Japanese Islands, forming the tips of the Boso Peninsula (Chiba Prefecture) and Miura Peninsula (Kanagawa Prefecture), as well as among the mountains of the lower Fossa Magna zone (Arai & Iwasaki 2014). On the east side of the Kanto

Mountains, a deep sedimentary basin had developed during Miocene rifting, and it was turned clockwise by the impact of the Izu Arc (Yanai *et al.* 2010) (cf. Figure 5.5: K).

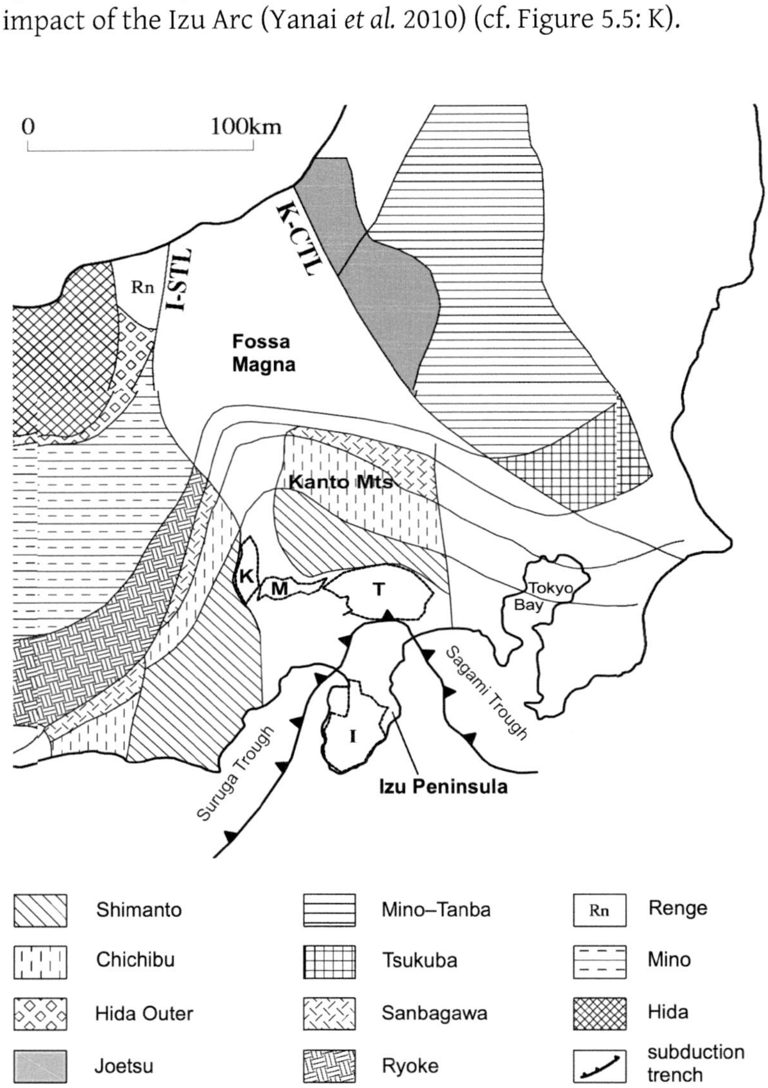

Figure 4.7 The Kanto Syntaxis

The intruding head of the Izu Arc bent once-parallel geological belts around itself. The previously accreted Kushigatayama/Koma (K), Misaka (M), Tanzawa (T), and Izu (I) blocks are in foremost position. The present plate boundary (toothed line) outlines the current forward edge of the intruding Izu block on the Philippine Plate, forming the Izu Peninsula. The Fossa Magna is shown bounded on the west by the Itoigawa-Shizuoka Tectonic Line (I-STL) and on the east by the Kashiwazaki-Choshi Tectonic Line (K-CTL).

The Fossa Magna

The Fossa Magna (Figure 4.7), named in 1885-86 by the pioneer geologist Edmund Naumann, who thought the long narrow valley transecting central Honshu from north to south along the Itoigawa-Shizuoka Tectonic Line (I-STL) was like a 'big trough' (i.e., *fossa magna*, in Latin).[2] It is a rifted graben

[2] See Yamashita (1995), written entirely as a response to Naumann's research.

structure resulting from the rotating away from each other of the SW and NE parts of Honshu during the opening of the Japan Sea Basin (Taira *et al.* 2016). The I-STL has three segments (Yanai *et al.* 2010):

- in the north, it forms a thrust fault (Figure 4.7: toothed line), with NE Japan slipping under SW Japan;
- in the central portion, it forms a strike-slip fault resulting in 12-km offset of the Median Tectonic Line (MTL) towards the north in the region of Lake Suwa (Yamashita 1995);
- in the south, there is an Active Fault zone due to the Izu arc collision; there the arc is overriding the western landmass (just opposite what is happening in the north).

The Itoigawa-Shizuoka Tectonic Line itself serves in a different way to demarcate SW and NE Japan: cultural differences including language dialects and food preferences are particularly marked across this border. The lat-long orientation of the two segments of Japan stemming from the Fossa Magna plus the differential in volcanic activity and regional climates have combined to produce contrasting lifestyles and histories from the prehistoric periods onward.

The eastern boundary of the Fossa Magna region is less definitive. Naumann hypothesized that it ran along the Chikuma River valley in Nagano Prefecture (Yamashita 1995: 31), but the Chikuma (called the Nagano River in Niigata Prefecture) is sourced in the Kanto Mountains and does not reach the eastern seaboard. Instead, a fault further east is postulated to constrain the syntaxis: the Kashiwazaki-Choshi tectonic line (K-CTL) (NCEC 1988: fig. 6. 2; Baba & Yoshida 2020: fig. 2) that parallels the eastern end of the MTL (cf. Figure 4.7). Some Fossa Magna scholars, however, are reluctant to accept this as the eastern boundary fault because doing so would include most of the Kanto region, thus going against Naumann's original definition of the Fossa Magna as *excluding* the Kanto Mountains and by inference anything east of them.

Takahashi (2006) has proposed that this eastern boundary – which he defines as the Tone River TL (TRTL)[3] instead of the K-CTL but running along the same course – is actually the joint between SW and NE Japan, based on the discontinuity of the Miocene andesitic Volcanic Front across this boundary. This contrasts with the predominant view that the TTL in Tohoku is the SW-NE divide (Isozaki *et al.* 2010). The two ends of the 18–16 Ma Volcanic Front, ending in the southwest at the I-STL and beginning in the northeast at the TRTL, are offset by 230 km (Takahashi 2006: fig. 4), implying that NE Japan rotated further than SW Japan. This makes sense if SW Japan had been doorstopped by the Izu Arc on the newly positioned Philippine Plate, causing the formation of a new southwestern Volcanic Front at 14 Ma in the Setouchi region.

Perspectives

In summary, what Naumann identified as the Fossa Magna – a narrow trough running through Honshu north to south along the I-STL – is now looked upon as a broad triangular area that relates to both the differential rifting of SW and NE Japan and to the deformation accommodating the indentation of the Izu Arc. The rearrangement of the basement geological belts by the Kanto Syntaxis was one of the last great tectonic events in the story of the opening of the Japan Sea Basin, and it is still ongoing today as the Izu Arc continues to push into Honshu. It is not the only volcanic edifice to affect the archipelago, however, as more volcanic seamounts, plateaus and ocean ridges are headed

[3] My designation to distinguish from TTL, the established Tanakura Tectonic Line; unfortunately TTL is also used for the Tonégawa Tectonic Line by some authors (e.g., Nakajima 2013).

towards Japan (cf. Figures 3.1, 6.10). A new era of movement and activity opened around 6 Ma with the stabilization of the Izu Arc position. Post-Miocene tectonics include compression and upifting of mountains, the re-activation of the MTL, and new extension zones forming within-arc basins, inciting modern volcanic activity – all to be examined in Chapters 5 and 6.

Figure & Table Sources

Figure 4.1 after (Okamura *et al.* 1995: figs. 1,4,15), redrawn by DAR
Figure 4.2 after (Yonekura *et al.* 2001: 297), extracted, modified and relabeled by DAR
Figure 4.3 compiled by GLB from (Otofuji *et al.* 1985: fig. 2; Otofuji 1996: fig. 5a; Yanai *et al.* 2010: fig. 3)
Figure 4.4 compiled by DAR from (Chigakudan 1995: fig. 3-1; Sugimura & Uyeda 1973: fig. 7.3); Setouchi volcano locations from (Kimura *et al.* 2005: fig. 3, 2914: fig. 13); author's photos from the Marusho quarry, Iwate Prefecture, Japan and Maritime Museum, Pusan, Korea
Figure 4.5 photos by author; artefacts from (Sugiyama 2020: unpg)
Figure 4.6 modified by GLB from Google Earth
Figure 4.7 modified by GLB from (Barnes 2007: fig. 9) with K,M,T,I locations inserted from (Koyama 1992: fig. 2)

BOX 4 photos by author
BOX 5 side-blow flake photo by author; sundial by I, Kenpei [https://commons.wikimedia.org/wiki/File:Nijyozan_medake2.jpg], CC-SA-3.0, cropped by GLB; reaping knives courtesy of Kagawa Prefectural Product Promotion Organization [www.kensanpin.org/wp-content/uploads/2019/ 09/timthumb-6-9.jpg]

Table 4.1 compiled from (Yoshida *et al.* 2014; Kimura *et al.* 2005; Nohda 2009; Nakajima, 2013)

References

ALTIS, Sungat (**1999**) "Interpretations of a Middle Miocene and late Quaternary steady dextral transpression in SW Japan and the opening tectonics for the Japan Sea." *Tectonophysics* 302: 257-285.

AMANO, Kazuo (**1989**) "Multiple collision tectonics in the South Fossa Magna, Central Japan." *Modern Geology Special Issue* 14: 315-330.

ARAI, Ryuta & Takaya IWASAKI (**2014**) "Crustal structure in the northwestern part of the Izu collision zone in central Japan." *Earth, Planets and Space* 66.21, 12pp. [DOI: 10.1186/1880-5981-66-21].

BABA, AK; T MATSUDA & T ITAYA *et al.* (**2007**) "New age constraints on counter-clockwise rotation of NE Japan." *Geophysical Journal International* 171: 1325-1341.

BABA, Kei & Takeyoshi YOSHIDA (**2020**) "Geological structures controlled the rupture process of the 2011 M9.0 Tohoku-Oki earthquake in the Northeast Japan Arc." *Earth, Planets and Space* 72.94: 14pp. [DOI: 10.1186/s40623-020-01212-3].

BARNES, Gina L (**2007**) *State formation in Japan: emergence of a 4th-century ruling elite.* London: Routledge.

---- (**2015**) *Archaeology of East Asia: the rise of civilization in China, Korea and Japan.* Oxford: Oxbow Books.

BLAKE, Stephen & Tom ARGLES (**2003**) *Growth and destruction: continental evolution at subduction zones.* Milton Keynes, UK: The Open University.

BOGATIKOV, Oleg (**2000**) *Magmatism and geodynamics: terrestrial magmatism throughout the Earth's history.* Amsterdam: Gordon and Breach Science Publishers.

BOGATIKOV, OA; EV SHARKOV & AV VESSELOVSKII *et al.* (**2009**) "Synchronism between development of Cenozoic volcanic arcs and back-arc basins: reasons and consequences." *Doklady Earth Sciences* 427A.6: 907-911.

CHEN, Shuping; Yuming TAN & Jiafu QI *et al.* (**2011**) "Cenozoic tectonic characteristics, evolution and geodynamics of Dongpu Sag, Bohai Bay Basin, China", pp. 143-168 in *New frontiers in tectonic research: general problems, sedimentary basins and island arcs*, ed. by E SHARKOV. Rijeka, HR/Shanghai: Intech.

Chigakudan (Chigaku Dantai Kenkyukai) (**1995**) *The development of the Japanese archipelago.* Tokyo: Tokai University Press (in Japanese).

FOULGER, Gillian R (**2010**) *Plates vs. plumes: a geological controversy.* Oxford, UK: Wiley-Blackwell.

HARRIS, Nigel & WHALLEY, John (**2001**) *Mountain building.* Milton Keynes, UK: Open University.

HASEGAWA, Shuichi; Seiko TSURUTA & Munekazu MAEDA (**2014**) "Preservation of sanukite, the highly sophisticated music instrument made of andesite", pp. 99-102 in *Engineering geology for society and territory, vol. 8*, ed. by G LOLLINO *et al.* Cham, CH: Springer.

HIGASHIMURA, T & T WARASHINA (**1975**) "Sourcing of sanukite stone implements by X-ray fluorescence analysis." *Journal of Archaeological Science* 2.3: 169-178.

---- (**1981**) "Provenances of sanukite implements in western Japan", pp. 121-130 in *Archaeological Studies of Pacific Stone Resources*, ed. by F LEACH & JM DAVIDSON. BAR International Series 104. Oxford: British Archaeological Reports.

HIRAHARA, Yuka; Jun-Ichi KIMURA & Ryoko SENDA *et al.* (**2015**) "Geochemical variations in Japan Sea back-arc basin basalts formed by high-temperature adiabatic melting of mantle metasomatized by sediment subduction components." *Geochemistry, Geophysics, Geosystems* 16: 1324-1347 [DOI: 1310.1002/2015GC005720].

HIRATA, Daiji; YAMASHITA Hiroyuki & Kazue SUZUKI *et al.* (**2010**) "Collision accretion tectonics of the proto-Izu–Mariana Arc: a review." *Chigaku Zasshi* 119.6: 1125-1160 (in Japanese with English title and abstract).

HIRATA, Naoshi; Boris Y KARP & Takashi YAMAGUCHI *et al.* (**1992**) "Oceanic crust in the Japan Basin of the Japan Sea by the 1990 Japan–USSR expedition." *Geophysical Research Letters* 19.20: 2027-2030.

HOSHI, Hiroyuki & Masakazu SANO (**2013**) "Paleomagnetic constraints on Miocene rotation in the central Japan Arc." *Island Arc* 22: 197-213.

ISOZAKI, Yukio; Kazumasa AOKI & Takaaki NAKAMA *et al.* (**2010**) "New insight into a subduction-related orogen: a reappraisal of the geotectonic framework and evolution of the Japanese Islands." *Gondwana Research* 18: 82-105.

ITAKI, Takuya (**2016**) "Transitional changes in microfossil assemblages in the Japan Sea from the Late Pliocene to Early Pleistocene related to global climatic and local tectonic events." *Progress in Earth and Planetary Science* 3.11: 21pp. [DOI: 10.1186/s40645-016-0087-4].

ITOH, Yasuto; Keiji TAKEMURA & Shigekazu KUSUMOTO (**2013**) "Neotectonic intra-arc basins within southwest Japan – conspicuous basin-forming process related to differential motion of crustal blocks", pp. 191-207 in *Mechanism of sedimentary basin formation - multidisciplinary approach on active plate margins*, ed. by Y ITOH. London: IntechOpen.

IWAMORI, Hikaru (**1991**) "Zonal structure of Cenozoic basalts related to mantle upwelling in southwest Japan." *Journal of Geophysical Research* 96.B4: 6157-6170.

JOLIVET, L; H SHIBUYA & M FOURNIER (**1995**) "Paleomagnetic rotations and the Japan Sea opening", pp. 355-369 in *Active margins and marginal basins of the western Pacific*, ed. by B TAYLOR & J NATLAND. Washington, DC: American Geophysical Union.

JOLIVET, Laurent; Kensaki TAMAKI & Marc FOURNIER (**1994**) "Japan Sea, opening history and mechanism: a synthesis." *Journal of Geophysical Research: Solid Earth* 99.B11: 22237-22259 [DOI: 10.1029/ 93JB03463].

KAWAI, Naoto; Tadashi NAKAJIMA & Kimio HIROOKA (1971) "The evolution of the island arc of Japan and the formation of granites in the circum-Pacific belt." *Journal of Geomagnetism and Geoelectricity* 23.3-4: 267-293.

KIMURA, Jun-ichi; James B GILL & Tomoyuki KUNIKIYO *et al.* (**2014**) "Diverse magmatic effects of subducting a hot slab in SW Japan: results from forward modeling." *Geochemistry Geophysics Geosystems* 15: 691-739 [DOI: 10.1002/2013GC005132].

KIMURA, Jun-Ichi; Robert J STERN & Takeyoshi YOSHIDA (**2005**) "Reinitiation of subduction and magmatic responses in SW Japan during Neogene time." *Geological Society of America Bulletin* 117.7/8: 969-986.

KOIZUMI, Itaru (**1992**) "Biostratigraphy and paleoceanography of the Japan Sea based on diatoms: ODP Leg 127", pp. 15-24 in *Pacific Neogene: environment, evolution and events*, ed. by R TSUCHI & JC INGLE Jr. University of Tokyo Press.

KOPTEV, Alexander; Evgueni BUROV & Eric CALAIS *et al.* (**2016**) "Contrasted continental rifting via plume-craton interaction: applications to Central East African Rift." *Geoscience Frontiers* 7.2: 221-236.

KOYAMA, Akira (**1990**) "Tectonic evolution of the northern Koma Mountains, southern Fossa Magna, central Japan." *Journal of Geosciences, Osaka City University* 33.1: 1-47.

KUSKY, Timothy M; Brian F WINDLEY & Mingguo ZHAI (**2007**) "Lithospheric thinning in eastern Asia; constraints, evolution, and tests of models", pp. 331-343 in *Mesozoic sub-continental lithospheric thinning under eastern Asia*, ed. by M ZHAI *et al.* London: The Geological Society of London.

LALLEMAND, S & L JOLIVET (**1985**) "Japan Sea: a pull-apart basin." *Earth and Planetary Science Letters* 76: 375-389.

LALLEMAND, Serge (**2016**) "Philippine Sea Plate inception, evolution, and consumption with special emphasis on the early stages of Izu-Bonin-Mariana subduction." *Progress in Earth and Planetary Science* 3: #15 [DOI: 10.1186/s40645-016-0085-6].

LISZEWSKA, KM; JC WHITE & R MACDONALD *et al.* (**2018**) "Compositional and thermodynamic variability in a stratified magma chamber: evidence from the Green Tuff Ignimbrite (Pantelleria, Italy)." *Journal of Petrology* 59.12: 2245-2272.

LIU, Jingao; Ronghua CAI & D Graham PEARSON *et al.* (**2019**) "Thinning and destruction of the lithospheric mantle root beneath the North China Craton: a review." *Earth-Science Reviews* 196: #102873, 18pp. [DOI: 10.1016/j.earscirev.2019.05.017].

MAHONY, SH; LM WALLACE & M MIYOSHI *et al.* (**2011**) "Volcano-tectonic interactions during rapid plate-boundary evolution in the Kyushu region, SW Japan." *GSA Bulletin* 123.11/12: 2201-2223 [DOI: 10.1130/B30408.1].

MARSAGLIA, Kathleen M (**1995**) "Interarc and backarc basins", pp. 299-330 in *Tectonics of sedimentary basins*, ed. by CJ BUSBY & RV INGERSOLL. Malden, MA: Blackwell Science Ltd.

MARUYAMA, Shigenori; Yukio ISOZAKI & Gaku KIMURA *et al.* (**1997**) "Paleogeographic maps of the Japanese Islands: plate tectonic synthesis from 750 Ma to the present." *The Island Arc* 6.1: 121-142.

Mindat (**n.d.**) "Taishu mine, Azu, Tsushima-Shimoshima island, Tsushima Islands, Nagasaki Prefecture, Kyushu Region, Japan" [www.mindat.org/loc-59047.html].

NAKAJIMA, Takeshi (**2013**) "Late Cenozoic tectonic events and intra-arc basin development in northeast Japan", pp. 153-189 in *Mechanism of sedimentary basin formation - multidisciplinary approach on active plate margins*, ed. by Y ITOH. London: IntechOpen.

NAKANISHI, Tetsuya & Eiji IZAWA (**2014**) "Evolution of silver-smelting technology of Japan in the middle of sixteenth century." *ISIJ international* 54.5: 1093-1097.

NCEC (Nihon no Chishitsu Editorial Committee) (**1988**) *Geology of Japan 4: Chubu region I*. Tokyo: Kyoritsu (in Japanese).

NIU, Yaoling; Yi LIU & Qiqi XUE *et al.* (**2015**) "Exotic origin of the Chinese continental shelf: new insights into the tectonic evolution of the western Pacific and eastern China since the Mesozoic." *Science Bulletin* 60.18: 1598-1616.

NOHDA, S (**2009**) "Formation of the Japan Sea basin: reassessment from Ar–Ar ages and Nd–Sr isotopic data of basement basalts of the Japan Sea and adjacent regions." *Journal of Asian Earth Sciences* 34: 599-609.

OGAWA, Yujiro (**1982**) "Tectonics of some forearc fold belts in and around the arc-arc crossing area in central Japan", pp. 49-62 in *Trench-forearc geology: sedimentation and tectonics on modern and ancient active plate margins*, ed. by JK LEGGETT. Oxford: Blackwell Scientific.

OKAMURA, Yukinobu; Mahito WATANABE & Rie MORIGIRI *et al.* (**1995**) "Rifting and basin inversion in the eastern margin of the Japan Sea." *The Island Arc* 4: 166-181.

OMORI, Soichi & Yukio ISOZAKI (**2011**) "Paleozoic Japan and the eastern extension of the collisional suture between the North and South China cratons." *Chigaku Zasshi* 120.1: 40-51 (in Japanese with English title and abstract).

ONO, Susumu (**1970**) *The origin of the Japanese language*. Tokyo: Kokusai Bunka Shinkokai (in English).

OTA, Setsuji (**1949**) "Studies on Oya-ishi: properties for building stone." *Chishitsugaku Zasshi* 55.64: 85-89 (in Japanese with English title and summary).

OTA, Setsuji & Toshio SUDO (**1950**) "Studies on 'Oya-ishi' Part 2: mineralogical composition." *Chishitsugaku Zasshi* 55.64: 650-651 (in Japanese with English title and summary).

OTOFUJI, Yo-ichiro (**1996**) "Large tectonic movement of the Japan Arc in late Cenozoic times inferred from paleomagnetism: review and synthesis." *The Island Arc* 5.3: 229-249.

OTOFUJI, Yo-Ichiro; Takaaki MATSUDA & Susumu NOHDA (**1985**) "Opening mode of the Japan Sea inferred from the palaeomagnetism of the Japan Arc." *Nature* 317: 603-604.

PETIT, Carole & Jacques DÉVERCHÈRE (**2006**) "Structure and evolution of the Baikal rift: a synthesis." *Geochemistry Geophysics Geosystems* 7.12 (Q11016): 26pp. [DOI: 10.1029/2006GC001265].

PHILPOTTS, Anthony R & Jay J AGUE (**2009**) *Principles of igneous and metamorphic petrology*. Cambridge University Press.

PICKERING, Kevin T; Michael B UNDERWOOD & Sanny SAITO *et al.* (**2013**) "Depositional architecture, provenance, and tectonic/eustatic modulation of Miocene submarine fans in the Shikoku Basin: results from Nankai Trough Seismogenic Zone Experiment." *Geochemistry Geophysics Geosystems* 14.6: 1722-1739.

ROSKOVEC, Bedrich; Ondrej SRAMEK & William F MCDONOUGH (**2018**) "Testing a proposed 'second continent' beneath eastern China using geoneutrino measurements." *arXiv* 1810.10914 [physics.geo-ph].

SHIKAZONO, Naotatsu (**1994**) "Hydrothermal alteration of green tuff belt, Japan: implications for the influence of seawater/volcanic rock interaction on the seawater chemistry at a back-arc basin." *The Island Arc* 3: 59-65.

---- (**2003**) *Geochemical and tectonic evolution of arc-backarc hydrothermal systems: implication for the origin of kuroko and epithermal vein-type mineralizations and the global geochemical cycle*. Amsterdam: Elsevier.

Smith, AJ (**1982**) "The Neogene to Recent geology of Japan and its surrounding seas." *Proceedings of The Geologists' Association* 93.2: 161-178.

Sueoka, S; Y Ikeda & K Kano *et al.* (**2017**) "Uplift and denudation history of the Akaishi Range, a thrust block formed by arc-arc collision in central Japan: insights from low-temperature thermo–chronometry and thermokinematic modeling." *Journal of Geophysical Research: Solid Earth* 122: 6787-6810 [DOI: 10.1002/2017JB014320].

Sugimura, Arata & Seiya Uyeda (**1973**) *Island arcs: Japan and its environs.* Developments in Geotectonics 3. Amsterdam: Elsevier (reprinted in 2013).

Sugiyama, Shinsaku (**2020**) "A witness to history: a photographic introduction to items from the collection." *Rekihaku Bulletin* 167. Abstract at [www.rekihaku.ac.jp/english/outline/publication/rekihaku/167/witness.html].

Taira, Asahiko (**1990**) *The Birth of the Japanese Islands.* Iwanami Shinsho 148. Tokyo: Iwanami (in Japanese).

Taira, Asahito; Y Ohara & S Wallis *et al.* (**2016**) "Geological evolution of Japan: an overview", pp. 1-24 in *The geology of Japan*, ed. by T Moreno *et al.* London: Geological Society of London.

Takahashi, Misaki (**2006**) "Tectonic boundary between Northeast and Southwest Japan Arcs during Japan Sea opening." *Chishitsugaku Zasshi* 112.1: 14-32 (in Japanese with English title and abstract).

Takahashi, Masaki & Kazuo Saito (**1997**) "Miocene intra-arc bending at an arc-arc collision zone, central Japan." *The Island Arc* 6: 168-182.

Takehiro, Fumiaki (**2013**) "Round-up on Inland Sea sanukite and andesite source areas." *Bulletin of the Institute for the Cultural Studies of the Seto Inland Sea* 41.1: 1-14.

Tamaki, Kensaku (**1995**) "Opening tectonics of the Japan Sea", pp. 407-420 in *Backarc basins: tectonics and magmatism*, ed. by B Taylor. New York: Plenum Press.

Tamaki, Kensaki; Kiyoshi Suyehiro & James Allan *et al.* (**1992**) "Tectonic synthesis and implications of Japan Sea ODP drilling." *Proceedings of the Ocean Drilling Program, Scientific Results* 127/128.2: 1333-1348.

Tapponnier, P; A Peltzer & AY Le Dain *et al.* (**1982**) "Propagating extrusion tectonics in Asia: new insights from simple experiments with plasticine." *Geology* 10.12: 611-616.

Tatsumi, Y (**1983**) "Generation of arc basalt magmas and thermal structure of the mantle wedge in subduction zones." *Journal of Geophysical Research* 88.B7: 5815-5825.

Tatsumi, Yoshiyuki & Kyoichi Ishizaka (**1982**) "Origin of high-magnesian andesites in the Setouchi volcanic belt, southwest Japan, I: petrographical and chemical characteristics." *Earth and Planetary Science Letters* 60: 293-304.

Treib, Marc (**1976**) "The Japanese storehouse." *The Journal of the Society of Architectural Historians* 35.2: 124-137.

Tsuchi, Ryuichi (**1992**) "Pacific Neogene climatic optimum and accelerated biotic evolution in time and space", pp. 237-250 in *Pacific Neogene: environment, evolution and events*, ed. by R Tsuchi & JC Ingle Jr. University of Tokyo Press.

Warashina, Tetsuo; Yoshimasa Kamaki & Takenobu Higashimura (**1978**) "Sourcing of sanukite implements by X-ray fluorescence analysis." *Journal of Archaeological Science* 5.3: 283-291.

Watanabe, Yasushi; Tetsuichi Takagi & Nobuyuki Kaneko (**2016**) "Mineral and hydrocarbon resources", pp. 431-456 in *The Geology of Japan*, ed. by T Moreno *et al.* London: Geological Society of London.

Wu, Jonny; John Suppe & Renqi Lu *et al.* (**2016**) "Philippine Sea and East Asian plate tectonics since 52 Ma constrained by new subducted slab reconstruction methods." *Journal of Geophysical Research: Solid Earth* 121: 4670-4741 [DOI: 10.1002/2016JB01].

Yamamoto, Shinji; Jun'ichi Nakajima & Akira Hasegawa *et al.* (**2009**) "Izu-Bonin arc subduction under the Honshu island, Japan: evidence from geological and seismological aspect." *Gondwana Research* 16: 572-580.

Yamashita, Noboru (ed.) (**1995**) *Fossa Magna.* Tokyo: Tokai University Press (in Japanese with English title).

Yanai, Shuichi; Kazumasa Aoki & Yoshimitsu Akahori (**2010**) "Opening of Japan Sea and major tectonic lines of Japan: MTL, TTL and Fossa Magna." *Chigaku Zasshi* 119.6: 1079-1124 (in Japanese with English title and abstract).

Yonekura, Nobuyuki; Sohei Kaizuka & Michio Nogami *et al.* (eds) (**2001**) *Geomorphology of the Japanese Islands, vol. 1: Introduction to Japanese geomorphology.* University of Tokyo Press.

Yoshida, Takeyoshi; Jun-ichi Kimura & Ryoichi Yamada *et al.* (**2014**) "Evolution of late Cenozoic magmatism and the crust–mantle structure in the NE Japan Arc", pp. 335-387 in *Orogenic andesites and crustal growth*, ed. by A Gómez-Tuena *et al.* London: Geological Society of London.

Zhai, Mingguo; Brian F Windley & Timothy M Kusky *et al.* (eds) (**2007**) *Mesozoic sub-continental lithospheric thinning under eastern Asia.* Geological Society Special Publication No. 280. London: The Geological Society of London.

Chapter 5

Making Japan's Mountains & Basins

This Chapter considers how to interpret the present-day geomorphology of Japan in terms of tectonic movements during the Pliocene and Pleistocene periods. Given that the forces outlined are still in progress, the landscape will also change in the future. For archaeological purposes, Japan's dichotomous division of the landscape into mountains and basins (including coastal basins) has exerted great influence on settlement and land use patterns over the millennia. Incremental change in landscape features due to tectonic activities may explain some features of past occupation over and above the effects of sea level change. Japan was not glaciated in the Pleistocene; instead, the activation of Quaternary volcanism, dealt with in Chapter 6, was a major factor in creating the land surface that we see today.

Setting the Stage

Plains, terraces & uplands

Japan is known as an extremely mountainous country, but other topographic units are crucially important for the human occupation of the land. Depending on how flat land is calculated, Japan consists of either 65% or 86% mountains versus 35% or 14% plains. It is the plains that host the majority of the Japanese population, though this has only been true for the last 2000 years since the inception of wet-rice agriculture in the Yayoi period.

The concept of 'plains' (*heiya*) in Japan (Table 5.1) is rather different than elsewhere and needs cultural interpretation. The 35% 'plains' figure *sensu lato* includes 11% foothills (*kyūryō*), 11% terraces (*daichi, dankyū*), and 13% flatland (alluvial fans *senjōchi* + alluvial plains *hanrangen* + river deltas *sankakusu*) (Yonekura *et al.* 2001: 200). The 14% 'plains' figure *sensu stricto* that one often encounters thus refers only to alluvial flatlands. Each of the major Plains (capitalized) in Japan have various combinations of these features.

According to Yonekura and colleagues, terraces (*daichi*) are flat landforms that are bordered by steep sides or scarps; they can include lava and pyroclastic flow terraces. However, terraces that occur along rivers or the seashore in stepped fashion are most often called *dankyū*. Another type of terrace is caused by earthquake activity, resulting in far more terraces in a river course than normal (Takahama 1997: 4, 184).

Table 5.1 Composition of some major plains

	Kanto Plains	**Echigo Plains**	**Nobi Plains**	**Osaka Plains**
major city	Tokyo	Niigata	Nagoya	Osaka
foothills	<30%	<30%	<30%	<30%
terraces	>30%	ca. nil	<30%	<30%
alluvium	>30%	>30%	>30%	>30%

It is significant that the coastal flatlands have developed only within the last 5000 years since the end of the Holocene Climatic Optimum.[1] Takahashi Manabu (2001) identified eleven stages of Holocene micro-geomorphological changes in plains development, resulting in three types of current alluvial plains. His F-type plains (the majority) exist in places where the Jomon Transgression during the Climatic Optimum (4000–2000 BC) did not extend far inland. Elsewhere the coast was defined by drowned valleys and rugged coastal topography (Umitsu 1991). In these interior plains, late alluvial fans cover much of the (buried) Pleistocene terrace surfaces, while Holocene I and II terraces and current

[1] This and the succeeding paragraph are based on Barnes (2010).

floodplains are clearly distinguishable. His L-type plains developed sandbar barriers during sea level regression, enclosing lagoons, while landward areas were occupied by two types of deltas. D-type plains developed after the formation of Holocene II terraces and are characterized by a particular delta formation, usually backed by relatively large inland basins where alluvial fans are particularly visible on Pleistocene terraces I and II.

All these plains have undergone constant agricultural and then urban modification over the last 3000 years, obscuring to a great extent previous landforms such as alluvial fans, floodplains, natural levees, backmarshes, lakes, sand dunes, lagoons, bays, tidal flats and deltas - each of which had different potential for human activities. When archaeological sites are found within those plains deposits, they cannot be interpreted vis-à-vis the surface topography but instead must be related to their contemporaneous microtopography of a previous age. Reconstruction of the Echigo Plains topographic succession in the Niigata Basin (cf. Figure 4.4), for example, has confirmed a palaeosurface at 3000 BC where an entire ancient barrier sand body, lagoon, backmarsh and delta have been buried between -12 m and -19 m depth. Middle-to-Late Jomon pottery recovered from within the lagoonal back marsh suggests fishing activities (Urabe & Takahama 2003; Urabe *et al.* 2004). In Chapter 13, we will look at reconstructed topography of the Nara Basin for its implications in settlement archaeology.

Inland from coastal regions one finds numerous mountain basins and river valleys that host terraces, alluvial fans, and alluvial bottomlands. These have of course been affected by Holocene alluviation, but many have much deeper and longer-term sedimentation histories, and the very existence of some is due to tectonic stresses. The contrast between lowlands and uplands is often stark, with some mountains rising almost vertically from the flat valley floors.

Changing Tectonic Regimes

The topographic relief of present-day Japan has been created under several post-rifting tectonic regimes that have resulted in regional folding and faulting, uplift and subsidence, throughout the archipelago. These all fall within the rubric of 'neotectonics', but this term has been differentially applied. Traditionally it is a term used to describe the activity of plate tectonics in the Pacific Rim region from the Miocene onwards, which, in Japan, would include rifting of the archipelago (Chigakudan 1981: 826). Pavlides & Mountrakis (1986) give a short introduction to definitions of neotectonics, which generally fall into two groups: those which limit themselves to particular time periods, and those which consider any geological activity as far back as necessary in order to explain Quaternary and ongoing present-day tectonics.

Within these guidelines, several processes of post-rifting neotectonics can be divided by stress regime: N–S compression, E–W compression, and latterly N–S extension behind the Ryukyu archipelago. The shift from extension tectonics during Japan Sea Basin rifting to compression tectonics thereafter took place throughout most of the archipelago with the exception of the opening of the Okinawa Trough (cf. Figure 6.11) in the far south; specific instances of tectonic stress and their outcomes are:

- the impact of the Izu Arc on central Honshu and the formation of the Fossa Magna, which have been discussed in Chapter 4
- N-S compression of SW Japan after the inception of Philippine Plate subduction, resulting in E–W trending folds between 7 and 5 Ma (Yonekura *et al.* 2001: 303)
- the beginning of E–W compression in NE Japan from the Pacific Plate, dated to 6.5 Ma by Nakajima (2013) that resulted in N–S trending folds in NE Japan from ca. 4 to 3 Ma and was fully developed by 2 Ma (Taira 2001)

• E–W compression of SW Japan from 3–2 Ma, resulting in the reactivation of the MTL as a right-lateral strike-slip fault (Toda 2016) and formation of NE-SW trending basins in the Inland Sea region (Altis 1999)

• a new extensional regime opening the Okinawa Trough around 9 Ma became more active at 1.9 Ma (Letouzey 1985), causing the rotation of southern Kyushu 30° anticlockwise and pulling open two new graben systems on the island.

• the entire archipelago was uplifted between the Early and Middle Pleistocene and then accelerated in late Middle Pleistocene, coincident with the beginning of volcanic activity (Shiba 2021)

• the formation of the Kinki Triangle outline under the E-W compressive regime with the major faults beginning movement in the Quaternary (Hujita 1978: 14), specifically the late Middle Pleistocene (Shiba 2021: 50)

• the wrenching of the Kinki Triangle in the Mid-Pleistocene to form the current N–S trending reverse faults and formation of N–S basins, associated with increased activity of the MTL from 800 ka (Sato *et al.* 2015)

The cause of this recent E–W compression that has affected all of Japan except Kyushu has been variously specified as the eastward movement of the incipient Amur Plate (Taira 2001) (Figure 5.1) or as a shift in the direction of subduction of the Philippine Plate (Kusumoto *et al.* 2001); however, Nakajima (2013) throws the net wider and attributes the effects to reciprocal changes in the Pacific and Antarctic Plates. Aspects of its various effects will be reviewed below, but before that, let us look at the current situation.

Current landmass movement

It is useful for comparison to view the stresses currently acting on the Japanese archipelago. The vector arrows in Figure 5.1 show a time-slice of direction and velocity of movement of landmasses in different areas in one limited time-slice as an example. While the Kurile microplate is moving south-westward, strengthening WNW movement in eastern Hokkaido; northern Japan down to Kobe City is moving westward under pressure from the Pacific Plate. However, Northern Kyushu and the western tip of Honshu are moving southwestward in opposition to the WNW move-ment of the Nankai microplate, which itself is moving together with the Philippine Plate via the MTL right-lateral strike-slip fault.

Figure 5.1 Current neotectonic activity in Japan

K-PR = Kyushu–Palau Ridge, MTL = Median Tectonic Line
I-STL = Itoigawa–Shizuoka Tectonic Line

A new tectonic line from Kobe to Niigata (N-KTL) has been mooted to continue into an incipient subduction zone along northwest Tohoku; movement on this line is right-lateral as on the MTL (Figure 5.1: purple half-arrows), resulting in a wrenching of Honshu to form the basins within the Kinki Triangle (orange triangle) (Itoh *et al.* 2013).

Regional landscape changes

Post-rifting uplift & subsidence

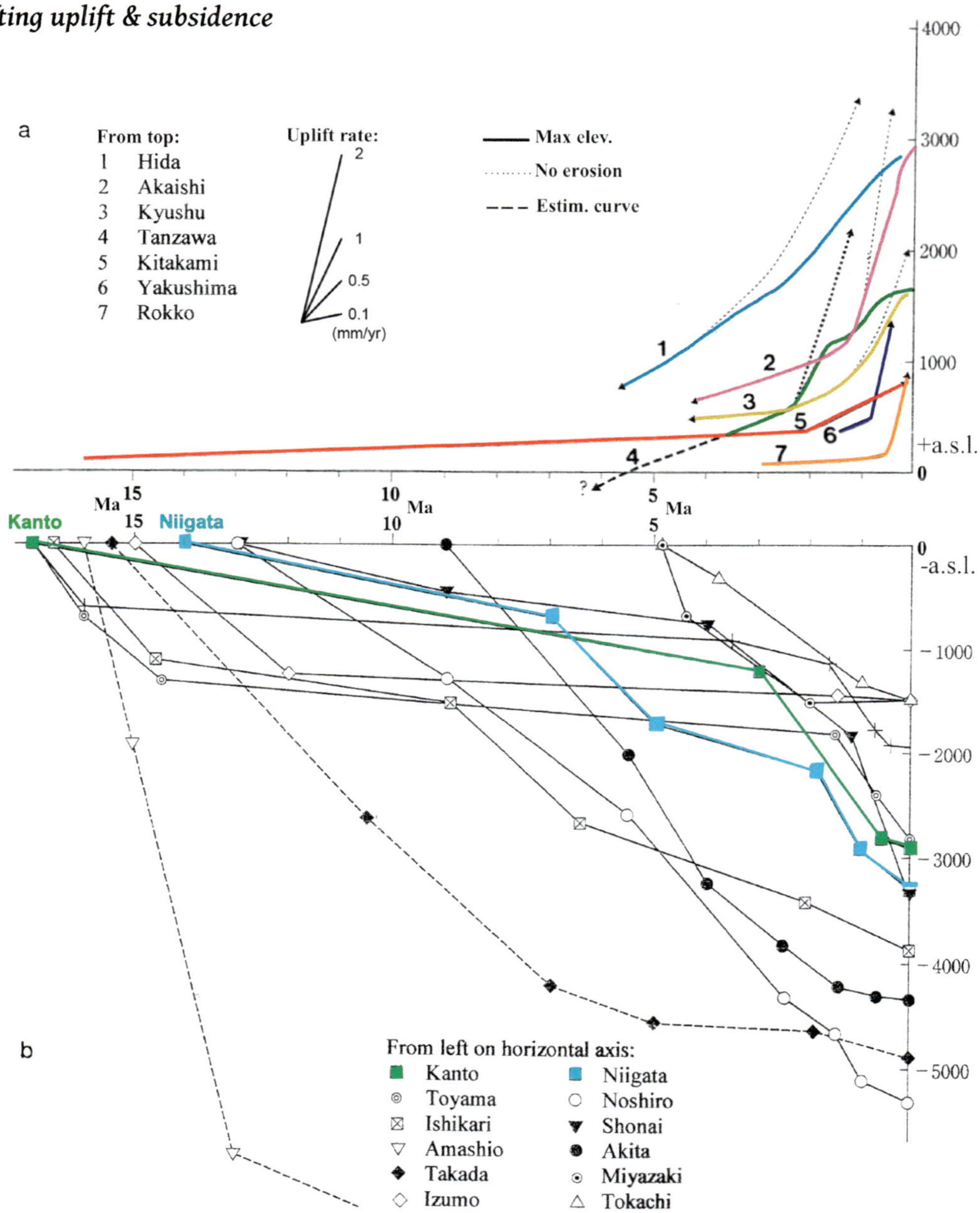

Figure 5.2 Early modelling of rifting/post-rifting progressive uplift and subsidence
a Uplift patterns of major mountain ranges from ca. 15 Ma; angle of line indicates rate
b Subsidence patterns for major sedimentary basins from ca. 17 Ma; Kanto and Niigata are emphasized here for discussion in the text

The stories of subsidence and uplift are intertwined with the rifting of the Japanese archipelago from the continent (Chapter 4). Figure 5.2 represents early modelling of these opposing processes, with uplift based on the assumption that low-relief areas of mountain ridges represent a planar surface after rifting and subsidence as evidenced in corings. Since then many more techniques have been developed for measuring uplift (Sueoka *et al.* 2016), but no overall assessment has yet been produced.

Both uplift and subsidence began around the time of cessation of rifting, although the upper half of Figure 5.2 suggests that the major stage of uplift commenced after 4 Ma. Basin formation and subsidence also commenced throughout Japan around the end of rifting, but different basins had very different trajectories of subsidence thereafter; nevertheless, the rate of subsidence generally increased after around 6 Ma. There are many more basins and mountain ranges that do not appear in this summary figure, and each has its own peculiar history – some of which are discussed below. Moreover, it is important to recognize that most of the major plains of Japan overlie deep tectonic basins such as the Echigo Plains in the Niigata Basin and the Kanto Plains in the Kanto Basin (cf. Figure 5.7). The Seto depression (BOX 6) – what may be called a 'megabasin' (Ingersoll & Busby 1995: i) – also holds the historically crucial basins of Osaka, Nara, and Kyoto.

The general processes of folding and faulting (cf. Figures 2.9, 2.10) are responsible for mountain and basin creation. Which ones were/are operative depend on local circumstances and both could change through time. These processes and the ultimate causes of directional movement are identified below as much as possible.

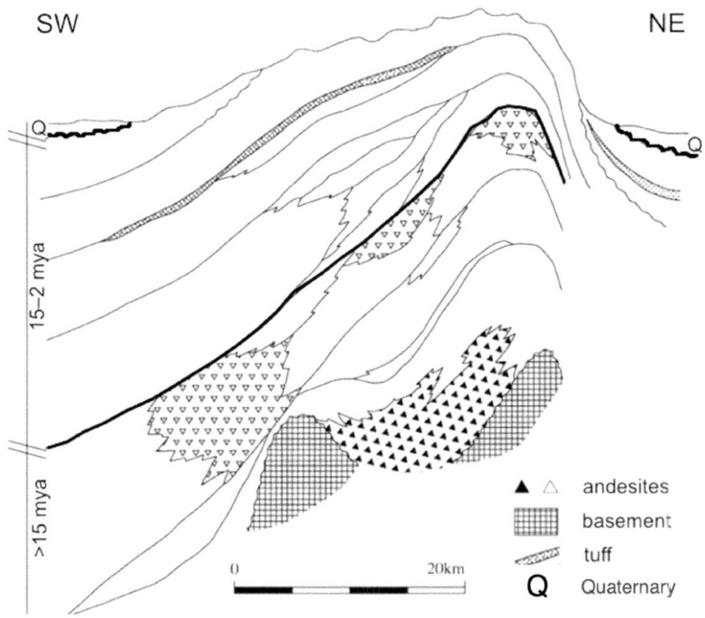

Figure 5.3 Folded Miocene–Pliocene hills in eastern Tohoku (location of transect shown in Figure 5.4)

Northeast Japan

Nakajima notes that an overall compressive regime, which began in the northeast from ca. 6.5 Ma, was signaled by a shift from Kurosawa marine deposits (Apx 9) to fluvial erosion products (conglomerates) in the basins, indicating lowered sea level and uplift of the Tohoku backbone Ou Range. Uplift increased after 3 Ma (Nakajima 2013: fig. 3), with intensi-fied compression resulting in areas of folding. As an example, Figure 5.3 illustrates a small ENE–SSW fold zone located on the eastern edge of Tohoku. The lower (Early Miocene) sandstone strata include andesitic deposits, while the upper sandstone strata (Late Miocene–Pliocene) are intercalated with tuff. These strata were pressured to form anticlines; Quaternary (Q, top left) sediments fill the valleys.

Contraction across NE Japan has even been calculated for a 5-year period by GPS monitoring; it particularly results in the uplift of the backbone Ou Range (Miura *et al.* 2004). TAKAHASHI Masaki gives the analogy that "NE Japan is contracted like a walnut caught in a nutcracker" (2017: fig. 5, caption), but he proposes that this is caused by the Philippine Plate moving to the northwest, not by Pacific Plate motion alone. He uses a 'scissor' metaphor to describe the changing direction of the Philippine Plate from its position *between* 15 and 3 Ma to that *after* 3 Ma (Takahashi 2006, 2017).

Tohoku fold belts

An example of coordinated tectonic basin and mountain range formation is found in Tohoku (Figures 5.4, 5.5). This has been identified as 'Basin & Range' (B&R) topography (Sueoka *et al.* 2016: fig. 3), but it seems that the term is used differently for Japan than elsewhere – based on folding rather than normal faulting. Neuendorf *et al.* (2005: 57) define 'basin & range' as a "regional geologic structure dominated by subparallel fault-block mountains spearated by sediment-filled basins; e.g., basin-range faulting characterized by movement on normal faults" as in the Basin-and-Range province of SW United States. Such structure is a function of extensional tectonics, still acting on the SW United States, whereas in Japan, the B&R structure is a product of compression tectonics which developed around 5–4 Ma after rifting (extension) ended, resulting in the folding wavelength and thereafter clearly involving reverse faulting (Koike *et al.* 2005: 313). The Tohoku B&R structure has a 50-km wavelength, but the axes of the basins in Tohoku are oriented in slightly different directions, NNW–SSE, and NNE–SSW, depending on

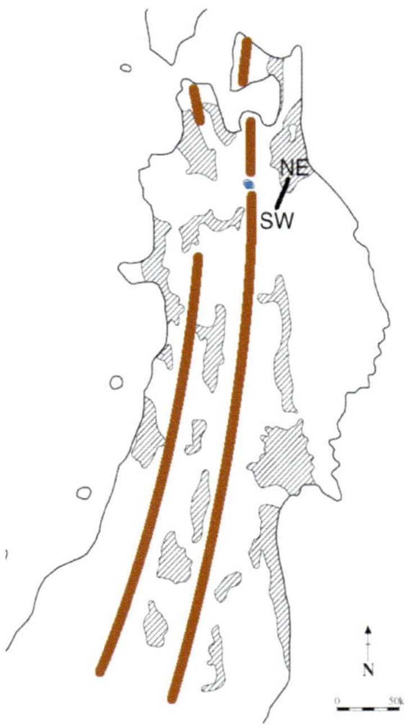

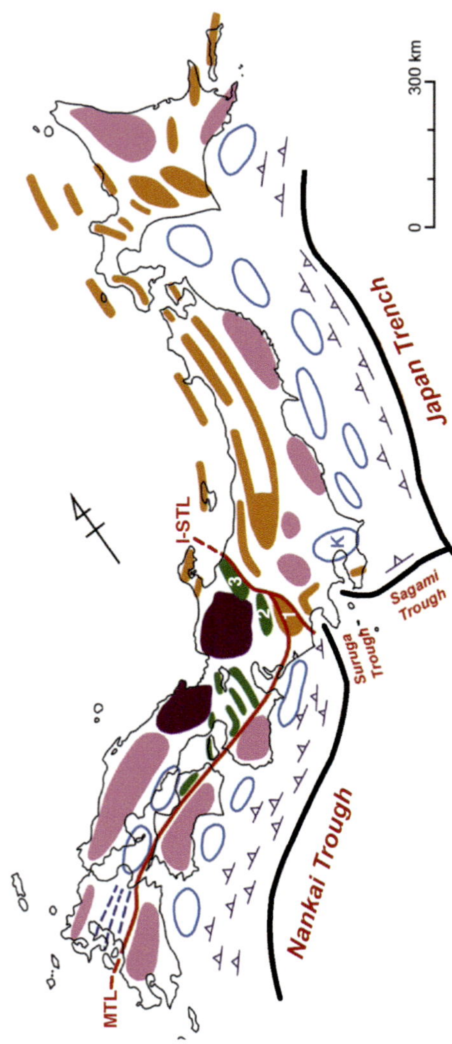

Figure 5.4 Above: Basins (hatched) between N–S trending mountain ranges in Tohoku, matched to Figure 5.5; NE–SW / transect shown in Figure 5.3

Figure 5.5 Right: Distributions of tectonically formed mountain ranges in Japan
Purple = folded, green = reverse (thrust) fault blocks, brown = basin & range folding/faulting, black = strike-slip fault blocks; blue ovals = fold basins, dashed lines = normal fault blocks, toothed lines = thrust faults of accretionary prisms; MTL = Median Tectonic Line, I-STL = Itoigawa–Shizuoka Tectonic Line, K = Kanto Basin

the direction and source of compression: the Kurile Forearc descending from the northeast or the Pacific Plate respectively. The western side of Tohoku is also a small fold belt, as can be seen from the emergence of the Nishiyama Hills in Figure 5.6, later riven by faults.

Green Tuff basins

Two of the economically important basins are the Akita and Niigata Basins in the Green Tuff Zone (Figure 4.4) – important because the sedimentary material deposited in them became catchments for Japan's limited oil and gas reserves. Figure 5.2 above shows the Niigata Basin as beginning to subside from 14 Ma, reaching its current depth at ca. -3300 m (measured to the base of the Nishiyama Formation under the Niigata Plain) (Figure 5.6). As the Japan Sea begins to close, we (or rather, our distant descendants!) may eventually see basin inversion – the folding or doming of their contents through lateral compression – as is already becoming evident in the Niigata Basin indicated by folding and faulting and enhanced by the formation of an incipient subduction zone running down the eastern edge of the Japan Sea Basin.

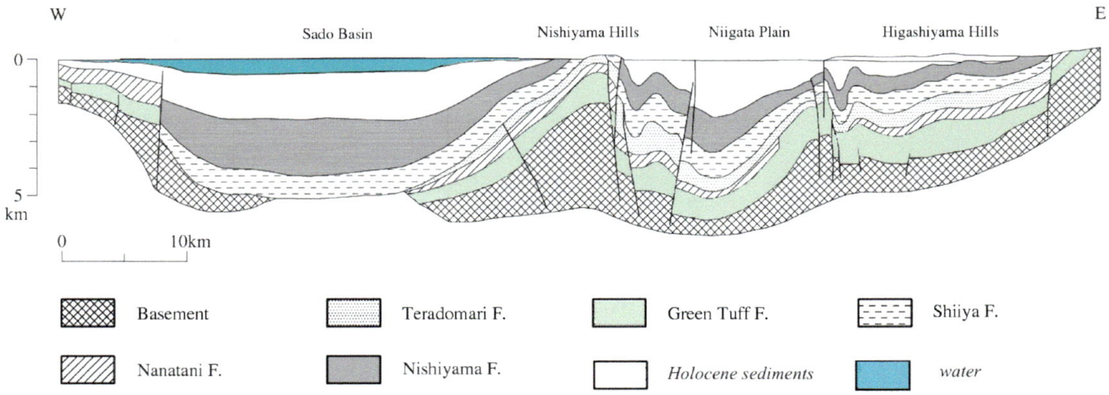

Figure 5.6 Cross-section of the Niigata basin. The Green Tuff Formation rests on continental basement rocks, followed by several sedimentary successions all of which have been folded and faulted by later tectonic activity. F. = Formation; sub-vertical lines indicate faults; the transect is taken from Sado Island in the west (cf. Figure 4.4)

In the year 2000, 2.2 million kilolitres of oil and 110 million cubic metres of natural gas were taken from Niigata's reserves, the largest in Japan (Tateishi & Kobayashi 2001: 75, 79). Most petroleum in this zone is found in the 'hard shales' (siliceous shale) and 'black shales' (dark grey mudstone) (Smith 1982: 163). In the Niigata gas fields, samples of porous and finely cracked rhyolitic (highly felsic) green tuff hosting natural gases have been brought up by coring from depths of 2.5 to 4.3 km (NPMRC 2000: plate 6.3), but the majority of productive strata lie at 2 km deep (Tateishi & Kobayashi 2001: 75).

Central Japan

Kanto Basin

Another important sedimentary basin is the Kanto Basin underlying the current Kanto Plains (Figure 5.7). Counter-intuitively, the basin is not parallel with today's Tokyo Bay but extends from NW–SE, as illustrated by the extent of Kazusa Group sedimentary deposition. Its current orientation E–W was likely determined by displacement by the Izu Arc impact from an original NE–SW orientation parallel to the off-shore trench (cf. Figure 5.5; note the angle of the Kanto Basin 'K' relative to others oriented parallel to the subduction trenches). In the mid-Pleistocene, the orientation changed again with the uplift of the Boso Peninsula seaboard, now represented by the direction of Tokyo Bay (Suzuki 2002). Kanto Basin then filled up with tephra deposits from Quaternary volcanoes and loess blown in from China, forming the Kanto loam of today's Kanto Plains as discussed in Chapter 6.

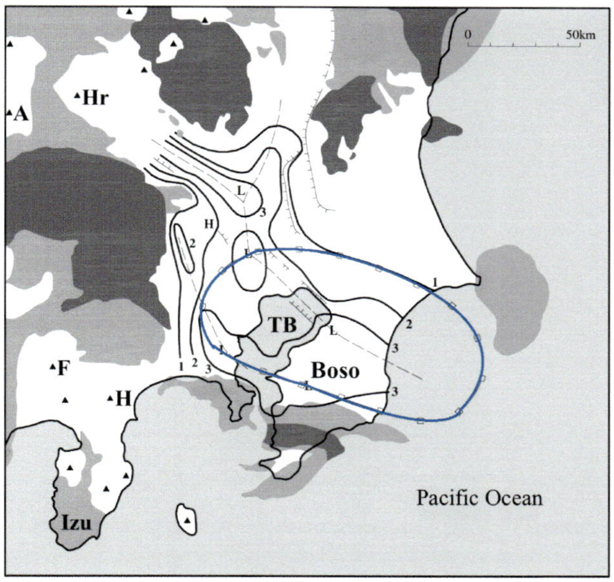

Figure 5.7 Plan of subsidence levels forming the Kanto Basin, crossing underneath Tokyo Bay and the Boso Peninsula. The Kazusa Group (blue line) comprises marine deposits from the Pliocene into Early Pleistocene.

Faulted mountains

Among Japan's mountains, the imbricated ranges of Akaishi, Kiso, and Hida in the southern Japan Alps (Figure 5.8) were impacted by the Izu Arc collisions, offsetting them to a NE–SW orientation. These are all parts of the Kanto Syntaxis, but their altitudes are due to further fault movement through compression. Both the Kiso Range and Hida Range have pop-up blocks, bordered by two reverse faults as the landmass was compressed (Figure 5.8: a, b). In the Hida Range (Figure 5.8: b), thrust faulting from the east along the I-STL has pushed up the mountains mainly towards the west.

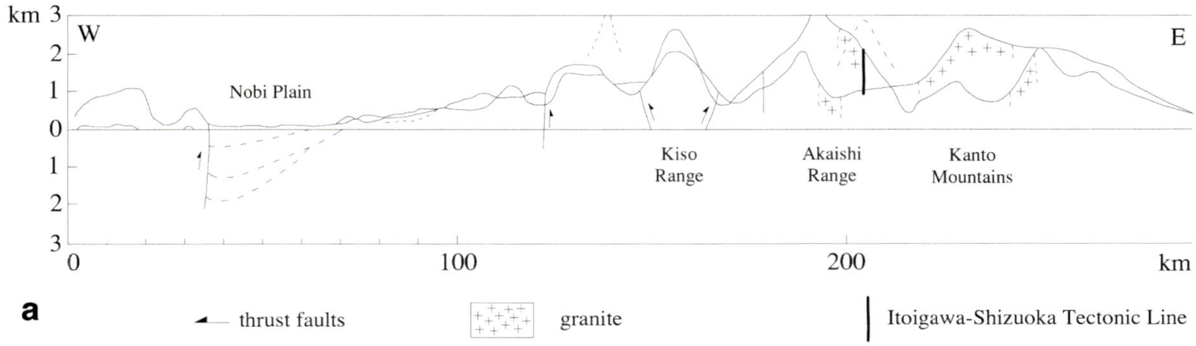

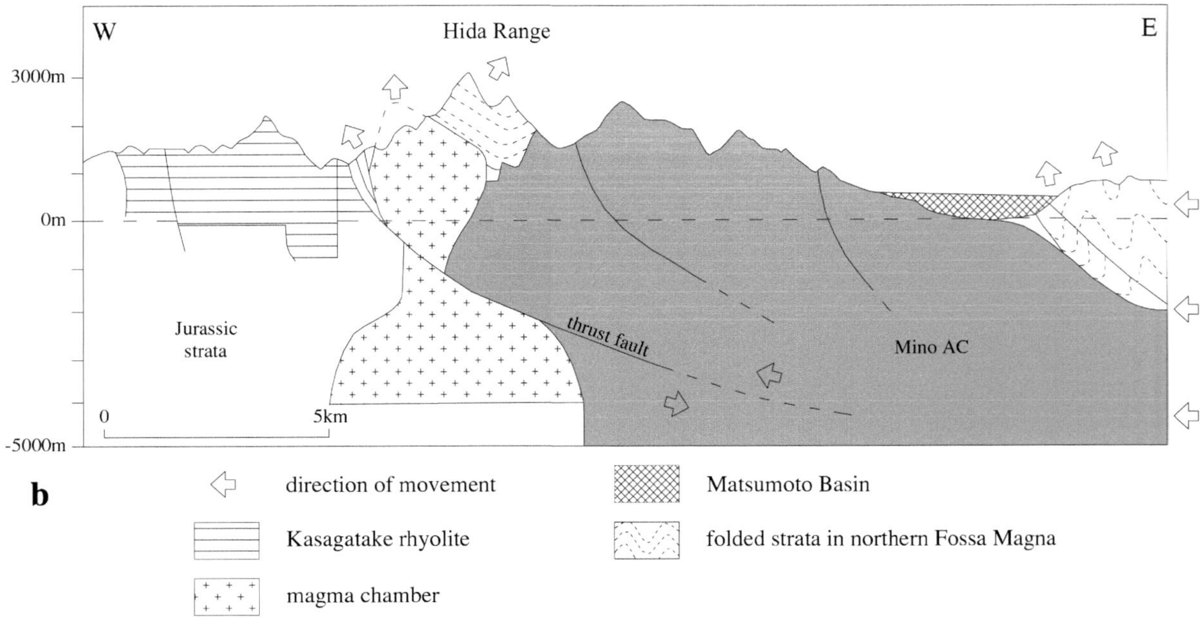

Figure 5.8 Faulted mountains of Kiso and Hida

a Horst-faulted mountains of the Kiso Range, **b** Thrust-faulting in the Hida Mountains

The Hida Range is the first among Japan's mountain ranges to show accelerated uplift (cf. Figure 5.2: a-1), starting around 6 Ma, after the Izu Arc began its greatest impact between 8 and 6 Ma (Mahony *et al.* 2011), and rising faster and higher than any others at an average rate of 1 m per 1000 years – currently resulting in eleven peaks over 3000 m (Machida *et al.* 2006: 169).

Southwest Japan

SW Japan & the Seto depression

In western Japan, the Cretaceous Ryoke Belt and Sanbagawa Metamorphic Belt (cf. Figure 3.1; Apx 9) have been exhumed along the Median Tectonic Line (MTL). Erosion of their original cover has 'unroofed' the Ryoke granites and metamorphic rocks, which now form the E-W trending mountains of Western Honshu and Shikoku. The MTL now forms the divider between the Inner and Outer Zones of SW Japan (Figure 5.9).

Southwestern Japan also became the focus of folding, but parallel to the MTL, as can be see in the undulations of Figure 5.9. The folding occurred post-rifting with the initiation of Philippine Plate subduction; the N-S compression regime that developed between 7 and 5 Ma resulted in folding on an E-W axis with a relatively long wavelength of 150 km; it uplifted the backbone ranges of Shikoku Island and western Honshu, while forming the extensive Seto depression between them (NCEC 1987: 243; Chigakudan 1996: 691; Yonekura *et al.* 2001: 303). This depression holds the current Inland Sea, but previously it extended farther inland to the east (BOX 6); the history of the Inland Sea is taken up in more detail in Chapter 13 for understanding the structure and contents of the Nara Basin.

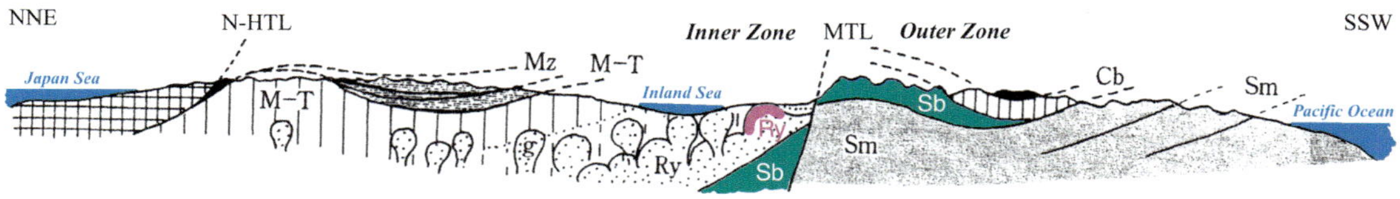

Figure 5.9 Transect ca. 300 km across western Honshu and Shikoku showing locations of Ryoke Belt exposure and Sanbagawa Metamorphic Belt faulted exhumation. Abbreviations:

		Accretionary complexes:	
MTL	Median Tectonic Line		
N-HTL	Nagato-Hida Tectonic Line	Mz	Maizuru
Sb	Sanbagawa MB	Cb	Chichibu
Ry	Ryoke Belt	Sm	Shimanto

In the mid-Pleistocene, the Philippine Plate changed subduction direction from NNW to NW, causing the formation of the Nankai microplate (Kagami *et al.* 1988; Altis 1999) and intensive movement in the Kinki Triangle (Huzita 1976) as the microplate forged westwards via right-lateral motion on the MTL.

The Kinki Triangle

The Kinki Triangle is bordered by the MTL on the south and on the northwest by Niigata-Kobe Tectonic Line; Itoh *et al.* (2013) propose that the eastern border is the little known Echizen–Shima Tectonic Line which runs from Wakasa Bay to Ise Bay, i.e., across the 'Waist of Honshu' (Apx 2: Fig. A). The tectonics of the Kinki Triangle were originally described as a product of time-transgressive activity along its boundary faults, with independent movement of blocks which either uplifted or subsided (Huzita 1962, 1968, 1969). Faults in SW Japan either have a NE–SW or NW–SE strike (cf. Altis 1999: fig. 3; Mitamura *et al.* 2016). The basins in the Kinki Triangle are bounded by reverse faults, such as those labeled F4 and F5 in Figure 5.10. Reverse faulting indicates compression, and indeed E-W compression is acknowledged as the overarching neotectonic regime from 3 Ma, superseding the N–S compression forming the Seto depression as seen in BOX 6.

So, how could compression cause basin formation? This 'paradox' (Itoh *et al.* 2013) resulted in the simultaneous deepening of the basins and rise of the surrounding mountains. This 'mechanical inconsistency' was explained through deformation modelling by Kusumoto *et al.* (2001). The important ingredients in the model are as follows:

- the MTL and its parallel echelon fault, the Arima-Takatsuki Tectonic Line (ATTL); these are right-lateral left-stepping faults, assumed to have vertical dip
- the secondary fault systems (F1-6) are reverse faults of assumed 80° dip
- movement on these boundary faults forming the Kinki Triangle pushed the triangle landmass east-wards in the south and westwards in the north, causing a twisting reaction within the landmass
- opposing movement on the boundary faults created normal faults between them which operate now as reverse faults;
- deformation of the land manifested itself as alternating pull-apart basins and pop-up blocks between these reverse faults, with the footwall side subsiding and the hanging wall side uplifting.

The researchers doing this modelling propose that all the faults formed simultaneously at the beginning of the Pleistocene. The Osaka Basin, oriented NE–SW, was judged to have formed first in the early Pleistocene, with the more strictly N–S oriented basins (Nara, Kyoto) forming after middle Pleistocene (see Chapter 13) in the Rokko Movement. With these new understandings, however, the older formulations of the Rokko Movement (cf. Chapter 3) have been superseded, and topography within the Kinki Triangle has become explainable in terms of the parameters of Kumoto's model. Moreover, it is not only the Kinki region but all of central Japan that is now recognized as having undergone a "Quaternary Uplift Movement" (Shiba 2021: 38).

For those who live in the Triangle, life is becoming more susceptible to tectonic movement along these fault lines, as demonstrated by the M_w6.9 Kobe Earthquake in 1995, the latter activated on the Nojima Fault extending into Awaji Island (Figure 5.10: F2) (see BOX 10, 12). In 2018, an M_w5.6 earthquake in northeast Osaka evidenced the active nature of the northern boundary of the Kinki Triangle (Hallo *et al.* 2019). This complex earthquake involved the intersection of two faults, one belonging to the NE–SW trending strike-slip ATTL zone and the other to the N–S Uemachi reverse fault system.

The Kinki Triangle provided Japan with its civilizational homelands: Kawachi and Nara Basins housed the first state-level society in Japan in the 5-7c AD and the Kyoto Basin thereafter from the 8c to 1868 (Figure 5.10: B3, B4).

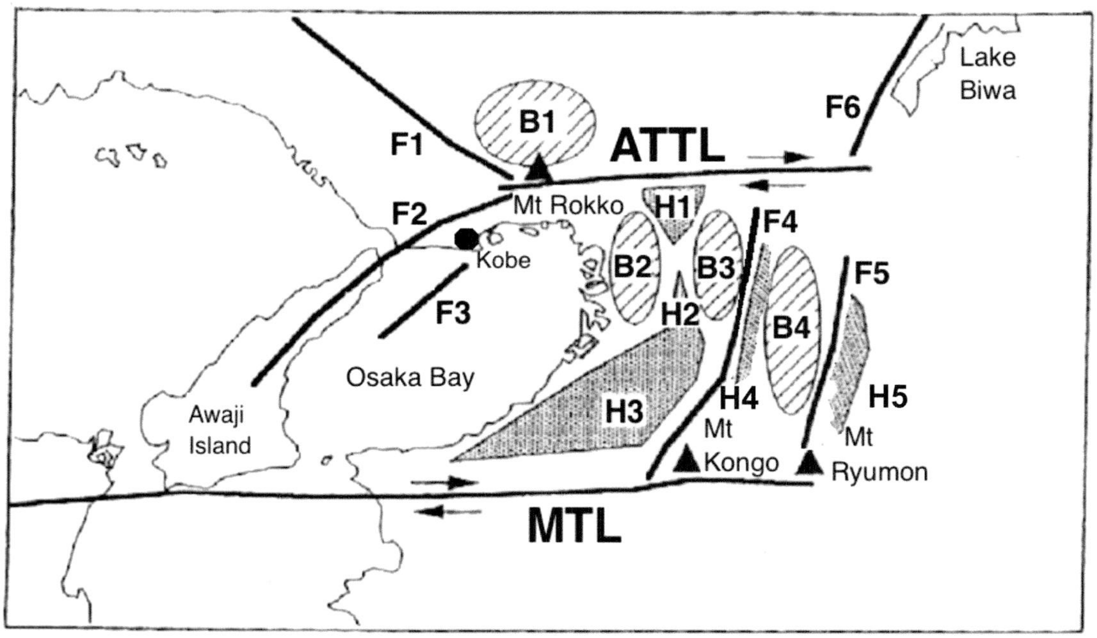

Figure 5.10 Tectonics in the Kinai region
ATTL = Arima-Takatsuki Tectonic Line
MTL = Median Tectonic Line
F = faults, B = basins, H = highlands

FAULTS — F1: Yamasaki Fault, F2: Rokko-Awaji Fault System, F3: Osaka Bay Fault, F4: Ikoma Fault System, F5: East Nara Fault System, F6: Hanaore Fault

BASINS — B1: Sanda Basin, B2: Osaka Basin, B3: Kawachi Basin, B4: Nara Basin

HILLS — H1: Senri Hills, H2: Uemachi Highland, H3: Senshu (Izumi) Hills; H4: Katsuragi-Nijo-Ikoma Mts; H5: Miwa-Kasuga Mts

BOX 6 The Inland Sea

The formation of the Inland Sea has been a major research focus, and two stages have been identified prior to its current extent: Inland Sea #1, and Inland Sea #2 – the latter also called the Palaeo-Inland Sea (Itihara 1993, 2001) and now termed the Second Setouchi Geologic Province (SHIBA Masahiro, pers. comm. 24 March 2022) because it was not always a 'sea'.

• Inland Sea #1 refers to the shallow marine and freshwater deposits laid diagonally across west-central Japan when it was submerged during Early to Mid-Miocene rifting. Marine deposits from this sea can be found today on plateaus and mountain tops standing at 400–1000 m msl throughout western Honshu, indicating they were uplifted soon after their deposition (Itihara 2001). In the faulted basins, such marine layers of Inland Sea #1 are preserved under subsequent sedimentary layers or on uplifted terraces.

• The 2nd Setouchi Geologic Province formed within the Seto depression made by Late Miocene folding. Its fill was initially composed of lacustrine and fluvial freshwater deposits. Those from the Pliocene were concentrated around Nagoya in the east between 5.3 and 3 Ma; these extended westwards through the Kinki at least to Shikoku by 1 Ma in the Pleistocene, with water drainage from east to west (Itihara 2001). Marine transgressions into the Nagoya region began ca. 1.8 Ma and in the Osaka region from ca. 1 Ma. The sediments of this Geologic Province are designated as the Second Setouchi Supergroup (below: right) and can be found in the mountain basins and along current shores within the extent of the "Palaeo-Inland Sea" illustrated below (Kataoka & Nagahashi

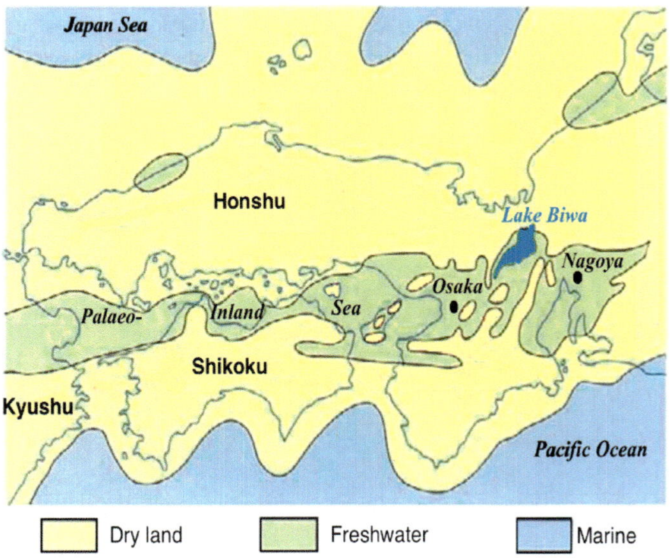

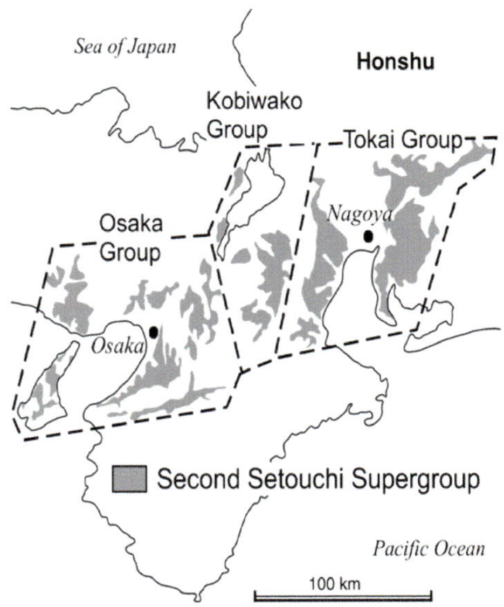

2012). In the far west, this geologic province cut across central Kyushu in the Beppu-Shimabara Trough, which continued into the Okinawa Trough (cf. Figure 5.11) – a graben rift zone in line with the MTL (Taira 2001: 111) – which may in time develop into a back-arc basin northwest of the Ryukyu Islands.

• The current Inland Sea #3 is totally marine and took its form with the increasing differentiation of uplands and lowlands during the activation of neotectonics in the region (Itihara 2001).

The composition of the Osaka Group will be examined more in detail in Chapter 13, with reference to the development of the Nara Basin adjacent to Osaka.

Kyushu rotation

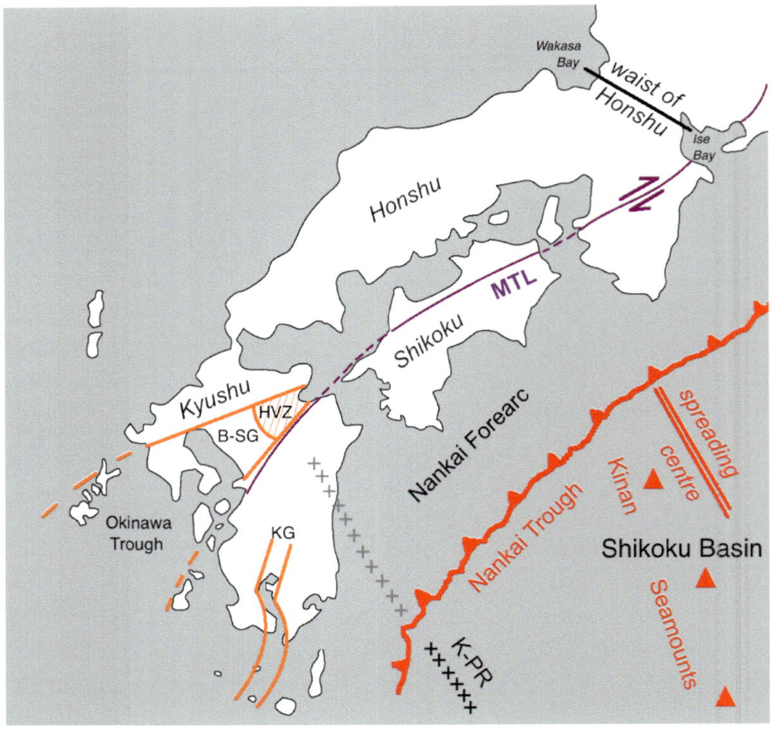

Figure 5.11 Kyushu neotectonics

HVZ = Hohi Volcanic Zone, B-SG = Beppu–Shimabara Graben, KG = Kagoshima Graben, K-PR = Kyushu-Palau Ridge

Kyushu is located at the crossroads of tectonic forces which are tearing the island apart (Figure 5.11). It is receiving the brunt of Nankai microplate (cf. Figure 5.1) movement to the WNW along the MTL at the same time the Kyushu-Palau Ridge (K-PR) is sub-ducting underneath it. Although the K-PR is a dead volcanic chain, unlike the Izu Arc, it is not clear what influence it has on the Hohi Volcanic Zone, which lies in its subduction path in northeastern Kyushu (Figure 5.11). Meanwhile, the Okinawa Trough is opening up on its northwest, in the first stage of back-arc basin formation behind the Ryukyu archipelago. These extensional processes have created the Beppu–Shimabara Graben (B-SG), forcing the southern part of the island to rotate 30° counter-clockwise and leading to the opening of a second graben in the south, the Kagoshima Graben. Much of the Quaternary volcanic activity in Japan, examined in Chapter 6, has occurred in the Kagoshima Graben and the Hohi Volcanic Zone at the head of the Beppu-Shimabara Graben.

Conclusions

It can be concluded that the Japanese Islands are not static but active in a very meaningful way. Although rifting from the continent ended millions of years ago, its ramifications are ongoing in the way the Philippine and Pacific Plates interact with the archipelago. More disruption can be expected through the subduction of several outlying seamount chains, plateaus, and the Shikoku Basin spreading centre not to mention the continuing impact of the Izu Arc on central Honshu. The effects of these tectonic activities manifest themselves in volcanic and earthquake activity in the short term; the long term results will be continued reorganization of the regional portions of the Japanese landmass. Though the basins and plains have held the cradle of Japanese civilization for protohistoric and historic times, these are also undergoing incremental change as subsidence and uplift continue to act on the archipelago.

Figure & Table Sources

Figure 5.1 after (Taira 2001: figs. 1, 4), redrawn by DAS, modified by GLB
Figure 5.2 after (Yonekura *et al.* 2001: fig. 10.1.6), redrawn by DAS, modified by GLB
Figure 5.3 after (NCEC 1989: fig. 5.14), redrawn by DAS, modified by GLB
Figure 5.4 after (NCEC 1989: fig. 8.2), redrawn by DAS, modified by GLB
Figure 5.5 after (Sueoka *et al.* 2016: fig. 3), modified by GLB

Figure 5.6 after (NCEC 1988: fig. 3.5), redrawn by DAS, modified by GLB
Figure 5.7 after (Chigakudan 1981: 196a; Kaizuka 1998: fig. 8.20), redrawn by DAS, modified by GLB
Figure 5.8 after (Machida 2006: figs. 1.1.2, 4.3.2b), redrawn by DAS, modified by GLB
Figure 5.9 after (Ota *et al.* 2004), redrawn by DAS, modified by GLB
Figure 5.10 after (Kusumoto *et al.* 2001 #9101: fig. 2), redrawn by DAS, modified by GLB to show mountains bordering Nara Basin
Figure 5.11 basemap by DAS, content compiled by GLB from (Sato *et al.* 2002: fig. 1B; Mahony *et al.* 2011: fig. 3; Chapman, 2009b: fig. 4.3)

BOX 6 left after (Itihara 2001: figs. 1.9, 1.10, 1.1); right after (Kataoka & Nagahashi 2012: fig. 1), bothmodified by GLB

Table 5.1 modified from (Yonekura *et al.* 2001: table 7.1.1)

References

ALTIS, Sungat (**1999**) "Interpretations of a Middle Miocene and late Quaternary steady dextral transpression in SW Japan and the opening tectonics for the Japan Sea." *Tectonophysics* 302: 257-285.

BARNES, Gina L (**2010**) "Landscape and subsistence in Japanese history", pp. 321-340 in *Landscapes and societies: selected cases*, ed. by P MARTINI & W CHESWORTH. Dordrecht: Springer.

CHAPMAN, Neil; Mick APTED & John BEAVAN *et al.* (**2009**) *Development of methodologies for the identification of volcanic and tectonic hazards to potential HLW repository sites in Japan: the Kyushu case study*. NUMO-TR-09-02. Tokyo: Nuclear Waste Management Organization of Japan (NUMO).

Chigakudan (ed.) (**1981**) *Dictionary of Earth Sciences*. Tokyo: Heibonsha (rev. ed.) (in Japanese).

HALLO, Miroslav; Ivo OPRSAL & Kimiyuki ASANO *et al.* (**2019**) "Seismotectonics of the 2018 northern Osaka M6.1 earthquake and its aftershocks: joint movements on strike-slip and reverse faults in inland Japan." *Earth, Planets and Space* 71: 21pp. [DOI: 10.1186/s40623-019-1016-8].

HUZITA, Kazuo (**1962**) "Tectonic development of the Median zone (Setouti) of Southwest Japan, since the Miocene." *Journal of Geoscience, Osaka City University* 6: 103-144 (in English).

---- (**1968**) "Rokko movements and its appearance." *The Quaternary Research* 7.4: 248-260 (in Japanese with English title and abstract).

---- (**1969**) "Tectonic development of Southwest Japan in the Quaternary period." *Journal of Geoscience, Osaka City University* 12: 53-70 (in English).

---- (**1976**) "Quaternary earthquake and tectonic activity in the Kinki region." *Chishitsu News* November 2, 1976: 10-20 (in Japanese) [www.gsj.jp/data/chishitsunews/76_11_02.pdf].

INGERSOLL, Raymond V & Cathy J BUSBY (**1995**) "Tectonics of sedimentary basins", pp. 1-52, in *Tectonics of sedimentary basins*, ed. by CJ BUSBY & R INGERSOLL. Malden, MA: Blackwell Science.

ITIHARA Minoru (ed.) (**1993**) *The Osaka Group*. Osaka: Sogensha (in Japanese with English title, table of contents & conclusion).

---- (**2001**) "Osaka Formation: Palaeo-Inland Sea river and lake systems." *Urban Kubota* 39: 2-13 (in Japanese).

ITOH, Yasuto; Keiji TAKEMURA & Shigekazu KUSUMOTO (**2013**) "Neotectonic intra-arc basins within southwest Japan – conspicuous basin-forming process related to differential motion of crustal blocks", pp. 191-207 in *Mechanism of sedimentary basin formation – multidisciplinary approach on active plate margins*, ed. by Y ITOH. London: IntechOpen (in English).

KAGAMI, Hideo; Hidekazu TOKUYAMA & Kazuto KODAMA *et al.* (**1988**) "Shortening of forearc wedge and breaking of oceanic crust in the subduction zone." *KAKEN* [https://kaken.nii.ac.jp/en/grant/KAKENHI-PROJECT-62490002].

KAIZUKA, Sohei (**1998**) *Developmental History Landform Studies*. Tokyo University Press (in Japanese).

KATAOKA, Kyoko S & Yoshitaka NAGAHASHI (**2012**) "A linkage between source volcanoes and alluvial basins, based on the facies and thickness variations of distal tephra beds of the Plio-Pleistocene Second Setouchi Supergroup, central Japan." *Chishitsu Zasshi* 118.3: 139-156 (in Japanese with English title and abstract).

KOIKE, Kazuyuki; Toshikazu TAMURA & Kiyotaka CHINZEI *et al.* (eds) (**2005**) *Geomorphology of Tohoku region*. Regional geomorphology of the Japanese Islands, vol. 3. University of Tokyo Press (in Japanese).

KUSUMOTO, Shigekazu; Yoichi FUKUDA & Keiji TAKEMURA *et al.* (**2001**) "Forming mechanism of the sedimentary basin at the termination of right-lateral left-stepping faults and tectonics around Osaka Bay." *Chigaku Zasshi* 110.1: 32-43 (in Japanese with English title and abstract).

LETOUZEY, Jean (**1985**) "Okinawa Trough genesis: structure and evolution of a backarc basin developed in a continent." *Marine and Petroleum Geology* 2: 111-130.

MACHIDA, Hiroshi; Tokihiko MATSUDA & Masatomo UMITSU *et al.* (**2006**) *Nihon no chikei 5: Chuubu*: Tokyo University Press (in Japanese).

MAHONY, SH; LM WALLACE & M MIYOSHI *et al.* (**2011**) "Volcano-tectonic interactions during rapid plate-boundary evolution in the Kyushu region, SW Japan." *GSA Bulletin* 123.11/12: 2201-2223 [DOI: 10.1130/B30408.1].

MITAMURA, Keisuke; Takamoto OKUDAIRA & Muneki MITAMURA (**2016**) "Meso-scale brittle deformation structures in and around the Ikoma fault zone." *Chishitsu Zasshi* 122.2: 61-74 (in Japanese with English title, summary, captions).

MIURA, Satoshi; Toshiya SATO & Akira HASEGAWA *et al.* (**2004**) "Strain concentration zone along the volcanic front derived by GPS observations in NE Japan arc." *Earth Planets Space* 56: 1347-1355.

NAKAJIMA, Takeshi (**2013**) "Late Cenozoic tectonic events and intra-arc basin development in northeast Japan", pp. 153-189 in *Mechanism of sedimentary basin formation – multidisciplinary approach on active plate margins*, ed. by Y Itoh. London: IntechOpen (in English).

NCEC (*Nihon no Chishitsu* Editorial Committee) (**1988**) *Geology of Japan 4: Chubu region I.* Tokyo: Kyoritsu (in Japanese).

---- (**1989**) *Geology of Japan 2: Tohoku region.* Tokyo: Kyoritsu.

NEUENDORF, Klaus KE; James P Jr MEHL & Julia A JACKSON (eds) (**2005**) *Glossary of geology*, 5th edition. Alexandria, VA: American Geosciences Institute.

NPMRC (Niigata Prefecture Map Revision Committee) (ed.) (**2000**) *Explanatory booklet for the Niigata prefecture geological maps.* Niigata: Niigata-ken Shōkō Rōdōbu, Shōkō Shinkōka (in Japanese).

OTA, Yoko; Toshiro NARUSE & Shingo TANAKA *et al.* (eds) (**2004**) *Regional geomorphology of the Japanese Islands, vol. 6: Kinki, Chugoku, Shikoku.* Tokyo University Press (in Japanese with English title).

PAVLIDES, S & D MOUNTRAKIS (**1986**) *Neotectonics: an introduction to recent geological structures.* Thessaloniki: University Studio Press.

SATO, H; S MACHIDA & S KANAYAMA *et al.* (**2002**) "Geochemical and isotopic characteristics of the Kinan Seamount Chain in the Shikoku Basin." *Geochemical Journal* 36: 519-526 (in English).

SATO, Hiroshi; Naoko KATO & Susumu ABE *et al.* (**2015**) "Reactivation of an old plate interface as a strike-slip fault in a slip-partitioned system: Median Tectonic Line, SW Japan." *Tectonophysics* 644-645: 58-67 (in English).

SHIBA, Masahiro (2021) "Characteristics of crustal uplift since the Pliocene in central Honshu, Japan, and sea level rise." *Chikyu Kagaku* 75: 37-55 (in Japanese with English title and abstract).

SMITH, AJ (**1982**) "The Neogene to Recent geology of Japan and its surrounding seas." *Proceedings of The Geologists' Association* 93.2: 161-178.

SUEOKA, Shigeru; Hiroyuki TSUTSUMI & Takahiro TAGAMI (**2016**) "New approach to resolve the amount of Quaternary uplift and associated denudation of the mountain ranges in the Japanese Islands." *Geoscience Frontiers* 7.2: 197-210 [DOI: 10.1016/j.gsf.2015.06.005].

SUZUKI, Hiroyoshi (**2002**) "Underground geological structure beneath the Kanto Plain, Japan." *Bōsai Kagaku Gijutsu Kenkyū Hōkoku* 63 (June): 1-19 (in Japanese with English title and abstract).

TAIRA, Asahiko (**2001**) "Tectonic evolution of the Japanese island arc system." *Annual Review of Earth and Planetary Sciences* 29: 109-134 (in English).

TAKAHAMA, Nobuyuki (ed.) (**1997**) *Research relating the establishment and destruction of sites to earth movements.* Hazards Research Institute, Niigata University.

TAKAHASHI, Manabu (**2001**) "Environmental archaeology for disaster risk management." *Ritsumeikan bungaku* 593: 582-569 (in Japanese).

TAKAHASHI, Masaki (**2006**) "Tectonic development of the Japanese Islands controlled by Philippine Sea Plate motion." *Chishitsu Zasshi* 115.1: 116-123 (in Japanese with English abstract).

---- (**2017**) "The cause of the east–west contraction of Northeast Japan." *Bulletin of the Geological Survey of Japan* 68.4: 155-161 (in English).

Tateishi, Masaaki & Iwao Kobayashi (**2001**) "Niigata's plains and hills and people's lifestyles", pp. 75-85 in *Ikiteiru chikyuu - Niigata kara kenshi suru Nihon retto no kakou, genzai, mirai*, ed. by Niigata University Public Lectures Committee. Niigata University (in Japanese).

Toda, S (**2016**) "Crustal earthquakes", pp. 371-408 in *The geology of Japan*, ed. by T Moreno *et al.* London: Geological Society of Japan.

Umitsu, M (**1991**) "Holocene sea-level changes and coastal evolution in Japan." *The Quaternary Research* 30.3: 187-196 (in English).

Urabe, A & N Takahama (**2003**) "Echigo Plains: reconstruction of the Middle Jomon palaeotopography in Nishi-Kanbara area", pp. 39-53 in *Zen-kindai no Niigata kasen kotsu to iseki ritchi no chiishiki-teki kenkyu*, ed. by S Kobayashi. Niigata University (in Japanese).

Urabe, A; N Takahama & H Yabe (**2004**) "Identification and characterization of a subsided barrier island in the Holocene alluvial plain, Niigata, central Japan." *Quaternary International* 155-156: 93-104 (in English).

Yonekura, Nobuyuki; Sohei Kaizuka & Michio Nogami *et al.* (eds) (**2001**) *Regional geomorphology of the Japanese Islands, vol. 1: Introduction to Japanese geomorphology.* TUniversity of Tokyo Press (in Japanese wth English title).

CHAPTER 6

Japan's Igneous Activity & Volcanic Arcs

As described above in the previous Chapters, the Japanese landmass has been subject to volcanic activity when part of the Cretaceous continental margin (the Ryoke Belt), as that landmass rifted from the continent in the Miocene (the Green Tuff Movement) and was subjected to Philippine Plate subduction (Setouchi volcanics). The Pliocene witnessed great caldera eruptions in the northeast, but the current volcanic arc configuration and activity was established within the Pleistocene from ca. 700,000 years ago. This Chapter includes references to those earlier activities but concentrates on the modern configuration, which has transformed the landscape with volcanic eruption products. That every inch of the archipelago has received volcanic ash fallout since this time gives great potential for tephrochronologic dating, and it affects the soils and agricultural productivity in many regions.

Japan as a Volcanic Archipelago

Japan is commonly thought of as a *volcanic island chain* (e.g., Nakada 2018), but in fact the present-day footprint of volcanoes themselves is very small compared to mountains of other compositions (Figure 6.1). The basement structure of Japan proved to be primarily pre-Miocene sedimentary accumulations that once formed the edge of the Eurasian continent (Accretionary Complexes), discussed in Chapter 3 as geotectonic 'belts'. However, throughout time, many of these AC have been subject to igneous intrusions or metamorphic transformation. Thus, the folded and faulted non-granites (cf. Figure 5.8) or metamorphic belts (MB) (cf. Figure 5.9).

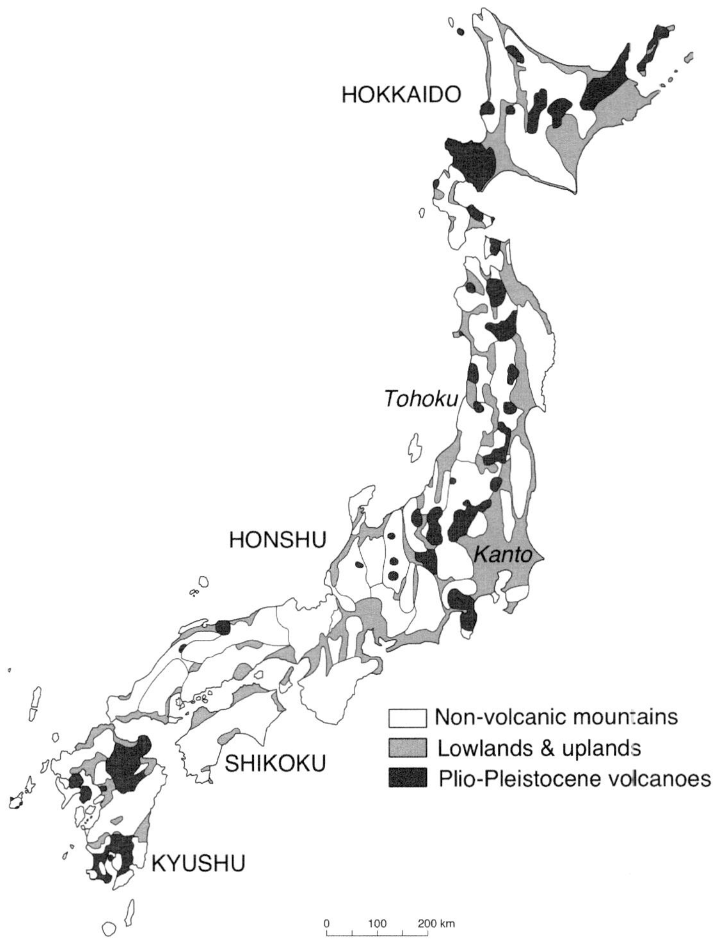

Figure 6.1 Mountains and plains of Japan
Volcanoes dating from 2.6 Ma in a landscape of folded and faulted mountains and low elevation 'plains' (lowlands & uplands)

For example, unroofed San'in and San'yo granites form the backbone ridge of southwestern Honshu, and the Ryoke granites form the mountains across northern Shikoku (cf. Figure 3.8), while the mountains formed by the Sanbagawa MB border Ryoke on the south (cf. Figure 5.9). Moreover, the AC often include sections of the ocean floor (ophiolites) and serpentinite mélanges, both of igneous origin. Finally, because of active volcanism in the rifted archipelago (cf. Chapter 4), every inch of the Japanese landmass has been covered in volcanic ash.

Thus, the igneous content of the Japanese landmass, as opposed to just the extant volcanoes, is much greater than illustrated here. It is more appropriate to distinguish between volcanism that affected the continental margin before rifting and the volcanic activities that took place after rifting when that continental margin was isolated as the Japanese archipelago. Moreover, it is necessary always to think of igneous rocks in their two forms: plutonic vs volcanic, and to remember that volcanic ash is considered a sediment forming sedimentary rocks. Finally, volcanics generally derive from magma chambers, barring exceptional circumstances, so that the presence of plutonic rocks (granites, gabbro: both are crystallized contents of former magma chambers) in regional geology usually indicates the formation of volcanoes or volcanic vents above those chambers, whose shapes – but perhaps not products (Nagahashi *et al.* 2000) – have been eroded and lost over time. This Chapter investigates the establishment of the modern Volcanic Fronts and the Quaternary activity of the different volcanic regions in Japan, but first, a look at pre-Miocene igneous activity.

Granitic rocks

Many believe that granites are probably the most diverse and complicated rocks in the geological record, so praise be to anyone tackling their genesis and composition. Several classification schemes are in use based on a variety of criteria, but it is most important to note that the IUGS defines 'granite' *sensu stricto* as having a specific mineral proportions (Apx 5: Fig. F), while *sensu lato* use includes others that are properly termed 'granitoids'. Because these distinctions are difficult to apply, the terms 'granite' or 'granitic' will be used here interchangeably to refer to both. The SIAM classification is based on source magma distinctions, while the tectonic setting can be used to classify granitic rocks into Orogenic, Transitional, or Anorogenic types (see Winter 2001: tables 18.3, 18.4; Allen 2001; Janousek *et al.* 2020).

Over 90% of Japan's granitic bodies were produced when the Japanese landmass existed as a continental volcanic arc at the edge of Eurasia; they can be grouped into small clusters dating 190–160 and 250–200 Ma, and large intrusions dating 140–130 Ma (Apx 10) – the last accounting for 80% of the total (Nakajima *et al.* 2016). From the start rifting which formed the Japanese Islands, granitic intrusions have continued from 20 Ma to the present, and granites form 30% of the current Japanese land surface; a map of these by age distribution throughout the archipelago can be seen in Nakajima *et al.* (2016: fig. 4). The granites in the San'in, San'yo, and Ryoke zones (Figure 3.8) are described as "a surface-exposed crustal cross-section of the Cretaceous Eurasian continental margin" and the granites are noted as becoming younger to the east (Nakajima *et al.* 2016: 254). A distinction is made by these authors between granites in Japan produced from slab dehydration processes (volcano-plutonic) (cf. Figure 2.5), and granites resulting from melting materials at high temperature and high pressure during extreme metamorphism occurring in the subduction zone (plutono-metamorphic) (cf. Figure 2.8). These designations are flagged in Appendix 10 to the extent available. These terms hark back to Harker (1889) – who conceptually divided the metamorphic T/P diagram (Figure 2.8) into quadrants. The two types of granites have different associated rocks – volcanic or metamorphic, respectively – but these associations can only be discerned in granites less than 280 million years old; earlier granites all have arc-type (volcano-plutonic) volcanic chemistry (Nakajima *et al.* 2016).

Early volcanic rocks

Chapter 4 dealt with the massive Green Tuff deposits, dating to ca. 14 Ma at the end of the rifting of the Japanese landmass from the continent, and andesitic volcanics in the Inland Sea region which marked the initiation of Philippine Plate subduction under SWJapan at the same time period. However, pre-Miocene volcanic deposits do exist in the Japanese geological record (cf. Apx 11). One of the largest ancient depositions of lava is the Nohi Rhyolite, preceded by calc-alkaline andesitic volcanism (cf. Apx 5: Fig. G) in the Mid-Cretaceous (ca. 100 Ma). The Late Cretaceous outpourings of Nohi Rhyolite, beginning in the Mino Belt (cf. Apx 6), consisted of 11 pyroclastic flows, 2 lava extrusions, and 2 rhyodacite intrusions (Sonehara & Harayama 2007), and today they fairly fill the eastern half of Gifu Prefecture in central Honshu. By their reckoning, the Nohi Rhyolite covers an area of 3500 km^2 at 1.5–2 kms thickness and has a total volume of ca. 5000–7000 km^3. The largest eruption sequence was NOHI-3 between 80 and 70 Ma, covering most of the total Nohi Rhyolite area. Mineral and chemical compositions of all these units are given in Sonehara & Harayama (2007: fig. 9; table 2).

In addition, Kimura *et al.* (2015) have identified 20 widespread tephra deposits from land-based volcanoes that were erupted prior to 2 Ma. Tuff, a sedimentary rock deriving from tephra is a common building material (cf. BOX 4), present and past, while two volcanic rocks in particular – sanukite (cf. BOX 5) and obsidian, discussed below – were critical resources for stone-tool making in the prehistoric periods.

In a ground-breaking archaeological study of Late Kofun-period stone coffins, (Wada 1986) analyzed and compared their types of stone and stylistic distributions. Four tuffs, one granite, and one andesite stone-types were identified, with tuff quarries identified by prefecture: Fukui, Gifu, Nara/Osaka border (Futakami tuff, Miocene), Kagawa (Washi-no-yama tuff, Miocene), Hyogo (Tatsuyama tuff, Cretaceous), and Kumamoto (Aso-Makado tuff, Cretaceous). Currently there is a revival of tuff stone-coffin use after centuries of cremation burial, providing much business to Hyogo Prefecture stonecutters (HJKK 2018).

Plio-Pleistocene Two-plate Subduction Regime

The current volcanic arcs within the island chain were fixed once the Izu Arc began to collide with Honshu (cf. Chapter 4) at the end of rifting. This led to the establishment of the two subduction regimes characterizing Quaternary volcanics in Japan (CCQVJ 1999, updated in JMA 2013 and 2017), but see Hone *et al.* (2007) and Chapman *et al.* (2009*a*: sec. 4.2).

Philippine vs Pacific Plate subduction

Northeast and southwest Japan had different rifting histories in the Miocene (24–6 Ma); and now in the Quaternary (2 Ma–present), their volcanic eruption patterns and chemistries also have been and still are completely different (Katayama *et al.* 2010). The Japanese Islands west of the Izu Peninsula are subject to the subduction activities of the Philippine Plate as well as to stress from the Pacific Plate pushing westwards (Takahashi & Saito 1997: 179), while NE Japan receives the full brunt of the Pacific Plate. These plates are moving at different directions, speeds and angles, and the ages of the subducting plates are radically different (Table 6.1). However, both NE and SW Japan show evidence of compression exerted by the oceanic plates on the newly positioned archipelago, and general uplift is thought to have been initiated by the increased rate and angle of subduction of the Pacific and Philippine Plates after 13 Ma (Yonekura *et al.* 2001: 309), with uplift increasing dramatically from ca. 3 Ma (cf. Figure 5.2) (Sueoka *et al.* 2016). These two plate regimes have led to the establishment of different volcanic Fronts in NE and SW Japan; however because these plates overlap in central and western Honshu, the magmatic chemistry there becomes much more complicated.

Table 6.1 Comparison of Pacific and Philippine Plates as currently subducting under Japan

kmd = km depth, Ma = million years ago, VF = Volcanic Front; th = thickness; icons: see Figure & Table Sources

Attribute	**Pacific Plate**	**Philippine Plate**	**Crust th**
Origination age	~192 Ma§	60 Ma#	
Plate area þ	ca. 130 x 10^6 km^2	ca. 9 x 10^6 km^2	
Angle of subducting slab†,‡	0–20 kmd: <10° dip 30–50 kmd: 15-25° dip >50 kmd: 30° dip	≈flat under Outer Zone, then 12° under Inner Zone, with sharp downturn at VF	
Direction of subduction	N40°W˜ (WNW)	N55°W^ (NW)	
Velocity*	8–9 cm/year	3–4 cm/year	
Volcanic Front position‡	100–120 km from trench	375 km from trench	
Kurile Arc *	8.2 cm/yr, 46.4° dip	--	~20 km
NE Japan Arc *	8.3 cm/yr, 28.7° dip	--	~36 km
Izu Arc *	6.1 cm/yr, 46.1° dip	--	~20 km
Ryukyu Arc»	--	north end 70° dip, south 40–50°	

Quaternary volcanic fronts

In accordance with the 'two-plate' regime, there are two Volcanic Fronts in Japan (Figure 6.2). Volcanic Fronts form at predictable distances from the subduction trench but depend on the angle and age of the subducting slab; the slab must be at least 100 km deep for dehydration to take place, producing fluids which lower the temperature of the overlying crust and cause rocks to melt into magma.

Table 6.1 indicates that very old parts of the Pacific Plate are subducting at a relatively steep angle in the northwestern Pacific. Thus, VF2–VF2′ sits fairly close (within 100–120 km horizontal distance) to the Japan Trench. In contrast, the Philippine Plate is young and subducting at a low angle, and VF1–VF1′ sits much further away (375 km) from the Nankai Trough. The age and angles of the subducting plates affect the amount of volcanism in the VF: NE Japan experiences "wholly intensive" volcanism due to the steep subduction of old Pacific Plate; SW Honshu has the weakest volcanism due to the gently sloped subduction of young Philippine Plate, and volcanism "becomes more vigorous" in the northern Ryukyu Arc where older Philippine Plate is being subducted "perpendicularly" (Uto & Tatsumi 1996: 256).

• VF2–VF2′ represents subduction of the Pacific Plate in the Japan–Izu Trenches, which stabilized around 700 ka in Tohoku. Three separate arcs comprise VF2–VF2′: the Kurile Arc extending into northeastern Hokkaido, the NE Japan Arc running from southwestern Hokkaido to central Honshu, and the Izu (IBM) Arc carrying on into the Bonin-Mariana Chain (Taira *et al.* 2016). The volcanics that have erupted along this front comprise the East Japan Volcanic Belt (Nakada 2018: fig. 3.2).

• VF1–VF1′ represents subduction of the Philippine Plate in the Nankai Trough in western Honshu and Kyushu from 7 Ma (Kimura *et al.* 2005: 975). VF1–VF1′ is divided into the SW Japan Arc, and the Ryukyu Arc – where the volcanics have been partially submerged due to crustal thinning and rifting behind the Ryukyu Archipelago (Taira 2001: 111). Volcanic activity along this front comprises the West Japan Volcanic Belt (Nakada 2018: fig. 3.2).

The major volcanoes in these Volcanic Fronts are discussed in Machida & Shirao (1998) chapters with photographs: Fuji and Hakone (ch. 1), Izu Oshima (ch. 2), Izu Peninsula (ch. 3), Asama (ch. 4), Towada and Hakkoda (ch. 5), SW Hokkaido volcanoes (ch. 6), and the Kyushu caldera volcanoes (ch. 7) as well as Paektu (Baekdu) (ch. 8), the Taupo zone in New Zealand (ch. 9), Hawaii (ch. 10), and Iceland (ch.11).

Zonation geochemistry

The location of the VF can change through time as the trench changes location or the subducting slab changes angle; the VF represents the point at which the slab reaches a depth of 110 km – the point where the dehydration of two important minerals (amphibole and chlorite) occurs, kickstarting the process of magma formation via fluids lowering the melting point of the mantle rock; a second depth marker at -180 km is the point of dehydration of the mineral phlogopite was also said to be operative in magma formation (Uto & Tatsumi 1996). These two dehydration points were viewed as responsible for the formation of the double rows of volcanoes that have formed in NE Japan.

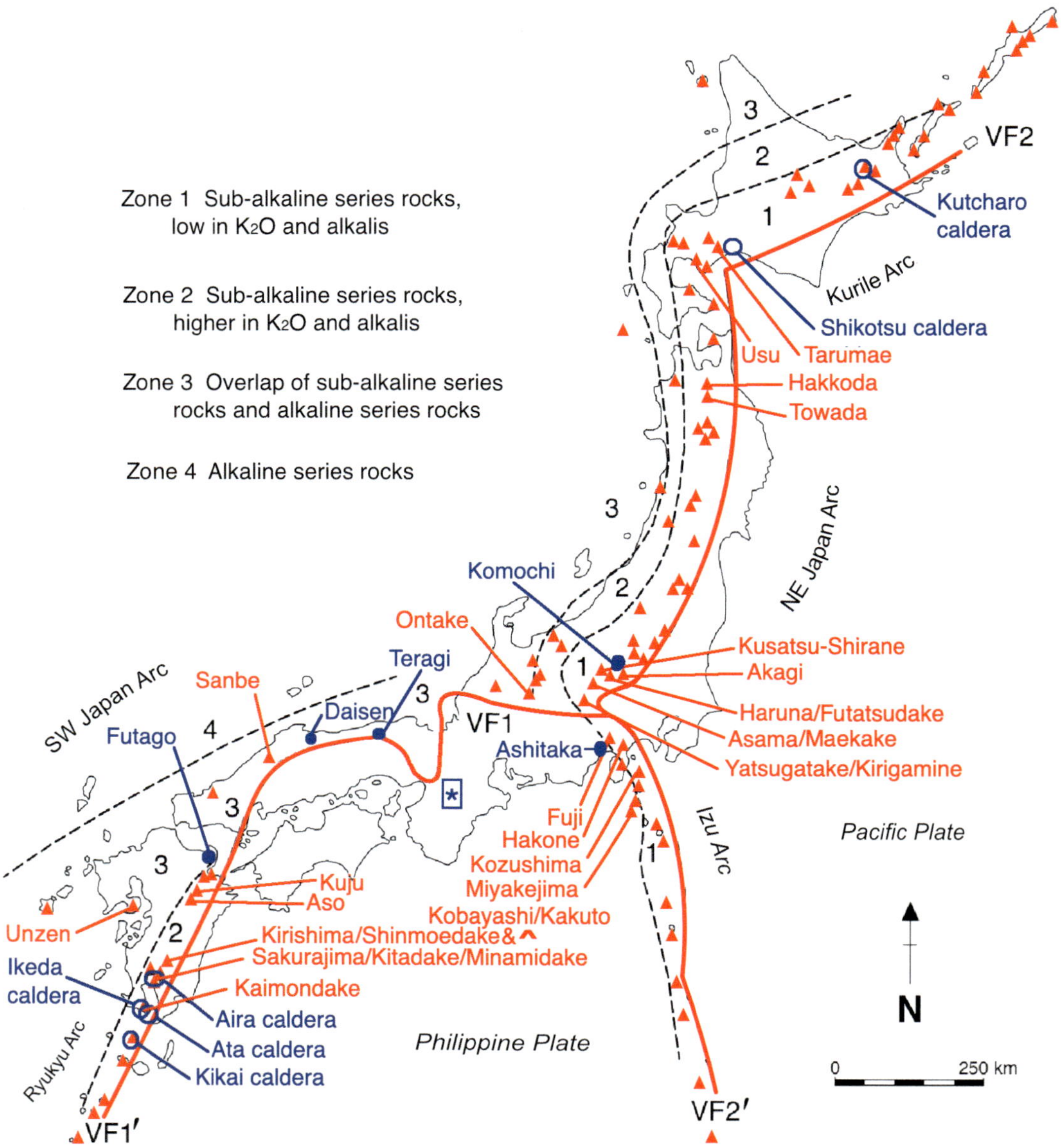

Figure 6.2 The Volcanic Fronts, geochemical zones, and Active Volcanoes in modern Japan
VF1–VF1′ = West Japan Volcanic Front, VF2–VF2′ = East Japan Volcanic Front
Named volcanoes occur in the text: triangles (red) = active; circles (blue) = inactive; open circles = caldera
* = see Figure 13.4 for Miocene volcanoes in Nara
See further Active Volcano names at [https://gbank.gsj.jp/volcano/Quat_Vol/act_map_e.html]

However, recent research on volcano distribution and chemistry in NE Japan has surpassed these simplistic ideas indicating a more complicated picture (see 'NE Japan Arc in Honshu' below).

Another aspect governing the geochemistry of the volcanic chains is the gradual change in composition from alkaline on the Japan Sea-side to calc-alkaline on the Pacific Ocean side. Zones 1 to 4 in Figure 6.2 indicate the changing geochemistry across the islands. Although these zones are coarse and do not reflect small regional variations, they illustrate how alkaline volcanics have continued in the back-arc region from Miocene Japan Sea Basin opening to the present (Uto & Tatsumi 1996). Mantle upwelling which began in the Miocene continues to feed volcanic eruptions in western Honshu, though the basic rock chemistry is more strongly alkaline now than during rifting (Tatsumi 1983; Iwamori 1991).

Finally, zonation is apparent in the occurrence of incompatible elements (such as Sn, Li, Rb, Sr), rare earth elements (REE), and radiogenic isotopic systems ($^{87}Sr/^{86}Sr$, $^{142}Nd/^{144}Nd$) – all of which have difficulty entering mineral lattices because of their size (too small, too large) or their electric charge (±) is wrong for the site (Katsui *et al.* 1978; Uto & Tatsumi 1996; Perkins *et al.* 2019: sec. 5.10.2). These increase from the forearc to the back-arc side of the islands in all volcanic rock compositions, i.e., of basaltic, andesitic, and rhyolitic, etc. (Apx 5: Table B).

Active volcanoes

In 1973 a Special Measurement Law for Active Volcanoes was introduced, and from the next year (1974), 5-year plans for predicting volcanic eruptions began In Japan. The first list of Active Volcanoes was made by the Committee for Predicting Volcanic Activity, formed in 1974 of university specialists and government workers in hazard prevention (Nakata 2005). At that time, 66 volcanoes were recorded as active within the last thousand years. The list was revised in 1999 (active within two thousand years), and in 2003 (active within ten thousand years) (Yamasato 2005). By 2006, 108 volcanoes were listed as active during the last 10,000 years (Aizawa 2006), and the current list has 111 entries (JMA 2017). Figure 6.2 illustrates their approximate locations, though some icons represent more than one volcano; 50 of these are being constantly monitored (GSJ n.d.). Since any volcano that has erupted within the Holocene is considered 'active', it is a potential hazard. Given the distribution of Active Volcanoes throughout the Japanese Islands, most of the resident population is at various levels of risk – in addition to the fact that 21 of 32 national parks in Japan are focussed on volcanoes (Nakada 2018).

The Active Volcano list does not include volcanoes that erupted prior to the Holocene, including those erupting in the Pleistocene and going back 14 million years ago into the Miocene (cf. Chapter 4). Though these earlier volcanoes are considered 'inactive' or even eroded, their products still exist in the landscape and have formed important sources of stone for artefact and architectural use from prehistoric into modern times. Moreover, eruptions prior to the 10 ka cut-off point for Active Volcanoes, such as Aira at 30,000 years ago, have been very disruptive to human society and cannot be ignored.

Volcanic Eruption Patterns

In general, the eruption styles of Active Volcanoes in Japan have been very explosive, as is common for subduction-zone volcanoes. Volcanic eruption style is commonly described via reference to outstanding and famous examples of historic eruptions, as in Figure 6.3. This scheme distinguishes

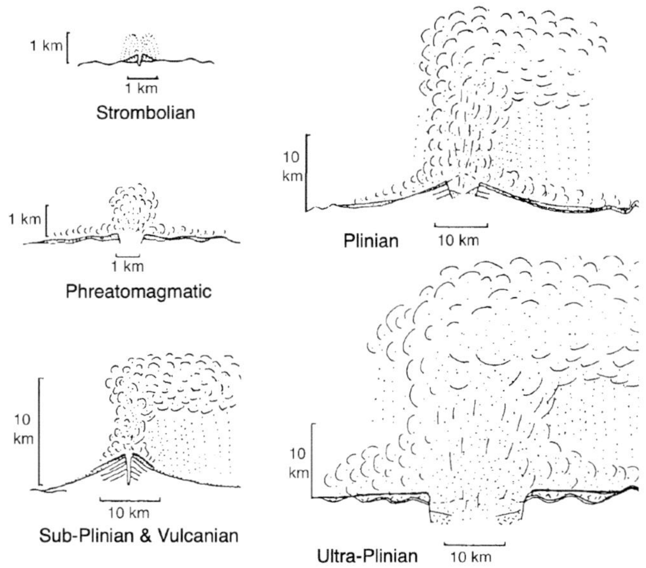

Figure 6.3 Volcanic eruption 'styles' 'phreatomagmatic' indicates magma mixing with water; 'phreatic' indicates release of steam and country-rock fragments without involving magma

between eruptions consisting mostly of flowing lava and extrusion of varying amounts and natures of tephra. However, "this classification scheme is not appropriate to describe poly-genetic volcanoes that typically display several different types of activity even in a single eruption. Eruption style on its own is thus not a very good way of classifying volcanoes as entities" (Hone *et al.* 2007: 204). For example, the nature of the Asama eruption of 1783 changed day by day (Figure 6.4), and the style varied between eruptions over time.

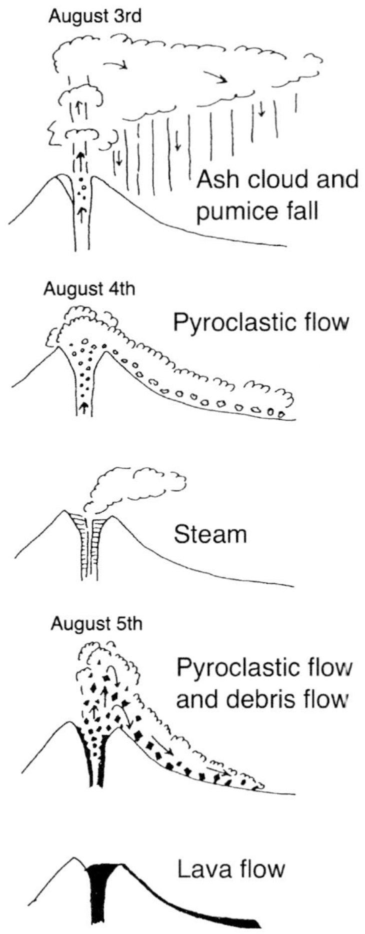

Figure 6.4 Different eruption styles of Mt Asama on the last three days of its eruption in 1783

The eruption 'style' is conditioned by two different processes:
1) the amount of gases (volatiles) in the magma and the form of their release; and 2) whether the magma encounters water when erupting. The latter is easiest to understand: if a volcano already contains a crater lake, the mixing of cold water with hot magma is like throwing water into hot oil, which every cook knows results in immediate explosion. The former is governed by the amount of silica in the magma (Apx 5: Table B) and is influenced by the *rate* of magma generation and *rate* of eruption (Yamamoto *et al.* 2018: 1). Because higher-silica magmas are increasingly viscous (sticky), they do not easily allow gas bubbles to form and escape, even as steam. When gases build up in high-silica magma, they tend to explode during decompression when reaching the surface, scattering the magma as individual crystals and glass (silica) bubble fragments, i.e., as tephra. An important caveat is that magma can 'evolve' through 'fractionation' (Fichte 2000): different elements combine to form minerals (the mineral fraction) in a certain order (Apx 5: Fig. B) with mafic minerals first then felsic with silica (quartz) being the last and often in excess. The silica fraction gathers with molasses-like viscosity, so the more silica that is generated, the more likely the eruption will be explosive. Magma renewal is cited as both the cause of changing magmatic geochemistry, which in turn affects eruption style, and the cause of

ongoing/increasing frequencies of eruptions rather than declining/decreasing frequencies (Yamamoto *et al.* 2018). These magma types influence both the energy of the eruption and the shape of the volcanic pile: most silica-poor (basaltic) magma is often emitted slowly in '*effusive* eruptions' of lava, forming shield volcanoes as in the Hawaiian Islands; their low viscosity allows gas to escape gradually. Silica-rich (rhyolitic) magmas tend generate '*explosive* eruptions' because their viscous nature does not allow continuous de-gassing; they can generate lightning during eruption (Van Eaton *et al.* 2022) and tend to form composite stratovolcanoes.

One product of high-silica rhyolitic (>70%wt SiO_2) magmatic eruptions is the volcanic glass, obsidian, having an amorphous mineral structure. Its existence contradicts the expectation that tephra is the only product of very silica-rich eruptions. As noted by Tripcevich (2007), the formation of obsidian is rare because a series of conditions must be met: extruded as a silica 'lava', it must already be devoid of excess water vapour and be immediately quenched (cooled) upon extrusion to prevent growth of a crystalline lattice; moreover, good quality obsidian is not usually extruded above ground but into rock fractures that contain no dust, ash, dirt, or rock particles. Obsidian hydration dating methods work because the volcanic glass is devoid of H_2O when erupted but then gradually rehydrates after deposition on the Earth's surface.

Kannari et al. state that geological sources of obsidian in Japan number more than 100 (2014: 49), and their map (fig. 4.2) shows around 50 named locations that were utilized for stone-working, and around 80 obsidian varieties have been identified.

The explosivity of volcanoes is measured by the VEI (Volcano Explosivity Index), ranging from 0 to 8 and based on the areal extent of tephra extruded, and also the DRE (Dense Rock Equivalent) as measured in km^3. Eruptions of VEI 7~8 are classed as 'super eruptions', the most prominent of which was the VEI 7 eruption of Mt Aso in Kyushu 87,000 years ago (Aoki 2008), emitting >600 km^3 DRE of tephra covering the entire archipelago in volcanic ash at levels 15 cm deep even in northwest Hokkaido (Nakada 2018: fig. 3.3) and resulting in a huge oval caldera 25 km NS x 18 km EW. Caldera volcanoes are usually caused by VEI 6~8 eruptions; 14 calderas have been formed in Japan within the last 150,000 years, calculated as one every 10,000 years or so (Nakada *et al.* 2018). Kimura *et al.* (2015) track 80 known tephras erupted from calderas from 5 Ma; only two occurred in the Holocene: Kikai and Towada, both of which caused regional destruction and widespread disruption of human society (Kuwahata 2019; Maruyama 2019) (cf. Chapter 8).

Yamamoto has collated eruption data on 56 individual volcanoes from the mid-Pleistocene (Yamamoto 2015), and a further analysis of 29 Quaternary volcanoes has revealed four eruption patterns, three for *cycles* and one for *volume*: predictable eruption cycles, non-predictable cycles, strictly periodic, and those with predictable eruption volumes (Yamamoto *et al.* 2018). Moreover, Yamamoto and colleagues (2018: fig. 2) found that volcanoes had increasing, decreasing, or fluctuating eruption patterns, whereby changes in magma generation led to magma increases, while stable magma plumbing systems led to magma decreases. A graph of DRE rates per 1000 years shows these different patterns: four of the highest DRE rates belong to volcanoes in the Izu Arc, while the fifth is Sakurajima in Kyushu.

This discussion leads to the point that few volcanoes can be encapsulated by a simple description: they change through time in their magma sources, geochemistry, eruption styles, and forms. The following section tours the different volcanic regions to examine how volcanic products differ across the Japanese Islands. Such differences, of course, are germane to the sourcing of artefacts made from volcanic rocks.

Regional Volcanoes

Japan is like other volcanic arcs with predominantly andesitic products (Apx 5: Table B), due to the melting of high-silica continental crust above the subducting slab (cf. Figure 2.5). However, basaltic volcanoes do exist and can form stratovolcanoes, not just shield volcanoes as in Hawaii. Each area of Japan has a distinctive repertoire of volcanic activities and products. This section briefly reviews the highlights, providing a general structure within which to investigate particular sources of volcanic materials and to understand the location and nature of current volcanic activity; more detailed histories, compositions, and analyses are in the references cited. The literature is full of reports on individual volcanoes and individual eruptions which cannot possibly be summarized here; see listings and descriptions on the JMA (2017) or GSJ (n.d.) websites. The following framework, however, allows these to be considered in the context of regional similarities and differences.

Hokkaido: Kurile & NE Japan arcs

Hokkaido has two clusters of Quaternary volcanic activity: that associated with the Kurile Arc in the northeast, and those volcanoes that are a continuation of the NE Japan Arc in the southwest. The island also incorporates the bend of VF2 as it transitions from the Kurile Arc in the north to the NE Japan Arc extending into Honshu (cf. Figure 6.2); the geochemical zonation is maintained around the bend (Table 6.2), as established early on (Katsui *et al.* 1978).

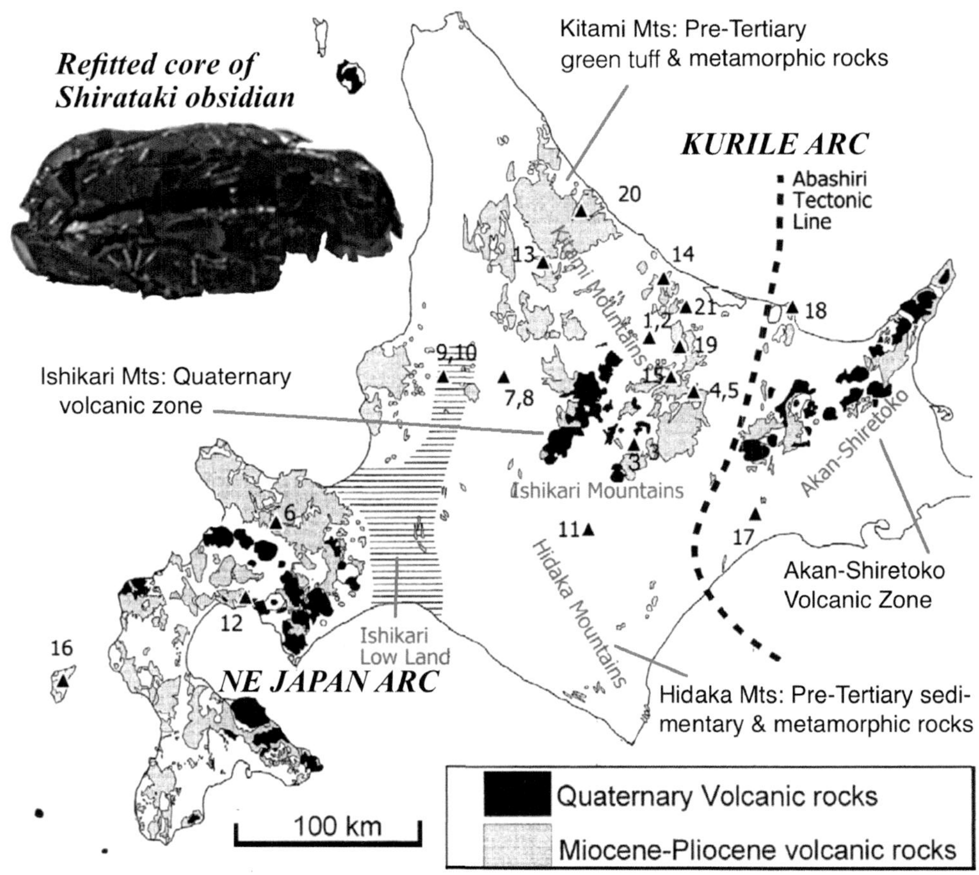

Figure 6.5 Volcanics in Hokkaido; inset: Shirataki obsidian
Triangles with numbers indicate volcano locations

Though high-silica, rhyolitic eruptions were rare (Table 6.2), prominent outcrops of obsidian occur in 19 localities in the Kurile Arc region while only two appear in the NE Japan Arc region in southwestern Hokkaido (Figure 6.5). Among these outcrops, obsidian from Shirataki is renowned for its bi-colouration: black and red (Figure 6.5: inset).

Table 6.2 Quaternary volcanic rocks in Hokkaido

	Basaltic	**Calc-alkaline**
Zone 1	mostly low alkali tholeiite: basalt, andesite	andesite, dacite, rhyolite (rare)
Zone 2	hi-Al basalt series: basalt, andesite	andesite, dacite, rhyolite (pumice)
Zone 3	alkali series: basalt	andesite

Volcanic products of the Late Miocene~Pliocene were extremely important to prehistoric inhabitants of Hokkaido and beyond for toolmaking; Shirataki obsidian was most widely traded – into Sakhalin and the Kurile Islands and into northern Honshu. Obsidian lava at Shirataki dates to 2.2 Ma and consists of ten flow units (Sano et al. 2013). Shirataki, Oketo, and Tokachi-Mitsumata were sourced by EDXRF (Hall & Kimura 2002), while these three resources and a fourth, Akaigawa, were sourced by INAA (Kuzmin et al. 2002; see also Izuho & Hirose 2010; Kuzmin et al. 2013). These four (nos. 1–6 in Figure 6.5) comprise the major obsidian resources for tool production, though Shirataki and Oketo are foremost; a further 17 primary sources have been tested, and all 21 were found to have unique geochemical fingerprints (Izuho & Sato 2007: table 1). These studies differ from those using Raman spectroscopy in the Pacific region (Kelloway et al. 2010).

NE Japan Arc in Honshu

Following the mid-Miocene Green Tuff Movement of effusive submarine volcanism and the establishment of the first andesitic volcanic chain in SW (cf. Figure 4.4), NE Japan experienced a new series of explosive eruptions occurring between 12 Ma and 1.7 Ma (from the Late Miocene into the Early Pleistocene) (Kimura *et al.* 2005: 974, 983). Calderas from great felsic eruptions between 8 and 1.7 Ma were left dotting the landscape of Tohoku (Figure 6.6) (Yoshida *et al.* 2014: 339). Lava and pyroclastic flow deposits from these calderas are still apparent in the regional geology. More than 40 calderas have been identified, most of them aligned down the centre of Tohoku, with diameters ranging from 1 to 20 km (NCEC 1989: 251). Sixteen volcanoes have formed within the last 50 ka, including 4 caldera eruptions; 11 were in southwestern Hokkaido and 5 in Tohoku (Chapman *et al.* 2009*a*: table 4.3, fig. 4.12).

Quaternary volcanism in the NE Japan Arc began around 700,000 years ago, with major Quaternary volcanoes arrayed in four chains (Figure 6.6: left), organized into two volcanic zones: three chains in the Nasu zone on the east and Chokai zone comprised of one chain on the west (Gates & Ritchie 2006: 127). A different view of the volcanic organization, however, has been offered in terms of mantle flow causing magmatic 'fingers' flowing west to east (opposite the subduction direction) which result in clusters of volcanoes along those fingers with incremental zonation of magma geochemistry (Figure 6.6: right) (Tamura 2002, 2003). Such 'fingers' have been confirmed by seismic tomography (Chapman *et al.* 2009*a*: fig. 4.2).

Hone *et al.* (2007) analyzed 59 of these Quaternary volcanoes and 129 volcanic 'edifices' in Tohoku, using the *Catalogue of Quaternary Volcanoes* (VSJ 1999, updated in JMA 2013, 2017), wherein 'volcano' was identified as "a single coherent magmatic system", and 'edifice' included the associated features of any particular volcano "where magma reached the surface", including "lava cones, lava flows, lava domes, maar deposits, collapse caldera, pyroclastic flow plateaus, volcanic necks" (Hone *et al.* 2007: 206, 207). They found that clusters 4 through 8 of volcanoes along the 'magmatic fingers' identified by Tamura had different compositions and histories, resulting from evolution through time and space in terms of magma types (basaltic, andesitic, rhyolitic, etc.). Their study identified four major groupings of volcanic attributes, each of which occurred in Tamura's clusters in varying proportions (Hone *et al.* 2007: table 4). These two analyses form the basis of what is currently known of the nature and distribution of the volcanic products in Tohoku and are important for geological sourcing of materials.

Volcanic eruptions have continued throughout the Holocene in Tohoku, with one of the most memorable being the 915 AD eruption of Towada (Box 7), causing much destruction and dislocation throughout northern Tohoku (Murakami 2019; Maruyama 2019).

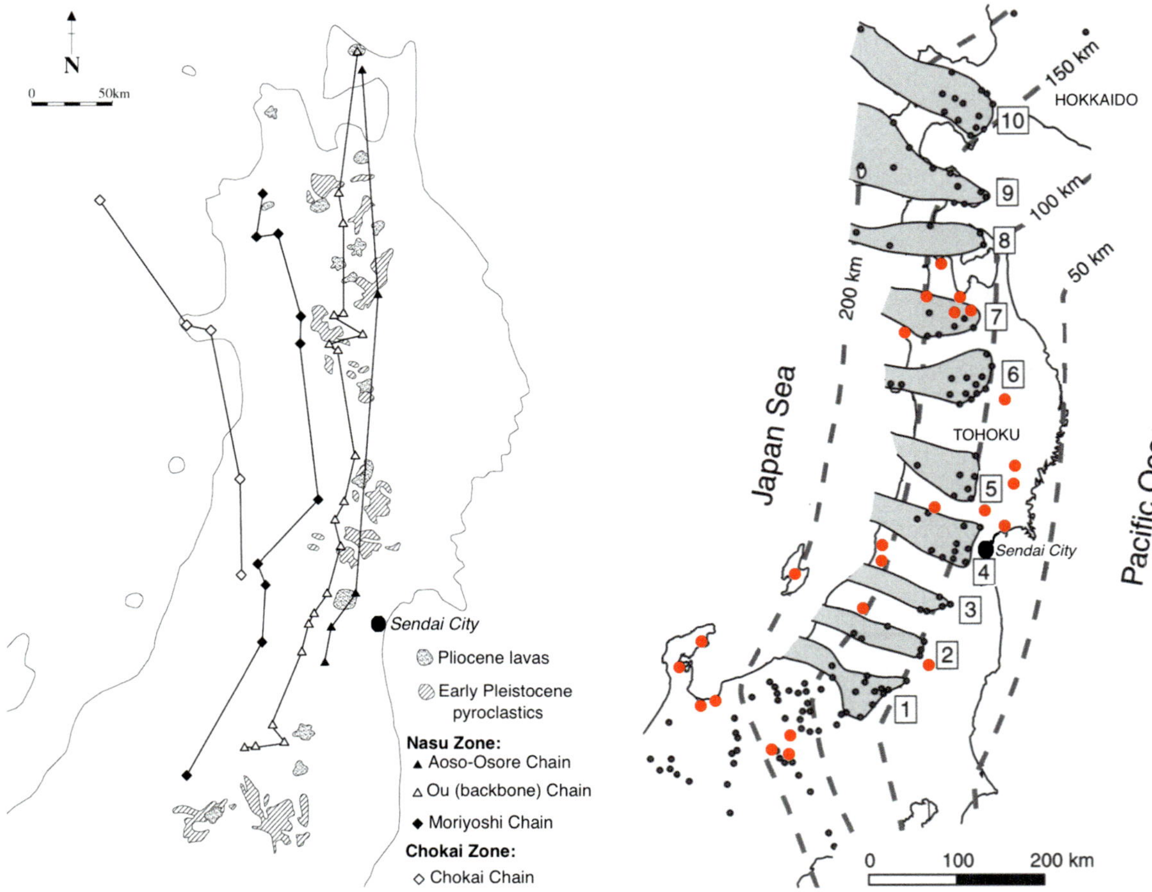

Figure 6.6 Volcanism in Tohoku, northeastern Japan

left: major volcanoes in four different volcanic chains, as presented in 1989
right: recent identification of 'magma fingers' flowing west to east, producing across-arc changing geochemistries; dashed lines indicate the depths of the downgoing Pacific Plate slab; volcanic clusters 1–10 are sited within the magma fingers which flow over basement highs, with low basement valleys forming the ca. 30–35 km gaps between clusters; red dots = obsidian resources

Obsidian resources are known throughout northeastern Japan (Figure 6.6: red dots), extending east to Sendai City and southwest into Fukui Prefecture; this distribution exceeds that of the Quaternary VF2 and thus, as in Hokkaido, Miocene through Pliocene products are also in evidence. Tsutsumi (2010: fig. 3.1) identifies 9 among the 24 major outcrops across this region as important for prehistoric tool manufacture.

BOX 7 Towada Caldera

Lake Towada fills an 11 km diameter caldera, the largest crater lake in Japan. The lake has become a major tourist area, drawing around three million people a year. The original size of the Towada volcano that erupted to form this crater is not known, but the crater is the same diameter as all of Mt St Helens. The eruptive history of this spot goes back 200,000 years (JMA 2017), varying over time between basaltic and rhyolitic products. Successive eruptions in 55, 36, and 15 ka built the current large caldera. Then between 15 and 12 ka, lava flows and eruptions continued in the southern part of the caldera. These later eruptions formed the 3 km diameter Nakanoumi crater, whose rim remnants comprise peninsulae within Lake Towada (one arm of the rim of this more recent crater can be seen in the upper right of this photo). In the upper left can be seen the dacite lava dome (Ogurayama) which grew on the Nakanoumi rim following an eruption 7600 years ago. As many as eight explosive eruptions have occurred since Nakanoumi crater formation until 915 AD, its last known eruption. This most recent phreatomagmatic eruption of VEI 5 in 915 from within Nakanoumi crater produced pumice and volcanic ash, pyroclastic flows and surge that – because of the water in the crater lake – immediately became a lahar that flowed out to the Japan Sea. This lahar acquired a folk tale telling of a hunter turned dragon residing in Lake Towada who fought explosively with a monk (the 915 AD eruption) and then flowed down the mountains to the sea (as a lahar), stopping to rest on its way (back-pooled lahar accumulations) (Kobayashi 2019).

The Nasu Volcanic Zone (Figure 6.6: left) extends through Tohoku into Central Japan. Kusatsu-Shirane, situated at its southern end in Gunma Prefecture (Figure 6.7), erupted suddenly and unexpectedly on 23 January 2018, resulting in one death.

Figure 6.7 The crater lake at Kusatsu-Shirane in 2007 before its eruption a decade later

Central Honshu

Central Honshu is the meeting point of the two lower arcs of FV2, the NE Japan Arc (a product of Pacific Plate subduction under the Okhotsk/NAmerican Plate) and the Izu Arc (a product of the Pacific Plate subducting under the Philippine Plate from the north). It is also the area where two oceanic plates subducting beneath Japan (Figure 6.8) literally overlap: the

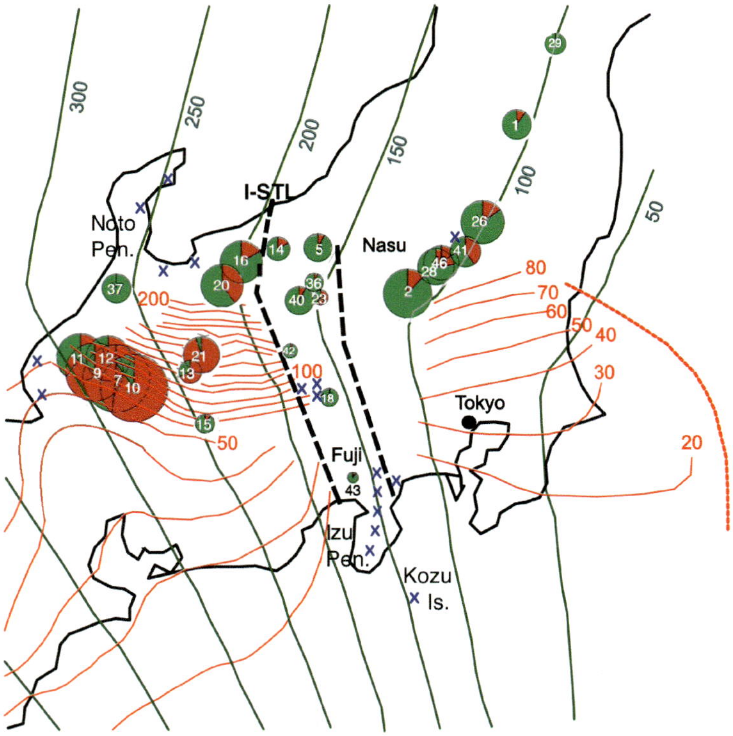

• pie charts = relative magma quantities from the Philippine Plate (red) or Pacific Plate (green)

• red contour lines = Philippine Plate (overlying), red dotted line = eastern limit

• green contour lines = Pacific Plate (underlying)

• blue x's = obsidian sources

• **black** dashed lines = Izu Arc subduction trajectory to NW, bordered on left by Itoigawa-Shizuoka TL

• volcano numbers = see table S2 in Nakamura *et al.* (2008) for names

Figure 6.8 Overlapping subducting plates under central Honshu

Philippine Plate subducting at an initially shallow level with a gentle slope then greatly steepening in the west, and the Pacific Plate subducting at a steep angle *underneath* the Philippine Plate (cf. Table 6.1). This complicated plate arrangement has ramifications for both magma formation and earthquake generation (Nakamura *et al.* 2008; Churikova 2008; Uchida *et al.* 2010; Nakamura *et al.* 2019).

Nakamura *et al.* (2008) tested the geochemistry of 28 Quaternary volcanoes in central Honshu, discovering that in this area of plate overlap, volcanoes could draw on magma generated by both downgoing plates – i.e., that the overriding Philippine Plate did not prohibit fluid flow from the underlying Pacific Plate. To discern how much of each volcano's magma was derived from dehydration of the respective plates, Pb-Sr-Nd isotopic ratios were calculated. The results show that magma from the Philippine Plate (Figure 6.8: red in pie charts) is concentrated in the western sector where the plate is between 50 and 150 km deep; nevertheless, magma from that plate is found as far north as southern Tohoku in small amounts. Conversely, magma from the Pacific Plate (green in pie charts) dominates through the Tohoku Nasu Volcanic Zone (cf. Figure 6.8: #29–2), into the North Kanto (Figure 6.8: #2) and across to the Noto Peninsula (Figure 6.8: #37). The central zone (Figure 6.8: dashed black lines) marks the northwestward subduction trajectory of the Izu Arc located on the Philippine Plate; in the southern sector, the plate is particularly dry along the arc, so the plate does not release much fluid. Therefore the volcanoes are lacking in magma from that plate, being provisioned primarily by the Pacific Plate underneath. Mt Fuji, however, is an exceptional example of mainly basaltic magma while most other arc volcanoes extrude andesitic–felsic magma/tephra. An extraordinary hypothesis has been mounted that Fuji is fed from a magma chamber of an Izu Arc volcano on the Philippine Plate that has been subducted while still active (seamounts are usually dead volcanoes when subducted) (Aoki *et al.* 2019). The extent of the Philippine Plate in the Kanto region as depicted here is challenged by Toda *et al.* (2008) but adopted by Uchida *et al.* (2016). Fuji's eruption is imminent but unpredictable, and the Japanese

Society for Civil Engineering has lamented the lack of sufficient hazard maps (Ieda & Inagaki 2003) (see Chapter 11).

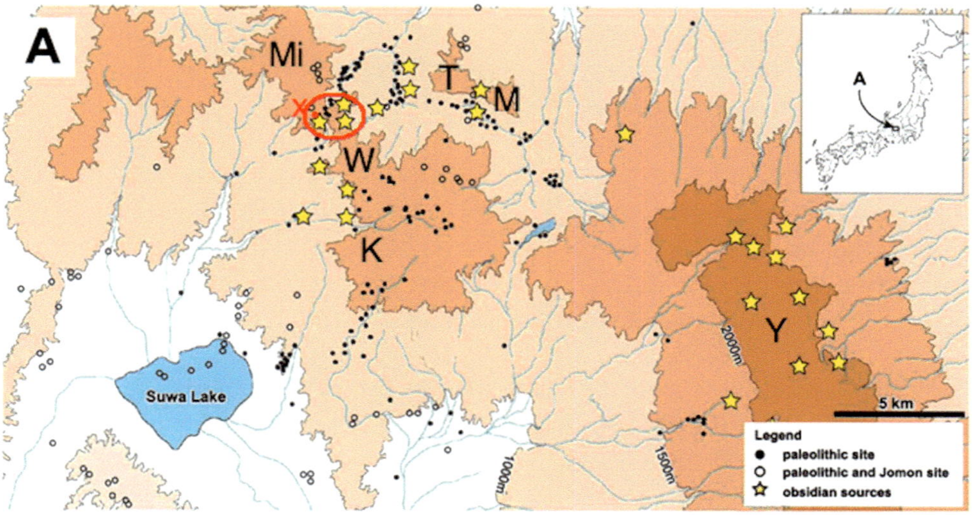

Figure 6.9 Obsidian resources and archaeological sites in Nagano Prefecture
X = Hoshikuzu-toge obsidian quarry; • COLS, red circle = Wada Pass area; Mountains: MI=Mitsumine, T=Takamatsu, M=Mushikura, W=Washigamine, K=Kirigamine, Y=Yatsugatake

In 2001, Meiji University established the Center for Obsidian and Lithic Studies (COLS) in Nagano Prefecture, following long-term excavation since 1984 at the Hoshikuso Pass obsidian quarries in conjunction with the Nagano Board of Education. 2006 saw the opening of a branch of the Meiji University Museum, the Hoshikuzu-no-sato Takayama Obsidian Experiential Museum; and from 2010 the Center has been involved in international collaborative research with scholars from Russia, the US, Korea, and Hungary.

Scores of Palaeolithic and Jomon sites have been discovered in the central mountains, with obsidian being traded out across Honshu over hundreds of kilometers. However, the Kozu Island source is notable for having been reached by boat during the Palaeolithic, as early as 38–27 ka soon after the occupation of the Japanese Islands around 42 ka, proving that seaworthy boats existed at that time (cf. Shimada 2012: 281).

Obsidian occurs in very limited areas in central Japan: around the Noto Peninsula on the Japan Sea coast; on the Izu Peninsula and Kozu Island (cf. Figure 6.8), as part of the Izu Arc volcanics; and in particular, in the central mountains near Ko-fuji (cf. Figure 6.8: #18). The last offers three major sources with outcrops on Mt Kirigamine, Mt Yatsugatake, and the Wada Pass area (red oval in Figure 6.9).

SW Japan Arc

Arc characteristics

Southwest Japan is subject to the subduction of the Philippine Plate, which moved into its current position by 15 Ma along a dextral strike-slip fault. The initiation of subduction of the plate into the Nankai Trough at that time caused the Setouchi andesite volcanics (cf. BOX 5) as well as considerable felsic volcanic activity in the Outer Zone of western Japan (see Nakada *et al.* 2016: fig. 5.1). However, the Philippine Plate has internal structural variations that have different regional impacts; it is divided into two sea basins: the Shikoku Basin and the West Philippine Basin, separated by the Kyushu–Palau Ridge (K-PR) (cf. Figure 5.11). The position of the K-PR fairly divides the Nankai Trough from the Ryukyu

Trench (Lallemand 2016: fig. 1); and while the young, hot ocean floor in the Shikoku Basin is subducting in the Nankai Trough under Shikoku and SW Honshu at a gentle angle, the older, colder West Philippine Basin floor is subducting under Kyushu at a steep dip (Mahony *et al.* 2011) – as can be seen in Figure 6.10. Tomographic surveys suggest that the K-P Ridge is subducting under Kyushu until 30 km depth, at which point it is hypothesized that the ridge detaches from the Philippine Plate and underplates the Eurasian Plate (Yamamoto *et al.* 2013).

The first products of Philippine Plate subduction were the Setouchi andesite volcanoes at 14 Ma (cf. Figure 4.4) and other felsic magmatism in the Outer Zone (Mahony *et al.* 2011: fig. 7). After a suspected hiatus between 10 and 6 Ma, reinitiation of subduction has since caused much volcanic activitiy in the western arm of the West Japan Volcanic Front (Figure 6.2: VF1) but less along the northern coast of western Honshu. These individual areas are discussed below.

As in central Honshu, southwest Japan is subject to the effects of the overlapping plate subduction, with the Pacific Plate at 200 km depth at the eastern end of VF1, descending to 600 km depth at the western tip of Honshu (Figure 6.10). Volcanoes in the SW Japan Arc can be assigned as either close to the Volcanic Front or in a back-arc position and show typical across-arc geochemical zoning (Shinjo *et al.* 2000). A systematic change in arc magmatism is noted from NE through central to SW Honshu (Nakamura *et al.* 2019), but the subduction of the Shikoku Basin on the Philippine Plate has produced more magma types than in VF2 (Kimura *et al.* 2014). Kimura and colleagues exclaim that the occurrence of 7 of the 8 possible magma types common in 'hot' subduction zones makes the SW Japan Arc unique in worldwide terms.

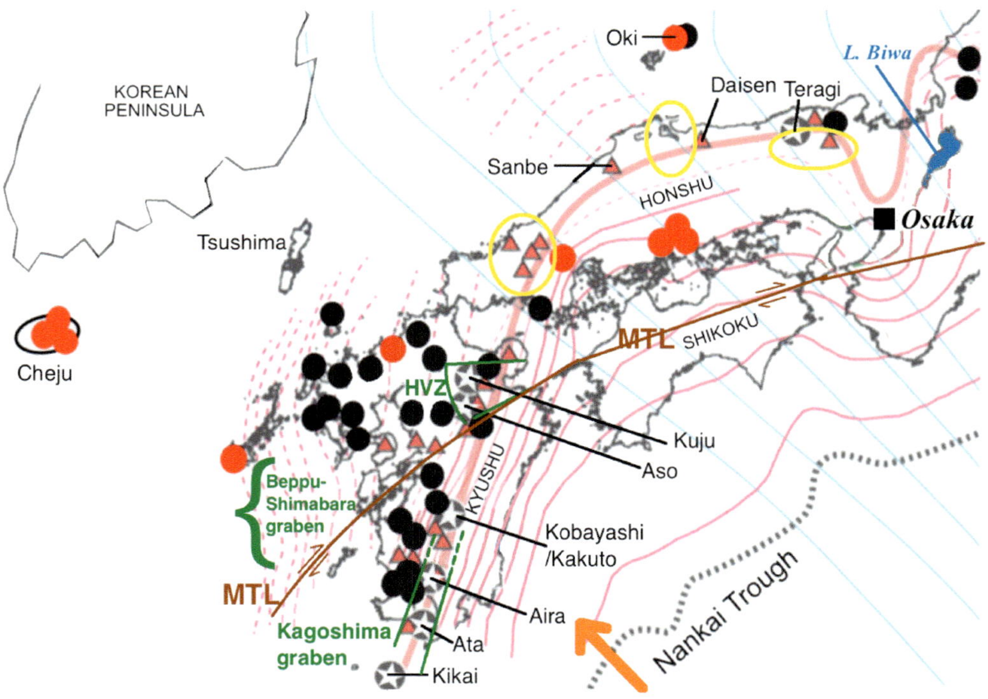

Figure 6.10 SW Japan Arc volcanics
Thick pink line = VF1; Thin pink line, solid & dashed = Philippine Plate subduction contours
Thin blue line = Pacific Plate subduction contours; Orange arrow = Kyushu-Palau Ridge;
Grey starred circles = named calderas; Red dots = plume magmas; Black dots = obsidian sources;
HVZ = Hohi Volcanic Zone; Pink triangles = volcanoes; Yellow ovals = monogenetic volcanic fields

The Pacific Plate can be seen to reach the Korean Peninsula in Figure 6.10 at a depth of 575 km; at that point the plate flattens out to the west because its further subduction is resisted by the greater viscosity of the Mantle Transition Zone (MTZ) between 410 and 660 km depth (Zhao 2015). Zhao explains how bending from the subduction angle to the stagnant flatness generates many earthquakes, causing further dehydration of hydrous minerals retained in the old Pacific slab. Just as fluids higher up in the subduction zone resulting from slab dehydration lower the melting point of crustal rock leading to magma formation and volcanic activity, the fluids released from the deeper level also lead to magma formation which is thought to be the source for the most active intraplate volcano in East Asia: Paektu/Baekdu (Mt Changbai).

Continuing plume activity

The northern coastal regions of western Honshu and northern Kyushu have been subject to mantle plume activity (Figure 6.10: red dots) that continued after Japan Sea Basin opening (cf. Chapter 4). Five volcanic systems are illustrated as active between 7 and 5 Ma, with one (Oki Island) continuing to 500 ka, while three new systems have erupted since 1.5 Ma (Kimura *et al.* 2014: fig. 3). Wet mantle plumes are responsible for producing OIB alkaline magmas in this region in different manifestations (Nakamura *et al.* 2019): spotty, zonal, and sandwiched. Several island hot spots occur in the Japan Sea and Korea Strait, particularly Oki and Cheju Islands, and plume products are seen from the Inland Sea region into northwestern Kyushu.

Northwestern Honshu coast

Although VF1 is illustrated as passing across the northwest coast of Honshu, in actuality, the Volcanic Front is underdeveloped there due to the fact that the Philippine Plate is subducting at a very shallow angle; since it does not reach the critical depth of ca. 110 km where the plate begins to dehydrate, it does not give off fluids for magma production in the continental crust above it (Mahony *et al.* 2011). Instead, the volcanics in this region are thought to derive either from a mantle plume under the subducting plate which is redirected northwards over the plate lip, resulting in alkalki basalts, or from the plume melting part of the plate itself, resulting in adakite volcanics (Connor *et al.* 2013: 9-10).

Adakites result from partial melting of altered basalt; their composition, described as having "peculiar geochemical features" by Yamamoto *et al.* (2018: 10), is featured in the two major andesitic volcanoes on the northwest coast, Daisen and Sanbe (cf. Figure 6.10). Adakites are characterized by "extreme light REE enrichment (e.g., La/Yb>9); very high Sr/Y ratios (e.g., Sr/Y>50), low yttrium and heavy REE concentrations (e.g., Y<20ppm, Yb <2ppm)" (Kelemen *et al.* 2005: 614). Kelemen *et al.* further note that there are no Mg values specific to adakites, which are often equated with sanukites (high-magnesium andesites) in the literature; however, they stress that not all adakites are high Mg# andesites, and not all high Mg# andesites are adakites. Five volcanic systems on the northern Honshu coast have produced adakites, most after 1.5 Ma; these are common to 'hot' subduction zones around the Pacific Rim (Kimura *et al.* 2014: 692), where young crust as in the Shikoku Basin is subducted.

Along the Japan Sea coast, both Daisen and Sanbe are known as 'polygenetic' volcanoes – those that erupt multiple times with changing magma chemistry. They contrast to 'monogenetic' volcanoes, which erupt only once briefly and therefore have a homogeneous magma chemistry. These traditional distinctions are being challenged by new research (e.g., Tchamabé *et al.* 2016); nevertheless, the north coast of western Honshu is known to have three separate monogenetic volcano fields comprised of ca. 10–40 volcanoes each (Figure 6.10: yellow ovals) (Connor *et al.* 2013: fig. 2.6).

Kyushu Island

Whole island concerns:

Mahony and colleagues divide Kyushu into four regions of volcanic activity (NW, SW, NE, and SE); the southeast has evidence of Miocene plutonic activity but no Quaternary activity. They chart regional production of five different magma types through four successive periods of varying volcanic intensity: 15–10, 10–6, 6–2, and 2–0 Ma (Mahony *et al.* 2011: tables 1-4; see their summary fig. 7). As expected, there is across-arc variation in rock geochemistry, with andesites produced close to the Volcanic Front, and alkali basalts produced in the back arc region.

A complicating factor in the distribution of Quaternary volcanoes in Kyushu is the presence of two zones of extension which can incite volcanism (cf. Figures 5.11, 6.10):

1) the segmentation of the island into north and south sectors by the Median Tectonic Line (MTL), with its associated Beppu-Shimabara Graben, passing through Kyushu NE–SW into the Okinawa Trough west of the Ryukyu Archipelago (Kamata 1998; Matsumoto *et al.* 2015; Mahony *et al.* 2011), and

2) the presence of the north-south Kagoshima Graben in the southern centre of the island, connecting with the Ryukyu Arc.

Both these graben areas have generated Quaternary caldera volcanoes and continue to feature active volcanism today. On the southeast side of the MTL is a shear zone that has potentially reduced shear stress on the MTL fault system (Matsumoto *et al.* 2015).

In addition to the above, the following tectonic events/processes have affected volcanism throughout Kyushu over the last 15 Ma (Mahony *et al.* 2011: 2201):

3) periods of slow subduction
4) changes in subduction direction
5) subduction of a ridge
6) shallow plate subduction
7) clockwise and anticlockwise vertical axis rotation of the forearc region
8) rollback of the subducting slab (and concomitant retreat of the trench)
9) back-arc and intra-arc rifting (including the above grabens)
10) changes in relative motion of adjacent tectonic plates.

All the above factors have contributed to the volcanism in Kyushu being far more complicated that in other areas of Japan; most of the current volcanic activity is occurring in the two grabens dissecting the island, enhanced by the subduction of the Kyushu-Palau Ridge (Chapman *et al.* 2009*b*).

Northwest Kyushu:

Most of the volcanic activity in this region occurred between 10 and 6 Ma; both ocean island basalt (OIB) and island arc basalt (IAB) were the greatest products (Mahony *et al.* 2011). Rhyolitic effusions during this time produced the Koshidake obsidian sources.

Obata and colleagues report that over 30 localities of obsidian are known on Kyushu Island, with Koshidake in the northwest having the best quality and therefore widely traded in the Jomon period – into the Ryukyus (Obata et al. 2010), and across the straits to the Korean Peninsula (Sugihara 2011).

Central Kyushu:

The MTL crossing Kyushu is referred to as the O-KTL by its geographical termini: from Oita Prefecture in the northeast to Kumamoto Prefecture in the south central. In the Pliocene the quadrant northeast of the O-KTL was an area of extension, the Hohi Volcanic Zone (HVZ) (Figures 5.11, 6.10), and volcanism occurred as lava erupting through fissures running E–W, most actively 5–4 Ma (Kamata 1998). The northern border of the Beppu-Shimabara Graben (Figure 6.10) runs along the northern edge of the HVZ, while the southern border is coincident with the O-KTL (Mahony *et al.* 2011: fig. 3). Much of current volcanic activity occurs within this graben among four Active Volcanoes: Aso, Futago, Kuju, and Unzen. Unzen is unusual in exhibiting characteristics of both arc and back-arc magmas (Chapman *et al.* 2009*b*). Volcanic activity began at Aso around 300 ka, forming the initial caldera and thereafter entailing 17 volcanic vents (AIST n.d.). Eruption of Aso-4 at 87 ka (Aoki 2008) created the pyroclastic flow plateaus that cover most of central-north Kyushu and the current Aso caldera, measuring 18 x 25 km (AIST n.d.). Among several post-caldera lava cones in the center, one continues to be active today: the Nakadake cone, which has eight 'craterlet' vents within its main crater (Hoshizumi & Miyabuchi 2012). There are 185 known eruptions of Aso confirmed for the Holocene (Smithsonian n.d.).

Southwest Kyushu:

As in Hokkaido discussed above, Kyushu Island entails two volcanic arcs: the southwestern portion of Kyushu actually belongs to the Ryukyu Arc, but it will be dealt with here as part of the island's structure. An interesting but perhaps counter-intuitive component in southern Kyushu tectonics is the effect of slab rollback on the Volcanic Front. Slab rollback occurs when the subducting plate begins to steepen – as can be seen in Figure 6.10 for southeastern Kyushu, forcing the bend (hinge) of the plate further oceanward during subduction and moving the trench oceanward; this was compounded by southern Kyushu rotating counter-clockwise ca. 30° around 2 Ma (Chapman *et al.* 2009*b*: 36). The two processes of slab rollback and rotation not only prompted extensional rifting (at the two grabens, Figure 5.11) but also resulted in migration oceanward of the critical depth of magma formation in the overriding slab. Thus, the Volcanic Front that was located along the southwestern coast of Kyushu between 6 and 1 Ma, resulting in 'flood andesites' (Shinjo *et al.* 2000) at 2 Ma before rollback, shifted the Volcanic Front eastwards to the Kagoshima Graben in the Quaternary (Mahony *et al.* 2011: figs. 8 & 9).

The Kagoshima Graben (Figure 6.10) runs north–south from the Kobayashi/Kakuto calderas in the Kirishima zone to the off-shore Kikai caldera; the volcanic chain continues into the Ryukyu Arc. The volcanoes in the Kagoshima Graben have been and continue to be some of the most volatile in the Japanese Islands:

- Aira Caldera forms the upper part of Kagoshima Bay in southern Kyushu; in the timespan of human occupation, it has erupted seven times (Nagaoka *et al.* 2001), with the most devastating at ca. 30 ka. However, a small parasite volcano, Sakurajima, has developed on its southern rim. Sakurajima is a composite stratovolcano comprised of two separate volcanoes, Kitadake and Minamidake, as well as hosting other vents (Iguchi & Nakada 2012). Periods of activity are known for Kitadake (26–24 ka, 13–5 ka), then activity shifted to Minamidake leading to four Plinian eruptions during historical times. The 1914 eruption extruded sufficient lava to connect the volcano to the Osumi Peninsula and left the Showa crater on the flank of Minamidake. The summit crater of Minamidake has been erupting almost daily since 1955, and activity recently resumed at the Showa Crater in 2008.
- Shinmoedake is a stratovolcano with a small crater in the Kirishima zone; it has erupted 84 known times during the Holocene (Smithsonian n.d.), with the 2011 event equaling the 1717 eruption size (Aizawa *et al.* 2014). Recent research has indicated that Active Volcanoes Shinmoedake and Sakurajima may share a magma chamber that underlies the northern portion of the graben; reciprocal inflation

and deflation episodes of the land surface over Aira and Kirishima were monitored by GPS before, during, and after the 2011 eruption of Shinmoedake, leading to a proposal of a common magma chamber (Brothelande *et al.* 2018).

• Ata Caldera lies under the lower part of Kagoshima Bay. Between 125 and 115 ka, it erupted several times, with more low-level activity between 60 and 50 ka, with its last eruption ca. 8 ka (Machida 2002: fig. 1b). The most significant Ata tephra was erupted between 110 and 90 ka; tephra fallout extended from the southern Ryukyus to the Kanto, but Miyazaki and Kagoshima Prefectures were subjected to widespread pyroclastic flows (Machida & Arai 1992: fig. 2.1-14). Kaimondake is a parasitic volcano of the Ata Caldera at the tip of the Satsuma Peninsula; between 2000 BC and 894 AD, there were four eruptions from Kaimondake (see Chapter 8).

• Kikai Caldera resides under the sea off the southwestern tip of Kyushu Island. Three super-eruptions are known: at 140, 95, and 7.3 ka, resulting in a double caldera (19 x 24 and 15 x 17 km) and one dome formation. Tatsumi *et al.* (2018) propose that both calderas were formed by the 7.3 ka eruption and that the dome is a post-eruption dome from that incident. The 7.3 ka eruption tephra is known as Akahoya tephra; if both calderas were responsible, their combined eruption extruded 140 km³ DRE of tephra. The Kikai Caldera is part of the Northern Ryukyu Volcanic Front.

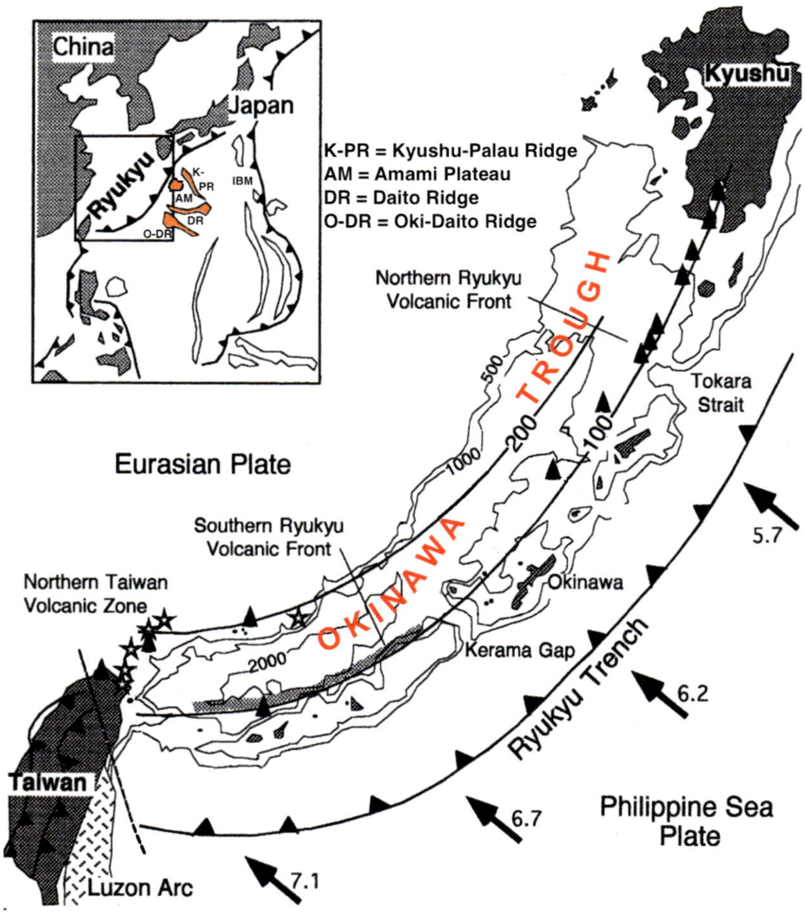

Figure 6.11 The Ryukyu Arc, stretching over 1200 km from southern Kyushu to Taiwan; arrows indicate plate speed cm/yr

Ryukyu Arc

In the above section, the volcanoes in the Kagoshima Graben were discussed in terms of the island of Kyushu; however, these volcanoes actually belong to the Ryukyu Arc. Various authors deal with them differently (Shinjo & Kato 2000; Shinjo *et al.* 2000; Nishimura *et al.* 2004). The dividing line between the SW Japan and Ryukyu volcanic arcs is placed *north* of the Kagoshima Graben within Kyushu Island.

The Ryukyu Arc itself is theoretically divided into three 'blocks' (north, central, south) which are divided by the Tokara Strait and the Kerama Gap (Figure 6.11), while volcanic activity is clustered in the Northern Ryukyu Volcanic Front, including southern Kyushu and crossing into the central block, but separately in the Southern Ryukyu Volcanic Front in the southern block (Shinjo *et al.* 2000; Shinjo & Kato 2000). Nishimura *et al.* (2004) found that these blocks moved independently, mainly in response to the opening of the Okinawa Trough behind the archipelago.

The Okinawa Trough is an incipient back-arc basin separating the Ryukyu Arc from the Eurasian continent, having begun rifting around 6 Ma (Mahony *et al.* 2011); as stated above, it continues from the Beppu-Shimabara Graben across central Kyushu. The trough is wide at the northeast end, ca. 200 km, but narrows down to 10 km at the southern end (Shinjo *et al.* 2000); the Okinawa Trough together with the Ryukyu Arc terminate at Taiwan, where the Luzon Arc is impacting. Two other points of impact are in the northern block, where the Kyushu-Palau Ridge is subducting under Kyushu Island, and the Amami Plateau impinging on the Ryukyu Archipelago (Figure 6.11: inset, orange landmasses).

Tephra

Definitions

Subduction zone volcanics tend to be more explosive than hot spot or rifting volcanics; this means that there is more tephra than lava generated. Ironically, the term 'tephra' originally meant 'ash' in Greek, and 'volcanic ash' was historically and mistakenly equated with wood ash. Even its formal term as a 'pyroclast' ('fire fragment') perhaps encourages this line of thinking. However, in a geological context, 'ash' is simply a measure of particle size: pyroclasts that are less than 2mm. Ash can further be divided into coarse ash (2 mm–0.06 mm) and fine ash (<0.06 mm) – the latter may also be called 'dust' (Lowe & Hunt 2001). Tephra is now used in geology as an umbrella term for aerial-borne particles of all size classes: ash (< 2 mm), lapilli (2–64 mm), then bombs and blocks (>64 mm). Lava bombs are ejected as fluid and solidify during flight (forming pointed ovoid shapes like an American or rugby football), while rock blocks are ejected as solids (Tucker 1991: table 10.1). Hot avalanches of volcanic ash, pumice, rock fragments, and gases (termed pyroclastic flows and pyroclastic surges) become 'ignimbrite' rocks upon solidifying; volcanic ash and other airborne deposits lithify to become tuffs and welded tuffs. These different tephra forms blanket the landscape in characteristic ways (Figure 6.12).

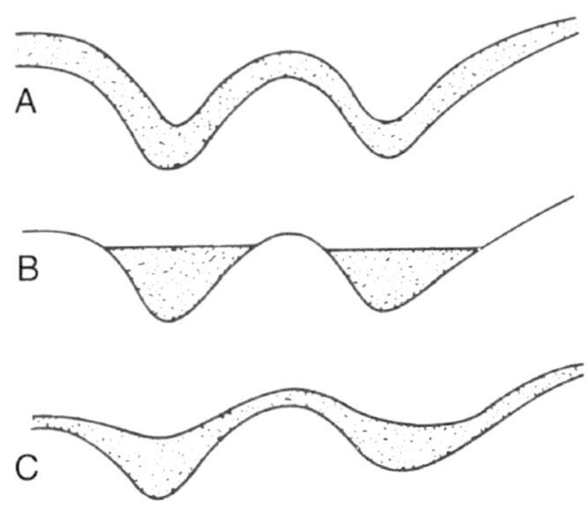

Figure 6.12 Tephra identification by deposit type

A Ash fallout blanketing
B Pyroclastic flow in hollows
C Pyroclastic surge draping the topography

Along with pyroclastic flows, lahars can be highly dangerous (Murakami 2019); these are water-saturated tephra flows that may start as landslides and then flow down hollows in the landscape – including river valleys where the tephra may be put into suspension in the water and be carried great distances. Lahar movement may coincide with deposition if the tephra is extremely wet; or rain-saturated tephra deposits may move at any time – even years – after their deposition. Barring the first possibility of wet tephra flowing upon deposition to form a primary deposit, by definition lahars are secondary deposits of tephra. They can consist of fine-grained mudflows of volcanic ash (the original meaning of *lahar* in Indonesian), or a lahar can be full of rocks and large-size tephra (then referred to as a blocky debris flow) (Vallance & Iverson 2015). Lahars and debris flows travel much slower than pyroclastic flows but can extend many kilometres.

Distributions

The regional extent of tephra deposition has been calculated for most highly explosive and important mid-to-late Quaternary eruptions in Japan, as presented in the *Atlas of tephra in and around Japan*, compiled by Machida & Arai (1992), first published in 1992, revised in 2003, 2011, and currently undergoing revision again. The three most devastating eruptions for human society were Aira (ca. 30 ka) and Kikai (7300 ka), both in southern Kyushu, and the Towada eruption in 915 AD (cf. BOX 7). Mapping of the tephra deposits by decreasing thickness over distance demonstrates that winds from the west/southwest dominate in Japan, so that explosive eruptions of tephra in northern Hokkaido do not affect the majority of Japanese citizens. However, seasonal changes of wind patterns such as winds from the northwest during winter or southeast in the summer (e.g., Sugihara 2001) and different wind directions during different eruptions of the same volcano may occur. These distributions are mapped in lobate isopachs (cf. Figures 6.14 a,b); however, animations of tephra fallout distribution from the 2010 eruption of Eyjafjallajökull in Iceland rather put paid to the idealistic view of the lobate interactions between tephra and wind (NMI 2010; djxatlanta 2010); this dissonance has yet to be incorporated into ongoing archaeological fieldwork.

Perhaps the largest eruption of tephra in Japan occurred around 87 ka from the Aso Volcano. Aso-4 tephra covered all of Japan except the northern tip of Hokkaido at depths greater than *15 cm* and buried northern Kyushu in pyroclastic flows (Machida & Arai 1992: 66-68). The second greatest eruption was from the Aira caldera at ca. 30 ka (Machida 2002: fig. 1b), causing an enormous pyroclastic flow, the 'Ito pumice', extending across southern Kyushu. The plateau that formed is 200–300 m deep, consisting of unsolidated pumice called *shirasu* ('white sand') (Figure 6.13). 500 km^3 DRE of Aira-Tanzawa tephra (AT or A-Tn) accompanying this eruption spread volcanic ash over all of the archipelago to half-way through Hokkaido (Machida & Arai 1992; Tatsumi *et al.* 2018) (Figure 6.14a). The third most devastating was the Kikai caldera eruption of 7300 BP (Akahoya tephra, K-ah), inundating south-western Japan with volcanic ash though its Koya pumice flow was mainly limited to the ocean (Figure 6.14b).

Figure 6.13 The *shirasu* (Ito pumice) of the 30 ka eruption of Aira Caldera

Names and abbreviations for naming Japanese tephras are historic rather than systematic. K-ah comes from the name of the volcano, Kikai, together with two letters from the word akahoya, meaning 'sea squirt', "a marine tunicate that has a body with orifices through which water flows into and out of a central pharynx" according to Google Translate. *Akahoya* (*Halocynthia roretzi*) is a food delicacy often called 'sea pineapple' because of bumps covering the surface; the tephra soils named after it are composed of red-coloured pumice granules. These *akahoya* tephrogenic soils are common across southern Kyushu and southern Shikoku. The AT tephra is more simply named for Aira, the caldera forming Kagoshima Bay in Kyushu, and Tanzawa, the Tanzawa Mountains in the Kanto region where it was first identified; AT is sometimes notated as A-Tn.

In the archaeological record, tephra deposits can be identified by type via their depositional patterns (cf. Figure 6.12), but this requires broadscale excavation and regional reconstructions. Moreover, tephra layers compress through time so that the amount revealed is always less than the original deposit. Explosive eruptions of tephra across the Japanese Islands have disrupted regional life several times in human history. Excavating archaeological sites buried under tephra has become a specialized sub-discipline in Japanese archaeology entitled 'tephroarchaeology' by Soda (2019). Southern Kyushu and Gunma Prefecture, northwest of Tokyo, have become the homelands of tephroarchaeology, with excavations revealing a host of detailed behaviour by both humans and animals that are not normally recovered in non-tephra areas. The results of many of these excavations are reviewed in Chapter 8.

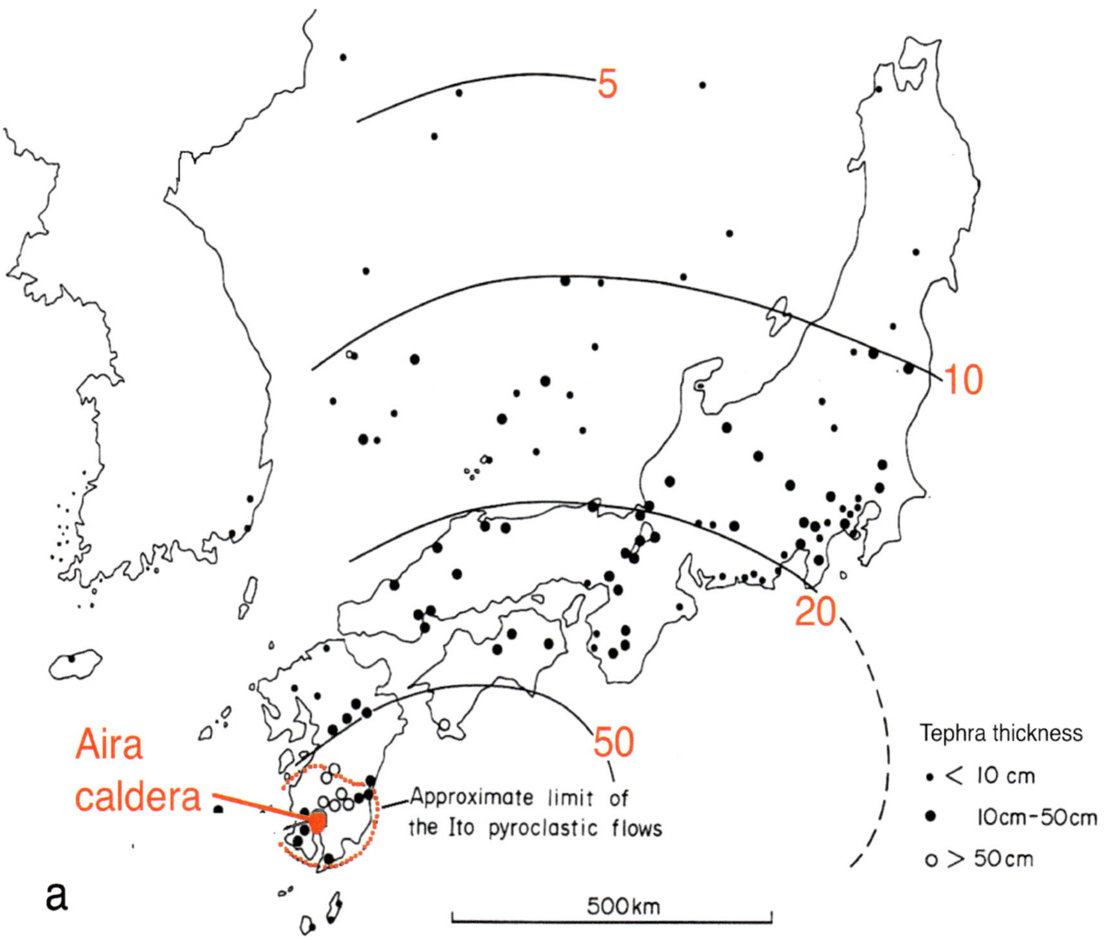

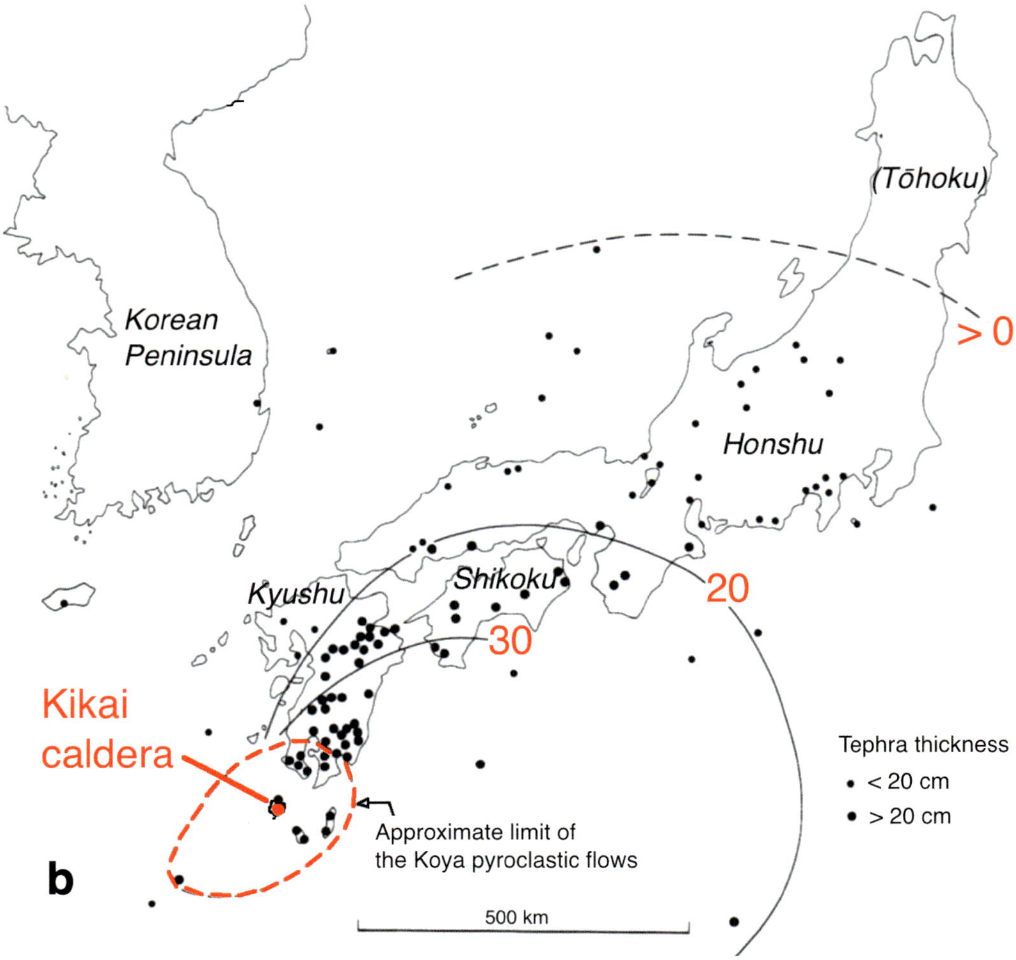

Figure 6.14 Isopach distributions of Aira (**a**) and Kikai (**b**) caldera tephra eruptions

Kanto loam

A particular type of sediment has developed in the Kanto region called the 'Kanto loam'. There have been some misunderstandings in the archaeological literature that the layers of loam are composed of volcanic ash, stemming from an 1881 suggestion that the "red earth was volcanic ash originating from volcanoes to the west" (Soda 2019: 25). However, the technical definition of a loam is that it consists of varying parts of sand, silt, and clay – measured in particle size, not genetic derivation. The Kanto loam consists of deep deposits of weathered Middle–Late Pleistocene loess (silt) and weathered tephra (clay) (Soda 2019). An early study of particle size of Kanto loam samples addressing this definitional problem gave the following distributions (Table 6.3), and it is acknowledged that the Kanto loam is not loam *sensu stricto* (KLRG 1961).

Only the range of particle size percentage is given in Table 6.3, as size did not correlate with the depth of the samples. Note that 'silt' is not represented in this analysis as these divisions do not correspond with the Wentworth scale (see Glossary), where the silt-size category occupies 0.62–0.004 mm. Reckoning by the Wentworth scale, there is even less 'sand' than shown – with that recorded falling in the 2~0.062 range (fine~very fine). Mineral types generally correspond with the type of tephra (basaltic from Fuji and Hakone in SKanto, andesitic from Asama, Haruna, and Akagi in NKanto). Also not shown here, unless included in the 'unidentified' category, is allophane – a transitional amorphous silica gel formed as tephra weathers to clay; it greatly inhibits plant growth and agriculture (see Chapter 7).

Table 6.3 Particle composition of North (Utsunomiya) and South (Tokyo) Kanto loams

	NKanto	**SKanto**
genesis	andesitic	basaltic
Composition (mm)		
sand/silt 2~0.2	% 4.4~17.9	% 1.4~7.4
0.2~0.1	5.5~7.5	2.4~6.6
0.1~0.02	5.6~22.9	3.0~16.7
0.02~0.002	28.7~42.6	39.8~47.6
clay (mm) 0.002~	21.3~42.8	21.1~50.4
minerals	relative proportions	
volcanic glass	*~***	*
plagioclase	*~****	**~***
hypersthene	**~***	*
amphibole	*	*
olivine	0	*~***
unidentified	*~**	**~***

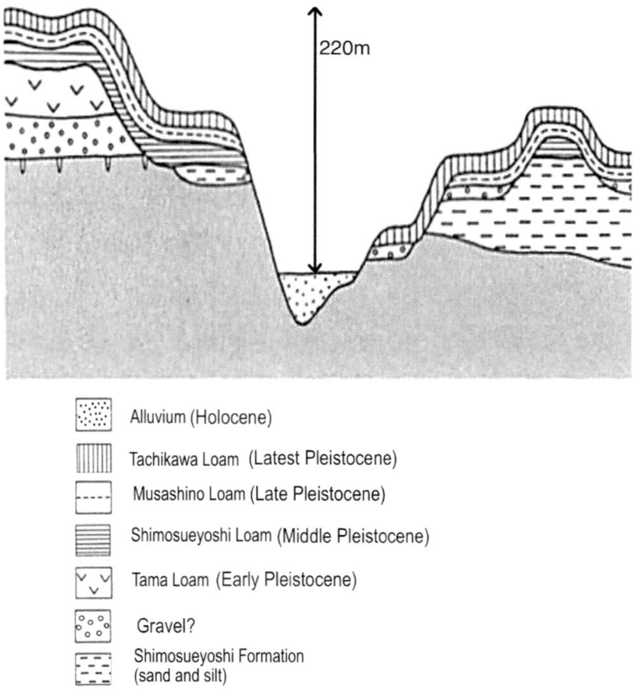

Figure 6.15 Kanto loam development and distribution

Thus, the Kanto loam has developed from three sources: *primary* ash fallout deposits from erupting regional volcanoes; aeolian (*secondary*) deposits of tephra blown in from primary deposits; and aeolian deposits of loess, mainly from the north China Mainland. Not until the 1980s was it discovered that the yellow haze *harugasumi* consisted of loess blown in from China. Although research on the Kanto loam began after its identification in the late 1800s, its stratigraphic layering was not determined until the late 1950s (KLRG 1961). Four loam strata were recognized, having formed during the Pleistocene; these can now be divided into different aged 'surfaces' (Suzuki 1972; Yoshikawa *et al.* 1981: fig. 6.3). The loams layers themselves are intercalated with palaeosols (indicating hiatuses in volcanic activity), pumice and scoria deposits from nearby volcanoes, as well as marker tephra from distant eruptions such as the Aira-Tanzawa (AT or A-Tn) ash fallout (Sugihara 1980).

The first Palaeolithic artefacts in Japan were discovered in 1946 by amateur archaeologist AIZAWA *Tadahiro in the Kanto loam near the foot of Mt Akagi in Gunma Prefecture (Aizawa 1957). Archaeologists previously believed that the loam layers were sterile, but Aizawa's discovery stimulated the formation of the Kanto Loam Research Group in 1953 and led to intensive research on the loam layers, compositions, and ages to provide context for these unexpected finds. The exposed terraces of the various layers of Kanto Loam are important to archaeologists because they contain many Palaeolithic sites dating up to ca. 40 ka (Yamaoka 2011; but see Arai 1971).*

Identification

Machida & Arai (1992) have outlined several steps leading to the identification of tephra separate from non-tephra sediments and the characterization of individual tephra-falls – first in the field and then in the lab (Table 6.4; see also Machida & Arai 2019). Proper field observations are extremely important, as they often provide defining characteristics for a tephra which cannot be known through lab analyses. This means that tephra samples should be taken by tephra specialists, not archaeologists, who can make a proper examination of in situ deposition.

Table 6.4 Analyses leading to tephra identification

Stratigraphic methods (mainly in the field)
Temporal and stratigraphic relationships of tephra stratum
Stratigraphic position
Dating by available radiometric means Dating by micro-fossils
Dating by cultural relationships
Documentation of tephra stratum characteristics Stratum thickness
Grain size
Colour
Fall unit
Materials of similar or different composition
Degree of weathering
Petrographic characteristics (mainly in the laboratory)
Rock level
Bubble form; crystal/glass ratio
Mineral composition (percentages of different minerals; heavy/light mineral ratio)
Chemical composition (whole rock [bulk], determined by XRF; wet chemical analysis)
Mineral level
Form (crystal habit; in case of volcanic glass, details of its shape type, colour, microcrystalline, or crystallite)
Refractive index (of e.g., volcanic glass, pyroxenes, amphiboles, feldspars)
State of magnetization (for strongly magnetized minerals, the Curie point, etc.)
Chemical composition (major and trace element percentages for volcanic glass and phenocrysts using EPMA, XRF, AA, ICP, INAA)
Other (e.g., lattice constant values of specific minerals; isotopic values of strontium-Sr in volcanic glass)

Laboratory procedures include optical and scanning electron microscopy, electron probe, and geochemical analyses (Lowe 2017: table 3). In Japan, the use of the refractive index of volcanic glass and minerals is an important tool for tephra characterization, combined with major and minor mineral identification (Arai, 1972; Machida 2002; Machida & Arai 2019); Nakamura (2016) additionally recommends the routine reporting of refractive index distribution and mineral proportions. Geochemical analyses with electron microprobe can reveal the chemical composition of individual mineral crystals and glass shards. As an example, the identification of a previously unknown tephra from Mt Samalas in Indonesia was made by correlating the geochemical composition of single flakes of volcanic glass from an Iceland core with the composition of pumice deposits near the Samalas volcano erupted in 1257; this correlation, however, has been taken as fact despite the tentative conclusions of the researchers (Lavigne *et al.* 2013; Lavigne & Guillet 2015). As Lane *et al.* caution (2011: 87), the chemical compositions of tephra erupted at different times from the same volcanic system may have similar compositions and "it seems therefore that composition alone is insufficient for the correlation of some widespread tephra layers: good stratigraphic information and/or robust dating control are also essential." Distinguishing different strata of tephra from a 'single' eruption that may undergo several stages, such as characterized for the Haruna Hr-FA tephra (Soda 2006), is therefore essential.

At archaeological sites in Gunma Prefecture affected by the Hr-FA tephra-fall, Soda identified 15 layers of the early 6c eruption, 7 layers of which were present at the Kanai Higashi-ura site and became crucial in understanding the relative devastation patterns by the evolving tephra deposition – from volcanic ash fallout, to pyroclastic surge, then to pyroclastic flow (Sugiyama 2019). Thus, despite the idea that a tephra deposition "can be considered instantaneous from a geological, evolutionary, and archaeological perspective" (Riede & Thastrup 2013: 1), there is a temporal sequence even within individual eruptions that can equate to the human timescale. The season and durations of tephra-falls at archaeological sites in Gunma Prefecture have been deduced from the preserved stage of the agricultural cycle and the overlay of tephra-filled footprints (Sakaguchi 2019).

Dating

The dating of tephra eruptions mainly goes by the name of 'tephrochronology'. Lowe distinguishes between broad and strict definitions – the 'broad' encompassing "all aspects of tephra studies and their application" (Lowe 2017: 4, fig. 2). The 'strict' definition is more of concern here: a tephra layer comprising an event-based distribution of sediment across the Earth's surface whose primary deposition provides a natural "stratigraphically fixed tie-point" that allows correlation between locations on a shared time-plane (Lowe 2017: 1). Correlation of distributions relies on accurate tephra characterizations or fingerprints, obtainable through such field observation and laboratory procedures as listed in Table 6.4. A particular stumbling block is in correlating near (proximal) and distant (distal, more than 800 km from the volcano) fallouts of the same eruption; if the volcanic ash includes heavy mineral crystals, these will be deposited closer to the volcano, with distal tephras composed almost entirely of magmatic bubble shards requiring glass chemistry analysis (Machida 2002). The transition of depositional style from proximal to distal, or even within the proximal area, is of changing character; Nakamura (2016) provides an insight through his table 1 recording various combinations of deposition: afa → pfl (from ash fallout to pyroclastic flow, or reverse), afa → pfa (from ash fallout to pumice fallout, or reverse), pfa → pfl (from pumice fallout to pyroclastic flow), or all three in various order.

Once identified, these key tephra layers each form contemporaneous marker beds over broad swaths of landscape, continuous or even discontinuous. Machida lists 31 major marker tephra emanating from 22 volcanoes, with the accompanying map (Figure 6.16) illustrating 13 maximum dispersals (Machida 2002: table 1, fig. 2). In these cases, one tephra type still may provide an "age-equivalent dating method" (Lowe 2017: 1) because it represents an "identical age" (an isochron) that occurs over a widespread area via the distribution of the tephra (Lowe 2011: 109).

Despite the 'chronology' in tephrochronology, the absolute age of the tephra may not be known and be dated only through association with cultural materials or relationship with other tephras in stratigraphic sequences. Such relative dates are useful in determining successions, and often it is only the tephra's association with archaeological materials that hint at its true date and even seasonality.

If the facilities and funds are available, a tephra's date may be assessed via numerous 'numerical' dating techniques (e.g., ^{14}C, Fission Track, K-Ar, U-series, Luminescence, ESR). Lowe discusses these under the term 'tephrochronometry' (Lowe 2017: 4, fig. 2) and lists those and other techniques such as dendrochronology in his table 5. Using different techniques on single tephras at different locations, however, may give a variety of dates; thus, Machida recommends that they be cross-checked against each other and with "strictly identified tephra sequences" (Machida 2002: 23). Lowe (2011) also warns of miscorrelations and erroneous age transfers among tephra deposits.

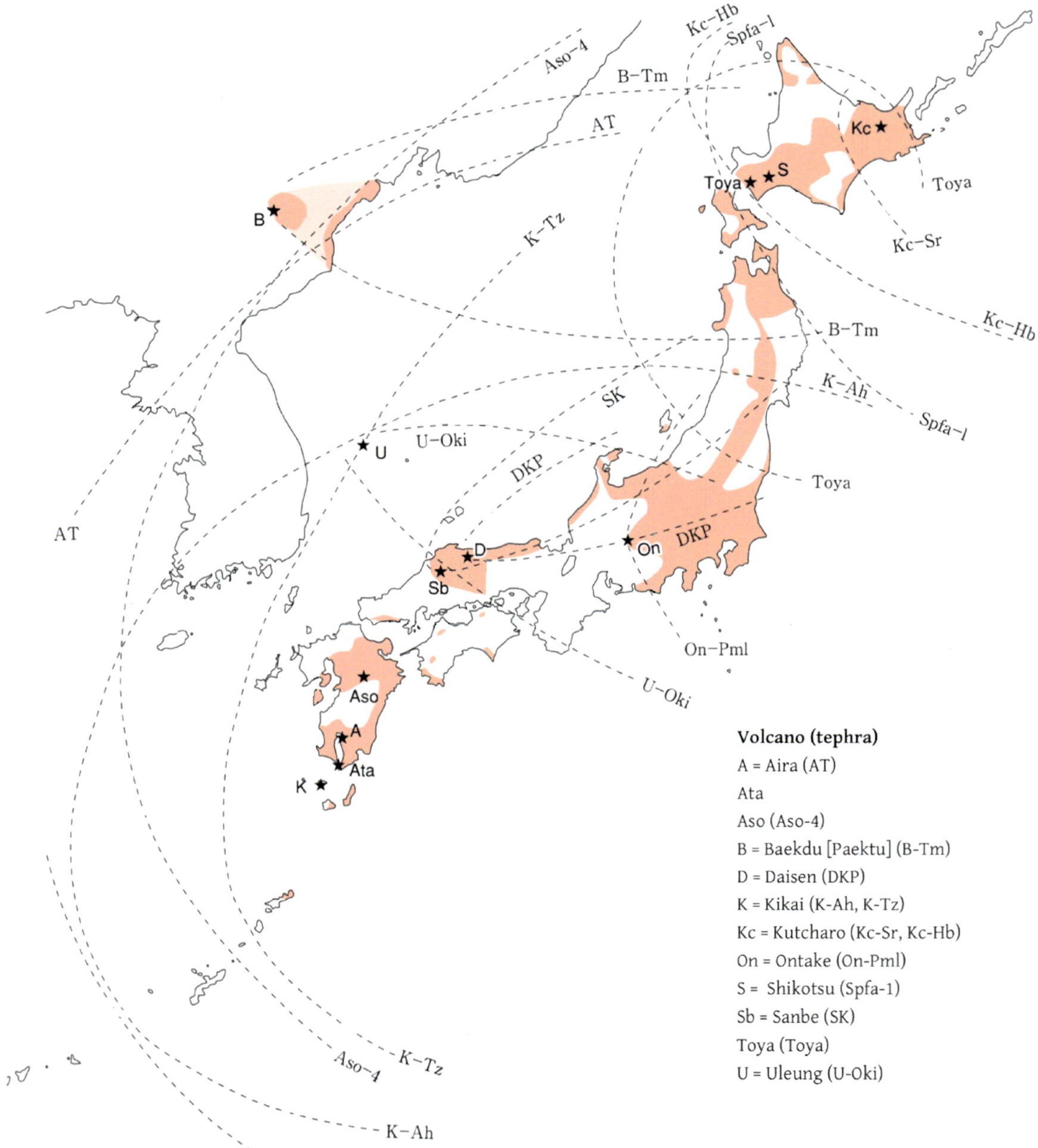

Figure 6.16 Major Late Pleistocene marker-tephra distribution maxima and volcanic soils (pink) that exceed 2 metres in thickness

In an infamous case of the misuse of tephrochronology, the amateur archaeologist FUJIMURA *Shinichi in Japan was caught on video by the Mainichi Newspaper in 2000 planting artefacts on a site in the early morning hours. After long investigation, it was discovered that he had been inserting Jomon-period lithic artefacts into Palaeolithic sites, burying them ever deeper under different tephra layers (Soda 2001) and thus increasing the depth of the Palaeolithic period in Japan from ca. 35,000 BP to 180,000 then 300,000 and finally 800,000 BP before he was caught. His activities over 25 years affected 186 archaeological sites, most of which were deemed by the Japanese Archaeological Association (JAA 2006) to be compromised, and the starting date of the Early Palaeolithic in Japan was brought back to 1970s levels, ca. 35,000-40,000 BP. Known as the "Palaeolithic Hoax" (Bleed 2000; Keally 2002, 2006, 2010; Kaner 2002; Hudson 2005), Fujimura's activities deeply rocked public faith in Palaeolithic archaeology, the discipline of archaeology, and academia in general in Japan; recovery is ongoing.*

Retrospection

Geological maps routinely list igneous rocks separately from periodized successions of sedimentary rocks. This is primarily due to the fact that igneous rocks tend to penetrate other rock successions and do not match the country rock chronology. Moreover, igneous rocks renew themselves continuously with changing geochemistry, creating their own palimpsests within the geological record. Thus, in this discussion, igneous rocks have been treated separately from the geological belts in which they occur rather in the formative sequences of Chapter 3.

The current Chapter has endeavoured first to briefly introduce pre-Plio-Pleistocene granitic and volcanic rocks that occur in the various geological belts of Japan and then to examine the Quaternary Volcanic Fronts and their products. By reviewing each region of Japan, a considerable variety has been evidenced in magma genesis, chemical compositions, and volcanic tephra and rock types relevant to palaeo-economics. Tephra is a category unto itself, which has the capacity to change the terrain and modify ecosystems across broad swaths of land. Eruptions in the Quaternary have had and continue to have a direct impact on people's lives. The basics set forth here enabling tephra research are operationalized in many archaeological studies and are important for agriculture past and present (cf. Chapter 7).

One problem that may be noted in coming to grips with the volcanic landscape of Japan is that volcanic activity has been occurring over a very long time, resulting in multitudes of volcanoes and volcanic vents with different names but often related through the magma plumbing system. It might be recalled that Sakurajima is parasitic on the Aira caldera rim, while Towada crater lake occupies two successive calderas in the same place. Kobayashi and Kakuto craters are so close together they form a figure-eight, while Yatsugatake is a collection of several volcanoes/vents. These overlaps and relationships often make it difficult to separate out "a" volcano from a volcanic field such as Kirishima – not all are as independent looking as Mt Fuji. But even Fuji overlies two predecessors in the same place, while Mt Usu is a collection of several stratovolcanoes each with their own name (Matsumoto & Nakagawa >2010).

Figure & Table Sources

Figure 6.1 modified by GLB after (Yonekura *et al.* 2001: fig. 1.3.2)
Figure 6.2 base map and zonation modified by GLB after (Miyashiro 1973: fig. 15.3), but VF2's east end trajectory after (Kimura *et al.* 2015: fig. 1); volcanoes after (GSJ n.d.); see [https://gbank.gsj.jp/ volcano/Quat_Vol/ act_map.html] for names in Japanese; click on names in map to see romanized spellings
Figure 6.3 modified by GLB after (Machida & Arai 1992: fig. 18)
Figure 6.4 modified by GLB after (Aramaki 1993: fig. 7)
Figure 6.5 map modified by GLB after (Izuho 2007: fig. 2); photo from display at Shirataki Geopark Museum

Figure 6.6 left: modified by GLB after (NCEC 1989: fig. 6.1); right: modified by GLB after (Tamura 2002: fig. 9), with red dots for obsidian after (Tsutsumi, 2010: fig. 3.1)
Figure 6.7 photo by author, 2007
Figure 6.8 compiled from (Nakamura *et al.* 2008: fig. 4) for western Philippine Plate contours and discs; (Uchida *et al.* 2010: fig. 7b) for eastern Philippine Plate contours & edge
Figure 6.9 modifed by GLB from Figure 2 at (COLS n.d., from Sugihara 2011); mountain locations from (Oyokawa 2013)
Figure 6.10 modified by GLB from (Kimura *et al.* 2015: fig. 1); plume products from (Nakamura *et al.* 2019: fig. 8a); obsidian locations from (Tsutsumi 2010: fig. 3.1); monogenetic volcanic fields from (Connor *et al.* 2013: fig. 2.6); K-PR location from (Chapman *et al.* 2009*b*: fig. 4.1 and Yamamoto *et al.* 2013: fig. 1)
Figure 6.11 modified by GLB from (Shinjo & Kato 2000: fig. 1)
Figure 6.12 modified by GLB from (Tucker 1991: fig 10.7)
Figure 6.13 By Ray_go (Own work) CC BY-SA 3.0 unported, via Wikimedia Commons [https://commons.wikimedia.org/wiki/File%3AShirasu_Cliff.jpg]
Figure 6.14a,b modified by GLB from (Machida & Arai 1992: figs. 2.1-2, 2.1-6)
Figure 6.15 modified by GLB from (Yoshikawa *et al.* 1981: fig. 6.1)
Figure 6.16 modified by GLB from (Machida 2002: fig. 2)

BOX 6 Photo by Soica2001 (Own work) [Public domain], via Wikimedia Commons [https://commons.wikimedia.org/wiki/File%3ALake_Towada_from_Ohanabe_2008.jpg]

Table 6.1 compiled from: (**§** Nakanishi & Ishihara 2015); (# Lallemand 2016); († Zhao 1997 – Pacific Plate); (‡ Isozaki *et al.* 2010: 84, fig. 5 – Philippine Plate); (^ Kelloway *et al.* 2010); (* Nakamura *et al.* 2019); (**þ** Blake & Argles 2003); (» Shinjo *et al.* 2000); (˜ Dominguez 1998); oldest extant age of Pacific Ocean is 167 Ma (Neall & Trewick 2008)
Table 6.2 compiled from (Katsui *et al.* 1978)
Table 6.3 compiled from (Tada *et al.* 1963)
Table 6.4 extracted from (Machida & Arai 2019)

References

AIZAWA, Koji (**2006**) "Present Situation of Monitoring, Prediction and Information for Volcanic Disaster Mitigation in Japan." *The 6th Joint Meeting of the UJNR Panel on Earthquake Research* 8–11 November 2006, Tokushima (PPT in English).

AIZAWA, Koki; Takao KOYAMA & Hideaki HASE *et al.* (**2014**) "Three-dimensional resistivity structure and magma plumbing system of the Kirishima Volcanoes as inferred from broadband magnetotelluric data." *Journal of Geophysical Research: Solid Earth* 119.1: 198-215 (in English) [DOI: 10.1002/2013JB010682].

AIZAWA, Tadahiro (**1957**) "The sequence of various stone industries found in 'Kanto Loam' beds near Mt. Akagi." *The Quaternary Research* 1: 17-22 (in Japanese with English title and abstract).

ALLEN, CM (**2001**) "25 years of I and S granites: Introduction." *Journal of Volcanology and Geothermal and Research* 48.4: 487-488.

AOKI, K (**2008**) "Revised age and distribution of ca. 87 ka Aso-4 tephra based on new evidence from the northwest Pacific Ocean." *Quaternary International* 178.1: 100-118 (in English).

AOKI, Yosuke; Kae TSUNEMATSU & Mitsuhiro YOSHIMOTO (**2019**) "Recent progress of geophysical and geological studies of Mt. Fuji Volcano, Japan." *Earth-Science Reviews* 194: 264-282 (in English).

ARAI, Fusao (**1971**) "Lithic artifact-bearing layers in the Kanto loam in North Kanto, Japan: problems on the so-called Early Palaeolithic in view of geology." *Daiyonki Kenkyū* 10.4: 317-335 (in Japanese with English title and abstract).

---- (1972) "Identification of particular tephras by means of refractive indices of orthopyroxenes and hornblendes." *Daiyonki Kenkyū* 11.4: 254-269 (in Japanese with English title and abstract).

ARAMAKI, Shigeo (**1993**) "Progressions in the 1783 eruption of Mt. Asama and their problems", pp. 83-110 in Arai (ed.) (1993) *Volcanic ash archaeology*. Tokyo: Kokoin Shoin (in Japanese).

BARNES, Gina L & Tsutomu SODA (eds) (**2019**) *TephroArchaeology in the North Pacific*. Oxford: Archaeopress.

Blake, Stephen & Tom Argles (**2003**) *Growth and destruction: continental evolution at subduction zones.* Milton Keynes: The Open University.

Bleed, Peter (**2000**) "Special Report: Digging out of the scandal" [www.t-net.ne.jp/~keally/ Hoax/ keally-bleed2.html] (dead link).

Brothelande, E; F Amelung & Z Yunjun [Zhang, Yunjun] *et al.* (**2018**) "Geodetic evidence for interconnectivity between Aira and Kirishima magmatic systems, Japan." *Scientific Reports* 2018.8: #9811, 9810pp. [DOI: 10.1038/s41598-018-28026-4].

CCQVJ (Committee for the Catalogue of Quaternary Volcanoes in Japan) (**1999**) *Catalogue of Quaternary Volcanoes in Japan, v. 1 (CD-R/OM).* Tokyo: Volcanological Society of Japan.

Chapman, Neil; Mick Apted & John Beavan *et al.* (**2009*a***) *Development of methodologies for the identification of volcanic and tectonic hazards to potential HLW repository sites in Japan: the Tohoku case study.* NUMO-TR-08-30. Tokyo: Nuclear Waste Management Organization of Japan (NUMO).

---- (**2009*b***) *Development of methodologies for the identification of volcanic and tectonic hazards to potential HLW repository sites in Japan: the Kyushu case study.* NUMO-TR-09-02. Tokyo: Nuclear Waste Management Organization of Japan (NUMO).

Churikova, Tatiana (**2008**) "Geochemistry: seeing through tectonic plates." *Nature Geoscience* 1: 350-351.

COLS (Center for Obsidian and Lithic Studies) (**2015**) "Home Page" [https://www.meiji.ac.jp/cols/english/].

---- (**n.d.**) "Maps" [www.meiji.ac.jp/cols/english/maps/maps.html].

Connor, Charles B; Laura Connor & Olivier Jaquet *et al.* (**2013**) *Spatial and temporal distribution of future volcanism in the Chugoku region.* NUMO-TR-13-03. Tokyo: Nuclear Waste Management Organization of Japan.

djxatlanta (**2010**) "Iceland's Eyjafjallajökull Volcanic Ash Plume [1080p]" [https://www.youtube.com/watch?v =HI1e7-sa84Y].

Dominguez, S; S Lallemand & J Malavieille *et al.* (**1998**) "Oblique subduction of the Gagua Ridge beneath the Ryukyu accretionary wedge system." *Marine Geophysical Researches* 20.5: 383-402.

Fichte, Lynn S (**2000**) "The evolution of igneous rocks and the significance of their tectonic associations" [https://csmgeo.csm.jmu.edu/geollab/Fichter/IgnRx/IgEvweb.html] (dead link).

Gates, Alexander E & David Ritchie (**2006**) *Encyclopedia of earthquakes and volcanoes.* New York: Infobase.

GSJ (Geological Survey of Japan, AIST) (**n.d.**) "Volcanoes of Japan" [https://gbank.gsj.jp/volcano/index_e.htm].

Hall, Mark & Hideaki Kimura (**2002**) "Quantitative EDXRF studies of obsidian sources in northern Hokkaido." *Journal of Archaeological Science* 29: 259-266.

Harker, Alfred (**1989**) "III.–Notes on the physics of metamorphism." *Geological Magazine* 6.1: 15-20.

Hone, DWE; SH Mahony & RSJ Sparks *et al.* (**2007**) "Cladistic analysis applied to the classification of volcanoes." *Bulletin of Volcanology* 70.2: 203-220.

Hoshizumi, H & Y Miyabuchi (**2012**) "084. Asosan." *JMA (Japan Meteorological Association)* [https://www.data.jma.go.jp/vois/data/tokyo/STOCK/souran_eng/volcanoes/084_asosan.pdf].

HJKK (Hōdenseki Jigyō Kyōdō Kumiai) (**2018**) "*Houdenseki* (Tatsuyama-ishi)" (in Japanese) [https://web.pref.hyogo.lg.jp/sr09/jibasan/31.html].

Hudson, Mark (**2005**) "For the people, by the people: postwar Japanese archaeology and the Early Paleolithic hoax." *Anthropological Science* 113.2: 131-139.

Ieda, Hitoshi & Hideki Inagaki (eds) (**2003**) *Preparing for volcanic eruption: when will Fuji-san erupt?* Tokyo: Doboku Gakkai (in Japanese).

Iguchi, M & S Nakada (**2012**) "JMA 090. Sakurajima." *JMA (Japan Meteorological Association)* (in English) [https://www.data.jma.go.jp/vois/data/tokyo/STOCK/souran_eng/volcanoes/090_sakurajima.pdf].

Isozaki, Yukio; Kazumasa Aoki & Takaaki Nakama *et al.* (**2010**) "New insight into a subduction-related orogen: a reappraisal of the geotectonic framework and evolution of the Japanese Islands." *Gondwana Research* 18: 82-105 (in English).

Iwamori, Hikaru (**1991**) "Zonal structure of Cenozoic basalts related to mantle upwelling in southwest Japan." *Journal of Geophysical Research* 96.B4: 6157-6170 (in English).

Izuho, Masami & Hiroyuki Sato (**2007**) "Archeological obsidian studies in Hokkaido, Japan: retrospect and prospects." *Bulletin of the Indo-Pacific Prehistory Association* 27: 114-121 (in English).

Izuho, Masami & Wataru Hirose (**2010**) "A review of archaeological obsidian studies on Hokkaido Island (Japan)", pp. 9-25 in *Crossing the straits: prehistoric obsidian source exploitation in the North Pacific Rim*, ed. by YV Kuzmin & MD Glascock. Oxford: Archaeopress (in English).

JAA (Japanese Archaeological Association) (**2006**) *Inspection of the Early and Middle Palaeolithic Problem*. Tokyo: Japanese Archaeological Association (in Japanese).

Janousek, V; B Bonin & WJ Collins *et al.* (eds) (**2020**) *Post-Archean granitic rocks: petrogenetic processes and tectonic environments*. GSL Special Publications 491. London: Geological Society of London.

JMA (Japan Meteorological Agency) (**2017**) "Information & interpretation of Active Volcanoes throughout the country" [https://www.data.jma.go.jp/vois/data/tokyo/open-data/data_index.html].

JMA (Japan Meteorological Agency) & VAJ (Volcanological Society of Japan) (**2013**) "National catalogue of the active volcanoes in Japan (the fourth edition, English version)." [https://www.data.jma.go.jp/svd/vois/data/tokyo/STOCK/souran_eng/menu.htm].

Kamata, Hiroki (**1998**) "Quaternary volcanic front at the junction of the South-west Japan Arc and the Ryukyu Arc." *Journal of Asian Earth Sciences* 16.1: 67-75 (in English).

Kaner, Simon (**2002**) "Trouble in the Japanese Lower and Middle Palaeolithic." *Before Farming* 2.4: 1-23.

Katayama, Ikuo; Ken-ichi Hirauchi & Junichi Nakajima (**2010**) "Variability of subduction processes beneath Japan." *Chigaku Zasshi* 119.2: 205-223 (in Japanese with English title and abstract).

Katsui, Y; Y Oba & S Ando *et al.* (**1978**) "Petrochemistry of the quaternary volcanic rocks of Hokkaido, north Japan." *Journal of the Faculty of Science, Hokkaido University, Series IV* 18.3: 449-484 (in English).

Keally, Charles T (**2002**) "Dirt and Japan's Early Palaeolithic hoax." *Sophia International Review* 24: 33-41 [www.t-net.ne.jp/~keally/Hoax/sir-dirt.html].

---- (**2006**) "Summary of the 'Final Report' on Japan's archaeological hoax." *Current Research in the Pleistocene* 23: 6-9.

---- (**2010**) "Index to materials on Japan's Early Palaeolithic hoax." *Japanese Archaeology* [www.t-net.ne.jp/~keally/Hoax/hoax.html].

Kelemen, PB; J Hanghøj & AR Greene (**2005**) "One view of the geochemistry of subduction-related magmatic arcs, with an emphasis on primitive andesite and lower crust", ch. 22 in *The crust*, ed. by RL Rudnick. Amsterdam: Elsiever.

Kelloway, Sarah J; Nina Kononenko & Robin Torrence *et al.* (**2010**) "Assessing the viability of portable Raman spectroscopy for determining the geological source of obsidian." *Vibrational Spectroscopy* 53.1: 88-96.

Kimura, Jun-ichi; James B Gill & Tomoyuki Kunikiyo *et al.* (**2014**) "Diverse magmatic effects of subducting a hot slab in SW Japan: results from forward modeling." *Geochemistry Geophysics Geosystems* 15: 691-739 (in English) [DOI: 10.1002/2013GC005132].

Kimura, Jun-Ichi; Yoshitaka Nagahashi & Yasufumi Satoguchi *et al.* (**2015**) "Origins of felsic magmas in Japanese subduction zone: geochemical characterizations of tephra from caldera-forming eruptions <5 Ma." *Geochemistry, Geophysics, Geosystems* 16: 2147-2174 (in English) [DOI: 10.1002/2015GC005854].

Kimura, Jun-Ichi; Robert J Stern & Takeyoshi Yoshida (**2005**) "Reinitiation of subduction and magmatic responses in SW Japan during Neogene time." *Geological Society of America Bulletin* 117.7/8: 969-986.

KLRG (Kanto Loam Research Group) (**1961**) "The Kanto loam and the Quaternary chronology of the Kanto district, Japan." *Earth Science: Journal of the Association for the Geological Collaboration in Japan* 54: 20-29 (in English).

Kobayashi, Masaru (**2019**) "Akita's three lakes: the legend of Hachiro Taro and the dragon", pp. 112 in Barnes & Soda (eds) (2019).

Kannari, T.; Nagai, M. and Sugihara, S. "The effectiveness of elemental intensity ratios for sourcing obsidian artefacts using energy dispersive X-ray fluorescence spectrometry: a case study from Japan", pp. 47- in *Methodological Issues for Characterisation and Provenance Studies of Obsidian in Northeast Asia*, ed. by A. Ono; Y.D. Kuzmin; M.D. Glascock and Y. Suda. Oxford: British Archaeological Reports.

Kuwahata, Mitsuhiro (**2019**) "The gigantic explosive eruption in Holocene epoch: response of hunter-gatherers to the disaster by the Kikai-Akahoya eruption." *Kikan Kokogaku* 146: 34-37 (in Japanese).

Kuzmin, Yaroslav V; MC Glascock & H Sato (**2002**) "Sources of archaeological obsidian on Sakhalin Island (Russian Far East)." *Journal of Archaeological Science* 29: 741-749.

Kuzmin, YV; MD Glascock & M Izuho (**2013**) "The geochemistry of the major sources of archaeological obsidian on Hokkaido Island (Japan): Shirataki and Oketo." *Archaeometry* 55.3: 355-369.

Lallemand, Serge (**2016**) "Philippine Sea Plate inception, evolution, and consumption with special emphasis on the early stages of Izu-Bonin-Mariana subduction." *Progress in Earth and Planetary Science* 3: #15 [DOI: 10. 1186/s40645-016-0085-6].

Lane, CS; SCE Blockley & J Mangarud *et al.* (**2011**) "Was the 12.1 ka Icelandic Vedde ash one of a kind?" *Quaternary Science Reviews* 33.6 February (2012): 87-99 [DOI: 10.1016/j.quascirev.2011.11.011].

Lavigne, Franck; Jean-Philippe Degeai & Jean-Christophe Komorowski *et al.* (**2013**) "Source of the great A.D. 1257 mystery eruption unveiled, Samalas volcano, Rinjani Volcanic Complex, Indonesia." PNAS 110.42: 16742-16747 [DOI: 10.1073/pnas.1307520110].

Lavigne, Franck & Sébastien Guillet (**2015**) "The unknown AD 1275 stratospheric eruption: climatic impacts in Europe and tentative volcanic source", pp. 63-74 in *Past vulnerability: volcanic eruptions and human vulnerability in traditional societies past and present*, ed. by F Riede. Aarhus, DK: Aarhus University Press.

Lowe, David J (**2011**) "Tephrochronology and its application: a review." *Quaternary Geochronology* 6: 107-153 [DOI: 10.1016/j.quageo.2010.08.003].

---- (**2017**) "Connecting, synchronising, and dating with tephras: principles and applications of tephrochronology in Quaternary research", pp. 269-300 in *14th Quaternary techniques short course - measuring change and reconstructing past environments*, ed. by MJ Vandergoes *et al.* Lower Hutt, NZ: National Isotope Centre, GNS Science.

Lowe, DJ & JB Hunt (**2001**) "A summary of terminology used in tephra-related studies", pp. 17-22 in *Tephras: chronology, archaeology*, ed. by ET Juvigné & J-P Raynal. Les dossiers de l'Archaéo-Logis No. 1. Goudet: CDERAD éditeur.

Machida, Hiroshi (**2002**) "Volcanoes and tephras in the Japan area." *Global Environmental Research* 6: 19-28 (in English).

Machida, Hiroshi & Fusao Arai (**1992**) *Atlas of tephra in and around Japan.* Tokyo: University of Tokyo Press, revised editions in 2003, 2011 (in Japanese with English title).

---- (**2019**) "Appendix E: The history of tephra characterization in Japan", translated by GL Barnes, pp. 305-316 in Barnes & Soda (eds) (2019).

Machida, Hiroshi & Motomaro Shirao (**1998**) *An illustrated natural history of volcanos.* University of Tokyo Press (in Japanese).

Mahony, SH; LM Wallace & M Miyoshi *et al.* (**2011**) "Volcano-tectonic interactions during rapid plate-boundary evolution in the Kyushu region, SW Japan." *GSA Bulletin* 123.11/12: 2201-2223 [DOI: 10.1130/B30408.1].

Maruyama, Koji (**2019**) "Volcanic disaster research using archaeological methods: 10th-century eruption and population movements in northern Tohoku, Japan", pp. 140-157 in Barnes & Soda (eds) (2019).

Matsumoto, A & M Nakagawa, (>**2010**) "15. Usuzan." [www.data.jma.go.jp/vois/data/tokyo/STOCK/souran_eng/volcanoes/015_usuzan.pdf]. JMA (Japan Meteorological Association).

Matsumoto, Satoshi; Shigeru Nakao & Takahiro Ohkura *et al.* (**2015**) "Spatial heterogeneities in tectonic stress in Kyushu, Japan and their relation to a major shear zone." *Earth, Planets and Space* 67.172: 9pp (in English) [DOI: 10.1186/s40623-015-0342-8].

Miyashiro, Akiho (**1973**) *Metamorphism and metamorphic belts.* London: George Unwin & Allen (translation of 1965 Japanese text).

Moreno, Teresa; Simon Wallis & Tomoko Kojima (eds) (**2016**) *The geology of Japan.* London: Geological Society of London.

Murakami, Yoshinao (**2019**) "Katakai-Ienoshita site, Akita Prefecture, buried by the lahar accompanying the 10th-century eruption of Mt Towada", pp. 97-113 in in Barnes & Soda (eds) (2019).

Nagahashi, Y; Y Satoguchi & S Yoshikawa (**2000**) "Correlation and stratigraphic eruption age of the pyroclastic flow deposits and wide spread volcanic ashes intercalated in the Pliocene-Pleistocene strata, central Japan." *Chishitsugaku Zasshi* 106: 51-69

Nagaoka, Shinji; Mitsuru Okuno & Fusao Arai (**2001**) "Tephrostratigraphy and eruptive history of the Aira caldera volcano during 100–30 ka, Kyushu, Japan." *Chishitsugaku Zasshi* 107.7: 432-450 (in Japanese with English title and abstract).

Nakada, Setsuya (**2018**) "Volcanic archipelago: volcanism as a geoheritage characteristic of Japan", pp. 19-28 in *Natural heritage of Japan: geoheritage, geoparks and geotourism*, ed. by A Chakraborty *et al.* Cham, CH: Springer International Publishing.

Nakada, Setsuya; Takahiro Yamamoto & Fukashi Maeno (**2016**) "Miocene-Holocene volcanism", pp. 273-308 in T Moreno *et al.* (eds) (2016).

Nakajima, Takashi; Masaki Takahashi & Teruyoshi Imaoka *et al.* (**2016**) "Granitic rocks", pp. 251-272 in T Moreno *et al.* (eds) (2016).

Nakamura, Hitomi; Hikaru Iwamori & Jun-ichi Kimura (**2008**) "Geochemical evidence for enhanced fluid flux due to overlapping subducting plates." *Nature Geoscience* 1(June): 380-385 [DOI: 10.1038/ngeo200].

Nakamura, Hitomi; Hikaru Iwamori & Mitsuhiro Nakagawa *et al.* (**2019**) "Geochemical mapping of slab-derived fluid and source mantle along Japan arcs." *Gondwana Research* 70: 36-49 (in English).

Nakamura, Yugo (**2016**) "Stratigraphy, distribution, and petrographic properties of Holocene tephras in Hokkaido, northern Japan." *Quaternary International* 397: 52-62 (in English).

Nakanishi, M & T Ishihara (**2015**) "Tectonic evolution of the Jurassic Pacific Plate." *American Geohysical Union, Fall Meeting 2015 abstracts*: V21A-3017 (in English) [https://ui.adsabs.harvard.edu/abs/2015AGUFM.V21A3017N/abstract].

Nakata, Setsuya (**2005**) "The forefront of volcanic disaster prevention software technology", pp. 130-137 in *Kazan funka ni sonaete*, ed. by *Doboku Gakkaishi* Editorial Committee. Tokyo: Agricultural Engineering Society (in Japanese).

NCEC (*Nihon no Chishitsu* Editorial Committee) (**1989**) *Geology of Japan 2: Tohoku region*. Tokyo: Kyoritsu (in Japanese).

Neall, Vincent E & Steven A Trewick (**2008**) "The age and origin of the Pacific islands: a geological overview." *Philosophical Transactions of the Royal Society B* 363: 3293-3308.

Nishimura, Sou; Manabu Hashimoto & Masataka Ando (**2004**) "A rigid block rotation model for the GPS derived velocity field along the Ryukyu arc." *Physics of the Earth and Planetary Interiors* 142.3/4: 185-203 (in English).

NMI (Norwegian Meteorological Institute) (**2010**) "Eyjafjallajökull's ash cloud." Modified and uploaded by J Crowe. [www.youtube.com/watch?v=K-4TB47N3Y&feature=emb_title] (dead link).

Obata, Hiroki; Isao Morimoto & Susumu Kakubuchi (**2010**) "Obsidian trade between sources on northwestern Kyushu Island and the Ryukyu Archipelago (Japan) during the Jomon period", pp. 57-73 in *Crossing the straits: prehistoric obsidian source exploitation in the North Pacific Rim*, ed. by YV Kuzmin & MD Glascock. Oxford: BAR Publishing.

Ono, Akira; Michael D Glascock & Yaroslav V Kuzmin *et al.* (eds) (**2014**) *Methodological issues for characterisation and provenance studies of obsidian in Northeast Asia*. BAR International Series 2620. Oxford: Archaeopress.

Oyokawa, Minoru *et al.* (**2013**) "Field survey of obsidian sources in the Kirigamine region of the central highlands, Nagano Prefecture, Japan: Wada-toge-nishi and Tsuchiyabashi-higashi obsidian sources." *Natural Resource Environment and Humans* 3.77-94 (In Japanese with English title and abstract).

Perkins, Dexter; Kevin R Henke & Adam C Simon *et al.* (**2019**) *Earth materials: components of a diverse planet*. Boca Raton, LA: CRC Press.

Riede, Felix & Mads Bakken Thastrup (**2013**) "Tephra, tephrochronology and archaeology – a (re-)view from Northern Europe." *Heritage Science* 1.15: 17pp. [www.heritagesciencejournal.com/content/1/1/15].

Sakaguchi, Hajime (**2019**) "Archaeological investigation of the seasonality and duration of the 6th-century eruptions from Mt Haruna", pp. 184-191 in Barnes & Soda (eds) (2019).

Sano, Kyohei; Keiji Wada & Hiroyuki Sato *et al.* (**2013**) "Geological and geochemical study in Shirataki obsidian lava complex, northern Hokkaido, Japan", pp. 93 in *5TH ARHEOINVEST SYMPOSIUM 'Stories Written in Stone' International Symposium on Chert and other Knappable Materials, Iași, 20-24 August 2013*, ed. by ON Crandell & V Cotiugă. Iași, RO: Editura Universității „Alexandru Ioan Cuza" din Iași (English abstract).

Shimada, Kazutaka (**2012**) "From gathering to mining: prehistoric human activities around obsidian sources in central Japan." *Archeometriai Mühely* 2012.4: 229-245 (in English with Hungarian title and abstract).

Shinjo, Ryuichi & Yuzo Kato (**2000**) "Geochemical constraints on the origin of bimodal magmatism at the Okinawa Trough, an incipient back-arc basin." *Lithos* 54: 117-137 (in English).

SHINJO, Ryuichi; Jon D WOODHEAD & Janet M JERGT (**2000**) "Geochemical variation within the northern Ryukyu Arc: magma source compositions and geodynamic implications." *Contributions to Mineralogy and Petrology* 140: 263-282 (in English).

Smithsonian (Smithsonian Institution National Museum of Natural History) (**n.d.**) "Global volcanism program" [http://volcano.si.edu/].

SODA, Tsutomu (**2001**) "Utilization and issues in volcanic ash chronology within Japanese archaeology." *Chishitsu to Chosa* 2001.3: 19-25 (in Japanese).

---- (**2006**) "The great eruptions of Mt Haruna in the Kofun period: searching for shifts in eruption styles from tephra", pp. 54-66 in *Haruna 30-nen monogatari*, ed. by Kamitsuke no Sato Hakubutsukan. Takasaki City: Kamitsuke no Sato Museum (in Japanese).

---- (**2019**) "Tephroarchaeology and its history in Japan", pp. 24-40 in BARNES & SODA (eds) (2019).

SONEHARA, T & S HARAYAMA (**2007**) "Petrology of the Nohi Rhyolite and its related granitoids: a Late Cretaceous large silicic igneous field in central Japan." *Journal of Volcanology and Geothermal Research* 167: 57-80 (in English).

SUEOKA, Shigeru; Hiroyuki TSUTSUMI & Takahiro TAGAMI (**2016**) "New approach to resolve the amount of Quaternary uplift and associated denudation of the mountain ranges in the Japanese Islands." *Geoscience Frontiers* 7.2: 197-210 (in English) [DOI: 10.1016/j.gsf.2015.06.005].

SUGIHARA, Shigeo (**1980**) "The present state and problems in Kanto tephrochronology research." *Sundai Shigaku* 50: 286-295 (in Japanese).

---- (**2001**) "Volcanic eruptions of Tenjouzan Volcano at Kozushima and Mukaiyama Volcano at Niijima, Izu Islands." *Chigaku Zasshi* 110.1: 94-105 (in Japanese with English title).

---- (**2011**) "Kyushu Koshidake, hypotheses on the source of obsidian artifacts from Hirazora and Suzuoke sites." *Sundai Shigaku* 142: 111-137 (in Japanese).

SUGIHARA, S (ed.) (**2011**) *Keiko-X sen bunseki souchi ni yoru kokuyoseki sei ibutsu no gensanchi suitei: Kiso deta shu* (2) [Sourcing obsidian artifacts using X-ray fluorescence analyzer. Data book 2], Tokyo: Meiji Daigaku Kobunkazai Kenkyusho [Cultural Properties Laboratory, Meiji University], 294pp.

SUGIYAMA, Hidehiro (**2019**) "Investigations into the Kofun period disasters caused by Mt Haruna eruptions", pp. 168-182 in BARNES & SODA (eds) (2019).

SUZUKI, Masao (**1972**) "Chronology of the Tachikawa loam as established by fission track and obsidian hydration dating." *The Quaternary Research* 11: 280-287 (in English).

TADA, Atsushi; Hajime TAKENAKA & Koichi SOMA *et al.* (**1963**) "On the characteristics of the particles of the Kanto Loam volcanic ash soil." *Transactions of The Japanese Society of Irrigation, Drainage and Rural Engineering (JSIDRE)* 7: 14-21 (in English).

TAIRA, Asahito; Y OHARA & SR WALLIS *et al.* (**2016**) "Geological evolution of Japan: an overview", pp. 1-24 in T MORENO *et al.* (eds) (2016).

TAKAHASHI, Masaki & Kazuo SAITO (**1997**) "Miocene intra-arc bending at an arc-arc collision zone, central Japan." *The Island Arc* 6: 168-182 (in English).

TAMURA, Y (**2002**) "Hot fingers in the mantle wedge: new insights into magma genesis in subduction zones." *Earth and Planetary Science Letters* 197: 105-106 (in English).

---- (**2003**) "Some geochemical constraints on hot fingers in the mantle wedge: evidence from NE Japan", pp. 105-106 in *Intra-oceanic subduction systems: tectonic and magmatic processes*, ed. by RD LARTER & PT LEAT. London: The Geological Society of London.

TATSUMI, Y (**1983**) "Generation of arc basalt magmas and thermal structure of the mantle wedge in subduction zones." *Journal of Geophysical Research* 88.B7: 5815-5825 (in English).

TATSUMI, Yoshiyuki; Keiko SUZUKI-KAMATA & Tetsuo MATSUNO *et al.* (**2018**) "Giant rhyolite lava dome formation after 7.3 ka supereruption at Kikai caldera, SW Japan." *Scientific Reports* 8.#2753: unp. [DOI: 10.1038/s41598-018-21066-w].

TCHAMABÉ, Boris Chako; Gabor KERESZTURI & Karoly NÉMETH *et al.* (**2016**) "How polygenetic are monogenetic volcanoes: case studies of some complex maar-diatreme volcanoes", pp. 355-389 in *Updates in volcanology: from volcano modelling to volcano geology*, ed. by K NÉMETH. Croatia: InTech.

TODA, Shinji; Ross S STEIN & Stephen H KIRBY *et al.* (**2008**) "A slab fragment wedged under Tokyo and its tectonic and seismic implications." *Nature Geoscience* 1: 771-776.

TRIPCEVICH, Nicholas (**2007**) *Quarries, caravans, and routes to complexity: prehispanic obsidian in the South-Central Andes*. Doctoral dissertation, Department of Anthropology, University of California Santa Barbara.

TSUTSUMI, T (**2010**) "Prehistoric procurement of obsidian from sources on Honshu Island (Japan)", pp. 27-55 in *Crossing the straits: prehistoric obsidian source exploitation in the North Pacific Rim*, ed. by YV KUZ'MIN & MD GLASCOCK. Oxford: Archaeopress.

TUCKER, Maruice E (**1991**) *Sedimentary petrology: an introduction to the origin of sedimentary rocks*. Oxford: Blackwell Scientific. 2nd Revised edition; 3rd edition 2009.

UCHIDA, Naoki; Youichi ASANO & Akira HASEGAWA (**2016**) "Acceleration of regional plate subduction beneath Kanto, Japan, after the 2011 Tohoku-oki earthquake." *Geophysical Research Letters* 43: 9002-9008 [DOI: 10.1002/2016GL070298].

UCHIDA, Naoki; Toru MATSUZAWA & Junichi NAKAJIMA *et al.* (**2010**) "Subduction of a wedge-shaped Philippine Sea plate beneath Kanto, central Japan, estimated from converted waves and small repeating earthquakes." *Journal of Geophysical Research* [DOI: 10.1029/2009JB006962].

UTO, Kozo & Yoshiyuki TATSUMI (**1996**) "Quaternary volcanism of the Japanese Islands." *The Island Arc* 5.3: 250-261 (in English).

VALLANCE, James W & Richard M IVERSON (**2015**) "Lahars and their deposits", pp. 649-664 in *The Encyclopedia of Volcanoes*, 2nd edition, ed. by H SIGGURDSSON. Elsiever Science.

VAN EATON, Alexa R; Cassandra M SMITH & Michael PAVOLONIS *et al.* (**2022**) "Eruption dynamics leading to a volcanic thunderstorm—the January 2020 eruption of Taal volcano, Philippines." *Geology* 50.4: 491-495 [DOI: 10.1130/G49490.1].

VSJ (Volcanological Society of Japan, Committee for Catalog of Quaternary Volcanoes in Japan) (**1999**) *Catalogue of Quaternary volcanoes.* Tokyo: The Volcanological Society of Japan.

WADA, Seigo (**1986**) "Political interpretations of stone coffin production in protohistoric Japan", pp. 349-374 in *Windows on the Japanese past*, ed. by R PEARSON *et al.* Ann Arbor: University of Michigan, Center for Japanese Studies.

WINTER, JD (**2001**) *An Introduction to igneous and metamorphic petrology*. Prentice Hall, NJ: Prentice Hall.

YAMAMOTO, Takahiro (**2015**) "Cumulative volume step-diagrams for eruptive magmas from major Quaternary volcanoes in Japamn." *GSJ Open-File Report, no. 613* [www.gsj.jp/researches/openfile/openfile2014/openfile0613.html].

YAMAMOTO, Takahiro; Takashi KUDO & Osamu ISIZUKA (**2018**) "Temporal variations in volumetric magma eruption rates of Quaternary volcanoes in Japan." *Earth, Planets and Space* 70.65: 12 pp. [DOI: 10.1186/s40623-018-0849-x].

YAMAMOTO, Yojiro; Koichiro OBANA & Tsutomu TAKAHASHI *et al.* (**2013**) "Imaging of the subducted Kyushu-Palau Ridge in the Hyuga-nada region,western Nankai Trough subduction zone." *Tectonophysics* 589: 90-102.

YAMAOKA, Takuya (**2011**) "Transitions in the Early Upper Palaeolithic: an examination of lithic assemblages on the Musashino Upland, Tokyo, Japan." *Asian Perspectives* 49.2 (Fall 2010): 251-279 (in English).

YAMASATO, Hitoshi (**2005**) "Classification of prior volcanoes and levels of volcanic activity", pp. 138-145 in *Kazan funka ni sonaete*, ed. by *Doboku Gakkaishi* Editorial Committee. Tokyo: Agricultural Engineering Society (in Japanese).

YONEKURA, Nobuyuki; Sohei KAIZUKA & Michio NOGAMI *et al.* (eds) (**2001**) *Introduction to Japanese geomorphology.* Regional geomorphology of the Japanese Islands, vol. 1. Tokyo: University of Tokyo Press (in Japanese).

YOSHIDA, Takeyoshi; Jun-ichi KIMURA & Ryoichi YAMADA *et al.* (**2014**) "Evolution of late Cenozoic magmatism and the crust–mantle structure in the NE Japan Arc", pp. 335-387 in *Orogenic andesites and crustal growth*, ed. by A GÓMEZ-TUENA *et al.* London: Geological Society.

YOSHIKAWA, Torao; Sohei KAIZUKA & Yoko OTA (**1981**) *The landforms of Japan*. Tokyo: University of Tokyo Press (in English).

ZHAO, Dapeng (**1997**) "Morphology of the subducting slab boundary in the northeastern Japan arc." *Physics of the Earth and Planetary Interiors* 102.1/2: 89-104 (in English).

---- (**2015**) "The 2011 Tohoku earthquake (Mw 9.0) sequence and subduction dynamics in Western Pacific and East Asia." *Journal of Asian Earth Sciences* 98: 28-49.

Chapter 7

Tephra-derived Soils of Japan in Comparative Context

Around the world, soils derived from volcanic ash are judged to be fertile and good for agriculture (e.g., Bellwood 1987; Fiantis *et al.* 2019). Japan, however, seems to be an anomaly in that tephra-derived soils have traditionally been avoided because they were classed as 'infertile' soils (*yasechi*). Granted, the major crop in Japan is wet-rice – which is generally grown in lowland regions where alluvial (redeposited) sediments are fashioned into paddy-fields. And if paddy-fields are located in upland regions, they still involve irrigation which brings in nutrients and silt to enrich the paddy-fields. But only the modern application of lime and phosphate have allowed the growing of crops with good yields in upland dry-field areas. There are several variables that determine land use patterns on tephrogenic soils, and soil chemistry plays a large part. This Chapter comprises a foray into sorting out some of these variables, challenging some of the myths about the fertility of volcanic ash soils, and hopefully laying the groundwork for further investigation of tephra-derived soil use in prehistory.

Tephrogenic Soils

Rocks and **tephra** weather to clays through two different processes: via simple water/rock interaction over time, *or* through the activity of plants (Velde & Meunier 2008). Whereas buried rocks (including tephra) may be transformed to clay over millennia via the first process (think granite → kaolinite), the surfaces of tephra strata are not only exposed to plant activity but to further sedimentation by loess, floodwater, or even subsequent tephra fallout, etc. Thus, the tephra becomes diluted with other materials including organic matter, and chemicals generated by those plants or deposited through sedimentation including aeolian **loess**. Nevertheless, the basic natures of the soils developing on tephra are influenced by the mineralogical content and glassy fragments of the tephra.

The term 'tephragenic' is used by some to describe volcanoes producing tephra, e.g., tephragenic volcanoes or tephragenic eruptions (Kittleman 1979: 6; Machida & Arai 1983; Hotes *et al.* 2004), or the products of explosive eruptions, e.g., tephragenic **clasts** (Ehlmann *et al.* 2017: 2511). The term is adopted here with the spelling modified to 'tephrogenic' to match the sense of 'originating in' (as in 'neurogenic'). Tephrogenic soils are thus those that develop on or include substantial amounts of tephra, but soils also may form on lava (James *et al.* 2000).

Andosols: what are they?

The major soil type formed from volcanic ash is variously called an **andosol**, or andisol, or andept; these terms all derive from the Japanese word *ando* meaning 'dark earth', but in Japanese such soils are more often known as *kurobokudo* 'black fluffy earth'. In this Chapter, the 'andosol' spelling is used because it retains the Japanese descriptive, *ando*. In contrast to less than 1% of the Earth's landmass covered by andosols, they are estimated to cover 17% of Japan's land surface (Omura 1993: 247) (Figure 7.1). The commonly accepted characteristics of andosols are listed in Table 7.1 (see also James *et al.* 2000: table 1).

Before proceeding further, it is useful to consider two dimensions of soil research: definition and classification. Every discipline has created its own definition of soil for its own purposes (Holliday 2004;

An earlier version of this Chapter was published in *TephroArchaeology in the North Pacific*, ed. by GL Barnes & T Soda (eds) (Oxford: Archaeopress, 2019).

Table 7.1 Andosol characteristics

* a high phosphate (P) retention of ≥85%
* low bulk density of ≤0.90 kg dm^{-3} (FAO) /
≤0.9 Mg m^{-3} @ 33 kPa water retention (USDA)
* $Al_{ox}\frac{1}{2}Fe_{ox}{}^{2}$ value of ≥2% (FAO) /
Al_{ox} $0.5Fe_{ox}$ content ≥20 g kg^{-1} (USDA)

- variable charge
- high water retention
- friable consistency
- excellent tilth
- a high humic content
- highly stable soil aggregates
- "an abundance of non-crystalline minerals such as allophane, imogolite and metal-humus complexes" (Imaya *et al.* 2010: 454)

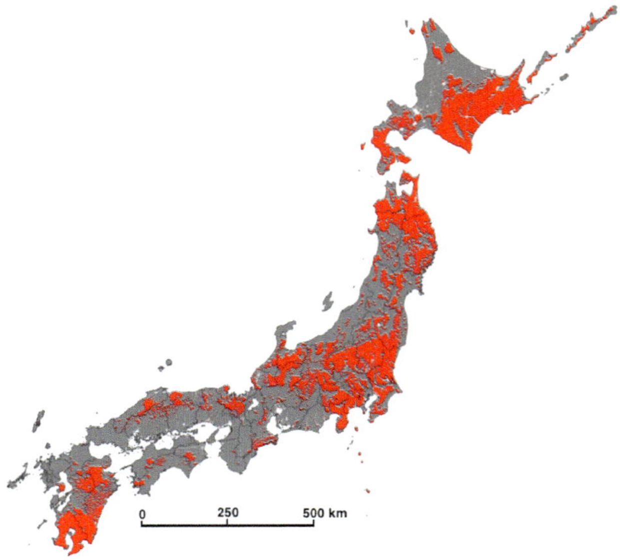

Figure 7.1 Distribution of andosols in Japan

Hartemink 2016). Soil scientists, however, develop classifications on one-metre deep sections from the surface where at least A and C horizons can be detected, even if a B horizon is not apparent (cf. Figure 7.9). A single sediment, such as freshly deposited tephra or alluvium or a paddy-field, is not a 'soil' according to this view. Moreover, any sediments that occur at levels of one metre or more are not included in these classifications.[1] In this Chapter, 'soil' is used in its pedological sense: sediments defined according to their genetic horizons (Schaetzl & Thompson 2015). However, Hosono & Sase (2015) have recently coined a term *kuroboku-dosō* (black 'fluffy soil' stratum), which does not refer to a soil profile but to a single layer (*sō*). This definition might be useful in archaeology if black andic layers are discovered, for example, as palaeosols buried more than one metre deep or when individual layers rather than profiles are discussed. However, there is a long-lasting disagreement between geologists and pedologists how the word *kurobokudo* should be used (Shirato 2015).

Most extant systems of soil classifications have been produced with an eye to agricultural potential, and soil scientists in several countries have created their own national classifications (Gerrard 2000: ch. 5). Takahashi & Shoji (2002) compare domestic classifications in Japan and New Zealand with worldwide classification schemes, illustrating many differences in scope and definition. The US Soil Taxonomy of the United States is often used as a referent (USDA Soil Survey Staff 1999); but the FAO-UN (2015) scheme integrates local soil types where feasible. The above terms for volcanic ash soils are used variously in the different schemes (see Retallack 2019: ch.5), for example 'andosol' in the FAO-UN scheme; 'andisol' in the USDA Soil Taxonomy, previously referred to as 'andept'; and 'andisol' in the New Zealand national classification (Kato 1983: 8). Most of these schemes are hierarchical, with global, regional, and local foci of increasingly fine distinctions, but commonalities do exist. The *starred items in Table 7.1 are given by the FAO-UN (2015: 63) as diagnostic properties of 'andic' soils that must co-occur; the USDA taxonomy gives some of these in different units of measurement (Imaya *et al.* 2010: 461). A useful comparator is given in Morisada *et al.* (2004: table 1).

[1] Based on a discussion of 'soil' definitions with OHKURA Toshiaki, NIAES (National Institute for Agro-Environmental Sciences, Tsukuba), pers. comm. 23 Oct 2018.

Japanese soil classifications have historically been developed within three categories: *ando* soils, paddy soils, and upland soils (Obara *et al.* 2015). Instead, Obara and associates have proposed a new classification using criteria that are defined quantitatively and qualitatively across all land categories (Figure 7.2). Andosols (Figure 7.2-D) are one of ten proposed 'Great Soil Groups'; andosols are further divided into six 'Soil Groups': regosolic, gleyed, wet, fulvic, non-allophanic, and allophanic andosols (Ibid.: table 2).[2] Among these, allophanic andosols "are the central concept of Andosol great group and are characterized by the dominance of allophane and imogolite formed by the weathering of volcanic ash under well-drained conditions and accumulated humus" (Obara *et al.* 2015: 222). Allophane and imogolite are both forms of alterite clay minerals which have active Al and Fe components; they will be discussed more fully below.

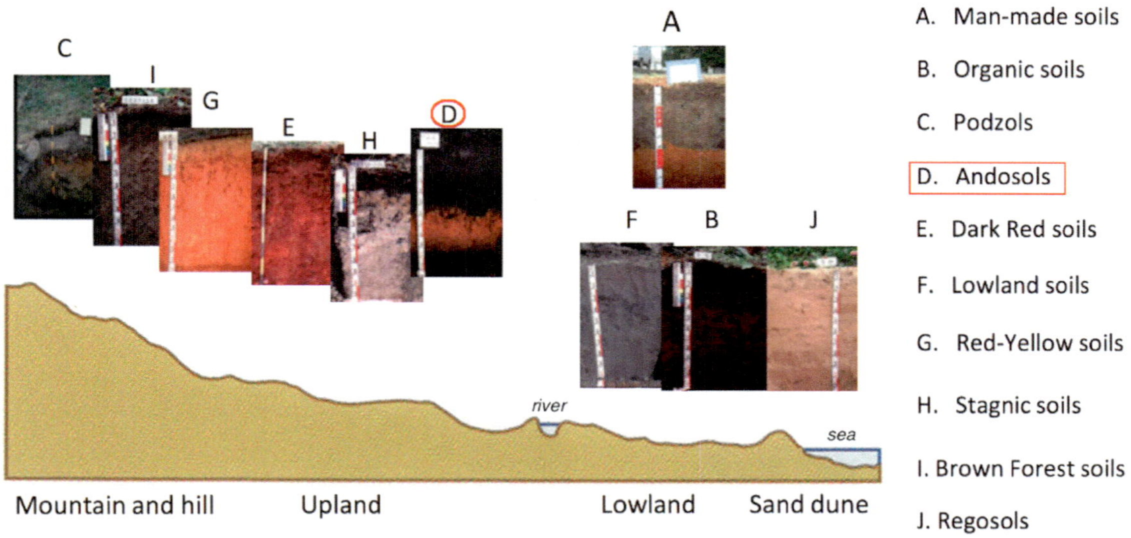

Figure 7.2 The proposed ten Great Soil Groups of Japan as distributed in different ecological zones © JIRCAS

Thus, we have a situation where allophanic andosols "are the central concept of Andosol Great Group" (Obara *et al.* 2015: 222), but they comprise only one of six andosol Groups. Among the other five, non-allophanic andosols, were only recognized in 1978 (Takahashi & Shoji 2002). Mapping these two types illustrates the dominance of allophanic over non-allophanic andosols (Figure 7.3). Moreover, tephra can be found in other Great Groups such as Brown Forest Soils. Beyond Japan, volcanic ash soils are present in several orders (Entisols, Ustivitrands, Spodosols, Mollisols, Alfisols, Ultisols, Vertisols, Inceptisols, Oxisols as seen in Ugolini & Dahlgren 2002: table 3; Dahlgren *et al.* 2004: table 1). Many of these are transformations to and from andosols with changing climatic conditions.

The essential differences between allophanic and non-allophanic andosols are presented in Table 7.2 below. These two types of andosols form under different pH conditions: allophanic andosols – which are characterized by allophane, imogolite, and ferrihydrite – preferably form under pH values >5.0; non-allophanic andosols are characterized by Al-humus complexes and opaline silica and are dominant under pH<5.0 conditions. Fe prefers formation of ferrihydrite than incorporation into Fe-humus

[2] Takahashi & Shoji (2002) name the fourth group 'forest' instead of 'fulvic', and the sixth group simply 'andosols' rather than 'allophanic andosols'. 'Fulvic' is sometimes misspelled as 'fuluvic', primarily in East Asia (e.g., Ge & Cheng 2001).

complexes because ferrihydrites are more stable (Burrell & Fields 1952; Ugolini & Dahlgren 2002). When aluminium is complexed with humus in low pH (acid) soils, it prevents Al from forming the alterite clays,[3] and the excess Si precipitates as opaline silica; this is termed the 'anti-allophane effect' and operates to produce non-allophanic andosols (Shoji *et al.* 1993: 113). The presence of excess Al, however, makes non-allophanic andosols very acid and contributes to Al toxicity.

Table 7.2 Comparison of the two major andosol groups

	Allophanic andosols	**Non-allophanic andosols**
Recognized in	1947	1978
% of Andosols	70%	30%
Termed by WRB (2014)	silandic (silica dominant) "typically gives a strongly acid to neutral soil reaction"	aluandic (aluminium dominant) "gives an extremely acid to acid reaction"
humus	ca. 3–11% of soil matter; Japan = 33%; low content of complexing organic compounds	ca. 4–22% of soil matter; Japan = 55%; rich in organic matter
Active Al/Fe	allophane imogolite ferrihydrite Al/Fe humus complexes	Al-humus complexes (with opaline silica by-product) can cause Al-toxicity
pH(H_2O)	pH 5.2–6.0 in humus-poor soil due to H from carbonic acid in water; pH 5.0–5.7 in humus-rich soil due to humus complexing	4.8–5.3; H from organic acids
base metal presence	base-rich andesitic basalt and basalt coloured glass	base-poor volcanics: rhyolite, dacite, andesite non-coloured glasses
clays	ca. 33%(wt) allophane / imogolite ca. 23%(wt) crystalline clays (mainly 2:1)	ca. 2%(wt) allophane / imogolite ca. 42%(wt) 2:1 crystalline clays
precipitation	<1000 mm/yr	>1000 mm/yr
tephra age	younger volcanic materials	older more weathered volcanics

It will be noted that many of the differences between the two andosols in Table 7.2 are mainly due to the age of the tephra. In the Non-allophanic andosols column, it can be seen that older tephra has had longer to be transformed into clays (cf. Figure 7.4; Apx 12), accounting for the decrease of the transitional clays allophane / imogolite (to 2%) and increase in 2:1 crystalline clays (42%) in contrast to clays from younger volcanic materials in the Allophonic andosols column. Precipitation rates can also increase weathering. The amount of humus in each also reflects time expended since the tephra deposition. A similarity that is not shown in Table 7.2, however, is that both andosols are highly deficient in P, the lack of which is one of the basic attributes of andic soil (cf. Table 7.1).

[3] The term 'alterite' follows usage by Bauer & Velde 2014: 130, indicating minerals altered through weathering. As discovered by Goldich (1938), minerals generally dissolve in the same order as formed as specified in the Bowen Reaction Series (see Apx 5: Fig. B).

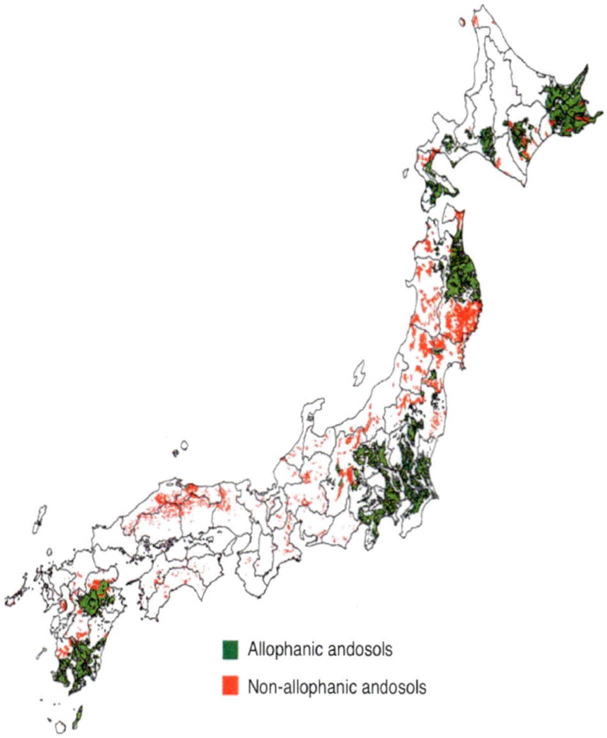

Figure 7.3 Distribution of allophanic and non-allophonic andosols in Japan

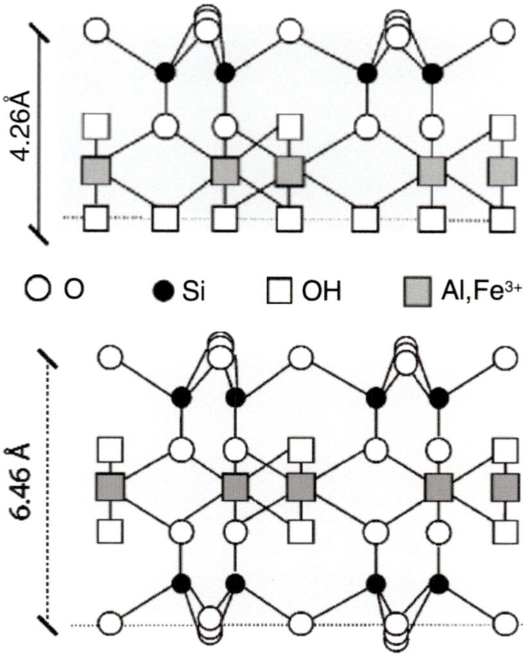

Figure 7.4 Layered phyllosilicates
Top: 1:1 clay structure
Bottom: 2:1 clay structure

Figure 7.3 clearly shows that allophanic andosols (in green) are correlated with active Quaternary volcano clusters, from left to right: southern Kyushu Island, Mt Aso in central Kyushu, the Kanto region, northeastern Tohoku, and areas of southern and eastern Hokkaido Island. According to McDaniel *et al.* (2012), non-allophanic soils (in red) comprise 25–30% of andosols. They name a crucial condition giving rise to allophanic andosols (2012: 23): frequent dustings with new volcanic ash, providing more Al for allophane construction. If loess from China is the dominant sedimentary deposition, the Al content becomes diluted and/or depleted, leading to complexing of organic matter instead, a trait of non-allophanic soils.

Tephra in other soil classes

Andosols are not the only soils that contain tephra. Two other Great Soil Groups in Japan have members with a significant tephra-derived component.

First, Regosols (Figure 7.2-J); compare:

> Volcanogenous *regosols*: "Soils in which the cumulative thickness of un-weathered pyroclastic materials with phosphate absorption coefficient of less than 300 and organic carbon contents of less than 3%, is 25 cm or more within 50 cm of the mineral soil surface" (Obara *et al.* 2015: 223).

> Regosolic *andosols*: "Soils belonging to the Andosol great group and with vitric [glassy] soil materials 25cm or more within 50cm of the mineral soil surface" (Obara *et al.* 2015: 222).

The general definition of regosols applies to any rock altered through weathering (producing alterites); it is potentially composed of three zones from bottom up: the structurally coherent altered rock, a structurally coherent but friable and porous zone, and a friable structurally incoherent zone that is usually unconsolidated but may be cemented (Velde & Meunier 2008: 115). Often termed an 'immature' soil, regosol does not contain any significant organic material and is composed mainly of minerals. When and how a volcanogenous regosol of pyroclastic materials becomes a regosolic andosol is not discussed, but a reasonable deduction is that it does so by the action of plants.

The tephrogenic soils of Indonesia may be offered here in comparison. Based on regosols, they are not very black (contain little humus) and they do not contain allophane – both conditions due to the young age of the tephra; but they do have andic properties (T OHKURA, pers. comm. 23 Oct 2018).

Second, Brown Forest Soils (BFS) (Figure 7.2-I) may contain tephra; these soils correspond to Udepts in the USDA system and to Cambisols and Stagnosols in the FAO-UN taxonomy. BFS cover 53% of Japan's land area (NDHG 2015: 22). According to Kubo (1982), they can form from mountain ridges down to the lowlands, in dry to wet conditions; six different BFS sub-types were devised to accommodate this variability. However, Imaya *et al.* (2010) argue that instead of the traditional classification of BFS by moisture and slope conditions, BFS can be better characterized by their varying 'andic soil properties', particularly the amount and vertical distribution of aluminium oxylate (Al_{ox}), plus gravel which represents bedrock weathering. The focus on aluminium oxalate, written variously as $(NH_4)_2C_2O_4$ *or* $C_2H_8N_2O_4$, is due to its role in the formation of allophane, which is a "unique clay mineral of volcanic ash soils" (Imaya *et al.* 2010: 461). They state that although 70% of forest soils in Japan are brown (BFS), contrasting with black forest soils (andosols), BFS may incorporate varying amounts of volcanic ash either through direct tephra fallout or soil creep, mixing with and changing their chemical compositions.

Implications

The above distinctions in soils were not apparent to prehistoric and historic peoples, who had to judge the fertility of soils merely on their apparent productivity (c.f. Lowe *et al.* 2014; Force 2015; Vogel *et al.* 2016). We now know that upland tephrogenic soils in Japan were generally avoided for three main reasons: the lack of phosphate-P, soil acidity, and aluminium-Al toxicity. Without knowing the causes of infertility, however, direct solutions – such as the modern addition of phosphorous and lime (which ameliorates acidity and therefore conceals Al toxicity) – were not feasible. Instead, indirect solutions such as short-term cultivation (swidden) or green manuring were undertaken to greater or lesser effect. The andic characteristics of, and colloidal clay presence in, a soil can today be tested by scientific means, but we may be limited in prehistory to judging land use practices today as a proxy.

Tephra Transformations

Tephra-derived soils have a particular weathering sequence that soil scientists classify as a progression from tephric (initial tephra weathering) → vitric (coarse, still containing large amounts of unweathered glass) → andic properties (having the properties listed in Table 7.1 above) (FAO-UN 2015: 62). The FAO-UN provide quantitative descriptions of soils with 'tephric' and 'vitric' properties where glass* (in Table 7.3) refers to volcanic glass, glassy aggregates, and other glass-coated primary minerals; in this case 'glass' indicates amorphous quartz. The transformation of newly deposited tephra into andic soils (andosols) takes time, with the negative aspects (P retention, acidity, and Al toxicity) gradually increased through the process.

• Tephric properties (tf) particularly belong to 'tephric material' (tephra of all sorts), but the tephra may be "reworked and mixed with material from other sources...tephric loess, tephric blown sand and volcanogenic alluvium" (Ibid.: 84).

• Vitric properties (vi) develop in tephra "which contain a limited amount of short-range-order minerals or organo-metallic complexes"; when tephric material acquires vitric properties through weathering, "it is then no longer regarded as tephric material" (Ibid.: 74, 84). Vitric properties are "closely linked with andic properties, into which they may eventually develop" (Ibid.: 74). Hence the tephric→vitric→andic sequence (andic as *starred in Table 7.1) (see Shoji *et al.* 1996 for precise specifications).

Table 7.3 Tephric & vitric properties of soils compared

	Tephric (tf)	**Vitric (vi)**
glass*	≥30% grains in the sand fraction	≥5% grains in the sand fraction
$Al_{ox}½Fe_{ox}$[2]	low	≥0.4%
phosphate retention	none	≥25%
carbon	none	<0.5% C_{py}/OC[3] or $C_i/C_{py}4$
organic C	none	<25% by mass

The transformation from tephra to andosol entails two different processes: rock weathering and soil formation; these are "independent open systems" (Velde & Meunier 2008: 114), though they may occur simultaneously; moreover, clays may form during both or either of these processes. Weathering of rock into clay takes "hundreds *of* thousands of years", whereas clay mineralization in the soil horizon stabilizes within "hundreds *to* thousands of years" (Ibid.: 242, emphasis added). Weathering is studied by geologists and soil by pedologists, and never the twain shall meet! Velde & Meunier propose that this is because "pedologists study the soil downward...while geologists begin their investigation upward from the unaltered rocks" (2008: 113). But Eash *et al.* (2016: 11) are clear that "microorganisms are involved in rock weathering from the start." Volcanic ash soils are exceptional because research into their formation has incorporated rock weathering – unlike research on soils of other lithologies (Ugolini & Dahlgren 2002).

Weathering has been mentioned above in terms of regosol zones, but the product of weathering – clay formation – will be investigated below. The weathering of tephra involves all three components of tephra – volcanic glass, mineral crystals and lithic fragments – which weather at different rates, providing inorganic substances (the elements and minerals) for clay formation. The treatment here is necessarily brief, doing little justice to the multitudes of work written on the subject (e.g., Ugolini & Zasoski 1979; Shoji *et al.* 1993; Yamanoi 1996; Sase & Hosono 1997; Ugolini & Dahlgren 2002; Dahlgren *et al.* 2004; Hosono & Sase 2015; Shindo & Nishimura 2016; and bibliographies therein).

From tephra to clay

As stated above, all rocks weather to clay over time due to the action of water – whether atmospheric, hydrothermal or oceanic water – upon silicate materials.[4] However, the weathering of tephra is unique in having intermediate stages where the amorphous clay mineral allophane and/or tubular-structure imogolite are some of the first products. As used by Velde & Meunier (2008: 3), the term 'clay' refers to phyllosilicates (Greek *phyllon* 'leaf'+ Latin *silic* 'flint') rather than being used as a size category. Despite

[4] This section is heavily based on Velde & Meunier 2008.

the etymology, however, they also include non-leafy aluminosilicate phases (e.g., imogolite and allophane) in the phyllosilicate / clay category.

Crystalline clays are generally divided into groups depending on the numbers and types of their 'sheets' as 2:1 and 1:1 clays (cf. Figure 7.4; Apx 12). One type of sheet is formed of silicate tetrahedra SiO_4^{4-} or aluminate tetrahedra AlO_4^{5-}, which are linked together by their oxygen atoms (cf. Apx 5: Table B); a second type is an octrahedral sheet containing metals, e.g., Al^{3+}, Fe^{3+}, Fe^{2+} or Mg^{2+}, bonded to O^{2-} (oxygen) or OH^- (hydroxyl). These electrically charged atoms are ions: cations are positively$^+$ charged while anions are negatively$^-$ charged, with the number (valency) indicating the capacity of the atom to form chemical bonds, as in the clay sheets. These sheets are then organized into layers, with 1:1 clays containing one each of a tetrahedral and an octrahedral sheet, and 2:1 clays in which one octrahedral sheet is sandwiched between two tetrahedral sheets (see Velde & Meunier 2008: ch. 1.1.1. for details). Kaolinite (as well as the serpentine mineral lizardite) are examples of 1:1 clays, while 2:1 clays include smectites, illites, micas, vermiculite, chlorite (and the metamorphic mineral talc). When the octrahedral sheet is found alone, it constitutes the non-clay mineral gibbsite (aluminium hydroxide); gibbsite is often the last product of the clay weathering cycle.

Numerous clays, by-products, and detritus are produced during the weathering process. Only a few are relevant here (Apx 12), but it is important to note that clay species are inherently unstable; depending on changes in their environment, a particular clay may undergo transformation into another type of clay in different clay group altogether. Thus, the information given in Appendix 12 does not represent all the varieties and possibilities that are inherent in clay minerals.

Weathering of tephra

Tephra is composed of three components: volcanic glass, mineral crystals, and rock fragments. All these are forms of or collections of minerals, and therefore mineral weathering is the most important process. However, volcanic glass has its own problems, which will be discussed separately.

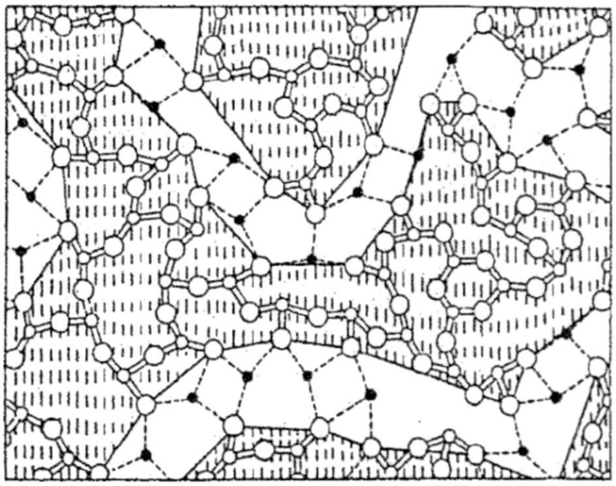

Figure 7.5 Proposed structure of vitreous glass

Minerals crystallize from magma melt in a given order, and the mineral species depends on the chemical composition of the magma (Apx 5: Figure B). The felsic minerals are the last to crystallize and they dominate intermediate and felsic tephras; in order of abundance, these felsic minerals in tephra are: noncoloured volcanic glass, plagioclase, other silica minerals (e.g., quartz–Q), and mica. Heavy mafic minerals are less common: coloured volcanic glass, hypersthene and other opaque minerals, augite, and hornblende (Shoji *et al.* 1993: 105). These various minerals weather in order from least stable (glass and olivine) to most stable: coloured → non-coloured glass = olivine ← plagioclase ← augite ← hypersthene ← hornblende ← ferromagnetic minerals (Ibid.: 112). The major elements provided by these rock-forming minerals (cf. Apx 5: Table A) are Si, Ca, K, Na, Mg, K, Al, Fe, and Ti – unsurprisingly the same elements tested for quantities in Quaternary tephras in Japan (Machida & Arai 2011). The rock-forming minerals also provide minor and trace elements. Minerals of non-volcanic sources may be present in tephra, deriving from aeolian dust (particularly quartz), or having undergone hydrothermal alteration. The products of this weathering are then available for clay formation.

Weathering of volcanic glass into crystalline clays

The weathering of tephra begins as it is exposed to volcanic gases during eruption; sulfuric, hydrochloric, and nitric acids condense on the glass particles in tephra which, after deposition, initiate chemical changes producing soluble compounds of Na^+ and Ca^{2+} with SO_4^{2-}, Cl^-, and NO^{3-} (Ugolini & Dahlgren 2002). Further interaction occurs between water and volcanic glass fragments. The glass particles are composed of so-called 'amorphous' silica: vitreous silica SiO_2 tetrahedra organized randomly (Figure 7.5) rather than in a crystalline structure like quartz. Ions of several types of impurities may also be incorporated into the glass (Al, boron-B, alkalis and alkaline earth elements), while the water itself most commonly contains sodium–Na ions (Icenhower & Dove 2000). Volcanic glasses in Japanese Quaternary tephras are characterized by quantities measured by EPMA of silicon–Si, titanium–Ti, aluminium–Al, calcium–Ca, manganese–Mn, magnesium–Mg, potassium–K, sodium–Na, iron–Fe and then presented as oxides (Machida & Arai 2011: app. 2). Upon the dissolution of volcanic glass, these various elements enter into the circulating solution from which clay minerals can form.

The dissolution of amorphous silica has been a puzzle, but it appears that the uneven surface structure of silica tetrahedra links provides sites for the detachment of individual tetrahedra (Dove *et al.* 2008). Moreover, recent research proposes that amorphous silica (greyed in Figure 7.5) may have greater surface area than previously imagined, with channels (white) formed through the glass matrix by links to ionic impurities (the dots •) (Anaf 2010). These channels are opened up through ion exchange during leaching, exposing internal silica tetrahedra as well as surface silica tetrahedra to hydrolysis. Previous studies on crystalline quartz dissolution have shown that rates can be increased to 50x by the presence of NaCl (sodium chloride, common salt) in water (Dove *et al.* 2008: 9904), and it is hypothesized that ions of other metals (K, Mg, Ca) in solution will also increase dissolution rates up to 26x (Icenhower & Dove 2000: 4202). The removal of these ions weakens the linkages between silica tetrahedra (Shoji *et al.* 1993: 110-111), allowing the formation of silicic acid $Si(OH)_4$ which is unstable in water and quickly combines to form other minerals such as clays, or is taken up by plant growth (Belton *et al.* 2010). Water can act on the glass fragments in three ways (Figure 7.6):[5]

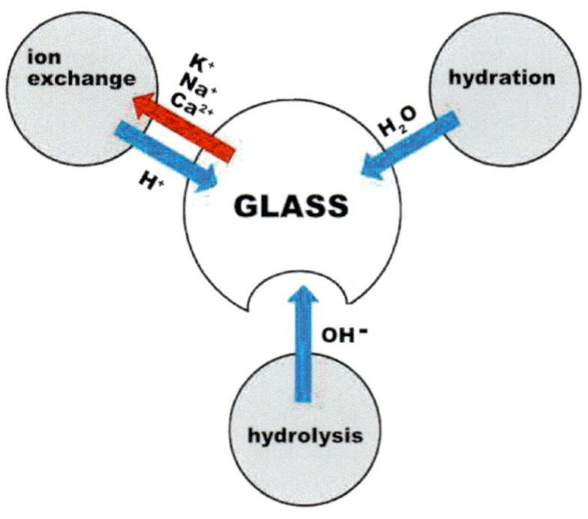

Figure 7.6 Corrosion of glass by water

• Rainwater and even morning dew acts on minerals via *ion exchange*, facilitated by carbonic acid H_2CO_3 (carbon dioxide dissolved in water); since Al is a large atom with a high charge (Al^{3+}), it is generally difficult to replace through exchange and is thus left behind.

• Water may be chemically bonded into the volcanic glass upon eruption, but further *hydration* of the glass is possible by absorption of water on fragment edges; this is the same

[5] This diagram should not be confused with the silica tetrahedron (cf. Apx 5: Table B)! For details on the weathering of volcanic glass, see Shoji *et al.* (1993: ch.5.2.4).

process as obsidian hydration, obsidian being a volcanic glass extruded like lava rather than explosively like tephra.

• *Hydrolysis* causes a chemical reaction between hydroxols (negatively charged OH molecules) in water and solid silica to form silicic acid: $SiO_2\ 2H_2O \rightarrow Si(OH)_4$ *or* H_4SiO_4, the major ingredient for building clays. James *et al.* (2000: 320) note that hydrolysis releases Si, Al, and Fe ions faster than they form crystals, and the ions are thus incorporated into short-range order and amorphous minerals and humic complexes.

Clay formation

Rock weathers to clay varieties as constrained by several variables: temperature/climate, rock types; and for igneous rocks – magma types (cf. Apx 5: Fig. E), chemical compositions, soil pH, etc. For felsic igneous rocks, the weathering of feldspars produces illite-mica and smectite – the clay of alluvial soils (T OHKURA pers. comm. 23 Oct 2018); further weathering transforms these into kaolinite and gibbsite (Figure 7.7 left). The major clay species of the smectite group is montmorillonite (Apx 12), but in Japan, montmorillonite is almost exclusively formed by hydrothermal alteration from rhyolite tephra in the Green Tuff zone of Miocene date (cf. Chapter 4) (Takeshi 1978). In this same region, other clay minerals, such as vermiculite and mica, as well as quartz are also present; their presence in non-allophanic andosols is attributed to aeolian deposition of loess from the continent on the windward side of Japan (Shoji *et al.* 1993: 122). It is notable that allophanic andosols do not have the vermiculite component (T. Ohkura, pers. comm. 23 Oct 2013).

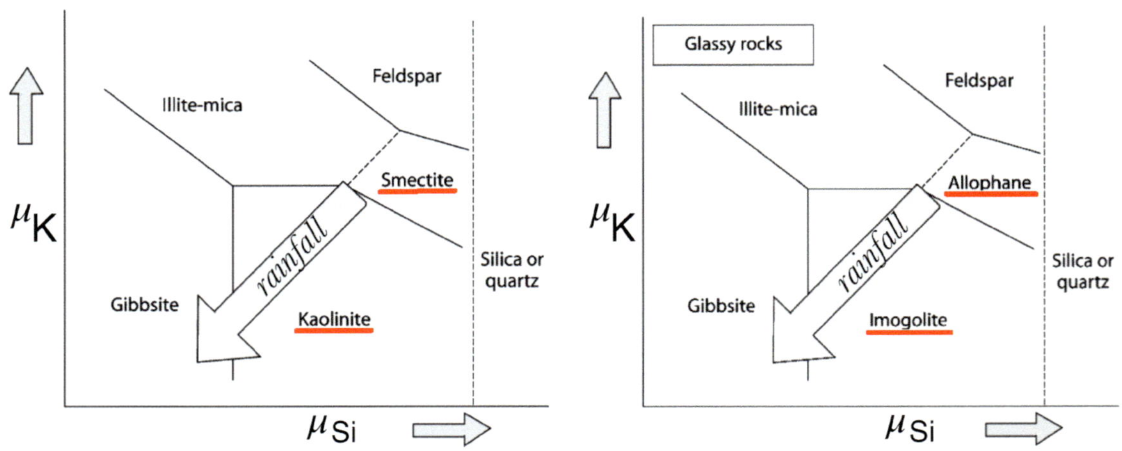

Figure 7.7 'Normal' clay successions (left) and glassy-rock clay successions (right)
NB: gibbsite, the final product, is not a clay

The rapid weathering of volcanic glass in Quaternary tephra produces more Al, Fe, and Si than can be quickly built into crystalline clay minerals (Ugolini & Dahlgren 2002); moreover, template minerals for clay formation (chlorite and mica) are generally absent in tephra (Shoji *et al.* 1993: 101-102). Consequently, so-called alteration (alterites) form instead:[6] the amorphous allophane and imogolite colloidal clays (Figure 7.7, right), plus ferrihydrite, Al/Fe-humus complexes, and opaline silica. Which of these form or dominate is determined by soil acidity; in acidic andosols, aluminum prefers combination with humus rather than forming the colloidal clays (Apx 12). Most alteration products stem from felsic minerals, while ferrihydrite is an iron hydroxide that forms from volcanic glass and

[6] Formation processes and conditions are quite complicated for each colloid; see Shoji *et al.* 1993: ch.5.3, 5.4.

minerals of mafic composition. Rather than participate in clay formation, ferrihydrite can crystalize as goethite and haematite when Fe is released from the humus complexes (James *et al.* 2000: 322).

Amorphous clays and alterite formation

Allophane and imogolite are formed of hydrous aluminosilicate colloids – that is, they comprise gels made of silicon and aluminium with water, and their structures are not detectable by X-rays. Some refer to them as short-range-order clays (SROCs) (Neall 2006), but Shoji *et al.* (1993: 101) prefer the term 'non-crystalline' for allophane and ferrihydrite as they "show no repeat of structural units in any of the three dimensions". A combination of infrared spectroscopy, thermal tests, and chemical analyses reveal the components of allophane to be hollow spherules 35–50 Å in diameter; imogolite forms tubes of even smaller scale: 6.4–21.4 Å. Allophane develops in volcanic ash, whereas imogolite develops on pumice – particularly fibrous pumice so that the structure of the fibers is retained in tubular crystalline structures through "incongruent precipitation" (T Ohkura, pers. comm. 23 Oct 2018).

Allophane is dependent on slightly higher pH and Si concentrations, while imogolite forms under lower pH and Si concentrations. As long as Si is available in higher pH value soils, colloid precipitates of AlSi will preferentially form over Al/Fe-humus complexes. Intermittent tephra airfall may interject new supplies of Al and Si, but when these elements decrease in availability as volcanic glass weathers out, the nature of the soil will change from andosol to other types. However, Harsh (2005) emphasizes that both allophane and imogolite are not confined to tephra-derived soils and can form in soils derived from a variety of different rocks (gneiss, loess, and other igneous and sedimentary rock), while allophane can occur in six of the twelve soil orders. The high water content of colloidal clays makes them dependent on humid conditions with constant moisture; thus, they are characteristic of areas with high rainfall, and if the soils dry out, they destabilize and convert to various clay minerals: imogolite→halloysite (hydrated kaolinite), and allophane→smectite. If further dessicated, the halloysite will lose its hydration and convert to kaolinite. More will be said about these colloidal clays in a section below as they relate to plant growth.

Alteration rates

Lest one think that the process is a quick one, the full transformation of tephra to clay takes thousands of years, and although the alteration products are seen as somewhat temporary or transitional, they can exist for similar time periods. The rate of tephra transformation has been documented to depend on the chemical composition of the parent magma (Apx 5: Fig. E) and morphology of the extruded tephra. In New Zealand, andesitic tephra has been found to decay to allophane within 300 years, but rhyolitic tephra took 3000 years to decay to allophane; then it took 7000 years for rhyolitic allophane to crystallize to halloysite, but andesitic allophane persisted for up to 100,000 years (Kirkman & McHardy 1980). Parfitt (2009) notes allophane ages of 250,000 years.

Turning tephra into soil

Plant activity

Only deeply buried rocks are affected by water interaction without influence from plants. Any rocks (including tephra) near the Earth's surface are subject, first, to colonization by bacteria and spore-producing life-forms (mosses, lichens, algae, and fungi). Thereafter, vegetative regeneration occurs by a variety of means, via wind-blown seeds or the "resprouting of buried branches"; plants may benefit by tapping into buried palaeosols for nutrients (Shoji & Takahashi 2002: 114). Lahars and debris flows provide exceptionally good bases for new growth as they often contain much previously developed soil, its micro-organisms, and plant roots and seeds. The rate of revegetation will depend on temperature and moisture regimes, with climate stress limiting recovery (del Moral & Grishin 1999).

A review of plant regeneration on volcanic rocks and sediments revealed complicated and idiosyncratic patterns of recovery around the world; an extreme example was provided for Krakatau: "Each island of the group was developing differently as a result of unique colonization events and subsequent disturbance" (del Moral & Grishin 1999: 145). In contrast, "Although successions in New Zealand are endlessly varied, some patterns emerge" (Wardle 1991: 554). The pioneer species colonizing each area depended on the current ecosystem flora, type of volcanic product both in form and chemistry, climate, precipitation, altitude, animal presence, etc. Thus, is it very difficult to generalize about or predict floral communities – each case needing investigation anew.

Nitrogen-N

The one constant, however, is that pioneer plants need nitrogen to grow. It is the building block of proteins and DNA, etc., and it is usually provided in soils from decaying plant and animal matter. However, if organic N is unavailable, as in fresh tephra cover, pioneer species find other ways to obtain the N they need. Several opportunities exist for obtaining N from rock material itself or assisted by bacteria or fungi.

It has long been known that atmospheric N_2 could be fixed (transformed into nitrate NO_3^-) by bacteria in the soil. These belong to plant growth promoting rhizobacteria (PGPR), from Greek *rhiza* 'roots', and are categorized by root proximity: around roots (rhizosphere), on roots (rhizoplane), in root tissue (endophytes), and in special root nodules (Gopalakrishnan *et al.* 2015). The last group includes the well-known legume-associated Rhizobia as well as Frankia, which form actinorhizal symbiosis within vascular plant nodules. While there are 1700 species of legumes (family *Fabaceae*) that are involved in rhizobia-legume symbiosis, actinorhizal plants number about 220 species and are mainly trees and shrubs in the Rosid I lineage of angiosperms (see Wall 2000: fig. 2). The other nitrogen-fixing bacteria "do not induce the formation of observable plant structures" (LMIL 2018: n.p.). One particular genus of nitrogen fixer bacteria, *Azospirillum*, is known to be associated with *Poaceae* species[7] (Steenhoudt & Vanderleyden 2000).

Among pioneer tree species, there is evidence of symbiosis with soil fungi of the EM (ectomycorrhizal) type. Like some PGPR, the fungi remain outside the plant root system, and many have long root sheaths themselves that bring up water to the plants in exchange for carbohydrates in 'obligate symbiosis'; they also can provide up to 86% of the nitrogen needed by host plants (Peay *et al.* 2007). Obase *et al.* (2007) note that endomycorrhizal fungi (AM) were also present, but they were not consistently associated with the woody pioneer species.

It is notable that N is not part of the standard formulae for 'rock-forming minerals' that are common in tephra (Apx 5: Table A). But these formulae do not take account of variations in the crystal lattice where other ions may be substituted in. Ammonium tetrahedron (NH_4) is found in decreasing amounts in biotite, muscovite, K-feldspar, and plagioclase; its distribution is "quite systematic, suggesting that NH_4 is one of the stable geochemical components in high temperature processes" (Honma & Itihara 1981: abstract). Research since the millennium has revealed that there is much nitrogen incorporated into bedrock and circulating thermal waters: sedimentary and metasedimentary rocks may contain up to 1000 mg N kg^{-1} organically bound nitrogen in ammonium-rich bedrock, or it may be available from

[7] Scientific names of all levels are italicized, following recommendations by Thines *et al.* 2020.

thermal waters (Holloway & Dahlgren 2002). This non-atmospheric nitrogen may be very important in supplying plants with the N needed for growth (Morford *et al.* 2011).

Plant Regeneration

As mentioned above, plant regeneration patterns on newly deposited volcanic products are very diverse; del Moral and Grishin (1999) emphasize the randomness of which species will be colonizers. They also make a distinction between survival and succession (not always clear from other reports); others distinguish between primary succession on a sterile sediment and secondary succession after a disaster (Brewer n.d.). Arnalds (2013) emphasises that the height of plants affects recovery. If plant life has not been completely extinguished, then recovery is possible via survivors, including those pushing through tephra from below or from root bundles caught in lahars; quick erosion of tephra often exposes underlying plant matter that is then able to grow again. In contrast, succession relies on transport by wind or birds of spores and seeds from nearby survivors. The importance of 'snags' – dead tree branches serving as bird perches – may allow the seeds of far distant trees and shrubs to be carried to new areas. And the geomorphology of the area may determine what species colonize which terrains.

Only a few examples can be offered here of case studies in plant regeneration on tephra or tephra-derived soils.

• A rapid recovery of plants on tephra is reported by Shoji *et al.* (1993: 45-48); case studies up to that point indicate recolonization by 'higher plants' on >50 cm tephra within 10 years but next-year recovery on tephra <50 cm thick. Seeds of such species were transported in by the wind and by animals.

• Shoji & Takahashi (2002) report on revegetation on Mt Usu in Hokkaido after the eruption in 2000. *Populus maximowiczii* (Japanese poplar, *doronoki*), *Betula platyphylla* var. japonica (Japanese white birch, *kabanoki*), and *Salix sachalinensis* (Sakhalin willow, *onoe yanagi*) were early colonizers. By 2004, 14 tree species were documented (Obase *et al.* 2007); and after 25 years, forests were established on extremely thin soils, ca. 10 cm, with regosolic properties persisting below.

• Dahlgren *et al.* (2004: 151) identify lupine (*Lupinus* spp.) as a nitrogen fixer and colonizer of Mt St Helens pyroclastic flows, creating "islands of soil fertility" by supplying nitrogen, capturing aeolian sediments, and incorporating organic matter. They also mention alder (*Alnus* spp.) as an N-fixer.

• In the Philippines, *Saccharum spontaneum* (wild cane) was able to colonize Mt Pinatubo tephra in the Philippines within seven years (Shoji & Takahashi 2002), and it remained the dominant grass among four types 15 years later when much tree growth (*Parasponia rugose*, belonging to the hemp family) had developed (Marler & del Moral 2013). The hemp trees and wild cane favoured flat terraces, while the minor grasses and ferns preferred talus slopes.

• In Kyushu, the evergreen shrub *Ardisia japonica* (coralberry, *yabukōji*) can form "a stable pioneer scrub community in southwest Japan that resists invasion [by trees] indefinitely" (del Moral & Grishin 1999).

• From New Zealand, is it known that beech forests supress the growth of grasses and establish thick stands of climax forest within 300–400 years (Wardle 1991: 556). Neall (2006) comments that forest litter improves soil fertility, but certain trees may have acidic litter, contributing to the acidification of the soil (Jobbágy & Jackson 2003). In northern Japan, fulvic andosols develop under beech forests (*buna*; *Fagus crenata, Fagus japonica*) (Shoji *et al.* 1993: 8) with an understory of bamboo grass. Shirakami-Sanchi is a "primordial Japanese beech forest that stretches over 1,300 square kilometers of mountainous land" in Aomori and Akita Prefectures (Uzawa & Kitchen 2012: 23); a portion has been designated as a UNESCO World Heritage Site to commemorate this indigenous forest type.

Studies of plant regeneration in clear-cut forest areas are also instructive:

• Clear-cut laurel forests growing on allophanic andosols in the Canary Islands were first colonized by the pioneer species – heather (*Erica arborea*) and the native firetree (*Morella faya*), the latter having the ability to fix nitrogen in the soil and therefore having the "capability to change entire ecosystems" (HEAR 2017 n.d.). Regrowth of an intermediate mature laurel forest occurred within 50 years (Arévado & Aboal 2015). This latter information is relevant to western Japan where the indigenous Holocene forest system was laurilignosa, though now replaced by secondary red pine.

• Another case of clear-cutting, however, provides a different perspective: andosols in northwestern North America were colonized by bracken, which acidified the soil and prohibited the regrowth of forest (Johnson-Maynard *et al.* 1998).

Such studies have yet to be carried out routinely at archaeological sites in tephra-affected areas to monitor the recovery of plant species. Dahlgren *et al.* introduce the important concept of "disturbance gradient" in assessing the amount of tephra cover ranging from "complete obliteration of the landscape near the mountain to a mere dusting of volcanic ash hundreds of kilometers from the vent" (2004: fig. 1, 115). Not all tephra fallout is damaging, as they note that dustings can rejuvenate soils and increase their fertility.

In Japan, the seven grasses of autumn (*aki no nanakusa*) are praised in poems of the 8c *Manyoshu* poetry anthology. These are mixed grasslands which take root in disturbed areas around settlements and could be seen until quite recently (Suga *et al.* 2011). Among the seven were *Miscanthus*, the only Poaceae, accompanied by a flowering shrub (*hagi, Lespedeza*) and several wildflowers. Such mixed grasslands should not be confused with those grasslands comprised of *Miscanthus* and *Pleioblastus*, which are major andolizers on volcanic soils. A study on volcanic islands, but not in Japan, revealed that plants with a greater amount of dry matter (mainly ground-hugging wildflowers) rather than shrubs and trees accumulated less P, K, S, sulfate, Mn, Fe, and Ni – indicating that they thrived despite having low P levels and avoided the others which inhibited photosynthesis (Kristin-Kirsty *et al.* 2018: 10). It would be interesting to know if *Miscanthus* and *Plioblastus* reacted the same way.

Andolization

Andolizer species

The progression tephric→vitric →andic is basic to this discussion, as we are particularly interested in andosols rather than other soil types. After initial colonization by bacteria, fungi, and the pioneer species, the transformation of tephric to andic volcanic ash soils begins with 'andolizer' species which convert volcanic ash to andosols; these include bamboo grasses, particularly *sasa* (*Sasa* spp.) and *nezasa* (*Pleioblastus variegatus* Makino) (cf. Stapleton 2006-2018); *Cortaderia* (pampas grass, of the subfamily *Arundinoideae*); *Miscanthus sinensis* (*susuki*, eulalia or Chinese silvergrass, sub-family *Panicoideae*) (Figure 7.8); and *Saccharum spontaneum* (wild cane, family *Poaceae*). In Japan, *Miscanthus* is the major andolizer in the cool-temperate zone, while *Miscanthus* and *Pleioblastus* colonize the warm-temperate zone (Shoji & Takahashi 2002; Nanzyo 2004). Andosols are not just formed on fresh tephra: bluejoint grass

(*Calamagrostis Canadensis*) has also been noted to be an andolizer by converting tephra-derived spodosols into andosols after forest fires (Shoji *et al.* 1993: 49).

The andolizers named above all belong to the *Poaceae* (previously, *Gramineae*) family of grasses. Because grasses need silica to form the phytoliths / plant opal in their plant parts, especially to keep themselves stiff and upright, the silica of dissolving tephra is a perfect source to exploit. Great amounts of plant opal from grasses (*Poaceae* phytoliths) are found in *kurobokudo* soils, indicating they develop under the colonization of grasses and herbs rather than forest. The primary contributors of these phytoliths are *Miscanthus* and *Pleioblastus* (bamboo) (Kubo 1982; Inoue *et al.* 2015). Such heavily humic soils can be radiocarbon-dated, providing an indirect age determination for the volcanic parent material (Chigaku-dan 1981: 313a). *Miscanthus* aids and abets the formation of colloidal clays by biocycling alkali / alkaline base metal ions, maintaining higher pH values and base-rich soils (Shoji *et al.* 1993: 131). Also, since grass parts (leaves, stalks, husks) contain much plant opal, returning phytoliths to the soil every year as the plants die keeps up the supply of silicon (Si) available for further colloid formation.

Figure 7.8 The *Miscanthus* fields of Sengokuhara, Japan

Andosol soil profiles

Soils are defined by having particular stratigraphic horizons (Figure 7.9, left), in general designated O through C; these are termed 'genetic horizons' relating to their formation, in contrast to 'diagnostic horizons' that are used in soil classification (Schaetzl & Thompson 2015: chs. 3, 8). The O horizon is comprised of dead plant matter (humus, see Apx 13: Table A); horizon A is topsoil; B is subsoil; and C is the regosolic substratum. Not shown are E (an eluviated stratum with loss of minerals) and R (bedrock). In the andosol class, the A horizon is much thicker than usual; compare the genetic sequence depths (Figure 7.9 left, in both centimetres and inches) with an andosol in Ibaraki Prefecture (right, in cm), where the A horizon reaches more than 1 m depth. Solleiro-Rebolledo & Cabadas-Báez present cases in

Figure 7.9 Soil profiles for genetic sequences (left) and an allophanic andosol (right) in equivalent scale

Mexico where tephra takes from <500 to 6000 years to be transformed into an andisol. For Japan, development of a deep A horizon depends on time lapsed and climate; one containing at least 10% humus and measuring 30 cm deep generally takes 1000 years to develop, and for a full ABC profile to develop takes 1500 years (Kato 1983: 20, 22). The dark humus layer is commonly referred to as *kurotsuchi* (black earth) and the underlying yellow-red sediments as *akatsuchi* (red earth). The upper dark layer

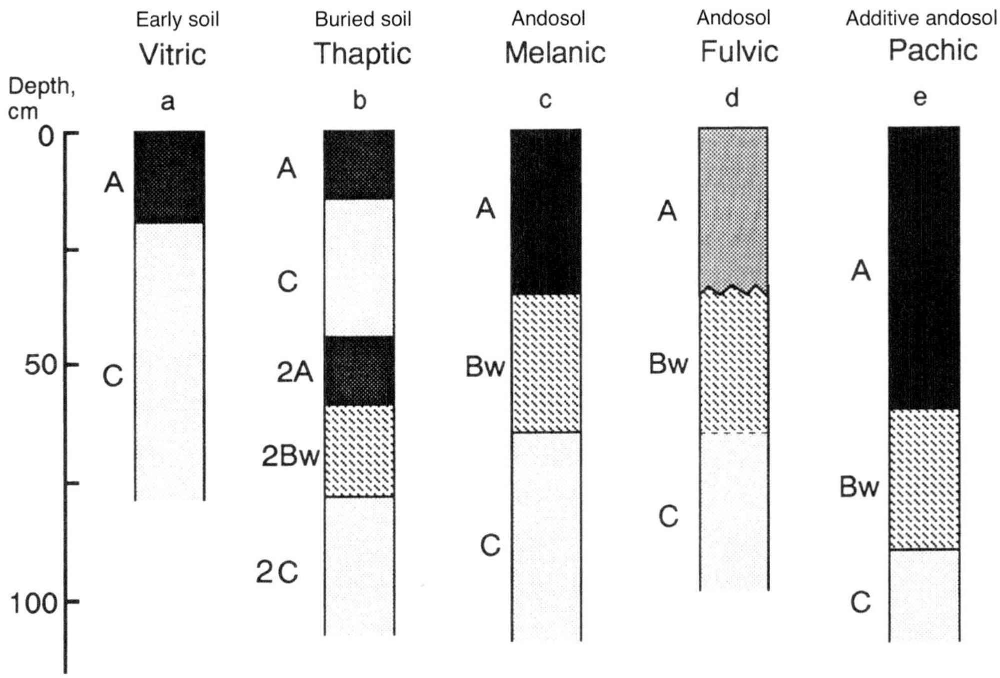

Figure 7.10 Select andosol profiles as found in Japan

has virtually the same composition as the underlying red layer, only differing in the addition of plant remains (T Noto & T Soda pers. comm. 9/2018).

Six major diagnosic soil profiles have been established for northeastern Japanese andosols (Figure 7.10), beginning with the 'vitric' stage (a) as designated in Table 7.3; the A horizon contains humus, while the C horizon is the weathering tephra. The 'buried' tephra column (b) illustrates the repetition of soil formation after repeated tephra fallout; the buried portion (indicated by the numeral '2') existed long enough for the development of an A horizon and a weathered B horizon (Bw), but the new tephra on top (C) has only developed an A horizon. The 2A layer may become a source of nutrients for deeply rooted plants. Sometimes, the periodic deposition of volcanic ash does not even allow the sequential A layers to develop melanic or fulvic profiles and remain regosolic andosols (Shoji & Takahashi 2002).

The melanic index in Table 7.4 is the humic / fulvic acid ratio in the organic fraction of the soil (USDA Soil Survey Staff 1999: 860). However, fluvic andosols can also be quite dark and can only be distinguished from melanic andosols by laboratory analysis determining the ratio of humic acids to fulvic acids (Honna *et al.* 1988). These humic substances are discussed more fully in Appendix 13 together with other nutrients. These colour-types of andosols can occur in either allophanic or non-allophanic andosols. The characteristics and productivity will vary with each combination.

Table 7.4 Properties of the two main diagnostic andosol horizons

	MELANIC ANDOSOLS	FULVIC ANDOSOLS
Maturity	andic (see Table 7.1)	andic (see Table 7.1)
Colour	both: Munsell colour value and **chroma of ≤ 2**, moist, *and* a **melanic index of < 1.7**	either or both: a. Munsell colour value or **chroma of > 2**, moist b. a **melanic index of ≥ 1.7***
Organic carbon	weighted average of ≥ 6% *soil organic carbon*, and ≥ 4% soil organic carbon in all parts; may contain 25–30% organic carbon	weighted average of ≥ 6% *soil organic carbon*, and ≥ 4% *soil organic carbon* in all parts
Thickness	combined thickness of ≥ 30 cm with ≤ 10 cm non-**melanic** material in between	combined thickness of ≥ 30 cm with ≤ 10 cm non-**fulvic** material in between
Humus	**humic acids (HA) A-type dominant** over fulvic acids	**fulvic acids (FA) dominant** over humic acids (mainly P-type)

Grasslands as 'pyromes'

International concerns

Although the natural succession of plants is from colonization of open ground by grasses and herbs to shrubby growth and the eventual invasion of forest, pioneer growth can last a long time: centuries for bracken (Johnson-Maynard *et al.* 1998), and thousands of years for grasses. Long-lived grasslands occur worldwide: North American prairies, South American plains, African savannahs, Eurasian steppes, and Australian bush. A variety of conditions have been named as maintaining these grasslands: climate, precipitation, soil-rooting depth, fire, and clearance activities; but tephra-derived soils are not a precondition for grassland growth.

One critical aspect of grasslands in general is maintenance of their boundaries with forests. Existing forests provide seeds and roots for colonization of areas of tephra fallout, promoting grassland recovery. Studies around the world report increasing encroachment of forest on grassland, though the rate may be decadal (Cuni-Sanchez *et al.* 2016). The reasons given are varied: increase in precipitation, increase in global CO_2 and other growth promoting factors, and fire. Many of these 'fires' are intentional: "in the Congo basin, it has been suggested that forest is expanding into savannas because of urban-migration and a consequent reduction in fire frequency" (Cuni-Sanchez *et al.* 2016: 2, 5). Conversely, in Brazil, fire is rejected as a single cause but is treated in conjunction with precipitation and tree-rooting depth (Langan *et al.* 2017).

Archibald *et al.* (2013) have developed the concept of a 'pyrome' that differs from 'biome', a plant community; they define pyrome as "regions of the globe that have similar fire frequencies, intensities, sizes, burned areas, and fire season lengths" (2013: fig. 2 caption). They use five criteria to establish five pyromes (Table 7.5: top half) and then classify different biomes as potentially hosting the different pyromes (Table 7.5: bottom half). A negative correlation is noted between fire frequency and intensity, reflecting the fact that fires occurring at frequent intervals do not allow the build-up of fuel to support intensive fires. Another factor is precipitation: despite the high productivity of high rainfall climates, the fuel is usually too wet to burn. With few exceptions, all the biomes are subject to all five pyromes but in varying percentages.

The ICS (blue) pyrome in Table 7.5 correlates with deforestation and agricultural regimes and is therefore indicative of human activity; its fires are "fairly small" with "intermediate return time" and have significant occurrence in all biomes except boreal forests. Grasslands that have a higher risk of fire other than in ICS are in FIL, FCS, RIL, and RCS pyromes to varying degrees. The colours denoted in Table 7.5 are keyed to a map (Archibald 2013: fig. 2) which has no datapoints for Japan. This is not to say, however, that Japan is not subject to fires; Zorn (2002) presents data that shows the number of forest fires has been decreasing from avg. ca. 4000 incidents per annum in the 1980s to avg. ca. 3300 in the 1990s, and between 2004 and 2013 declining from ca. 2600 to ca. 2000 incidents (FDMA 2014: fig. 1-4-1). The Fire and Disaster Management Agency specify early spring as forest fire season. In 1933 there was an interesting article published about natural causes of seasonal forest fires initiated simultaneously across the archipelago by approaching cold weather fronts (Editorial 1933), but it seems more likely that the FDMA is correct in assessing that the fires are due to weather being relatively dry with strong winds in early spring and also that people are more likely to be in the mountains collecting wild vegetables or taking hikes. Data on anthropogenic fires is contradictory. Zorn presents data from the Forestry Agency Japan dated to 2000 that indicate ca. 70% of forest fires were human-induced, but they also quote a 1987 study that states 99% of forest fires in Japan were caused by humans; the FDMA states that anthropogenic fires are very predominant. In any case, the majority of fires seem to be human induced; however, what were the fires set for? That is the conundrum of *kurobokudo* in Japan.

Table 7.5 Characteristics of the five pyromes identified by model-based expectation maximizing clustering

Characteristic	FIL (yellow)	FCS (orange)	RIL (green)	RCS (purple)	ICS (blue)
Mean burned area, %	14 (8–36)	9 (3–17)	1 (0–2)	0 (0–0.5)	0 (0–1)
Estimated FRI, y	3 (1–4)	1 (1–2)	>50	>50	12 (6–19)
Max FRP, MW	473 (350–660)	197 (156–253)	476 (283–844)	187 (108–334)	224 (143–352)
Max fire size, km^2	414 (155–1437)	25 (15–43)	83 (38–214)	4 (2–9)	9 (5–17)
Length of fire season, mo	4 (3–4)	3 (3–4)	2 (1–2)	1 (0–1)	3 (3–4)
Tropical moist broadleaf forests	2	10	5	**28**	**56**
Tropical dry broadleaf forests	3	**21**	6	16	**53**
Tropical coniferous forests	0	10	13	17	**59**
Temperate mixed forests	1	2	11	**41**	**45**
Temperate coniferous forests	2	0	**26**	**45**	**26**
Boreal forests	1	0	**46**	**47**	6
Tropical grasslands and shrublands	**29**	**33**	9	6	**23**
Temperate grasslands and shrublands	10	3	**21**	**29**	**37**
Flooded grasslands	**46**	17	8	5	**24**
Montane grasslands	6	15	13	**31**	**36**
Mediterranean vegetation	1	3	**25**	**36**	**35**
Xeric vegetation	18	1	**31**	**24**	**26**

The median and 25th to 75th quantiles of each fire characteristic are reported (parentheses), as are the percentages of 12 World Wildlife Fund biome classes that fall into each pyrome (values greater than 20% are bold)

Fire characteristics: F = frequent, I = intense (FIL, RIL)/intermediate (ICS), C = cool, L = large, S = small

Kurobokudo as a pyrome

Deep black soils around the world may or may not be related to fire. For example, plaggen soils in Europe are known to have been formed by the addition of heather or grass topsoil slabs to the fields, thus building up deep dark humic anthropogenic layers (Blume & Leinwether 2004), and loess in China has been modified by adding earth mixed with manure containing much humic matter (Gong 1994). On the other hand, the deep black *terra preta* soils of the Amazon clearly contain great amounts of biochar; Glaser *et al.* (2001: 39-40) propose that rather than burning forest to create this biochar, it was added as hearth residue from the "low-heat smouldering domestic fires commonly used by the native population

Figure 7.11 Two bamboo types
above: *sasa* (*Sasa* spp.)
below: *nezasa (Pleioblastus variegatus Makino)*

for cooking and heating". *Terra preta* soils are limited in distribution and have now been proven to be midden deposits, containing household refuse, pottery, plant and animal remains as well as hearth debris (Schmidt *et al.* 2014). The *terra preta* soils are anthrosols, not andosols; they are not andic and did not develop on tephra.

Above we saw that andosols in Japan occur in two general colours, melanic and fulvic (Table 7.4); these correlate with different biomes: deep melanic profiles are assessed to form under grasslands (mainly by *Miscanthus sinensis*), but fulvic profiles form under beech forests (by *Fagus crenata*, Japanese beech *buna*) (Shoji *et al.* 1993: 45-47). Both of these have high organic matter contents, but the difference in colour is attributed to the presence in melanic andosols of minute pieces of charred plant matter which supply black humic acids. Thus, melanic andosol formation hinges on the burning of plant matter.

Since grass phytoliths are very prominent in *kurobokudo*, and since *Miscanthus* is a known andolizer, it has been assumed that the humic content (cf. Apx 13: Table A) of melanic andosols in Japan mainly comes from *Miscanthus*, a C4 plant. However, further investigation of carbon stable isotopic ratios ($\delta^{13}C$) in Japanese andosols revealed that only half of the plant material is contributed by C4 plants (Hiradate *et al.* 2004); C3 plants such as woody plants and trees among others contribute the other half. C4 plants comprise only 3% of all terrestrial plants, but 79% of C4 plants belong to grass or sedge families (Bareja 2013); thus, a 50% contribution to melanic andosols by C4 plants is quite substantial. Nevertheless, C4 plants like *Miscanthus* have been found *inessential* to the formation of andosols (Takahashi & Dahlgren 2016), though it seems that Japanese *kurobokudo* do mainly form under grasses that have been burned.

It is a mistake, however, to think that even maintained grasslands were comprised only of *Miscanthus* (Inoue *et al.* 2015); two varieties of bamboo have also been prominent: ground-hugging bamboo grass (*sasa*) and the taller dwarf bamboo *nezasa* (Figure 7.11). The phytoliths of these bamboos and *Miscanthus* are found in *kurobokudo* in varying quantities; moreover, the humic acids in *kurobokudo* match these plants (Suga *et al.* 2011: 113).

The picture of andosol formation is rapidly changing, but one important ingredient for melanic andosols is recognized to be FIRE. However, some scholars are deeply sceptical that fire is the *only* cause (T OHKURA, pers. comm. 23 Oct 2018). Nevertheless, research in North America demonstrates that melanic andosols result from frequent forest fires in Alaska and northern California's Cascade Range (Takahashi *et al.* 1994); and Dahlgren speculates that charcoal, from the Native American management of grassland over thousands of years, contributes to the dark colour of prairie soils (pers. comm. 17 Feb 2018). Schmidt *et al.* (2002) have discovered black carbon in chernozem soils of central Europe, challenging the traditional understanding of chernozem development and suggesting fire as an inalienable formative component. Finally, the biochar found in Amazonian *terra preta* soils, though not from grassland fires but charcoal added to the soil, is still an important component in the formation of those black soils.

Because of the high humic acid content of melanic andosols, the humic material itself can be radiocarbon dated; contra Yamanoi's (1996) conclusion of a Holocene date, these black andosols can date back to 30,000 years ago, though most of them are clustered between 5000 and 1000 years ago (Takahashi & Dahlgren 2016). But now, *kurobokudo* have been found together with Palaeolithic stone tools well below the AT marker tephra of 30,000 years ago. Hosono & Sase (2015) report that on the flanks of Mt Ashitaka, just south of Mt Fuji, six sequential palaeosols (of the Ashitaka loam) have been recovered,[8] with the earliest associated with stone tools at 40,000 BP. This date is close to that currently known for the peopling of Japan (Barnes 2015: 65); thus, Hosono and Sase conclude that as soon as Palaeolthic peoples occupied the islands, they began to transform the landscapes using fire. They underwrite this by noting that grasslands in New Zealand also started to appear only after colonization by Polynesian peoples (the current Maori). However, archaeologists in Japan are very reluctant to accept the interpretation that all *kurobokudo* soils are due to anthropogenic interference (NOTO Takeshi, pers. comm. 1 Oct 2018).

Searching for charcoal in bog stratigraphies and taking charcoal fragments as evidence of firing the landscape have become the major foci in environmental reconstruction studies. Sasaki & Takahara (2012) found fire evidence at 900 BP and and then continuously from 2500 BP in the Tanba Highlands, while Inoue *et al.* (2012) documented fire at 7000 BP and then continuously from 1500/1000 BP in the Soni Plateau, Nara Prefecture. At Mt Aso volcano in Kyushu, charcoal was found continuously in the local stratigraphy from at least 11,000 years ago in both brown and black soils; macroscopic charcoal concentrations more than doubled during the Climatic Optimum (6000–4000 BP) and in recent times (Kawano *et al.* 2012: fig. 4). These charcoal results correlate with phytolith studies, which indicate the establishment of grasslands variously comprised of *Miscanthus*, *Pleioblastus*, and *Zoisa* (lawn grass).

These studies should be taken in context of even longer charcoal histories. Corings in Lake Biwa have recovered small peaks of macroscopic charcoal between 92 and 87 ka, as well as at 75 ka (Inoue *et al.* 2001: fig. 3). These occurred far earlier than the human occupation of the archipelago after 40 ka. A similar peak occurred between 30 and 25 ka, but these peaks are dwarfed by charcoal concentrations of all sizes from 18 ka. It is clear that natural fires occurred in the Pleistocene. One of the problems in working with charcoal is distinguishing natural fires from anthropogenic fires – in this case particularly the 30–25 ka peak. In either case, however, the result of fires is the establishment of grasslands, either temporary or continuous.

Japanese grassland formation & continuity

Hosono & Sase (2015: 325) report that three environmental conditions are essential for grasslands to develop in Japan:

- Annual average temperatures greater than 0°C
- Warmth index (WI) >30~35°C
- A wet environment with a Köppen index of K>18

[8] These would be *kuroboku-dosō* rather than *kurobokudo*, as explained above.

Figure 7.12 The annual celebratory firing of Mt Wakakusa behind Kofukuji Temple in Nara City

It follows that for *kurobokudo* to have been formed in the latter half of MIS 3 (which lasted from 60 ka to 24 ka), [9] there must have been warmer humid interludes. *Kurobokudo* have been found in Kyushu by 30 ka; however, conditions were not favourable in Hokkaido until the Holocene. Grassland development on the Kanto Plains was first by colonization of new tephra by *Miscanthus*, the andolizer species. If these plants are not subject to fire renewal, the land is gradually reforested; the shade kills the *Miscanthus*, whence bamboo grass species colonize the understory. If the forest is cut, bamboo grass remains as the ground cover. Thus, the dwarf bamboo, *nezasa*, has been the main grassland species of the Kanto region (NOTO Takeshi, pers. comm. 1 Oct 2018; see also Inoue *et al.* 2015).

Two external factors in the longevity of melanic andosols are 1) the natural rejuvenation of volcanic ash layers through repeated eruptions (cf. Figure 7.10-e), with constant recreations of grasslands, 2) and/or the potential impact of human activities. If grasslands are not maintained (either naturally or artificially – as indicated in the signboard of Figure 7.8: "managed land") – they are subject to forest succession. In a humid climate like Japan, succession from grassland to forest is a certainty unless humans intervene. In Kyushu, it is reported that shrubs such as *Lespedeza bicolor* and *Hydrangea paniculata* dominated meadows within six years after after cessation of anthropogenic burning (Miyabuchi *et al.* 2012). Suga *et al.* (2011: 114) note that northern grasslands that became disused at the beginning of the 20c have been colonized by white birch (*shirakaba, Betula platyphylla*) forest within the century.

Today, grasslands account for less than 1% of Japan's land surface (as opposed to two-thirds being forested). It is well-known in Japan that the current grasslands such as Sengokuhara in Kanagawa Prefecture (Figure 7.8), Wakakusa-yama in Nara Prefecture (Figure 7.12), or the Akiyoshi karst landscape of Yamaguchi Prefecture (cf. BOX 3) are maintained by annual firing. Even the name *wakakusa* means 'young grass'. From land-use maps of the Meiji period (1868–1912), Ogura (2011) notes that grasslands were more widespread than today and that there is documentary evidence of their management by firing them in early spring; this is a traditional practice called *hi-ire* ('setting fires').

Fire, however, is only one method of maintaining grasslands. Suga *et al.* name two other methods: cutting and grazing (2011: 101-102, 108). In the Kofun period (5c AD) grasslands were maintained for fodder for horses and cows after these were imported from the continent (Shimomura 2011). Finally, harvesting grasses for green fertilizer and thatch for roofing dates back to at least 800 AD, intensifying around 1500 AD, with the acreage then increasing exponentially. Miyabuchi *et al.* (2012) note that the

[9] MIS = Marine Isotope Stage

long historic continuity of grasslands around Mt Aso, comprising the largest area of grasslands in Japan at 220 km^2, has resulted from mowing and grazing in addition to burning the fields every year in early spring.

***Kurobokudo* and prehistoric humans**

The areal coverage of *kurobokudo* in the late Pleistocene and early Holocene was limited by the small population densities of hunter-gatherers. Hosono & Sase (2015) note that two expansions of *medake* bamboo (*Pleioblastus simonii*) have been found in the lower Tachikawa loam in the southern Kanto district which may correlate with the formation of *kurobokudo* via human activities. Most *kurobokudo* date to Initial and Early Jomon (10,550–3520 BC). This concurs with the proposal by Suga *et al.* (2011) that intentionally fired specialized grasslands date back to the Jomon period. However, Hosono & Sase puzzle over the lack of such soils at the height of occupation in Middle Jomon (3520–2470 BC) at the Sannai Maruyama site in Aomori Prefecture, concluding that the village was built on grasslands whose *kurobokudo* soils were disturbed by large-scale urban earthworks; but in the Late Jomon (2470–1250 BC), the Sannai Maruyama site area reverted to grasslands, building up another thick layer of *kurobokudo*; the area was used for hunting and cutting thatch through the 10c AD.

But what were these grasslands used for in the prehistoric period? In the mid-1960s a hypothesis for Jomon agriculture was proposed for the area around Mt Aso in Kyushu based on tool types (Shimomura 2011: 133), and in the 1980s, it was hypothesized that the Jomon grew buckwheat in mountain swidden fields based on charcoal and buckwheat pollen found together (Tsukada *et al.* 1986; Barnes 1986*b*). Buckwheat grains have been recovered from several prehistoric sites, but they could be contaminations from above and are not fully accepted with clear dating (Nakayama 2010). Buckwheat (*soba*, *Fagopyrum esculentum*) is not a grass but belongs to the *Polygonaceae*. It is native to the Himalayan region of Yunnan Province in China, having been domesticated in the San Jiang region bordering Yunnan, Sichuan and Tibet; molecular DNA studies indicate it spread from southern to northern China, through the Korean Peninsula to the Japanese Islands (Gondola & Papp 2010). Its arrival date is unknown, but it may have accompanied dry-rice arriving in the Late Jomon period. Nevertheless, it is rarely found in flotation samples despite it having a very distinctive shape of grain (Gary Crawford, pers. comm. 21 Oct 2018). By the 9c, it is mentioned in government documents encouraging planting, so we know that it formed one of the non-rice grain staples for peasant farmers (von Verschuer 2016: 25).

Currently, there is no firm evidence of swidden in the Jomon period, though burned-field (*yakihata*) cultivation was known in Japan up into the 1990s. Such fields were part of the *satoyama* ('village mountain') complex (Barnes 2010; Berglund 2014), whereby hillslopes behind villages were managed for various resources. Some villagers are now trying to revive the practice, for example at Itsuki Village in Kumamoto Prefecture, where one field was fired for the third time in 2012; the fired field was then sown with buckwheat for the first year, then millets and barnyard grass in the second, with beans and soybeans in the third.[10] Research on southern Kyushu has revealed a pattern of firing bamboo forest rather than trees. The first year after firing, foxtail millet (*awa*, *Setaria italica*) was sown, then the second year, bamboo roots were harvested; thereafter the bamboo forest was allowed to grow back, ready for firing again in ten years' time (Kawano 2008). This pattern does match the build-up of *Poaceae* phytoliths in the deep *kurobokudo* soils but not of *Miscanthus* per se.

[10] See the two videos from Itsuki Village, first the firing of the fields then sowing buckwheat: [www.youtube.com/watch?v=pRF4GQVuhuU], [www.youtube.com/watch?v=o8WmdA8j1s4]

Other prehistoric uses of grasslands have been suggested. Suga *et al.* (2011) propose that the firing of grasslands from the Late Palaeolithic period in Japan related to hunting practices to keep forest growth down as the climate warmed. This would probably also obtain in the Jomon periods when deer and boar were the major terrestrial protein; deer are not forest animals but like open clearings near the forest edge where they can graze and then escape to tree cover. Their preference for parkland is exemplified today by the Nara deer park in Japan, the Raby Castle deer park in England, and on the golf course of Estes Park, Colorado. Boar and deer continued to be hunted in the Medieval period (despite Buddhist prohibitions against eating the meat of four-legged animals), and Nakazawa (2011) has specified 'rough meadows' (*kōya*) as the hunting grounds.

It was mentioned above that *Miscanthus* was an important roofing material in historic times, with *Miscanthus* fields maintained for cutting thatch. However, it has not been emphasized enough that the basic residence of prehistoric times (and existing through the first millennium AD) was the pit-house, thatched from roof tip to ground. With growing populations and the need to thatch houses every generation, the demand for *Miscanthus* must have been high. Much more research is needed on these possibilities.

Andosol Productivity

> Volcanic ash soils or Andisols are among the most productive soils in the world.... In contrast, volcanic ash soils or Kurobokudo in Japan were originally regarded as impoverished soils because of high amounts of active aluminium, very low contents of available phosphate, low concentration of exchangeable bases and strong acidity (Shoji *et al.* 1993: 209).

How can both of these statements be true? We have seen that there are at least six types of andosols alone, with a clear distinction between allophanic and non-allophanic andosols, plus other soil types that incorporate volcanic ash. Moreover, there is tremendous variation in the chemical and mineral compositions of tephra-derived soils, with some clay products and by-products affecting those compositions. Agricultural productivity is said to depend "primarily on the degree or intensity of pedogenic development", which in itself is governed by soil composition, climate, precipitation, and temperature, etc. (Dahlgren *et al.* 2004: 115, 122) Thus, it is difficult to generalize, as the soil of any location has its own characteristics, and deficiencies must be identified for mitigation.

Andosol properties

The properties of tephra-derived soils have been described in detail by several authors (e.g., Ugolini & Zasoski 1979; Shoji *et al.* 1993; Shoji & Takahashi 2002; Takahashi & Shoji 2002; Ugolini & Dahlgren 2002; Dahlgren, Saigusa & Ugolini 2004; Lowe & Palmer 2005; Parfitt 2009; Takahashi & Dahlgren 2016; Taylor *et al.* 2016). Complicated chemical reactions are involved in producing these properties and conditions; here we can only present those properties inherent in andosols (Table 7.6) without reference to climate, tephra age, and composition – factors which will further affect fertility. Appendix 13 addresses some of the issues concerning nutrients. Table 7.6 does not distinguish between allophanic and non-allophanic andosols, where some of the greatest differences in fertility arise; this division is the most important for soil management and depends on the electric charges of materials in each type of andosol (Dahlgren *et al.* 2004). Within the tephric→vitric→andic sequence discussed above (cf. Table 7.3), there is a gradation of productivity.

Table 7.6 Andosol properties

	Advantages	Disadvantages	Mitigation
Water supply	Good water retention if pumaceous	Strongly hydrated → low bearing capacity & stickiness = waterlogging	Drainage
Colloid clays	High water-holding capacity; plant available	Can dehydrate	
Texture	Highly friable → good rootability; stable soil aggregates	If too porous → water loss	add bentonite /montmorillonite
Base metal supply	Fertile if basic or andesitic	Leaching of base metals; low exchangeable base-holding capacity leading to soil acidity	liming to add Ca; add gypsum or phosphogypsum to increase subsoil Ca
Phosphate	Major nutrient, from apatite	Active Al and Fe lead to P fixation; adsorb P from solution thus high P retention at >85%	add lime, gypsum, silica, green manuring, and P fertilizer
Potassium	Major nutrient, from weathered glass	Depleted in allophanic andosols; used for 2:1 clays in non-allophanic andosols	add potassium-K fertilizers
Nitrogen supply	small pool of easily larger pool of slow release	Gradual decrease of N with continued glass weathering	green manuring
Toxicity		potential Al, Na, and Io toxicity	liming, green manuring

Shoji *et al.* (1996) describe these progressive levels of productivity (Table 7.7, with my interpolations). Vitr-andic transitional andosols, which have not completely acquired andic properties, are often lowest in productivity because colloids which hold water and nutrients have not yet developed thoroughly; conversely, PO_4 is generally available for the same reason. Aluandic or Non-allophanic soils have variable productivity; the deficiency of P needs to be overcome with fertilizers, then non-allophanic andosols may be very productive as long as Al-toxicity does not develop. The main drawback of allophanic andosols is the low availability of P, which again can be mitigated with fertilization. For tropical andosols, Shoji *et al.* describe other problems and solutions.

These authors emphasize that the non-crystalline contents of andosols (allophane, imogolite, Al/Fe-humus complexes, ferrihydrite) are major determinants of the productivity of andosols, as they have "variable charge, high water-holding capacity, high phosphate retention, low bulk density, high friability, weak stickiness, formation of stable soil aggregates, etc." (Shoji *et al.* 1996: 606). High phosphate retention is obviously a negative andosol property, but what about colloid clays and humus complexes?

As we have learned, allophane and imogolite are "mutual solid solutions of silica, alumina, and water and minor amounts of bases" (Parfitt 2009, quoting Ross & Kerr), and the ratios vary from Si-rich to Al-rich (Al:Si = 4:1 max). Water molecules may occupy the internal spaces of allophane spheres or be chemically adsorbed to the AlOH component, accounting for their water retention properties.

Allophane is a kind of immature clay, and it can comprise up to 60% of an andosol (w/w); it is thus useful in pottery production. Allophane heated to high temperatures (900–1000°C) will begin to form the mineral mullite from the aluminium content (Parfitt 2009); and mullite is the hallmark of stoneware pottery, in contrast to earthenware which will melt at those temperatures due to the fluxing agency of the iron content (Barnes 1992/2001). Thus, areas that have volcanic soils with considerable Al content are prime areas for stoneware and porcelain production as well, since kaolin clay is dehydrated halloysite.

Because the colloidal clays have variable surface charges (both positive and negative), they can bind individual ions, such as those of phosphate, as well as humic substances. Organic matter that is bound to allophane or imogolite decomposes only very slowly (Parfitt 2009); thus, although the soil looks fertile because of its black humic content, much of that is bound to the colloidal clays where the humic and fulvic acids are not available for plant use. Dahlgren *et al.* (2004: 146) describe one way that allophane and other non-crystalline colloids in soil hinder organic matter decomposition: their high adsorption of P deprives micro-organisms from using it themselves so that they cannot act to decompose organic matter; the authors also note that carbon is also 'protected', with decreased rates of mineralization from C to CO_2. Phosphorus which is bound into these complexes is thus said to be 'retained' and not available for plant nutrition.

Al–humus complexes form preferentially in acidic soils, leading to the formation of non-allophanic andosols. These complexes perform the same function as the non-crystalline clays: adsorbing various metal ions and humic substances. In doing so, they stabilize the soil organic matter and 'protect' it from microbial decomposition, thus denying its use by plants (Takahashi & Dahlgren 2016). Andosols characterized by a predominance of Al–humus complexes exhibit the 'anti-allophane' effect of competition for Al between the humus and allophane, meaning that more Al-humus complexes will form rather than allophane. Non-allophanic soils have low base concentrations, high exchangeable Al content, and strong acidity – all of which contribute to Al-toxicity (Dahlgren *et al.* 2004).

Table 7.7 Comparisons of productivity among andosol types common in Japan

	Vitr-andic	**Alu-andic = Non-allophanic**	**Sil-andic = Allophanic**
Productivity level	low	variable	high when fresh
Colloid clays	not yet fully developed	fully developed	fully developed
Phosphorus	not yet bound to colloids ∴ plentiful	strong fixation capacity	strong fixation capacity leading to lower productivity
Humus retention	low	high, bound to Al	high, bound to colloid clays
Al		potentially toxic from 2:1 clays	adsorbed by humic materials
Base metals		low exchangeable base content	high nutrient retention but decreases with weathering
Acidity		increases with weathering	

The fact that there is a continuum between allophane/imogolite formation and Al-humus complexing means that local conditions will determine which direction and how much of each form. These different products have different dissolution rates, with allophane/imogolite dissolving more slowly than Al-humus complexes, with the ratios tending towards equilibrium (Dahlgren *et al.* 2004: fig. 9). However, both can be broken down by tilling, when the micro- and macro-aggregates of soil particles stuck together via these colloids are broken apart and the SOM (soil organic matter) is exposed to microbial action; dehydration may also promote microbial action which cannot take place while soil particles are saturated by water (Takahashi & Dahlgren 2016: 113), but with the loss of water-holding colloids, soil particle aggregates become unstable and lead to higher bulk density and less friability.

General cropping

Different crops have different nutritional requirements so that one may grow in a regime that is unsuitable for another; both Neall (2006) and Uchida (2000) discuss specific crops grown on andosols in terms of deficiencies and some mitigation procedures, while Shoji *et al.* give a brief survey of worldwide agricultural tephras (1993: ch. 8.2). Today, *Miscanthus* stands in Japan are artificially maintained and form only a minority of growth on andosols. By recent accounts, 34% of Japanese andosols are dry-fields (*hatake*), 28% are in paddy (*suiden*), 22% is pasture, and 17% is orchard (NARO/NIAES 2010). These statistics indicate that at least 78% of these 'impoverished soils' have been overcome by intensive mitigation and are producing crops for human consumption.

Neall (2006) gives a general overview of what crops grow best volcanic ash soils of different countries and what deficiencies might develop due to local conditions. For example, iodine toxicity is common in northern and central Honshu in newly constructed paddies. Mg and S may limit grass growth on andesitic or rhyolitic soils; lack of mafic mineral elements may lead to nutritional deficiencies (Cu, Ca, Iodine, or Mb and F toxicity). A 'boron–B' deficiency is noted for lettuce in Japan.

Surprisingly, some plants such as tea and buckwheat (in addition to hydrangea!) actually accumulate Al in chelated form; internal chelation is accomplished by various organic complexes (e.g., citrate, oxalate) and acts to detoxify absorbed Al (Yokel 2002). These plants may be said to be Al-tolerant, and it is no surprise that they are major crops grown on tephra-derived soils in Japan. Takahashi & Dahlgren (2016) report that higher levels of aqueous Al^3 in non-allophanic soils can protect against root rot in common bean and potato crops. Other Al-tolerant plants grown in Asia include wheat (*Triticum aestivum*, grown in Hokkaido), sickle senna (*Cassia tora*), soybean (*Glycine max*), and taro (*Colocasia esculenta*), while western Al-tolerant crops include rye, oats, maize, rapeseed, and snapbean (*Phaseolus vulgaris*) (Ma *et al.* 2001). These crops mainly exude organic acids around the root tips which then chelate Al and protect the roots. The tree, *Paraserianthes falcataria* (batay, Peacock's plume) indigenous to island Southeast Asia, is a pioneer species that is not only Al-tolerant but can fix nitrogen (Krisnawati *et al.* 2011).

Interestingly, flooding the soil in paddy-fields dissolves P from its bound forms and makes it available for plant use – but this does not help in upland dry-fields. One crop that is tolerant of low P levels is the sweet potato; as an introduced crop to Japan, it was farmed extensively in the volcanic soils of southern Kyushu to the extent that it is known by its regional name *satsuma-imo* (potato from the Satsuma Peninsula). Tephra-derived soils, if sufficiently fertilized and fine-grained at depth, are excellent for long root crops such as the large *daikon* radish (mooli) and edible burdock (*gobō*) in Japan. The soils are light and fluffy, amenable to extensive cultivation of buckwheat (Figure 7.13). Mulberry bushes have been a standard crop in volcanic areas for the raising of silkworms (Box 8).

Box 8 The Japanese Silk Industry on Volcanic Soils

In his novel, *The Gate*, Natsume SOSEKI wrote of Yamanashi Prefecture, but his words can equally well have been applied to Gunma (1910: 121, 123, brackets added):

> "*This gentleman has come all the way from Yamanashi prefecture to sell his cloth in Tokyo.*"...Mrs. Sakai explained that the soil of the pedlar's native region was of volcanic rock and unsuitable for raising rice, or even millet so that the natives in desperation had planted mulberry trees and taken to raising silkworms... [when he had] turned all his goods into cash [he went] back again in his little village of volcanic rock which lay to the north of Mount Fuji.

Gunma Prefecture became a major producer of silk from the late 19c, the silkworms fed with mulberry leaves grown on the volcanic soils of Mt Haruna and Mt Akagi (Noto & Barnes 2019). This area was favoured by the innovative agricultural expert FUNATSU Denjibei, a colleague of the German agricultural technician Max FESCA, who came to Japan at the age of 36 and stayed for 12 years (1882–1894) as a lecturer at the Komaba Agricultural School in Tokyo. Funatsu learned of German agricultural methods and instituted many of Albrecht THAER's "Principles of Rational Agriculture". He devoted much effort to instituting high-yield mulberry crops for silkworm fodder (Tanaka 2013), and he was instrumental in increasing silk production in Gunma using western technology. The Tomioka Silk Mill in Gunma, pictured below, was the first silk-reeling factory built in Japan, established in 1872. By the 1930s, Japanese raw silk exports were three times that of China, accounting for an 80% share of the world market (Ma 2005). Mulberry trees were the major crop grown on Gunma's volcanic flanks from Funatsu's time into the 1980s when the silk industry in Japan collapsed. Today Kiryu City is virtually the only large silk producer in the prefecture, while the Tomioka Silk Mill was designated a UNESCO World Heritage Site in 2014.

More and more, 'precision agriculture' in either low-tech or hi-tech forms is being practiced. Low-tech involves recognising spatial differences in soil development (using lobate tephra fallout patterns as in Figures 6.13, 6.15 and micro-climates) and applying specific fertilisers to micro-environments. Hi-tech relies on instrumentation such as "sensors, robots, GPS, mapping tools and data-analytics software to customize the care that plants receive" (Ling & Bextine 2017: n.p.).

Figure 7.13 Ploughed fluffy andosol field ready for sowing buckwheat used to make *soba* noodles, a Gunma specialty

Summary

This review has not been comprehensive but indicative of what happens to tephra after its deposition on the Earth's surface, especially under the climatic conditions characteristic of Japan. Thus, the focus has been on andosol formation and use, but the story will be quite different for tephra in arid regions or the tropics. In the least, an example has been provided where volcanic soils do *not* have "excellent physical and chemical properties for plant growth" (Cronin *et al.* 2008).

Lesson 1: The soil types and clays formed from tephra vary depending on parent material, climate, altitude/slope, temperature, precipitation rates, and vegetation, etc. Thus, it is difficult to make a blanket statement about tephra-derived soils, and doubly difficult because their characteristics modify over time depending on changes in all the above variables. Investigation into prehistoric land use patterns must take account of or reconstruct the relevant variables in the time slice being dealt with, which may even have changed seasonally.

Lesson 2: Not all andosols are created equal, though they must qualify as 'andic' (having the starred diagnostics in Table 7.1) to be considered an andosol. Within this category, however, there are differences in colour (melanic vs fulvic) that reflect their genesis under the influence of burned grasses and forest cover, respectively; there are also differences in the formation of non-crystalline clays and Al-humus complexes, the former being typical of allophanic andosols and the latter of non-allophanic andosols. This is the basic difference in tephras that determines fertility and aluminium toxicity levels.

Lesson 3: We do not know enough about plant regeneration after tephra eruptions, particularly how humans dealt with the changes in soil characteristics and plant communities. The gross categories of responses to volcanic eruptions (extinction – abandonment – making do – recolonization) need finer evaluations in the latter two categories. More studies of off-site macro- and micro-botanical remains are needed to understand the timing of plant recolonization of tephra and the potential human use of the plants available or how crops are re-initiated in areas covered by tephra.

As for the *kurobokudo* soils of Japan, these melanic andosols are now known to have formed under grasslands which were generally maintained by periodic firing. The fact that they first appear at the same time as colonization of the archipelago ca. 40,000 years ago speaks volumes for the possibility of human manipulation of the environment using fire. *Kurobokudo* soils can belong to either allophanic or non-allophanic andosols, with different problems for agriculture in each. Melanic andosols have developed under grassland, but if the grass is not continuously maintained by firing, the land is soon subject to forest encroachment.

We know a considerable amount about current cultivation practices in overcoming some of the deficiencies of andosols, but as yet we know almost nothing about relative problems in using tephra-derived soils versus non-tephrogenic soils in prehistory. The special conditions of wet-rice agriculture have been the focus of much archaeological work in Japan (Barnes 1986*a*, 1990), with paddy-field and irrigation complexes excavated at many sites. What we do not yet know is how tephra-covered uplands were integrated into the agricultural regime. Statistics for 2007 show that the amount of land devoted to paddy-fields (2.53 million ha) was slightly more than dry-fields (2.12 million ha); nevertheless, paddy-fields only contributed 22% of agricultural output in the form of rice, while dry-fields contributed 25% in the form of other grains, vegetables, and fruits (MAFF 2008; MLIT 2008). If this is the case in a fully mature wet-rice agricultural system, then how were uplands exploited in pre- and proto-historic Yayoi and Kofun period times while wet-rice technology was being developed and expanded? At several archaeological sites, ridge-and-furrow plots have been recovered from underneath tephra cover (Sakaguchi 2019; Kuwahata 2019; Noto & Barnes 2019). More detailed analysis is needed of the soils from these plots and the plant remains therein to assess dry-field agriculture in times of paddy-field reliance.

Before agriculture, we have the Palaeolithic and Jomon peoples who seem to have managed their environments with fire in order to exploit the plants and animals that inhabited grasslands. The Jomon are known horticulturalists who domesticated several species of plants, including the herb Perilla (*shiso*) and *hie* barnyard grass (*Echinochloa crus-galli* → *E. utilis/esculenta*), during their long habitation of the archipelago (Crawford 2011); middens have been proposed as the locus of initial domestication (Matsui & Kanehara 2006). But it is also suspected that late Jomon peoples also cultivated buckwheat and dry rice, crops imported from the continent; where and how did they manage the soil and field systems for these crops? There are many questions to answer regarding the use of tephra-derived soils in prehistory, not least how grasslands were used for hunting.

Figure & Table Sources

Figure 7.1 after (NARO [NIAES] 2010), Creative Commons Attribution License 4.0, modified by GLB: Ryukyus eliminated, new scale font

Figure 7.2 after (Obara *et al.* 2015: fig. 2), modified by GLB. © National Research and Development Agency, Japan International Research Center for Agricultural Sciences, with permission

Figure 7.3 after (Saigusa 2007: fig. 7), modified by GLB

Figure 7.4 extracted from (Velde & Meunier 2008: fig. 1.4) and modified by GLB

Figure 7.5 after (Lebedev 1921 via Anaf 2010: fig. 1), modified by GLB, with permission of CeROart

Figure 7.6 after (Anaf 2010: fig. 9), modified by GLB, with permission of CeROart

Figure 7.7 compiled and modified from (Velde & Meunier 2008: figs. 6.3, 6.4)

Figure 7.8 courtesy of Denice Cabanban with permission
Figure 7.9 left: Soil profile by US Department of Agriculture [Public domain], via Wikimedia Commons, [https://commons.wikimedia.org/wiki/File%3ASoil_profile.png]
Figure 7.9 right: Andosol profile by NARO (NIAES) under Creative Commons Attribution License 4.0, [https://twitter.com/i/web/status/840121713526374400]; modified by GLB: cropped, pixel enhanced
Figure 7.10 after (Shoji *et al.* 1993: fig. 2.1), modified by GLB
Figure 7.11 by 名古屋太郎 CC BY-SA 3.0, via Wikimedia Commons [https://commons.wikimedia.org/wiki/File%3A090124_wakakusa_yamayaki.jpg]
Figure 7.12 above: *sasa* By Sten CC BY-SA 3.0 via Wikimedia Commons, [https://commons.wikimedia.org/wiki/File%3ASasa-palmata-winter.JPG]
Figure 7.12 below: *nezasa* By harum.koh from Kobe city, Japan (Pleioblastus chino var. viridis) [https://commons.wikimedia.org/wiki/File%3APleioblastus_chino_var._viridis_(22800900376).jpg] CC BY-SA 2.0, via Wikimedia Commons
Figure 7.13 author's photo March 2001

BOX 9 Meiji-period woodblock print, archived in the National Diet Library, Japan. Public domain.

Table 7.1 compiled from (Obara *et al.* 2015; Imaya 2010; FAO-UN 2006: 42; Takahashi & Dahlgren 2016)
Table 7.2 compiled from (Shoji *et al.* 1993; Ugolini & Dahlgren 2002; WRB 2014; FAO-UN 2015: quotes from p. 62); Takahashi & Dahlgren 2016
Table 7.3 compiled from (FAO-UN 2015: 74, 84)
Table 7.4 extracted from (FAO-UN 2015: 35-6, 39-40; McDaniel *et al.* 2012: 19)
Table 7.5 from (Archibald *et al.* 2013: table 1), with permission
Table 7.6 compiled from (FAO-UN 2006: 70-71; Shoji *et al.* 1993: ch. 5)
Table 7.7 compiled from (Shoji *et al.* 1996).

References

in English unless otherwise noted

ARÉVALO, José Ramón & Jesús Ramón ABOAL (**2015**) "Species turnover during secondary succession in a laurel forest stand 50 years after clearcutting." *Forest Systems* 24.1: 5pp.
ANAF, Willemien (**2010**) "Study on the formation of heterogeneous structures in leached layers during the corrosion process of glass." *CeROArt EGG* 1: 22pp. + references [http://journals.openedition.org/ceroart/1561].
ARCHIBALD, Sally; Caroline ER LEHMANN & Jose L GÉOMEZ-DANS *et al.* (**2013**) "Defining pyromes and global syndromes of fire regimes." PNAS 110.16: 6442-6447.
ARNALDS, Olafur (**2013**) "The influence of volcanic tephra (ash) on ecosystems." *Advances in Agronomy* 121: 331-380.
BAREJA, Ben G (**2013**) "Plant types: II. C4 plants, examples, and C4 families." *Crops Review* [www.cropsreview.com/c4-plants.html].
BARNES, Gina L (**1986*a***) "Paddy field archaeology." *Journal of Field Archaeology* 13.4: 371-379.
---- (**1986*b***) "Japanese agricultural beginnings." *Nature* 322.6080: 595-596.
---- (**1990**) "Paddy field soils." *World Archaeology* 22.1: 1-17.
---- (**1992**) "The development of stoneware technology in southern Korea", pp. 197-208 in *Pacific Northeast Asia in Prehistory*, ed. by CM AIKENS & SN RHEE. Washington State University Press [republished as Chapter 3 in *State formation in Korea* by GL BARNES (Curzon/Routledge 2001)].
---- (**2010**) "Landscape and subsistence in Japanese history", pp. 321-340 in *Landscapes and societies: selected cases*, ed. by P MARTINI & W CHESWORTH. Dordrecht: Springer.
---- (**2015**) *Archaeology of East Asia: the rise of civilization in China, Korea, and Japan.* Oxford: Oxbow.
BARNES, Gina L & Tsutomu SODA (eds) (**2019**) *TephroArchaeology in the North Pacific.* Oxford: Archaeopress.
BAUER, Andreas & Bruce D VELDE (**2014**) *Geochemistry at the Earth's surface: movement of chemical elements.*

Heidelberg: Springer.
BELLWOOD, Peter (**1987**) "The prehistory of island Southeast Asia: a multidisciplinary review of recent research." *Journal of World Prehistory* 1.2: 171-224.
BELTON, David J; Olivier DESCHAUME & Siddharth V PATWARDHAN *et al.* (**2010**) "A solution study of silica condensation and speciation with relevance to in vitro investigations of biosilicification." *Journal of Physical Chemistry* B 114.31: 9947-9955.
BERGLUND, Björn E *et al.* (**2014**) "Traditional farming landscapes for sustainable living in Scandinavia and Japan: global revival through the Satoyama Initiative." *AMBIO: a Journal of the Human Environment* 43.5: 559-578.
BLUME, Hans-Peter & Peter LEINWEBER (**2004**) "Plaggen soils: landscape history, properties, and classification." *Journal of Plant Nutrition and Soil Science* 167: 319-327.
BREWER, A (**n.d.**) "Tropical grassland / savanna." *Biology B3 Tincher* [https://tropicalgrasslandsavannabrewerb3.weebly.com/succession.html].
Chigakudan (ed.) (**1981**). *Dictionary of Earth sciences.* Tokyo: Heibonsha (rev. ed.) (in Japanese).
CRAWFORD, Gary W (**2011**). "Advances in understanding early agriculture in Japan." *Current Anthropology* 52.4: S331-S345.
CRONIN, Shane J; Karoly NEMETH & Vincent E NEALL (**2008**) "Volcanism and archaeology", pp. 2185-2196 in *Encyclopedia of archaeology*, ed. by DM PEARSALL. Amsterdam: Elsevier.
CUNI-SANCHEZ, Aida; Lee JT WHITE & Kathryn J JEFFERY *et al.* (**2016**) "African savanna-forest boundary dynamics: a 20-year study." *PLOS* 11.6: e0156934.
DAHLGREN, RA; M SAIGUSA & FC UGOLINI (**2004**) "The nature, properties and management of volcanic soils." *Advances in Agronomy* 82: 113-182.
DEL MORAL, Roger & Sergei Yu GRISHIN (**1999**) "Volcanic disturbances and ecosystem recovery", pp. 137-160 in *Ecosystems of disturbed ground*, ed. by LR WALKER. Amsterdam: Elsevier.
DOVE, Patricia M; Nizhou HAN & Adam F WALLACE *et al.* (**2008**) "Kinetics of amorphous silica dissolution and the paradox of the silica polymorphs." *PNAS* 105.29: 9903-9908.
EASH, Neal S; Thomas J SAUER & Deb O'DELL *et al.* (**2016**) *Soil science simplified.* Hoboken, NJ: Wiley Blackwell.
EBINGER, John E (**1962**) "Validity of the grass species *Digitaria adscendens.*" *Brittonia* 14.3: 248-253.
Editorial (**1933**) "Forest fires in Japan." *Nature* 1933 (February 4): 161.
FAO-UN (**2006**) *World reference base for soil resources 2006: a framework for international classification, correlation and communication.* Rome: Food and Agriculture Organization of the United Nations.
---- (2015) *World reference base for soil resources 2014: international soil classification system for naming soils and creating legends for soil maps, update 2015.* Rome: Food and Agriculture Organization of the United Nations.
FDMA (Fire and Disaster Management Agency) (**2014**) "4 Wildland fire policies." *Heisei 26 Nenban Shobō Hakusho* [www.fdma.go.jp/html/hakusho/h26/h26/index3.html] (in Japanese).
FIANTIS, Dian; Frisa Irawan GINTING & Gusnidar *et al.* (**2019**) "Volcanic ash, insecurity for the people but securing fertile soil for the future." *Sustainability* 11.#3072: 19pp. [DOI: 10.3390/su11113072].
FORCE, Eric R (**2015**) "Geologic aspects of ancient Villanovan settlement distributions in central Italy." *Catena* 125: 162-168.
GE, Hong guang & Li hua CHENG (**2001**) "Study on separation of fulvic acid in peat coal." *Journal of Yanan University* (Natural Science Edition 2001.03.) [http://en.cnki.com.cn/Article_en/CJFDTOTAL-YSZR200103021.htm].
GERRARD, John (**2000**) *Fundamentals of soils.* London: Routledge.
GLASER, Bruno; Ludwig HAUMAIER & Georg GUGGENBERGER *et al.* (**2001**) "The 'terra preta' phenomenon: a model for sustainable agriculture". *Naturwissenschaften* 88: 37-41.
GOLDICH, Samuel S (**1938**) "A study in rock weathering." *Journal of Geology* 46.1: 17-58.
GONDOLA, István & Peter P PAPP (**2010**) "Origin, geographical distribution and phylogenic relationships of common buckwheat (*Fagopyrum esculentum* Moench.)." *The European Journal of Plant Science and Biotechnology* 6.2: 1-10.
GONG, Zitong (**1994**) "Formation and classifcation of anthrosols." *Transactions of the 15th World Congress of Soil Science, Acapulco, Mexico* 6A: 120-128.
GOPALAKRISHNAN, Subramaniam; Arumugan SATHYA & Rajendran VIJAYABARATHI *et al.* (**2015**) "Plant growth promoting rhizobia: challenges and opportunities." *3 Biotech* 5.4: 355-377.
HARSH, J (**2005**) "Amorphous materials", in *Encyclopedia of soils in the environment*, ed. by D HILLEL. Amsterdam:

Academic Press; abstract by *Science Direct* [www.sciencedirect.com/topics/agricultural-and-biological-sciences/imogolite].

HARTEMINK, Alfred E (**2016**) "The definition of soil since the early 1800s." *Advances in Agronomy* 137: 73-126 [DOI: 10.1016/bs.agron.2015.12.001].

HEAR (Hawaiian Ecosystems at Risk project) (**2017**) "*Morella faya* (Myricaceae)" [www.hear.org/species/morella_faya/].

HIRADATE, Syntaro; Toshie NAKADAI & Haruo SHINDO *et al.* (**2004**) "Carbon source of humic substances in some Japanese volcanic ash soils determined by carbon stable isotopic ratio, δ^{13}C." *Geoderma* 119: 133-141.

HOLLIDAY, Vance T (**2004**) *Soils in Archaeological Research.* Oxford Scholarship Online 2020 [DOI: 10.1093/oso/9780195149654.001.0001].

HOLLOWAY, JM & RA DAHLGREN (**2002**) "Nitrogen in rock: occurrences and biogeochemical implications." *Global biogeochemical cycles* 16.4: 1-65.

HONMA, Hiroji & Yuuko ITIHARA (**1981**) "Distribution of ammonium in minerals of metamorphic and granitic rocks." *Geochimica et Cosmochimica Acta* 45.6: 983-988.

HONNA, T; S YAMAMOTO & K MATSUI (**1988**) "A simple procedure to determine melanic index that is useful for differentiating melanic from fulvic andisols." *Pedologist* 32: 69-78 (in Japanese with English title and abstract).

HOSONO, Mamoru & Takashi SASE (**2015**) "Formation history of *kuroboku-doso*: a trial interpretation from the perspective of the anthropogenic ecosystem." *The Quaternary Research - Daiyonki Kenkyū* 54.5: 323-339 (in Japanese).

ICENHOWER, Jonathan P & Patricia M DOVE (**2000**) "The dissolution kinetics of amorphous silica into sodium chloride solutions: effects of temperature and ionic strength." *Geochimica et Cosmochimica Acta* 54.2: 4193-4203.

IMAYA, Akihiro; Shuichiro YOSHINAGA & Yoshiyuki INAGAKI *et al.* (**2010**) "Proposal for advanced classification of brown forest soils in Japan with reference to the degree of volcanic ash additions." *Soil Science and Plant Nutrition* 56.3: 454-465.

INOUE, Jun; Ryo NISHIMURA & Hikaru TAKAHARA (**2012**) "A 7500-year history of intentional fires and changing vegetation on the Soni Plateau, Central Japan, reconstructed from macroscopic charcoal and pollen records within mire sediment." *Quaternary International* 254: 12-17 [DOI: 10.1016/j.quaint.2010.08.012].

JAMES, Peter; David CHESTER & Angus DUNCAN (2000) "Volcanic soils: their nature and significance for archaeology", pp. 317-338 in *The archaeology of geological catastrophes*, ed. by WG MCGUIRE *et al.* GSL Special Publications 171. London: Geological Society of London.

JOBBÁGY, Esteban G & Robert B JACKSON (**2003**) "Patterns and mechanisms of soil acidification in the conversion of grasslands to forests." *Biogeochemistry* 64: 205-229.

JOHNSON-MAYNARD, JL; PA MCDANIEL & DE FERGUSON *et al.* (**1998**) "Changes in soil solution chemistry of andisols following invasion by bracken fern." *Soil Science* 163.10: 814-821.

KATO, Yoshiro (**1983**) "Formative mechanisms of tephra-derived soil", pp. 5-30 in *Tephra-derived soil - formation, character and typology*, ed. by Nihon Dojō Hiryō Gakkai. Tokyo: Hakuyūsha (in Japanese).

KAWANO, Kazuaki (**2008**) "Bamboo swidden: agriculture able to sustain forest growth and eating the forest." *Sato Project News Letter* 43.12 December [www.chikyu.ac.jp/sato-project/news_letter/news_letter_43.pdf] (in Japanese).

KIRKMAN, JH & WJ MCHARDY (**1980**) "A comparative study of the morphology, chemical composition and weathering of rhyolitic and andesitic glass." *Clay Minerals* 15: 165-173.

KRISNAWATI, Haruni; Eveliina VARIS & Marrit KALLIO *et al.* (**2011**) *Paraserianthes falcataria (L.) Nielsen: ecology, silviculture and productivity.* Bogor, Indonesia: Center for International Forestry Research.

KUBO, Tetsushige (**1982**) "Classification of forest soils in Japan." *Journal of Agricultural Research Quarterly* 16.3: 212-217.

KUWAHATA, Mitsuhiro (**2019**) "Restoration of agricultural assets after volcanic disasters in southwest Japan", pp. 192-201 in BARNES & SODA (eds) (2019).

LANGAN, Liam; Steven I HIGGINS & Simon SCHEITER (**2017**) "Climate-biomes, pedo-biomes or pyro-biomes: which world view explains the tropical forest-savanna boundary in South America?" *Journal of Biogeography* 44: 2319-2330.

Lebedev AA (**1921**) "Polymorphism and annealing of glass." *Transactions of the Optical Institute, Petrograd* 2: 1-20.

Ling, Geoffrey & Blake Bextine (**2017**) "Precision farming increases crop yields." *Scientific American* 26 June 2017 [www.scientificamerican.com/article/precision-farming/].

LMIL (Legume-Microbe Interactions Laboratory) (**2018**) "Plant molecular biology symbiotic nitrogen fixation: associative nitrogen fixation." [http://staceylab.missouri.edu/bradyrhizobium-japonicum-genetics-and-genomics/].

Lowe, DJ & DJ Palmer (**2005**) "Andisols of New Zealand and Australia." *Journal of Integrated Field Science* 2: 39-65.

Lowe, David J; Megan R Balks & Nadia Laubscher (**2014**) "Once despised now desired: innovative land use and management of multilayered pumice soils in the Taupo and Galatea areas, central North Island, New Zealand." 2014 Society of Soil Science "*Soil Science for Future Generations*" conference field trip notes.

Ma, Debin (**2005**) "Between cottage and factory: the evolution of Chinese and Japanese silk-reeling industries in the latter half of the nineteenth century." *Journal of the Asia Pacific Economy* 10.2: 195-213.

Ma, Jian Feng; Peter R Ryan & Emmanuel Delhaize (**2001**) "Aluminium tolerance in plants and the complexing role of organic acids." *Trends in Plant Science* 6.6: 273-278.

Machida, Hiroshi & Fusao Arai (**2011**) *Atlas of tephra in and around Japan.* Tokyo University Press (in Japanese with English title).

MAFF (Ministry of Agriculture, Forestry and Fisheries) (**2008**) *White paper on foodstuffs, agricultural industry, and farming villages.* Tokyo: Norin Suisansho (in Japanese).

Marler, Thomas & Roger del Moral (**2013**) "Primary succession in Mount Pinatubo." *Communicative & Integrative Biology* 6.6: e25924.

Matsui, Akira & Masa'aki Kanehara (**2006**) "The question of prehistoric plant husbandry during the Jomon period in Japan." *World Archaeology* 39.2: 259-273.

McDaniel, PA; DJ Lowe & O Arnalds *et al.* (**2012**) "Andisols", 52pp. in *Handbook of soil sciences, vol. 1: Properties and processes*, ed. by PM Huang *et al.* Boca Raton, FL: CRC Press.

MLIT (Ministry of Land, Infrastructure and Transport) (**2008**) *Land white paper.* Tokyo: Kokodu Kotsusho (in Japanese).

Morford, Scott L; Benjamin Z Houlton & Randy A Dahlgren (**2011**) "Increased forest ecosystem carbon and nitrogen storage from nitrogen rich bedrock." *Nature* 477: 78-81.

Morisada, Kazuhito; Kenji Ono & Hidesato Kanomata (**2004**) "Organic carbon stock in forest soils in Japan." *Geoderma* 119: 21-33.

Nakazawa, Katsuaki (**2011**) "Hunting and fields", pp. 201-225 in M Yumoto *et al.* (in Japanese).

Nakayama, Seiji (**2010**) *Botanical archaeology and the origins of Japanese agriculture.* Tokyo: Dōseisha (in Japanese).

Nanzyo, Masami (**2004**) *Utilization of apatite in volcanic ash using functions of plants.* Sendai: Tōhoku Daigaku Dagakuin Nōgaku Kenkyūka (in Japanese & English).

NARO/NIAES (National Agriculture and Food Research Organization) (**2010**) "*Kurobokudo.*" Nihon Dojō Inbentorii (in Japanese) [http://soil-inventory.dc.affrc.go.jp/explain/D.html].

Neall, VE (**2006**) "Volcanic soils", pp. 23-45 in *Land use, land cover and soil sciences*, Vol. VII in Encylopedia of Life Support Systems, ed. by W Verheye. Paris: Developed under the Auspices of the UNESCO, Eolss Publishers, Paris, France [http://www.eolss.net].

NDHG (Nihon Dojō Hiryō Gakkai) ed. (**2015**) *Secret of the soil.* Tokyo: Asakura (in Japanese with English title).

Noto, Takeshi & Gina L Barnes (**2019**) "Farming tephrogenic soils in Gunma: before and after volcanic eruptions", pp. 234-260 in Barnes & Soda (eds) (2019).

Obara, Hiroshi; Yuji Maejima & Kazunori Kohyama *et al.* (**2015**) "Outline of the Comprehensive Soil Classification System of Japan – First Approximation." *Japan Agricultural Research Quarterly* 49.3: 217-226.

Obase, Keisuke; Yutaka Tamai & Takashi Yajima *et al.* (**2007**) "Mycorrhizal associations in woody plant species at the Mt. Usu volcano, Japan." *Mycorrhiza* 17: 209-215.

Ogura, Jun'ichi (**2011**) "Column 2: Micro-charcoal and *kurobokudo*", pp. 137 in M. Yumoto *et al.* (in Japanese)

Ōmura, Kazuo (**1993**) "Civil engineering technology suited to volcanic sediments", pp. 247-261 in *Volcanic ash archaeology*, ed. by Fusao Arai. Tokyo: Kokon Shoin (in Japanese).

Parfitt, RL (**2009**) "Allophane and imogolite: role in soil biogeochemical processes." *Clay Minerals* 44: 135-155.

Peay, Kabir G; Thomas D Bruns & Peter G Kennedy *et al.* (**2007**) "A strong species–area relationship for eukaryotic soil microbes: island size matters for ectomycorrhizal fungi." *Ecology Letters* 10: 470489.

RETALLACK, Gregory J (**2019**) *Soils of the past: an introduction to palaeopedology*. 3rd edition. Chichester, UK: John Wiley & Sons Ltd.

SAIGUSA, Masahiko (**2007**) "The university life captivated with the studies on the non-allophanic andosols and field science." *Hiryō Kagaku* 29: 1-62 (in Japanese).

SAKAGUCHI, Hajime (**2019**) "Archaeological investigation of the seasonality and duration of the 6th-century eruptions from Mt Haruna", pp. 184-191 in BARNES & SODA (eds) (2019).

SASE, Takashi; HOSONO, Mamoru & MIURA, Hideki *et al.* (**1997**) "Comment on 'Geological investigation on the origin of the black soil, distributed in Japan' by T. YAMANOI, 1996." *The Journal of the Geological Society of Japan* 103.7: 692-695 (in Japanese).

SCHAETZL, Randall & Michael L THOMPSON (**2015**) *Soils: genesis and geomorphology*. Cambridge University Press.

SCHMIDT, Michael W; Jan O SKJEMSTAD, & Cornelia JÄGER *et al.* (**2002**) "Carbon isotope geochemistry and nanomorphology of soil black carbon: black chernozemic soils in central Europe originate from ancient biomass burning." *Global Biogeochemical Cycles* 16.4: 70-1~70-8.

SHIMOMURA, Satoshi (**2011**) "Archaeological sites of the Yayoi–Kofun periods and the world of grasslands viewed from those sites", pp. 143-168 in YUMOTO *et al.* (in Japanese).

SHINDO, Haruo & Syusaku NISHIMURA (**2016**) "Pyrogenic organic matter in Japanese andosols: occurrence, transformation, and function", pp. 29-62 in *Agricultural and environmental applications of biochar: advances and barriers*, ed. by M GUO *et al.* Madison, WI: SSSA.

SHIRATO, Yasuhito (**2015**) "Review: *Japanese soils: Japanese soils: black soils and Jomon culture illuminated by geology*, by T. Yamanoi." *Nihon Dojō-Hiryōgaku Zasshi* 86.3: 281 (in Japanese).

SHOJI, Sadao; Masami NANZYO & Randy A DAHLGREN (**1993**) *Volcanic ash soils: genesis, properties and utilization*. Tokyo: Elsevier.

SHOJI, Sadao; Masami NANZYO & Randy A DAHLGREN *et al.* (**1996**) "Evaluation and proposed revisions of criteria for andosols in the *World Reference Base for Soil Resources*." *Soil Science* 161.9: 604-615.

SHOJI, Sadao & Tadashi TAKAHASHI (**2002**) "Environmental and agricultural significance of volcanic ash soils." *Global Environmental Research* 6: 113-135.

SOSEKI, Natsume (**1910**) *The Gate*. Translated by Francis MATHY. London: Peter Owen, 1972.

STAPLETON, Chris (**2006-2018**) *Bamboo identification* [http://bamboo-identification.co.uk/index.html].

STEENHOUDT, Oda & Jos VANDERLEYDEN (**2000**) "Azospirillum, a free-living nitrogen-fixing bacterium closely associated with grasses: genetic, biochemical and ecological aspects." *FEMS Microbiology Reviews* 24: 487-506.

SUGA, Takeshi; Atsushi USHIMARU & Hiroyuki TANAKA (**2011**) "The history of grasslands in the Japanese Archipelago and their plant and insect populations", pp. 101-122 in YUMOTO *et al.* (in Japanese).

TAKAHASHI, T & S SHOJI (**2002**) "Distribution and classification of volcanic ash soils." *NASA Global and Environmental Research* 6: 83-97.

TAKAHASHI, Tadashi & Randy A DAHLGREN (**2016**) "Nature, properties and function of aluminum–humus complexes in volcanic soils." *Geoderma* 263: 110-121.

TAKAHASHI, Tadashi; Randy A DAHLGREN & Takashi SASE (**1994**) "Formation of melanic epipedons under forest vegetation in the xeric moisture regime of northern California." *Soil Science and Plant Nutrition* 40.4: 617-628.

TAKESHI, Hideo (**1978**) "Smectites", pp. 221-242 in *Clays and clay minerals of Japan*, ed. by T SUDO & S SHIMODA. Tokyo & Amsterdam: Kodansha & Elsevier.

TAYLOR, Matthew D; David J LOWE & Peter HARDI *et al.* (**2016**) "Comparing volcanic glass shards in unfertilised and fertilised Andisols derived from rhyolitic tephras, New Zealand: evidence for accelerated weathering and implications for land management." *Geoderma* 271: 91-98.

THINES, Marco; Takayuki AOKI & Pedro W CROUS *et al.* (2020) "Setting scientific names at all taxonomic ranks in italics facilitates their quick recognition in scientific papers." *IMA Fungus* 11: #25, 5pp. [DOI: 10.1186/s43008-020-00048-6].

TSUKADA, Matsuo; Shinya SUGITA & Yoko TSUKADA (**1986**) "Oldest primitive agriculture and vegetational environments in Japan." *Nature* 322: 632-634.

UCHIDA, R (**2000**) "Essential nutrients for plant growth: nutrient functions and deficiency symptoms", pp. 31-55 in *Plant nutrient management in Hawaii's soils, approaches for tropical and subtropical agriculture*, ed. by JA SILVA & R UCHIDA. Honolulu, HI: College of Tropical Agriculture and Human Resources, University of Hawaii at Manoa.

Ugolini, FC & RA Dahlgren (**2002**) "Soil development in volcanic ash." *Global Environmental Research* 6.2: 69-81.
Ugolini, FC & RJ Zasoski (**1979**) "Soils derived from tephra", pp. 83-124 in *Volcanic activity and human ecology*, ed. by PD Sheets & DK Grayson. New York: Academic Press.
USDA Soil Survey Staff (**1999**) *Soil taxonomy: a basic system of soil classification for making and interpreting soil surveys.* Washington, DC: US Department of Agriculture Natural Resources Conservation Service.
Uzawa, Akihiko & Minoru Kitchen (**2012**) "Surrounded by a Japanese beech forest: Shirakami-Sanchi and northern Tohoku." *Niponica* 7: 22-23.
Velde, Bruce & Alain Meunier (**2008**) *The origin of clay minerals in soils and weathered rocks.* Berlin: Springer.
Vogel, Sebastian; Michael Märker & Ivano Rellini *et al.* (**2016**) "From a stratigraphic sequence to a landscape evolution model: Late Pleistocene and Holocene volcanism, soil formation and land use in the shade of Mount Vesuvius (Italy)." *Quaternary International* 394: 155-179.
von Verschuer, Charlotte (**2016**) *Rice, agriculture, and the food supply in premodern Japan.* London, Taylor & Francis (Kindle edition, translated and updated from the original French edition).
Wall, Louis Gabriel (**2000**) "The actinorhizal symbiosis." *Journal of Plant Growth Regulation* 19.2: 167-182.
Wardle, Peter (**1991**) *Vegetation of New Zealand.* Cambridge University Press.
Yamanoi, Tohru (**1996**) "Geological investigation on the origin of the black soil, distributed in Japan." *The Journal of the Geological Society of Japan* 102.6: 526-544 (in Japanese with English title and abstract).
Yokel, RA (**2002**) "Alumium chelation principles and recent advances." *Coordination Chemistry Reviews* 228.2: 97-113.
Yumoto, M; Sato, H & K Iinuma (**2011**) *Environmental history of fields and plains.* Tokyo: Bun-ichi (in Japanese).
Zorn, Tobias; Kouji Nakayama & Osamu Hashiramoto (**2002**) "Forest fire situation in Japan." *IFFN* 26: 54-60.

Part I

Reflections

In reciting the compilation of the Japanese landmass, we have touched on two very important topics: the relative importance of magmatism and accretion tectonics in creating the Japanese landmass,[1] and the episodic nature of the formation of that landmass. Two other issues deserve our attention: the differing views on 'continental collision' in relation to Japan, and new research by Japanese geologists on 'second continent formation'.

The Relative Importance of Magmatism and Accretion Tectonics

Japan has long been characterized as a volcanic island arc – with the emphasis on 'volcanic' – implying that its major constituents are igneous rocks. Now, with the majority of its basement rocks classified as sedimentary (albeit with igneous underplating) and with Accretionary Complexes (AC) recognized as the main components of the Japanese basement geology, the importance of accretion tectonics cannot be overstated – for these views totally overturn previous theories of the formation of the Japanese Islands. However, one cannot have sedimentary rocks without sediments; and in the case of Japan, the majority of sediments in AC are erosional products of former igneous rocks. Except for the Ryoke Belt, containing a granite batholith, the records of former magmatic activity have been erased by erosion, and the importance of those processes is hidden in the succeeding sedimentary complexes. It is now clear that both magmatism and accretion are equally important in Japanese basement formation, and they both result from subduction and are linked by erosion – as part of the ongoing 'Rock Cycle' (p. 28 figure).

Although alien bodies such as volcanic seamounts and ocean floor slices (ophiolites) have been incorporated into the Japanese landmass, they are by far minor components of the geological setting. However, a distinction must be made between accretion tectonics as the means of forming the dominant geological belts (cf. Apx 6) that comprise the Japanese Islands and the chain of causation inherent in the Rock Cycle that allowed accretion processes to occur. For accretion to take place, rather than the oceanic plate eroding the continental edge (which is quite common), the subduction trench must have a sufficient amount of eroded sediments from the landmass itself to be bulldozed up again into an accretionary prism (which is quite rare.) Erosion is stimulated by uplift; thus, the creation of mountains leads to their erosion and the eroded products contribute to the formation of Accretionary Complexes – sedimentary units later subject to metamorphism. Thus, the Japanese landmass is neither a pure volcanic island chain implying only the presence of volcanic rocks, nor has it always been an archipelago, with most materials accreted to the continental edge before rifting only 16 million years ago.

Episodic Formation of the Japanese Landmass

The worldwide presence of Accretionary Complexes has been recognized as both uncommonly formed and rare in survival (see Kasahara *et al.* 2010: fig. 2). In contrast, tectonic erosion by the subducting oceanic plate is the more common process: the subducting plate grinds away at the continental edge,

[1] My use of 'accretion' here refers specifically to the formation of Accretionary Complexes; see the discussion in the second section here.

eroding what AC might have existed in the past. Thus, those AC that remain in Japan are exemplary for the study of subduction zone processes and continental landmass increase.

Accretionary processes were not gradual and ongoing but intermittent and drastic in nature. Wakita *et al.* (2018) identify six stages of AC formation between the Cambrian and Palaeogene (541–23 Ma). Except for the very early example (South Kitakami Terrane), which is composed of an ophiolite and forearc sediments, the majority of AC consist of small amounts of basalt and much chert (both of these from the ocean floor), and some trench fill (Ibid.: fig. 3). The ages and P/T nature of metamorphic transformations of the various AC in these several stages are laid out in their figs. 3 and 4.

AC formation decreased primarily during times of oblique subduction of an oceanic plate, and their edge could be eroded tectonically by a subducting plate. They increased particularly during times of uplift which exacerbated land-based erosional processes, making available more terrestrial sediments to the subduction trench to be bulldozed back up on land. Uplift may be caused by the subduction of an oceanic ridge or very young oceanic crust, or the exhumation of a metamorphosed AC. An extraordinary example of the effects of exhumation is the case of the ultra high-pressure metamorphic unit formed during the collision of the North China and South China continental blocks at ca. 250 Ma (cf. Figure 3.5). It is estimated that this unit was metamorphosed at the 'diamond-level' in the deep crust, and when it was exhumed, it displaced three million cubic kilometres of terrestrial materials from the suture zone. These materials were then washed into the offshore subduction trench, only to be subsequently accreted as the Jurassic AC; the volume of eroded materials was equivalent to a land area of ca. 144 km (90 miles) cubed (Isozaki 1997*b*: 46)!

Based on these cycles of uplift, igneous activity, erosion and accretion, three episodes of crust-building (orogeny) in the Japanese area have been proposed by different authors (Maruyama 1997: 112):

- the Abean Orogeny at 450 Ma, named after an early historic clan, the Abe, in the Tohoku district (Minato *et al.* 1979),
- the Akiyoshi Orogeny at 180–250 Ma, identified by Kobayashi (1941) and linked with the incorporation of the Akiyoshi limestones (cf. BOX 3), but rejected today as a single process,
- and the Sakawa Orogeny at 100 Ma (Kobayashi 1941).

These differ from Maruyama's own datings of 450, 320, 250–180 Ma (1997: 112), which in themselves do not correspond to the dating of oceanic ridge subduction at 300, 220, 180? and 75 Ma in his figure 12. Nevertheless, Maruyama estimated that an oceanic ridge is subducted every 100 million years or so and represents the climax of an orogenic cycle. Hirsch & Ishida (2002: 4) report on this cycle as consisting of the formation of "an accretionary complex (AC), tectonically interlayered regional metamorphic belts and posterior granitoid intrusions", with the latter deemed most important because they provide the granitic sediments for accretion (Hirsch & Ishida 2002: 4). The granites are produced especially during ridge subduction as the ridge, containing considerable water, disrupts the mantle wedge and magma is formed from ridge dehydration.

Though the above orogenic names have been given here for correlation with the existing literature, there is a shift among Japanese geologists not to use these old names since they are associated with out-dated concepts of geosyncline theory – but the dates remain important. Thus, there is a tendency of late to speak of 'events' rather than 'orogenies'. Nevertheless, within worldwide physical geology, there is increasing recognition that the main feature of some named orogenies (like the Sakawa above) is not mountain-building *per se* but rather addition to continental landmass increase via accretionary tectonics and production of granitic rocks. Maruyama (1997) terms this 'cordilleran-type orogeny' – in contrast to the 'collision-type orogeny' that has produced the Himalayas, for example.

Confusion over Collision

Nappes and thrust zones

The use of the Alpine terms 'nappe' (and 'klippe') in relation to accretionary tectonics has caused much confusion, not only about the mechanics involved in accretion but in the appropriateness of the terms themselves to the accretionary context. 'Nappe' in its most basic meaning simply implies a sheet covering something, like a tablecloth, from which the term is derived in French; Isozaki used it to describe the Permian–Triassic AC (Akiyoshi, Sangun, Maizuru Belts) "as sub-horizontal nappes, forming large-scale imbricated bodies" (Isozaki 1997*a*: 5). 'Nappe' has also been applied to a pile of Accretionary Complexes in eastern Hokkaido (Ishiwatari *et al.* 2016). This mirrors the downward younging polarity of sediment accumulation within the accretionary prism, stacked more like napkins than tablecloths (see Figure 2.5: forearc ridge). Downward younging is the reverse of normal stratigraphy, which is upward younging; the former is achieved by underplating of units from below with younger materials.

One definition of nappe offered is "thrust sheets which have moved more than about 10 km relative to the footwall" (Ramsay & Huber 1987: 521), but it does not distinguish between over-thrusting and under-thrusting – a crucial difference. Overthrust faulting commonly happens in continent-continent collision but rarely in accretionary contexts – though possibly responsible for emplacing ophiolites (Ishiwatari *et al.* 2016). For ACs, the contents are *underthrust*, as the subducting plate embeds material into the continental edge in a downward younging polarity.

Isozaki (1997*a*: fig. 3) used the term 'nappe' to describe the positioning of the Permian-Triassic AC complex (Akiyoshi, Sangun, and Maizuru Belts) as they lay *over* a younger Jurassic AC (Mino-Tanba). This figure has been revised in Isozaki *et al.* (2010: fig. 8) so that the sequence of Permian–Triassic–Jurassic ACs are younging from west to east, each underplating the earlier AC. The faults separating them are underthrust faults – having the essential nature of the subduction underthrusting of an oceanic plate – accommodating the angular stacking of the ACs as they accumulate under the forearc region and extend the continental edge. In conclusion, Isozaki and colleagues justify their application of 'nappe' to ACs: "Detailed field mapping of ACs coupled with OPS analysis clarified that all AC units in Japan occur as thin subhorizontal fault-bounded geologic bodies, i.e., nappe, and that on the whole they show a clear downward (and oceanward) younging polarity" (Isozaki *et al.* 2010: 86).

Charvet (2013: 92) has mounted a criticism that "the first and main difficulty" with the Pacific-Miyashiro Orogeny model (Miyashiro 1973; Maruyama 1997) is that it is based on ridge subduction causing nappe (AC) emplacement, arguing that ridge subduction causes tectonic erosion, not accretion. This may be true in the majority of cases, but in Japan, there is clear evidence that ridge subduction is followed several million years later by AC formation (Isozaki *et al.* 2010: fig. 3). Isozaki and colleagues do not deny the reality of tectonic erosion at the continental edge, as is currently happening along northeastern Japan, but they argue that the subduction of a hot oceanic ridge will cause both uplift and magma production at the continental edge. The uplift and increased volcanic activity will result in a higher rate of erosion, which supplies sediments to the trench. It is these sediments that are then bulldozed back up into the continental edge to form an AC, long after the ridge has been subducted.

Accretion vs collision

In arguing for tectonic erosion rather than accretion as the main result of subduction, Charvet instead proposes a different process resulting in the formation of ACs. He insists (2013: 94) that "the arrival at the trench of a small block with a lighter crust, and therefore positive buoyancy, like a mature arc or micro-continent is able to induce a collision producing permanent compressive structures". He offers the North Shimanto Belt as an example of a 'continent-type block', which collided and underthrust along the BTL between 80 and 60 Ma. However, he does not state where this 'Shimanto block' came from and does not define what 'continent-like' means in terms of the North Shimanto Belt. Isozaki *et al.* (2010: 86) identify the North Shimanto Belt as an AC formed 80 Ma in the lower crust of a forearc zone; it partially underwent high pressure (deep) metamorphism between 70–60 Ma after which it came to be positioned under the main islands along the BTL, a previous subduction plane. Thus, it was not an allochthonous micro-continent that came from somewhere else.

Part of the problem here is whether the 'docking' of the North Shimanto Belt is viewed as accretion or collision. Is this perhaps a false dichotomy? Cloos (1993: 734) suggests that the "terms collision and collider should be reserved for bodies whose subduction caused or would cause some kind of rearrangement in the pattern of plate motion, generally the initiation of a new subduction zone and the creation of mountains". This is said to happen in continent-continent collision, which does not occur in "peripheral orogens," such as the Japanese landmass, permanently facing the open ocean at the continental edge and not being approached by another continental mass (Nance & Murphy 1994: 51).

However, island arcs and other oceanic bodies do, in a sense, collide with a continental edge in peripheral contexts. Orogenic zones around the world are filled with accreted island arcs, such as in the CAOB and western coast of North America – traditional 'exotic terrane' accretions (Howell 1995). One such event, the accretion of the South Kitakami Terrane is known for Japan. Otherwise, recognized instances of collision within the Japanese Islands include two collision sutures: Izu Arc with Honshu (Figure 4.7) and the Kurile microplate with Hokkaido (Figure 5.1) (Isozaki *et al.* 2010: 95), while only two instances of continent-continent collision affecting Japan are identified: the major collision of the North and South China Blocks at 230 Ma that left geological remnants in Japan (Chapter 3: Triassic) and the current minor colliding of SW and NE Japan along the Itoigawa-Shizuoka Tectonic Line (cf. Figure 3.6) to form the southern Japan Alps (Isozaki *et al.* 2010: 83, 93).

In sum, the very mechanisms that Charvet proposes for causing nappe (AC) formation (arc-arc and continent-continent collision) are already recognized as instances of collision but do NOT result in AC, and those that are not recognized as collision can be accommodated under the umbrella of accretionary tectonics – where things do collide but are accreted without rearrangement of plate motion or initiation of a new subduction zone or formation of ACs. If other 'small blocks' envisioned by Charvet have arrived to cause AC formation, then where are these blocks now? Where did they come from? The South Shimanto Belt is currently in the process of AC formation, as ascertained by the ODP drilling project (Taira *et al.* 1992), and certainly not marked by the arrival of a small continent. It seems the issue of terminology surrounding 'collision', such as discussed in Garzanti (2008) must be revisited vis-à-vis the mathematical calculation of stress and strain in fault systems within structural geology, and how much deformation is required to label an event a 'collision' – major or minor – as opposed to a 'meeting' or 'docking' or indeed 'accretion'. In their review of the problems, van Staal & Zagorevski clearly exclude from the concept of 'collision' the growth of an accretionary wedge through incorporation of "small bodies of sedimentary and/or volcanic rocks" (2020: 103); however, they propose that the accretion of island arcs, continental ribbons, and the like are a type of 'soft collision', in contrast to 'hard collision' when full subduction of an oceanic plate brings two continental masses into contact and collision (2020: 104). Maruyama & Safonova (2019) offer an interesting discussion over 'collision' that puts many aspects in historical context.

A New Paradigm: Second Continent Formation

In 2010, Japanese geologists announced to the world that they had identified the driving force that makes the Pacific Ring of Fire so tectonically active (Kasahara *et al.* 2010). This is the phenomenon of 'second continent' formation consisting of granitic materials at the Mantle Transition Zone (MTZ), between 400 and 660 km depth (figure below). These radioactive granitic materials brought into the mantle by the subducting oceanic slab, through tectonic erosion and island arc subduction, are thought to contain enough radioactive heat to power the formation of mantle plumes (Kawai *et al.* 2013). Such second continents are distinguished from 'slab graveyards', where ancient subducted cold oceanic slabs were downwelled to lie between 2700 and 2900 km depth at the mantle-core boundary (Voosen 2016; Klein *et al.* 2017; Niu 2018). Maruyama *et al.* (2007) first hypothesized that this downwelling of cold slab material caused a reciprocal upwelling of hotter mantle material which triggered the development of hot plumes, rifting of back-arc basins, and splitting off of microplates. However, Kawai and colleagues focus not on the cold slab but on the hot granitic material brought down into the mantle by that slab. Both findings link interaction between the subducting slab materials and the overlying continental plates in a kind of boomerang effect of plume generation. The second continent model joins the long stream of Japanese contributions to Plate Tectonics theory (cf. BOX 1).

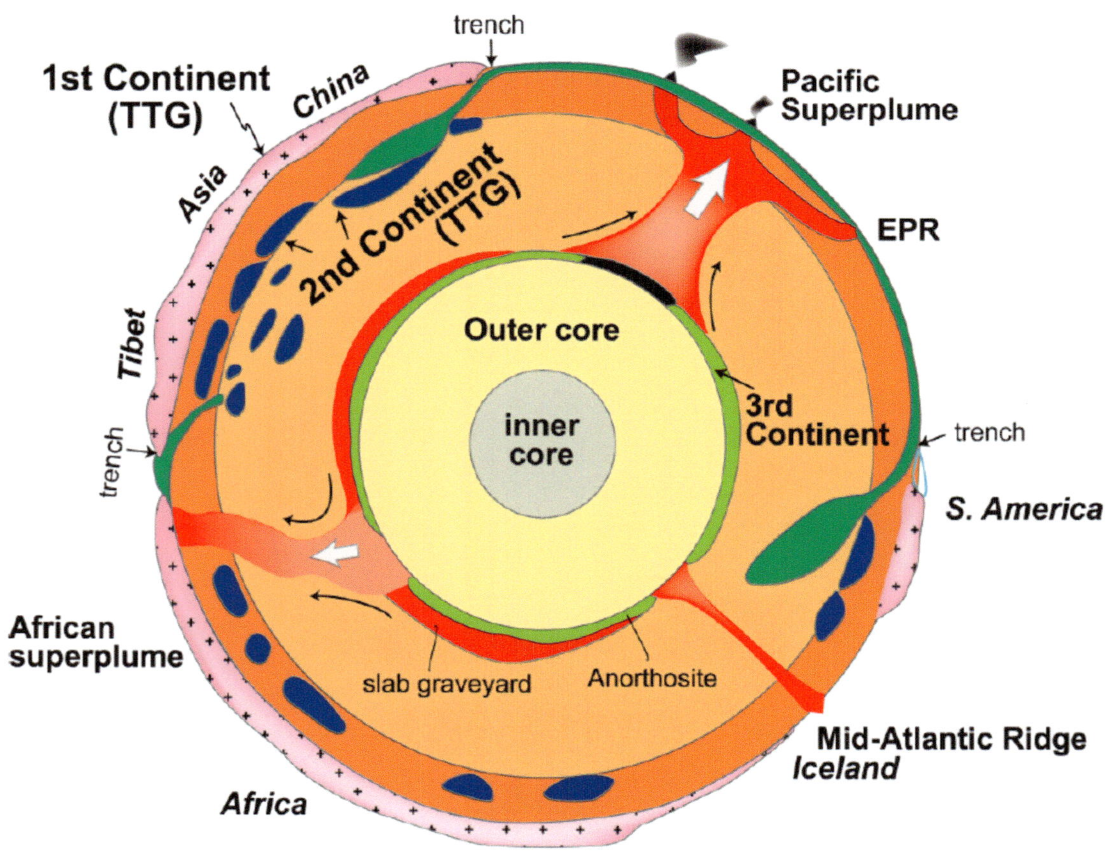

Illustration courtesy of Inna Safonova as discussed in (Maruyama *et al.* 2007; Safanova & Maruyama 2014)

This new research on second continents depends on the physical process of tectonic erosion, rather than adding to continental edges through accretion. As shown in the above figure of a slice through the Earth, in tectonic erosion, the subducting plate grinds away at the continental edge, carrying fragments of the granitic continental batholiths produced via previous subduction zone magmatism down into the mantle; these materials join sea floor basalts and island arc granites that have been subducted. But because granitic material is less dense and therefore more buoyant than mantle rock, it equilibrates in the MTZ forming 'second continents', while oceanic slab, comprised of heavy basaltic rocks, is drawn down to the slab graveyard at the core-mantle boundary. Kawai and colleagues (2013) estimate that the amount of granitic material in second continents is *six* times the amount currently found in the current Earth's crust, after potentially several billion years of subduction (cf. Azuma *et al.* 2017).

The heat from naturally radioactive elements in granite in second continents causes the formation of magma superplumes in the overlying mantle, which rise to break up the earth's surface crust. These are much larger than the magma extrusions produced through any individual volcano, including hot spot volcanoes. The many rifted basins in the East Asian region, such as Lake Baikal and the South China, Philippine, Bohai and Japan Sea basins, are thought to be caused by plume activity, now potentially linked to generation in the second continent region. These plumes are, however, smaller than the superplumes originating from the core-mantle boundary, which are thought to break up supercontinents (Maruyama *et al.* 2007; Safanova & Maruyama 2014).

Dividing Northeast and Southwest Japan

Due to the positioning of the Izu Arc, we now have two measures of 'northeast' and 'southwest' in Japanese geology. The pre-Miocene geological belts have a continuous distribution throughout Kyushu, Shikoku and Honshu up to the Tanakura Tectonic Line (TTL) (Figure 3.1a). Some of these belts occur in fragments in northern Tohoku and across Hokkaido, but they have been disrupted through serious faulting and rearrangement during back-arc basin opening. The TTL in Tōhoku is considered the geological division between NE and SW Japan for the basic *pre-Miocene* geological belt structure.

However, the collisional emplacement of the Izu Arc, at a point within the continuous belt distributions in the SW, has set a new dividing line *for volcanic activity* between NE and SW Japan in the Late Miocene and Quaternary, as discussed in Chapter 6. This variance in volcanic activity is determined by the different plates subducting under northeast and southwest Japan (Table 6.1). The directional referents of NE and SW thus have chronological implications which need to be clarified in discussions, and references to NE and SW in the geological literature must be deciphered as to whether these distinctions refer to the pre-Miocene geological belts or the later volcanic arrays. These concerns go beyond the pedantic: the Quaternary tectonic plate arrangement is, of course, of utmost relevance to volcanic and earthquake hazard planning.

Japan as a Subduction Zone Product

The hard rock geology of Japan is fairly equally divided, as discussed above, between the results of magmatic and accretionary (thereafter metamorphic) activities. The early accretion of materials on the then East Asian continental coastline of the Panthalassa Ocean formed the basement rocks of Japan since the Jurassic period. These have been intruded over time by plutonic and volcanic magmatic products – in the continental setting before rifting ca. 16 Ma (Chapter 3), and after rifting of the archipelago in the Setouchi Volcanic Zone (Chapter 4) and in the Quaternary Volcanic Fronts (Chapter 6). Both sedimentary accretion and magmatic activities are aspects of subduction of the succession of oceanic plates drawn under the landmass.

Quaternary volcanics began in the Middle Pleistocene, caused by the ongoing subduction of the Pacific and Philippine Plates underneath NE and SW Japan (as divided by the Izu Arc), respectively. The difference between pre-Quaternary and Quaternary remains are that tephras of the former reside in the landscape only as varieties of tuff, while unconsolidated Quaternary tephra on the surface has contributed to volcanic soil formation (Chapter 7). Among the world's tephrogenic soils, which are mainly very fertile, Japan's volcanic soils are distinctly *not* fertile due to the formation of allophane and other amorphous clay gels. In contrast to the small footprint of Quaternary volcanoes (Figure 6.1), non-volcanic mountains have been folded and faulted upwards by compression between the oceanic plates moving westwards and the East Asian continent's Amur Plate moving eastward (Figure 5.1). Several critical faults differentiating independent landmasses are termed Tectonic Lines; these landmasses have been moved and/or reorganized through time by plate activity (Chapter 3). Continuing compression into the Holocene has given rise to the concept of Active Faults occurring on land and on the surrounding continental shelf. Movements on these faults are the sources of great earthquakes that out-perform subduction zone earthquakes that are generated by the oceanic slabs as they are drawn down into the Earth's mantle. Thus, earthquakes and the tsunami they give rise to are also aspects of the subduction zone siting of the Japanese Islands.

Conclusions

Japan is one of the most active landmasses in the world and is constantly changing shape, size, and height on a rapid geological timescale. To occupants and visitors, earthquakes and volcanic eruptions are the day-to-day evidence of such changes that must be worried about. But a knowledge of how the islands came into being and what kind of stresses they are currently subjected to puts these events into a long-term context that not only aids geohazard planning but illustrates magnificent aspects of Earth history and operational mechanics. Earth is a living planet with its own rules of operation and regularities. However much we see it as 'steady-state', Japan has neither always been nor will always be the same mountainous island chain as exists in today's slice of time.

References

AZUMA, S; S YAMAMOTO & H ICHIKAWA *et al.* (**2017**) "Why primordial continents were recycled to the deep: role of subduction erosion." *Geoscience Frontiers* 8.2: 337-346.

CHARVET, Jacques (**2013**) "Late Paleozoic-Mesozoic tectonic evolution of SW Japan: a review – reappraisal of the accretionary orogeny and revalidation of the collisional model." *Journal of Asian Earth Sciences* 72: 88-101.

CLOOS, Mark (**1993**) "Lithospheric buoyancy and collisional orogenesis: subduction of oceanic plateaus, continental margins, island arcs, spreading ridges, and seamounts." *Geological Society of America Bulletin* 105: 715-737.

GARZANTI, Eduardo (**2008**) "Comment on 'When and where did India and Asia collide?' by Johnathan C. Aitchison, Jason R. Ali, and Aileen M. David'." *Journal of Geophysical Research (Solid Earth)* 113.B04411: 4pp.

HIRSCH, F & K ISHIDA (**2002**) " Izanami Plateau: Japan's Triassic low latitude carbonates." *Eclogae Geologicae Helvetiae* 95: 43-55.

HOWELL, David G (**1995**) *Principles of terrane analysis: new applications for global tectonics.* 2nd edition. London: Chapman & Hall.

ISHIWATARI, Akira; Kazuhito OZAWA & Shoji ARAI *et al.* (**2016**) "Ophiolites and ultramafic rocks", pp. 223-250 in *The geology of Japan*, ed. by T MORENO *et al.* London: Geological Society of London.

ISOZAKI, Yukio (**1997*a***) "Contrasting two types of orogen in Permo-Triassic Japan: accretionary versus collisional." *The Island Arc* 6.1: 2-24 (in English).

---- (**1997*b***) "Jurassic accretion tectonics of Japan." *The Island Arc* 6.1: 25-51 (in English).

ISOZAKI, Yukio; Kazumasa AOKI & Takaaki NAKAMA *et al.* (**2010**) "New insight into a subduction-related orogen: a reappraisal of the geotectonic framework and evolution of the Japanese Islands." *Gondwana Research* 18: 82-105.

Kasahara, Junzo; Osamu Sano & Nobuo Geshi *et al.* (**2010**) "Overview of a Special Issue on 'Geotectonic Evolution of the Japanese Islands under New Paradigms of the Next Generation (Part I-IIII)'." *Chigaku Zasshi* 119.6: 947-958 (in English).

Kawai, Kenji; Shinji Yamamoto & Taku Tsuchiya *et al.* (**2013**) "The second continent: existence of granitic continental materials around the bottom of the mantle transition zone." *Geoscience Frontiers* 4: 1-6 [DOI: 10.1016/j.gsf.2012.08.003].

Klein, Benjamin Z; Oliver Jagoutz & Mark D Behn (**2017**) "Archean crustal compositions promote full mantle convection." *Earth and Planetary Science Letters* 474: 516-526.

Kobayashi, Teiichi (**1941**) "The Sakawa orogenic cycle and its bearing on the origin of the Japanese Islands." *Journal of the Faculty of Science, Imperial University of Tokyo* Sec. 2.5: 219-578.

Maruyama, Shigenori (**1997**) "Pacific-type orogeny revisited: Miyashiro-type orogeny proposed." *The Island Arc* 6.1: 91-120 (in English).

Maruyama, Shigenori & Inna Safonova (**2019**) *Orogeny and mantle dynamics: role of tectonic erosion and second continent in the mantle transition zone.* Novosibirsk: Novosibirsk State University.

Maruyama, S; M Santosh & D Zhao (**2007**) "Superplume, supercontinent, and post-perovskite: mantle dynamics and anti-plate tectonics on the core–mantle boundary." *Gondwana Research* 11: 7-37 (in English).

Minato, Masao; Mitsuo Hunahashi & Jun Watanabe *et al.* (eds) (**1979**) *Variscan geohistory of northern Japan: the Abean orogeny.* Tokyo: Tokai University Press (in English).

Miyashiro, Akiho (**1973**) "Paired and unpaired metamorphic belts." *Tectonophysics* 17: 241-154.

Nance, R Damian & J Brendan Murphy (**1994**) "Orogenic style and the configuration of supercontinents." *Canadian Society of Petroleum Geologists,* Memoire 17: 49-65.

Niu, Yaoling (**2018**) "Origin of the LLSVPs at the base of the mantle is a consequence of plate tectonics: a petrological and geochemical perspective." *Geoscience Frontiers* 9: 1265-1278.

Ramsay, John G & Martin I Huber (**1987**) *The techniques of modern structural geology, vol. 2: folds and fractures*: Academic Press.

Safonova, Inna & Shigenori Maruyama (**2014**) "Asia: a frontier for a future supercontinent Amasia." *International Geology Review* 56.9: 1051-1071 [DOI: 10.1080/00206814.2014.915586].

Santosh, M (**2013**) "Evolution of continents, cratons and supercontinents: building the habitable Earth." *Current Science* 104.7: 871-879.

Taira, A; I Hill & J Firth *et al.* (**1992**) "Sediment deformation and hydrogeology of the Nankai Trough accretionary prism: synthesis of shipboard results of ODP Leg 131." *Earth & Planetary Science Letters* 109: 431-450.

van Staal, Cees & Alexandre Zagorevski (**2020**) "Accretion, soft and hard collision: similarities, differences and an application from the Newfoundland Appalachian orogen." *Geoscience Canada* 47: 103-118.

Voosen, Paul (**2016**) "Graveyard of cold slabs mapped in Earth's mantle." *Science* 354.6315: 954-955.

Wakita, Koji; Takanori Nakagawa & Masahiro Sakata *et al.* (**2018**) "Phanerozoic accretionary history of Japan and the western Pacific margin." *Geological Magazine* [DOI: 10.1017/S0016756818000742].

Part II

The Tectonic Archaeologies of Japan

The Japanese Islands currently rest on four different tectonic plates and are bordered on the east and south by two different subduction zones. As Chester's title (2008) exclaims, subduction zones are the "birthplace of earthquakes, volcanoes, and tsunamis". These three products of Plate Tectonic processes affect the archaeology of Japan in interrelated ways, but separate sub-disciplines of archaeologies have grown up to deal with the major consequences of each.

This portion of the book introduces these different sub-disciplines by reporting the work of Japanese archaeologists. Earthquake Archaeology, elsewhere known as archaeoseismology, is perhaps the oldest named sub-discipline, dating to the mid-1980s when the Geological Survey of Japan began mapping Active Faults across the landscape, noting their effects on monumental mounded tombs dating to the Kofun period in the mid-1st millennium AD. TephroArchaeology, elsewhere known as archaeological volcanology, dates from an edited compilation of excavations (Arai 1993) in volcanic ash that had been conducted for many years in Gunma Prefecture. In contrast, Tsunami Archaeology has arisen after the 2011 Tohoku-oki Earthquake and Tsunami, with evidence of many palaeotsunami being increasingly recognized at archaeological sites in Japan. Thus, these are new endeavours that complement traditional archaeological practices and concerns.

The effects and evidence of Earth processes on archaeological sites are not immediately evident to archaeologists trained as cultural historians. The educated eye, however, can identify occurrences of past tectonic-related events and assess their impact on both physical sites and features as well as contemporaneous cultural reactions and practices. Moreover, archaeologists can contribute to the geological sciences by helping to date "birthplace" events, in Chester's words, that feed back into understanding Earth processes on a refined human-scale timeline.

Understanding human reactions to such events in the past is also of interest to current bureaucratic concerns in Japan. For both the 9c Kaimondake eruption in Kyushu and the 18c Maekake eruption in northern Kanto, the reactions of the bureaucrats in the central government are being studied by historians for insights to their political concerns and duties, including the extension of welfare or tax relief to survivors and regional inhabitants. Hopefully these studies may yield ideas that can be applied to modern situations.

Tectonic archaeologists indeed have many functions and responsibilities to those both within and without their sub-disciplines. The penultimate chapter in this section reviews the inter-relationships among and between the geohazard events themselves, and between themselves and Disaster Archaeology, heritage concerns, and future preparedness. Prager's review (2020) of aspects of our "dangerous earth" suggests that by teasing out what we do *not* know about Earth processes, we can guide research to greater understanding in future.

Arai, F (ed.) (**1993**) *Volcanic ash archaeology*. Tokyo: Kokoin Shoin (in Japanese).

Chester, R (**2008**) *Furnace of creation, cradle of destruction: a journey to the birthplace of earthquakes, volcanoes, and tsunamis*. New York: Amacom.

Prager, E (**2020**) *Dangerous Earth: what we wish we knew about volcanoes, hurricanes, climate change, earthquakes, and more*. University of Chicago Press.

Chapter 8

TephroArchaeology

Volcanic eruptions are usually spectacular acts of Nature, regardless of their size or extent. What we think of as normal volcanic activity is geographically limited to those places where molten magma is able to work its way from the Earth's mantle to the surface – at mid-ocean ridges, hot spots, rifting zones, and back-arc basins – as well as forming from melted crustal rock (cf. Chapter 2.4.3). In these situations, magma is extruded as lava or tephra, but there are further effects that we rarely think of: the greenhouse gases that accompany eruptions, or deadly gases from the African lakes Kivu in the Albertine Rift, and Nyos and Monoun in Cameroon, or the warming of meteoric water to be discharged as hot springs. Volcanic Archaeology can encompass all of these, but in Japan, the greatest effects of volcanic activity come from the eruption of tephra – particularly volcanic ash. The archaeological investigation of volcanic effects is therefore specifically termed *kazanbai kōkogaku* (lit. 'volcanic ash archaeology'), best translated as TephroArchaeology. This chapter outlines the development and nature of TephroArchaeology in Japan for comparison with Volcanic Archaeology in other areas of the world such as Mesoamerica and Italy.

A Brief Comparison

Digging in tephra is not an activity exclusive to Japanese archaeologists. Ever since the discoveries of the Roman cities of Pompeii and Herculaneum in 1762, the preservation potential of tephra has been recognized; but it was with the seminal volume of Sheets & Grayson in 1979 that 'archaeological volcanology' became a specific focus (Riede *et al.* 2020). Sheets has gone on to do extensive work at the site of Cerén in El Salvador and the Arenal region of Costa Rica (1983, 1992, 2006; Sheets & McKee 1994), while others have collected many case studies from various countries where people were living 'under the shadow' of a volcano (Grattan & Torrence 2007; Riede 2016*a*, 2016*b* and references therein). Riede, working on Lachaar See marr eruptions in the middle of Europe (Riede 2017; Riede *et al.* 2011, 2017), and Elson, documenting Sunset Crater eruptions in Arizona (2007), demonstrate that volcanism associated with Plate Tectonics does occur in intraplate continental settings - not just in subduction zone settings - and must not be overlooked by archaeologists.

Japanese efforts to develop a 'Volcanic Ash Archaeology' were begun by Arai (1993) and introduced to the English language audience by Shimoyama (2002*a*), Machida (2002), and Machida & Sugiyama (2002). The Japanese field, newly designated in English as 'tephroarchaeology' (Soda 2016), was then given greater scope in Barnes & Soda (2019). These additions to the international repertoire of Volcanic Archaeology enlarge and enhance the data available on human responses over time - while also providing geologists with detailed data on volcano behaviour in the guise of Archaeological Volcanology.

This Chapter is based on earlier material published in Barnes (2014) "Vulnerable Japan: the volcanic setting of life in the archipelago", pp. 21-42 in *Environment and Society in the Japanese Islands*, ed. by Philip Brown and Bruce Batten. Corvallis, OR: Oregon State University Press; and material in *TephroArchaeology in the North Pacific*, ed. by GL Barnes & T Soda. Oxford: Archaeopress (2019).

TephroArchaeology in Japan

Development of the field

TephroArchaeology is a translation by Soda (2016) of the Japanese term *kazanbai kōkogaku* (lit. 'volcanic ash archaeology'). He chose the English term specifically to parallel 'tephrochronology', though he did not use the idiosyncratic capitalization used here to distinguish it from other such terms as 'tephrostratigraphy'.

The Japanese term was first used in 1988 but not developed until the publication of the definitive book on the subject, edited by ARAI Fusao (1993), Professor at Gunma University. Nevertheless, the topic has a longer history dating back even to the Edo period (1603–1868) when SUGAE Masumi documented houses and objects buried in lahar sediments eroding from the banks of a river in northern Honshu (Soda 2019). Sugae's drawings (see Murakami 2019: fig. 2) provide poignant examples how whole standing houses made of organic matter can be buried by tephra and rot in place; Murakami's modern excavations at the Katakai-Ienoshita site provide scientifically recovered examples of this phenomenon, which was fairly widespread in Yoneshiro River drainage in Akita Prefecture (cf. Apx 2: Fig. A).

SODA Tsutomu (2019) recounts the historiography of the field's development within Japan, with seminal events until Arai's publication being:

- the determination of the deep Kanto Loam layers (cf. Figure 6.15) as volcanogenic in nature – even if later determined to be secondary deposits
- the excavations in Kyushu, Kanto, and Hokkaido (cf. Apx 2: Fig. A) in the 1920s and '30s inspired by the eminent archaeologist HAMADA Kosaku's visit to Pompeii
- the discovery in 1946 of Palaeolithic artefacts in the Kanto Loam by an amateur archaeologist AIZAWA Tadahiro, initiating Palaeolithic studies in Japan (the loam was previously believed to have been sterile)
- the introduction of tephrochronology into Japan in 1948 via Thorvarinsson's work
- regional correlation of tephra layers facilitated by intensive post-war economic development and construction projects
- development of the refractive index method by Arai in 1972 (cf. Arai & Machida 2019) for identifying individual volcanic glass and minerals in different tephra eruptions
- the mapping of tephra distributions and the establishment of marker tephra by Machida & Arai in the late 1970s that gave rise to their *Atlas of Tephra* in the Japanese Islands (1992, 2003, 2011, and currently under revision)
- the beginnings of volcanic 'Disaster Archaeology' in the late 1970s and early '80s
- AMS radiocarbon dating and other numerical dating methods began to be deployed from 1980 in relation to tephra layers
- and 1983, when "tephrochronology began to be seriously integrated into the discipline of archaeology" (Soda 2019: 28)
- the publication of the first collection of papers on Volcanic Ash Archaeology, edited by Arai (1993); the table of contents is given in Soda (2019: 30), with other references therein.

Soda's review continues with discussions of the status of volcanogenic soils (as presented here in Chapter 7), the tephrochronologist's view of the Palaeolithic Hoax of 2000 (mentioned at the end of Chapter 6), and new research topics.

The use of tephrochronology in archaeology grew out of Arai and Machida's work in the 1970s (Arai 1972; Machida 1977*a,b*) and their application of it to prehistory (Arai 1979; Machida 1984). The dating technique, detailed in Arai & Machida 2019, has contributed to understanding the dates of individual sites and site stratigraphy (e.g., Morisaki *et al.* 2018) as well as correlative relationships of sites across the landscape (e.g., Moriwaki *et al.* 2016).

That Japan should have played a major role in formalizing these tephra-based archaeological sub-disciplines is not surprising, given the geographical prominence of volcanoes in that country. Active Volcanoes in Japan (cf. Chapter 6) account for 10% of the world's total, up to 110 in number depending on the source reference and definition. Although the footprint of these volcanoes is relatively small compared to the mountains formed by folding and faulting (cf. Figure 6.1), the entire archipelago has, at one time or another, been subjected to tephra fallout. Why this is so is conditioned by Japan riding on the edge of a subduction zone, as described in Chapter 2. It is therefore worthwhile reviewing those conditions to understand more of Japan's vulnerability to volcanic eruptions. The volcanoes mentioned below are located in Figure 6.2.

Why 'tephro'?

Subduction zone volcanoes, as seen around the Pacific Rim of Fire and in the Mediterranean, are inherently more explosive than hot spot volcanoes such as in Hawaii. Hot spot volcanoes are generally formed by a magma plume drawing directly from the mantle, which is extruded as basaltic lava flows or spatter as in fire fountains. In contrast, subduction zone volcanoes are formed by a complex process accompanying the drawing down of an oceanic slab underneath already-existing continental crust. Water dehydrated from the down-going slab lowers the melting point of the continental crust above it to form magma and then dissolves into the magma as a gas (water vapour). As the magma rises towards Earth's surface through a volcanic vent, the gas normally escapes from the magma, as has been happening at the Sakurajima volcano in Kyushu for the last half century; but when this does not happen fast enough because of increasing magma viscosity, gas pressure builds up inside the magma and causes it to decompress so quickly upon reaching normal air pressure at Earth's surface that it explodes in cooled fragments (ash, pumice, etc.) which may be extruded into the atmosphere or roll down the volcanic flanks as pyroclastic flows or pyroclastic surges. Such eruptions are usually classed as various degrees of Plinean-type eruptions (cf. Figure 6.3), and their products are generally high in silica content (i.e., andesitic to rhyolitic in composition; see Apx 5: Fig. F) with very little lava extrusion.

The 'tephro' in TephroArchaeology simply reflects the fact that the dominant volcanic product found in archaeological excavations in Japan is some form of tephra (see Chapter 6: tephra definitions), whether it is aerially ejected ash-sized particles, pumice nodules, blocks and lava bombs (because these are aerially distributed, they are included in tephra as opposed to flowing lava), or whether it is secondarily deposited by wind or in lahars. All of these forms of tephra can potentially be generated by a single eruption of a specific volcano, or a close succession of eruptions from that volcano, so that it is important to be able to identify and distinguish products in the archaeological record as to source and timing. For example, Soda (1996) identified 15 tephra strata dating to the early 6c eruption of Mt Haruna in Gunma Prefecture, seven of which were present at the Kanai Higashi-ura site (Sugiyama 2019: 168); three stages have been identified as volcanic ash fallout, pyroclastic surge, then pyroclastic flow. A second eruption of Mt Haruna in the mid-6c generated 36 tephra units in quick succession (Soda 1993: fig. 5), accompanied by a lahar. Gunma Prefecture has served as an excellent laboratory in which to monitor volcanic activity through the centuries and assess human reactions to the difficulties brought on by living at the foot of an Active Volcano (Horaguchi 2019).

Hundreds if not thousands of sites preserved by tephra have been excavated in Japan, including off-site fields and canal constructions. Periodic tephra fallout has not only affected long-term agricultural strategies (cf. Chapter 7) but led to landscape blight that not only cost lives but interrupted settlement and cropping patterns. These changes are under intense scrutiny as archaeologists analyze human responses to volcanic eruptions through the millennia. In the historic period, both the 9c Kaimondake eruption in Kyushu (Fujino & Kobayashi 1997) and the 18c Maekake eruption in northern Kanto (Aramaki 1993), among others, have been documented in writing; these texts recording the reactions of the bureaucrats at the time are being studied by historians for insights to their political concerns and duties, including the extension of welfare or tax relief to survivors and regional inhabitants.

Homelands of TephroArchaeology

Three areas were mentioned above where pre-war excavations in tephra were carried out: Hokkaido, Kanto, and Kyushu. Two of these have become the homelands of intensive tephroarchaeological research: Gunma Prefecture in the Kanto region, and Kagoshima/Miyazaki Prefectures in southern Kyushu. These two areas correspond to some of the most active volcanic areas where there are high present-day population densities. Recently, work in Tohoku has become an important third area for tephroarchaeological research.

1) The Kanto region has three factors contributing to its volcanic activity: it is located at the bend in the VF2 Volcanic Front and also where VF2 intersects the VF1 Volcanic Front, and finally it is where the Izu Arc is impacting into mid-Honshu (cf. Figure 6.2). The Kanto Loam (cf. Figure 6.15), incorporating much secondary deposit of tephra from the surrounding volcanoes, is the foundation of the Tokyo Metropolitan area; potential volcanism is thus of great concern.

2) Southern Kyushu, on the other hand, is the area where subduction of the Philippine Plate steepens to allow the extensive magma production that has been extruded in several large caldera explosions in the past that affected western Japan (cf. Figure 6.10). This situation is feeding the most continuously active volcano today, Sakurajima, and possibly Shinmoedake as well as others.

3) The northern part of Tohoku has been subject to large volcanic eruptions from Mt Towada, on the southern border of Akita Prefecture, and Mt Paektu, on the border between North Korea and China. Tephra from these eruptions were only 60 years apart, providing interesting chronological control on people's reactions to the different events.

In the following sections, examples from these three areas will be discussed thematically, examining population movements, extraordinary examples of data preservation, settlement and field systems, and recovery attempts.

Artefact Distributions & Population Recovery in Kyushu

Humans are calculated to have occupied the Japanese archipelago from 42,000 BP. By doing so, they wandered into a land that was far more tectonically active than whence they had come: the Korean Peninsula and/or the Primorye coast of Siberia — both inland continental regions not generally subject to volcanic activity (but see Pratt 2019 and Liu *et al.* 2015 on Mt Paektu [Baekdu, Changbai] on the border between North Korea and China, located in Figure 6.16: B).

Table 8.1 Archaeologically relevant volcanic eruptions in Kyushu

Volcano	Tephra	Date cal. BP	VEI	m^3
Aira	Aira-Tn, AT	ca. 30,000	7	450
"	A-Ito pumice	"		
Sakurajima	Sakurajima 11, Sz-11	ca. 8000	5	1–10
Kikai	Akahoya, K-Ah	ca. 7280	7	170
Kirishima	Goike, Kr-M	ca. 4600	5	1–10
"	Ushinosune, Kr-UsA	ca. 7600–7100	3	0.01–.01

Populations in western Japan were particularly subject to volcanic activity. In the Late Pleistocene and Early Holocene, successive eruptions in Kyushu (Table 8.1) affected not only the local regions but well beyond. Population movements are typically monitored by changing artefact repertoires – particularly stone tools and ceramics in these early periods – before and after tephra deposition.

Aira eruption 30,000 BP: Palaeolithic consequences

Fairly soon after Palaeolithic peoples' arrival, a large eruption of the Aira volcano of southern Kyushu (cf. Figure 6.14a) resulted in a caldera large enough to now form the upper part of Kagoshima Bay (Soda 2019: fig. 5). The Aira eruption at ca. 30,000 cal. BP[1] extruded more than 450 km^3 of material: A-Ito pumice, called *shirasu* ('white sand') from Miyazaki Prefecture southwards (cf. Figure 6.13); even today, the *shirasu* pumice, often hundreds of metres deep, "is a soil totally unsuited for agriculture, and for that matter, anything else" (Ibusuki-shi Kyōi 1996*b*: 62). Aira-Tn (AT) volcanic ash is distributed as far as northern Tohoku, while all of Japan west of Nagoya City was covered with 20 cm of ash (cf. Figure 6.14a) (Machida & Arai 1992: 60-65).

In north Kyushu, morphological evidence indicates that, immediately following the Aira eruption, small group(s) of Palaeolithic peoples from the Korean Peninsula introduced a specific type of tanged HS (*hakuhen sentōki*) spearhead/knife into northwestern Kyushu, and it spread rapidly throughout the island (Soda *et al.* 2010; Morisaki 2015). Fujiki (2019) documents that this new form of spearhead/knife was accompanied by a new knife form in southeastern Kyushu, and both were soon distributed across the new landscape reformed by the pyroclastic flows. More than half of artefacts recovered on the pyroclastic flow terraces were these new-form HS spearhead/knifes. The rapid recovery of the flora due to bamboo and *Miscanthus* growth – both of which were good forage for deer (Fujiki 2019: 25) – perhaps made good hunting grounds, and the HS lithics were indeed hunting tools. However, the presence of this new artefact type was short-lived for no apparent reason.

The Aira eruption prevented use of the south Kyushu obsidian quarry, causing a shift in raw material use. The pre-AT Palaeolithic assemblage of southeastern Kyushu consisted of scrapers made of obsidian, but other stone tools were made of hornfels; after the AT deposition, survivors on the southeastern coast of Kyushu made all their scrapers and knives of hornfels. While acknowledging initial total destruction in the pyroclastic flow areas, Fujiki (2019) argues that population decline in fringe areas

[1] Initial datings put the eruption at ca. 22–23,000 bp; this date has since been calibrated to ca. 30,000 BP (Smith *et al.* 2013).

was a result of environmental damage. The full environmental recovery time after AT deposition is estimated at between 1000 and 1500 years (Matsumoto 2019: table 1).

In the Kanto region, the deposition of AT ash was followed by rapid cultural change and technological diversity. Evidence from the Nozawa site, which exists as 10 cultural units within 13 natural strata of the Tachikawa Loam, indicates that the shift from an early core and flake technology (Phase I) to projectile point and knife-shaped tools (Phase II) began just after the deposition of the AT ash; moreover, a change is also noted in the site structure at that time (Oda 1993: 210).

Not only do these findings enable the comparison of Palaeolithic assemblages throughout Japan from below and above the AT ash layer, an important marker tephra, they also indicate that volcanic eruptions may have been severe enough to truncate certain lifestyles in a region which were replaced at a later date with other cultural forms. However, Shimoyama (2002*a*) cautioned that the correlation of volcanic activity and cultural change should not be mistaken for volcanic *causation* of cultural change. Moreover, in proximal areas where tephra layers may be 2 to 3 m deep (or 100 m in the case of Kagoshima *shirasu*, cf. Figure 6.13), it is often difficult to dig below the tephra to obtain a clear idea of what the cultural status was before the eruption to allow comparison with after the volcanic event.

Kikai eruption 7280 BP: Jomon consequences

In 7280 cal. BP, the Kikai volcano extruded more than 170 km^3 of volcanic material in another VEI 7 eruption in southern Kyushu (Machida & Arai 1992: 46, 56-57), again disrupting life in southwestern Japan. Pyroclastic flows inundated areas within a 50–150 km radius, killing almost everything, while Kikai airfall pumice (K-KyP) was deposited up to 25 cm thick across southern Kagoshima Prefecture. Isopachs show 30 cm of Kikai-Akahoya tephra (K-Ah) deposited in southern Kyushu, decreasing to 20 cms across Shikoku and the Inland Sea region (Figure 6.14b).

Early researchers assumed the Jomon pottery types used prior to the Kikai eruption completely disappeared in Kyushu, to be replaced after a hiatus of 900 years or so by a different type, Todoroki-style ceramics, derived from the north and west (Machida & Sugiyama 2002: 322-323; Machida 2002; but see Hall 2006). This interpretation has been rejected by Kuwahata's research (2016); he meticulously tracks ceramic changes through several districts and across several tephra fallout events. His results show that the continuation or not of ceramic styles across the marker tephra divide is quite variable (2016: fig. 27). While most pottery types with good AMS ^{14}C date ranges existed only between marker tephras, some began before or continued beyond the tephra deposit. The most important finding concerns the homogeneous Todoroki-A style of ceramics, which shared a design tradition with eastern Honshu; it was well established *before* the eruption and points to active interaction across the western archipelago already before the eruption; Kuwahata interprets the proliferation of design elements after the eruption within the same tradition, Todoroki-B, as indicating a florescence of the tradition due to more intensive interaction across western Japan.

Moreover, Kuwahata (2019*b*) is careful to monitor different reactions of hunter-gatherer communities within increasing distances from the volcano. He divides up the disaster area into four areas (à la Kuwahata 2019*a*; see Chapter 11 herein) with different outcomes (Kuwahata 2019*b*, 2016: table 7). The figures below for tephra thicknesses are excavated thicknesses; originally, they would have been much thicker; Kuwahata acknowledges that they represent only 50–80% of what had originally been deposited (2019*b*: 38).

- the proximal Area-A of pyroclastic flows: everything was decimated, and the environment took ca. 300~1000 years to recover; not suitable for humans.
- the near Area-B, tephra fallout >30 cm: the area was plagued by lahars and debris flows; initial movement out of the area and re-settlement was slow, and a lack of wooden implements in the sites may indicate forest destruction; difficult to adapt to.
- the far Area-C, covered by 10~30 cm of tephra: although ceramic types continued uninterrupted, there may have been populations movement eastwards, accounting for the sudden increase in sites around Tokyo Bay.
- the distal Area-D, <10 cm tephra: virtually no effects

Cases where settlement continuity or discontinuity through a tephra deposit have been found are discussed by Kuwahata (2019*b*):

- Jogao site, where stone clusters numbered 23 before the Sakurajima-11 eruption (10–30 cm tephra depth) and 9 clusters afterwards in almost the same distribution; this is interpreted as continuous occupation albeit with lower density.
- After the Kirishima Goike tephra deposit (1–2 m depth), ceramics appeared in the area after decades to 170 years; but houses were not built until between 150–310 years afterwards.
- The Kirishima Ushinosune tephra deposit is unusual in that the eruption continued over 500 years (7600–7100 BP). Ceramics are known from below the tephra, but a new type of pottery that appeared after the initial eruption is only present in very low frequencies in the area of >10 cm tephra thickness. This situation obtained for 300 years and is interpreted as people avoiding the area due to continuous eruption potential.

Coastal areas were affected not only by primary tephra cover but its erosion and deposition in shallow coastal seas. Shellfish communities in the Ise Bay region were buried by tephra, causing some people to move eastwards in search of new fishing grounds (Oda 1993: 221). They took with them a new type of pottery that has been distinguished not only by style but by chemical analyses (Ikeya 2017). Ikeya also reports that arrowhead use fell, indicating decline of terrestrial boar and deer in the ash-covered landscape. Not only did Tokai-region pottery spread into the Kanto region, Kansai-region pottery even spread into the Izu Islands. Uchiyama (2021) similarly interprets the spread of Sobata pottery, succeeding the Todoroki tradition, from north Kyushu coasts across to the Korean Peninsula and down into the Ryukyu Islands, indicative of the development of social networks to help with recovery more than a millennium after the Kikai eruption.

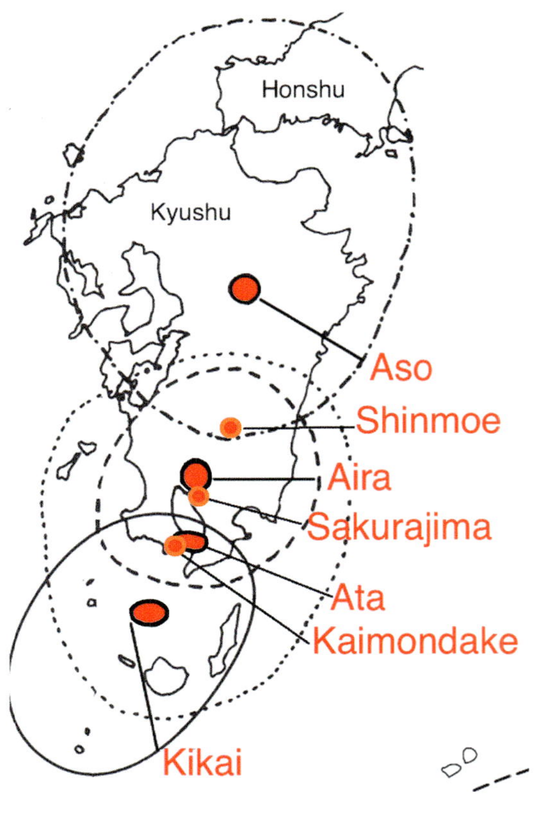

Figure 8.1 The volcanoes of Kyushu and their pyroclastic flow extents

Environmental recovery

The extent of pyroclastic flows across Kyushu from the major eruptions of Aso, Aira, Ata, and Kikai (Figure 8.1) indicate that the topography of the entire island has been determined by the laying down of tephra – sometimes accumulating to hundreds of metres thick (cf. Figure 6.13). Unlike volcanic ash fallout, which blankets the landscape fairly evenly, pyroclastic flows and surges tend to fill hollows and valleys (cf. Figure 6.12). This generally means that streambeds are filled and rivers are dammed or diverted; terraces, plateaus, and hills are created anew. At the heart of understanding environmental recovery is local plant response to widespread and deep tephra cover; at issue is how severely biomes are damaged and how quickly they can recover through plants survival or quick reseeding (see Chapter 7: Plant Regeneration). Andolizer plant types such as *Miscanthus* help form soils on tephra, while beech trees form the climax forests on tephra. Grasslands are known to have developed and maintained themselves continuously around Mt Aso in north-central Kyushu for the last 30,000 years (Miyabuchi & Sugiyama 2015). Over this time period for southern Kyushu, Moriwaki *et al.* (2016) have reconstructed a tephrostratigraphic framework that will allow for time-parallel reconstructions of cultural and environmental conditions.

Kuwahata has discussed recovery times in several publications (e.g., 2014). Comparison of environmental conditions prior to the Sakurajima-11 eruption (ca. 8000 BP) and before the Kikai Akahoya eruption (ca. 7280 BP) revealed grassland at the earlier time and laurelignosa forests later; instead of assigning this difference to recovery rates, Kuwahata interprets the difference as evidence of warming of the climate to the Holocene Climatic Optimum (2019*b*). Thus, for the early Holocene, there are additional factors to consider in assessing floral types in relation to tephra cover. Nevertheless, within the 100-km radius of Kikai Akahoya (K-Ah) pumice fallout (30–70 cm deep), it was found that there were no artefacts in the soil between that eruption ca. 7300 BP and the eruption of the Ikeda caldera tephra at 6400 BP, while plants recovered were limited to bamboo, grasses, ferns, and twigs 1–2 cm in diameter (Kuwahata 2016: 117). On the northern edge of pumice distribution, the laurelignosa forest was estimated to have taken 300–400 years to recover (2016: 118).

Volcanic eruptions affect the natural fauna and flora, settlement patterns, and subsistence regimes as well as changing the landscape itself (e.g., Takahama 1997; Moriya 1983). If the flora is destroyed, the fauna will not recolonize the area, and hunting and gathering efforts must be carried out elsewhere. There have been few studies as yet, especially in Japan, which calculate how fast biomes are recreated on tephra given the specific species available in an area (e.g., Suga *et al.* 2011; Hotes *et al.* 2004; Shoji & Takahashi 2002; Obase *et al.* 2007; del Moral & Grishin 1999). Such studies are necessary for understanding hunter-gatherer responses to volcanic eruptions – a field far less investigated than the effects on more complex societies (Uchiyama 2021).

Villages & Households

Northwest Kanto Plain: Mt Haruna & Mt Asama

The upper Kanto Plain in Gunma Prefecture is surrounded by several volcanoes, two of them historically active. Mt Haruna and Mt Asama have erupted several times since the 3c, affecting settlements and lifestyles on the plain and in the mountain valleys (Table 8.2). Archaeological site investigations reveal volcanic ash cover, pumice layers, pyroclastic flows and surges, and lahars and debris flows – especially in the 6c. Often destructive at the time of their occurrence, they preserve a level of detail that is not found in non-tephroarchaeological investigations.

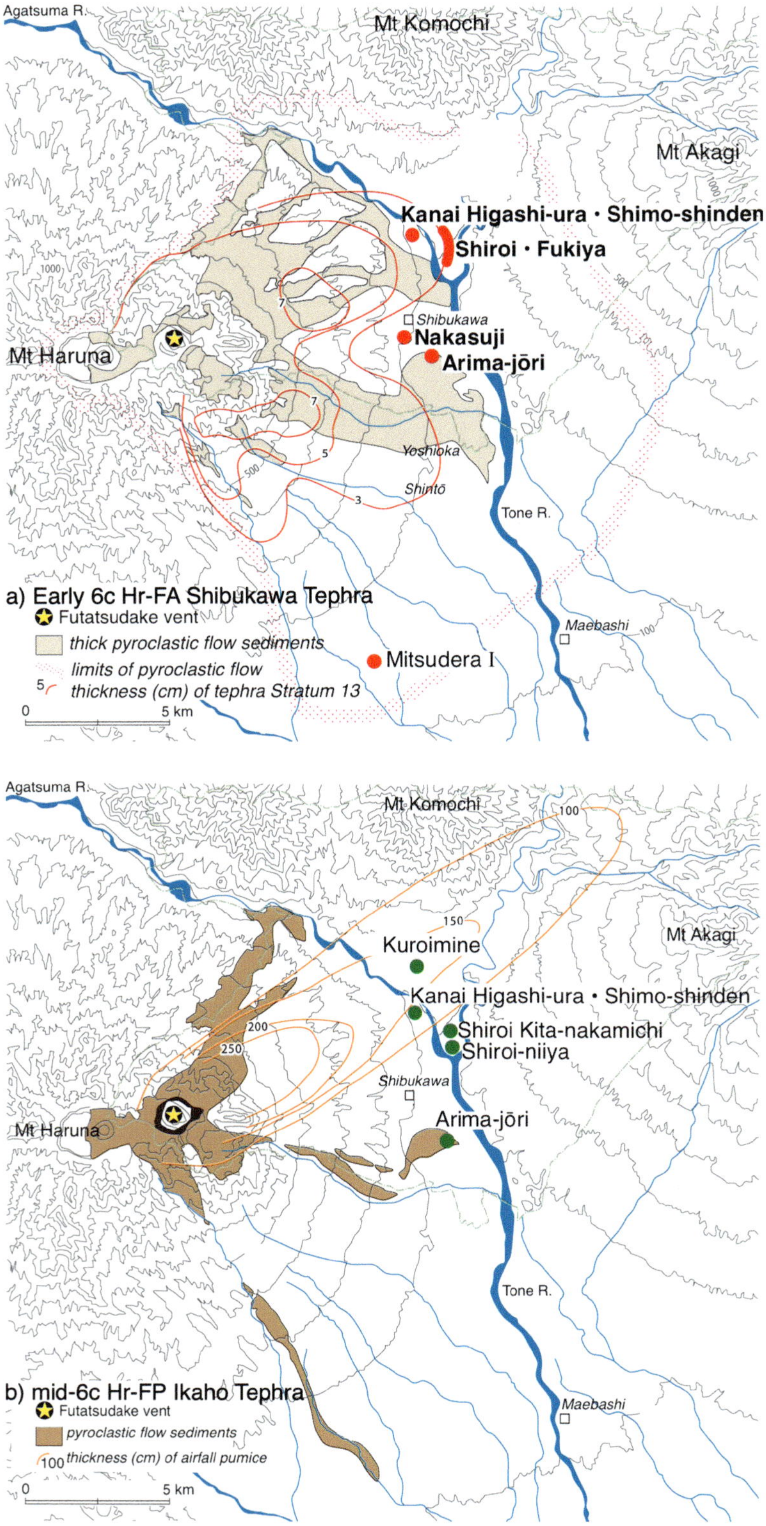

Figure 8.2 Mt Haruna eruptions from the Futatsudake vent

Mt Haruna (Hr) erupted twice in early prehistoric times (13,000–9000 BP, and 3700–3400 BP) (JMA ≥2012*a*); these would have affected Jomon life in the area. In the mid-first millennium, three eruptions occurred in the Kofun period (400–500, 489–490,[2] and 525–550 AD) (JMA ≥2012*a*). The first eruption in the 5c was relatively mild, spewing the Arima Ash (Hr-AA) in a thin layer across the landscape. The early 6c eruption from the Futatsudake vent was more serious, resulting in a lahar and the Futatsudake Ash (Hr-FA), known locally as the Shibukawa Tephra (Figure 8.2a). But the mid-6c eruption was most severe (Soda 1990), resulting again in a lahar and distributing a 2+ metre-thick pumice layer, the Futatsudake Pumice (Hr-FP) known locally as the Ikaho Tephra, across the landscape to the northeast (Figure 8.2b).

Asama (As), located on the border between Gunma and Nagano Prefectures, erupted in the 3c, 1108, and 1873. The last one occurred over a three-month period from May to August 1873, and eye-witness accounts establish the many variations in eruption pattern – though modern examination has found contradictions with the texts (Aramaki 1993). Figure 6.4 documenting the Asama eruption of 1783 exemplifies the changes that can occur during one single eruption, from generation of tephra to steam, pyroclastic and

[2] This eruption has been reported as between 500 and 525 AD (e.g., Soda 1993: 166; Barnes & Soda 2019).

debris flows, and lava extrusion - and these only over the last three days of the three-month period. When the tephra flow entered the river north of the volcano, lahars were formed that brought even large boulders down into the plains area (BOX 9).

BOX 9 Boulders Transported by Lahars

There are several very large boulders deposited from Mt Asama by lahars in Gunma and Nagano Prefectures. At least two in Gunma have been designated as National Natural Monuments. These are andesitic welded tuff, light enough to be carried by the lahars from Mt Asama north around the base of Mt Haruna and into the plains down to Takasaki City. The lahars laid down up to 10m of sediments in which these boulders are embedded, with more of their bulk underground.

The 'Asama rock' at Kanashima, Gunma (above); one of many brought ca. 60 km from Mt Asama in the 1873 lahar and deposited in Shibukawa City. Designated in 1952, it measures 9.5 x 15 m and stands 4.5 m high (Shibukawa-shi 2020). The lahar inundated 127 houses with the loss of 113 lives, and buried 54 ha of paddy fields.

Another large rock (left), 70 m in circumference and 10 m high, was brought 75 km to Maebashi City by a lahar 24,000 years ago from Mt Asama. It is preserved on the grounds of the Iwagami Inari Shrine and was designated a National Natural Monument in 1938 (Maebashi-shi Kyōi 2016).

As with sand strata covering a coastal site that might be a tsunami deposition (cf. Chapter 10), mud deposited over a site may be due to a lahar. Lahars can occur concurrently with the eruption of volcanic ash if it is raining at the time or if a pyroclastic flow enters a river course and the tephra is put into suspension; or heavy rainfall can wash volcanic sediments off slopes long after an eruption.

Lahar, aka 'mud flow' damage, is often attributed to flooding. The source of the mud is easily checked by microscopic examination of the mud to see if it is composed primarily of volcanic glass particles/shards and crystals that make up fine-grained tephra. The lahar deposits can range from centimetres to metres thick, and the blocks of lava or tephra that they can carry miles overland are similar to glacial erratics and need recognizing as such.

Table 8.2 Major sites in Gunma affected by tephra cover

Periods: Kofun 250–710 AD; Heian 794–1833; Edo 1603–1858

Site	Location	Period	Date AD	Tephra	Excavated Thickness
Dodo	Takasaki	Kofun Heian	4c early 6c mid-6c 1108	As-C pumice fallout Hr-FA ash fallout, lahar, flood Hr-FP lahar, flood As-B airfall pumice	10–20 cm 3–8 cm 60–100 cm 20–30 cm
Hamagawa-nagamachi	Takasaki	Kofun Heian	early 6c mid-6c 1108	Hr-FA ash fallout, lahar, flood Hr-FP lahar, flood As-B pumice fallout	50–100 cm 120–140 cm 20–30 cm
Shimo-Shiba sites	Takasaki	Kofun Heian	early 6c mid-6c 1108	Hr-FA ash fallout, lahar, flood Hr-FP lahar, flood As-B pumice fallout	163 cm 187 cm 15–20 cm
Kanai sites	Shibukawa	Kofun	early 6c mid-6c	Hr-FA ashfall, pyroclastic flow/surge Hr-FP pumice fallout	30 cm 200 cm
Nakasuji	Shibukawa	Kofun	early 6c	Hr-FA ash fallout, pyroclastic flow	72 cm
Kuroimine	Shibukawa	Kofun	early 6c mid-6c	Hr-FA ash fallout, pyroclastic flow Hr-FP pumice fallout	30 cm 200 cm
Shiroi sites	Shibukawa	Kofun	early 6c mid-6c	Hr-FA ash fallout, pyroclastic flow Hr-FP pumice fallout	8–40 cm 40–190 cm
Kanbara	Tsumagoi	Edo	1783	As-A pyroclastic flow/debris avalanche	650 cm
Higashimiya	Naganohara	Edo	1783	As-A pumice fallout, flood	100 cm

Nishigumi site: a farmyard buried in pumice

The Nishigumi site (Figure 8.3), revealed in broad-scale excavation, had diverse architecture, with one large pit-house and a small storehouse outside a fenced compound which contained several other structures. A stable with stalls contrasts with a courtyard structure, while in the centre of the compound are several surface structures, three round and three squarish. One small and one large storehouse stands inside the compound. The path out of the fenced area to the south leads directly to fenced plots of piled ridge & furrow gardens, with another ploughed ridge & furrow area to the west.

Short mounded plots are still used in Japan today as kitchen gardens for growing vegetables, though dry-field grains could also be sown in ridge and furrow areas. The village seemed prosperous and inhabited over several building episodes, indicated by overlap in some of the features. Horses might have been raised here, as this region of Kanto is known for its horses from the Kofun into Medieval times. Unlike the Kuroimine site, the houses at Nishigumi were devoid of pottery and tools, indicating they may have fled prior to the pumice inundation (Ishii 1990: 56).

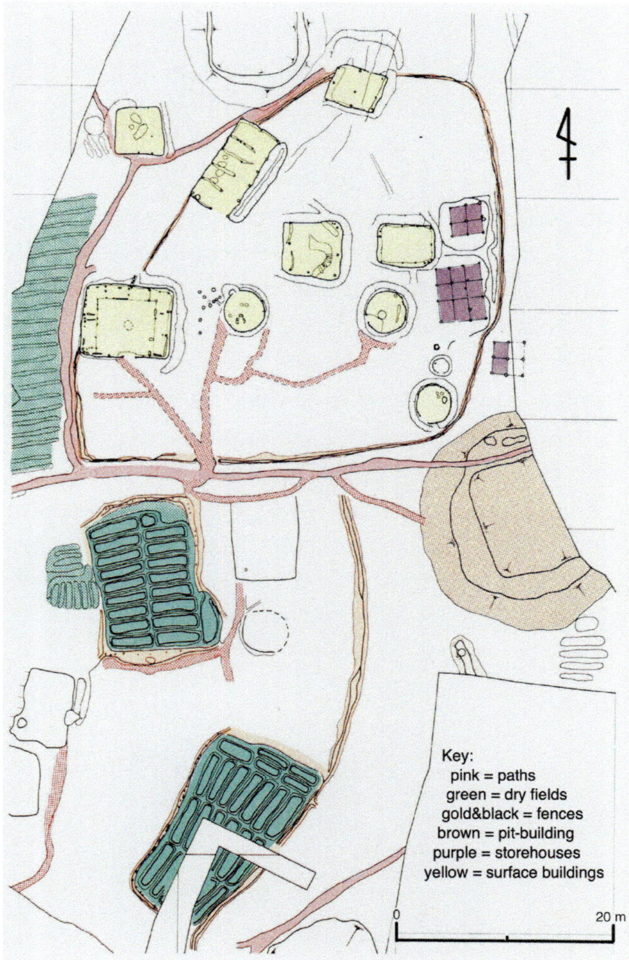

Figure 8.3 Housing complex, stables, and dry-fields at Nishigumi site, Gunma

Kuroimine site: a village buried in pumice

Kuroimine was one of the earliest excavations in tephra and is known as the 'Japanese Pompeii'. This is not due to the finding of body casts as at Pompeii but to the extraordinary uncovering of an entire village with its houses, outbuildings, fields, and paths; it is now designated as a National Historic Site (Tsude 1992). The village was located on a river terrace directly on the axis of the mid-6c Hr-FP Ikaho pumice fallout. The initial eruption laid 20 cm of volcanic ash over the site. However, this was followed by the fallout of 170 cm of pumice (Table 8.3: Stages 3,4) delivered in a short interval. Being so close to the source volcano, it also received periodical bombardment by rock fragments and large pumice nodules up to ca. 30 cm in diameter. Careful excavation revealed five stages of destruction linked to the nature of tephra fallout (Table 8.3).

Table 8.3 Mid-6c tephra fallout in five stages and their consequences at Kuroimine site

Stages	Pumice nodule sizes, deposition thickness	House and person damage
1	5 mm dm grey pumice	no outstanding damage
2	1 cm dm pumice, up to 1 m thick layer	houses damaged but not destroyed
3	10 cm avg ~ 30 cm dm pumice, 150-160 cm thick layer without pause	damage to weak points in houses, some fires; could cause wounds bone-deep if hit outside
4	1 cm dm pumice	damage to house walls and central collapse
5	volcanic ash and heavy rain > lahar	washed out small buildings

The broadscale excavation of Kuroimine village allowed understanding and reconstruction of Late Kofun rural life in the region. Not only was the organization of different structural functions (houses, granaries, fences, paths, etc.) recovered but social differences therein. Large houses had corrals for livestock as well as granaries while smaller houses were limited to the latter, illustrating that social stratification in the mid-6c had reached to the lowest settlement level. In addition, several interesting aspects of human reactions to the eruption have been noted in the Kuroimine excavation (Ishii 1990). The time of day of the eruption was deduced from the state of houses indoors: indicated by dead hearth fires, household tools untouched, and lack of agricultural equipment in the houses all suggested people were out in the fields working – thus during the day. An indication of survivors was the digging out of a tomb entrance filled with pumice so that a body could be deposited therein. Two other people are known not to have survived; when pumice flowed into their house, they used shovel and hoe to dig and pile dirt from the floor onto the pumice flow to help stop its progress and cool it; in the hole in the floor, they buried their precious possessions – a clamshell filled with red face pigment (*benigara*) and a bundle of adlay millet (*Coix lacryma-jobi*). They then took refuge in a corner and dug out dirt from the pit-house wall to put under their feet. All that was left of them was a lump of 'lard' and a necklace of rock crystal, amber, and glass beads (Ishii 1990: 56).

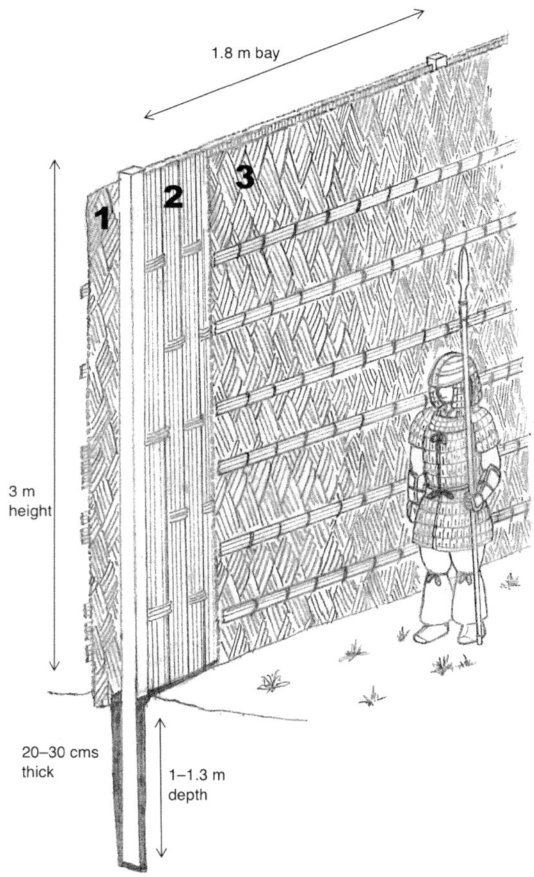

Figure 8.4 A three-layered fence at Kanai Shimo-shinden site, Gunma

The area of Kuroimine was not re-colonized until the 12c.

The Kanai sites: elite attempts at escape

The Kanai sites were most recently excavated and give great insight to elite households that can be compared to the Kuroimine village data. The sites are located on the lip of an alluvial fan formed by stream flow off Mt Haruna; the fan ends in a 20 m scarp that was cut orthogonally to the stream by the main river. Two areas were excavated: Kanai Higashi-ura and Kanai Shimo-shinden, summarized below from Sugiyama (2019).

Kanai Shimo-shinden consisted of a fenced enclosure containing two pit-houses with hearths and two pillared buildings; outside the fence stood another small pit-building and two more pillared buildings. Two of the three pillared buildings (one within the enclosure, two without) had underfloor supports indicating they were storehouses. The site was seemingly abandoned sometime before the eruption of Hr-FA tephra: the roof had been removed from the large pit-house so that the initial ash fallout covered both its interior and exterior. The pillars supporting the roof, however, remained but were pushed off vertical and carbonized by the succeeding pyroclastic flow. The fence had also been knocked down and carbonized, but enough organic matter was left to provide a rare example of a Kofun-period fence (Figure 8.4). Standing 3 m in height and 20–30 cm thick, it consisted of inner and outer panels of diagonally woven bamboo and reeds sandwiching bundles of reeds. The panels, 1.8 m wide, were reinforced with six horizontal reed bundles spanning the width. The elite nature of the site is known from the enclosure itself, the presence of multiple storehouses, ceremonial spots within the enclosure yielding different ritual items, armour /weaponry, and a cache of deer antlers ready to be made into knife hilts, etc.

Kanai Higashi-ura site is renowned for the discovery of an armoured man and a bejeweled woman and infant lying separately in a ditch covered with pyroclastic flow deposits; on the slope above were three horses and two youth skeletons. The man is the first to have been discovered actually wearing armour, rather than finding armour as grave goods in mounded tombs. Most poignantly, while the woman seemed to be fleeing downhill, the man in armour had turned towards the volcano, knelt down and took off his helmet, turning it around to face him on the ground and smoothing out the cheek-flaps to the side; he then rested his forehead on the helmet with his arms alongside, waiting for the inevitable. Excavators of the site believe this was an act of supplication to the volcano.

The armoured man has the skeletal features of a continental immigrant, while the woman had indigenous heritage. Strontium isotope analyses indicate that both had grown up somewhere in western Japan, but the child was locally born and bred. This couple may have been the occupants of a villa excavated at Kanai Higashi-ura, consisting of a large pit-house containing specialized weaving equipment, out-buildings, garden ridges mounded for planting, and a ritual area yielding around 700 small earthenware bowls and some larger pots, up to 10,000 talc disc beads, 290 other beads of mostly glass, 180 iron objects, and 160 small soapstone imitations of objects like bronze mirrors. The bowls had collected layers of volcanic ash interspersed with black sediments, indicating time lapse between eruptions.

The individuals described above may have lost their lives in the pyroclastic flow, but it seems others had escaped. Trails of footprints in wet ash led in and out of houses and finally accompanied horses off site. None of the footprints or hoofprints were rushed, indicating a calm exodus after the ash fallout before the pyroclastic flow, though it is not known if the procession was caught elsewhere by the eruption.

The area was not re-colonized until the 17c.

Mitsudera site: an abandoned housestead

The Mitsudera site, southeast of Mt Haruna in Gunma Prefecture (Figure 8.5), is another site that was covered with 78 cm of mud by the lahar that accompanied the Hr-FP pyroclastic flow. The lahar blocked river drainages, and many locations were flooded. Although far thinner than at Moto-Soja Kitakawa site, the 78 cm of mud caused the site to be abandoned (Soda 1990: 108, 1993: 149). Mitsudera is an extremely important site

Figure 8.5 The Mitsudera elite moated compound

in the history of Kofun-period archaeology as one of the first elite housesteads to have been excavated in detail (Shiraishi 1992). The site consisted of an elite house compound with internal fenced enclosures, surrounded by a moat, 30–40 m wide and 3.5 m deep. The moat was fed by waters diverted from the nearby river, which formed a perfect path for the lahar to overwhelm the site.

The area was never recolonized but turned into paddy-fields.

Kanbara Kannondo Temple: last refuge

The Asama eruption of 1783 lasted almost three months, with the tephra expelled to the southeast (cf. Figure 6.4). Following the largest eruption one evening, villagers to the southeast evacuated themselves. But ironically, on the same day the volcano sent debris avalanches and pyroclastic flows cascading down the mountainside to the north. Kanbara, located at the northern foot of the volcano, housed a simple temple to the Kannon goddess, built of organic materials on a small hill but accessed by a long series of stone steps. The debris flow covered the lower steps to form a new land surface. In 1981, a trench excavation uncovered the steps to find the skeletal remains of two females who had been escaping to the temple for shelter (see Soda 2019: fig. 5). Where the lahar entered the river northwest of Mt Haruna, it destroyed 127 out of 168 houses in Kawashima village and 54 ha of agricultural fields (Shibukawa-shi Kyōi 2020). The damage here can be compared to the lahar damage from Mt Towada in northern Tohoku, discussed below.

The temple still stands, but the lower stone staircase is buried by lahar sediments.

Northern Tohoku: Mt Towada & Mt Paektu

Katakai-Ienoshita site: the dragon lahar

It is in Akita Prefecture, however, that the most severe lahar damage has been explored. The Towada volcano (cf. BOX 7), which erupted in 915 AD in the middle of northern Tohoku, sent a lahar of

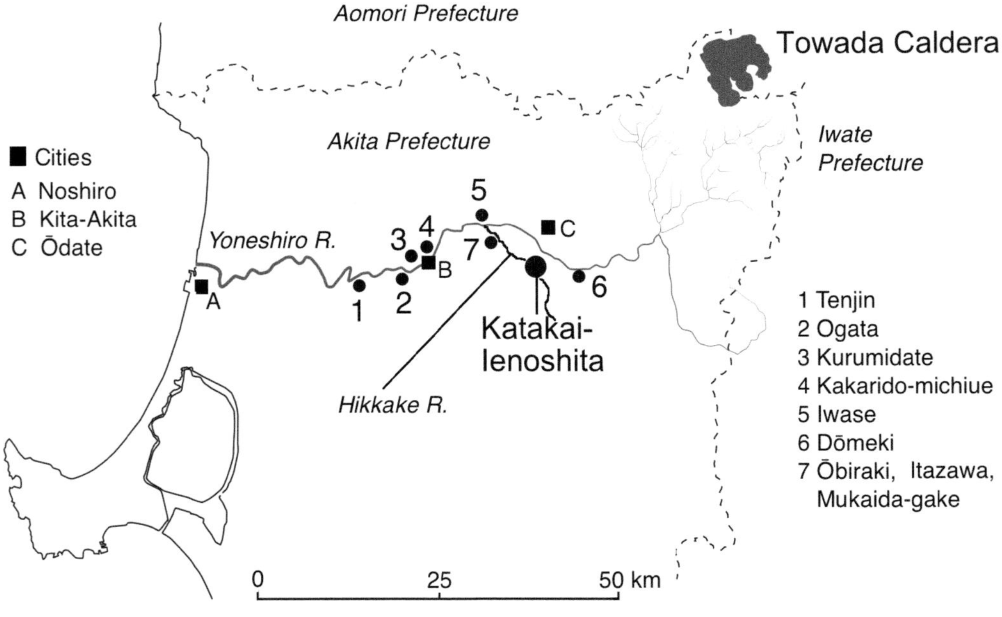

Figure 8.6 Path of the Towada lahar down the Yoneshiro River drainage and locations of buried sites

remobilized tephra down the Yoneshiro River to the Japan Sea (Figure 8.6). It has been characterized in folklore as a dragon (see Kobayashi 2019). There are several points along the river where the lahar was obstructed and back-flowed up tributary valleys; these were interpreted as ponds where the dragon rested before continuing downstream. At least ten sites in the Yoneshiro drainage have been confirmed to have had standing buildings buried by the lahar (Murakami 2019); one such was drawn by SUGAE Masumi at the turn of the 18c as mentioned above.

Trench excavations at Katakai-Ienoshita site by Murakami (2019) of the Akita Archaeological Center revealed a village with at least 13 pit-buildings, a pillared building, board fences, pits and burned earth areas; adjacent to the residences were rice paddies, ditches, and streambed. The site was first covered by pumice fallout of 5–10 cm, then the lahar mixed with pumice to form layers ranging from 10 to 250 cm thick. The distribution of tephra in the houses revealed differential use: some houses had been previously abandoned and backfilled with their roofs deconstructed; these were covered only by the lahar. Houses without roofs but with no anthropogenic fill bore pumice fallout on their floors and then lahar sediments. Houses with roofs had been currently occupied, with household ceramics being swept into the lahar. In one of the first discoveries of its kind, a hearth flue from a pit-building entered an adjacent pillared building. This resembles a traditional farmhouse with an earthen-floored kitchen area and raised tatami living area, but with an underfloor heating system? Portions of the paddy-fields had tephra removed before the lahar, indicated by many footprints in the area.

Since the lahar is water-saturated tephra, it is commonly assumed that the houses were preserved intact; in fact, most of the organic material was recovered only as stains or hollow pockets in the lahar. Those pieces of construction material that reached groundwater or were subject to water seepage, however, could be recovered. Stains, in particular since they were preserved in three-dimensional space, provide information on building practices that do not survive in normal acidic Japanese soil.

Northern population migrations

The Towada and Paektu tephras, erupted in 915 and 946 respectively, affected populations in northern Tohoku above the 40th parallel (cf. Maruyama 2019: fig. 3). Maruyama, in an ingenious study, investigated tephra layers from these two eruptions that had been deposited in pit-houses and subsequently excavated archaeologically (Figure 8.7). By contrasting the area and extent of tephra

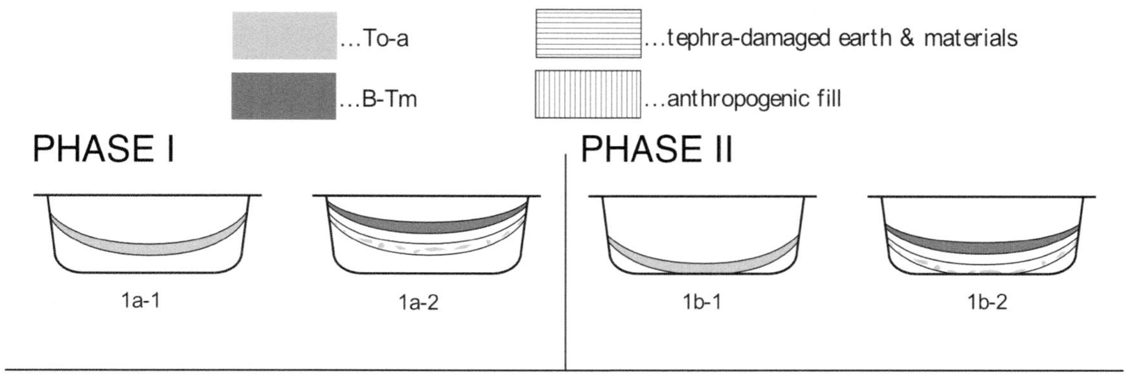

Figure 8.7 Tohoku pit-house fills with different tephra stratigraphies: 4 examples out of 21 combinations of solid and blocky Towada-a (To-a), and solid Baekdu-Tomakomai (B-Tm) tephra. Whether the tephra touched the floor and/or filled the corners assists in determining the chronology of occupation.

layers and other sedimentary patterns within the houses, he was able to discern the chronological construction, use, abandonment, and rebuilding of the houses before, between, and after these two eruptions. By expanding the study to monitor which sites within several local areas revealed decreases in population and which sites had increases in population, and by tracking ceramic change, he was able to reconstruct population movements between areas via "previously unknown ceramic types spreading into new areas" (Maruyama 2019: 154).

The findings are significant. At this time period, the northern part of Honshu was the frontier between the Nara state and indigenous farmers referred to as *emishi*, meaning those who lived beyond state jurisdiction. Maruyama found that in the southern region, there was centripetal movement of *emishi* (and their ceramics) into areas where the state "maintained a presence" and was appealed to for "assistance in recovery and reconstruction" (2019: 155). However, north of the nominal border, where the *emishi* had less interaction with the state to begin with, the population moved further north, presumably to call on kin and trade networks to aid recovery.

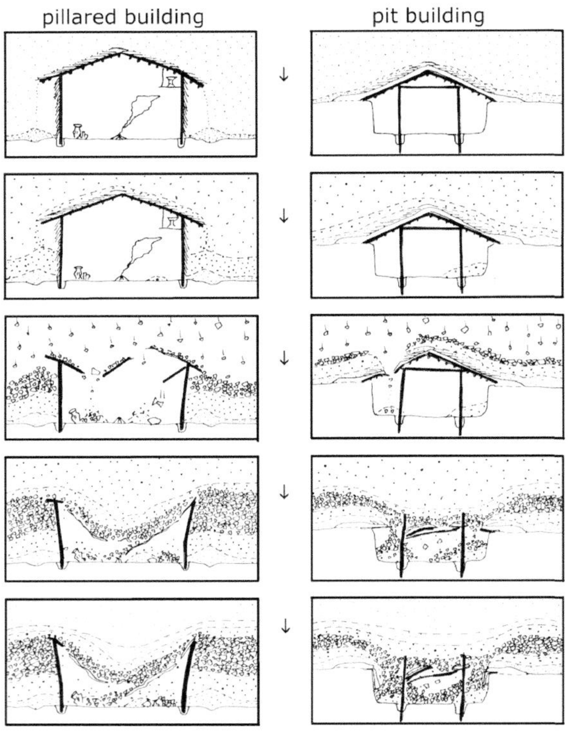

Figure 8.8 Five hypothesized stages, from top to bottom, of destruction from tephra fallout on pillared buildings (left) and pit-buildings (right) at Hashimure-gawa site, Kyushu

Southern Kyushu: Mt Kaimondake

Hashimure-gawa: destruction of a house

The earliest eye-witness record of a volcanic eruption in Japan comes from the Hashimure-gawa site at the southern tip of Kyushu, on the Satsuma Peninsula of Kagoshima Prefecture. The site stands on a base layer of ignimbrite from the final pyroclastic flow eruption of the Ikeda caldera (6400 cal. BP) (Inakura *et al.* 2015), but it sports a succession of tephra layers from Mt Kaimondake thereafter. Kaimondake has erupted at least seventeen times in the postglacial period and four times in historical periods (JMA ≥2012*b*), most significantly in Late Jomon (4500 and 3500 cal. BP); five times in the Middle Yayoi period (2400–2000 cal. BP); three times in the Kofun period; and twice between 874 and 885 AD. The largest was the Jomon eruption which affected the lower halves of the Osumi and Satsuma Peninsulae of Kagoshima Prefecture (Ibusuki-shi Kyōi 1996*a*: 7). A sequence of time slices of the representative periods of Japanese archaeology are thus preserved at Hashimure-gawa by these successive tephra covers.

The Hashimure-gawa site was first excavated in 1918–1919 by HAMADA Kosaku (1921), then Professor of Archaeology at Kyoto University, and thereafter 30 times until 1996 (Ibusuki-shi Kyōi 1996*a*: table 1). Designated as a National Historic Site in 1924, its coverage was increased in 1995. The site itself is larger, covering more than 10 ha, and has yielded remains from the Jomon through the Heian periods (Ibusuki-shi Kyōi 1991; 1996*a*). Shimoyama (2002*b*) notes that 20 cm of tephra fallout destroyed fields but not houses; the fields were abandoned but the houses were cleaned out and reused. But, after a subsequent

30 cm tephra fallout and much damage to village structures, the occupants apparently left voluntarily. Features recovered from under tephra layers at Hashimure-gawa include Kofun-period pit-buildings, path, canals, harrow marks, pottery dumps and the course of the ancient river, including ramps leading up to the riverbank. The Heian-period remains include ridge and furrow fields, cleanly covered by tephra; paths; bridge ramps; shell dumps; and embedded-pillar buildings. One of the new features recognized from this excavation consists of the remains of uprooted trees in the archaeological record.

This site is unusual as the 10c *Nihon Sandai Jitsuroku* records the destruction of the Heian-period settlement from scoria and ash fallout on the 4th day of the 3rd month in 874.[3] Most extraordinarily, the eye-witness account could be correlated with the sequence of the collapse of the buildings in archaeological terms (cf. Figure 8.8); among such buildings are some that tilted but did not collapse, preserving an air pocket inside. The Heian-period structures destroyed in the Kaimondake tephra fall have been reconstructed at the Hashimure-gawa site in Ibusuki City, accompanied by a site museum, the Hashimuregawa CoCCo Museum.

Among the Hashimure-gawa remains, the Kofun-period village was especially important for illuminating the lifestyles of the ethnic minority, the Hayato peoples, living in southern Kyushu (Shimoyama 1996). Although many Hayato were recruited to act as guards at the Nara Palace, as recorded in the *Nihon Shoki* (Aston 1972), the occupants of the Osumi Peninsula resisted being incorporated into the growing Nara state until they were defeated in a rebellion lasting from 720 to 721.

Preserved Field Systems

Paddy-fields

Some of the most eye-opening discoveries of the late 20c involved the excavation of intact field systems under substantial tephra layers as at the Dodo site (Figure 8.9). Beginning in 1978, with the discovery of mid-4c paddy-fields at Hidaka site, scores of field remains have been excavated from under three major proto-historic tephra layers: As-C from Asama in the

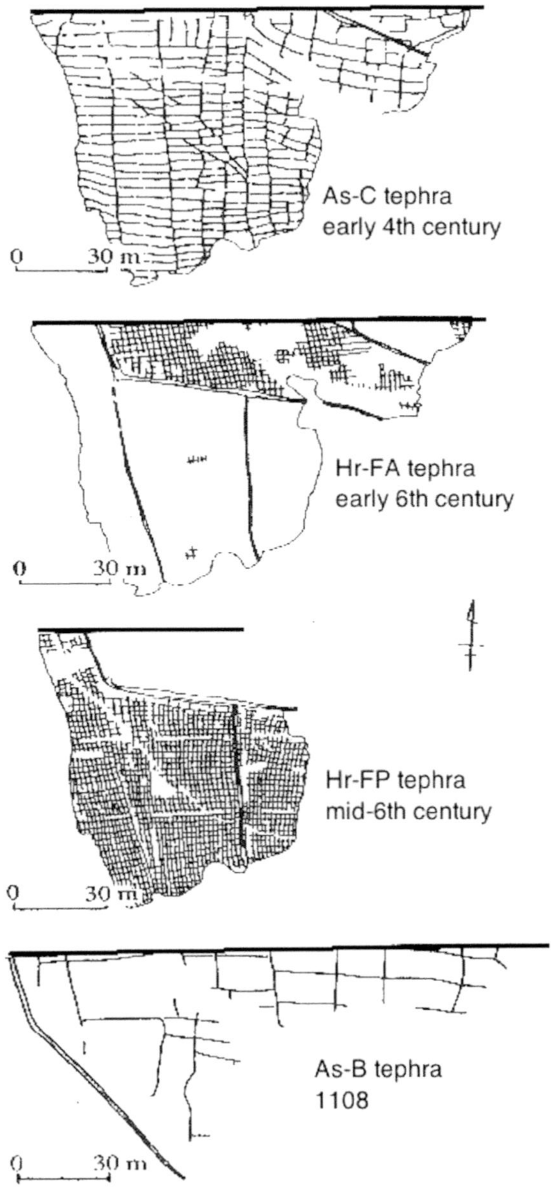

Figure 8.9 Stratified layers of different field sizes, orientation, and use under different tephra falls at Dodo site, Gunma

[3] The date according to the lunar calendar; the corresponding solar calendar date was 25 March 874.

mid-4c, Hr-FA from Haruna at the beginning of the 6c, and Hr-FP from Haruna in the mid-6c. Most of the fields have been paddy-fields for growing rice, but many are dry-fields (*hatake*) or garden plots as at Kuroimine and Kanai Higashi-ura.

One of the surprising aspects of these early field systems is that the paddy-fields are much smaller than late historic examples. Generally rectangular shaped measuring 1.5 m x 2 m, they are barely large enough to be worked by two people at a time. Moreover, the fields are separated by slightly mounded ridges which are far less substantial than historic bunds, the latter serving as raised paths between fields. In setting the field grid, ridges for the long sides of the rectangles were made first, dividing the land area into columns; then the length of the rectangular field was determined by making the short ridges, complete with small openings for water distribution. Irrigation water was thus passed through the columns of fields. It is estimated that at some sites, the entire field area was leveled after harvest, and every year the ridges and field pattern were created anew – in the least, it seems that the horizontal ridges were dug out and the fields were redug within their columnar layout.

These facts indicate that the protohistoric field system was not as permanent as the *jōri* field system – a one-hectare, square field grid instituted in the Nara–Heian periods and still in evidence in some places today – and that more labour was invested on an annual basis for digging up and resetting the field layout.

Horse hoofprints

In addition to the horse hoofprints at the Kanai sites in Gunma, the Shiroi site yielded around 40,000 hoofprints from under the tephra, allowing investigators to estimate the size and kind of horse used in the early 6c. The whither height (125–135 cm) and hoofprint size match the native medium-size Kiso horse the best (Inoue & Sakaguchi 1997).

Moto-Soja Kitakawa site: building a small dam

At the Moto-Soja Kitakawa site in Maebashi City, the first eruption of Hr-FA volcanic ash is known to have been light and disregarded by those working in the area to construct a dam across a river to divert water into a canal (Sakaguchi 2019: 186-7). The dam stratigraphy has near its base two fall units of Hr-FA tephra: S_1 and S_7. There was no earth fill between these tephra layers, so it is known that they occurred close together and the workers did not add soil to the rampart between these layers. However, on top of these tephra layers, the dam rampart was completed with soil incorporating blocks of S_1 and S_7, indicating both that the initial tephra fallouts had stopped and there was enough time to complete the dam. Further upstream, footprints suggest that the worker(s) were not idle between the S_1 and S_7 fallouts which must have occurred a few days apart. Footprints were filled with both tephra fall units, and some overlapped (see Sakaguchi 2019: fig. 6). As they were positioned near a 'quarry' for earth fill, it is understood the workers were obtaining more soil for the dam rampart. Unfortunately, all this work was then quickly buried in a lahar; Sakaguchi (2019: 186) estimates there were "at least several hours after the S_7 fallout in which the farmer(s) then worked mounding the dam and diverting water before the lahar struck." The lahar consisted of around 5 layers of different colours and was between 2.5 and 3.5 m thick.

A thin layer of paddy soil on top of the Hr-FA lahar indicates efforts to re-initiate agriculture in the mid-6c. However, a thin 1-centimetre layer (I_{19}) of Hr-FP pumice was laid on the fields, preserving plough marks. After this tephra fallout, bund construction and ploughing continued, mixing up the tephra into the bund structures. Then a lahar struck and laid 5 cm of mud across the fields. Some of this first lahar contents and earlier tephra were incorporated into the bunds, indicating "efforts at continuing construction or re-construction" between tephra depositions (Sakaguchi 2019: 188). It is estimated that this period of recovery lasted between a few days and a week. Further footprint evidence shows them filled with I_{19} tephra, first lahar sediments, and more footprints filled with sediments from a second lahar, 30 cm thick. Thus, this site escaped the worst of the pumice deposition, but the lahar still brought an end to activity there.

Field restoration & land use changes

Depending on the depth of lahar and pumice deposition, restoration of fields may or may not have been pursued. If the situation could not be rectified immediately, or if it was too late in the season, the tephra devastated chances for the year's harvest. Moreover, it made good arable unusable in the short term. Thus, there is some evidence for the vigorous restoration or reclaiming of fields. In Gunma, it has been noted that the historic human response to tephra-fall, except for instances of exceptionally thick cover, has been to bravely recolonize the land and continue previous activities (Soda 1990: 111). This is especially apparent where several stratigraphic layers of paddy-fields, successively covered by tephra, have been excavated. In some cases where the tephra fall was light enough, farmers reused surviving paddy bunds to recreate field areas. The time lapse between field destruction and revitalization has been estimated as minimal, and even in areas of heavy tephra fall, such as the 2 metre-thick pumice cover, house pits were dug directly into the upper pumice surface before any soil formation could take place.

The Dodo site in Gunma is the earliest indication that restoration work was undertaken (see Noto & Barnes 2019: 250-52, fig. 10). Late Yayoi paddy-fields had been covered by Asama (As-C) tephra deposited in the late 3c. In the Kofun period, longitudinal bunds were retained in the paddy system before and after light tephra fallout. Noto surmises that for the bunds to have been re-used, they must have been visible and restored within three years of the tephra fallout, otherwise they would have eroded away and become invisible. The main problem for the Kofun-period farmers was how to make paddy-fields (which need to retain water during the growing season) on top of pumice (which is very porous). Noto suggests that soil was brought down from the uplands, either manually or by adding the soil to the irrigation water upstream. This technique, called 'flowing-water guest soil' (*ryūsui kyakudo*), was used at the turn of the 6c and also in modern times (Noto & Barnes 2019: 253).

At the early 6c site of Shimo-Shiba Tenjin in Gunma, it was found that permanent dry-fields of ridge-and-furrow (*une-unema*) construction had been restored by digging the furrows deeper and mounding the dug-out soil onto the tops of the ridges (Noto & Barnes 2019: 254).

Cases have also been noted where land use has changed after a tephra fallout. Phytolith analysis on samples in Gunma Prefecture taken from immediately above and below the three historical tephra layers – As-C (first half of the 4c), Hr-FA (early 6c), and As-B (dated to 1108) – indicate that wet-rice was cultivated only from the 5c in the Maebashi City area (Inoue & Sakaguchi 1996). And yet, after paddy-fields on the Maebashi Terrace were buried by the Asama-B tephra in 1108, the area was re-planted in dryland crops (Soda 1990: 58). This may have been due to political rather than agricultural concerns. Paddy-fields around the provincial capital were restored soon after the Asama-B tephra fallout in order to comply with rice-production and taxation inspections by the central government. However, paddies in peripheral areas beyond the inspection route were converted to dry-fields, as they required much less labour (Noto & Barnes 2019: 254).

The lahar from the 1783 Asama eruption destroyed 70% of crops and 30% of houses in its flow path. Rocks in the lahar made it difficult to construct paddy-fields. The solution was a technique called 'turn the earth over' (*tenchi-gaeshi*). As Noto describes it (2019: 255):

> ...the soil was renewed from underneath by digging successive parallel trenches down into the buried paddy soil and piling the sediments from digging one trench into an already dug and empty trench beside it. In this way, the top lahar mud was deposited into the trenches first, and the paddy soil from underneath, deposited last, became the new topsoil. This technique resembles that described above for the Kofun and Heian Periods but is totally different in intent: the former involved digging trenches to unearth previous topsoil for mounding on top of tephra to make new ridges, leaving the trenches to serve as the new furrows; the latter sought to bury tephra in newly-dug trenches, which when filled became the new ridges and the space between them became the new furrows.

Compare these efforts with 'flipping' buried soil horizons in New Zealand (Lowe *et al.* 2014: 5). Noto also discussed techniques for clearing tephra from dry-fields (2019: 255):

> Excavations have also uncovered dry fields...which received considerable pumice fallout (15 cm excavated thickness, probably 20–25 original depth). These dry-fields were reclaimed by digging the pumice into the existing soil. Various digging patterns (zig-zags, spirals) across the field extents are evident...some farmers scraped pumice from their field surfaces and piled it up in field corners as *haikaki-yama* 'mountains of raked tephra'.... There was also the option of removing the pumice altogether and depositing it in another place.

In Kyushu, Kuwahata (2019*c*) has assessed the various responses of farmers to the tephra deposited in the Miyakonojo Basin by the eruption of Sakurajima in 1471. Fields at four archaeological sites had been abandoned, but at another five sites, there were efforts to restore the fields for cropping. Within the 30-cm isopach of remaining tephra thickness (perhaps 60–300 cm at the time), only dry-fields were found, and all were abandoned. However, within the 10-cm isopach, both paddies and dry-fields were restored.

Seasonality of eruptions

The Ofuro site in Takasaki City offered up an excellent example by which to judge the seasonality of the Hr-FA eruption (Harada & Noto 1984). Broadscale excavation under tephra cover revealed four fields which were at different stages of paddy-field construction. First was a field without separating bunds but mud trampled by footprints. The second had some longitudinal bunds emplaced, while the third had both longitudinal bunds and short cross-ridges in place with footprints across the plot surface. These all contrasted to the fourth field of eroded remnants of last year's paddies, with their separating bunds worn down and the field surfaces rough. These fields in different stages of preparation indicated that the season of tephra fallout was June, before the paddies were planted with rice, preserving them in their unfinished states. Also, at the Moto-Soja Kitakawa site, work on paddy-field construction was truncated by the 10 cm tephra fallout of combined Hr-FA S_1 and S_7 fall units, indicating eruption from May into June (Sakaguchi 2019: 188).

At the Arima-jori site in Shibukawa City, a similar situation was preserved by the Hr-FP tephra (Harada & Noto 1984). Three different stages of preparation were ascertained (Sakaguchi 2019: 189):

1) surfaces with no bunds at all but areas of deep footprints,
2) areas where the longitudinal bunds had been started but not completed, and
3) levelled field surfaces where stirring of the paddy soils was taking place within completed surrounding bunds.

Then at the Kuroimine site, jars of beans and rice from the previous year's harvest were found (Iishi 1990: 56). Both these circumstances indicate that the mid-6c eruption also occurred in the paddy-field preparation season of early summer.

Interdisciplinary Contributions

The on-site cooperation of archaeologists and volcanologists is of benefit to both disciplines, as shown by Soda's ability to identify 15 different phases of eruption from stratigraphic ditch fills at archaeological sites. The identification of multiple fall units and lahar units enabled the assessment of human activities before, during, and after these occurrences, providing insights on how local people of certain periods reacted to volcanic eruptions. Tephra that looks all the same to an archaeologist might have many more stories to tell through different eyes.

Examples of interdisciplinary discoveries come from the Kanai-Higashi site in Gunma (Sugiyama 2019):

- Sensitive excavation and specialist analyses of tephra in several of the thousands of impact traces have enabled volcanologists to reconstruct the exact direction, strength, and damage of the pyroclastic flow. Wedge-shaped linear indentations in the ground were assessed to be impact structures of rocks carried by the pyroclastic flow, all oriented due east without following topographic hollows that pyroclastic flows usually fill.
- Also, the western paving of rocks on a tomb mound were blasted off.
- House pillars were blown over. The dynamics of the latter, including mechanical characteristics of the soil in which they were embedded, indicate that the pyroclastic flow was moving at 108 km/hour, having an impact equalling 250–500 kg of material.
- Tephra grain-size analysis revealed that the skeletal remains at Kanai Higashi-ura were preserved due to the compact tephra forming a 'capillary barrier' that prevented moisture percolating through the layers and kept the bones dry.

An interesting use of archaeological techniques to reveal a buried forest also illustrates how careful stratigraphic excavation can contribute to the understanding of volcanic processes. In a valley 5 km from Mt Sanbe in Shimane Prefecture, standing tree trunks were discovered during a construction project (Sawada & Sampei *et al.* 2000; Sawada & Nakamura *et al.* 2000). Excavation removed up to 17 m of tephra and alluvial deposition to uncover a buried forest (Figure 8.10). Analyses focussed on determining the temperatures of past pyroclastic and debris flows via carbonization of materials assessed by H/C thermometry.

The beauty of the excavated trees was that carbonization could be tested at intervals along the trunks (Figure 8.10: A-5, A-9) as well as on carbonized material within the pyroclastic flows. Based on the revealed temperatures, the stratigraphy was revised to identify a second pyroclastic flow, and between the two, a potential interval of lower temperatures when floral growth might have resumed. The excavated area was converted into an underground museum where one can descend 15 m down to view the logs and standing tree trunks, one measuring 2 m in diameter. The figure below illustrates how valleys can be filled up with tephra over short periods of time, illustrating how completely the landscape was altered for human occupation in the mid-4th millennium BC.

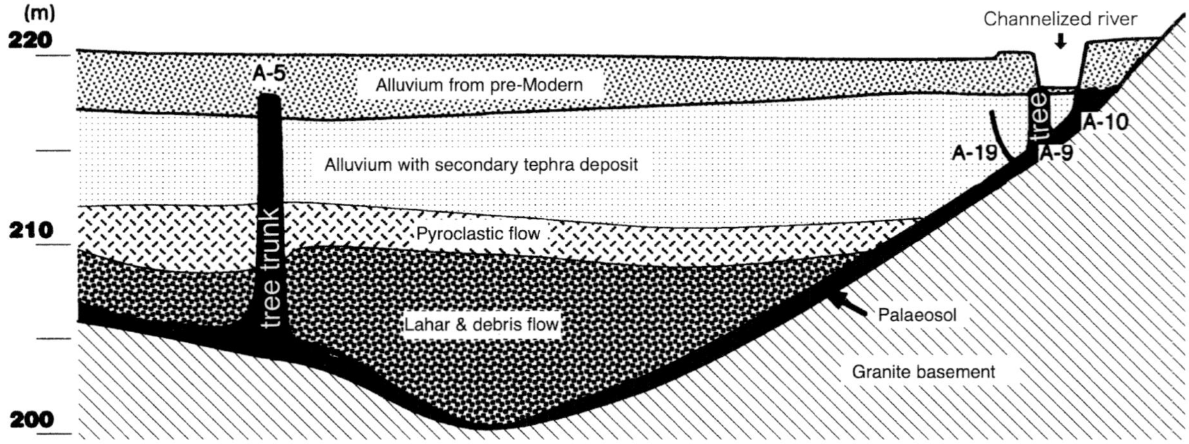

Figure 8.10 Cross-section of the Sanbe forest buried by tephra

The relationship between archaeology and history is also important. Many eruptions do not have the privilege of being recorded in writing for posterity, but there are several documents that have been coordinated with archaeological finds. The first eye-witness account of a volcanic eruption was recorded in Kyushu in 874 AD, as discussed above for the Hashimure-gawa site. More detailed observations are available from 1783 during the 3-month long eruption of Mt Maekake, the latest volcanic edifice of Mt Asama (As) in Gunma Prefecture (Aramaki 1993). It was tracked by eyewitnesses who itemized debris falls and flows and their destruction of houses, landscapes, and people. These eyewitness accounts bring back the horror of those people trusting enough to live next to an active Quaternary volcano. One of the interesting issues arising from the comparison of the documentary and archaeological records of the Maekake eruption are their contradictions and omissions (Aramaki 1993: 91, 105). For example, no mention is made of the final extrusion of lava; all eyes were on the incredible damage done by the pyroclastic flows and the flaming scoria fallouts which set houses on fire up to 10 km distant. The eyewitness accounts are also at variance with the succession of tephra fallouts and flows and their timings as documented in the field. Thus, textual accounts need to be correlated with archaeological data, in regard to both volcanic and tsunami disasters as mentioned in Chapter 11.

Conclusions

This chapter has outlined the development of the field of TephroArchaeology in Japan and some of its results. More detailed explanations of why tephra is so prevalent in Japan (as in any subduction zone) have extended the information in Chapters 2 and 6. Two homelands of TephroArchaeology are Gunma Prefecture in central Honshu and southern Kyushu Island in southwestern Japan; however, more work is being undertaken recently in Tohoku, northern Honshu. Excavations reveal an entire range of settlement and field systems from prehistoric into the historic periods, while analyses track recovery attempts and population movements.

Surviving of volcanic eruptions may entail population movements, increases in regional interaction, or changes in material culture and subsistence patterns due to resource limitations. Human recolonization of devastated areas after volcanic eruptions in prehistoric periods is usually attested by artefact distributions, but this may take several hundred years due to environmental recovery times. Following the AT tephra deposition, changes in Palaeolithic technology on Kyushu Island point both to the introduction of new stone tool forms by incoming groups from the Korean Peninsula in the northwest and a shift in local raw material use by survivors in the southeast. Standard interpretations of Jomon reactions to the Akahoya tephra deposition have been overturned by recent detailed research monitoring ceramic change before, during, and after successive tephra depositions. Reactions to several other prehistoric eruptions in Kyushu give insight to varied human adaptations.

One of the most confounding issues in assessing prehistoric human behaviour following volcanic eruptions is an as-yet inadequate understanding and monitoring of environmental recovery. Although the principles of floral succession are known (as described in Chapter 7), on-the-ground explications of biotic recovery following specific tephra fallouts are few and far between. This needs much further research in order to assess the constraints on human reactions and areal recovery.

Reactions to volcanic eruptions changed in the protohistoric and early historic period, due to the development of political structures and more sophisticated technology that can mitigate the worst effects and also require maintenance of infrastructure. Work carried out in Gunma Prefecture has first identified multiple fall units and lahar units and then assessed human activities before, during, and after these occurrences. These studies have provided insights on how work in the fields continued through light volcanic ash cover and even lahar inundations, seen through paddy-field and dam construction and footprint evidence. Instances from the Kofun through Heian periods as well as more recent history illustrate a developing capacity of the state in encouraging field recovery and aiding population resettlement.

A number of excavated Kofun-period sites were described, having been exceptionally preserved by tephra from the 6c Mt Haruna eruptions in Gunma Prefecture. These give great insight into the organization of settlements during the period of state formation. Both farming villages and elite compounds have been revealed, while off-site excavations of field systems show opportunistic use for diverse ends: paddy-field, pasture, and dry-field cropping. Poignant examples of people's behaviour prior to their demise in pyroclastic flows and pumice fallouts have been recovered through careful excavation, and rare examples of biologically degradable artefacts open windows on military and ritual provisions.

Historical instances of volcanic disasters range from house collapse records in Kyushu to standing house preservation in Tohoku. Differential population movement in eruption areas in 10c Tohoku have been attributed to varying degrees of interaction with the early state. Field restoration often depended on taxation requirements, and ingenious methods of shifting tephra from fields were devised. The depth of tephra fallout often determined whether and how fields could be restored to productivity.

TephroArchaeology is the creation of Japanese archaeologists who have found themselves digging in tephra. Their meticulous attention to detail has unearthed a world of data that is rarely found in other areas. As Robin Torrence stated at the TephroArchaeology Forum at the WAC8 meeting in Kyoto in 2016,

> *The data in Japan is so incredibly rich and the archaeology is so fantastic that there is huge potential here within one cultural area for us all to get together and map out some of these differences. I think it could set a standard for us who have much less data to compare.*

Figure & Table Sources

Figure 8.1 modified by GLB from (Machida *et al.* 1986: fig. 19)
Figure 8.2 modified by GLB after (Soda 2003: figs. 52, 53)
Figure 8.3 modified by GLB after (Komochi-mura Kyōi 1990: fig. 5)
Figure 8.4 modified by GLB after (Hara 2017: appendix, p. 7) as incorporated into (Sugiyama 2019)
Figure 8.5 courtesy of the Kamitsukeno-sato Museum of Archaeology, Takasaki City, Japan
Figure 8.6 after (Murakami 2019: fig. 1)
Figure 8.7 after (Maruyama 2019: fig. 2)
Figure 8.8 modified from (Soda 1993: fig. 7)
Figure 8.9 modified from (Saitoh 1999: 26, fig. 1)
Figure 8.10 modified from (Sawada & Nakamura *et al.* 2000: fig. 3)

Box 9 author's photos

Table 8.1 compiled from (Ishii 1990: 55)
Table 8.2 modified by GLB from (Horaguchi 2019: table 1)

References

ARAI, Fusao (**1972**) "Identification of particular tephras by means of refractive indices of orthopyroxenes and hornblendes." *Daiyonki Kenkyu* 11.4: 254-269 (in Japanese with English title and abstract).

---- (**1979**) "Key tephra layers for the Jomon onwards in northwestern Kanto." *Kōkogaku Jānaru* 157: 41-52 (in Japanese).

---- (ed.) (**1993**) *Volcanic ash archaeology.* Tokyo: Kokoin Shoin (in Japanese).

ARAI, Fusao & Hiroshi MACHIDA (**2019**) "Appendix E: The history of tephra characterization in Japan", pp. 305-316 in Barnes & Soda (eds) (2019).

ARAMAKI, Shigeo (**1993**) "Progressions in the 1783 eruption of Mt. Asama and their problems", pp. 83-110 in ARAI (ed.) (1993) (in Japanese).

ASTON, WG (**1896/1972**) *Nihongi: chronicles of Japan from the earliest times to A.D. 697.* Tokyo: Charles E. Tuttle Company, reprint 1972.

BARNES, Gina L & Tsutomu SODA (eds) (**2019**) *TephroArchaeology in the North Pacific.* Oxford: Archaeopress.

DEL MORAL, Roger & Sergei Yu GRISHIN (**1999**) "Volcanic disturbances and ecosystem recovery", pp. 137-160 in *Ecosystems of disturbed ground*, ed. by LR WALKER. Amsterdam: Elsevier.

ELSON, Mark D; MH ORT & KC ANDERSON *et al.* (**2007**) "Living with the volcano: the 11th century AD eruption of Sunset Crater", pp. 107-132 in *Under the shadow: cultural and environmental responses to volcanic eruptions*, ed. by J GRATTAN & R TORRENCE. Walnut Creek, CA: Left Coast Press.

FUJIKI, Satoshi (**2019**) "The great eruptions of Aira caldera and the people." *Kikan Kokogaku* 146: 22-25 (in Japanese).

FUJINO, N & T KOBAYASHI (**1997**) "Eruptive history of Kaimondake volcano, southern Kyushu, Japan." *Kazan* 42.3: 195-211 (in Japanese).

GRATTAN, John & Robin TORRENCE (eds) (**2007**) *Living under the shadow: cultural impacts of volcanic eruptions.* Walnut Creek, CA: Left Coast.

HALL, Mark (**2006**) "Tougher than you think: a comment on Grattan's 'Aspects of Armageddon: an exploration of the role of volcanic eruptions in human history and civilization'." *Quaternary International* 183: 143-144.

HAMADA, Kōsaku (**1921**) "A prehistoric site at Ibusuki in the province of Satsuma and the pottery found in it." *Report upon Archaeological Research, Department of Literature, Kyoto Imperial University* 6: 11-15 (in Japanese with English title and summary).

HARA, Masanobu (**2017**) "The ritual features and fence structure at the Kanai Shimo-shinden site", pp. 4-9 in *Kanai Shimo-shinden iseki no nazo ni idomu*, ed. by Gunma Archaeological Research Foundation. Shibukawa City: GARF (in Japanese).

HARADA, Tsunehiro & Takeshi NOTO (**1984**) "The seasonality of volcanic disasters." *Gunma Kenritsu Rekishi Hakubutsukan Kiyō* 5.1: 1-22 (in Japanese).

Horaguchi, Masashi (**2019**) "TephroArchaeology in the Gunma region: an overview", pp. 158-166 in Barnes & Soda (eds) (2019).

Hotes, S; P Poschlod & H Takahashi *et al.* (**2004**) "Effects of tephra deposition on mire vegetation: a field experiment in Hokkaido Japan." *Journal of Ecology* 92: 624-634.

Ibusuki-shi Kyōi (**1991**) *Hashimuregawa site. Ibusuki-shi Maizō Bunkazai Hakkutsu Chōsa Hōkokusho* 9.

---- (**1996*a***) *Hashimuregawa site XI*. Kagoshima: Ibusuki City Board of Education.

---- (**1996*b***) *The Archaeological Museum of Ibusuki: COCCO Hasimure*. Ibusuki City: Ibusuki-shi Kyōiku Iinkai (in Japanese with English title).

Inakura, Hirohito; Hideto Naruo & Mitsuru Okuno *et al.* (**2015**) "Eruptive history of Ikeda caldera, southern Kyushu, Japan." *AGU Fall Meeting Abstracts* [https://agu.confex.com/agu/fm15/webprogram/Paper74749.html].

Inoue, Masame & Hajime Sakaguchi (**1997**) "Estimating the withers height of the ancient Japanese horse from hoof prints." *Anthropozoologica* 25/26: 119-130.

Ishii, Katsumi (**1990**) "Kuroimine village disaster", pp. 55-57 in *Kazan funka to Kuroimine mura no kurashi*, ed. by Gunma-ken Hakubutsukan. Takasaki City: Gunma Prefectural Museum (in Japanese).

JMA (Japan Meteorological Association) (**≥2012*a***) "43. Harunasan" [www.data.jma.go.jp › tokyo › STOCK › souran › main › 43_Harunasan].

---- (**≥2012*b***) "92. Kaimondake" [www.data.jma.go.jp › STOCK › volcanoes.]

Kobayashi, Masaru (**2019**) "Akita's three lakes: the legend of Hachiro Taro and the dragon", p. 112 in Barnes & Soda (eds) (2019).

Komochi-mura Kyōi (**1990**) *Nishikumi site interim excavation report*. Kitagunma-gun: Komochi-mura Board of Education.

Kuwahata, Mitsuhiro (**2014**) "How did human beings correspond [sic] to the volcanic disasters in southern Kyushu?" *Abstracts of the Japan Botanical Society*, 29th meeting: 12-17.

---- (**2016**) *The influence of large volcanic eruptions on populations: focussing on the Kikai Akahoya eruption in southwestern Japan*. Tokyo: Yuzankaku (in Japanese with English title).

---- (**2019*a***) "Volcanic disaster archaeology: comments on methodological prospects and issues", pp. 41-46 in Barnes & Soda (eds) (2019).

---- (**2019*b***) "Correspondences between hunting-gathering peoples and damage caused by the Kikai Akahoya eruption." *Kikan Kōkogaku* 146: 34-37 (in Japanese).

---- (**2019*c***) "Restoration of agricultural assets after volcanic disasters in southwest Japan", pp. 92-101 in Barnes & Soda (eds) (2019).

Liu, Jia-qi *et al.* (**2015**) "Geological background and geodynamic mechanism of Mt Changbai volcanoes on the China-Korean border." *Lithos* 236/237: 46-73.

Lowe, David J; Megan R Balks & Nadia Laubscher (**2014**) "Once despised now desired: innovative land use and management of multilayered pumice soils in the Taupo and Galatea areas, central North Island, New Zealand." 2014 Society of Soil Science "Soil Science for Future Generations" conference field trip notes.

Machida, Hiroshi (**1977*a***) *The Testimony of Volcanic Ash*. Tokyo: Sōjushōbo (in Japanese).

---- (**1977*b***) "Tephrochronology", pp. 59-68 in *Nihon no Daiyonki Kenkyū—sono hatten to genjō*, ed. by Nihon Daiyonki Kenkyūkai: Tokyo University Press (in Japanese).

---- (**1984**) "The significance of explosive volcanism in the prehistory of Japan." *Geological Survey of Japan, Report* 263: 301-313 (in English).

---- (**2002**) "Impact of tephra forming eruptions on human beings and the environment." *Global Environmental Research* 6: 61-68.

Machida, Hiroshi & Fusao Arai (**1992**) *Atlas of tephra in and around Japan*. University of Tokyo Press, revised editions in 2003, 2011 (in Japanese; English title).

Machida, Hiroshi; Fusao Arai & Hiroshi Moriwaki (**1986**) *Knowledge of geological layers: investigation the Quaternary*. Tokyo: Tokyo Bijutsu (in Japanese).

Machida, Hiroshi & Shinji Sugiyama (**2002**) "The impact of the Kikai-Akahoya explosive eruptions on human societies", pp. 313-325 in Torrence & Grattan (eds) (2002).

Maebashi-shi Kyōi (**2016**) *National Natural Monument: maintenance report for Iwagami 'flying rock'*. Maebashi-shi: Maebashi City Board of Education (in Japanese).

MARUYAMA, Koji (**2019**) "Volcanic disaster research using archaeological methods: 10th-century eruption and population movements in northern Tohoku, Japan", pp. 140-157 in BARNES & SODA (eds) (2019).

MATSUMOTO, Shigeru (**2019**) "Kirishima Kobayashi pumice: mobile society and volcanic damage." *Kikkan Kōkogaku* 146: 26-29 (in Japanese).

MIYABUCHI, Yasuo & Shinji SUGIYAMA (**2015**) "90,000-year phytolith records from caldera rim to western foot of Aso Volcano, Japan: implications for vegetation history since catastrophic eruption." *Quaternary International* 397 (2016): 392-403.

MORISAKI, Kazuki (**2015**) "Appearance of *hakuhen-sentoki* (HS points) and second modern human migration into Kyushu, Japan", pp. 376-388 in *Emergence and diversity of modern human behavior in Paleolithic Asia*, ed. by Y KAIFU *et al.* College Station, TX: Texas A&M University Press.

MORISAKI, Kazuki; Masami IZUHO & Hiroyuki SATO (**2018**) "Human adaptive responses to environmental change during the Pleistocene-Holocene transition in the Japanese Archipelago", pp. 91-122 in *Lithic technological organization and paleoenvironment change*, ed. by E ROBINSON & F SELLET. New York: Springer International Publishing AG.

MORIWAKI, Hiroshi; Naoko NAKAMURA & Toshio NAGASAKO *et al.* (**2016**) "The role of tephras in developing a high-precision chronostratigraphy for palaeoenvironmental reconstruction and archaeology in southern Kyushu, Japan, since 30,000 cal. BP: an integration." *Quaternary International* 397 (2016): 79-92.

MORIYA, Ichio (**1983**) *Volcanic landforms of Japan*. Tokyo University Press (in Japanese).

MURAKAMI, Yoshinao (**2019**) "Katakai-Ienoshita site, Akita Prefecture, buried by the lahar accompany-ing the 10th-century eruption of Mt Towada", pp. 97-113 in Barnes & Soda (eds) (2019).

NOTO, Takeshi & Gina L BARNES (**2019**) "Farming tephrogenic soils in Gunma: before and after volcanic eruptions", pp. 234-260 in Barnes & Soda (eds) (2019).

OBASE, Keisuke; Yutaka TAMAI & Takashi YAJIMA *et al.* (**2007**) "Mycorrhizal associations in woody plant species at the Mt. Usu volcano, Japan." *Mycorrhiza* 17: 209-215.

ODA, Shizuo (**1993**) "Volcanic disasters in the Palaeolithic and Jomon periods", pp. 207-224 in F ARAI (ed.) (1993) (in Japanese).

PRATT, Keith (**2019**) "Portrait of a volcano: the paradox of Paektu (Changbaishan)", pp. 114-139 in Barnes & Soda (eds) (2019).

RIEDE, Felix (ed.) (**2016*a***) Volcanic activity and human society. Quaternary International 394: 1-5 [DOI: 10.1016/j.quaint.2015.08.090].

---- (ed.) (**2016*b***) Volcanic activity and human society. *Quaternary International* 394 (Thematic issue).

---- (**2017**) *Splendid Isolation: the eruption of the Laacher See volcano and southern Scandinavian Late Glacial hunter-gatherers.* Aarhus University Press.

RIEDE, Felix; Gina L BARNES & Mark D ELSON *et al.* (**2020**) "Prospects and pitfalls in integrating volcanology and archaeology: a review." *Journal of Volcanology and Geothermal and Research* 401: #106977.

RIEDE, F; O BAZELY & AJ NEWTON *et al.* (**2011**) "A Laacher See-eruption supplement to Tephrabase: investigating distal tephra fallout dynamics." *Quaternary International* 246: 134-144.

RIEDE, F; F SAUER & C HOGGARD (**2018**) "Rockshelters and the impact of the Laacher See eruption on Late Pleistocene foragers." *Antiquity* 92: e2.

SAKAGUCHI, Hajime (**2019**) "Archaeological investigation of the seasonality and duration of the 6th-century eruptions from Mt Haruna", pp. 184-191 in Barnes & Soda (eds) (2019).

SAWADA, Y; Y SAMPEI & M HYODO *et al.* (**2000**) "Estimation of emplacement temperatures of pyroclastic flows using H/C ratios of carbonized wood." *Journal of Volcanology and Geothermal Research* 104: 1-20.

SAWADA, Yoshihiro; Tadashi NAKAMURA & Chikako KASHINE *et al.* (**2000**) "Estimation of emplacement temperatures of past pyroclastic flow using in situ carbonized wood: an example from the Sanbe-Azukibara buried forest, Shimane Prefecture." *Geoscience Report, Shimane University* 19: 17-26 (in Japanese with English title and abstract).

SHEETS, Payson (ed.) (**1983**) *Archeology and volcanism in Central America: the Zapotitan Valley of El Salvador.* Austin, TX: University of Texas Press.

---- (1992) *The Ceren site: a prehistoric village buried by volcanic ash in Central America.* Fort Worth, TX: Harcourt, Brace, Jovanovich.

---- (2006) *The Ceren site: an ancient village in Central America buried by volcanic ash.* Belmont, CA: Wadsworth Publishing.

Sheets, Payson D & Donald K Grayson (eds) (**1979**) *Volcanic activity and human ecology.* New York: Academic Press.

Sheets, Payson & Brian R McKee (eds) (**1994**) *Archaeology, volcanism and remote sensing in the Arenal Region, Costa Rica.* Austin, TX: University of Texas Press.

Shibukawa-shi (**2020**) "The Asama stone in Kanashima" [www.city.shibukawa.lg.jp/kankou/kankou/midokoro/history/shiteibunkazai/p000295.html#kanasimanoasamaisi] (in Japanese).

Shimoyama, Satoru (**1996**) "1992 excavation results; 1993 excavation results", pp. 65-66, 108 in *Hashimuregawa 11*, ed. by Ibusuki-shi Kyōi. Ibusuki City: Ibusuki City Board of Education (in Japanese).

---- (**2002*a***) "Basic characteristics of disasters", pp. 19-27 in Torrance & Grattan (eds) (2002).

---- (**2002*b***) "Volcanic disasters and archaeological sites in southern Kyushu, Japan", pp. 326-341 in Torrence & Grattan (eds.) (2002).

Shiraishi, Taichiro (**1992**) "Mitsudera", pp. 220-222 in *Ancient Japan*, ed. by R Pearson. Washington DC: Smithsonian Institution.

Shoji, Sadao & Tadashi Takahashi (**2002**) "Environmental and agricultural significance of volcanic ash soils." *Global Environmental Research* 6: 113-135.

Smith, Victoria C; Richard A Staff & Simon PE Blockley *et al.* (**2013**) "Identification and correlation of visible tephras in the Lake Suigetsu SG06 sedimentary archive, Japan: chronostratigraphic markers for synchronising of east Asian/west Pacific palaeoclimatic records across the last 150 ka." *Quaternary Science Reviews* 67: 121-137.

Soda, Tsutomu (**1990**) "Natural history and climate of Gunma Prefecture", pp. 39-129 in *Gunma Prefectural history, vol. 1: Prehistorical to ancient periods, no. 1]*, ed. by Gunma Prefectural History Editorial Committee. Maebashi City: Gunma Prefectural Government (in Japanese).

---- (**1993**) "The eruptions of Mt. Haruna Futatsudake in the Kofun period", pp. 128-150 in Arai (ed.) (**1993**) (in Japanese).

---- (**1996**) "Explosive activities of Haruna volcano and their impact on human life in the sixth century AD." *Geographical Reports of Tokyo Metropolitan University* 31: 37-52 (in English).

---- (**2003**) "Environmental changes following the last glacial period", pp. 101-127 in *Shinpen Takasaki shishi*, ed. by Takasaki City History Compilation Committee. Takasaki City: Takasaki Shiyakusho (in Japanese).

---- (**2016**) "Tephroarchaeology and its history in Japan." Paper presented at WAC8, 2 Sept 2016, Kyoto, Japan (in English).

---- (**2019**) "Tephroarchaeology and its history in Japan", pp. 24-40 in Barnes & Soda (eds) (2019).

Soda, Tsutomu; Masami Izuho & Hiroyuki Sato (**2010**) "Human adaptation to the environmental change caused by the gigantic AT eruption (28–30 ka) of the Ito caldera in south Kyushu, Japan." *Abstracts of International Field Conference and Workshop on Tephrochronology, Volcanism and Human Activity*, Kirishima City, Kagoshima, Japan (May 9–17, 2010): 82.

Suga, Takeshi; Atsushi Ushimaru & Hiroyuki Tanaka (**2011**) "The history of grasslands in the Japanese Archipelago and their plant and insect populations", pp. 101-122 in *Environmental history of fields and plains*, ed. by M Yumoto *et al.* Tokyo: Bun-ichi (in Japanese).

Sugiyama, Hidehiro (2019) "Investigations into the Kofun period disasters caused by Mt Haruna eruptions", pp. 168-182 in Barnes & Soda (eds) (2019).

Takahama, Nobuyuki (ed.) (**1997**) *Research relating the establishment and destruction of sites to earth movements.* Niigata: Niigata University Hazards Research Institute (in Japanese).

Torrence, Robin & John Grattan (eds) (**2002**) *Natural disasters and cultural change.* London: Routledge.

Tsude, Hiroshi (**1992**) "Kuroimine", pp. 223-225 in *Ancient Japan*, ed. by R Pearson. Washington, DC: Smithsonian Institution.

Uchiyama, Junzo (**2021**) "After a super volcanic eruption: a new project on social-ecological impacts of the Kikai-Akahoya disaster, 7,300 years ago" (in English) [www.sainsbury-institute.org/info/after-a-super-volcanic-eruption-a-new-project-on-social-ecological-impacts-of-the-kikai-akahoya-disaster-7300-years-ago?fbclid=IwAR3mmbm6viTHyXsdf0aM7g4aglIIa9l-tVuTAgch58utYHMotxQxATrrbaE].

Chapter 9

Earthquake Archaeology

Earthquake Archaeology developed in Japan simultaneously with that in the Mediterranean in the mid-1980s. By 1996, evidence of earthquake occurrence had been documented at 378 sites throughout the archipelago. The main features identified include various results of liquefaction, faults, landslips, and surface cracking. This evidence differs greatly from the standard Mediterranean focus on building damage, and the reasons for the very different natures of archaeoseismology in these world regions are explained herein. This Chapter first recounts the development of this new subfield in Japan,[1] inspired by the interest of geomorphologist SANGAWA Akira. It explains why buildings per se are not a prominent component of Japanese Earthquake Archaeology and how sediment deformation, in its more geological sense, is used instead to glean information about past earthquakes, as illustrated by recent advances in identifying soft-sediment deformation structures by geoarchaeologist MATSUDA Jun-ichirō. The evidence of earthquake activity at archaeological sites can be matched with earthquakes caused by either active fault movement or subduction. The historical record of earthquake occurrence, already documented back to 599 AD, is extended into the prehistorical record through relative dating of artefacts and features on archaeological sites. Both the identification and the dating of the archaeological evidence of earthquakes can be challenged, though the 'territorial approach' gives the data a significance that is not achieved through analysis of single sites.

The New Subdiscipline

Sangawa's creation

Named *jishin kōkogaku* ('Earthquake Archaeology') in the mid-1980s, this field has been developed in Japan primarily through the initiative of one geomorphologist with the Geological Survey of Japan, SANGAWA Akira, who participated in the compilation of the national survey of Active Faults and the rendering of these onto 1:50,000 scale maps (RGAFJ 1992).

A chance encounter in his student days with the 5c Kondayama mounded tomb, which shows evidence of dike offset along a fault line (Figure 9.1; see also Figure 9.7), led Sangawa to develop the idea that even human-made constructions in Japan can give evidence of past earthquake activity. This may sound odd to Mediterranean archaeologists, who deal with earthquake damage to buildings and site destruction layers as a matter of course (cf. Galadini *et al.* 2006), but the novelty of this idea in Japan was illustrative of the difference between Earthquake Archaeology in Japan and archaeoseismology elsewhere – a difference discussed in detail below.

Sangawa's ideas were introduced to the archaeological community in 1987 through the Kodaigaku Kenkyukai (Paleology Research Group) at Doshisha University in Kyoto (Sangawa 1988). Primarily through that publication, archaeologists became aware that past earthquake activity can be seen in excavations and that many unexplained features in some sites were, in fact, the unrecognized effects of earthquakes. Such discoveries are continuously being published in the journal *Kodaigaku Kenkyu*, and Sangawa has expanded upon his original views in 1995, 2001*a*, and 2007, in addition to many individual

[1] An earlier version of this chapter was published as "Earthquake Archaeology in Japan: an overview" in *Ancient Earthquakes*, ed. by M SINTUBIN; IS STEWART; TM NIEMI, & E. ALTUNEL. GSA Special Paper 471 (2010: 81-96). Boulder, CO: Geological Society of America [DOI: 10.1130/2010.2471(08)].

articles – including a personalized excursion through the progress of his Earthquake Archaeology research (Sangawa 2009).

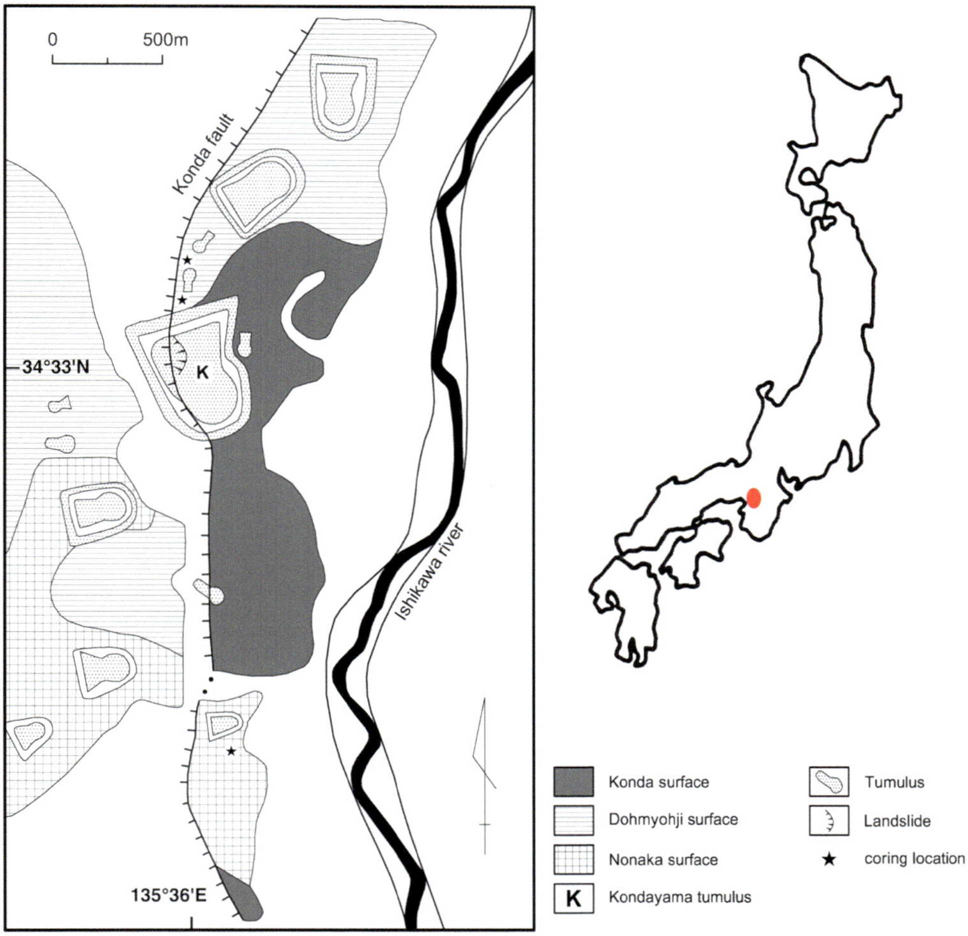

Figure 9.1 The Kondayama Tomb and Konda fault

Attributed to the protohistoric sovereign Ojin, the Kondayama Tomb is situated in a late 4–5c AD cluster of keyhole-shaped mounded tombs, established by early Japanese ruling elites, on a Pleistocene terrace in southeastern Osaka. Some tombs are moated with additional greenbelts or dikes, which exhibited displacement along the Konda fault.

The Athens conference

As Sangawa was formulating the new field of Earthquake Archaeology in Japan, archaeological approaches to earthquake damage were also developing overseas, particularly in the Mediterranean (Rapp 1986: 365). Archaeological sites with building damage and destruction layers were well known, but it was not until the late 1980s that a systematic approach to the data began to be constructed. In 1991, an international conference in Athens "brought together around a hundred specialists from several countries...to exchange ideas and discuss the problems of identification and study of ancient earthquakes from the complementary standpoints of their social, cultural, historical and physical effect" (Stiros & Jones 1996: 1). Several of these conference papers were published in the volume entitled *Archaeoseismology* (Stiros & Jones 1996).

The Athens conference discussed and defined the terms seismic archaeology, archaeological seismicity, earthquake geology, and paleoseismology. It seems that archaeoseismology, which was initially presented in 1971 by Ambraseys (Niemi 2017), has been generally adopted as the name of the field since it interfaces with archaeology. Sangawa then preferred the term seismic archaeology, but for the purposes of this article I have chosen to translate *jishin kokogaku* directly as Earthquake Archaeology (as Sangawa also does now) and will use this term only with reference to Japan to avoid confusion. In Japan, Earthquake Archaeology is considered a branch of archaeology, which itself is considered a branch of history. Its sister discipline, historical seismology (Ishibashi 2004), has concentrated on the documentary evidence for earthquakes in Japan, beginning with the *Nihon Shoki*, the earliest chronicles of 720 AD. In contrast, the editors of the compilation of papers in the special issue of the *Journal of Seismology* (Galadini *et al.* 2006) treat archaeoseismology as a developing branch of seismology, emphasizing the refinement of quantitative seismological analyses. The Stiros & Jones volume (1996) stressed that this new subdiscipline, whatever it is called, needs full interdisciplinary status to accomplish the goals of both neotectonic geology and archaeology in terms of identification and documentation of past seismic events, reconstruction of seismic cycles, assessment of earthquake damage on cultural properties, and methods of prevention and conservation in earthquake-prone areas. Caputo & Helly (2008) have contrasted several of these new subdisciplines – but not always to the best informative level, as we shall see in the following discussion.

Japan and the Mediterranean compared

Earthquake Archaeology in Japan accomplishes only a few of Stiros & Jones' objectives. Until 1995, it was a rather narrowly defined field, confined to what is discovered at archaeological sites rather than including damage to cultural properties in general – the latter assigned to Disaster Archaeology (see Chapter 11). However, given the reduced rate of excavation in today's economic retrenchment, publicly employed archaeologists (more than 7000 at the peak time in the late 1980s) are being asked to address the built environment as well, assessing extant cultural properties. Still, there is a split between what is seen as earthquake damage to standing buildings and earthquake activity in the archaeological record.

For reasons detailed below, building damage is not a major concern in Japanese Earthquake Archaeology – unlike archaeoseismology elsewhere (see photos in Sintubin 2011; Rodríguez-Pascua *et al.* 2011). The types of earthquake evidence at archaeological sites are totally different in the Mediterranean than in Japan: in the former area, buildings and layers of destroyed cultural remains (destruction layers) are of concern, while in the latter, sediment deformation is the focus. Around 75% of the sites assessed in the DCAN volume (1996) bore evidence of liquefaction (see also Garrett *et al.* 2016), while fissures and faults were reported at far lower levels of incidence (Table 9.1).

Table 9.1 Predominant types of earthquake damage at archaeological sites in the Mediterranean compared with Japan

Mediterranean	Japan
1. Displacement	1. Fault slippage (*dansō*)
2. Building damage	2. Landslides (*jisuberi*)
3. Building deformation	3. Cracks (*kiretsu*), fissures (*jiwari*)
4. Destruction layers	4. Sand eruptions (*funsa*)

Niemi (2017) discusses two more quantitative measures for determining earthquake potential and damage: the Archaeoseismic Quality Factor (AQF) and the Earthquake Archaeological Effects (EAE). But see criticisms of such a quantitative approach in Forlin & Gerrad (2017).

Buildings in Japan

Throughout the ages, wood architecture has been the traditional building structure – until the British introduced brick buildings after 1868, many of which promptly fell down in the 1891 Nobi Earthquake (Clancey 2006). Stone had an early use in tombs but was not utilized on a large scale until the construction of Medieval castles – and even then applied only to the castle base and walls with traditional architecture within. Since wooden architecture is vulnerable to the high levels of precipitation and humidity in Japan, it has been customary to renew such buildings every generation. This is best demonstrated at the national Ise Shrine, derived from a granary structure (Figure 9.2) of the Yayoi period (Apx 2: Table B), where two compounds alternate hosting the main shrine building as it is renewed every 20 years. This has barred it from UNESCO World Heritage status since the shrine, however ancient and important, is periodically rebuilt rather than maintained from ancient times.

Figure 9.2 A raised storehouse of Late Yayoi agriculturalists, ca. 1–2c AD, reproduced at the Toro site, Shizuoka Prefecture. Total height of storehouse: 4.3 m; floor: 1.45 m above ground; floorspace: 4 × 2.5 m. This type of granary structure is thought to have been transitioned into a form of Shinto shrine in the 6c in competition with incoming Chinese-style temple architecture with the adoption of Buddhism.

Damage to traditional and monumental architecture

Given their vulnerability to climate and also to fire, wooden buildings rarely survive in Japan's archaeological record; thus, Japanese archaeology is, perforce, mainly post-hole archaeology. Prehistoric buildings were mostly pit-houses constructed of organic materials such as wood, thatch, reed mats, and wattle-and-daub. These latter materials occasionally survive due to being water-logged; these lead to a greater understanding of the buildings' superstructure and account for the many reproductions at archaeological sites throughout Japan (cf. Figure 9.2).

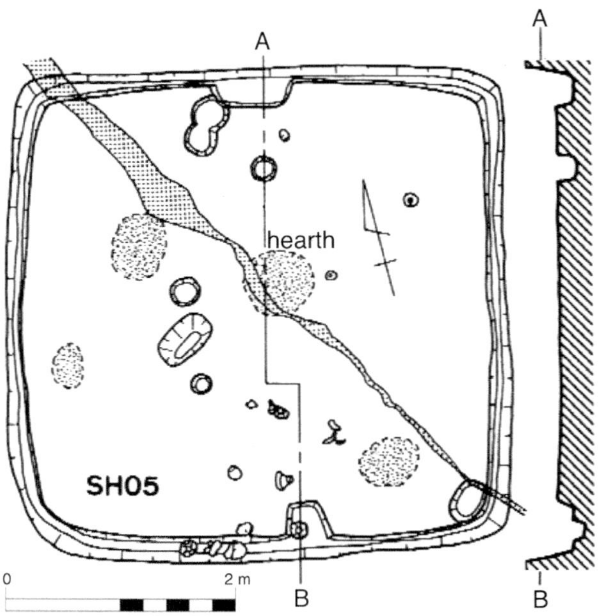

Figure 9.3 Remains of an Early Kofun-period pit-building

However, most prehistoric buildings have not survived except for their house pits and post-hole impressions (Figure 9.3). The floor of this pit-building was marred by various pits and circles of burned earth. A sand dike can be seen to cut diagonally through the house floor and through one of the burned-earth circles, giving a *terminus post quem* date of later than the 4c. It has been assigned to the earthquake of 701 AD (DCAN 1996).

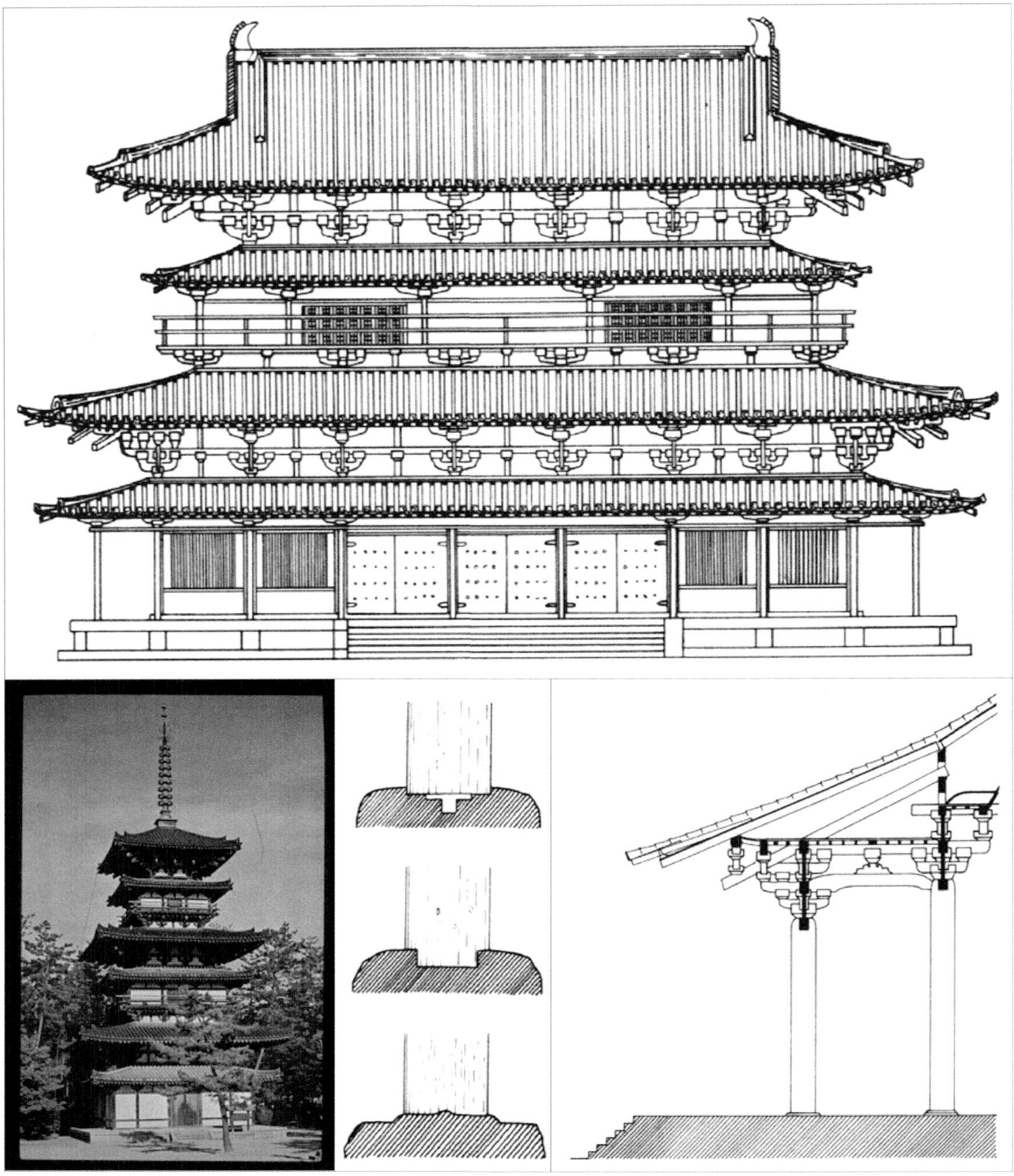

Figure 9.4 Chinese-style architecture with foundation platforms, foundation stones, load-bearing pillars, wattle-and-daub walls, heavy roof bracketing, and ceramic roof tiles

From the 6c to 8c, Chinese-style palatial architecture was introduced, first via Buddhist temple architecture (Figure 9.4), and then through state administrative architecture. The buildings themselves were similar to the indigenous architecture, using load-bearing wooden posts and wattle-and-daub walls; however, the roofs were laid with ceramic tiles, and the posts were set on foundation stones, while the whole structure was underlain by a pounded-earth foundation platform.

Figure 9.4 illustrates Chinese-style architecture in temples built in Nara after the introduction of Buddhism in the mid-6c. At the top is a scale drawing of the Main Hall at Yakushiji Temple, dating to the late 7c, while one of its two pagodas, built in 698, is shown in the lower left photo. The central pillars of pagodas are free-standing on carved foundation stones; a variety of such base-isolation systems are shown in the lower centre. The heavy roofs covered with ceramic tiles were supported by elaborate and heavy bracketing systems as at the Main Hall of Toshodaiji Temple post-dating 759, shown from the front in the top drawing and in profile in the lower right. Chinese-style buildings also used stone as facing material for the foundation platforms, for eaves-drip drainage facilities around the structure, and for the foundation stones themselves. Some of these buildings were paved with ceramic floor tiles; however, bricks, whether sun-baked or fired, were unknown until the 19c.

Tombs can also be considered as monumental architecture, and Caputo & Helly (2008) include them in their 'buildings' category while bundling 'buildings' into their 'artefacts' category. Nevertheless, these authors misrepresent the Japanese archaeological record (as well as Japan's seismic instrumentation record), indicating a 500-year depth in Japan for artefacts (buildings?) – only back to the 16c (Caputo & Helly 2008: fig. 3). On the contrary, mounded tombs (artefacts or buildings in Caputo & Helly's terms) have been built in Japan over several centuries, dating back to the 1c BC, but mostly between 100 and 710 AD.

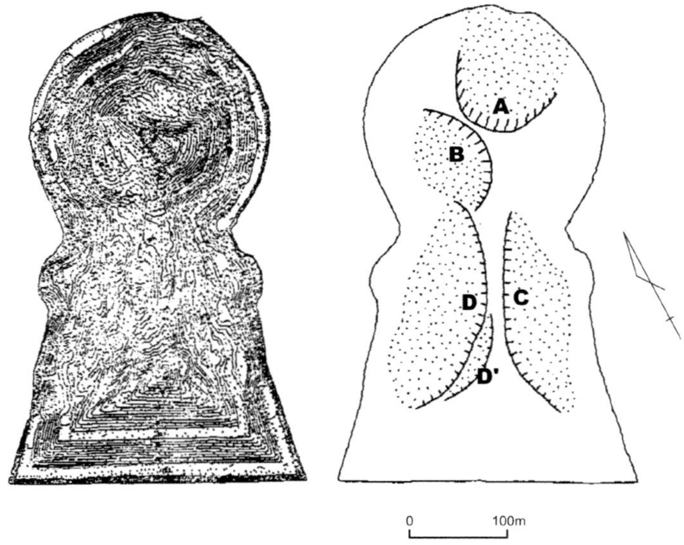

Figure 9.5 Small landslips on the Oyama Tomb, five estimated occurrences (A–D′)

Tombs of the Kofun period often had burial chambers of dry-wall stone construction and coffins carved of tuff. The mounds are particularly subject to landslips, fissures, and fault displacement, which can disrupt both their internal stone chamber structures and their external surfaces (Figure 9.5). These deformations are particularly difficult to date except with a *terminus post quem* date as occurring after tomb construction; since the tombs are rarely reused, there are usually no later cultural materials to sandwich the seismic event. However, recent assessment of landslides on two tombs (see Figure 9.7: #8, 25) located within the Arima-Takatsuki/Rokko fault zone (cf. Figure 5.10) were determined to have sustained these landslips due to liquefaction of the tomb foundation in the 1596 Keicho-Fushimi Earthquake (Kamai & Sangawa 2011).

Many building foundations and facilities such as house pits, pounded-earth platforms, and stone-lined drainage canals will also show fault displacement, fissures, and crack damage, but the buildings themselves have often collapsed and/or were burned in the fires that accompanied the earthquakes; rain and rot then dispersed the remains. Tombs exhibit fault displacement, as in the seminal Kondayama case (cf. Figure 9.1), and minor landslips such as at Oyama (Figure 9.5) and Imashiro-zuka tombs (Sangawa 2001*b*).

Survival of traditional architecture

While prehistoric indigenous architecture rarely survives in Japan due to the several environmental and fire destruction and renewal practices, Chinese-style architecture had a better chance of being maintained throughout the ages. There are no better built wooden buildings that those in the Chinese style, present throughout greater East Asia. Rapp (1986: 368) noted that "well built wooden structures will flex under the stress of strong earth vibrations" but that "in most structures the locations where components are joined together are the weakest points." Despite Rapp's assessment, it is exactly their joinery that makes Chinese-style architecture flexible. Two aspects of this style (cf. Figure 9.4) allowed buildings to survive earthquakes if they avoided fire: their mortice and tenon joinery construction and their foundation-stone/platform setting. The former allowed a building to flex and shake with the tremors, then come to rest in its original state. The latter – load-bearing pillars set on stones without any other fixtures – allowed the pillars to move separately from the foundation platform, comprising a sophisticated and ingenious ancient base-isolation system (see Okubo 2016: fig. 3). These were not always fool-proof, as several temples were noted to have collapsed in the Nobi Earthquake of 1891 (Clancy 2006). But they were good enough for Japan to have provided humanity with the world's oldest wooden buildings, dating back to the 7c.

Though both China and Japan can be considered "seismic cultures" (Stiros 1995: 726), there is as yet no evidence that these building techniques, dating to the late 2nd millennium BC, were active responses to earthquake damage. They might, however, be viewed as the result of 'natural selection', in that buildings of such structure survived more often than others. However, we *do* know that the Chinese were sensitive to earthquakes: the first ever seismograph was invented in China – by ZHANG Heng, a Han Dynasty court mathematician who operated it between 132 and 139 AD (Pajak 2005). His bronze seismograph has been re-created from fragments buried in his tomb (Xinhua News Agency, 13 June 2005).

Japan has the oldest extant wooden buildings in the world, and most are Chinese-style architecture. Due to the ravages of fire, many original buildings have been replaced through time, but even these replacements are often several hundred years old. In 1964, MUTO Kiyoshi, professor of structural engineering at Tokyo University, ostensibly modeled his innovative flexible steel-frame lattice on the Kan'eiji Temple pagoda, one of the few buildings in Tokyo to survive the 1923 Kanto Earthquake intact (NHK 2002). The structure of a pagoda, with a central pillar set into a foundation stone and the wooden multistoried structure arranged around it with bracketing, enabled the building to shake independently around the central pillar (cf. Figure 9.4; see also Okubo 2016: fig. 1). Muto's flexible steel-frame lattice was first used in the Kasumigaseki Building, the first skyscraper to be built in Japan, completed in 1968.

Thus, Japan has a long history of wooden architecture, and those wooden buildings that have survived through the ages are generally shrines or Chinese-style temple buildings still in use. Because they were important for political or religious purposes, they were usually repaired relatively quickly if they had sustained earthquake damage. Most other wooden buildings, whether collapsed by earthquakes or not, succumbed to rot or fire and have disappeared from the archaeological record, unless fragments are preserved by water-logging. Thus, Earthquake Archaeology will primarily recover data from building foundation remains but not superstructure. Japanese archaeologists do not generally work on standing buildings, which are assessed within the Cultural Heritage sector, particularly in conjunction with designations as National Treasures or Historic Sites (see Barnes 1990). The UNESCO World Heritage Site program was inspired by Japan's preservation efforts, dating back to the late 1800s, in designating important cultural properties for protection and maintenance. Unlike World Heritage Sites in many other parts of the world, however, Japan's *historic* sites are still living buildings – even if not currently occupied as at several palaces – rather than ruins.

Earthquake Records

No country is more likely to be subjected to earthquakes than Japan. Two types of earthquakes have been identified as repeatedly affecting the archipelago and surrounding continental shelf: subduction earthquakes and Active Fault earthquakes (cf. Chapter 2.4.2.1). As the archipelago sits on a conjunction of at least four plates in a subduction zone (cf. Figure 2.3), subduction earthquakes are caused by the descent of the Pacific plate in the northeast and the Philippine plate in the southwest (cf. Table 6.1). Most subduction earthquakes to date are recorded for the Pacific seaboard, while earthquakes on the Japan Sea side have so far been assigned to land-based Active Fault activity. However, there is an incipient subduction zone developing along the edge of the Japan Sea in the northwest as the back-arc basin begins to close (cf. Chapter 5). The 1995 Kobe earthquake might have been related to faults connecting Kobe and Niigata (cf. Figure 5.1), in what Itoh et al. (2013) call the Niigata-Kobe Tectonic Zone; but the immediate cause was movement on the Nojima Fault, an Active Fault crossing through western Honshu into Awaji Island in the Inland Sea. This earthquake is discussed further below.

Active Faults, defined as rock fractures caused by pressure with vertical and/or lateral movement of rock bodies against each other, occur only in the upper 20 km of brittle crust. In Japan, Active Faults are created by the archipelago itself being squeezed between the continental Amur plate in the west (part of the Eurasian plate) and the oceanic plates to the east (cf. Figure 5.1). The continental plate is moving eastward at up to 1 cm/yr, while the oceanic plates are moving westward at 2–5 cm/yr (cf. Table 6.1) (Taira 2001: 112, 114). The stress loading on Japan, caught in between, is much higher than for subduction earthquakes, so that Active Fault earthquakes are both more infrequent and much stronger than ones originating in subduction of a plate. Active Faults are said not to produce earthquakes of anything less than magnitude 6.5 (Scholz 1990), notated here as M6.5.

Between 1975 and 1979, the Active Fault Survey Group (RGAFJ 1992) documented thousands of faults across the archipelago and on surrounding continental shelves as known through surface and submarine features such as horizontal displacements, fault scarps, etc. These faults are ranked into three 'Certainty' groups with reference to the likelihood of their having been active in the present Quaternary period: I > 90% probability; II > ~50% probability; and III lower probability of activity (cf. Chapter 2.6.3.2). Some 80% of Japanese earthquake epicenters of ≥M6.5 in the last century have occurred on or within 5 km of a mapped Active Fault (RGAFJ 1992: 37), and maximum magnitudes have been predicted for Active Fault earthquakes likely to occur in specific fault zones within Japan (RGAFJ 1992: fig. 5.1).

Earthquakes are noted in many historical sources but are, of course, undifferentiated according to type. The earliest believable historical mention of an earthquake in Japan, attributed to 599 AD, occurs in the *Nihon Shoki* chronicle, which was compiled in 720 AD (Aston 1972). The oldest actual listing of earthquakes dates to 900 AD, documenting 700 earthquakes before 887 AD (Ishibashi 2004: 340, 344), while Matsu'ura (2017) documents further collections. The current collations of 'historical earthquakes' – defined by Ishibashi as those occurring before seismological instrumentation was developed in Japan in the early 1800s – run to 25 volumes and are cited in Ishibashi (2004), while Usami *et al.* have updated their listings (2013).

The first modern seismographs were developed in Japan between 1880 and 1883 through the collaboration of four men (J.A. Ewing, T. Gray, S. Sekiya, and J. Milne), and from 1884, Gray-Milne seismographs began to collect data in Japan and Britain (Utsu 2003). Thus, instrumentation records in Japan run back ca. 125 years and are not limited, as implied by Caputo & Helly (2008), to the modern seismological network covering Japan today. The modern network, however, gives immediate information on current earthquakes: within 4 or 5 minutes of occurrence, the location of the epicenter, depth, and magnitude of the earthquake are posted on the website of the Japan Meteorological Agency [www.jma.go.jp/en/quake/].

One of the more interesting aspects of Earthquake Archaeology is the ability of scientists today to collate these historical and instrumentation records of earthquakes and differentiate their causes, as described in the next two sections.

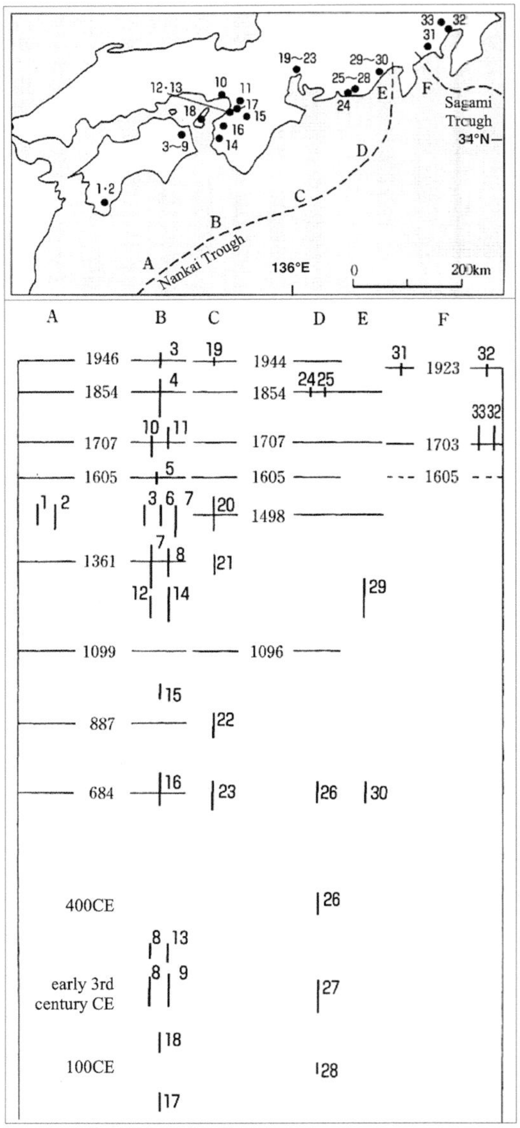

Figure 9.6 Subduction earthquakes of southwestern Japan from the historical and archaeological records

Earthquake Types & Archaeological Correlations

Subduction earthquake damage

Figure 9.6, compiled by Sangawa (2001*a*), correlates earthquake evidence sets recovered from 33 archaeological excavations with major subduction earthquakes as recorded in the historical documents after 684 AD. The map in Figure 9.6 shows the archaeological sites with earthquake damage; the offshore subduction troughs (Nankai to the left, Sagami to the right) are indicated by dashed lines. The troughs are divided into regions (A–F) of earthquake activity. The chart below this map shows the dates of the earthquakes on horizontal lines that indicate their regions of occurrence; the short vertical lines indicate the site (numbers keyed to the map) and the date range of its earthquake evidence. Site damage occurs in clusters of sites located in the Tokai (1–18), Nankai (19–30), and Kanto regions (31–33). Earthquakes in the Nankai and Tokai regions are generated in Nankai Trough rupture zones A–B and C–E, respectively, by Philippine Plate subduction under southwestern Japan. Pre-instrumental earthquakes are increasingly being found to have caused tsunami along the eastern seaboard (Fujino *et al.* 2018; Shimada 2019). Kanto earthquakes are generated northeast of the Sagami Trough (F) by Pacific Plate subduction. The last Great Kanto Earthquake was in 1923, and a future one in this region is apparently being delayed by the wedging of a broken piece of Pacific Plate in the subduction channel between the continental plate and oceanic plates (Toda *et al.* 2008; Miller 2008).

Two patterns are apparent in the horizontal date lines of Figure 9.6, indicating the regional extent of historically documented earthquakes. First, earthquakes occur simultaneously or nearly so in rupture zones A–E, as would be expected by the subduction of a single plate; however, not all zones are always attested in the documentary or archaeological records. Second, the time depths for the different zones are quite different, with far fewer earthquakes recorded in regions D–F before 1498. This has to do *not* with actual frequency but was affected by the location of the ancient capitals where literacy rates were higher: areas of old literacy documented more incidents. The capitals were located in the Kinai (around sites 11–14, 15, 17) until the late 12c, and then in Kanto thereafter.

The archaeological dates indicated by vertical date ranges were obtained from excavated stratigraphic and artifactual associations with damage assigned to earthquakes. Several sets of relative dates predate historical records at sites 8, 9, 13, 17, 18, and 26–28. These sites were assigned dates according to their cultural contents, and at least at site 8, it appears that earthquakes struck twice. The problems of relative dating of earthquakes, as well as of assigning observed damage to earthquake causes, are discussed more fully below.

Active Fault earthquake damage

The 1596 Keicho-Fushimi Earthquake in the Kinai region (Figure 9.7) was copiously documented in the historical record, and several Active Faults are proven through trench excavations on the fault lines to have caused the widespread damage (Sangawa 2001*a*: 109). Such excavations have been carried out in the last few decades as part of another new discipline, earthquake geology (cf. Caputo & Pavlides 2008).

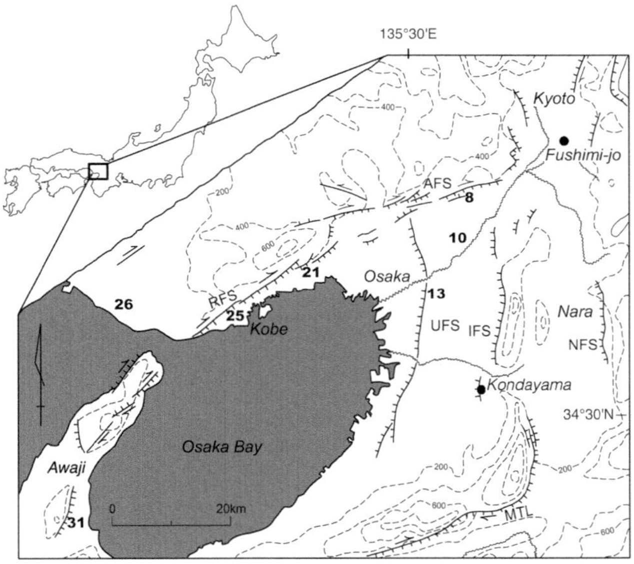

Figure 9.7 Active faults in the Kinai region and archaeological sites (described below) exhibiting damage from the 1596 Keicho-Fushimi Earthquake. Dated (right) with reference to the Medieval (1185–1603) and Edo (1605–1868) periods

RFS Rokko Fault System
AFS Arima-Takatsuki Fault System
UFS Uemachi Fault System
IFS Ikoma Fault System
NFS Nara (Toen) Fault System
MTL Median Tectonic Line

Site number KEY:

8. Imashiro-zuka Tomb: landslip of mound surface into the inner moat, covered with sediment dated by radiocarbon to early 6c; 1596 liquefaction-induced landslip

10. Tamakushi site: sand liquefaction dike cut Medieval stratum, then was overlaid by Edo stratum

13. Osaka Castle, outer moat: a landslide archaeologically dated 1583–1598

21. Ashiya ruined temple: a late 7c temple through which ran a 1m-wide fissure, which collected sand and dirt clods from a rainy period, then was filled with roof tiles made only in the late 16c. Contemporaneous documents describe how a typhoon hit five days after the 1596 earthquake, filling the fissure with sand and roof tiles. The earthquake destroyed the temple, and it was abandoned at that time.

25. Nishi-motomezuka Tomb: sand in moat liquefied to intrude the Medieval stratum above, then was overlaid by Edo stratum; 1596 base liquefaction caused landslips

26. Hyogo-no-tsu site: sand boils existed just underneath a fire layer dated to 1596

31. Shimo-naizen site: a sand dike cutting through a stratum containing Medieval artifacts, then covered with an Edo stratum.

Geological excavations took place in the Arima-Takatsuki tectonic zone (cf. Chapter 5) and included the Nodao and Higashiura faults, both on Awaji Island. Furthermore, 31 archaeological excavations in the Kinai region have revealed damage that may be correlated with the 1596 event, several of which are detailed in the figure key.

Movement on Active Faults generates the majority of shallow earthquakes in the archipelagic crust. Much current research is involved in monitoring crustal movement as manifested in faulting and shallow earthquake formation, and such movement is displayed by maps of vector arrows as in Figure 5.1. These vectors show that most of the archipelago is moving westwards — in other words, being compressed against the Eurasian landmass by the movement westwards of the Pacific Plate. Thus, earthquakes caused by crustal faulting in Japan are mainly due to east-west compression (Toda 2016: 375). Studies have shown that such shallow-crust earthquakes are particularly known "to have high damage potential, disproportionate to their magnitudes" (Chatterjee *et al.* 2018: abstract).

Identifying and dating earthquake damage

As seen in the key to Figure 9.7, relative dating by archaeological artefacts is rather loose, and many assumptions are involved in choosing to assign cause of the damage to the Keicho-Fushimi 1596 Active Fault earthquake. The subduction earthquake nearest in time, the 1605 Keicho Earthquake (M7.9) (cf. Figure 9.6, zones A–D), is the best competitor, but Sangawa (2001*a*) states that its intensity inland would have been too weak to cause the observed liquefaction damage (seen at sites 10, 25, 26, 31). Liquefaction is acknowledged to occur at not less than M5 and to become common at M5.5–6 (Obermeier 1996: 331); the land-based 1596 Keicho-Fushimi Earthquake has been assessed at M7.5 (Kanaori & Kawakami 1996), and the hypocenter(s) of the various faults was located much closer to the area of damage than was the 1605 Keicho earthquake. Thus, the late Medieval deformation in archaeological sites is assigned to the nearby Active Fault activity causing the 1596 Keicho-Fushimi Earthquake rather than being assigned to distant subduction activity causing the 1605 Keicho earthquake. For reference, the 1995 Kobe Earthquake, caused by the Nojima fault in line with the Arima-Takatsuki Tectonic Line, measured M7.3 and caused extensive liquefaction in the reclaimed land areas near Kobe as well as widespread damage in the Kinai region of Osaka, Nara, and Kyoto - a scope similar to the 1596 Keicho-Fushimi Earthquake (cf. Figure 9.7).

A significant difference in recurrence time characterizes the two types of earthquakes: subduction (interplate) earthquakes recur within 100-year cycles, while Active Fault (intraplate, plate boundary-related) earthquakes recur in periodicities of more than 1000 years (RGAFJ 1992: 41). The main consumers of earthquake archaeology data in Japan (and abroad, Silva *et al.* 2011) are paleoseismologists elucidating earthquake cyclicity. A very large combined Tokai-Nankai earthquake is expected before the mid-21c; it promises to affect all A–E rupture zones with great damage (Sangawa 2001*a*: 83-84). Thus, the severity and extent of the 2011 Tohoku-oki Earthquake, discussed below, surprised most everyone since attention had been focussed on the southwest, not the northeast.

In filling out the cyclicity chart of past earthquakes to refine expectations of future earthquakes, the main problems in Japan (as elsewhere) relate to identifying damage or deformation at archaeological sites that can be *unequivocally* assigned to an earthquake of a specific type and not some other cause, and to date them sufficiently.

Earthquake Evidence in Sediments

Liquefaction features

Soil liquefaction was first recognized in Japan in the 1948 Fukui Earthquake; studies of it increased after the 1964 Niigata Earthquake (Yasuda 2005: 152), when apartment buildings fell over sideways in softened sediments with relatively little damage. Liquefaction occurs only in sediments (e.g., silt, sand, gravel) saturated with water and usually requires a relatively impermeable layer (e.g., of clay) somewhere above them which prevents the water from dispersing as pressure builds (Obermeier 1996: 334-5). Ground shaking increases the pressure of the water, which fills the pore spaces between the sediment grains, and in extreme cases, it puts the sediment particles into suspension. Ishihara & Cubrinovski (2005: 10) stated that "the motion in the direction of maximum intensity is the key component of the ground motion that effectively controls the development of pore water pressures and consequent ground deformation." Sand eruptions are the usual result of subsurface liquefaction, as sediments (not just sand but silt and gravels) are pushed out to the surface with the pressurized groundwater forming sand dikes, blows, or boils (Figure 9.8). Usually, the ground surface becomes depressed and cracked as water is expelled from the lower layers (Obermeier 1996: 335).

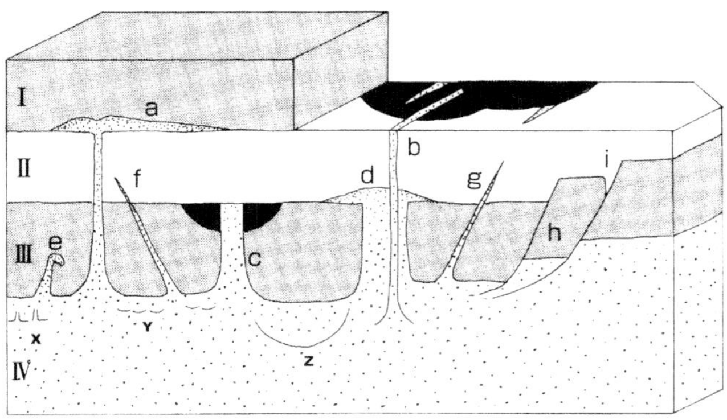

Figure 9.8 Liquefaction and faulted structures revealed in archaeological sites

Liquefaction features can take numerous forms (a–f). Dikes (b) characteristically cut through cultural layers, or through features such as pits and ditches; however, the cut-and-fill relations help date the earthquake occurrence rather than contributing greatly to an understanding of destruction to a built environment.

Key:
I–IV: strata
black areas: cultural features
a: *sand boil* later than strata **II** and **III** but earlier than stratum **I**
b: *sand dike* cutting through earlier *sand boil* **d** into preexisting cultural feature
c: *sand eruption* cutting through existing stratum **III** and cultural feature, but truncated by surface razing before stratum **II** was laid
d: *sand boil* cutting through existing stratum **III** before stratum **II** was laid
e: deflected *sand eruption*
f: fizzled *sand eruption*
g: *sand dike* cutting through existing strata **II** and **III** but cut by later cultural features
h, i: normal listric fault slips
x: *pillar* structures
y: *dish* structures
z: disturbed area

Obermeier also described the phenomenon of lateral spreading as an adjunct to liquefaction, happening on level ground of less than 5% incline. When an underlying stratum liquefies, the capping stratum fissures or breaks into segments, spreading laterally, especially when the cap is faced on one side by an abrupt change of slope like a terrace scarp or river bank (Obermeier 1996: 335-7). Although he states that the liquefied layer itself does not undergo a change in volume (Obermeier 1996: 333-4), Ishihara *et al.* postulated that dilation of the overlying ground mass must have occurred, based on data from the 1995 Kobe Earthquake lateral spreading, and that "it appears highly likely that the liquefied soil was sucked into the [land]masses instead of being ejected to the surface" (1997: 23, brackets added).

The year 1985 was the year that liquefaction features were first recognized at Japanese archaeological sites (Takahama *et al.* 2000: 157), but as far as is known, lateral spreading, with its surface fissures indicative of subsurface liquefaction, has not yet been identified in the archaeological record and has not been methodologically addressed, though its existence is well attested in modern earthquakes (Ishihara *et al.* 1997; Kuribayashi & Tatsuoka 1975; Nagase *et al.* 2006). Galadini *et al.* (2006: 400) include liquefaction and lateral spreading as 'off-site paleoseismological' effects that may damage buildings.

One of the most useful aspects of Earthquake Archaeology is instructing archaeologists how to distinguish sand dikes by their widening at the bottom and connection to the liquefied sand layer underneath, so that these features are not mistaken for canals or ditches. Another diagnostic comes from direction of grain-size fining: sand dike sediments *fine upward*, distinguishing them from fissures propagating downwards, which fill in from above and *fine downward.*

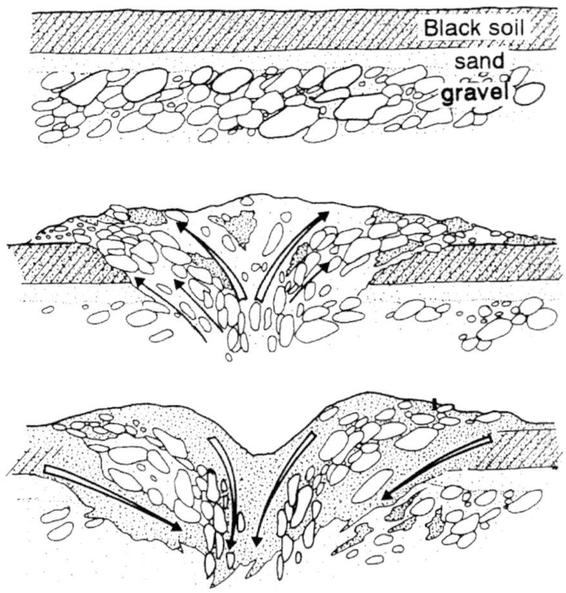

Figure 9.9 Liquefaction draw-in of cultural materials from overlying cap

Top: Black cultural layer
Middle: Sand and gravel erupts to surface
Bottom: Erupted material and cultural layer are partially drawn back into the liquefaction eruption channel

Takahama *et al.* (2000) have also elucidated the 'draw-in' phenomenon of liquefaction, where at the end of a sand eruption, surface materials (including potential cultural layers) and previously extruded sand/cobbles are sucked back into the top of the eruption path (Figure 9.9). Often, the surface layer contains cultural materials that can be used to relative-date the eruption. Or conversely, sand-blows can bring underlying artefacts to the surface (Gaur *et al.* 2005). Again, careful excavation will reveal the connection with the disturbed sand layer below, so that the feature - if circular - is not mistaken for a post-hole.

Shaking of saturated sediments can have quite opposite effects in the sedimentary record, as seen at two sites which illustrate explosive liquefaction and mild sediment disturbance, respectively.

The most extraordinary case of liquefaction in the Japanese archaeological record is at the Izumida site in Fukui Prefecture (Figure 9.10), dated to the Late Yayoi (ca. 100 AD) (Tomiyama 2009). Here, a large vein of cobbles up to 20 cm in diameter breached the surface; Yayoi people were apparently so surprised at this that they embedded a 35-cm-tall standing stone at the head of the eruption and provided it with a votive jar.

Figure 9.10 Liquefaction eruption of ca. 20-cm-diameter cobbles and its subsequent cultural marking with a standing stone as a sacred site, Late Yayoi Izumida site, Fukui Prefecture. The cobbles were derived from a lower layer, seen in the lower level of the trench wall.

In contrast, at the Nishi-Sanso/Yakumo-Higashi site in Osaka, only mild liquefaction features have been recovered in section (Figure 9.11). Dish and pillar structures are identifiable in the upper sand stratum; dike intrusions strike upward, some obviously abortive, while two small efforts at sill formation are visible.

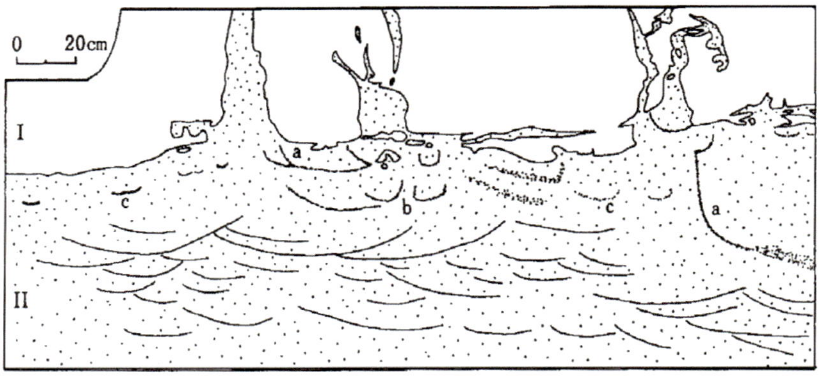

Figure 9.11 Liquefaction structures at Nishi-Sanso/Yakumo-Higashi site, Osaka

Eruptions of sand from stratum II into stratum I:
a—disturbed area, **b**—pillar structures, **c**—dish structures

In sum, sand and/or cobble eruptions and the phenomenon of lateral spreading are two immediate and drastic effects of liquefaction that can be incontrovertibly assigned an earthquake cause. No other Earth motions have the shaking severity necessary to raise pore pressure to such explosive levels.

Obermeier (1996: 334-337) has also implied that the mild sand dikes, sills, and dish-and-pillar structures are known results of liquefaction and that liquefaction itself is a result of earthquake activity. If we accept this, we may assume that liquefaction features at an archaeological site, if properly identified, are evidence of past earthquake events. The advantage of finding these features at sites rather than in the great geological outdoors is the association of cultural materials that may aid in dating the events. However, others argue that other mechanisms may cause mild deformation of soft sediments and that it cannot be assumed a priori that these derive from earthquake activity.

Soft-sediment deformation structures

Fine-grained (mud, clay) sedimentation is characteristic of low-energy depositional environments such as ocean floors and lake bottoms. The muddy deposits that collect there can be disturbed and distorted either during or after sedimentation by several processes – not only by seismic shaking but by "loading during rapid sedimentation, localized artesian conditions, or slumping" (Obermeier 1996: 332). The deformation structures formed under these pressures are patterned and have been given evocative names – some overlapping with liquefaction features: ball-and-pillow, flame, water-escape structures, neptunian dikes, load casts, slumps, and convolutions. These comprise plastic deformations of the soft sediments, which then lithify over time; they can be seen either in rock formations going back to Palaeoproterozoic times or in modern unlithified sediments. The challenge is to prove which ones were caused by earthquakes rather than by some other disturbance. As Guidoboni emphasized (1995: 10-11), structural damage and landscape changes cannot be assumed to result from earthquakes but must be proven so.

Obermeier (1996: 332) noted that the geological literature is 'replete' with references to seismites – such patterned disturbances that are assumed to have been caused by earthquake shaking, with some experimental work demonstrating their replication (Moretti *et al.* 1999). The first decade of the millennium saw a concerted effort to assess critically the formation processes of soft-sediment deformation structures (Ettensohn *et al.* 2002; Shiki *et al.* 2000), and the term 'seismite' has somewhat fallen out of favour as an initial descriptor. Even after sufficient investigation, a suspected seismite might be prefaced with different degrees of inclusion in that category as certain, probable, likely, or possible (Greb *et al.* 2002). Several authors offer criteria by which to determine an earthquake cause for deformed sediments (e.g., Rossetti & Góes 2000; Moretti 2000; Moretti & Sabato 2007; Greb *et al.* 2002; Greb & Dever 2002; Davies *et al.* 2004; Bowman *et al.* 2004; Mazumder *et al.* 2006). Among these sources, six specify two important criteria: exclusion of normal processes, and correlative horizons (i.e., lateral continuity) over wide areas. Four speak of the importance of regional abundance of data, so that earthquake attribution does not rely on a single site or single observation. Five specify the obvious criteria of being located in a tectonically active zone, having other tectonic evidence nearby, and occurring in conjunction with deposition (syn-sedimentary evidence). Three emphasize that the deformed strata should be discrete, occurring sandwiched between other undeformed strata or strata of different origins. Two note that such deformed strata should recur, possibly rhythmically or cyclically. Two emphasize the importance of deformation immediately post-depositional while still unconsolidated. Finally, individual sources cite further interesting criteria: the necessity to define a core zone of activity; the presence of complex upward deformation; the presence of long periods of quiescence; size of deformation; and the similarity of deformed patterns to modern cases or experimental results.

Thus, we cannot immediately discuss the presence of seismites in Japanese archaeological sites without first proving their seismic origins; instead, we must first talk about soft-sediment deformation structures. The criteria for judgment should be the same as proposed already for other areas outside Japan, and it is notable that the initial forays in this direction within Japan very much take the regional or 'territorial' approach common to most of the aforementioned authors, as well as Rapp (1986), Obermeier (1996), and Galadini *et al.* (2006). Damage at any one site or location may have been caused by local conditions, but an earthquake usually has broader geographical repercussions. This approach simply relies on recovering similar evidence of damage from a wide area (several sites' worth) at a certain date, so that through repetition of data, the conclusion that the damage was caused by an earthquake and not a more localized situation is reinforced. A limitation of this approach, however, is chronological uncertainty that each instance of damage was caused by the same earthquake; there is always the potential of collapse of different events into one, resulting in the exaggeration of event size (Galadini *et al.* 2006: 407-408).

Matsuda Jun-ichiro's research in the eastern Osaka Basin, with its attention to regional occurrences of soft-sediment deformation structures in the late Holocene archaeological record, is an important new direction in Japanese Earthquake Archaeology. A trained geomorphologist and sedimentologist working as a geoarchaeologist, Matsuda has focused on disturbance layers in the stratigraphic succession of alluvial and cultural deposits from postglacial Jomon to recent times. His publication (2000) synthesizes the sequences from six archaeological sites clustered in an area 7 km E–W by 4 km N–S in the southern Osaka Plains (cf. Figure 9.7, just east of site 13). The deformation layers are demonstrated to occur in a tectonically active zone at multiple, separated site intervals, and are territorially correlative (Figure 9.12). These data appear to fulfill the territorial approach criterion for assessing seismogenic status; their repetitive nature at intervals of several centuries, occurring between undisturbed layers of parallel laminated strata, also argues for occasional large disturbances separated by periods of quiescence rather than routine storm or slumping conditions. For these reasons, the deformations identified by Matsuda may be considered to be seismites.

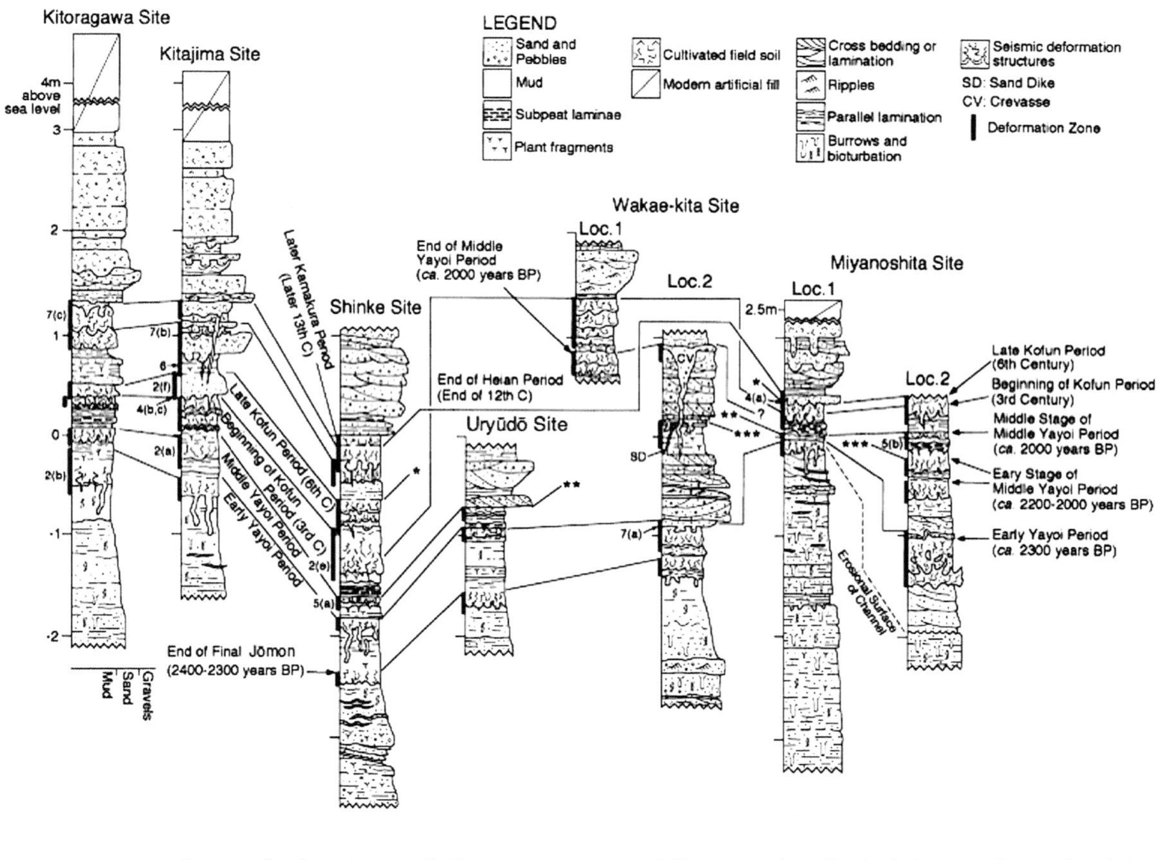

Figure 9.12 Correlation of soft-sediment deformation zones at different archaeological sites on the Osaka Plains

At the Kitoragawa site (Figure 9.12: far left; Figure 9.13), the evidence of soft-sediment deformation structures in layers relatively dated by cultural materials can be correlated with six or seven historically known earthquakes, all assessed by Usami at magnitudes between M7 and M8.4 (Matsuda 2000: fig. 11). This seems to go against Obermeier's statement that "soft-sediment deformations often form at such low levels of seismic shaking that the shaking poses no hazard" (1996: 332). Several of the named earthquakes were Nankai subduction earthquakes, which have already been shown to have affected archaeological sites in the eastern Inland Sea region in and around Osaka (cf. Figure 9.6). The unaddressed question is, what was the level of intensity of ground shaking at these various sites to create greater (liquefaction) or lesser (soft-sediment deformation) disturbance features?

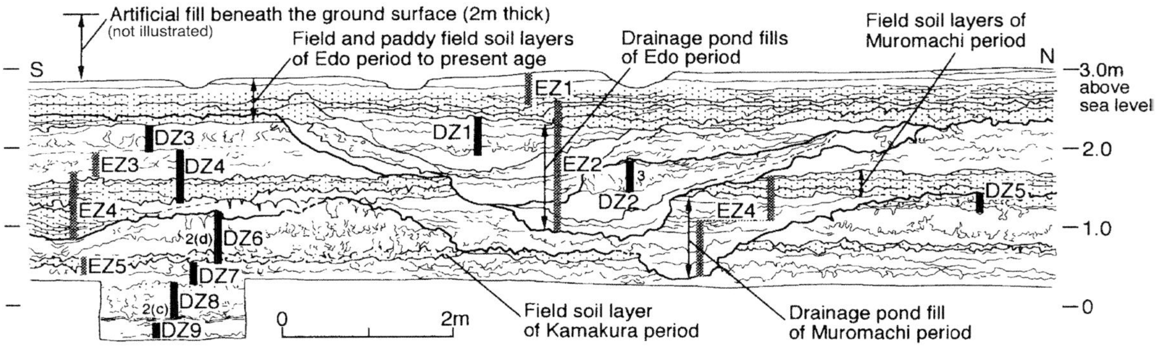

Figure 9.13 Soft-sediment deformation zones (DZ) and anthropogenic zones (EZ) at Kitoragawa site

The features investigated by Matsuda were all generated in muddy strata submerged under standing water at the time of deformation: in deltaic back marshes, small shallow lakes, or paddy-fields. They appeared in the plans and sections of archaeological trenches; Matsuda cut out blocks and returned them to the laboratory for further analysis, including X-ray radiographs of 1-cm-thick slices (Figure 9.14). He identified several kinds of features in stratigraphic succession, described, from bottom to top: downward fissures and microfaults at the bottom, load structures then plumose patterns, and homogenization at the top. The extensive use of radiographs in addition to sketch drawings of the structures answers Obermeier's lament concerning liquefaction that "the deformational effects...have rarely been illustrated and discussed in vertical section, the view most useful for paleoliquefaction studies" (1996: 332).

Several earthquakes were identified through these deformation patterns that are not attested in the historical record: one in early Heian, one in Late Kofun, and at least one in the first half of Yayoi. Prehistoric subduction earthquakes are also attested at other archaeological sites, as shown in Figure 9.6, though the evidence is quite different. It remains to correlate the soft-sediment deformation structures with other dated sites to construct a potential prehistoric series of earthquakes. Soft-sediment deformation studies have broadened the earthquake data-gathering capacity of archaeological excavation, but few specialists are available to do them. Japanese geoarchaeologists have unknowingly responded to Rapp's original goal of paying more attention to the sediments in archaeoseismic excavations (1986: 370), and this methodological aspect of Japanese Earthquake Archaeology is well worth exporting, since many other countries are also subject to post-hole archaeology without substantial building remains.

The 1995 Kobe Earthquake (officially called the Hanshin-Awaji Dai-shinsai in Japanese) brought to the fore this developing subdiscipline of Japanese archaeology, as it galvanized archaeological action on a multi-faceted front (BOX 10). After the Kobe disaster, young archaeologists in the Kinai region around Osaka, Nara, and Kyoto (cf. Figure 9.7) joined together to form the Disaster Concerned Archaeologists' Network (Maibun Kankei Kyuen Renraku Kaigi; see e.g., DCAN 1996) to assist with communications related to cultural properties damaged in the earthquake. A newsletter edited by OKAMURA Katsuyuki knitted the group together until they achieved nation-wide results.

One of the DCAN network's outstanding contributions to the scholarship of Earthquake Archaeology was the organization of a conference jointly sponsored with the Buried Cultural Properties Research Group (Maizō Bunkazai Kenkyūkai). Papers given at the conference revealed evidence of earthquakes in the archaeological record. Over 150 scholars from around Japan gathered near Kobe just 19 months after the earthquake to present evidence on 378 sites throughout Japan that were assessed as having sustained previous earthquake damage. Their published report runs to 826 pages and contains a master list of the sites covered (DCAN 1996). Their purpose in this compilation, however, was more than academic: they saw that if the cycle of earthquakes could be established through time in different regions, they could help contribute to the future safety of society.

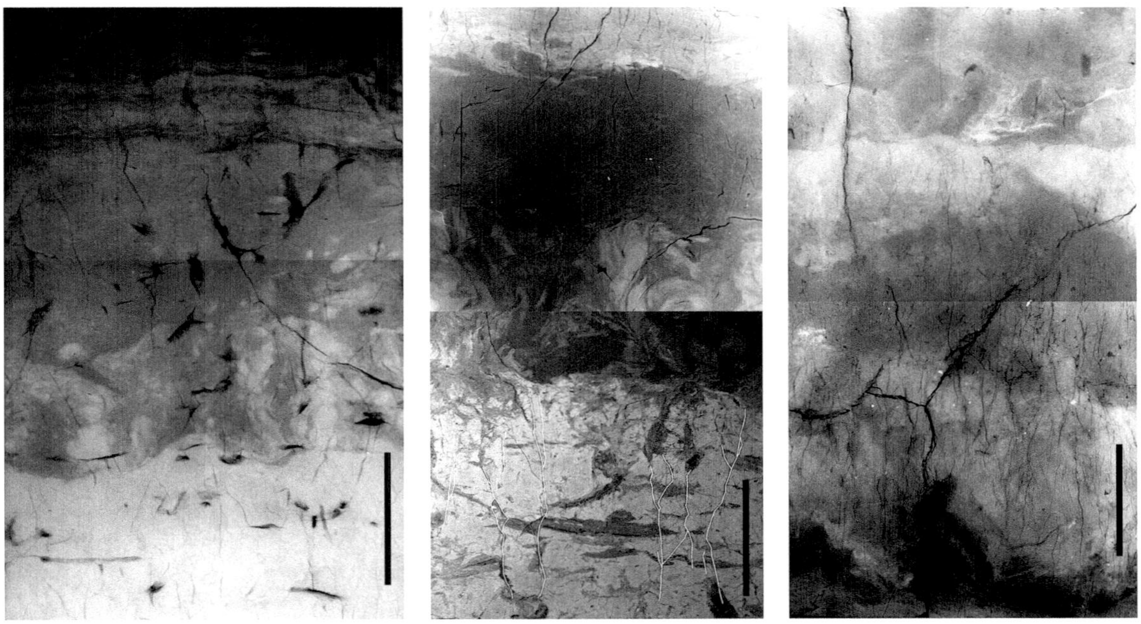

Figure 9.14 Radiographs of soft-sediment deformation structures
Scale bar = 10 cm

Left (Kitajima site): Deformation zone characterized by, from the top down, liquidized deformation unit with laminae of a weak load structure from a younger deformation, a hydroplastic deformation unit with plumose pattern of light and white intermixed and depressions at its base, and a lower, white hydroplastic deformation unit.

Middle (Kitoragawa site): Two deformation zones. The top light-colored layer is a young deformation zone. The older zone begins with the dark liquidized deformation unit having an undulated base in response to downward pressure. This is underlain by a hydroplastic deformation unit with twisted plumose pattern; the lower area of microfaults (white vertical fracture lines) indicates a brittle deformation unit. Lower half of this image is a normal photograph taken in daylight.

Right (Kitoragawa site): Two deformation zones with load structure. The darker liquidized homogenized band through the center marks the top of the older deformation zone; the lowermost peaty mud bed is deformed into blocky shapes. The light-colored layers of the younger deformation zone are separated horizontally into a hydroplastic deformation unit with plumose patterns above and a hydroplastic deformation unit below having downward lobes that involve the older homogenized layer.

The 1995 Kobe Earthquake also involved archaeologists in cultural preservation and heritage activities unrelated to tourism as well as actual excavation. Okamura (2013) reports that 250 ha of central Kobe City, entailing 280 archaeological sites, were devastated in the earthquake; within one month, management policies were put into place to provide excavation prior to the re-establishment of essential utilities or demolition/reconstruction of buildings. A precedent was set by the establishment of a rescue excavation team that brought in 121 regional archaeologists from up to 40 prefectures; they accomplished 1600 pre-construction investigations and 540 site excavations. The latter attracted attention from and fulfilled the curiosities of the populace via site open days and school study sessions. However, Okamura notes that the major disadvantage was the rotation of excavators, leaving behind unpublished data by departing staff; he recommends combined teams of local and seconded archaeologists, with the former in charge for the longer term and responsible for publication.

BOX 10 Archaeological Contributions to Understanding the Kobe Earthquake

Comparisons between soil types underlying Kobe City and palaeo-geomorphology led to insights on earthquake intensity. Hayashi *et al.* (1999) reported that the worst-hit areas in Kobe (Intensity VII on the JMA scale, cf. Table 2.2; Apx 7) correlate with the outcrop of the Pleistocene Osaka Group of marine and lacustrine sediments forming low hills (for comparison see Chapter 13). However, Uda *et al.* (1997) note that the area of destruction was limited to the Holocene alluvial strip along the coast (Figure A below). Within this zone, the direction of collapse of buildings – regardless of age, structure, or material – could only be explained by relative movement and twisting on a right-lateral strike-slip fault. Takahashi (1998, 2001) gives further detailed geomorphological evidence that the most severe shaking occurred on ground that was lagoonal between 7400–6300 years ago in Middle Jomon times (Figure B), relying on archaeological relative dating. Figure C illustrates coincidences between structural collapse and stream courses.

A) Disaster map of Hyogo Prefecture coastal region:

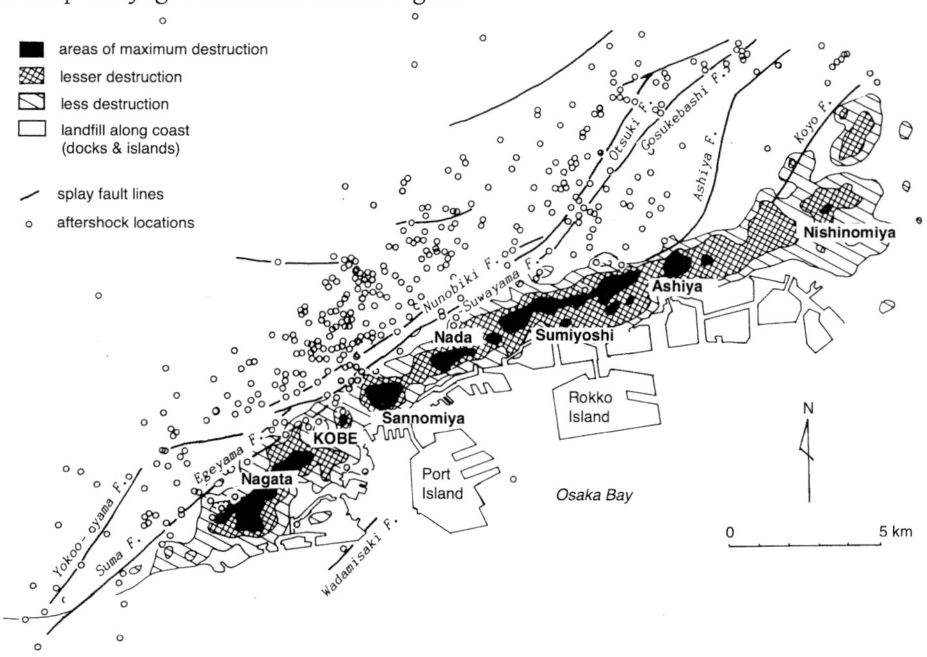

[cont'd]

B) Geomorphological reconstruction of limited section of that coastline:

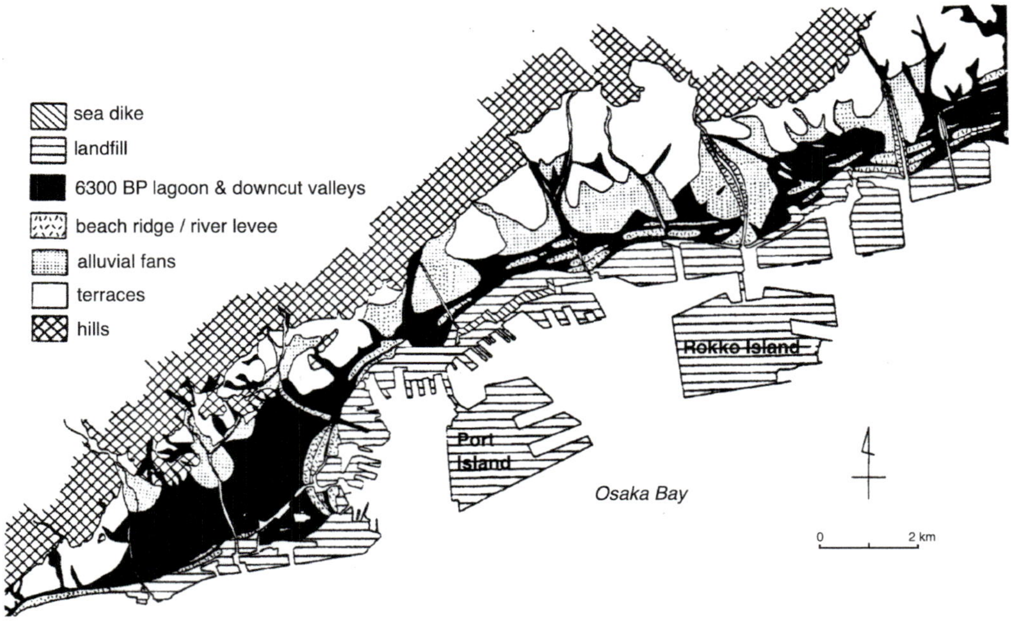

C) Deaths and structural collapse locations:
Moreover, in Nishinomiya City, locations of collapsed railway (Shinkansen) bridges and houses where people had died particularly correlated with buried braided stream courses that developed from the Final Jomon period.

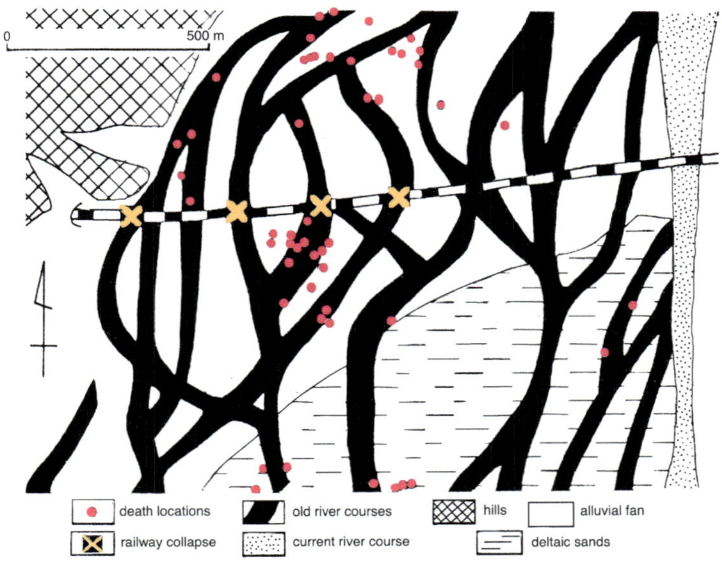

In sum, the effects of the earthquake differed in two geomorphological zones: alluvium (affected by strong ground motion) and landfill (affected by liquefaction). The kinds of destruction consisted primarily of structural damage and collapse in the former, but in the latter, sediment eruption, settling, and spreading that affected structures built on those sediments.

Concluding Remarks

Earthquake archaeology in Japan arose simultaneously in the mid-1980s with archaeoseismology in the Mediterranean, but these have taken off in quite different directions. Having few building structures in which to assess earthquake damage, the actual sediments at Japanese archaeological sites and their tectonic disruption have come under examination. Geomorphology and sedimentology rather than paleoseismology or historical seismology has been the main disciplinary inspiration; thus, it is the identification of morphological features that becomes the foremost task. Archaeologists are being encouraged to recognize features such as fault displacement, fissures, and liquefaction eruptions, while a few geoarchaeology specialists are engaged in sediment analysis for soft-sediment deformation patterns. Much of the earthquake damage recovered in archaeological excavations in Japan is geological damage to soils, sediments, and strata.

The main consumers of Earthquake Archaeology in Japan have been palaeoseismologists interested in establishing recurrence rates. Since such rates are different for subduction and active fault earthquakes, the challenge has been to divide the archaeological evidence to support one or the other series. This relies both on accurate dating and on the definition of broad areas of co-seismic damage. Although relative dating by association with cultural artefacts or layers is not as precise as documentary or absolute dating, the Japanese ceramic chronology is the most refined in the world, with generational (20-year) spans of earthenware types and 5-year spans of stoneware types in the prehistoric and early historic periods in the best-case examples. This is better than establishing 'seismic periods' (Caputo & Pavlides 2008: 3) or spans of a 'few decades' (Galadini *et al.* 2006: 407).

That relative dates are imprecise may irritate instrumentalists used to dealing with precise measurements, but any time human behaviour enters the research picture, we must deal with the probabilities rather than certainties. Rather than ignore or discard the datings provided by archaeoseismologists, these should be taken as supporting data for working hypotheses about past earthquake occurrence. Earthquake Archaeology provides a finer framework for discovering and dating past earthquakes than do trenching activities undertaken by earthquake geologists, and so the additional data should be welcomed.

Moreover, extension of research interests beyond earthquake damage to former civilizations (Caputo & Helly 2008) by including worldwide Paleolithic and Neolithic sites should enormously expand the data available. As Noller & Lightfoot (1997) did for prehistoric sites in America, Japanese archaeologists have demonstrated that any earthquake evidence that occurs in conjunction with cultural materials can be relatively dated, regardless of period or level of social development and whether or not those 'artefacts' suffered 'damage'. A more thorough treatment of approaches to archaeoseismology worldwide, however, would need to take account of Japanese-language sources and even the increasing body of English-language materials on the Japanese situation, including the depth of historical information about earthquakes in Japanese texts (cf. Ishibashi 2004) and the instrumentation records that go back to the invention and deployment of the seismograph in Japan and England in the 1880s (Clancey 2006; Utsu 2003).

Earthquake Archaeology in Japan has demonstrated that sites of all ages are repositories of earthquake damage, once archaeologists learn to read the excavated traces. Also, damage need not be restricted to buildings but can affect the actual sediments at archaeological sites. The inclusion of the full range of Japanese historic and archaeological earthquake data cannot but enhance the discipline of archaeoseismology, as would information from other tectonically active regions of the world where post-hole archaeology is the norm.

Figure & Table Sources

9.1 courtesy of A. Sangawa, redrawn by Durham Archaeological Services and modified by GLB
9.2 photo by author
9.3 modified by GLB after (DCAN 1996: 759)
9.4 drawings compiled from (Kidder 1972: figs. 45, 52, 56) and author's photo
9.5 courtesy of A. Sangawa, modified by Durham Archaeological Services
9.6 after (Sangawa 2001*a*: figs. 64 and 65); modified by Durham Archaeological Services
9.7 after (Sangawa, 2001*a*: 88), redrawn by Durham Archaeological Services; descriptions extracted from (Sangawa 2001*a*: 110–115; with added information from (Kamai & Sangawa 2011)
9.8 modified by GLB from (Sangawa 2001*a*: 52)
9.9 after (Takahama *et al.* 2000: fig. 9), modified by GLB
9.10 photo courtesy of A. Sangawa
9.11 after (Sangawa 1999: fig. 8), modified by GLB
9.12 modified from (Matsuda 2000: fig. 9), modified by GLB
9.13 after (Matsuda 2000: fig. 10), modified by GLB
9.14 photograph courtesy of J.-i. Matsuda; caption based on Matsuda (2000), with permission

BOX 10-A after (Uda 1997: fig. 3), modified by GLB
B after (Takahashi 2001: fig. 22), modified by GLB
C after (Takahashi 2001: fig. 24), modified by GLB

References

ASTON, William George (transl.) (**1972** [1896]) *Nihongi: chronicles of Japan from from the earliest times to AD 697*. Tokyo & Rutland, VT: Tuttle.

BARNES, Gina L (**1990**) "The origin of bureaucratic archaeology in Japan." *Journal of the Hong Kong Archaeological Society* 12: 183-196.

BOWMAN, D; A KORJENKOV & N PORAT (**2004**) "Late-Pleistocene seismites from Lake Issyk-Kul, the Tien Shan range, Kyrghyzstan." *Sedimentary Geology* 163: 211-228 [DOI: 10.1016/S0037-0738(03)00194-5].

CAPUTO, R & B HELLY (**2008**) "The use of distinct disciplines to investigate past earthquakes." *Tectonophysics* 453: 7-19 [DOI: 10.1016/j.tecto.2007.05.007].

CAPUTO, R & SB PAVLIDES (**2008**) "Earthquake geology: methods and applications." *Tectonophysics* 453: 1-6 [DOI: 10.1016/j.tecto.2008.01.007].

CHATTERJEE, A; T NATARAJAN & K RAJENDRAN (**2018**) "On the source characteristics of inter and intraplate earthquakes in the Japanese Trench." *AGU Fall Meeting Abstracts* [adsabs.harvard.edu] (dead link).

CLANCEY, Gregory (**2006**) *Earthquake nation: the cultural politics of Japanese seismicity, 1868-1930*. Berkeley: University of California Press.

DAVIES, NS; P TURNER & IJ SANSOM (**2004**) "Soft-sediment deformation structures in the Late Silurian Stubdal Formation: the result of seismic triggering. *Norwegian Journal of Geology* 85: 233-243.

DCAN (Disaster Concerned Archaeologists' Network) (**1996**) *Excavated evidence of earthquakes*. Osaka: Maibun Kankei Kyūen Renraku Kaigi (in Japanese).

ETTENSOHN, FR; N RAST & CE BRETT (eds) (**2002**) *Ancient Seismites*. GSA Special Paper No. 359. Boulder, CO: Geological Society of America.

FORLIN, Paolo & Christopher M GERRARD (**2017**) "The archaeology of earthquakes: the application of adaptive cycles to seismically-affected communities in late medieval Europe." *Quaternary International* 446: 95-108 [DOI: 10.1016/j.quaint.2017.06.030].

FUJINO, Shigehiro; Haruo KIMURA & Junko KOMATSUBARA *et al.* (**2018**) "Stratigraphic evidence of historical and prehistoric tsunamis on the Pacific coast of central Japan: implications for the variable recurrence of tsunamis in the Nankai Trough." *Quaternary Science Reviews* 201: 147-161 [DOI: 10.1016/j.quascirev.2018.09.026].

GALADINI, F; K-G HINZEN & S STIROS (**2006**) "Archaeoseismology: methodological issues and procedure." *Journal of Seismology* 10: 395-414 [DOI: 10.1007/s10950-006-9027-x].

GARRETT, Ed; Osamu FUJIWARA & Philip GARRETT *et al.* (**2016**) "A systematic review of geological evidence for Holocene earthquakes and tsunamis along the Nankai-Suruga Trough, Japan." *Earth-Science Reviews* 159: 337-357 [DOI: 10.1016/j.earscirev.2016.06.011].

GAUR, Aniruddh Singh & KH Vora SUNDARESH (**2005**) *Archaeology of Bet Dwarka Island: an excavation report.* New Delhi: Aryan Books International.

GREB, SF & GR DEVER Jr (**2002**) "Critical evaluation of possible seismites: examples from the Carboniferous of the Appalachian basin", pp. 109-125 in *Ancient Seismites*, ed. by FR ETTENSOHN *et al.* GSA Special Paper No. 359. Boulder, CO: Geological Society of America.

GREB, SF; F ETTISOHN & F OBERMEIER (**2002**) "Developing a classification scheme for seismites." *Geological Society of America Abstracts with Programs* 34.2: A-102. Boulder, CO: GSA.

GUIDOBONI, E (**1995**) "Archaeology and historical seismology: the need for collaboration in the Mediterranean area", pp. 7-13 in STIROS & JONES (eds) (1995).

HAYASHI, Yasuhiro; Kazuo TAMURA & Masafumi MORI *et al.* (**1999**) "Simulation analyses of buildings damaged in the 1995 Kobe, Japan, earthquake considering soil-structure interaction." *Earthquake Engineering and Structural Dynamics* 28: 371-91.

ISHIBASHI, K (**2004**) "Status of historical seismology in Japan." *Annals of Geophysics* 47.2/3: 339-368.

ISHIHARA, K & M CUBRINOVSKI (**2005**) "Characteristics of ground motion in liquefied deposits during earthquakes." *Journal of Earthquake Engineering* 9 (Special Issue 1): 1-15 [DOI: 10.1142 /S1363246905002304].

ISHIHARA, K; K YOSHIDA & M KATO (**1997**) "Characteristics of lateral spreading in liquefied deposits during the 1995 Hanshin-Awaji earthquake." *Journal of Earthquake Engineering* 1.1: 23-55 [DOI: 10.1142/ S1363246997000039].

ITOH, Yasuto; Keiji TAKEMURA & Shigekazu KUSUMOTO (2013) "Neotectonic intra-arc basins within southwest Japan — conspicuous basin-forming process related to differential motion of crustal blocks", pp. 191-207 in *Mechanism of sedimentary basin formation - multidisciplinary approach on active plate margins*, ed. by Y. Itoh. London: IntechOpen.

KAMAI, Toshitaka & Akira SANGAWA (**2011**) "Landslides on ancient embankments in the Kinki district (Japan): Strong motion seismoscope of the 1596 Keichou-Fushimi earthquake." *Quaternary International* 242.90-105.

KANAORI, Y & S KAWAKAMI (**1996**) "The 1995 7.2 magnitude Kobe earthquake and the Arima-Takatsuki tectonic line: implications of the seismic risk for central Japan." *Engineering Geology* 43.2-3: 135-150 [DOI: 10.1016/0013-7952(96)00056-7].

KIDDER, JE Jr (**1972**) *Early Buddhist Japan.* London: Thames & Hudson.

KURIBAYASHI, E & F TATSUOKA (**1975**) "Brief review of liquefaction during earthquakes in Japan." *Soil and Foundation* 15.4: 81-92.

MATSUDA, J-i (**2000**) "Seismic deformation structures of the post-2300 BP muddy sediments in Kawachi lowland plain, Osaka, Japan." *Sedimentary Geology* 135: 99-116 [DOI: 10.1016/S0037-0738(00)00066-X].

MATSU'URA, Ritsuko S (**2017**) "A short history of Japanese historical seismology: past and the present." *Geoscience Letters* 4.3: 15 pp.

MAZUMDER, R; AJ VAN LOON & M ARIMA (**2006**) "Soft-sediment deformation structures in the Earth's oldest seismites." *Sedimentary Geology* 186: 19-26 [DOI: 10.1016/j.sedgeo.2005.12.002].

MILLER, M (**2008**) "Breaking the slag." *Nature Geoscience* 1: 730-731 [DOI: 10.1038/ngeo341].

MORETTI, M (**2000**) "The interpretation of soft-sediment deformation structures as seismites." *European Geophysical Society, 25th general assembly (1029-7006) 01/01/2000 abstracts.*

MORETTI, M; P ALFARO & O CASELLES *et al.* (**1999**) "Modelling seismites with a digital shaking table." *Tectonophysics* 304: 369-383.

MORETTI, Massimo & L SABATO (**2007**) "Recognition of trigger mechanisms for soft-sediment deformation in the Pleistocene lacustrine deposits of the Sant Arcangelo Basin (Southern Italy): Seismic shock vs. overloading." *Sedimentary Geology* 196.1-4: 31-45.

NAGASE, H; K ZEN & A HIROOKA *et al.* (**2006**) "Zoning for liquefaction and damage to port and harbor facilities and others during the 2005 Fukuoka-ken Seiho-Oki earthquake." *Soil and Foundation* 46.6: 805-816.

NIEMI, Tina M (**2017**) "Archaeoseismology", pp. 47-57 in *Encyclopedia of geoarchaeology*, ed. by Allan S. GILBERT. Dordrecht: Springer.
NHK (Japan Broadcasting Company) (**2002**) "Kasumigaseki building—endlessly fighting towards skyscrapers: revolutionary technologies for the Japanese earthquake-prone archipelago." Tokyo: NHK Sofutowea (DVD) (in Japanese).
NOLLER, JS & KG LIGHTFOOT (**1997**) "An archaeoseismic approach and method for the study of active strike-slip faults." *Geoarchaeology* 12: 117-135.
OBERMEIER, F (**1996**) "Using liquefaction-induced features for paleoseismic analysis", pp. 331-396 in *Paleoseismology*, ed. by JC MCCALPIN. San Diego, CA: Academic Press.
OKAMURA, Katsuyuki; Atsushi FUJISAWA & Yasuhisa KONDO *et al.* (**2013**) "The Great East Japan Earthquake and cultural heritage: towards an archaeology of disaster." *Antiquity* 87.335: 258-269.
OKUBO, T (**2016**) "Traditional wisdom for disaster mitigation in history of Japanese architectures and historic cities." *Journal of Cultural Heritage* 20: 715-724.
PAJAK, J (**2005**) "Signal processing in the 'Zhang Heng Seismograph' for remote sensing of impending earthquakes", in *Proceedings of the 1st International Conference on Sensing Technology*, 21-23 November 2005, ed. by GS GUPTA *et al.* Palmerston North, New Zealand.
RAPP, G Jr (**1986**) "Assessing archaeological evidence for seismic catastrophes." *Geoarchaeology* 1: 365-379 [DOI: 10.1002/gea.3340010403].
RGAFJ (Research Group for Active Faults in Japan) (**1992**) *Maps of Active Faults in Japan with an Explanatory Text.* Tokyo: University of Tokyo Press (English condensed version of 1991 full publication in Japanese).
RODRÍGUEZ-PASCUA, MA; R PÉREZ-LÓPEZ & JL GINER-ROBLES *et al.* (**2011**) "A comprehensive classification of Earthquake Archaeological Effects (EAE) in archaeoseismology: application to ancient remains of Roman and Mesoamerican cultures." *Quaternary International* 242: 20-30.
ROSSETTI, DF & AM GÓES (**2000**) "Deciphering the sedimentological imprint of paleoseismic events: an example from the Aptian Codó Formation, northern Brazil." *Sedimentary Geology* 135: 137-156 [DOI: 10.1016/ S0037-0738(00)00068-3].
SANGAWA, A (**1988**) "Evidence of earthquakes recognized as objects of archaeological research." *Kodaigaku Kenkyū* 116: 1-16 (in Japanese with English title).
---- (**1995**)" Searching for ancient earthquakes in archaeological data", pp. 107-124 in *Kojishin o Saguru*, ed. by Y ŌTA & K SHIMAZAKI. Tokyo: Kokon Shoin (in Japanese).
---- (**1999**) "Palaeoliquefaction features at archaeological sites in Japan." *Chigaku Zasshi* 108.4: 391-398 [DOI: 10.5026/jgeography.108.4_391] (in Japanese with English summary).
---- (**2001*a***) Earthquakes: a history of catfish activities. Tokyo: Daikōsha (in Japanese).
---- (**2001*b***) "Landslip traces recognized on the Imashiro-zuka tomb." *Nihon Bunkazai Kagakukai Dai 18-kai Taikai Kenkyū Happyō Yōshi* 2001: 24-25 (in Japanese).
---- (**2007**) *Earthquakes in Japanese History*. Tokyo: Chūō Kōron Shinsha (in Japanese).
---- (**2009**) "A study of paleoearthquakes at archeological sites: a new interdisciplinary area between paleoseismology and archeology. *Synthesiology*, English edition, 2: 84-94.
SHIKI, T; MB CITA & DS GORSLINE (**2000**) "Sedimentary features of seismites, seismo-turbidites and tsunamiites: an introduction." *Sedimentary Geology* 135.1-4: vii-ix [DOI: 10.1016/S0037-0738(00)00058-0].
SCHOLZ, Christopher H (**1990**) *The mechanics of earthquakes and faulting*. Cambridge University Press.
SHIMADA, Yumi; Shigehiro FUJINO & Yuki SAWAI *et al.* (**2019**) "Geological record of prehistoric tsunamis in Mugi town, facing the Nankai Trough, western Japan." *Progress in Earth and Planetary Science* 6.3: 15 pp. [DOI: 10.1186/s40645-019-0279-].
SILVA, PG; M SINTUBIN & Klaus REICHERTER (**2011**) "Guest editorial: New advances in studies of earthquake archaeology and palaeoseismology." *Quaternary International* 242: 1-3.
SINTUBIN, Manuel (**2011**) "Archaeoseismology: past, present and future." *Quaternary International* 242.4-10.
STIROS, SC (**1995**) "Archaeological evidence of antiseismic constructions in antiquity." *Annali di Geofisica* 35: 725-736.
STIROS, S & RE JONES (**1996**) *Archaeoseismology*. Fitch Laboratory Occasional Paper No. 7. Athens: British School at Athens.

TAIRA, A (**2001**) "Tectonic evolution of the Japanese island arc system." *Annual Review of Earth and Planetary Sciences* 29: 109-134 [DOI: 10.1146/ annurev.earth.29.1.109].

TAKAHAMA, N; T ŌTSUKA & B BRAHMANTYO (**2000**) "A new phenomenon in ancient liquefaction, the draw-in process, its final stage." *Sedimentary Geology* 135: 157-165 [DOI: 10.1016/S0037-0738(00)00069-5].

TAKAHASHI, Manabu (**1998**) "Geo-environmental analysis of the 1995 Hyōgo-ken Nanbu earthquake disaster." *The Memoirs of the Geological Society of Japan* 51: 127-134 (in Japanese).

---- (**2001**) "Environmental archaeology for disaster risk management." *Ritsumeikan Bungaku* 593: 582-569 (in Japanese with English title and abstract).

TODA, S (**2016**) "Crustal earthquakes", pp. 371-408 in *The geology of Japan*, ed. by T MORENO *et al.* London: Geological Society of London.

TODA, S; RS STEIN & SH KIRBY *et al.* (**2008**) "A slab fragment wedged under Tokyo and its tectonic and seismic implications." *Nature Geoscience* 1: 771-776 [DOI: 10.1038/ngeo318].

TOMIYAMA, M (**2009**) "The Izumida location at the Hayashi-Fujishima site." *Fukui-ken Maizō Bunkzai Chōsa Hōkoku* 106: 87, 89 (in Japanese).

UDA, Tsuyoshi (ed.) (**1997**) "Research relating to surface movement and archaeological site formation/ destruction", unpg. in *Earthquake fault inferred from the severe disaster belt in the Kobe area.* Niigata: Mombusho Kagku Kenkyūhi (No. 06455010) (in Japanese).

USAMI, T; H ISHII & T IMAMURA *et al.* (**2013**) *Materials for comprehensive list of destructive earthquakes in Japan 599-2012.* University of Tokyo Press (in Japanese).

UTSU, T (**2003**) "Historical development of seismology in Japan." *Swiss National Centennial Report to the International Association of Seismology and Physics of the Earth's Interior*, Ch. 79.33 Japan, Part 2: 19 pp.

YASUDA, S (**2005**) "Survey of recent remediation techniques in Japan, and future applications." *Journal of Earthquake Engineering* 9.1: 151-186 [DOI: 10.1142/S1363246905002213].

Chapter 10

Tsunami Archaeology

This chapter provides an introduction to the nature of tsunami waves and their causes, resources available for studying past tsunami worldwide, and difficulties in identifying tsunami sediments.[1] Seventeen case studies of sites where tsunami deposits have been investigated throughout the Japanese and Ryukyu archipelagos are presented. The most important archaeological evidence of previous tsunami and their effects has been discovered in sites on the Sendai Plain, Miyagi Prefecture, which was also inundated in the 2011 Tohoku-oki Tsunami. Middle Yayoi-period paddy-fields which had been developed on the alluvium were covered a by tsunami sand sheet around 100 BC and thereafter abandoned for four centuries. The deprivation of coastal rice-growing may have led to shifts in the subsistence system towards exploitation of the highlands – a reversion to previous Jomon ways of life. The backmarsh and beach ridge areas were only re-developed when different technologies and social structures were introduced in the Kofun period.

Introduction

Tsunami damage to archaeological sites in Japan has been recognized since the 1980s, but the Great Tohoku-oki Earthquake & Tsunami of 11 March 2011 stimulated geologists and archaeologists to find evidence of previous tsunami in Japan and elsewhere (Goff & Witter *et al.* 2020), investigate the responses of earlier inhabitants to tsunami, and assess the probability of future occurrences. Excavated sites on the Sendai Plain in northeastern Japan, partially inundated in this recent tsunami, have been crucial in this endeavor, with recovered data at times contradicting historical sources. These investigations have contributed to understanding why many Middle Yayoi settlements on the Sendai Plain were abandoned, not to be reoccupied for 400 years.

The events of 2011 not only stimulated the formulation of *tsunami kokogaku* (Tsunami Archaeology) but also a new field of Tsunami Geology (Goto *et al.* 2014). After the March 2011 earthquake and tsunami (referred to as 3/11 in Japan, spoken as *san-ichi-ichi*, lit. 'three-one-one'), the Tohoku-oki Tsunami Joint Survey Group was formed, involving more than 300 Japanese and international geologists and engineers (Sato 2015). Although their purposes differ, geologists and archaeologists often work on the same sites or in collaboration to identify palaeotsunami (e.g., Ogura 2016). However, mere identification is not enough: geologists strive to derive the "physical parameters such as wave height, flow velocity, inundation extent, sediment sources, modes of transport and patterns of sediment dispersion, among others" (Costa & Andrade 2020: 1190). Especially important is recognizing backwash and how that changes the signature of tsunami deposits; the destructive nature of backwash must also be taken into consideration. For example, the Rikuzen-Takata City Museum was built with a solid wall facing the sea, but the back-flowing seawater burst through the building from the opposite side, 'sweeping away' artefacts and killing all six museum staff (Kaner *et al.* 2011). The return flow of the tsunami can be more powerful than the leading tsunami wave (Koshimura & Shuto 2016).

In early studies, fining-up units were assigned to different tsunami waves - this was too simplistic; moreover, multiple tsunami waves may leave no stratigraphic evidence, while other tsunami leave several deposits depending on different stages and flow characteristics of the tsunami (Switzer *et al.* 2012). Deposit characteristics depend on "tsunami height and period, local bathymetry & topography, sediment source, vegetation, and other roughness elements" (Switzer *et al.* 2012: 71). Fujiwara (2015) has written a

[1] An earlier version of this chapter was published as "The search for tsunami evidence in the geological and archaeological records, with a focus on Japan." *Asian Perspectives* 56.2: 132-165 (2017).

comprehensive book on various aspects of tsunami worldwide, including erosional patterns as well as sedimentation; palaeobiological materials in tsunami sediments; various models of deposition; and other common concerns.

While geologists aim at understanding the distribution and nature of tsunami deposits, archaeologists are more concerned, once tsunami activity has been recognized, to investigate past human responses to those tsunami. The event then becomes a potential explanatory factor for understanding subsequent human activities (or lack thereof) in the area. As such, tsunami deposits are similar to tephra deposits (cf. Chapter 6) in affecting the daily and calendrical routines across defined regions.

The New Field of Tsunami Archaeology

In contrast with inland earthquake zones, tsunami usually occur at long intervals and their effects are usually limited to narrow coastal plains. Before the advent of vulnerable modern infrastructure, the numbers of people and villages in coastal areas were often matched or exceeded by inland populations, so that entire cultures and civilizations were not at risk of total destruction – though any loss of human life is a tragedy. Following Shimoyama (2002), Tsunami Archaeology addresses how to define and assess disasters resulting from tsunami in the stratigraphic record. This is not an easy task, because the scientific parameters of tsunami are not yet well-understood in the archaeological community. For this reason, the definition, causes, measurements, and actions of tsunami are briefly set out below.

Figure 10.1 Major geographical locations mentioned

The cases discussed in the previous section demonstrate that human behaviour is visible in tsunami events. These case studies form a jumping-off point for further analyses – which must *not* start with the question "What happened?" but with "What happened *next*?" Was the tsunami a simple occurrence with limited effect or was it an event that changed the course of history (Sewell 2005)? Did people simply recover and go back to normal? Or were their social, political, or subsistence systems changed by this disruption of normal life? Goff & Nunn (2016) propose that tsunami is one type of environmental agent forcing cultural change. Several questions for investigating such change in Japan are offered in the conclusions here.

Japan is a good place to start a discussion on Tsunami Archaeology. The entire Pacific seaboard of Japan is vulnerable to Pacific Ocean tsunami (Minoura & Nakaya 1999; Kitamura *et al.* 2020). The effects of these tsunami sometimes intrude into the Inland Sea, though the waves developed there have slowed down and do not strike orthogonally. Even the Japan Sea has a record of tsunami (Minoura & Nakaya 1991; Hamasaki *et al.* 2012; Urabe 2017). For example, the 1983 Nihonkai-Chubu Tsunami scattered the concrete tetrapods that lined the shore for erosion control onto the land (Imamura n.d.). The coastal plains (totaling no more than 14% of the archipelago's land area, cf. Table 5.1) have provided the majority of flat land for housing the population in Japan since the early centuries BC. These plains are at the mercy of tsunami activity. The effect of tsunami on modern society was well illustrated in 2011, when the devastation to Japan's Sanriku coast in Tohoku had global repercussions (Kaner *et al.* 2011) (Figure 10.1). In pre-industrial society, damage was not as widespread. Nevertheless, archaeological excavations have revealed that economies suffered and settlements were disrupted, with presumed loss of life.

As was illustrated by the 2004 Indian Ocean Tsunami, Japan is not the only country affected by destructive tsunami. Coastal areas around Europe, the Pacific Rim, and the Indian Ocean are all at risk. Geologists have been studying tsunami deposits for some time (see references in Dominey-Howes *et al.* 2006 and Matsumoto *et al.* 2016; also see Fujiwara 2015; Shiki *et al.* 2008). But attention to tsunami deposits in Japanese archaeological sites only began in the late 1980s (Goto *et al.* 2014; Saino 2015b) and on the northwestern coast of North America in the 1990s (Goff & Bobrowsky *et al.* 2020). Such archaeological investigations have become more common after the turn of the millennium. Around the same time, scholars in New Zealand began dealing with tsunami deposits in archaeological sites (Goff *et al.* 2001, 2021; Goff & McFadgen 2001, 2003; McFadgen & Goff 2007). New Zealand archaeologists deal not only with individual site histories but also with general methodology. The New Zealand studies thus form an important body of comparative material for Japan, though the historical depths and complexities of the archaeological records in the two countries are quite different.

This chapter is a first attempt to map out the incipient field of Tsunami Archaeology from a Japanese viewpoint. The few geological and archaeological investigations that have been conducted are examined and some conclusions on human behavior are drawn from these case studies. These investigations are not comprehensive but are indicative of the kinds of research being carried out today.

Defining and Measuring Tsunami

Definition

Tsunami is a Japanese word combining two roots, *tsu* (harbour) and *nami* (wave). The word first appeared written in modern characters in a 1612 document titled *Sunpuki* (*Records of the Sunpu Castle*), which recorded a tsunami in 1611 (Cartwright & Nakamura 2008). Atwater and colleagues (2005: 47) relate that the word entered the English language via a story about a tsunami titled "The Living God", written in 1897 by Japan-resident Irish author Lafcadio Hearn. Goff *et al.* (2016) give an even deeper history of the term and its correlates. Though often given the plural suffix '-s' in English texts, Japanese words do not take a plural form; tsunami should be treated as a collective noun to mean one or many. English speakers sometimes refer to tsunami as 'tidal waves' (Kaner 2011); although they can interact with tides, tsunami comprise a different phenomenon.

Causes

Tsunami are generated by a variety of sometimes interrelated processes including volcanoes, earthquakes, and landslides (Fritz *et al.* 2009; Lockridge 1990; Pendick 1998; Goff & Bobrowsky *et al.* 2020; Goff & Terry 2016) (see Figure 10.2). Lockridge (1990) notes that even the eruption of underwater gases may cause tsunami in lakes and warns that landslides occur repeatedly in certain locations, while Goff & Dudley (2021) cite methane hydrate releases. Reicherter (2015: 1) notes that tsunami caused by earthquakes "have small wave heights offshore and very long wavelengths" while those caused by mass movements of earth or ice, on land or underwater, "have very high waves and very short wavelengths close to the origin." He states that 85% of tsunami are tectonically generated (i.e., subduction earthquakes, plate movements], and the remaining causes are 5% from volcanic activity, 5% from landslides, and 5%

other causes (2015: 1). Although the causes of tsunami that are *not* generated by earthquakes are difficult to decipher, data from the National Oceanic and Atmospheric Administration (NOAA) have enabled researchers to estimate percentages of tsunami that arose from each geologic process over the past 4000 years (Goto 2015*b*) (Table 10.1). These figures differ somewhat from those published by UNESCO *et al.* (2018), noted in parentheses in the table below. Japan accounts for 22% of confirmed tsunami sources, with the South Pacific Islands accounting for 15%; seven other areas account for 7% or less each (UNESCO *et al.* 2018).

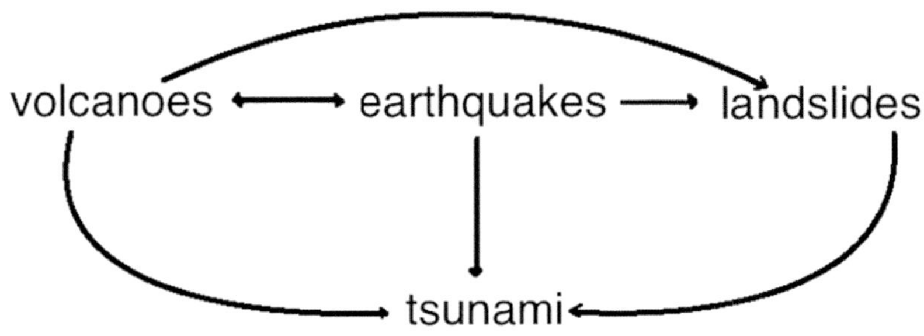

Figure 10.2 Relationships between different tectonic activities and their possible repercussions. Note that most arrows are unidirectional, but volcanic eruptions and earthquakes are inter-sensitive and can cause each other. Tsunami have three possible tectonic causes plus two others.

Table 10.1 Multiple causes of tsunami

%	Cause	Description
76.2 (81)	earthquakes	tectonic deformation (a sudden drop or rise) in the ocean floor and the water column above it
10.5 (4)	undetermined	–
7.6 (10)	landslides, (6%) seismic and (4%) nonseismic	start on land into slide into the ocean, or underwater lahars or debris flows from cone collapse during a volcanic eruption, or continental shelf sediments that are dislodged by an earthquake: 3.7% by earthquakes and landslides, 3.5% by landslides alone [including submarine slides and slumps], 0.4% by volcanoes and landslides [lava, pyroclastic flows, lahars, debris flows, caldera collapse]
5.0 (5)	volcanic eruptions	including submarine
0.7	volcanoes and earthquakes	ca. 22% of volcanic eruptions are accompanied by earthquakes
–	impacts	asteroid, bolide or comet
–	explosions	modern explosives; airwave disturbances from volcanic eruptions

Measurements

Multidimensional measurements of tsunami waves include depth, height, distance, speed, and number of waves. Further measurements are taken once the tsunami hits land, including inundation level and distance. Finally, the force and destructive powers of tsunami are measured in various ways, with nonseismogenic tsunami often holding records (Table 10.2). For archaeological work, the measurements on land are particularly important. Inundation level and distance should be clearly distinguished from run-up (maximum height from sea level obtained on land). The run-up is usually the most impressive part of the tsunami, but it is often transient and localized. For example, the 2011 Tsunami caused by the Great Tohoku-oki Earthquake reached a run-up height of 40 m at Ofunato

(Dunbar *et al.* 2013: 117). A similar run-up height of 38 m was attained by the 1896 Meiji Sanriku Tsunami (Satake 2007: 487).

All these measurements are somewhat determined by the topography. If a tsunami wave enters the mouth of a river leading into a steep-sided valley, it will be funneled up to higher levels with less horizontal extent. If a wave strikes a beach bordered by dunes covered with stands of trees (common in Japan), the speed of the wave will be reduced. The reason tsunami were originally named 'harbour waves' is that they attain greater height when they are funneled into a harbor setting and resonate off enclosing land surfaces. Tsunami heights can be enhanced by other factors such as coincidence with high tides, spring (as opposed to neap) tides, or storm surges. These same factors can also increase the extent of inundation. Mean tide ranges vary greatly by location. Areas with low mean tide ranges may be less subject to tsunami inundation than areas with high mean tide ranges. If a tsunami coincides with a low tide, the extent of inundation is lessened, but if the tide is exceptionally high, the results of a tsunami can be devastating (Roberts 2005).

Table 10.2 Ways of measuring tsunami and some of the worldwide record holders of large events

Measurement Criterion	Description
Wave measurement	**Depth** from sea floor to sea surface (changing as the wave passes over various sea floor structures)
	Distance between wave peaks (which shortens when approaching land)
	Height of wave from trough to peak (double amplitude, Satake 2007: 490); amplitude increases when approaching land
	Speed of travel
	Number of waves in wave train
Measurement on land	**Inundation level** is the height of the wave measured from tide level at that time; flow depth is depth of wave on land
	Inundation distance is how far the waves reach inland*
	Run-up of a tsunami is measured as the maximum height from sea level attained on land, which may exceed that of the inundation level*
Worldwide record holders	**Oldest** recognized tsunami occurred 3.5 billion years ago, covering all of the small continental plates that existed at that time except for the highest mountain tops (Glickson 2004)
	Biggest modern run-up in height (see below) reached by the tsunami wave on land was 524 m in Lituya Bay, Alaska in 1958 when an earthquake dislodged 30 million m^3 (90 million tons) of rock forming a landslide into the bay and a run-up on the opposite hill spur (Fritz *et al.* 2009); trees were stripped along the shoreline up to 520 m (Satake 2007: 487)
	Greatest loss of life in Indian Ocean tsunami of 2004, taking 350,000 lives (Athukorala & Resosudarmo 2005)
	Largest magnitude was recorded for the 1960 Chilean earthquake at M9.5; it affected the entire Pacific seaboard of Honshu and Shikoku
	Worldwide, about 25% of deaths in volcanic eruptions are due to tsunami (Satake 2007: 497)

* Both of these are conditioned by the topography: as the waves are squeezed into bays and river drainages, they get taller and therefore travel longer distances than they would have if they had exhausted themselves on open flat land.

Actions

A basic misunderstanding of the nature of tsunami waves has been perpetrated by the frequent reproduction of Hokusai's great wave woodblock print to signify a tsunami, when in fact it depicts a windstorm wave with a breaking crest (Cox 2001). Cox (2001) and Doughton (2005) speculate that this imagery may have contributed to many recent tsunami deaths in Indonesia and Thailand because people did not recognize the tsunami wave form. Instead of waves continually crashing onto a beach, tsunami are more likely to form a wall of water in a shallow coastal sea, then the wall moves straight inland to flood the land.[2] To understand how this happens, a review of wave generation is in order.

Because most tsunami are caused by deformation of the ocean floor and the water column above it, tsunami waves can be thousands of feet deep in the ocean, down to the ocean floor (Pendick 1998). They can travel up to 710 km per hour in a 4000 m deep ocean (Satake 2007: 502). The waves caused by such deformation spread in concentric circles, much like the ripples caused by throwing a stone into a pond, though much deeper. As with major earthquakes, which are preceded by tremors and may have aftershocks, major tsunami rarely consist of only one wave but rather a series of waves in a wave train. In the open ocean, tsunami wave trains may have crests several miles apart. Such wave trains hit land one after another. As they approach shallow water near landfall, the wave crests rise to form a wall of water; they may break at the shoreline but do not rear up in the open sea like storm waves. Each tsunami wave in the train hits areas of the coast that have different characteristics. The nearshore seafloor morphology and the sedimentary environment and dynamics of the seabed are important (Switzer *et al.* 2012). Moreover, whether a wave impacts the coast as a trough or a peak depends on whether the seabed moved up or down in the earthquake (McFadgen 2007: 25).

It is therefore difficult to generalize measurements for any single tsunami event because the results are different for each wave in each area. Rather unusually, the rise of the ocean floor during the Great Tohoku-oki Earthquake of 2011 did not result in successive wave formation on the landward side. Instead, DART (Deep-ocean Assessment and Reporting of Tsunamis) data from monitoring the ocean bottom using pressure recorders and satellite altimetry show that the waves spread seaward through the Pacific (Fine *et al.* 2013).

Tsunami Sedimentary Evidence

Inundation processes

It appears that only tsunami waves greater than 5 m high generate a sedimentary signature – in some cases, tsunami do not leave any deposits at all; the waves may scour the surface rather than leave anything behind (McFadgen 2007: 35). In either case, an erosive, sharp, or abrupt base to tsunami deposits is common, and this can be helpful in identifying tsunami activity (Costa & Andrade 2020). Where tsunami do leave deposits, they are "highly variable" (Satake 2007: 492), due to reworking and redeposition of sediments from different ocean floor, beach, and sand dune areas with local characteristics. The general geological consensus is that tsunami waves characteristically leave behind sheets of sand, gravel, or pebbles on inundated land (Costa *et al.* 2015; McFadgen & Goff 2007; Satake 2007: 489); Reicherter (2015) states that 'hard (rocky) coasts' are characterized by boulder deposits. For 'soft coasts', Reicherter gives eight characteristics of sand sheets: sheetlike geometry, landward thinning, landward fining, fining upwards or downwards, multiple sheets, marine materials, rip-up clasts and partially 'armoured' mud balls, and marine geochemical signatures (2015: 5). The multiple sheets may have sediments graded in different directions (up, down), while massive deposits may have no grading at all; each and every one

[2] For a graphic comparison, see the *Tsunami!* website (DESS n.d.).

may have erosional bases (Switzer *et al.* 2012: 62). Marine shelf sediments may also be laid down in freshwater lakes near the coast (Furumura *et al.* 2011; Sawai *et al.* 2008) or lagoons (King *et al.* 2017). Unfortunately, post-depositional reworking – including erosion and other physical processes such as liquefaction or frost heaving, chemical processes, bioturbation, and anthropogenic activities – can reduce the deposits and destroy structures therein. By eroding coastal dunes and depositing sediments in lagoons, tsunami may alter shoreline topography extensively (Switzer *et al.* 2012).

Boulders (particularly of coral), trees, and other debris (including human remains) can be left by either tsunami (Davies 2002; Goff & Terry 2016: fig. 5) or storms. Boulders can be imbricated by strong tsunami waves – evidence for a Bristol Channel tsunami in 1607 (Roberts 2005).[3] Some researchers have utilized physical attributes such as wave height, boulder weight, and elevation of deposit to distinguish boulders deposited by tsunami from those deposited by storms (Goff *et al.* 2012; see also Nandasena *et al.* 2011; Kitamura *et al.* 2017). An equation yielding a 'transport figure' (Tf) based on Weight (tons) x Distance moved from a cliff (metres) x Vertical cliff height (metres) can be used to distinguish tsunami boulders from storm boulders, as the latter have an upper limit of Tf=2,000 (Reicherter 2015). Goto *et al.* (2010) determined that storm-deposited boulders on Ishigaki Island weighed less than 47 tons, while *tsunami-ishi* (tsunami 'rocks') weighed up to 216 tons; these were also distributed in different places. Since tsunami have long-period wavelengths lasting much longer (up to an hour) than individual storm waves which have short-period wavelengths (M. Brewer in Anon. 2020), Watanabe and colleagues (2016) devised a numerical scheme using initial wave height and transport distance as determined by wave period to investigate boulders left by the 1771 Yaeyama Tsunami on Ishigaki Island (see Case Studies below). Costa & Andrade (2020: 1195) note studies that suggest "tsunamis produce unorganized boulder deposits whereas storms are capable of organizing boulders in lines and clusters;" however, the distributions of hundreds of tsunami boulders on Ishigaki Island do not show these organizational differences (cf. Goto *et al.* 2010).

If the earthquake was proximal, there is a tendency to assess tsunami-affected land as having been downwarped by the earthquake (Atwater *et al.* 2005); subsidence due to volcanic eruption is also known (Lockridge 1990). Downwarping or subsidence increases tsunami hazard, as these processes allow waves to inundate more land. There are, however, cases of co-seismic uplift rather than downwarping (Kitamura & Kobayashi 2014; McFadgen & Goff 2007). Moreover, subsidence and uplift can occur simultaneously on different parts of the fault causing the earthquake (Satake 2007: 486). These uplifts and downwarps relax between earthquakes, making it difficult to assess sea levels before the tsunami occurred.

Sometimes immense amounts of sand are deposited along the shoreline to form a new beach. Where does the tsunami sand come from? It is common knowledge among coastal peoples that the sand on a beach may decrease or disappear entirely during winter storms. Heavy waves can pick up dune and beach sand and carry it out to sea. Tsunami wave trains then pick up sand deposits – at fair-weather wave depths, the intertidal zone, beach, and even dunes – put the sand into suspension and then lay some of the sand down during inundation when tsunami waves cover previously dry land surfaces. There is even evidence that tsunami can raise sediments from the continental shelf and deposit them on land (Switzer *et al.* 2012: 71). The 2011 Tohoku-oki Tsunami waves picked up sand from on-land beach ridges and sand ejected during earthquake liquefaction events (cf. Figure 9.8) in the paddy-fields (Goto *et al.* 2011). Sand can also be deposited secondarily from dunes stripped of cover by tsunami waves and then eroded by the wind (McFadgen 2007: 37, 234). The draining of tsunami water can also deposit 'runout' sediments and create

[3] For an intimate visualization, see the BBC documentary about this possible tsunami in the Bristol Channel in 1608 (Roberts 2005).

run off channels; see the whole sequence from pre-tsunami to final tsunami stages in Switzer *et al.* (2012: fig. 9).

Because the waves of tsunami usually form 'wave trains', hitting the coast one after another, each has the potential to erode, reorganize, and finally cover the earlier deposits. Thus, archaeological investigation must be alerted to the potential of recurrent sand sheets covering sites, not just one sand sheet. Moreover, backwash sediments must be distinguished from landward deposition.

Recent work has demonstrated that the extent of sand deposits does not correlate with the extent of inundation which, like run-up extent, is demarcated by wrack line deposits (Reicherter 2015). The 2011 Tsunami laid down terminal sand deposits under 2 m of water, representing only 60% of the extent of inundation, and a 60% rule of thumb has been devised. This specifies that the proportion of the inundation distance covered by the sand sheet is about 60%. Beyond that, mud was deposited between 60 to 95% of the inundation distance, and beyond that, marine water left chemical constituents that could be retrieved for analysis only for several months after the tsunami waters had abated (Chagué-Goff *et al.* 2012*b*). In wetland peat deposits, tsunami signatures include smaller grain-sizes of sand, increased organic matter, and higher concentrations of iron or sulfur (McFadgen 2007: 35). Thus, archaeologists must pay attention to more than sand layers in sites if they want to understand the devastating effects of marine water incursions.

Before attempting to identify tsunami deposits in archaeological sites, it behooves archaeologists to also evaluate the likelihood of preservation. Richerter (2015) notes that preservation is promoted by areas that undergo subsidence and sedimentation rather than uplift and erosion; he names lagoons and marshy flatlands that are particularly good for preservation – which is good news for investigating ancient paddy-fields in coastal areas.

Identifying previous tsunami deposits

Since 2000, there has been considerable focus in the geological literature on problems in identifying tsunami deposits in the sedimentary record (Fujiwara 2015; Scheffers & Kelletat 2003; Shuto & Fujima 2009; Tappin 2007). Tsunami sediments are fairly easily distinguishable from other types of sediments deposited on a plain that derive from seasonal riverine flooding (alluvium), flash floods, debris flows, volcanic tephra fallout, and – in Japan – seasonal loess deposits carried from the Asian continent (Saino 2012*a*: 225; Yamada *et al.* 2016). However, distinguishing tsunami deposits from aeolian dune erosion (McFadgen 2007: 37) or storm deposits is a great challenge (cf. Marriner *et al.* 2017 vs Vött *et al.* 2018). There are, nevertheless, biological remains such as pollen, diatom, foram, shell, even animal skeletons (Goff & Chagué-Goff 2009), as well as stratigraphic and chemical signatures that separate them (Chagué-Goff *et al.* 2012*b*; Morton *et al.* 2007*a*). Goff & McFadgen (2003) compiled a list of characteristics that can be used to differentiate tsunami deposits from other sorts of deposits (Table 10.3) – but the potential for tsunami signatures to vary by event makes generalization difficult (Chagué-Goff *et al.* 2011, 2012*b*; Costa *et al.* 2015; Dominey-Howes *et al.* 2006) – and the list is being constantly updated (e.g., Chagué-Goff *et al.* 2011; 2012*b*; Goff & Bobrowsky *et al.* 2020 – ranging from 11 to 30 listed proxies). Meanwhile, Cain *et al.* (2019) have added that the discovery of mass cemeteries near coastal areas, including Kamakura, without evidence of warfare or violence may indicate tsunami deaths (see also Minami *et al.* 2012).

One important distributional aspect is that sand is deposited differently by tsunami and storm waves (Morton *et al.* 2007*b*) (Figure 10.3). Tsunami sands are widespread and drape the existing topography, while storm wave sand tends to settle in hollows in the limited area of breaking waves. However, tsunami 'backwash' also tends to follow channels (Costa & Andrade 2020). Tsunami deposits, high wave deposits, and alluvial deposits all have different areas of geographical distribution that can be used as clues in distinguishing them from storm deposits. Because tsunami deposits are difficult to distinguish from storm deposits at *any one site*, areal comparisons must be made. Davies (2002: 40) notes that tsunami sand deposited on 'residual soils' is easier to distinguish than tsunami sand deposited on 'older sand' layers, so examining deposits in several contexts is useful.

Table 10.3 Some diagnostic characteristics for distinguishing tsunami deposits from others (see Goff & Bobrowsky *et al.* 2020: table 1 for a more extensive list)

1) particle/grain-sizes range from boulders to coarse sand to fine mud
2) sediments generally become finer inland and finer upward within the deposit
3) each wave in a wave train can potentially form a distinct sedimentary unit
4) distinct upper and lower subunits (run-up and backwash) can often be identified
5) lower contact is unconformable or erosional
6) can contain intraclasts of reworked material
7) often associated with loading structures at base of deposit
8) marked changes in foraminifera (and other marine microfossils) assemblages
9) pollen concentrations are often diluted
10) increases in Na, S, Cl, Ca, Mg due to seawater inundation and/or high shell content
11) shells and shell-rich units are often present
12) often associated with buried vascular plant material and/or buried soil
13) shell, wood, and less dense debris often found "rafted" near top of sequence
14) often associated with reworked archaeological remains
15) known local or distant tsunamigenic sources can be postulated or identified

Minoura and colleagues (2001: 84) note that storms on the Tohoku coast are generally erosive and "do not produce regionally extensive deposits of marine sand on the flood plain". Thus, a regional (or territorial) approach – similar to that being developed in Earthquake Archaeology (Chapter 9) – will help in distinguishing storm from tsunami deposits (McFadgen & Goff 2007; Obermeier 1996). This approach hinges on the stipulation that evidence from one site is not enough to propose sedimentary changes, so

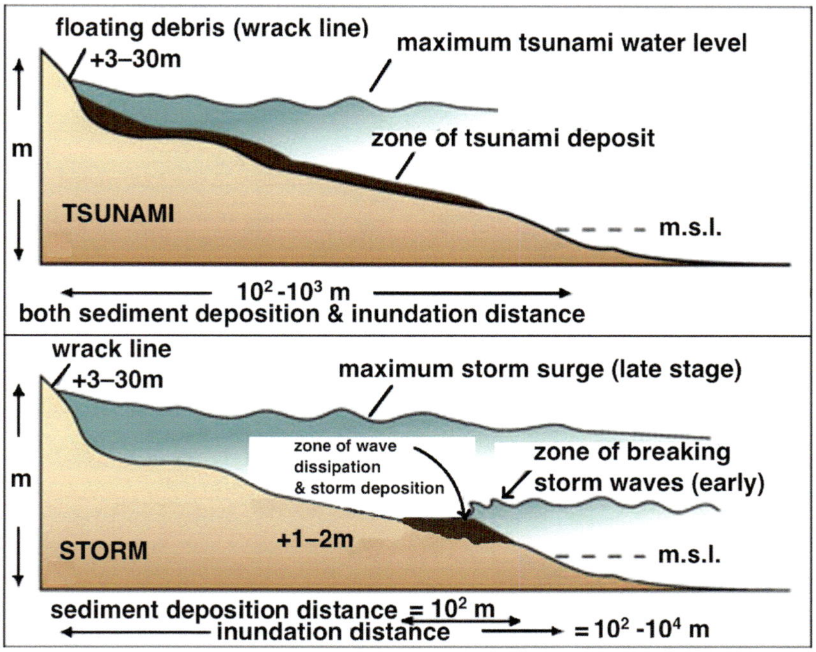

Figure 10.3 Comparison of tsunami and storm wave sand deposits

corroborating effects must be documented. Excavations on the Sendai Plain in Miyagi Prefecture (Figure 10.1) have confirmed the utility of broad-scale excavation and regional investigation in identifying tsunami deposits (Saino 2015*b*: 245; 2021).

Archaeologists in Japan have developed limited criteria for investigating tsunami deposits. The following information is stated as necessary to determine whether a certain deposit was made by a tsunami: grain-size characteristics, age of deposit, distribution and nature of deposit, and the palaeotopography including the position of the ancient shoreline (Matsumoto 2015: 4). Saino (2015*b*: 243) outlines the following five steps for investigating tsunami [brackets added]:

- Geologically distinguish tsunami deposits from alluvial and high wave [storm/typhoon] deposits
- Date tsunami deposits and cultural damage with both archaeological and geological dating techniques
- Reconstruct contemporaneous topography and coastline, in order to measure the distance of archaeological sites from the sea
- Estimate the spread of tsunami deposits and inundation distance
- Determine the cause of the tsunami, particularly if there was a proximate earthquake at the same time

Historical documents that give the date and even the time of tsunami waves provide precise dating (Table 10.4). As Saino (2012*a*, 2013, 2015*a*: 7) cautions, however, it is important to confirm that these records are accurate with reference to archaeological finds. The discovery of tsunami deposits in archaeological sites is especially advantageous since cultural materials found in layers above and below tsunami sands can provide relative bracket dates (e.g., Goff & Lamarche *et al.* 2011), as can tephra dates. In off-site cases where such information is absent, there are several other methods of determining the date. One is using dendrochronology that can specify a certain calendar year depending on whether a tree died with tsunami inundation or lived briefly thereafter. Absolute dating methods include palaeomagnetism, AMS ^{14}C, ^{210}Pb, ^{137}Cs, ^{36}Cl, OSL, and tephrochronology (Reicherter 2015; Rixhon *et al.* 2018). For radiocarbon dates taken from organic matter in sediments above, below, or even within tsunami deposits, close attention is needed to identify possible inbuilt (or pre-sample) age in case the organic matter is considerably older than the tsunami event (Gavin 2001; McFadgen 1982).

Inbuilt ages can significantly distort dates, especially during the last millennium. In Japan, the sedimentary column often contains tephra from near or distant volcanoes; identifying and dating the tephra can provide *terminus post quem* or *terminus ante quem* dates for tsunami deposits. However, tephra may also be redeposited in a tsunami, forming a continuous layer that is much later than its eruption date (Yamada *et al.* 2019). A newly suggested dating tool is the presence of microplastics. Dilek *et al.* (2021) note that microplastics are contaminated with the radioactive caesium isotope ^{137}Cs from nuclear testing; both plastic production and ^{137}Cs date from the 1960s so that recent tsunami sediments containing microplastics can be differentiated from pre-1960 tsunami deposits (see also Kawamura et al. 2021). Ishizawa *et al.* (2020) discuss the various pros and cons of dating methods and their applicability to fine-sediment deposits, while Rixhon *et al.* (2018) have reviewed dating methods for boulders.

Cyclical Tsunami and Recovery

Despite the terrible destruction by the 2011 Tohoku-oki Tsunami to the Sanriku coast northeast of Sendai City in Miyagi Prefecture, the Sendai Plain has become a showcase for studying cyclical tsunami activity

(Goto *et al.* 2015; Saino 2015*b*). The identification of stratigraphic layering of tsunami deposits on the Sendai Plain has revealed recurring mega-tsunami and accompanying earthquakes. Recovered deposits dated to 100 BC, AD 869, and 2011 suggest that large tsunami strike the Sendai coast roughly every thousand years (Table 10.4). The first documented tsunami struck during the Middle Yayoi period around 100 BC, causing the abandonment of several village sites, which were not recolonized for another 400 years. The second one struck in the traditional reign date of Jogan 11 (869 AD) during the Heian period (AD 794–1185). At that time, most damaged villages continued to be occupied, although some paddy-fields were abandoned. A growing number of scientists consider the 869 Jogan Tsunami a comparable forerunner to the March 2011 event. The 2011 Tohoku-oki Tsunami not only partially covered the Sendai airport, but also forced the survivors of at least six coastal villages on the Sendai Plain to be rehoused in a new town farther from the coast.

Table 10.4 Dates of tsunami discussed in the text

* reign dates of specific Emperors

Date	**Designation**	**Minimum area affected**
2011	Tohoku Tsunami	NE Japan coast
2004	Sumatra-Andaman Tsunami	Indian Ocean Basin
1983	Nihonkai-Chubu Tsunami	NW Japan coast
1960	Chile Tsunami	NE Japan coast
1958	Lituya Bay Tsunami	Lituya Bay, Alaska
1896	Meiji Sanriku Tsunami	NE Japan coast
1833	Krakatau Tsunami	Sumatra/Java, Indonesia
1771	Yaeyama Tsunami	S Ryukyu Islands
1707	Hoei 4* Tsunami	SW Japan
1700	Orphan Tsunami	NE Japan coast & NW America
1611	Keicho 16* Tsunami	NE Japan coast
1361	Nankai Tsunami	Kyushu Island
869	Jogan 11* Tsunami	NE Japan coast
684	Nankai Tsunami	Kyushu
100 BC	Middle Yayoi Tsunami	NE Japan coast
6000–3000 BC	four successive tsunami	Kanto region, Japan
ca. 7300 BC	Akahoya Tsunami	Kyushu

The modern and Middle Yayoi earthquakes and tsunami are assessed at similar magnitudes and extents, but the 869 Jogan Earthquake was estimated at lesser strength (M8.3) (Minoura *et al.* 2001). It is notable that abandonment of affected areas has been a favoured option for the stronger events, but Jogan-period people seem to have been able to weather the damage and were resilient in continuing their settlement of the plains (Saino 2015*a*: 6–7). The difference between the Yayoi and Jogan responses may have been due to differences in organizational complexity. The Yayoi peoples in this area were subsistence agriculturalists living in presumably self-sufficient villages without a more regionally integrated political context. This means that they had to depend on neighbours and relatives in dire times.

By the Heian period, the expanding state had established a provincial center, Tagajo Kokufu, and a branch of the national Buddhist temple network, Mutsu Kokubunji, at the northern edge of the Sendai Plain. Thus, not only was the 869 Jogan Tsunami and its 1000 death toll recorded by the local scribes, but help might have been available from government facilities to assist farmers suffering from the destruction of agricultural land. In rice-growing areas, the addition of seawater salts to paddy soils, not to mention any sand cover, is detrimental to crops. Salt seeps farther into sandy soils than clay soils but leaches out more quickly. It is estimated that soils are unusable for at least a year after a tsunami, depending on the local precipitation rates (McFadgen 2007: 233). Since paddy fields in Japan are underlain by a hardpan that prevents water from draining through the soil (to keep the crop inundated during the growing season), neither saltwater nor freshwater will leach out of paddy soils quickly.

Changes in settlement location after a tsunami often result in a change in subsistence, since coastal resources were damaged or are no longer available. After the 2011 Tohoku-oki Tsunami in Japan, there was much discussion about moving coastal towns in the Sanriku area to upland locations; and the presence of historic markers of previous tsunami run-up or inundation levels, which had been disregarded in the interest of modern urbanization, became a sensational focus of the media (e.g., Fackler 2011). The need to establish new communities and facilities on higher ground stimulated an unusual level of archaeological excavation prior to construction (cf. Chapter 11). Despite rehousing efforts, however, many people preferred to return and rebuild their old settlements on the coast rather than move elsewhere. Building complete 'new towns' was another alternative. For example, the survivors of six destroyed villages were resettled in the new town of Tama-ura Nishi, which was built only 3 km from the coast even though the inundation distance for the 2011 Tohoku-oki Tsunami had reached 5 km inland in places (Ishibashi 2015). Sako Architects suggested raising communities atop 40-metre-high rubble cushions (Gilhooly 2012). At Rikuzen-Takata, an area that bore heavy destruction from the 2011 Tsunami, authorities have been stripping nearby mountains to provide landfill to raise parts of the town up on foundations high enough to avoid future inundations, although this practice is endangering archaeological resources (Guttenfelder & Hoshiko 2016). Such resources were not available to prehistoric peoples, so the archaeological record instead speaks of site abandonment as the primary response to destructive tsunami.

Earlier tsunami sediments up to 3800 years old have been found in the stratigraphy of the Sendai Plain (Goto *et al.* 2011). Minoura and colleagues (2001) dated a lower sand layer to ca. 910–670 BC, which falls within the Tohoku Final Jomon phase (Apx 2: Table A). Jomon settlements in east and northeast Japan were commonly located on high ground, giving pause to one archaeologist, MATSUI Akira, who "often wondered why almost all the shell midden sites in the Kanto and Tohoku regions were located only on top of small hills, with very few sites being found on the low-lying land near the coast. In the course of our rescue activities, I have been made painfully aware of the reason for this" (quoted in Kaner *et al.* 2011: unpg.). Jomon residents were likely aware that tsunami periodically washed over the coast.

Owing to high sea levels between 4000 and 2000 BC during the Holocene Climatic Optimum, and plains formation thereafter, it is unlikely that tsunami remains will be found for any time earlier than 2000 BC, but tsunami evidence in Oita Prefecture, Kyushu, has been assigned to the Kikai Caldera eruption in 7300 BC (Fujiwara *et al.* 2010; Fujiwara 2015: ch. 6.1.3.1). The effects of this 'Akahoya tsunami' on available natural resources could have been disastrous for later Jomon and other historical populations such as the Ainu who lived close to nature (Fukasawa 2015; see also King *et al.* 2018, 2020). As Goff & McFadgen recount for New Zealand (2001: 159), terrestrial, lacustrine, and marine food sources would have been lost and vegetation destroyed. Lakes and lagoons would have dried up, sediment would have infilled streams and rivers, and shellfish beds would have been smothered. Settlements and gardens at the bases of cliffs would have been vulnerable to landslides. Travel either on foot or by canoe would have been seriously impeded in the short term because of landslips, fissuring, and the destruction of vegetation. Nevertheless, as discovered after the 2004 Indian Ocean Tsunami, beaches can recover naturally by landward migration of sandbars from near-shore zones; moreover, beaches located near tidal inlets recovered faster if not uniformly than beaches distant to tidal inlets, while tidal inlets themselves took several decades to recover (Wong 2012).

The plains of Japan, including Sendai, are subject to other natural disasters such as typhoons, earthquakes, floods, landslides, lahars, and volcanic eruptions. The effects of tsunami on the inhabitants both past and present must be separated from these other causes in order to assess the risk people face in future from sea incursions and salt-water inundation and the widespread destruction they cause.

Tsunami Excavations in Japan: Case Studies

Until recent excavation by geologists or archaeologists, information about past tsunami has been limited to what could be observed by contemporaries. Although entire coastlines may have been subjected to tsunami wave trains, they were only documented in those areas where damage was recorded by literate people. As Goto (2015*b*) states and Atwater and colleagues (2005) discovered, the Japanese records are extremely important for understanding Pacific Ocean tsunami, both emanating from Japan and arriving in Japan. Japan has a written record of tsunami that extends back 1300 years, with the first historical mention of tsunami in AD 869 (Table 10.4). This is three times older than Mesoamerican and South American records and far longer ago than records from Kuril/Kamchatka, North America and Hawai'i, or New Zealand/Polynesia (Goto 2015*b*: 10). Tsunami deposits in Japan have been documented back to the prehistoric period to 7300 cal. BC (Fujiwara *et al.* 2010; Maeno *et al.* 2006; Garrett *et al.* 2016; Fujino *et al.* 2018; Shimada *et al.* 2019), but few of these are associated with archaeological remains as yet.

The 2011 Tsunami stimulated geologists to investigate palaeotsunami occurrence in greater depth using coring, geo-slicing, and test-trenching. Although the 2011 Tsunami brought wide attention to past tsunami, excavations of tsunami deposits had been conducted before this. It has become a "matter of profound regret" that archaeologists did not publicize their findings widely enough to influence public policy for tsunami preparedness (Okamura *et al.* 2013: 267). The recent 2011 Tsunami has been compared above to the 869 Jogan Tsunami and to the Middle Yayoi Tsunami in 100 BC. Case studies of these and other tsunami are presented below, with the most recent first and progressing backwards in time.

2011 Tohoku Tsunami

Sendai Plain, Miyagi Prefecture (Chagué-Goff *et al.* 2012*a*)

An international team surveyed a transect across the Sendai Plain from coast to Sendai Airport to "document physical evidence of tsunami flow characteristics – such as debris in trees, high-water stains on buildings, and sedimentary deposits" (Jaffe *et al.* 2011: 2). The transect site was chosen based on research on the 869 Jogan Tsunami that had also affected the Sendai Plain (Chagué-Goff *et al.* 2012*b*; Jaffe *et al.* 2011). Jaffe *et al.* (2011: 4) caution, however, that "the relative magnitudes of the Jogan and 2011 tsunamis [sic] cannot be inferred from the thickness of each deposit, which is affected by local topography as well as sediment grain-size and tsunami flow speed." The deposits were assessed according to local topography, microtopography, flow directions, and water levels. Coring equipment (the geo-slicer) was used to assess sediment thickness, grain-size, and sedimentary structures.

Along this transect, 2011 Tsunami sand deposits extended 2.9 km inland, but mud deposits were found 4.65 km inland. However, markers of marine water such as sulfur (S) and chlorine (Cl) were found 4.85 km inland, close to the maximum known inundation of more than 5 km; S and Cl existed farther inland than any tsunamigenic sedimentation but decreased over time via leaching by rainwater. Other elements such as strontium-Sr and rubidium-Rb seem to be more persistent, since they were found in the lower deposits of the 869 Jogan Tsunami (see below).

Minami Soma-shi, Fukushima Prefecture

After the 2011 Tsunami damaged the Fukushima nuclear reactor ponds, the area around the reactor was evacuated. Since then, only basic activities such as removing rubble, repairing roads and dikes, and cutting grass in paddy-fields have been carried out, so the tsunami deposits have been left undisturbed. Kaner (2015) gives a first-hand review of the effects on archaeology in the Fukushima exclusion zone.

Ogura (2016) and his colleagues conducted a trenching project over an area 1.6 km E–W x 1.4 km N–S of paddy-fields adjoining the Odaka River east of the Odaka JR (Japan Railways) station. This land had reportedly subsided about 1 m after the earthquake and collected water after a heavy rain. Thirty test trenches, each 50–80 cm per side and 20–30 cm deep, were cut at 50 m intervals. Of these 30 trenches, 22 were cut on an E–W transect 1135 m long, that is, along the axis of the incoming waves, and the other 8 were cut on three shorter orthogonal transects. The trenches were dug through the tsunami deposits to reach the level of previously cultivated paddy-fields, though some trenches went slightly deeper.

Four basic strata were identified in this area, where the tsunami waves had been 10 m high:

Layer 0 = recent deposits occurring after the tsunami
Layer 1 = tsunami deposits, divided into three layers:
1*a* <11 cm thick, 1*b*-upper <19 cm thick, 1*b*-lower <12 cm thick
Layers 2 and 3 = previous agricultural and natural deposits

Layer 1 consisted of fine gray sand and fine gravel with clumps of mud: in the order they were laid down (bottom to top), 1*b*-lower was lumpy, while 1*b*-upper was laid down in fine laminae in regular fashion. Layer 1*a* consisted of gray silt and very fine sand laid in a regular structure; there were tiny clumps of mud in only one trench in this layer. In Trench 1 nearest the sea, the lower stratigraphy was scoured out and large mud clumps could be seen. Ogura (2016) suggests that the mud clumps may have come from the scoured area and from the rampart of the dike that borders the sea.

The thickness of Layers 1*a* and 1*b* varied widely among the trenches, from virtually 0 cm to 30 cm thick. The thickest layers were laid east of two road ramparts, when the waves washed over the fields until they hit the road ramparts. In a third area, the thickest deposits were on the western side of a road rampart. Apparently, the waves bypassed the ramparts using a canal that then acted as a funnel to deposit more sand inland than might otherwise have occurred.

The paddy-fields were crossed by three roads and a river, all of which were accompanied by artificial earthworks such as bunds and dikes. These earthworks seem to have been gutted and mixed with ocean sand for deposit farther inland. Since there were no houses or other structures present in the area, no cultural materials were found in the tsunami deposits. But given the situation with the earthworks, it could be imagined that *had there been* built structures seaside, it is likely their remains would have been mixed into the tsunami deposits, as were seashells from the beach.

Ogura (pers. comm., 7 December 2015) explained that the laminar sands in Layer 1*a* consisted of dark grains at the bottom of the layer with lighter-colored grains toward the top. The grains all tended to be the same size, but the dark grains consisted of volcanic crystals such as pyroxenes and hornblendes. He hypothesized that these crystals were volcanic erosion products that had previously been deposited on the seabed, then were redeposited by the tsunami waves on land.

Takaose Site, Iwanuma City, Miyagi Prefecture

This site stretches from the third set of beach ridges into a former backmarsh and is known to have been affected by tsunami over the last five centuries (Goto 2015a: 3; Kawamata 2015; Saino 2015b: 249-251, table 1). Excavations in 2013 revealed that modern paddy-fields (Layer 1) had been covered by 2011 Tsunami sands. Two lower sand layers are thought to be remains of earlier tsunami: Layer 4 (5 cm thick) from a tsunami caused by an earthquake in 1611 AD (Keicho 16 in traditional dating) and Layer 8 (20 cm thick) from the 869 Jogan Tsunami. Towada-a tephra dated to 915 AD were situated between these layers.

Hasunuma Site, Kujukuri-cho, Chiba Prefecture

The 2011 Tsunami struck orthogonally on the Chiba coast at Kujukuri, approximately 60 km east of Tokyo and much farther south than is generally recognized. A survey team headed by the Geological Survey of Japan (AIST) visited the site every month for five months after the tsunami, establishing three transects perpendicular to the coastline, taking 29 samples for grain-size analysis, and running geochemical analyses (Matsumoto *et al.* 2016; Shinozaki *et al.* 2016). Tsunami deposits were a maximum of 35 cm thick. Several sedimentary structures and sedimentary units were identified therein. Both landward thinning of deposits and fining of grain-size have previously been cited as characteristics of tsunami deposits (cf. Table 10.3). The research at the Hasunuma site confirmed that tsunami deposits tend to thin out landwards but did not confirm that grain-size becomes finer.

1771 AD Yaeyama Tsunami in the Ryukyu Islands

Miyako Island, Okinawa Prefecture

Eight villages were established at the outlet of a small river on Miyako Island, six lined up along the coast and two behind on the river delta. All eight villages were affected by a tsunami in 1771. They were all abandoned, and three new villages were established on higher ground. At two of these new villages (Sunakawa and Tomori), rituals (*nāpai*) were created to ward off future tsunami. A stele was also erected at Maeyama on Miyako Island to commemorate the lives lost in the tsunami (Kugai 2015).

Excavations at the Tomori Motojima site revealed that the tsunami deposits covered over burials dating to the 11–12c AD (Kugai 2015: 8-9). The tsunami had also disrupted stone pavements, which had possibly been a road, and had left a large coral (*Porites* sp.) boulder on Maiba Beach, Bora, Miyako Island.

Ishigaki Island, Okinawa Prefecture

The 1771 Tsunami also hit the southern coast of Ishigaki Island, killing over 10,000 people, or one-third of the inhabitants. Despite this, most settlements there have recovered and carry on to this day. The tsunami followed the M8 Yaeyama Earthquake in 1771 occurring in the southern Ryukyu Trench, though the tsunami's direct cause is debated (Goto *et al.* 2010; Yamada *et al.* 2019; Okamura *et al.* 2018). Inundation varied locally on Ishigaki from 800–1000 m (Goto *et al.* 2010). Recent excavations have revealed cracks made by earthquakes at tomb and shellmound sites which were filled with white sand containing coral – interpreted as tsunami sediments (Yamamoto 2012).

The maximum run-up height on Ishigaki Island was 32.7 m (Kugai 2015: 8-9). A coral boulder named 'Bari-ishi' commemorates the event. This is one of five boulders emplaced by the 1771 Tsunami on the east Ishigaki coast; they were designated a Natural Monument in 2013.[4] However, there are far more boulders

[4] The situation of the 1998 Tsunami in Papua New Guinea (Davies 2002; Tappin *et al.* 2008) is a modern comparator for what might have happened to lowland settlements in Japan.

along the Pacific coast than these; Goto and colleagues (2010) measured 626 boulders in only two short sections of the coast. These were geographically separated into two groups that were differentially distributed along the shoreline (382 in number) and along the fringing reef crest (numbering 244); those on the shoreline consisted of *Porites* non-reef corals from the moat between the reef and land, while those on the reef crest were corals that had been dislodged from the seaward reef slope. The size and weights of these were significantly different, indicating the relative force of tsunami waves depositing boulders on the shoreline versus storm waves – in this case typhoons, only piling rocks up on the reef. A similar situation was discovered on Kudaka Island nearby (Goto *et al.* 2009).

1707 AD Tsunami from Philippine Plate subduction earthquakes

Shimizu Plain, Shizuoka Prefecture

Kitamura & Kobayashi (2014) undertook corings and a trench excavation to assess the recent tsunami deposits on the Shimizu Plain. Instead, they identified evidence of four historical earthquakes and tsunami, already known from historical documents. The team documented co-seismic uplift with the 1854 earthquake and three sand layers that may date as far back as 1498 or as recently as 1707. They also found evidence of four prehistoric tsunami dating between 6000 and 3000 years ago in the bay sediments.

Old Kobe Foreigners' Residence Site, Kobe, Hyogo Prefecture

Masuda & Taniguchi's (2011) urban rescue excavation at 3.5 m above mean sea level (AMSL) near Kobe City Hall revealed the remains of bricks from the Meiji-period Foreigners' Residence area. The site was located in the former beach lowlands just seaward of the alluvial fan of the granite Rokko Mountains. Underneath the brick layer was 2.5 m of sand and clay deposits. Between 1.7 and 2.0 m AMSL was a 30 cm thick sand stratum thought to be a tsunami deposit. It was exposed in a trench more than 30 m long. Below this stratum extended a series of deposits 215 cm thick of alternating sand and mud/clay layers.

The layers in the lower stratum were formed by repeating alternating units of mud/clay to sand, coarsening upward, with the mud/clay gradually changing to sand. The sand grains were poorly sorted, being of every size, and the larger the sand grains, the thicker the layer. This sedimentation pattern is characteristic of a floodplain, where internal lamination is poorly developed and there is much mud suspended in the flood waters. This stratum is thus interpreted as a floodplain located just off the tip of the alluvial fans built by the rivers running down from the Rokko Mountains. This area was most highly damaged in the 1995 Kobe Earthquake (BOX 10, BOX 12).

One of the key factors in interpreting the upper stratum as a tsunami deposit was its completely different structure from the lower stratum. The upper stratum exhibited parallel lamination, including layers of fine-grained sand and fine gravel derived from rotten granite (*masado*, regolith); the gravel-sized grains were either granite pieces or feldspar crystals (cf. Apx 5: Fig. F, Table A). The laminations are easily distinguished from the layers below. Further analysis of the upper stratum vertical profile revealed that the grains in alternating sand layers were imbricated in different directions, indicating a change in direction of the water flow. These data are interpreted as deposits resulting from the tsunami wave reversals: first incoming and then outgoing (i.e., backwash). The upper stratum is also distinguished from normal wave deposits on a beach by its thickness. Turbulence from normal waves lasts perhaps 10 seconds, while one tsunami wave can cause turbulence lasting from a few to more than 10 minutes. The successions in the upper stratum indicate it was formed from three or four waves. Grain-size analysis also revealed several differences between the two strata. The upper stratum had consistently strong sorting for large grains (0–2ϕ), but the lower stratum had more layers with finer grains (2–4ϕ).

Radiocarbon dating of a piece of wood sampled from the topmost sand layer in the lower stratum gave a date of 340±20 BP (redated by OxCal, the calibrated date range is 1539–1635 AD).[5] Three earthquakes were caused by the subduction of the Philippine Plate during or after this time: in 1605, 1707, and 1854. The 1605 earthquake is thought to have been too weak to cause a tsunami, and the 1605 historical date falls within the dating range of the wood chip *below* the tsunami deposits. Conversely, the 1854 earthquake caused co-seismic uplift, not flooding, so the 1707 earthquake was most likely responsible for the tsunami.

Note that the tsunami layers documented in Kobe show that the Inland Sea region (cf. Figure 10.1) is not immune to tsunami from earthquakes generated in the Pacific. Predictions for the next big earthquake on the Philippine Plate suggest tsunami waves could reach up to 5 m high in the Inland Sea area (DRI 2014). Also note that even though the epicenter of an earthquake in 1995 was very near the Kobe shoreline, it did not cause a tsunami. Instead, tension pulled apart port facilities, and lateral spreading plunged many sea walls and platforms into the sea. This may have been because the 1995 Kobe Earthquake was generated by an Active Fault, not plate subduction (cf. Chapter 9).

Ryujin Lake, Kyushu

This lake, just south of Oita City, was inundated by the 1707 Hoei Tsunami accompanying the Nankai Trough earthquake in southwestern Japan. Furumura *et al.* (2011) used numerical modeling to demonstrate that the extent of tsunami inundation required a longer source rupture distance than originally expected. Moreover, in addition to the regular Nankai Trough earthquake cycle of 100–150 years (cf. Chapter 9), they identified a 'hyper-earthquake cycle' of every 300–500 years. Tsunami generated by hyper-earthquakes left deposits in 684, 1361, and 1707.

869 AD (Jogan 11) Heian-period tsunami

At the most general level, historical records, eyewitness observations, local traditions, and legends are important sources for identifying and describing tsunami (Goto 2015*b*: 11; McFadgen & Goff 2007). Indeed, the usefulness of ethnographic and historical records was amply demonstrated in the search for evidence of the 1700 'orphan' tsunami, a tsunami of unknown origin that affected the northwest coast of North America. Japanese historic records revealed the source to have been a Japanese earthquake (Atwater *et al.* 2005). Another earthquake and tsunami were reported for May 5, Jogan 11 (AD 869), in a historical document compiled in AD 901 entitled *Nihon Sandai Jitsuroku* (The True History of Three Periods in Japan) (Sakamoto 1991). Saino (2015*b*: 250) cautions, however, that documentary evidence for the 869 Jogan Tsunami damage at Tagajo might be exaggerated. The written record cannot be taken at face value but must be evaluated in light of archaeological evidence. The following discussion therefore deals only with archaeological research on the 869 Jogan Tsunami.

According to Saino (2012*a*, 2012*b*, 2013, 2015*a*: 7), tsunami deposits dating to 869 have been discovered at two archaeological sites (Numamukai and Shimo-masuda Iizuka, Figure 10.4: A, B) on the Sendai Plain; three others are suspected to have tsunami deposits (Fujita Shinden and Shimoiida in Sendai City, Figure 10.4: a, b; and Takaose in Iwanuma City, Figure 10.4: c). These are all paddy-field sites that exhibit the trend of the times toward exploiting coastal lowlands while inhabiting sites located slightly inland. Only some of the paddy-fields covered by tsunami deposits were abandoned, and seven habitation sites remained inhabited without discontinuity. This is in great contrast to Middle Yayoi times (discussed below).

[5] [https://c14.arch.ox.ac.uk/oxcal.html]; see Reimer *et al.* 2013.

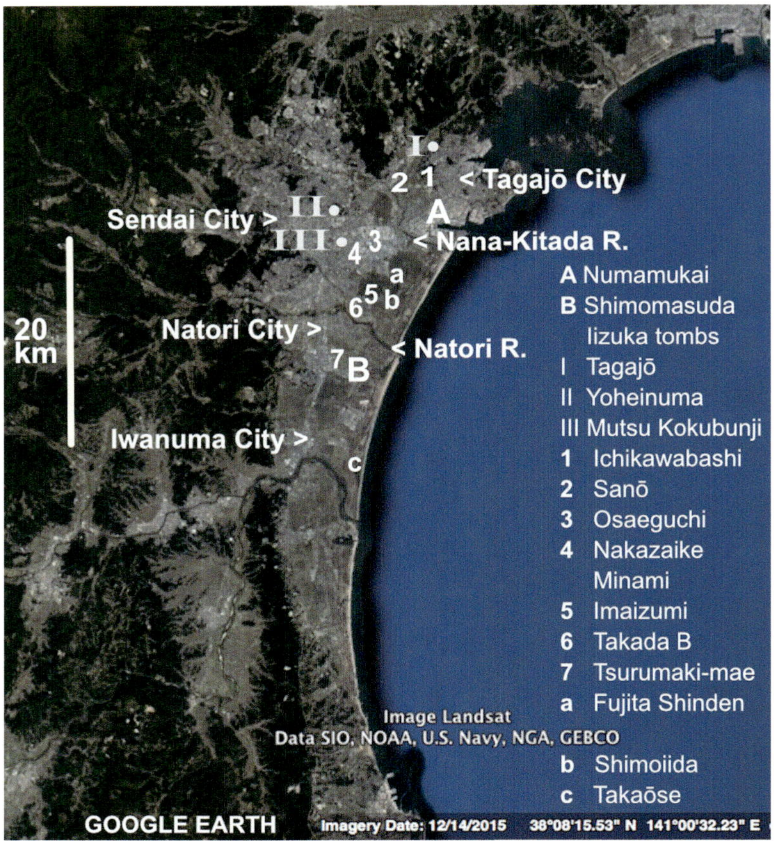

Figure 10.4 Heian occupation of Sendai Plain
attested tsunami remains: A, B; possible tsunami remains: a, b, c
village continuity: 1–7; important historical sites (dots):
I – provincial centre, II – kiln site, III – provincial temple.

Numamukai Site, Miyagino-ku, Sendai City, Miyagi Prefecture

The Numamukai site was excavated 34 times between 1996 and 2004 (Saino 2012*c*; 2015*a*: 7; 2015*b*: 244; 2017). It extends from the beach ridge system into a prior backmarsh area. Layer 2-b in the beach ridge area and Layer 6 in the old backmarsh have been identified as deposits from the 869 Jogan Tsunami. This archaeological evidence thus corresponds with the *Nihon Sandai Jitsuroku* documentation for a tsunami in 869 AD. The sand is only about 1 cm thick in these areas. It is one of four natural layers sandwiched between paddy-fields (Layer 9a) dating to the end of Nara/early Heian period (end of the 8c) and Towada-a tephra (Layer 4, dating to 915 AD). No corresponding tsunami sand layers were discovered at the Sano or Ichikawabashi sites northwest of Numamukai.

Shimomasuda Iizuka Tomb Cluster, Natori City, Miyagi Prefecture

At Shimomasuda Iizuka, Heian-period paddy-field remains (Layer 7) extending more than 1000 m^2 were covered with tsunami deposits (Layer 6) and thereafter abandoned (Goto 2015*a*; Saino 2015*b*: 248). The fields were dated by Towada-a tephra (915 AD) in Layer 4, by late 9c Sué ware occurring in Layer 6 of an irrigation canal, and by early 9c Haji ware occurring in layers below Layer 6. Thus, it is proposed that these tsunami deposits were also the result of the 869 AD earthquake. The site was located about 1.5 km from the contemporaneous shoreline, so, using the 60% rule discussed above, the tsunami inundation distance is calculated at 2.5 km.

Middle Yayoi-period tsunami ca. 100 BC

The tsunami that struck the Sendai Plain affected many sites in the region. Particular work has been carried out in the Natori River drainage whose outlet is situated in the upper middle of the Sendai Plain (cf. Figure 10.4). Archaeological investigations at individual sites (Figure 10.5) are discussed first, with a regional overview of settlement changes in the Natori River drainage in BOX 11 and Table 10.6.

Kutsukata Site, Wakabayashi-ku, Sendai City, Miyagi Prefecture (Figure 10.5: 1)

Kutsukata is the best example of Tsunami Archaeology in Japan, with Goto (2015*a*: 3) and Saino (2012*a*, 2012*b*, 2012*c*, 2012*d*, 2015*b*: 244-245) reporting evidence for the 100 BC tsunami at this site. Paddy-fields covering 20 ha and dating to the middle of Middle Yayoi were discovered during a 2006 excavation in the Sendai coastal lowlands. Situated 2.5 km from the contemporaneous shoreline, the ancient fields (Layer 6a1) were covered with a thin (2–5 cm) layer of sand (Layer 5b).

The sand was similar to two samples taken from beach dunes on the coast. Grain-size analysis revealed that the structure of the grain-size distribution matched the dune sand but not the riverine flood deposits (alluvium) on the plains. All grains fell into the 'medium sand' range of 0.25–0.5 mm, but the alluvium had a greater size range, as shown by its greater standard deviation value in Table 10.5.

As seen in histogram form in Figure 10.6, all distributions approximated normal distribution. The alluvial flood deposits on the plain had a broad grain-size distribution (i.e., large standard deviation) (Figure 10.6: B), while the Layer 5b sand grains (Figure 10.6: A) were confined to a tight size range similar to the samples of coastal sand (Figure 10.6: C, D). This clearly illustrates that the sand deposited on the plain was derived from the coast and not river flooding.

The sand at Layer 5b of the Kutsukata site was thus traced across the plain back to the beach dunes, 0.5–1.0 km away. Matsumoto (2015) comments that sand carried across this distance could not have been carried by a storm surge. He concluded that the deposits on the paddy-fields were most likely beach/dune sand that had been washed over the fields by a tsunami.

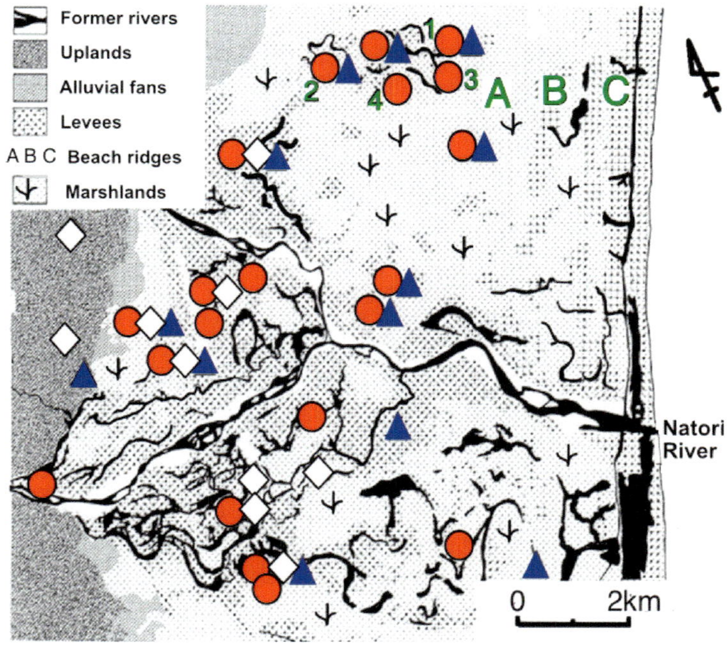

Figure 10.5 Yayoi–Kofun occupation of the Natori River drainage on the Sendai Plain. Beach ridges **A**, **B**, **C** (from oldest to youngest) parallel the shore. Site occupation symbols are organized left to right horizontally to indicate changes through time at any one site:

circles (•) represent Yayoi phases I–IIIa (before tsunami)

lozenges (◊) represent Yayoi IIIb–V (after tsunami)

triangles (▲) represent Early Kofun

Four sites at the top of the survey area are discussed in detail in the text: **1** Kutsukata, **2** Nakazaike-Minami, **3** Arai-Hirose, and **4** Arai-Minami. See BOX 11 for further descriptions.

There is strong evidence that the beach ridges (A,B,C above) along the coast correlate with previous earthquakes (Goff & Sugawara 2014; McFadgen 2007: 24).

Excavations of the 2011 tsunami deposits around Kutsukata can be compared with the Middle Yayoi Tsunami deposits to estimate inundation distances for the latter. The 2011 tsunami had an inundation distance of 4 km in this area. Based on the principle that the sand deposits ending at 2.5 km from the contemporaneous shoreline represent 60% of inundation distance, the Middle Yayoi Tsunami is calculated to have had an inundation distance of more than 4.2 km. Another comparison can be made between the relative amounts of mud in the deposits. The 2011 tsunami deposited 60–70% of the sand within the first 2.3 km or so, while 25–40% of the sediments on the landward side were mud distributed over the last 2 km of inundation. Therefore, the 2011 tsunami deposited most of the mud at its leading edge. By contrast, the excavated Yayoi deposits were mostly sand, which suggests that the leading edge of the tsunami lay beyond the excavated area.

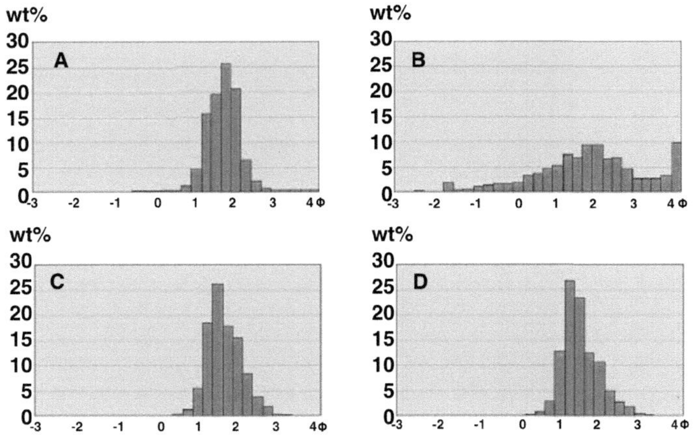

Figure 10.6 Comparison of sand grain-sizes

A: sand from Layer 5b at Kutsukata site
B: riverine alluvial deposits at Kutsukata site
C and D: samples of coastal sand

Table 10.5 Comparative sand grain-sizes at Kutsukata site

	Kutsukata	Dunes	Alluvium
mean phi (ϕ)	1.83	1.64, 1.79	1.92
standard deviation σ	0.46	0.46, 0.43	1.19

The tsunami was dated by radiocarbon from samples just above and below the sand deposit. Both came out as 2060±30 BP (IAAA-62310, IAA-62311). These dates were uncalibrated in the report; however, calibrating them with OxCal gives 170 BC–AD 4, which falls within the late Middle Yayoi period. Once covered by tsunami sands, the fields were abandoned; there was no evidence for recovery or reuse.

Arai-Minami, Wakabayashi-ku, Sendai City, Miyagi Prefecture (Figure 10.5: 4)

The Arai-Minami site is situated in the previous backmarsh area behind the first set of beach ridges (Figure 10.5: 4). Paddy-fields (Layer 6) covering 18 ha were excavated, revealing that tsunami sands (Layer 5b) had been deposited on them; the fields were subsequently abandoned (Saino 2015*b*: 245-246). The tsunami deposits cover the eastern (seaward) side of the site and were supplied from that direction. These paddy and tsunami deposits are contemporaneous with those at Kutsukata site.

Arai-Hirose Site, Wakabayashi-ku, Sendai City, Miyagi Prefecture (Figure 10.5: 3)

The Arai-Hirose site (Figure 10.5: 3) is adjacent to the Kutsukata site discussed above (Saino 2015*a*: 6). The ca. 100 BC earthquake caused a crack to develop in the site sediments; the crack then filled with tsunami deposits. One stone tool was discovered within the crack among the tsunami deposits.

Nakazaike-Minami Site, Wakabayashi-ku, Sendai City, Miyagi Prefecture (Figure 10.5: 2)

The Nakazaike-Minami site is situated in the previous backmarsh area of the Sendai Plain on an alluvial levee and former river course (Figure 10.5: 2). Excavation of the site yielded mid-Middle Yayoi pottery from two river course fill episodes (Layers 14, 15) (Saino 2015*b*: 246-248). The side of the riverbank also bore two long (7–10 m) cracks caused by the ca. 100 BC earthquake; the cracks propagated through Layer 15. Analysis of diatoms found in the fill layers revealed that marine diatoms were concentrated in Layers 15–17 downward. This area was 3.5 km from the contemporaneous shoreline, so the diatoms are interpreted as having been deposited during the tsunami inundation. No marine diatoms were found in Layer 14 though it contained Yayoi pottery. This suggests that the earthquake and tsunami that occurred in Layer 15 did not completely wipe out habitation in this area.

Nakasuji Site, Yamamoto-cho, Miyagi Prefecture

Nakasuji is located on a low hill spur of the Abekuma Mountains at 7–10 m elevation. This site is unusual in that paddy-fields were found on higher ground, while the other sites discussed above were located in backmarsh areas. The paddy-fields (Layer 6) were covered with tsunami deposits (Layers 5b,c) and thereafter abandoned (Saino 2015*b*: 248-249). Mid-Middle-Yayoi pottery can be compared with ^{14}C dates on material just above (2020±30 BP) and below (2070±30 BP) the tsunami deposits. Using IntCal 13 (Reimer *et al.* 2013) via OxCal 4.2, these radiocarbon dates respectively yield 107 BC–AD 59, and 174–19 BC calibrated dates.

Table 10.6 Differential settlement across the Sendai Plain through time
+ marks denote presence of dated artefacts, circles denote two density levels
(see BOX 11 for further discussion)

		Jomon	Yayoi before tsunami				Yayoi after tsunami							Kofun
Area	Phase	late Final	I	II	IIIa1	IIIa2	IIIb	IV	Va	Vb	Vc	Vd	Ve	Early
Nana-Kitada R.	drainage:													
Hills/terraces			+	+	⊙	+		+						⊙
Alluvial levees		+	+	+	◉	◉								◉
Beach ridges		⊙	+	⊙	+	+	+	+	+	+	+	+	+	◉
Natori R.	drainage:													
Hills/terraces			+	⊙		+				⊙	+		◉	⊙
Alluvial fan	Hirose R. lft. bank	⊙	+	+	⊙	◉	+	⊙	+	+	+			◉
	backmarsh	⊙	⊙	⊙	⊙	◉	⊙	◉	◉	⊙	⊙	⊙	⊙	◉
	Natori R. rt. bank	+	⊙	◉	⊙	+	+	+	⊙	+	+	+	+	◉
Alluvial levees	Natori R. lft. bank	+	+	⊙	◉	◉	+		+		+			◉
	Natori R. rt. bank				+	+					+			◉
Beach ridges	Natori R. rt. bank		+		+	+								⊙

Middle Holocene tsunami after Kikai Akahoya eruption 7300 cal. BP

Yoko-o Site, Oita City, Oita Prefecture

Evidence of Kikai Tsunami sediments has been proposed for the Yoko-o Shellmound site in Oita City, Oita Prefecture, Kyushu (Fujiwara *et al.* 2010). This tsunami is assigned to the eruption of the Akahoya (K-Ah) tephra from the Kikai caldera off the tip of southern Kyushu, which covered the southern tip of the island (Fujiwara *et al.* 2010; Maeno & Imamura 2007; Maeno *et al.* 2006; Fujiwara 2015: ch.6.1.3). The shellmound site bore 60 cm of volcanic ash within the 30–20 cm isopachs (cf. Figure 6.14b).

BOX 11 Surviving (or not) the Middle Yayoi Tsunami

Changing settlement patterns have been tracked across the Natori River drainage, with site occupation symbols located on horizontal transects to indicate temporal trajectories (Figure 10.5). According to Saino (2015*b*: 250-251), Yayoi sites on the Sendai Plain had been occupied from Yayoi I to IIIa phases (Figure 10.5: circles); when the tsunami in Middle Yayoi IIIa inundated the paddy-fields with sand sheets, many of the occupation sites were abandoned. However, some of the settlements survived into the Yayoi IIIb phase (Figure 10.5: circles overlapped by lozenges) and new settlements were established farther inland (Figure 10.5: lozenges). The backmarsh and beach ridge areas were not recolonized until 400 years later, during the Early Kofun period (250–400 AD) (Figure 10.5: triangles).

In Figure 10.5, it can be seen that of 27 excavated sites, 20 were occupied in Middle Yayoi (circles); 13 of the 20 were abandoned, but 6 were reoccupied in the Kofun period (circles + triangles). Two of the sites continued from Middle Yayoi into Late Yayoi and then were abandoned (circles + lozenges). Among the 27 sites, only 4 sites were occupied straight through from Middle Yayoi to Kofun (circles + lozenges + triangles), while 3 were established in new locations in Late Yayoi (independent lozenges) and 3 in Kofun (independent triangles).

Interestingly, the three sites newly established in Late Yayoi did not last into the Kofun. Two of them were in the highlands, suggesting perhaps a change in economic activities to hunting or forestry or maybe they were established for defense purposes. The other Late Yayoi settlements were positioned on the alluvial fan where dry crops might have become more important than rice growing in the narrow river flats. In any case, it is clear that settlement retreated from the beach ridges and marshlands, where fishing and rice growing were practiced. It is unknown whether this was because the villages' populations had been destroyed or whether they had retreated and not sought to return.

Table 10.6 provides a closer view of the relationship of site to topography. The table is divided horizontally into two river catchments, which show slightly different settlement strategies. The vertical division between Yayoi IIIa2 and IIIb marks the timing of the tsunami at 100 BC; another vertical division partitions the Yayoi from the Kofun period. Before the tsunami, Yayoi settlement focused on slightly elevated areas (alluvial fans and levees); after the tsunami, the Nana-Kitada River drainage was virtually abandoned, while in the Natori River drainage, settlement withdrew to the alluvial fan. Only in the Kofun period were all ecozones reoccupied, with settlement focused heavily and consistently throughout the lowlands.

Conclusions

Tsunami Archaeology is just in its beginning stages. Cooperation is required amongst archaeologists and geologists to identify tsunami deposits in archaeological sites and off-site localities. Great strides have been made in differentiating tsunami deposits from storm deposits, with a wide variety of geochemical, sedimentary, and distributional evidence being taken into account. The recognition of tsunami sands and gravels in conjunction with cultural materials at archaeological sites aids geological dating of tsunami cycles.

While this first phase of intensive geological work on identification is being conducted, archaeologists are beginning to assess the impacts on habitable areas and resources, with the aim of understanding human responses to tsunami through time. In Japan, the Sendai Plain is the locus of the most advanced investigations. Considerable differences are noted there among peoples of different prehistoric and historic periods. While abandonment was apparent during the Middle Yayoi and presumably for as yet undocumented earlier periods, later historic occupants and modern peoples seem to have been able to overcome the devastation from tsunami and recover their fields to grow crops. When fields could not be recovered or resources were too depleted, tsunami survivors have moved their settlements or pursued different economic activities.

Researchers following Shimoyama's (2002) parameters for investigating disasters in the archaeological record are still struggling to identify tsunami deposits and the causes of tsunami among multiple possibilities (i.e., earthquakes, landslides, volcanic eruptions). Such research supplies data about Shimoyama's first two parameters, initiation and immediate causes. For archaeological sites where tsunami deposits have been confirmed, the limited areal extent of excavation confines observations to the stratigraphic sequence and precludes wider understanding of the extent of damage and human reactions. In order to address Shimoyama's last parameter (reactions to tsunami), a regional approach such as Saino's, which can track population movements and subsistence changes, is necessary.

For an intimate view of reactions to tsunami, it would be instructive to assess the lifestyles of people who occupied those sites on the Sendai Plain that survived through the tsunami or were established afterwards (Figure 10.5). Did they show a change in organization or subsistence economies? Can migrants be traced within the region or beyond? What happened to village sizes and structure? Did inhabitants stop growing rice in the lowlands and turn to fishing and hunting in the uplands? These questions go beyond the identification and documentation of tsunami strata and excavation of destroyed settlements. Future research will certainly illuminate reactions of previous inhabitants to the constant but unpredictable threat of tsunami along the "hostile shores", using McFadgen's (2007) phrase, of the Japanese archipelago.

Being limited to coastal fringes, tsunami damage seldom leads to collapse of a society or civilization, though the socio-economic status of the affected society is crucial to the nature of human response. Force & McFadgen (2012) emphasize the knock-on effects of a tsunami in their study of the Maori. With the loss of the elderly and coastal craftspeople, depletion of the cultural knowledge base led to reduction in technological sophistication and retraction of trade networks, while the loss of women and children stimulated a state of warfare between communities. On the other end of the socio-economic scale, the Fukushima nuclear disaster triggered by the 2011 Tohoku-oki Tsunami illustrated how modern technology is vulnerable and non-local effects are considerable. A massive tsunami attacking Tokyo directly would be extremely destructive. Whether or not tsunami can be considered an agent forcing cultural change (Force 2015; Goff & Nunn 2016) or an 'event' in Sewell's (2005) terms, archaeology as a discipline has moved beyond merely rejecting environmental determinism to really considering the effects of environmental changes on society. Tectonic archaeology informs on the reasons, locations, and character of natural disasters such as tsunami that can be traced to plate movement, earthquakes, and landslides into the ocean. Disaster archaeology picks up where the tectonic explanations leave off to deal with the human reactions to such natural events.

Figure & Table Sources

Figure 10.1 by author
Figure 10.2 modified by GLB from (Lockridge 1990: 404, fig. 1)
Figure 10.3 modified by GLB from (Morton *et al.* 2007*b*, USGS permitted)
Figure 10.4 after (Saino 2015*b*: 252: fig. 7); courtesy of Google Earth maps
Figure 10.5 after (Saino 2015*b*: 251: fig. 6) modified by GLB with permission
Figure 10.6 after (Saino 2015*b*:243, fig. 3; created by H. Matsumoto), with permission

Table 10.1 compiled from (Goto 2015*b*; Lockridge 1990; Satake 2007)
Table 10.2 constructed by author

Table 10.3 after (Goff & McFadgen 2003: 612, table 2)
Table 10.4 compiled by author
Table 10.5 compiled from (Matsumoto 2015)
Table 10.6 after (Saino 2012*a*: 233, table 1).

References

Anon. (**2020**) "A network of state-of-the-art tsunami buoys is being deployed from New Zealand up into the Pacific to keep communities safer." *NIWA Taihoro Nukurangi* [https://niwa.co.nz/ news/new-tsunami-monitoring-system-for-pacific].

ARTHUKORALA, Prema-chandra & Budy P RESOSUDARMO (**2005**) "The Indian Ocean tsunami: Economic impact, disaster management and lessons." *Asian Economic Papers* 4.1: 1-39.

ATWATER, Brian F; Satoko MUSUMI-ROKKAKU & Kenji SATAKE *et al.* (**2005**) *The orphan tsunami of 1700: Japanese clues to a parent earthquake in North America.* Reston, VA: USGS & University of Washington Press.

CAIN, Genevieve; James GOFF & Bruce MCFADGEN (**2019**) "Prehistoric coastal mass burials: did death come in waves?" *Journal of Archaeological Method and Theory* 26: 714-754 [DOI: 10.1007/s10816-018-9386-y].

CARTWRIGHT, Julyan HE & Hisami NAKAMURA (**2008**) "Tsunami: a history of the term and of scientific understanding of the phenomenon in Japanese and Western culture." *Notes and Records of the Royal Society* 62: 151-166.

CHAGUÉ-GOFF, Catherine; Kazuhisa GOTO & James GOFF *et al.* (eds) (**2012*a***) *The 2011 Tohoku-oki tsunami.* Special Issue of *Sedimentary Geology* 282.

CHAGUÉ-GOFF, C; A ANDREW & W SZCZUCINSKI *et al.* (**2012*b***) "Geochemical signatures up to the maximum inundation of the 2011 Tohoku-oki tsunami – implications for the 869 AD Jogan and other palaeotsunamis." *Sedimentary Geology* 282: 65-77.

CHAGUÉ-GOFF, Catherine; Jean-Luc SCHNEIDER & James R GOFF *et al.* (**2011**) "Expanding the proxy toolkit to help identify past events — lessons from the 2004 Indian Ocean Tsunami and the 2009 South Pacific Tsunami." *Earth Science Reviews* 107: 107-122.

COSTA, Pedro JM & César ANDRADE (**2020**) "Tsunami deposits: present knowledge and future challenges." *Sedimentology* 67.3: 1189-1206 [DOI: 10.1111/sed.12724].

COSTA, Pedro JM; Cesar ANDRADE & Cascalho JOAO *et al.* (**2015**) "Onshore tsunami sediment transport mechanisms inferred from heavy mineral assemblages." *The Holocene* 25.5: 795-809.

COX, Doak (**2001**) "The inappropriate tsunami icon." *Science of tsunami hazards* 19.2: 87-92.

DAVIES, Hugh (**2002**) "Tsunamis and the coastal communities of Papua New Guinea", pp. 28-42 in TORRENCE & GRATTAN (eds) (2002).

DESS (**n.d.**) (Department of Earth and Space Sciences) "Tsunami!" Seattle, WA: College of the Environment, University of Washington [https://earthweb.ess.washington.edu/ tsunami/toc.html].

DILEK, Yildirim; Yujiro OGAWA & Yasukuni OKUBO (**2021**) "Characterization of modern and historical seismic-tsunamic events and their global-societal impacts", pp. 1-22 in *Characterization of modern and historical seismic-tsunamic events, and their global-societal impacts*, ed. by Yildirim DILEK *et al.* London: Geological Society of London [DOI: 10.1144/SP501-2021-17].

DOMINEY-HOWES, Dale TM; Geoff S HUMPHREYS & Paul P HESSE (**2006**) "Tsunami and palaeotsunami depositional signatures and their potential value in understanding the late-Holocene tsunami record." *The Holocene* 16: 1095-1107.

DOUGHTON, Sandi (**2005**) "Signs' tsunami icon misleads." *The Seattle Times 17 Jan'05* [www.seattle times.com/seattle-news/signs-tsunami-icon-misleads/].

DRI (**2014**) *1.17 Hanshin-Awaji Great Earthquake after 20 years: messages to the future, to the world.* Exhibition 2014.1.7–2015.6.28 at The Great Hanshin-Awaji Earthquake Memorial. Kobe: Disaster Reduction and Human Renovation Institution (DRI).

DUNBAR, P *et al.* (**2013**) "NOAA's historical tsunami event database: raw and processed water level data, and model output relevant to the 11 March 2011 Tohoku, Japan earthquake and tsunami", pp. 113-127 in *Tsunami events and lessons learned: environmental and societal significance*, ed. by YA KONTAR *et al.* Basel: SpringerNature.

FACKLER, Martin (**2011**) "Tsunami warnings, written in stone." *The New York Times* (20 April 2011) [www.nytimes. com/2011/04/21/world/asia/21stones.html?_r=2].

FINE, Isaac V; Evgueni A KULIKOV & Josef Y CHERNIAWSKY (**2013**) "Japan's 2011 tsunami: characteristics of wave propagation from observations and numerical modelling." *Pure and Applied Geophysics* 170.6: 1295-1307 [DOI: 10.1007/s00024-012-0555-8].

FORCE, Eric R (**2015**) *Impact of tectonic activity on ancient civilizations: recurrent shakeups, tenacity, resilence, and change.* Lanham, MD: Lexington Books.

FORCE, Eric R & Bruce G McFadden (**2012**) "Influences of active tectonism on human development: a review and neolithic example", pp. 1195-1202 in *Climates, Landscapes, and Civilizations*, ed. by L GIOSAN *et al.* Washington, DC & Hoboken, NJ: American Geophysical Union & Wiley-Blackwell Geopress.

FRITZ, Hermann M; Fahad MOHAMMED & Jeseon YOO (**2009**) "Lituya bay landslide impact generated mega-tsunami 50th anniversary." *Pure and Applied Geophysics* 166.1-2: 153-175.

FUJINO, Shigehiro; Haruo KIMURA & Junko KOMATSUBARA *et al.* (**2018**) "Stratigraphic evidence of historical and prehistoric tsunamis on the Pacific coast of central Japan: implications for the variable recurrence of tsunamis in the Nankai Trough." *Quaternary Science Reviews* 201: 147-161 [DOI: 10.1016/j.quascirev.2018.09.026].

FUJIWARA, Osamu (**2015**) *The science of tsunami deposits.* Tokyo University Press (in Japanese).

FUJIWARA, Osamu; Hiroshi MACHIDA & Jun-ichi SHIOCHI (**2010**) "Tsunami deposit from the 7,300 cal BP Akahoya eruption preserved in the Yokoo midden, north Kyushu, West Japan." *Daiyonki Kenkyu* 49.1: 23-33 (in Japanese).

FUKASAWA, Yuriko (**2015**) "Ainu oral tradition about tsunami", pp. 16 in *Symposium: traces and experiences of past tsunami disasters in the Pacific Rim, and the succession of knowledge*, ed. by Cultural Property Section, Sendai City Board of Education. Sendai: Sendai-shi Kyōi (in Japanese and English).

FURUMURA, T; K IMAI & T MAEDA (**2011**) "A revised tsunami source model for the 1707 Hoei earthquake and simulation of tsunami inundation of Ryujin Lake, Kyushu, Japan." *Journal of Geophysical Research* 116: B02308.

GARRETT, Ed; Osamu FUJIWARA & Philip GARRETT *et al.* (**2016**) "A systematic review of geological evidence for Holocene earthquakes and tsunamis along the Nankai-Suruga Trough, Japan." *Earth-Science Reviews* 159: 337-357 [DOI: 10.1016/j.earscirev.2016.06.011].

GAVIN, Daniel G (**2001**) "Estimation of inbuilt age in radiocarbon ages of soil charcoal for fire history studies." *Radiocarbon* 43.1: 27-44.

GILHOOLY, Robert (**2012**) "Islands on land could make towns tsunami-proof." *New Scientist* 2852 (15 February 2012) [www.newscientist.com/article/mg21328525-800-islands-on-land-could- make-towns-tsunami-proof].

GLICKSON, Andrew Y (**2004**) "Early Precambrian asteroid impact-triggered tsunami: excavated seabed, debris flows, exotic boulders, and turbulence features associated with 3.47–2.47 Ga-old asteroid impact fallout units, Pilbara Craton, Western Australia." *Astrobiology* 4.1: 19-50.

GOFF, J; P BOBROWSKY & D HUNTLEY *et al.* (**2020**) "Palaeotsunamis along Canada's Pacific coast." *Quaternary Science Reviews* 237: #106309 [DOI: 10.1016/j.quascirev.2020.106309].

GOFF, JR & C CHAGUÉ-GOFF (**2009**) "Brief communication: Cetaceans and tsunamis – whatever remains, however improbable, must be the truth?" *Natural Hazards and Earth System Sciences* 9: 855-857.

GOFF, J; C CHAGUÉ-GOFF & S NICHOL (**2001**) "Palaeotsunami deposits: a New Zealand perspective." *Sedimentary Geology* 143.1-6.

GOFF, James R; Catherine CHAGUÉ-GOFF & James P TERRY (**2012**) "The value of a Pacific-wide tsunami database to risk reduction: putting theory into practice", pp. 209-220 in TERRY & GOFF (eds) (2012).

GOFF, James & Walter DUDLEY (2021) *Tsunami: the world's greatest waves.* Oxford: Oxford University Press.

GOFF, James; Yuichi EBINA & Kazuhisa GOTO *et al.* (**2016**) "Defining tsunamis: Yoda strikes back?" *Earth-Science Reviews* 159: 271-274 [DOI: 10.1016/j.earscirev.2016.06.003].

GOFF, James; Geoffroy LAMARCHE & Bernard PELLETIER *et al.* (**2011**) "Predecessors to the 2009 South Pacific tsunami and the Wallis and Futuna archipelago." *Earth-Science Reviews* 107: 91-106 [DOI: 10.1016/j.earscirev.2010.11.003].

GOFF, JR & BG MCFADGEN (**2001**) "Catastrophic seismic-related events and impact on prehistoric human occupation, coastal New Zealand." *Antiquity* 75(287): 155-162.

---- (**2003**) "Large earthquakes and the abandonment of prehistoric coastal settlements in 15th Century New Zealand." *Geoarchaeology* 18: 609-623.

GOFF, James; Bruce MCFADGEN & Nick MARRINER (**2021**) "Landscape archaeology—the value of context to archaeological interpretation: a case study from Waitore, New Zealand." *Geoarchaeology an International Journal* 36: 768-779 [DOI: 10.1002/gea.21864].

GOFF, James & Patrick D NUNN (**2016**) "Rapid societal change as a proxy for regional environmental forcing: evidence and explanations for Pacific island societies in the 14–15th centuries." *The Island Arc* 25(2015): 305-315.

GOFF, James & Daisuke SUGAWARA (**2014**) "Seismic-driving of sand beach ridge formation in northern Honshu, Japan?" *Marine Geology* 358: 138-149.

GOFF, James & James P TERRY (**2016**) "Tsunamigenic slope failures: the Pacific Islands 'blind spot'?" *Landslides* 13.6: 1535-1543.

GOFF, J; R WITTER & J Terry *et al.* (**2020**) "Palaeotsunamis in the Sino-Pacific region." *Earth-Science Reviews* 210: #103352 [DOI: 10.1016/j.earscirev.2020.103352].

GOTO, Kazuhisa (**2015*a***) "Mechanisms of near- and far-field tsunamis", pp. 2-3 in *Symposium: traces and experiences of past tsunami disasters in the Pacific Rim, and the succession of knowledge*, ed. by Cultural Property Section, Sendai City Board of Education. Sendai: Sendai-shi Kyōi (in Japanese and English).

---- (**2015*b***) "Tsunami evidence in the Pacific", pp. 10-11 in *Symposium: traces and experiences of past tsunami disasters in the Pacific Rim, and the succession of knowledge*, ed. by Cultural Property Section, Sendai City Board of Education. Sendai: Sendai-shi Kyōi (in Japanese and English).

GOTO, Kazuhisa; Catherine CHAGUÉ-GOFF & Shigehiro FUJINO *et al.* (**2011**) "New insights of tsunami hazard from the 2011 Tohoku-oki event." *Marine Geology* 290: 46-50.

GOTO, Kazuhisa; Shigehiro FUJINO & Daisuke SUGAWARA *et al.* (**2014**) "The current situation of tsunami geology under new policies for disaster countermeasures in Japan." *Episodes* 37.4: 258-264.

GOTO, Kazuhisa; Kunimasa MIYAGI & Hideki KAWAMATA *et al.* (**2010**) "Discrimination of boulders deposited by tsunamis and storm waves at Ishigaki Island, Japan." *Marine Geology* 269: 34-45 [DOI: 10.1016/j.margeo.2009.12.004].

GOTO, Kazuhisa; Kiyohiro OKADA & Fumihiko IMAMURA (**2009**) "Characteristics and hydrodynamics of boulders transported by storm waves at Kudaka Island, Japan." *Marine Geology* 262: 14-24 [DOI: 10.1016/j.margeo.2009.03.001].

GOTO, Tomoko; Kenji SATAKE & Toshihiko SUGAI *et al.* (**2015**) "Historical tsunami and storm deposits during the last five centuries on the Sanriku coast, Japan." *Marine Geology* 367.May: 105-117.

GUTTENFELDER, David & Eugene HOSHIKO (**2016**) "Cities flattened by Japan tsunami literally rise." *AP Images Spotlight (blog)* 9 March 2016 [apimagesblog.com/blog/2016/03/09/cities-flattened-by-japan-tsunami-literally-rise].

HAMASAKI, Shinji; Yuko OKAZAKI & Shoichi SHIMOYAMA *et al.* (**2012**) "Evidence of tsunami and high tides at Kajikurihama site, Shimonoseki City, Yamaguchi Prefecture." *Kokogaku Kenkyu* 59.2: 1-5 (in Japanese).

IMAMURA, Fumihiko (**n.d.**) "Recent tsunami research in Japan." Powerpoint presentation [www.fzk.uni-hannover.de/fileadmin/institut/Veranstaltungen/DFG-RT/DFG-pdf/2.1.1_ imamura.pdf] (dead link).

ISHIBASHI, Hideaki (**2015**) "Tsunami survivors open new town on Miyagi farmland." *Asahi Shimbun* 20 July 2015 [fwww.fukushima-is-still-news.com/2015/07/new-town-in-miyagi-pref.html].

ISHIZAWA, Takashi; Kazuhisa GOTO & Yusuke YOKOYAMA *et al.* (**2020**) "Datiing tsunami deposits: present knowledge and challenges." *Earth-Science Reviews* 200: #102971, 102911 [DOI: 10.1016/ j.earscirev.2019.102971].

JAFFE, Bruce; Bruce RICHMOND & Helen GIBBONS (**2011**) "International team studies tsunami deposits in Japan to improve understanding and mitigation of tsunami hazards." *Sound Waves* June [soundwaves.usgs.gov/2011/06/].

KANER, Simon (**2011**) "Tidal wave: the day Japan shook." *Current World Archaeology* 49: 22-24.

---- (**2015**) "Visiting Fukushima: archaeology in a nuclear exclusion zone." *Current World Archaeology* 70: 46-51.

KANER, Simon; Junko HABU & Akira MATSUI (**2011**) "Rescuing archaeology affected by the Japanese earthquake and tsunami." *Antiquity* 85(335): 258-269 [www.antiquity.ac.uk/projgall/kaner329/].

KAWAMATA, Takahisa (**2015**) "The Takaōse site, Iwanuma City, Miyagi Prefecture", in *Hakkutsu sareta Nihon retto 2015: shin-hakken kōko sokuhō*, ed. by Agency for Cultural Affairs. Tokyo: Kyōdo Tsūshin [http://archaeology.jp/sites/2015/takaose.htm].

KAWAMURA, K; K OGIRI & T TOYOFUKU *et al.* (**2021**) "Tsunami-triggered dispersal and deposition of microplastics in marine environments and their use in dating recent turbidite deposits", pp. 381-390 in *Characterization of modern and historical seismic-tsunamic events, and their global-societal impacts*, ed. by Y DILEK *et al.* GSL Special Publication No. 501. London: Geological Society of London.

KING, DN; JR GOFF & C CHAGUÉ-GOFF *et al.* (**2017**) "Reciting the layers: evidence for past tsunamis at Mataora-Wairau Lagoon, Aotearoa-New Zealand." *Marine Geology* 389: 1-16 [DOI: 10.1016/j. margeo.2017.05.001].

KING, Darren N; Wendy S SHAW & Peter N MEIHANA *et al.* (**2018**) "Māori oral histories and the impact of tsunamis in Aotearoa-New Zealand." *Natural Hazards and Earth System Sciences* 18: 907-919 [DOI: 10.5194/nhess-18-907-2018].

KING, Darren N; Maurice MANAWATU & Wendy S SHAW (**2020**) "Comparing and combining ethnographic records with active Māori histories to provide insights on tsunami hazard." *Quaternary Research* 95: 43-55 [DOI: 10.1017/qua.2019.84].

Kitamura, Akihisa; Takafumi Imai & Yosuke Miyairi *et al.* (**2017**) "Radiocarbon dating of coastal boulders from Kouzushima and Miyake islands off Tokyo Metropolitan Area, Japan: implications for coastal hazard risk." *Quaternary International* 456: 28-38.

Kitamura, Akihisa & Konatsu Kobayashi (**2014**) "Geologic evidence for prehistoric tsunamis and coseismic uplift during the ad 1854 Ansei-Tokai earthquake in Holocene sediments on the Shimizu Plain, central Japan." *The Holocene* 24.7: 814-827.

Kitamura, Akihisa; Kazuyoshi Yamada & Daisuke Sugawara *et al.* (**2020**) "Tsunamis and submarine landslides in Suruga Bay, central Japan, caused by Nankai Suruga Trough megathrust earthquakes during the last 5000 years." *Quaternary Science Reviews* 245: #106527, 23 pp. [DOI: 10.1016/ j.quascirev.2020.106527].

Koshimura, Shunichi & Nobuo Shuto (**2015**) "Response to the 2011 Great East Japan Earthquake and Tsunami disaster." *Philosophical Transactions of the Royal Society of London A: Mathematical, Physical and Engineering Sciences* 373.2053 [DOI: 10.1098/rsta.2014.0373].

Kugai, Mitsugu (**2015**) "The tsunami trace and disaster in Miyako and Yayaema Islands", pp. 8-9 in *Symposium: traces and experiences of past tsunami disasters in the Pacific Rim, and the succession of knowledge*, ed. by Cultural Property Section, Sendai City Board of Education. Sendai: Sendai-shi Kyōi [in Japanese and English].

Lockridge, Patricia A (**1990**) "Nonseismic phenomena in the generation and augmentation of tsunamis." *Natural Hazards and Earth System Sciences* 3: 403-412.

Maeno, F & F Imamura (**2007**) "Numerical investigations of tsunamis generated by pyroclastic flows from the Kikai caldera, Japan." *Geophysical Research Letters* 34: L23303.

Maeno, Fukashi; Fumihiko Imamura & Hiromitsu Taniguchi (**2006**) "Numerical simulation of tsunamis generated by caldera collapse during the 7.3 ka Kikai eruption, Kyushu, Japan." *Earth Planets Space* 58: 1013-1024.

Marriner, N; D Kaniewski & C Morhange *et al.* (**2017**) "Tsunamis in the geological record: making waves with a cautionary tale from the Mediterranean." *Science Advances* 3: e1700485, 12 pp.

Masuda, Fujio & K Taniguchi (**2011**) "Edo-period turbulence (tsunami?) sediments investigated at the Old Kobe Foreigners' Residence site", pp. 9-15 in *Old Kobe Foreigners' Residence site: excavation report*, ed. by Kobe City Board of Education. Kobe: Kobe-shi Kyōi.

Matsumoto, Dan; Yuki Sawai & Koichiro Tanigawa *et al.* (**2016**) "Tsunami deposit associated with the 2011 Tohoku-oki tsunami in the Hasunuma site of the Kujukuri coastal plain, Japan." *The Island Arc* 25: 369-385.

Matsumoto, Hideaki (**2015**) "Characteristics of tsunami deposit about 2000 years ago and some geological sections of tsunami deposit AD869 in the Sendai Plain", pp. 4-5 in *Symposium: traces and experiences of past tsunami disasters in the Pacific Rim, and the succession of knowledge*, ed. by Cultural Property Section, Sendai City Board of Education. Sendai: Sendai-shi Kyōi (in Japanese and English).

Matsumoto, Dan; Yuki Sawai & Koichiro Tanigawa *et al.* (**2016**) "Tsunami deposit associated with the 2011 Tohoku-oki tsunami in the Hasunuma site of the Kujukuri coastal plain, Japan." *The Island Arc* 25: 369-385.

McFadgen, Bruce G (**1982**) "Dating New Zealand archaeology by radiocarbon." *New Zealand Journal of Science* 25: 379-392.

---- (**2007**) *Hostile shores: catastrophic events in prehistoric New Zealand and their impact on coastal Maori communities.* Auckland University Press.

McFadgen, BG & JR Goff (**2007**) "Tsunamis in the New Zealand archaeological record." *Sedimentary Geology* 200.3-4: 263-274.

Minami, M; T Nakamura & T Nagaoka *et al.* (**2012**) "^{14}C dating human skeletons from medieval archaeological sites in Kamakura, Japan: were they victims of Nitta Yoshisada's attack?" *Radiocarbon* 54.3/4: 599-613.

Minoura, K; F Imamura & D Sugawara *et al.* (**2001**) "The 869 Jogan tsunami deposit and recurrence interval of large-scale tsunami on the Pacific coast of northeast Japan." *Journal of Natural Disaster Science* 23: 83-88.

Minoura, Koji & Shyu Nakaya (**1991**) "Traces of tsunami preserved in inter-tidal lacustrine and marsh deposits: some examples from Northeast Japan." *The Journal of Geology* 99.2: 265-287.

Morton, RA; Guy Gelfenbaum & Bruce E Jaffe (**2007*a***) "Physical criteria for distinguishing sandy tsunami and storm deposits using modern examples." *Sedimentary Geology* 200.3-4: 184-207.

Morton, R.A; Guy Gelfenbaum & Bruce E Jaffe *et al.* (**2007*b***) "Distinguishing tsunami from storm deposits in the geologic record." *Sound Waves* (November) [soundwaves.usgs.gov/2007/11/].

Nandasena, NAK; R Paris & N Tanaka (**2011**) "Reassessment of hydrodynamic equations: minimum flow velocity to initiate boulder transport by high energy events (storms, tsunamis)." *Marine Geology* 281: 70-84.

Obermeier, Stephen F (**1996**) "Using liquefaction-induced features for paleoseismic analysis", pp. 331-396 in *Paleoseismology*, ed. by JP McCalpin. San Diego, CA: Academic Press [a more detailed version of this chapter was also published in *Engineering Geology* 44: 1-76 (1996)].

OGURA, Tetsuya (**2016**) "The condition of deposition of sediments by the 3.11 tsunami in Minami-Soma City, on the lower reaches of the Odaka River, Fukushima Prefecture." *Osaka Bunkazai Kenkyusho Kiyo* 2016(March): 1-12 (in Japanese).

OKAMURA, Katsuyuki; Atsushi FUJISAWA & Yasuhisa KONDO *et al.* (**2013**) "The Great East Japan Earthquake and cultural heritage: towards an archaeology of disaster." *Antiquity* 87(335): 258-269.

OKAMURA, Yukinobu; Azusa NISHIZAWA & Yushiro FUJII *et al.* (**2018**) "Accretionary prism collapse: a new hypothesis on the source of the 1771 giant tsunami in the Ryukyu Arc, SW Japan." *Scientific Reports* 8: 13620, 11 pp.

PENDICK, Daniel (**1998**) "Waves of destruction: tsunami – remembrance of waves past." *PBS Online: The Savage Earth* [www.pbs.org/wnet/savageearth/tsunami/html/sidebar2.html].

REICHERTER, Klaus R (**2015**) "Tsunamis as paleoseismic indicators," 12 pp. in *Encyclopdia of Earthquake Engineering*, ed. by E PATELLI *et al.* New York, NY: SpringerReference [DOI: 10.1007/978-3-642-36197-5_26-1].

REIMER, Paula *et al.* (**2013**) "INTCAL13 and MARINE13 radiocarbon age calibration curves 0–50,000 years cal BP." *Radiocarbon* 55.4: 1869-1887.

RIXHON, G; SM MAY & M ENGLE *et al.* (**2018**) "Multiple dating approach (^{14}C, ^{230}Th/U and ^{36}Cl) of tsunami-transported reef-top boulders on Bonaire (Leeward Antilles): current achievements and challenges." *Marine Geology* 396: 100-113.

ROBERTS, Sam (Director) (**2005**) "The killer wave of 1607." *BBC Timewatch* season 24, episode 5 (1 April 2005) [www.bbc.co.uk/programmes/b0078w1y].

SAINO, Hirohiko (**2012*a***) "The central-northern Sendai Plains in the Yayoi-period, and evidence of the Heian-period tsunami and settlement dynamics." Heisei 19 nendo ~ Heisei 23 nendo MonbuKagakusho shiritsu daigaku gakujutsu kodoka suishin jigyo 'open research center seibi jigyo' Tohoku chiho ni okeru kankyo, seigyo, jijutsu ni kansuru rekishi dotaiteki sogo kenkyu. *Kenkyū Seika Hōkokusho* 1: 225-258 (in Japanese).

---- (**2012*b***) "Evidence of agricultural disasters on the Sendai Plains: from Yayoi to Pre-modern periods", pp. 186-215 in *Collected writings on Tohoku history*. Tokyo: Seibundō Shinkōsha (in Japanese).

---- (**2012*c***) "Kutsukata site and Numamukai site in Sendai City." *Kōkogaku Kenkyu* 58.4: 114-115 (in Japanese).

---- (**2012*d***) "Excavated evidence of tsunami in Sendai plain." *Kōkogaku Kenkyu* 58.4: 6-11 (in Japanese).

---- (**2013**) "Natural disaster evidence research and ancient documents on the Jōgan 11 earthquake in Mutsunokuni", pp. 1-21 in *Shishi Sendai*. Sendai: Kyōiku-kyoku Bunkazaika (in Japanese).

---- (**2015*a***) "Tsunami disasters of Yayoi period and Heian period in the Sendai Plain", pp. 6-7 in *Symposium: traces and experiences of past tsunami disasters in the Pacific Rim, and the succession of knowledge*, ed. by Cultural Property Section, Sendai City Board of Education. Sendai: Sendai-shi Kyōi (in Japanese and English).

---- (**2015*b***) "Tsunami evidence left in archaeological sites on the Sendai Plains", pp. 240-254 in *Iwanuma Shishi*. Iwanami City: Board of Education (in Japanese).

---- (**2017**) *Investigate method of Tsunami disaster traces: cooperation with concerned many fields*. Tokyo: Dōseisha (in Japanese with English title).

---- (**2021**) *Learning about tsunami disasters from the East Japan Earthquake and archaeological sites*. Tokyo: Dōseisha (in Japanese).

SAKAMOTO, Taro (**1991**) *The six national histories of Japan*, trans. by John S BROWNLEE. Vancouver & Tokyo: UBC Press & University of Tokyo Press.

SATAKE, Kenji (**2007**) "Tsunamis", pp. 484-510 in *Treatise on geophysics, vol. 4: Earthquake seismology*, ed. by H KANAMORI. Amsterdam: Elsevier.

SATO, Shinji (**2015**) "Characteristics of the 2011 Tohoku tsunami and introduction of two level tsunamis for tsunami disaster mitigation." *Proceedings of the Japan Academy* Series B 91: 262-272.

SAWAI, Yuki; Yushiro FUJII & Osamu FUJIWARA *et al.* (**2008**) "Marine incursions of the past 1500 years and evidence of tsunamis at Suijin-numa, a coastal lake facing the Japan Trench." *The Holocene* 18.4: 517-528.

SCHEFFERS, Anja M & Dieter H KELLETAT (**2003**) "Sedimentologic and geomorphologic tsunami imprints worldwide—a review." *Earth-Science Reviews* 63.1-2(Oct): 83-92.

SEWELL, William Henry (**2005**) *Logics of history: social theory and social transformation*. University of Chicago Press.

SHIKI, Tsunemasa; Yoshinobu TSUJI & Koji MINOURA *et al.* (eds) (**2008**) *Tsunamiites: features and implications*. Amsterdam: Elsevier.

SHIMADA, Yumi; Shigehiro FUJINO & Yuki SAWAI *et al.* (**2019**) "Geological record of prehistoric tsunamis in Mugi town, facing the Nankai Trough, western Japan." *Progress in Earth and Planetary Science* 6.3: 15 pp. [DOI: 10.1186/s40645-019-0279-9].

SHIMOYAMA, Satoru (**2002**) "Basic characteristics of disasters", pp. 19-27 in R TORRENCE & J GRATTAN (eds) (2002).

SHINOZAKI, Tetsuya; Yuki SAWAI & Junko HARA *et al.* (**2016**) "Geochemical characteristics of deposits from the 2011 Tohoku-oki tsunami at Hasunuma, Kujukuri coastal plain, Japan." *The Island Arc* 25: 350-368.

SHUTO, Nobuo & Koji FUJIMA (**2009**) "A short history of tsunami research and countermeasures in Japan." *Proceedings of the Japan Academy* Series B 85.8: 267-275.

SWITZER, Adam D; S SRINIVASALU & N THANGADURAI *et al.* (**2012**) "Bedding structures in Indian tsunami deposits that produce clues to the dynamics of tsunami inundation", pp. 61-77 in TERRY & GOFF (eds) (2012).

TAPPIN, David R (**2007**) "Sedimentary features of tsunami deposits — their origin, recognition and discrimination: an introduction." *Sedimentary Geology* 200.3-4: 151-154.

TAPPIN, DR; P Watts & ST GRILLIA (**2008**) "The Papua New Guinea tsunami of 17 July 1998: anatomy of a catastrophic event." *Natural Hazards and Earth System Sciences* 8: 243-266.

TERRY, JP & J GOFF (eds) (**2012**) *Natural Hazards in the Asia-Pacific Region: Recent Advances and Emerging Concepts.* GSL Special Publications 361. London: The Geological Society of London.

TORRENCE R & J GRATTAN (eds) *Natural disasters and cultural change*. London: Routledge.

UNESCO *et al.* (**2018**) *Tsunami sources 1610 B.C. to A.D. 2017 from earthquakes, volcanic eruptions, landslides, and other causes.* Map by UNESCO/IOC-ITIC-NOAA-ICSU-WDS, available from the Tsunami Information Center [http://itic.ioc-unesco.org/index.php?option=com_content& view=article&id=1613& Itemid=2892].

URABE, Atsushi (**2017**) "Reconstruction of tsunami history based on event deposits in the Niigata area, eastern coast of the Sea of Japan." *Quaternary International* 456: 53-68.

VÖTT, A; HJ BRUINS & M GAWEHN *et al.* (**2018**) "Publicity waves based on manipulated geoscientific data suggesting climatic trigger for majority of tsunami findings in the Mediterranean – response to 'Tsunamis in the geological record: making waves with a cautionary tale from the Mediterranean' by Marriner *et al.* (2017)." *Zeitschrift für Geomorphologie,* May 2018, 39 pp. [DOI: 10.1127/zfg_suppl/2018/0547].

WATANABE, Masashi; Kazuhisa GOTO & Fumihiko IMAMURA *et al.* (**2016**) "Numerical identification of tsunami boulders and estimation of local tsunami size at Ibaruma reef of Ishigaki Island, Japan." *Island Arc* 25(2015): 316-332.

WONG, Poh Poh (2012) "Impacts, recovery and resilience of Thai tourist coasts to the 2004 Indian Ocean Tsunami", pp. 127-138 in Terry & Goff (eds) (2012).

YAMADA, Masaki; Shigehiro FUJINO & James GOFF *et al.* (**2016**) "Large-scale erosion and overbank deposition caused by the July 2013 flood of the Abu River, Yamaguchi City, Japan." *The Island Arc* 25: 386-399.

YAMADA, Masaki; Shigehiro FUJINO & Takashi CHIBA *et al.* (**2019**) "Redeposition of volcaniclastic sediments by a tsunami 4600 years ago at Kushima City, south-eastern Kyushu, Japan." *Sedimentology* 67 (2020): 1354-1372.

YAMAMOTO, Masaaki (**2012**) "Evidence of tsunami found at archaeological sites: examination of the sites around Karadake, Ishigaki Island." *Kōkogaku Kenkyu* 59.2(234): 5-8 (in Japanese).

Chapter 11

The Inter-relatedness of Tectonics & Hazard Research

Tectonic Archaeology is an umbrella promoting an integrated view of the causes or sources of many of the phenomena seen on archaeological sites. As proposed in Chapter 1 and detailed in Chapters 2–7, it begins with deep Earth processes and deals with their surface manifestations. However, it is apparent that many tectonic 'geohazards' can cause other types of non-tectonic hazards such as floods and fire. Disaster Archaeology deals with damage from all types of hazards, thus overlapping with the concerns of Tectonic Archaeology. This chapter addresses geohazards and their inter-relatedness, with description of efforts in the modern world to mitigate their damage. The closing section discusses archaeologists' involvement in two dimensions of Disaster Archaeology: investigating geohazard damage in the archaeological record and dealing with the aftermath of such damage in contemporary life.

The Hazards of Living in Japan

The fact remains...
that when it comes to the danger of natural disasters,
complacency is the norm.

The above quote appeared in the newspaper *Nikkei Weekly* (Anon. 2011: 5) soon after the March 2011 Tohoku-oki Earthquake, but the same complaint about complacency is seen in Fukushima & Ishihara's 2005 study concerning the eruption of Sakurajima, discussed below. Such an attitude has long been attributed to a deep-seated sense of fatalism in the Japanese character that can be traced back almost to the beginnings of their written history in the 8c (see review by Plath 1966). Plath, however, argues that the Japanese as well as Westerners "will opt for some lesser form of determinism that can be reconciled with free will, reflective-choice, effort-optimism, or ethical responsibility" and that fatalism in this minor form "is no bar to progress" (1966: 162, 169). Therefore, living with thousands of years of volcanic eruptions, tsunami, and earthquakes, not to mention typhoons, floods, landslides, and even drought have probably inured the inhabitants of the archipelago to an acceptance of the inevitable, equivalent to the feeling of *que sera sera*. However, population levels have increased to the extent that many more people worldwide are living 'under the shadow' of high-risk disaster potential, as in many other parts of the world past and present (Balmuth *et al.* 2005; Grattan & Torrence 2007; Okay 2018). Accordingly, efforts are increasingly being made to protect Japanese citizens periodically exposed to such natural hazards not only through identification of hazards, protection from disasters, and mitigation of their effects but also to aid in rapid response and long-term recovery.

The three types of natural hazards covered in the previous chapters are, in fact, inextricably interlinked with each other (cf. Figure 10.2) and with further natural hazards. Thus, it is perhaps unrealistic to separate out the products and activities of each, though that is what is being done here. Ideally, it is imperative for archaeologists and policy-makers dealing with each of these activities to be acquainted with the other related activities. The most obvious relationship is tsunami after earthquake, keeping in mind that tsunami can also be caused by non-tectonic events (Chapter 10, Nagata 2018). Earthquakes themselves are of two types with different periodicities (Chapter 9), but any of them can cause a variety of damage including building collapse and fires, disrupting roads and tracks and in modern times infrastructural supply lines, offsetting rivers to cause flooding, and landslides. Volcanic hazards also involve both direct and indirect damage (cf. Chapter 6): via primary lava flows, pyroclastic flows, base surges, mudflows (lahars), debris avalanches, tephra deposition, and volcanic gas emissions, and secondary landslides, lightning, forest fires, floods, earthquakes, and tsunami.

The Central Disaster Management Council (CDMC) of Japan was established under the Cabinet Office in 1962. Its remit is entailed in the Disaster Countermeasures Basic Act of 1961, and amended periodically; it "addresses all of the disaster phases of prevention, mitigation and preparedness, emergency response as well as recovery and reconstruction with roles and responsibilities among the national and local governments clearly defined; it is stipulated that the relevant entities of the public and private sectors are to cooperate in implementing various disaster countermeasures" (Cabinet Office 2015). In the Japanese language, 'hazard' and 'disaster' have not traditionally been differentiated (Goto *et al.* 2014: 258); the word *kiken,* translated as hazard, conflates risk and danger. Defining hazard as a 'potential source of harm', however, is necessary to quantify that potential and the thresholds of harm. This is what the risk equation addresses.

The potential risk for geohazards is represented by the equation Rs = H x E x V (Culshaw 2010), where Risk (Rs) depends not only on the Hazard (H) itself but also on the Element (E) at risk and its Vulnerability (V). Elements can be people, property, or infrastructure. Risk entails "both physical and social components," while vulnerability can be defined as "the magnitude of loss potential" (Warrick 1979: 174-5); Riede (2016), however, notes that vulnerability can really only be known *post hoc.* Losses are defined by numbers of fatalities, destruction of property, disruption of livelihoods, and economic consequences of all these. A composite hazard map for Japan and the adjacent continent was published in 2007 by the United Nations (OCHA ROAP 2007) that provides risk assessment for insurance purposes for several types of natural hazards (Figure 11.1). This figure is ordered from left to right as affecting property in the country at increasing levels of risk.

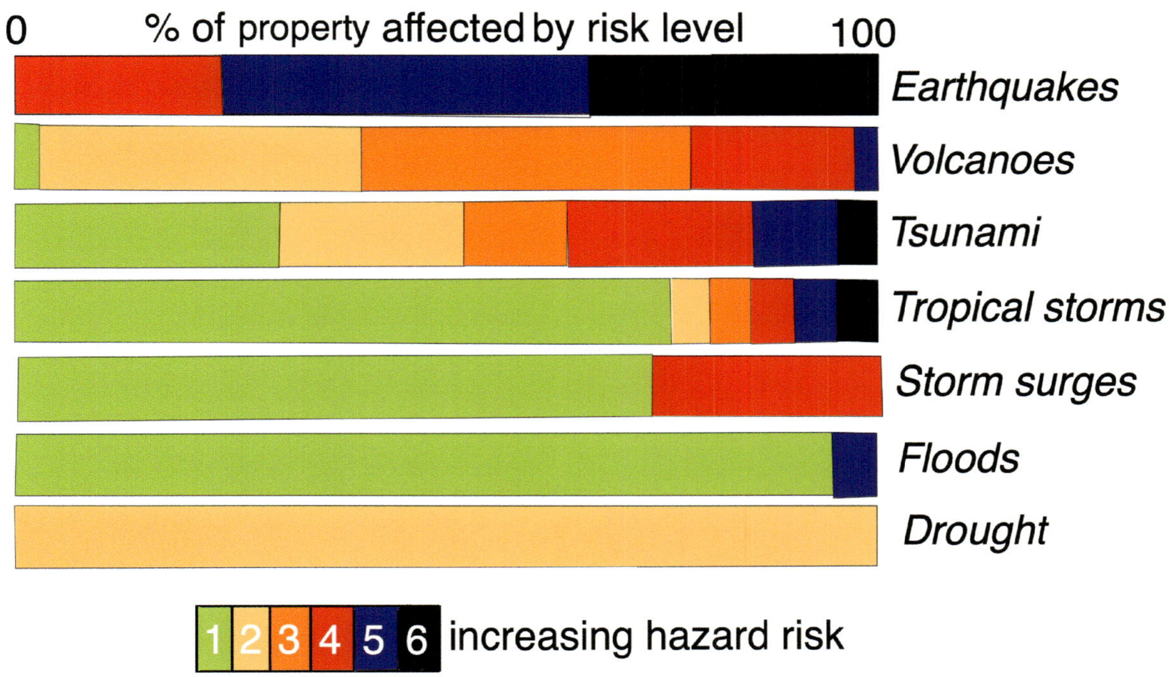

Figure 11.1 Proportions of hazard risks in Japan from 1 to 6, calculated for insurance purposes

Figure 11.1 reveals that:

- Despite the high precipitation rates in Japan, drought may affect the entire country at a low level of 2.
- Storm surges only affect a small proportion of the country, ca. 4% presumably along coastlines, but the risk there is high at level 5.
- Just over a quarter of property in Japan is at risk from volcanic eruptions, but the risk in those areas is high at level 4.
- The risk of tsunami is limited to just under 30% of the Japanese land area, presumably along the coasts, but five levels of risk are equally distributed within that area.
- Flooding significantly affects ca. 70% of properties at increasingly severe levels.
- Hurricanes (aka typhoons) affect 98% of property at mid-levels of risk, with the coasts presumably hit hardest.
- Earthquake risk is the winner, affecting all of Japan at high levels: 24% at level 4, 43% at level 5, and 33% at level 6; the risk of ground-shaking through most of Japan is put at 8–12 on the Modified Mercalli Scale, with only northern Hokkaido at level 7 (i.e., JMA intensity 5-strong ~7, and 5-weak: see Table 2.6; Apx 7).

Monitored over a 30-year period (1980–2010), the major geohazards (earthquakes, tsunami, and volcanic activities) were found to occur ca. ≤100/year, but other types of hazards (extreme temperatures, droughts, fires, floods, storms, etc.) have quadrupled in frequency (Dilek *et al.* 2021: fig. 1). The authors attribute these increases to "the increasing effects of global warming, increased greenhouse emissions, and rising ocean water temperatures during the past 30 years" (2021: fig. 1 caption). Whether the levels 30 years ago were already much higher than in the (prehistoric) past is a problem to which archaeologists can contribute information.

Risk is highly dependent on exposure; this is usually based on population levels. Volcanoes which erupt in northeastern Hokkaido, far from any centres of habitation and whose ash-clouds are blown northeast out to sea, form very little risk because few people live near the volcano or in the path of tephra dispersal. Risk from volcanic hazard is thus spread unevenly throughout the islands. The eruption of Mt Fuji or Sakurajima would pose a high risk of fatalities because of the high number of Elements (E) at risk and their vulnerability. Historical examples of the highest fatalities have been the 29,025 persons who perished in the pyroclastic flow of the 1902 eruption of Mt Pelée over the town of St. Pierre on the island of Martinique (Bullard 1979), or the 830,000 deaths in the 1556 Shaanxi Earthquake (Ritchie 2018), or the 228,899+ deaths from the 2004 Boxing Day Tsunami in the Indian Ocean (NCEI n.d.). However, fatalities are not the only subject for quantification of risk. The 2011 Tohoku-oki Earthquake and Tsunami did not result in as many fatalities (<20,000, Vervaeck & Daniell 2011) as they could have done precisely because of several factors: the lack of large cities on the Sanriku Coast area of direct impact, and the presence of sea barriers, warning systems, and population preparedness – albeit less than sufficient. Though the population levels were low in coastal Tohoku, tsunami damage to infrastructure is estimated at USD $365 billion (Dilek *et al.* 2021: fig. 1). What would the fatalities and damage be if a tsunami hit a large city directly?

The Japan Meteorological Agency [JMA.go.jp] monitors many hazards for the whole country, including volcanoes and earthquakes (JMA 2020*a*). A 'real-time risk map' (JMA 2020*b*) that updates every 10 minutes is available for landslides, inundations, and floods; these are mapped in colours indicating risk level on a zoomable topographic map. Earthquakes are recorded with associated tsunami risk. The JMA website is fully presented in Japanese but has several English pages – essential for foreigners living in

Japan. In addition, the Japanese Geospatial Information Authority maintains composite maps for weather-related hazards, including landslides (MLITTJ n.d.). Several of these geohazards will be reviewed below in terms of JMA monitoring and preparedness. This, however, only gives a national overview, while many local administrative units undertake to alert their citizens about hazard risk and help their preparedness (e.g., Koshimura & Shuto 2015).

Volcanic Hazards

Prediction & monitoring

In 1974 the Ministry of Education, Culture, Sports, Science and Technology (currently MEXT) compiled the Volcanic Eruption Prediction Plan, and from 1989 a research project on the prediction of a volcanic eruption was started within the National Research Institute for Earth Science and Disaster Prevention (NIED). From 1994 a series of 24-hour monitoring stations at 26 Active Volcanoes was installed; now 50 of the currently recognized 111 Active Volcanoes are monitored, among which 25 are intensively monitored (Aoi *et al.* 2020). The monitors feed their data on volcanic activity into four Volcanic Observation and Information Centers (VOIC) administered by the Japan Meteorological Agency (JMA). Regional centres are placed in Sapporo, Sendai, Tokyo and Fukuoka– near the main clusters of active volcanic chains but with regional responsibilities that cover areas at risk. NIED itself runs 55 Volcano Observation Network (V-net) stations on 16 Active Volcanoes (Aoi *et al.* 2020). The monitoring stations are mainly seismic, with surface-level seismometers and GNSS receivers (Aoi *et al.* 2020) recording earth tremors or small earthquakes that indicate the movement of magma within the volcano; but GPS, infrasonic microphones, tiltmeters, and high-sensitive visual cameras are also deployed (Aizawa 2006).

Satellite GPS can monitor changes in ground level movement, tracking expansion or contraction of the volcano. For example, tiltmeter readings revealed that the Miyakejima magma moved to the northwest during eruption and that the island itself contracted (Ukawa *et al.* 2000). This is a sophisticated replacement for the visual tracking of volcano expansion, of which the most famous low-tech incident occurred in 1944–1945. Then, the Usu volcano in Hokkaido rose 300 m in 20 months before it erupted, as documented by an office worker near the volcano tracing its outline on a piece of paper taped to a window (Nunn 1994: fig. 3.15).

Three other new technologies are being used in Japan (Nakata 2005): laser profiling (LIDAR) was employed for the first time on Mt Usu and Miyakejima and has been used in mapping sections of Mt Fuji that are densely forested. And, in addition to COSPEC (correlation spectrometer), a large heavy instrument which has traditionally been employed in measuring SO_2 (sulfur dioxide) levels in volcanic ash-clouds, a newer smaller lightweight instrument called Mini-DOAS (differential optical absorption spectroscopy) is making it possible to monitor emissions other than SO_2. Finally, improvements in radar have also allowed volcanoes to be 'seen' through ash-clouds. The technology is called SAR (synthetic aperture radar), mounted on aircraft, satellites or space vehicles. Its antenna picks up variable signals from an object as the craft moves over it; an image is then synthesized (or, more appropriately, simulated) from the successive signals.

The Tokyo Volcanic Ash Advisory Center (Tokyo VAAC 2021) is one of nine world-wide regional centres established in 1993 under the framework for the International Airways Volcano Watch for the express purpose of notifying airline companies, civil aviation authorities and the like about volcanic-ash cloud hazards. Any ash-cloud detected by satellite imagery is monitored at six-hour intervals unless more rapid change demands closer tracking. The Tokyo VAAC, administered by the JMA, co-operates with the

Anchorage VAAC to the northeast, the Washington VAAC covering the South Pacific, and Darwin VAAC to the south in monitoring volcanic activity along the western Pacific Rim.

Twenty-four hour advisories are published on the Tokyo VAAC homepage for volcanoes located from Kamchatka in the north to the northern half of the Philippines. For any eruption of ash within this area, data are provided on the eruption time, height of the ash plume, while forecasts in graphic format and satellite images show movement of the volcanic ash-cloud at six-hour intervals at three altitudes (from the surface up to 55,000 feet); however, its effect on aviation (via the aviation colour code) is not specified in the Tokyo VAAC presentations. The online database goes back three months or so, and the online archive begins from 2006. Monthly reports describe volcanic eruptions and unrest (including gas and steam plumes, seismicity and number of tremors, and ground movements determined by GPS measurements).

In 2003, a Committee for Predicting Volcanic Eruptions, organized under the auspices of the JMA, undertook a ranking of Active Volcanoes in Japan as to their likelihood of eruption (Yamasato 2005). Three ranks were established:

A – Especially High Possibility of eruption: volcanoes that had erupted more than 5 times in the last 100 years or more than 10 times in the last 10,000 years
B – High Possibility of eruption: volcanoes that had erupted more than once in the last 100 years, or more than 7 times in the last 10,000 years
C – Low Probability of eruption: others among the active volcano list, excluding A and B and also volcanoes in the northern Kurile chain or submarine volcanoes.

Volcanic eruptions are potentially very highly probable in Hokkaido, the Izu Islands (Ito/Ogasawara), and Kyushu; eruptions in Tohoku and Kanto are less probable. See Nakamura *et al.* (2008: table 1) for some of the volcanoes classed into these categories before 2006.

For informing the public of potential eruptions, the JMA runs a volcanic warning and forecast system with three levels of warning/forecast and five levels of alert based on the feed from the VOIC centers (Table 11.1) (JMA n.d.*a*). According to Aizawa, volcanic alerts are issued "when volcanic activities are extremely intensified and urgent countermeasures to prevent/mitigate damage to human lives are required"; in contrast, advisories are issued "when unusual volcanic phenomena are observed and preparation for disaster prevention/mitigation is required" (Aizawa 2006: unpg.). Nevertheless, complaints had been received about the warning levels: that the statements themselves were hard to understand and that there are no cut-off dates; the Committee urged the JMA to adopt the more internationally used colour-coded triangles to no avail at that time (Yamasato 2005). Today, the JMA uses a modified color scheme similar to that of USGS (Table 11.1).

However, the Tokyo Metropolitan Government publishes a 64-page disaster guide (DPD 2020) that only mentions volcanic hazards on 2 pages (11, 44) and these deal with Active Volcanoes in the Izu Island chain; the potential eruption of Fuji and impact on Tokyo residents (cf. Figure 11.2) goes unremarked upon.

The prediction network was instrumental in dealing with the eruptions of both Usu and Miyakejima in 2000. Nakata reports that for the first time, the real-time movement of magma in the chambers could be monitored, allowing eruption predictions, but that the system was still deficient in predicting how the magma would erupt and how long the eruption would last (Nakata 2005: 132-133). He was optimistic that especially with volcanoes such as these, which erupt every 20 to 50 years, close scrutiny of multiple eruptions will help define and refine predictions of where and when they might erupt again. On the other hand, he warned that it is extremely unwise to try to predict how the volcano might erupt – predict its 'eruption scenario' as he calls it – since so many variables can produce an enormous range of results (Chapter 6), making a wrong prediction quite dangerous.

Table 11.1 Coloured triangle system for volcanic activity warnings

USGS	triangles	JMA	Alert	triangles
Normal	Green	Potential for increased activity	1	White
Advisory	Yellow	Do not approach the crater	2	Yellow
Watch	Orange (with eye)	Do not approach the volcano	3	Orange
Warning	Orange (with exclamation point)	Prepare to evacuate	4	Red
Watch	Red (with eye)			
Warning	Red (with exclamation point)	Evacuate	5	Purple
Unassigned	White			

The 1 September 2004 eruption of Asama further illustrates the working of the observation network (Aizawa 2006). A month before eruption, the volcanic glow was captured by high-sensitive video camera, while the hourly number of earth tremors increased dramatically on August 31st. The next day, the observation level was increased from Level 3 Advisory to a Level 2 Alert, and the volcano erupted at 8.02 pm that evening. Volcanic bombs were thrown 2 km from the crater. Eruptions continued in late September, and another large eruption occurred in February 2009 when again the observation level was increased from 3 to 2.

Much effort has been expended on monitoring volcanic eruptions in the past, as the risk factor rises with denser human occupation near the 111 Active Volcanoes (Figure 6.2). However, in the last decade, several factors combine to cause a decline in volcanologists both in university courses and in employment. Shozawa (2014) gives a damning report on the seeming indifference of the government to ensure volcanoes are properly monitored, with public education and acknowledgement of risk also lagging behind. He reported (2014: unpg.): "It's as if the volcanoes notice our absence." This sentiment was voiced by researchers at the Volcanological Society of Japan's convention, held in Fukuoka City, Fukuoka Prefecture, in November 2014. In Kagoshima Prefecture, Mount Shinmoe erupted in 2011 after the University of Tokyo shifted to unmanned observation, while a pyroclastic eruption occurred on Kuchinoerabu Island in August 2014 following the retirement of a Kyoto University professor who had monitored volcanoes in Kagoshima for decades. The eruption of Mount Ontake in September also came after an authority on the volcano retired from his professorship at Nagoya University.

The decline in teaching capacity and job opportunities for volcanologists in Japan does not sit well with the expected increase in volcanic activity in the future (Fujii & Yamasato 2015).

Mitigation

Patterns of ash fallout are dependent on the prevailing winds, which historically tend to blow from the west across Japan. Each volcano has its historical set of lobate ash-fall distributions, mapped by isopachs, through various eruptions (compare the red lines in Figure 8.2). In the 2009 Asama eruption, the tephra fallout extended 140 km to the southeast (including Tokyo), while in the 2004 eruption, it was blown 250 km northeast near to Sendai City. In terms of human settlement, at least people downwind can be warned of impending ash fallout, but surprisingly, there is not much public information on how this can be mitigated.

The hazard map for Kamakura City (**2022**), for example, addresses several hazards including flooding, landslides, earthquakes, and tsunami - but still not volcanoes, even though the city is located approximately 75 km directly east of Mt Fuji. The volcano last erupted in 1707, but more than 75 eruptions are documented for the last 2200 years (Wolman 2006). Perhaps the citizenry are under the illusion that "it can't happen here", but the hazard map for Mt Fuji shows that not only Kamakura but all of Tokyo are at risk (Figure 11.2), and revised calculations for the first time in 17 years predict greater and more rapid dispersal of eruption products (Anon. 2021); twice as much lava and four times the amount of pyroclastic flow are expected (Kyodo 2021). Yamamoto & Nakada (2015: 355) explain that because Fuji is sitting within a collision zone at the intersection of two volcanic arcs in the "most active fault system in Japan" (cf. Figures 6.2, 6.8), they predict an M8 eruption "in the near future within several hundred years".

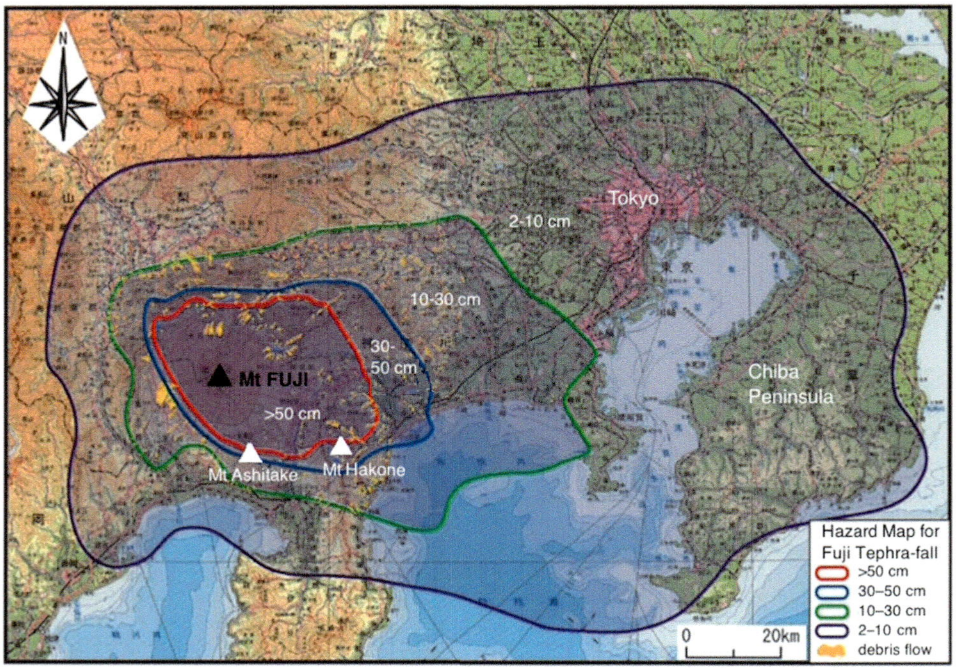

Figure 11.2 Volcanic hazard map for Mt Fuji and the southern Kanto region

In southern Kyushu, Sakurajima is Japan's most active volcano and has been in "a period of frequent intermittent moderate explosive eruptions" since 1955 (Smithsonian n.d.). Fukushima & Ishihara note a measure of complacency among the citizens of Kagoshima, attributed to the increasing sophistication of volcanic eruption prediction technology; they observe that (2005: 1):

> improvements in such facilities tends [sic] to decrease disaster prevention awareness of local people... residents may become complacent, feeling as though they live in an entirely safe environment and forgetting about previous disasters. In the long term, and with the

disasters being distant memories, stories of disasters are not passed on to younger generations.

Their solution is to establish 'eco-museums' in areas affected by natural disasters, to bring the local environment to citizens' attention and keep attention honed to the possibilities of reoccurrence. They trialed their ideas during a postcard exhibition of the 1914 eruption of Sakurajima, which was viewed by about 3,500 people; 80% of their surveyed viewers said their awareness for disaster prevention had increased, even though many were not interested in the subject.

Meanwhile, people living close to constantly erupting volcanoes such as Sakurajima are blasé, while those surprised by the sudden eruption of, for example, Shinmoedake in 2011 or Ontake in 2016, will resume their lifestyles close to the volcano when eruption 'ceases'. In Japan, where there are so many active volcanoes close to prime flatlands for occupation, agriculture, and industry, the eruption risk to person and property is somehow perceived as less than earthquake or landslide risk, with people willing to co-exist with volcanoes.

The OCHA ROAP map (2006) of Holocene eruptions until 2002 indicates that southwestern Hokkaido, the Japan Alps down into the Izu Arc, and Kyushu into the northern Ryukyu Arc have been particularly vulnerable, with VEI 2 to 4 level eruptions. Nakamura *et al.* (2008: 297) reported that by 2008, 60 volcanic hazard maps had been published; they offer criticisms of these on many fronts and note that "no local governments have yet conducted assessments of volcanic risk analyses"; the authors argued that what is really needed is "real-time hazard maps with probability tree algorithms for forecasting volcanic events", but also they argued that prediction maps for specialists' use should be separate from those for disaster management focussing on people's needs. The National Research Institute for Earth Science and Disaster (NIED 2016) established a website database and CD compilation of volcanic hazard maps and other reference materials, but it has not been updated in five years. That website, though still extant, has been superseded by the "Overlapping Hazard Map" with local search facilities (MLITTJ n.d.) and the real-time reporting by JMA [www.jma.go.jp/bosai/#lang=en&pattern= earthquake_volcano& area_type=japan&area_code=010000].

Fatalities

A study of documented volcanic fatalities around the world shows that pyroclastic flows have claimed the most lives (about 29% of total fatalities) (Simkin *et al.* 2001). These flows occur when a cloud of hot ash, too dense and heavy to be cast into the atmosphere, instead rolls down the volcanic flanks at high speed, carrying with it large rocks and anything else entrapped within. In 1991, the eruption of the Fugen-dake lava dome in the Unzen volcanic complex of northwest Kyushu generated a pyroclastic flow, moving at up to 200 km/hr; it took 14 lives including those of several professional volcanologists (demonstrating inabilities to predict the volcanic eruption style). Next in line in Simkin *et al.*'s study for claiming lives come indirect causes of volcanic eruptions (landslides due to volcanic tremors, or gas emissions that damage proximal flora and cause widespread climate changes resulting in famine, etc.). An example of the latter is the Mt Laki eruption in Iceland in 1783, notable for producing 14 km^3 of lava (OSU 2021) and releasing 0.5 billion tons of toxic gasses, including 122 Mt of SO_2, which killed over 60% of their cattle – which ate fluorine-contaminated grass – and resulted in the death of one-quarter of the then population of Iceland by famine due to crop failure from acid rain (Thordarson *et al.* 1996; Thordarson & Self 2003; OSU 2021). The Laki eruption was exceeded by the Eldgjá eruptions from AD 934, emitting 219 Mt of SO_2 and causing drought and a locust plague in China (Fei & Zhou 2016).

Volcanoes and water are a deadly mixture. Those with crater lakes or snow and ice cover are likely to produce explosive eruptions (Figure 6.3) and lahars, as evidenced by the Dragon Lahar of Mt Towada in 915 (Kobayashi 2019) which buried several villages along the Yoneshiro river course (Murakami 2019). Volcanoes situated near the coast can also cause tsunami. For example, the previous eruption of Mt Unzen in 1792 is one of seven worldwide eruption events killing more than 10,000 from all causes since the year 1500 (Simpkin *et al.* 2001: 255). The eruption was followed by the Fugen-dake lava dome collapse and debris flow into Ariake Bay. The landslide is estimated to have been 30 m thick at its front and containing 150 million m^3 of sediments traveling at 100 m/s (Miyamoto 2010). The landslide triggered tsunami along 77 km of coastline around the Ariake Sea, which resulted in approximately 15,000 deaths on the opposite shores from the volcano (Shiki *et al.* 2008: 167). Another example is the eruption of the Krakatau volcano in Indonesia in 1883 which caused tsunami responsible for more than 36,000 deaths (Simkin & Fiske 1983). The eruption generated tsunami by several processes: "landslide, pyroclastic flow entering seawater, caldera collapse, shock wave produced by large explosion" (Paris *et al.* 2014). Of all the natural hazards, volcanoes offer the greatest range of multi-dimensional disasters; within these, the water-related tsunami and lahars together with pyroclastic flows are the most intractable in terms of dealing with their occurrence and aftermath (Simkin *et al.* 2001).

Graphs of cumulative fatalities and of fatal volcanic eruptions both show upward curves, nearly exponential, between the 14c and 20c (Simkin *et al.* 2001). The authors reason this is not because volcanic eruptions are becoming more frequent or devastating, but because more and more people are living near volcanoes and becoming vulnerable to eruptions. The biennial conference "Cities on Volcanoes", organized by the International Association of Volcanology and Chemistry of the Earth's Interior (IAVCEI), moves around the world to volcanic sites with large populations past or present to emphasize the risks involved. Simkin and colleagues note that "the most common killer in fatal eruptions is tephra, which includes all fragmental material thrown from volcanoes, killing mainly by collapse of ash-covered roofs (the most deaths) or by projectile impact (by far the largest number of fatal tephra accidents)" (2001: 255). They argue that improved building codes and public education to keep people away from volcanoes suspected of eruption are the most efficient ways to lower the potential death rate in volcanic eruptions. However, people are extremely tenacious, as exhibited during eruptions of Mt Merapi in 2020 in Indonesia where evacuees refused to take up new accommodations and many travelled back to their farms every day from refugee camps regardless of the risk (Mei & Lavigne 2012).

Earthquake Hazards

Earthquake types & magnitudes

The largest earthquakes are usually from Active Faults, caused by stress within the Earth's crust rather than oceanic slab subduction. As seen in Chapter 4, the collision of the Indian Continent into the Eurasian Continent is squeezing continental material north of the Himalayas out to the east in escape tectonics. Thus, Japan is caught between the eastward movement of the Eurasian continent and the westward movement of the oceanic plates, giving rise to considerable compression faulting of the archipelago's landmass. Active Faults are where pressure builds up over the centuries and is released as an earthquake about every 1000 years on any one fault. The Kobe Earthquake of 1995 (BOX 10, 12) was one such Active Fault earthquake on the Nojima Fault; another was the great Keicho Fushimi Earthquake of 1596 (Lin *et al.* 1998), which caused much damage in the Kinai region. Active Fault earthquakes are seldom less than M6.5.

In contrast to Active Fault earthquakes, subduction zone earthquakes can occur at a variety of strengths, and plates can even subduct aseismically (Platt *et al.* 2018). Very large subduction earthquakes occur when topographic highs such as seamounts and ridges on the oceanic plate obstruct smooth passage (Zhao 2015). Blockage is what caused the 2004 Indian Ocean Earthquake, when the down-going plate broke free after being stuck for some time, resulting in a quake with a magnitude of 9.2 (CIT Tectonics Observatory n.d.). A similar scenario is proposed for the Kanto earthquake series. The Great Kanto Earthquake of 1923 was ca. M8.2, with the greatest number of fatalities in Japanese history at 103,000 lives lost (Dilek *et al.* 2021). That was the most recent of a series of large earthquakes affecting Tokyo, with earlier ones in 1855 (the Ansei-Edo Earthquake, M7.2) and 1703 (the Genroku Earthquake, M8.2); a repeat performance – already overdue – is expected within this century. Toda *et al.* (2008) have proposed that the delay is due to a broken piece of the Pacific Plate stuck under the Kanto region, obstructing the smooth subduction of the rest of the plate.

The great Tohoku-oki Earthquake in 2011 was unexpected, both in its timing and extent as well as in its severity at M_w9.0; it was also accompanied by many aftershocks, at least seven greater than M7, hundreds between M6 and M7 and thousands less than M6 (Zhao 2015). Zhao has presented evidence that the larger earthquakes occurred in low-velocity zones along the Tohoku coast, as identified by seismic tomography, interleaved with high-velocity zones – a situation Zhao says resembles the hot magma fingers feeding volcanic activity in Tohoku (cf. Figure 6.6) as proposed by Tamura (2002). Far more epicentres of the many earthquakes in the Tohoku-oki series were located within the low-V zones or at their borders with the high-V zones (Zhao 2015: fig. 7). This did not prevent earthquakes in the high-V zones, but many there are of the 'slow earthquake' variety facilitated by fluids derived from the magma underlying these zones. Dehydration and fluid release is the main factor in magma formation that leads to subduction zone volcanoes (cf. Chapter 2.4.3), and it is now recognized that "fluids may be involved in nucleation and rupture processes of all types of earthquakes" (Zhao 2015: 26).

Earthquakes also occur in conjunction with volcanoes. There are two types of tremors which are usually less than M5: short (volcano-tectonic, VTs) and long-period earthquakes (LPs, or long tremors) (PNSN n.d.). LPs are induced by movement of magma into the chamber; these can be used, among other indications, to predict eruptions (PNSN n.d.). Short tremors can occur with or without the direct involvement of magma; without magma, the tremors are induced by faulting around the volcanic edifice due to weak rock structures and changing stresses in volcanic inflation or deflation (PNSN n.d.). The movement of magma into dykes, and such VT-seismicity, often occurs in 'swarms' (Roman & Cashman 2006; White *et al.* 2019). Such swarms generally occur in compressive regimes, e.g., Mt Usu, but also occur in extensional grabens such as the Beppu-Shimabara Graben (B-SG) containing Mt Unzen, in Japan (cf. Figure 5.11).

Finally, earthquakes are known be a response to changing environmental stress conditions as mentioned in Chapter 2.4.2. For example, tides can trigger earthquakes, as can increases in atmospheric pressure during winter – thought to affect subduction earthquakes of the Philippine Plate under SW Japan (Heki 2003). However, the greatest influence among these stress conditions in Japan is snow load leading to crustal deformation. Heki has histogrammed by season the numbers of M≥7.0 earthquakes occurring in the 'snow country'[1] region of northwest Japan; earthquakes there are more common in the snow-melt spring–summer season than autumn–winter (Heki 2003: fig. 4; Toyota & Takada 2015). For

[1] "Snow Country" is the title of Kawabata Yasunari's novel (published in Japanese 1935–1937) that took place in Niigata Prefecture; it was published in English translation by Edward Seidensticker in 1957 and was one of three books that earned Kawabata a Nobel Prize in Literature in 1968.

the northern coast of southwest Honshu (San'in district), a spring/autumn bimodal patterning in earthquakes that correlates with snow melt and heavy rain has been identified. The melting of snow releases pressure on the Earth's surface causing isostatic rebound, while precipitation from both snow melt in the spring and rain in the June monsoon season and August typhoon season may percolate down to the "seismogenic zone, at depths around 8 km...weakening fault strength" (Ueda & Kato 2019).

These various causes of earthquakes indicate that people living in different areas of Japan are subject to different kinds of earthquake risk – risk that may be calculated on a seasonal or yearly basis, or even by century or millennium. Without instrumentation, earthquake prediction was conducted by observing omens or portents in the weather, atmospheric conditions, changes in groundwater levels, and abnormal animal behaviour (Clancey 2006: 151-154; Tajima & Hayashida 2018). With instrumentation (since 1876 in Japan), the expectations for earthquake prediction have increased – but still remain elusive as illustrated by the surprise 2011 Tohoku-oki Earthquake (Kagan & Jackson 2013).

Monitoring earthquakes

Until the 1995 Kobe Earthquake (cf. BOX 10, 12), monitoring was only carried out on the areas thought most at risk, but since the 2011 Tohoku-oki Earthquake, Aoi *et al.* state that (2020: 1, emphasis added):

> policies in Japan aimed at disaster prevention were modified drastically to perform uniformly spaced earthquake observations all over Japan on land, as well as earthquake and tsunami observations on seafloors, in acknowledgement that earthquakes could occur *anywhere in the country at any time.*

NIED has been designated overall responsibility for observing earthquakes, which is conducted through the High Sensitivity Seismograph Network Japan (Hi-net, 800 stations) and the Kyoshin Network (K-NET) and Kiban Kyoshin Nework (KiK-net) with 1700 stations combined, and Full Range Seismograph Network of Japan (F-net, 73 stations); the sensors used are velocity seismometers, tiltmeters, strong motion accelerometers, broadband seismometers, and pressure gauges (Aoi *et al.* 2020). Many of the instruments are placed in deep boreholes up to 4 km deep, to reach stable pre-Tertiary basement rocks (cf. Chapter 3). However, few are yet placed on the ocean floor around Japan, and this is the next step to completing the coverage (Zhao 2015). This confluence of these 4235 stations is called the 'big-net'; operational since 2007, it now detects 20 to 30 earthquakes a day and updates every second (Tajima & Hayashida 2018: fig. 3, p. 8). This network is augmented by industry-specific monitoring networks, such as those for the Shinkansen Bullet Train, semi-conductor factories, and chemical plants – producing a sum of >10,000 ground-motion stations (Wartman *et al.* 2013).

Earthquake warning systems

In its reporting of earthquake occurrences, the JMA (2020*a*) will post information on epicentre location, hypocentre depth, magnitude, and intensity within two minutes of occurrence.[2] Also displayed are all the earthquakes in the last 30 days according to magnitude and intensity.[3] The JMA does not distinguish between the varieties of earthquakes listed above, but those that pose a tsunami risk are flagged. Japan's 'big-net' Earthquake Early Warning (EEW) system cannot predict earthquakes, but once they start to happen warnings can be sent out to the public via various media: TV, radio, and texts. However, Tajima & Hayashida note there are proximal (grey) areas to the epicentre (10–30 km, 3 to 9 seconds after an earthquake) where ground-shaking can start before an alert is given (2018: fig. 1). Alerts are given on

[2] [www.jma.go.jp/bosai/map.html#8/33.973/132.849/&elem=int&contents=earthquake_map&lang=en]
[3] [www.jma.go.jp/bosai/map.html#3/44.126/131.473/&elem=hypo&contents=earthquake_map&lang=en]

the estimated level of ground-shaking intensity, 5-lower (cf. Table 2.2, Apx 7), not the magnitude of the earthquake – as there is little relationship between magnitude and casualties/damage. The latter depend on the level of preparedness of those receiving the alerts. Building codes in Japan have been revised in 1981 and 2000 after major earthquakes, while power and gas cuts can now be made automatically. Tajima & Hayashida document a shift in thinking about earthquake prediction (2018: 5, emphasis in the original):

> We used to think that if an earthquake were predicted, earthquake hazards could be mitigated. Then, we asked ourselves: how shall we respond to an earthquake prediction if it is announced minutes, hours or days in advance?... But we have achieved the stage where we can receive an alert from an EEW system seconds before a strong quake hits a site, and explore how to mitigate earthquake damage using the information.... We explore: what can we do effectively in seconds before strong ground shaking?

In conclusion they note, first, that the EEW system may not be able to issue alerts within the 'grey areas' fast enough for people to rely on them, and second, that EEW systems and their incremental improvement are independent of preparedness in terms of base isolation systems, seismically engineered buildings, private front-alarming systems, and clear public guidance. Lacking these preparations, alerts are hardly better than going without.

Earthquake & volcano interactions

Researchers have recently become aware of volcanic activity that has been stimulated by earthquakes. Ground-shaking can change the physical orientation of rocks and affect magma chamber dimensions, feed-in and outlet paths. The magma itself can undergo changes such as "dilatation that introduces gas bubble growth in magma, resulting in a buoyancy force that causes the magma to erupt, or the contraction that squeezes magma in the crust, causing it to rise to the ground surface" (Nishimura 2017: 7750). Nishimura discovered that such stimulation only occurred with earthquakes >M_w7.5 and only affected volcanoes within a 200 km radius of the epicentre; moreover, it operated on volcanoes that were already 'ready' for eruption. This is a wholly different phenomenon than volcanic earthquakes that occur spontaneously prior to eruption, as discussed below. It is now acknowledged that the three major eruptions of Mt Fuji followed known subduction earthquakes, the most infamous being the 1703 Genroku–Kanto earthquake followed by the 1707 Hoei eruption of Fuji (Mori & Ogawa 2021).

The 'seismic staircase'

Based on New Zealand evidence, a chain reaction of events begun by seismic activity called the "seismic staircase" has been identified. This involves the chronological progression of several of the processes detailed below, first and foremost being tsunami, and landslides and rock avalanches. From that point, "vegetation disturbance, rapid coastal dune building, river aggradation, and abandonment of prehistoric coastal settlements" take place in succession (Goff & McFadgen 2002: 2229). A graphic illustration of these processes in Goff & Knight *et al.* (2016) postulates immediate or temporary abandonment (as indicated at Kutsukata site in northeast Japan, discussed in Chapter 10; Saino 2012) or ensuing social disruption. These authors further explain how this succession can be anthropogenically disrupted by damming and channelling rivers, using the Sendai Plain as an example.

Tsunami hazards

Soda (2008) details relationships between earthquake and tsunami evidence at archaeological sites. Between 75% and 81% of tsunami are caused by earthquakes, so-called 'seismic tsunami' (cf. Table 10.1;

averaged below as 80% according to Wartman 2013). However, monitoring itself does not determine the cause, and the cause does not necessarily determine the severity of tsunami. Thus, tsunami hazards are dealt with here *en toto*.

Major tsunami occurrences

The last few hundred years have given us the most information on tsunami disasters, with palaeotsunami research gaining ground (see the review by Goff & Witter *et al.* 2020). The worst historical tsunami in Japan occurred in conjunction with the 1792 eruption of Mt Mayu in the Unzen Volcanic Complex of western Kyushu, Japan, as described above, resulting in ca. 15,000 fatalities. It provided a clear link between tsunami and volcanic eruptions. The 1960 Chile earthquake and tsunami, despite its distant origin, devastated Japan. This tsunami provided data on house destruction, which is partial when tsunami height is less than 2 m but significant when more than 4 m (Satake 2007: 492). The 2011 Tohoku-oki Tsunami accompanied the M9.0 Tohoku-oki Earthquake; it inundated about 500 km^2 (Jaffe *et al.* 2011) and took approximately 18,500 lives (NOAA 2015). Buildings which were constructed to modern earthquake standards generally survived the earthquake, but unfortunately most wooden structures were washed away in the following tsunami. The average height of the tsunami in the same area was 10 m but reached 33 m in places – comparable to the earthquake and Meiji Sanriku Tsunami in 1896 when waves reached 38 m in height and caused 27,000 deaths (USGS Earthquake Hazards Program 2013). In addition to actual damage to properties, running to trillions of yen, the world's increased vulnerability to what Warrick (1979) called "systemic disruption" was well demonstrated as the economic and nuclear effects of the 2011 double hit rippled outwards.

Although the Tohoku-oki Tsunami was at first thought to have been caused exclusively by the earthquake, evidence has been found of a submarine landslide caused by the earthquake that in turn caused the high run-up (Tappin *et al.* 2014; Tappin 2015). This phenomenon was first recognized for the 1998 Papua New Guinea Tsunami (Synolakis *et al.* 2002; Tappin *et al.* 2008). 'Submarine slumping' had also been suggested for the less destructive Meiji Sanriku Tsunami of 1986 (Satake 2007: 495). In light of these findings, recent publications have revisited historic tsunami, especially those whose dimensions cannot be reproduced with standard modelling, to propose that many were exacerbated by similar mass movements (e.g., Heidarzadeh *et al.* 2019; Pakoksung *et al.* 2019; Schambacha *et al.* 2020).

Tsunami databases

Goff & Terry (2012) report that several historical tsunami databases have been compiled for different countries including Indonesia, Greece, Italy, Australia, China, and even New Caledonia – with New Zealand's (PTDB 2017) perhaps being most relevant to Japan. Several of these databases have deep records, such as Japan's back to 684 AD and Chile's back to 1562 AD. They are also developing a database for the PICs (Pacific Island Countries, defined as those east of the subduction trenches of the western Pacific), which have been neglected in the past (see their table 1 for database architecture, and table 2 for their proxies). Meanwhile, two 'historical conventional tsunami databases' have been compiled, integrating records on a worldwide basis (Joseph 2011: 69):

1) The "Basic Pacific Tsunami Catalog and Database" was begun in 1997 as a joint project between the International Union of Geodesy and Geophysics Tsunami Commission (IUGG-TC) and the International Coordination Group for the Tsunami Warning System in the Pacific (ICG/ITSU). The database was created and maintained by the Novosibirsk Tsunami Laboratory (NTL) in Russia,[4] providing an offline, stand-alone

[4] The *Tsunami Laboratory* website is maintained by the Institute of Computational Mathematics and Mathematical Geophysics in Novosibirsk, Russia [http://tsun.sscc.ru/WinITDB.htm]. The National Geophysical Data Center

tool working on a Windows platform (WinITDB) which could be distributed on CD-ROM entitled the "Integrated Tsunami Database for the World Ocean".[5] The features on this database were last updated in 2005, but the content is now available online for the Atlantic, Mediterranean, and Pacific (the Pacific data only go back to 47 BC) via the International Tsunami Information Center (ITIC) website. In 2007, data on Japan in this database were assessed by the Japan Meteorological Agency (JMA), producing 2500 additional records and 1200 corrections (Igarashi 2007).

2) The "Global Historical Tsunami Database, 2100 BC to present" was compiled by the National Geophysical Data Center/World Data Service (NGDC/WDS) now housed in NOAA.[6] This is actually a combination of two databases for historical tsunami. The first covers more than 2400 tsunami source events, and the second provides approximately 20,000 tsunami run-up measurements for the Atlantic, Indian, and Pacific oceans and for the Mediterranean and Caribbean seas. These databases can be searched online or downloaded in full. This tsunami database compiled by the National Geophysical Data Center (NGDC)/NOAA, has been available since 2004; however, it is perennially incomplete and new additions are constantly being made, though there is some time lag (Goff & Terry 2012).

Another database only covering the Japan region is available from Tohoku University. The 'Japan Tsunami Trace Database' was compiled and is maintained by the Disaster Control Research Center (DCRC) in the Graduate School of Engineering of Tohoku University and the Japan Nuclear Energy Safety Organization (JNES).[7] Only 60 tsunami events from 1596 AD onward are included in this database. However, it also compiles data drawn from historic documents such as field surveys, research reports, and coastal policies that describe 25,664 'traces' (*konseki*) of tsunami. Traces are marks left on buildings, trees, rocks, and so on, that indicate run-up heights and inundation distances (Iwabuchi *et al.* 2012). The 3367 traces that are dated before 1887 are categorized as 'historic tsunami' traces, while the 22,297 traces dated after 1888 (for which seismic data are available) are called 'modern tsunami' traces (Iwabuchi *et al.* 2012). Of the modern tsunami traces, 15,079 date since the Chilean Tsunami of 1960.

Although some aspects of the database website, such as the section on tsunami deposits, are still in the build stage, data began to be released in 2010. This database has the unusual feature of 'reliability' assessments, carried out by a committee of experts (Satake 2007: 488). Different scales of reliability of the collected data in categories A–D (from reliable to increasingly uncertain), X (unreliable), and Z (needs more investigation) were established for pre- and post-1960 events, based on the tsunami from Chile of that year (Iwabuchi *et al.* 2012: I_1327, table 1 and I_1329, table 8). From 2007, the committee has conducted quarterly reviews of the reliability of both the documentary/field evidence for tsunami and the database system. Despite considerable documentation of tsunami in the last 400 years, very few mentions are reliable. Published measurements show a trend toward historic tsunami trace reliability in categories A–B at 14% and modern trace reliability in categories A–B at 40%. Such low reliability leaves the field of tsunami research wide open for geologists and archaeologists to further assess tsunami deposits.

(NGDC) is now part of NOAA (National Oceanic and Atmospheric Administration) National Centers for Environmental Information; the website can be accessed at [www.ngdc.noaa.gov/].

[5] "Integrated Tsunami Database for the World Ocean (WinITDB/WLD)" can be accessed from the *Tsunami Laboratory* website, [http://tsun.sscc.ru/WinITDB.htm].

[6] The National Geophysical Data Center (NGDC), now known as NCEI (National Centers for Environmental Information), the NOAA (National Oceanic and Atmospheric Administration) National Centers for Environmental Information. The database can be accessed at [DOI: 10.7289/V5PN93H7].

[7] The "Japan Tsunami Trace database" is at [https://tsunami-db.irides.tohoku.ac.jp/] (in Japanese and English), with the March 2016 *Tsunami Trace Database Operation Manual 7.2 Edition* available for download at URL: [tsunami-db.irides.tohoku.ac.jp/tsunami/manual_ippanEN.pdf] (in English).

Tsunami warning & prediction

The Chile Tsunami in 1960 stimulated research into historical tsunami and resulted in the establishment in 1965 of the Pacific Tsunami Warning System (PTWS) (UNESCO/IOC-NOAA 2021). In 2001, the Deep-Ocean Assessment and Reporting of Tsunamis (DART) system was established then upgraded and expanded in 2008 (NOAA 2020). The latter operates via a pressure recorder anchored to the sea bottom which transmits information on seafloor temperature and water pressure to a satellite via a surface buoy; the satellite then relays the information to Tsunami Warning Centers (TWC). DART began with 6 buoy stations and now consists of a network of 39 stations at critical points in the world's oceans; these have interactive communication facilities so the pressure recorders can be reset in anticipation of a tsunami. Real-time data can be accessed through NOAA's web page. DART is augmented by additional monitoring systems in several countries: the United States, Australia, Chile, India, and Japan (Lancaster 2020). In Japan, the JMA earthquake and tsunami warnings operate in tandem, while the JMA hosts the Northwest Pacific Tsunami Advisory Center (NWPTAC) which covers an area from Kamchatka through Japan and South Korea, the Philippines, and Papua New Guinea but not the South China Sea, serviced by the SCSTAC (JMA n.d.*b*).

Modelling future tsunami force levels and destructive potential has traditionally been based on information about historical earthquakes and tsunami (Furumura *et al.* 2011; Maeno & Imamura 2007; Maeno *et al.* 2006; Minoura *et al.* 2001; Satake 2007; Sato 2015; Shuto & Fujima 2009). Goto *et al.* (2014: 258), however, emphasize that such modelling should be "based on seismology, irrespective of past occurrence." Simulated data can be used to forecast the scale and nature of future events, as well as for estimating the strength of previous earthquakes and tsunami to account for past inundation and damage. Those areas of the world that are particularly at risk of tsunami appear in several maps curated by the Tsunami Information Center.[8] In particular, the map "Tsunami Sources 1610 B.C. to A.D. 2017 from Earthquakes, Volcanic Eruptions, Landslides, and Other Causes" (UNESCO *et al.* 2018) can alert archaeologists working in those areas as to what to possibly expect in their excavations. Their excavations can contribute to improving risk assessment and means of mitigating destruction from future tsunami as well as understanding past reactions and recovery from palaeotsunami.

Although these tsunami databases are very useful, particularly for archaeologists investigating past events, these historical data are not used in current modelling of future earthquakes and tsunami. It is now CDMC policy that the maximum possible strengths of each be calculated regardless of the attested historic levels or geological assessments of size/strength; a prediction for a Nankai Trough earthquake in SW Japan made in 2012 estimated 320,000 fatalities and ¥220 trillion damage – far beyond past occurrences (Goto *et al.* 2014). Goto and colleagues, however, are still in favour of Tsunami Geology investigating past events for discovering and evaluating both seismogenic and non-seismic tsunami; they have constructed a database on tsunami sedimentary deposits to complement the Japan Tsunami Trace Database at Tohoku University. Archaeologists would be able to contribute to this database through their site excavations.

Volcano & tsunami inter-relationships

In the section above, the relationship between volcanoes and tsunami was introduced, but there is a further facet of this relationship that is being developed by researchers. Coastal sedimentary deposits

[8] Several maps of historical tsunami maps and other data are available from the Tsunami Information Center [http://itic.ioc-unesco.org/index.php?option=com_content&view=article&id=1613&Itemid=2892].

are being examined for both volcanic materials and tsunami remains; Paris *et al.* (2014: 21) have identified five different coastal sedimentary facies for one volcanic eruption:

1) bioclastic sands deposited by tsunami
2) pumiceous tsunami sands
3) rounded pumice lapilli from pyroclastic fallouts reworked by tsunami
4) pumiceous ash fallout deposits (primary or reworked)
5) pyroclastic surge deposits overlying the uppermost tsunami deposits

Such information is feeding into understanding not only the tsunami itself but also the volcanic eruption with which it is linked. Thus, archaeological projects which recover tsunami sediments can equally be useful in exploring this link.

Landslides

The most feared

Despite Japan being located in a subduction zone with its many attendant hazards, Japanese people are allegedly most afraid of landslides. Not volcanic eruptions, not earthquakes, not tsunami, but landslides! Part of this is due to the long return time of tectonic activities, while landslides are comparably frequent. The other part is due to the country being 86% mountainous, including hilly flanks (cf. Table 5.1), where much alluvial flatland directly abuts acutely sloped ground. Serey *et al.* (2019: 1153) comment for other mountainous areas such as the Andes that "landslides represent perhaps the most frequent geological hazard present in mountainous environments due to the geological, geomorphological and geotechnical characteristics of steep upland landscapes". Because the mountains are often covered with unconsolidated sediments from recent subduction zone processes (i.e., tephra), the slopes may be unstable. Landslides of precarious sediments can be caused by earthquake tremors, including those associated with volcanic eruptions, as well as heavy rains. This precipitation causes a high rate of erosion of the mountain slopes – particularly those that are undergoing rapid uplift such as the Hida Range (cf. Figure 5.8a) – as well as the acidification of volcanic soils due to alkali and alkaline elements being leached away (see Chapter 7).

Average annual precipitation in Japan ranges from 944 to 4060 mm, but it usually exceeds 1020 mm. This rate puts Japan on a par with South China, the Congo, and Brazil but is not as high as Indonesia. Japan has two rainy seasons a year: a monsoon season in early summer and a typhoon season in late summer (although climate warming is extending these). In winter, humidity can drop as low as 50% on Japan's eastern seaboard, opposite 'snow country' on the west; but in summer, the humidity throughout the lower islands is often 98% even when it is not raining. Not only can heavy rain saturate sediments, making them easier to slip, snowpack melt can also lubricate fractures to cause tremors, as mentioned above. Thus, Japan is in the unenviable situation of being subject both to high summer precipitation and humidity rates throughout the islands as well as the winter dichotomy of heavy snow in the north-northwest and forest fires in the east.

Landslide generalities

There are several ways to classify landslides (Table 11.2). The BGS and USGS sources give simplistic explanations with illustrations of a variety of types (Highland & Johnson 2004; BGS 2021). Highland & Johnson name five geological causes and nine morphological causes for landslides as well as eight anthropogenic causes; among these are three that contribute to the most damage around the world: seismic, volcanic, and water-related causes. Earthquakes can cause landslides through simple shaking

Table 11.2 Landslide types as presented by the USGS

TYPE OF MOVEMENT		TYPE OF MATERIAL		
		BEDROCK	ENGINEERING SOILS	
			Predominantly coarse	Predominantly fine
FALLS		Rock fall	Debris fall	Earth fall
TOPPLES		Rock topple	Debris topple	Earth topple
SLIDES	ROTATIONAL	Rock slide	Debris slide	Earth slide
	TRANSLATIONAL			
LATERAL SPREADS		Rock spread	Debris spread	Earth spread
FLOWS		Rock flow	Debris flow	Earth flow
		(deep creep)	(soil creep)	
COMPLEX		Combination of two or more principal types of movement		

or through soil dilation that allows water to infiltrate; volcanic eruptions in conjunction with water, ice, or snow can cause debris flows and lahars; finally, landslides and flooding are highly related due to the slides potentially being caused by high precipitation rates and then blocking rivers, contributing to backwater and downstream flooding (Highland & Johnson 2004).

Keefer (1984) first identified 14 kinds of earthquake-induced landslides which ranged from hard rock falls to lateral spreading during liquefaction. It is such co-seismic landslides – those that occur simultaneously with earthquakes – that are of interest here. It must be remembered, however, that volcanic activity also involves earthquakes, one event of which led to one of the worst world-wide disasters in Japan in 1792 with the landslide of Unzen volcano that caused the tsunami in Ariake Bay, Kyushu (Miyamoto 2010). Keefer documents that weak shaking will cause 'internally disrupted landslides', referring to disintegration of internal structure in soil or rock – in contrast to 'coherent' slides (see also Wartman *et al.* 2013: table 2). Coherent slides require stronger shaking and lateral spreading, flows are caused by strong shaking, while rock and/or soil avalanches need the strongest shaking (Keefer 1984: 406). He cites data from the United States on what magnitudes cause which kinds of landslides (Ibid.: 409), with agreement that a minimum of M4 is needed to cause landslides, and M5 is needed for soil liquefaction leading to lateral spreads. In conclusion, Keefer lists 10 types of materials (rocks, soils) that are susceptible to certain types of landslides.

Among co-seismic landslides, Serey *et al.* further distinguish between those that are triggered by "shallow crustal earthquakes" (presumably on Active Faults) or by "megathrust earthquakes" on the subduction plane (2019: 1161). Both are exacerbated by steep slopes and unstable surface deposits. The range of damage is fairly proximal in shallow-crust earthquakes, but deep focal earthquakes on the subduction plane are capable of causing landslides at great distances between 100 and 500 km from the epicentre, as known in Central Asia (Havenith *et al.* 2016; see also Keefer 1984: fig. 2).

In the Andes, which are most comparable to Japan, co-seismic landslides occur in three types of deposits: Quaternary sediments, Palaeogene-Neogene volcanic and volcano-sedimentary (tephra) rocks (42% of landslides), and Cenozoic intrusive rocks (Serey *et al.* 2019: fig. 5). In the Andes, 52% of landslides occurred on slopes between 20° and 40°, as documented also in Kashmir and southwestern China, illustrating the risk in mountainous regions (Serey *et al.* 2019). Serey *et al.* (2020) model the different kinds of landslides caused by shallow-crust and subduction earthquakes that can occur in glacial valley or fluvial valley settings (Serey *et al.* 2020: figs. 3~6, table 2). Many of these slides reach the valley floor where there might be settlements, and they have the potential to dam river courses causing floods and to disrupt transport routes through the mountains. Lest one think such damage is trivial, consider the 2008 Wenchuan Earthquake in southwestern China which triggered up to 200,000 landslides and 257 landslide dams along the front slopes of the Sichuan Basin just northwest of Chengdu City (Havenith *et al.* 2016: fig. 1, table 4: 71). Three factors were named as the combination causing such damage: steep slopes, Active Faults, and a humid climate; furthermore, if the fault had a surface rupture of ca. 300 km and was located in the mountains, landslide numbers increased. And yet, landslides may themselves form new locales for settlement, as noted in the Kurile Islands (Fitzhugh 2012).

Landslides in Japan

Landslides are such a concern that The Japan Landslide Society (https://japan.landslide-soc.org/) was established in 1963; it now has member units located across Japan in vulnerable areas: Hokkaido, Tohoku, Niigata, Kanto, Chubu, Kansai, and Kyushu. Within their research remit are cliff collapses, lahars, and rock falls, etc. as well as sedimentary landslides. In 2002, an International Consortium on Landslides was formed in Kyoto, which now publishes the journal *Landslides* (Springer), and The Japan Landslide Society participates in their activities while publishing the *Landslide Society Journal*. Oguchi *et al.* (2001: 5) attribute hillslope failures and landslides to a combination of "steep watersheds and frequent heavy storms...these processes tend to be activated when rainfall intensity exceeds 50 mm/h or 100–200 mm/day" - conditions that recur about every ten years. They further note that vegetation does not anchor the soil but is swept away with the hillside collapse.

In assessing the 3477 landslides that occurred after the 2011 Tohoku-oki Earthquake in Japan, it was found that 80% of them took place in Quaternary soil and Neogene rock units (cf. Figure 5.3); and the most common types were "(1) disrupted landslides in Neogene sedimentary rocks and (2) lateral spreading in Quaternary sediments" (Wartman *et al.* 2013: 1519). However, steepness was not a factor in Tohoku as in the Andes: lateral spreading in low-relief plains contributed to most of the landslide erosion by large mass movements (Ibid.: 1517-18). Wartman *et al.* are meticulous in their descriptions of the evolution and topography of the Tohoku landscape (cf. Part I herein) in order to make judgments about landslide risk. Interestingly, the highest Peak Ground Acceleration in the earthquake was 0.80 g ±1σ (0.31–1.29 g) in the Ou backbone range of Tohoku west of the 2011 epicentre, comprised of tuff, volcanic breccia, sandstone, mudstone, and conglomerate; this acceleration was much higher than in the Kanto unconsolidated Quaternary sediments (cf. Figure 6.15) at 0.69g ±1σ (0.28–1.10g), albeit they were much farther away (Wartman *et al.* 2013: table 1). The authors also noted that since the 2011 earthquake took place in March in the dry season, the accompanying landslides "typically occurred in unsaturated materials, [and] flow-type failures in saturated masses were virtually nonexistent" (Wartman et al 2013: 1507). Thus, the season as well as the type of land surface affects the nature of the landslides.

The most spectacular series of landslides in Japan (Figure 11.3) occurred in southern Hokkaido in September 2018 after the passage of a tropical typhoon was followed by the M_w6.6 Iburi Earthquake causing JMA 7-level shaking (cf. Table 2.2) (Kameda *et al.* 2019; Zhang *et al.* 2019). An inventory of landslide locations resulted in 5625 identifications within 46.3 km^2, accounting for 36 deaths in a relatively unpopulated area; the earthquake itself caused 41 deaths, 691 injured, 394 houses destroyed and 1061 damaged (Zhang *et al.* 2019). The landslides occurred between 7 and 30 km north of the epicentre, mainly in surface tephra (Normile 2018; Ito *et al.* 2020: fig. 3). One cause of these landslides

Figure 11.3 Multiple landslides in southern Hokkaido following the Iburi Earthquake in September 2018; landslide scarps are brown among the green mountains, fields, and golf courses

was a distinctive example of liquefaction; Kameda and colleagues (2019) report how halloysite clay derived from the underlying pumice became liquefied in the ground-shaking, causing a fluidized landslide; such landslides are viewed as exceptionally dangerous because of their rapid deployment and long reach.

Halloysite is a weathering product of tephra (cf. Chapter 7) and "has been identified as one of the principal clay minerals on the failure surfaces of volcanic landslides in Japan, Indonesia, and New Zealand" (Kameda *et al.* 2019: unpg.). However, the role of volcanic soils cannot be taken for granted. Two years earlier in 2016, after the passage of four tropical typhoons and an earthquake, it was found that fossil periglacial soils were the weakness that allowed multiple landslides in the Tokachi Range (Furuichi *et al.* 2018). Landslides on the coast of Lake Biwa, mobilized by liquefaction during four different earthquakes since the 14c, have carried parts of three villages into the lake (Anon. 2012).

In a radical new understanding, it is now recognized that, in addition to dams being vulnerable to landslides, the water in the dammed reservoir itself can incite earthquakes and soil instability that lead to landslides (Masano 2010). This is due to both saturation of the lower reaches of the slopes and pressure that the water exerts on the subsurface. Settlements on the surrounding slopes are vulnerable to the landslides, and if the dam collapses as well, the 'tsunami' that floods downstream is always devastating. Japanese construction agencies are now involved in 'landslide protection' engineering, but if the composition of the surrounding mountain slopes is landslide prone – such as volcanic tuff which absorbs water – then such mitigations are useless. Masano reports that the Landslide Prevention Law defines a landslide as "land that slides because of underground water or another reason"; the emphasis on groundwater here is significant, but as primary factors contributing to landslides, she refers to geological fragility regarding "geological formations that slide, bedrock that disintegrates easily, and other unstable geological features" (2010: 7 of 11). As an appeasement, the Toné River Drainage Erosion Control Office has established the Yuzurihara Landslide Museum on the Kanna River in Fujioka City, Gunma.

One overwhelming conclusion about co-seismic landslides is that they do not occur singly or in isolation. The number of landslides per earthquake can range from tens to thousands and can encompass regions up to 500 km from the epicenter. This implies that the landscape is constantly changing through time via mechanisms faster than simple erosion. The presence of landslide soil, plants, and structures may remain in the sedimentary record, especially in narrow valleys. Archaeological research has only begun to identify landslips on tombs (cf. Figure 9.1 and Tanaka *et al.* 2020), while landslide formations in site stratigraphies have yet to be generally recognized (but see Gerrard *et al.* 2021). This means that we are perhaps missing an important factor that has influenced settlement patterns and communication routes in the past. In addition to potentially causing death and destruction, they can radically change the landscape for which new land uses and travel routes need to be established – similar to conditions caused by tephra fallout (Oetelaar 2019).

Earthquakes >M6 can trigger slope failures and debris flows (Oguchi *et al.* 2001: 9). While co-seismic landslides are easily identified, there is reason to believe that earthquakes can cause slope instability that can be activated up to 200 years later, often by heavy rainfall (Goff & Sugawara 2014). Moreover Goff & Sugawara make a case for the formation of beach ridges by sediment load, based on data from the Sendai Plain (cf. Figure 10.5), that can take decades to centuries after landslides occur for those sediments to make their way down to the coast. These authors have made a convincing case for beach ridge formation on the Sendai Plain after the particularly large 869 Jogan and 1611 Keicho Earthquakes (2014: 145).

Typhoons

Severe tropical storms are called different things in different regions; the following usages are approved in the BBC and the Met Office (King 2019): typhoons in the western Pacific, cyclones in the South Pacific and Indian Ocean, and hurricanes in the North Atlantic and NE Pacific. Such severe tropical storms are far stronger than normal storms and wreak unheard-of destruction, as evidenced by Hurricane Katrina in 2005. Without itemizing the toll on human life and the economy, the primary damage by these storms comes from ocean surges, high waves, strong winds, and intense rainfall; secondarily, they cause flooding, landslides, erosion of coasts and beaches, debris flows, mud flows, and wind-blown debris and falling objects, as well as sheer impact of waves and debris (Heidarzadeh 2021: fig. 14).

In Japan, the JMA is responsible for monitoring storms in the West North Pacific; there, typhoons are "defined by central pressure falling below or equal to 970 hPa" (Feng & Terry 2012: 188), and as having winds greater than 119 km/hr, with three classes of increasing speeds: typhoon, very strong typhoon, and violent typhoon (>194 km/hr) (Heidarzadeh 2021: table 1). Storms that do not meet these criteria are classed as tropical storms. Most typhoons affecting Japan begin in the area around the Philippines; tracking their paths and wind speeds has revealed two types: those following straight paths and those that follow sinuous paths – the latter being far more dangerous because their track cannot be predicted (Feng & Terry 2012). An interesting finding of Feng & Terry has been that rainfall is not closely linked to typhoon ferocity. The OCHA ROAP hazard map (2007) uses the Saffir-Simpson scale of storm intensity to identify areas of Japan vulnerable to different storm intensities:

- Level 5 at >250 km/hr: Okinawa and the southern Izu Islands (Iwojima)
- Level 4 at 210–249 km/hr: southern Kyushu and northern Ryukyus, and northern Izu Islands
- Level 3 at 178–209 km/hr: central and northern Kyushu, south and southwestern Shikoku, and the Izu Peninsula – in addition to inland swaths across western Honshu and into the Kinki region
- Level 2 at 154–177 km/hr: Kanto and central Tohoku and most of the rest of western Honshu
- Level 1 at 118–153 km/hr: Hokkaido and Tohoku coasts and a swath across western Honshu

These are very strange patterns, with tongues of differential intensities sometimes following coasts and currents but also crossing land in unexpected places. Despite the 'low level' of intensity in Level 1, this is still typhoon-like wind, wreaking much more damage than lesser tropical storms.

Work has begun on documenting typhoon watermarks, inundation distances and run-up heights, river flooding, and debris scatters (e.g., Heidarzadeh *et al.* 2021), which is essential in defining the differences between tsunami and storm evidence. For comparison, data collected by Heidarzadeh and colleagues for the Typhoon Lionrock in 2016, which unusually made landfall at the north edge of Sendai Bay in Tohoku, gave run-up heights between 1.7 and 4.3 m, and inundation distances between 16 and 56 m – all on the Sanriku Coast affected by the 2011 Tohoku Tsunami; considerable mud was deposited from river flooding. It is also notable that their study documented that the highest surges occurred at pressures of 980 hPa despite lower pressures that followed <970 hPa. It is 'storms' such as these that must be distinguished from evidence of past tsunami, as discussed in Chapter 10, since they consist not only of wind and rain but of considerable wave surges as well. One factor that may be useful for this distinction is that the typhoon season in the West North Pacific lasts from May to November (Feng & Terry 2012), while tsunami can occur at any time of the year; seasonal biological materials therefore may assist in the differentiation.

Heidarzadeh and colleagues note that the major damage from successive typhoons often varies from inland flooding, wind-blown debris, to mudslides; Etienne (2012: 27) adds to this list: "landslides on steep weathered slopes, tree fall, vegetation burned by sea spray, coral reef erosion and debris production, sandy island migration, beach erosion at shoreline, washover accumulations inland". This variety makes it difficult to plan for mitigation compared to inundation and debris flows of tsunami. Karlsrud (2006) estimates that if typhoons occur simultaneously with tsunami, inundation levels would be higher; however, the odds are low for this co-incidence, and since beaches would be depopulated during typhoons, he suggests that fatalities might be lower than otherwise.

Tectonics & Disaster Archaeology

It is apparent from the above sections that tectonogenic hazards are not only interwoven with each other but also interrelated with non-tectonic causes of disasters. In essence, this means that although archaeologists may specialize or mainly deal in one kind of tectonic hazard or another, they have to be aware of all the other possible events and impacts associated with the main cause. Thus, in an archaeological site, one might find evidence of an earthquake, volcanic and tsunami deposits, and landslides all in one place. These have to be recognized and sorted – to understand not only their different effects but also their potential time differences. And it is significant that they are all evidenced by different types of 'dirt' of the kind traditionally scraped away by archaeologists in order to get at the artefacts (cf. Chapter 1). The combination of different hazards, their evidence and effects, thus leads to the archaeology of disasters in general.

Becoming Disaster Archaeology

Since its beginning as a named discipline near the turn of the millennium, growing out of the earlier pursued 'hazard research' (e.g., Sheets 1980; Van Buren 2001; see Sheets & Cooper 2012 for a history), Disaster Archaeology has encompassed investigations into multitudes of disasters in the past as well as in the present day. Research into disasters, however, is not exclusively the preserve of archaeology – more than 30 disciplines have been cited as involved. Liritzis (2019: 1310) emphasizes that "from a geological perspective, these events are not catastrophes but merely the regular history of the earth and the geodynamics of its evolution"; however, Sheets & Cooper (2012) propose that 'natural disasters' are rarely fully 'natural'. Whereas humans have previously been seen only as 'victims', the disastrous death tolls and damage levels that follow these natural processes are now recognized as being conditioned by human involvement. Five levels of involvement are apparent: simple ignorance, wilful ignorance, inadequate preparedness, traditional solutions, and lately, the exertion of technological control.

Disaster Archaeology has five strands:

1) The more traditional strand has focussed on the impact of natural disasters on society and human life at the time of occurrence. Scores of case studies document the who, what, when, and where of past disasters, and they tend to be framed in cultural historical studies that attempt to capture what life was like in the past (in the timeslice preserved by tephra, for example), or they try to assess impact on and recovery of the social system at the time (in terms of human behaviour patterns of abandonment, recolonization, etc.). However, investigations into the 'why' have evolved from scientific explanations of the natural processes involved to exploration of how the structure of social systems contribute to the 'why' of the disaster. Egan & Sheets (2018: 1) note that mainly since the millennium has it been recognized that "social processes are fundamental components of disasters": they condition the responses to disaster and therefore vary according to the structure of society. Impacts cannot be fully understood minus articulation with socio-political-economic organization, belief and value systems.

2) Secondly, early in this millennium, forensic archaeology as developed by the late Richard Gould (2004, 2007, 2011) brought archaeological methods to information recovery on human-caused disasters beginning with the 2001 attacks on the World Trade Center in New York. Expanding on previous work concentrating on human remains and/or crime scenes (e.g., Hunter & Cox 2005), Gould extended his interests to all material culture found at disaster sites, and his textbook is now used in Disaster Archaeology courses (e.g., Beisaw 2016).

3) Thirdly, such studies that previously were confined to past events are now being utilized in service to modern society: what can previous disasters tell us about human adaptation capabilities, strategies, and resilience – and how can these inform present-day policies of risk reduction and disaster mitigation? The latter was brought to fore by the Eagle Hill meeting in 2009 of participants from "a dozen disciplines and nations worldwide", which sought to "to make a well-understood past serve to create a more genuinely sustainable future and increase human resilience in the face of both gradual and sudden change" (McGovern 2012: vii, viii). This third strand combines research on the past with current policy-making. Dilek *et al.* (2021: 17 of 22) note that there was a shift in thinking from 2000 "from a focus on disaster management [reactive] to a focus on disaster-risk management [preventative]" (interpretive brackets inserted), and that it was necessary to move away from centralized disaster-management

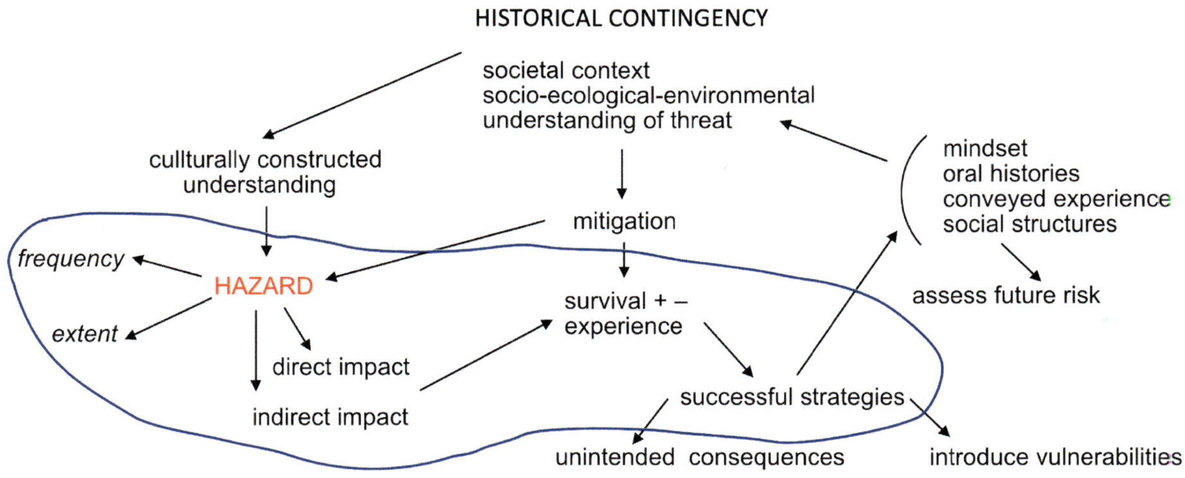

Figure 11.4 Scheme showing interrelationships between society and nature; enclosed area is the traditional arena of Disaster Archaeology, with more recent concerns added outside it

systems to empowering communities to ensure their own preparedness. The papers in Cooper & Sheets (2012) led the way in modelling the relationships between past and present (Figure 11.4).

Perhaps one of the most succinct statements in that volume was by Ben Fitzhugh (2012: 42), concluding from his studies of egalitarian societies that even in highly developed societies:

> ...hazard planning has to include capacity building for decentralized response systems. Families, households, and local communities need to have the ability to respond creatively, with decision decentralization supported by higher governmental institutions so responses can scale with capacity. This also requires systems for rapid and decentralized information sharing.

4) Scale is another important concern in studying disasters. Wholesale abandonment of sites and human demise are often treated under 'collapse' or 'catastrophe' theory (Diamond 2011; Keys 1999), but societies often meet and rise above periodic damage and can even thrive in disaster-prone areas. Eric Force (2015) hypothesizes that earthquakes occurring in current or fossil subduction zones (collision zones) have encouraged societies located within the vicinity to reinvent themselves over and over through time, while societies located in seismically inactive regions tended to be culturally stable for much longer periods. He cites this as one reason the great Mediterranean and hypo-Himalayan societies became foci of civilization as they were located on the tectonic borders of the African/Eurasian/Indian plates. These borders, defined by large extensive fault zones, provided trade routes and fertile valleys for settlement, while the repetition of devastating earthquakes forced recovery and innovation. Such positive long-term effects must be set against immediate destruction, whether they may be creation of new land by pyroclastic flows, additions of new soil through landslides and tsunami deposits, or rerouting of rivers to new areas for colonization, etc.

5) Archaeologists conducting research on past disasters or in areas of 'chronic disasters' are finding it increasingly important to collaborate with local peoples, not just administrative entities (e.g., Ryzewski & Cherry 2012). Archaeological finds may or may not be closely entwined with local identities,

but either way negotiations entail educational programmes, outreach, inclusive participation and more. Ryzewski & Cherry draw a distinction between the sites and artefacts that constitute cultural heritage to an archaeologist and intangible heritage such as food, festivals, and music that are more meaningful to local inhabitants – who often want to forget a colonial past represented by archaeological finds. However, in Japan, ostensibly a country of 'chronic disasters', the past is revered, and efforts are made to keep previous disasters in 'living memory' by engaging local populations.

Disaster Archaeology in Japan

The cross-over from archaeological knowledge to present preparedness stimulated the reviews above of the monitoring and warning systems in place in Japan for individual tectonogenic hazards. However, there is a movement in Japan to combine the three interrelated named subdisciplines of Tectonic Archaeology as described above into a more encompassing Disaster Archaeology (*saigai kōkogaku*) that goes beyond tectonic causes to include all sorts of natural disasters (Okamura 2015; Okamura *et al.* 2013; Saino 2012*b*). A Disaster Archaeology database compiling information from published site reports has been developed at the Nara Research Institute for Cultural Properties (Nara Bunkazai Kenkyūjo, abbreviated as Nabunken) (Okamura 2015: 251). In Japan as elsewhere, disaster archaeologists hope to prevent worse losses in the future by understanding past disasters. Current research there tends to focus on past and present human reactions in order to predict and prepare for future disasters (Okamura *et al.* 2013). Traditional solutions to many types of disasters are also being collated to highlight how simple but low-cost efforts are effective in accepting disasters rather than trying to conquer them (Okubo 2016).

The efforts at database construction by Nabunken close the circle on Shimoyama's attempts in the late 1990s and early 2000s to promote the field of Disaster Archaeology and to present on the archaeology of natural disasters at World Archaeology Congresses (WAC4 in 1999, WAC5 in 2003) (Shimoyama 1997, 2002a, 2002b; Torrence & Grattan 2002). Shimoyama (2002a) set out parameters for defining and assessing the nature and extent of disasters – whether caused by humans or natural processes – and assessing human reactions to them. He defined a disaster as an event that directly or indirectly has human victims, while avoiding the point that when plant and animal communities are subject to disasters, they can have long-term, possibly delayed, effects on human communities. Although he admitted that human reactions to similar events can be astoundingly variable, Shimoyama emphasized that "it is valuable and important to establish a number of elements that are common to all cases and to propose a standard framework for their analysis" (2002a: 26). He specified six topics that need addressing: initiation, immediate causes, local conditions, damages, assessment, and actions; these were echoed in Sheets & Cooper's specification of hazards, past impacts, and mitigation, that need attention – though they extended their concern to future risks (2012: 6).

Meanwhile, Schlanger *et al.* (2016) distinguish 'disaster-led archaeology' from 'developer-led archaeology'. Although both these approaches fall into the realm of 'rescue' archaeology (more recently termed 'preventive' archaeology, Schlanger 2012), for Japan, disaster-led archaeology references the dramatic increase in archaeological excavations necessitated by recovery and re-building efforts following major disasters. Despite the Cultural Properties Law in Japan requiring archaeological investigation of any ground-disturbing construction, public or private (see Barnes 1990), the regulations were eased after major disasters. Re-building after the 1995 Kobe Earthquake required intensive work over three years, resulting in "as many archaeological discoveries...as would normally be made in 20 years" (Okamura *et al.* 2013: 263). Moreover, after the 2011 Tohoku-oki Earthquake and Tsunami, 'Rescuing archaeology' (Kaner *et al.* 2011) came to have a double meaning that brought a new dimension to Disaster Archaeology. The reconstitution of knowledge held in disarrayed and damaged exhibits and storage facilities (Figure 11.5) was an enormous task and poignant in the sense that it was a job twice done: original collection and cataloguing, and then after the disaster, sorting, re-conserving, and cataloguing.

The 2011 Tohoku-oki earthquake and tsunami

As did the 1995 Kobe Earthquake, the event of 2011 in northern Japan, formally known as the Great East Japan Earthquake, radically transformed the scope of Disaster Archaeology. A look at the intensity map (cf. Figure 2.6) reveals that almost all of the northeastern coast endured shaking intensity of 6 or above (cf. Apx 7). The tsunami that followed was incredibly destructive, as everyone now knows, with 20,000 dead or missing. Ten years have passed since that event, but the losses are still raw for many people in northeastern Japan. It vividly illustrated just how vulnerable Japan is to repeated natural disasters.

Most affected were flat alluvial coastal plains, now housing most of Japan's current population, despite tsunami 'stone markers' set on high ground to warn about previous tragedies. One reads "Remember the massive tsunami that caused disaster.... Do not build homes on sites lower than where you stand.... Bear this in mind even after many years have passed" (Anon. 2011). Though such disasters happen perhaps only once in a lifetime, any loss of life is tragic; and this time was perhaps the first time that such a localized event had such wide-ranging ramifications in complex economic systems.

In addition to the devastation of several towns and deaths of their inhabitants in the tsunami, many standing cultural properties such as castles, shrines and temples, as well as museums, archives, archaeological research labs, and artefact storage facilities bore great damage, from the earthquake tremors or from the tsunami water or both, with objects and records saturated as well as being dispersed, buried in sand and mud or debris, and often crushed. The amount of knowledge residing in these cultural properties was precious. As with the formation of DCAN after the Kobe Earthquake, archaeologists, historians, and "specialists in cultural informatics" formed a new organization called

Figure 11.5 Inside the disturbed archaeological storage unit at Nobiru, Higashi-Matsushima City, after the 2011 Tohoku-oki Tsunami

CEDACH (Consortium for Earthquake-Damaged Cultural Heritage) for the "documentation and restoration of endangered cultural heritage" (Okamura *et al.* 2013: 265). Government archaeological units across Japan also responded immediately by seconding members to travel to Tohoku and assist with cleaning up, preserving, and reorganizing the materials that had been affected (Figure 11.5). CEDACH's remit, however, goes beyond this immediate response, with the founders hoping to develop a system of disaster recovery that can be shared throughout the world (Kondo *et al.* 2011; Kondo *et al.* 2013; Fujimoto *et al.* 2011).

The tragedy is that earthquake and tsunami damage along the northeastern coast was already well known among archaeologists and historians working in the area. However, as lamented by Atsushi Fujisawa (in Okamura *et al.* 2013), these research results were not translated into policy. Nor were the historical stele warnings ('do not build your houses below this mark') along the coast heeded. Earthquake archaeologists are adamant that knowledge of prior disasters must be integrated into public planning policy, and the new sub-discipline of Tsunami Archaeology in Japan is active in documenting prior occurrences (Chapter 10).

Finally, there has developed a strong opinion that housing and businesses should be moved away from the Tohoku coast to the uplands. The hills supported large Jomon populations (cf. Apx 2: Table A), and sites are densely distributed in the region. To build new communities in the uplands, the areas needed to be archaeologically excavated according to current laws. As with the Kobe 1995 Earthquake, the tsunami indirectly has caused a great resurgence of archaeological activity in uplands eastern Tohoku, potentially leading to a much greater understanding of Jomon livelihoods in the region. Concomitantly, archaeological excavations on the alluvial flatlands are yielding more information about the Yayoi inhabitants who farmed rice on the coastal plains until devastation by tsunami (cf. Chapter 10). In this way, Japanese archaeologists have not only come to be responsible for rescuing and preserving cultural materials in cases of major disasters but also have the opportunity for more intensive fieldwork associated with reconstruction.

Organizational outcomes

Positive outcomes of a disaster, such as the increased excavation and knowledge gained after earthquakes/tsunami, are also noted for the Cumberland River flooding in 2010 in Tennessee, which provided an unwelcome opportunity to survey and collect data from over 100 eroded sites along the riverbank (Peres & Deter-Wolf 2018). The necessary collaborative efforts, such as needed at short notice in both Japan and Tennessee, led to the establishment in Japan of two organizations: the Disaster Concerned Archaeologists' Network (DCAN) after the 1995 Kobe Earthquake as described in Chapter 9, and the Consortium for Earthquake-Damaged Cultural Heritage (CEDACH) after the 2011 Tohoku-oki Earthquake. However, as testimony to the difficulty of keeping a past disaster forefront in everyone's mind, Kondo *et al.* noted in 2013 that even two years after the 2011 earthquake, the need for remote support supplied by CEDACH was diminishing.[9] Instead of reactive collaboration on the spot, Peres & Deter-Wolf's suggestions for pre-disaster planning are even more imperative: "build the necessary relationships now, have a methodology and strike force ready now, follow best practices for mitigation and data collection, and share information with the public – they want and need to know what is happening to our collective history".

[9] And indeed, internet searches for their website and activities in 2021 were fruitless, although the GIS system and e-Learning facilities they established may still be in use.

Archaeologists bear the burden of preserving the memory of natural disasters for future generations. As stated by Fujisawa (in Okamura *et al.* 2013: 262), "It seems that it was only archaeologists who could have made it known to the general public that tsunamis of the scale of the 2011 disaster have inundated the Sendai plain in the past". In future, they can do this by contributing their knowledge of past events to local governments and particularly museums, playing a role in community empowerment by clarifying when and how badly the area had been involved in previous disasters. Museum exhibits, whether temporary or permanent, and whole museums that memorialize specific outstanding disasters in local

BOX 12 1995 Kobe Earthquake Disaster Monuments

Kobe Port liquefaction monument walkway with explanation boards. The lateral spreading is seen in the canal with lightposts leaning in opposite directions.

Fallen and overturned chimney at Moegi House, Kobe, preserved *in situ* "to remind future generations of the power of nature".

Catfish sculpture: a visual reminder that the roiling fish are traditionally thought to cause earthquakes.

Beware of earthquakes!

history are increasingly present. These complement heritage monuments which are preserved as relics of the disaster or constructed in memory of those disasters (BOX 12).

Disaster museums and disaster monuments such as the Peace Dome in Hiroshima, or the Tsunami Boats in Banda Aceh, Sumatra, or the Hikkaduwa railcar conserved at the Community Tsunami Museum in Peraliya, Sri Lanka (Goff & Dudley 2021), or the earthquake damage in Kobe (BOX 10, 12) comprise an alternative narrative of cultural heritage (Rico 2014). They do not pre-date the disaster as does most designated cultural heritage; moreover, Rico makes the point that they do not exist in order to restore the past nor do they represent anything but local concerns and memories, defying current concepts of 'at-risk heritage' which involve Western-based assessments of value and conservation. Moreover, keeping the disaster visually present is one way to keep residents alert to the possibilities of future occurrences, and many of the disaster museums have research sections that work to prevent similar disasters in future.

Prospective

It is clear that the tectonic archaeologies are integrally related to each other and overlap with many other lines of enquiry into the past. It is equally clear that the past is increasingly looked to for strategies and solutions that may be applicable to the future. Efforts to make archaeology 'relevant' to modern-day concerns now go far beyond the need to justify the time and investment devoted to it, as there is a real world need for understanding long-term processes of which people experience only a small slice in time. With great transformations happening to the natural world via our agency, human societies are already dealing with change on an unprecedented timescale.

Tectonic Archaeology promotes greater knowledge of deep Earth systems and their ability to modify the Earth's surface and even the atmosphere and hydrosphere. Currently focusing on the identification and recording of disasters in the archaeological record, it needs also to move on to a clear understanding of how damages were conditioned by social systems, how human reactions were either creative or constrained by those systems, and how the latter reconstituted themselves. The tension between tenacious occupation of hazardous land and the willingness and/or ability to relocate has repercussions far beyond Tectonic Archaeology, but some of the best case-studies may well be from the past.

Into the future, slogans adopted for climate change studies, such as 'Informing the Future by Understanding the Past' and 'Searching the Past for Clues to the Future' (Cooper 2012), are equally applicable to the investigation of human resilience and adaptation to disasters. Nevertheless, Kintigh *et al.* (2014: 18-19) emphasize that "The challenge is to move from case or regional studies to larger-scale comparative research, and to learn how to make generalizable statements about how people make choices that draw on universal biases in cognition (and in fact to study the evolution of those biases)." Torrence & Doelman (2007: 43) characterize the focus on individual events and losses thereof as the 'insurance perspective'; they recommend instead taking a long-term view of "material culture, social exchange, conceptions of place, and patterns of land use" to ascertain coping strategies.

As archaeologists in Japan lamented, it was only they who could have warned of impending tsunami based on knowledge of the scale of previous tsunami deposits and destruction in the region. How to feed archaeological knowledge into public policy is perhaps one of the most perplexing issues that we have today, since no points are given in university systems for work beyond the required publication of research results.

Figure & Table Sources

Figure 11.1 Hazard risk redrawn from Munich Reinsurance Company, Geo Risks Research Department (OCHA ROAP 2007)

Figure 11.2 By 富士山防災協議会 (the Mt. Fuji Volcano Disaster Management Conference); Miya.m [sic] added the URL and the information to the Map and uploaded under the permission of the Cabinet Office of the Japanese Government; [https://commons.wikimedia.org/wiki/ File%3APredicative_map_of_Mt.Fuji_volcanic-ash-fall.jpg] CC-BY-2.5, via Wikimedia Commons

Figure 11.3 NASA Earth Observatory images by Lauren Dauphin, using Landsat 8 – OLI data from the U.S. Geological Survey, as published in (Voiland 2018), public domain; NASA photograph on 15 September 2018 downloaded from [https://earthobservatory.nasa.gov/images/92832/landslides-in-hokkaido]

Figure 11.4 abstracted from text in (Sheets & Cooper 2012; McGovern 2012; Redman 2012)

Figure 11.5 courtesy of FUJISAWA Atsuji; originally published in (Okamura *et al.* 2013)

BOX 12 photos by author and David W. Hughes

Table 11.1 compiled from (USGS n.d.; JMA n.d.*a*)

Table 11.2 excerpted from (Highland & Johnson 2004: fig. 2), USGS public domain

References

AIZAWA, Koji (**2006**) "Present situation of monitoring, reduction and information for volcanic disaster mitigation in Japan." The 6th Joint Meeting of the UJNR Panel on Earthquake Research 8–11 November 2006, Tokushima (PPT in English) [http://cais.gsi.go.jp/UJNR/6th/orally/O08_ Present.pdf] (dead link).

Anon. (**2011**) "So much for the myth of safety." *Nikkei Weekly* Special Magazine Issue Summer: 5.

Anon. (**2012**) "The causes of three villages at the bottom of Lake Biwa were landslides." *Isabōnet* [https://isabou.net/Convenience/aviso/news_20120223.asp] (in Japanese).

Anon. (**2021**) "Fujisan lava extrusion earlier: eruption map revised after 17 years" [www.jiji.com/jc/article?k=2021032600795&g=pol&p=20210326ax02S&rel=pv].

AOI, Shin; Youichi ASANO & Takashi KUNUGI *et al.* (**2020**) "MOWLAS: NIED observation network for earthquake, tsunami and volcano." *Earth, Planets and Space* 72.126: 31pp. [DOI: 10.1186/s40623-020-01250-x].

Balmuth, Miriam S; David K Chester, & Patricia A Johnston, (**2005**) *Cultural responses to the volcanic landscape.* Boston, MA: Archaeological Institute of America.

BARNES, Gina L (**1990**) "The origin of bureaucratic archaeology in Japan." *Journal of the Hong Kong Archaeological Society* 12: 183-196 [www.academia.edu/4206136/The_origins_of_bureaucratic_archaeology_in_Japan].

BARNES, Gina L & Tsutomu SODA (eds) (**2019**) *TephroArchaeology in the North Pacific.* Oxford: Archaeopress.

BEISAW, April M (**2016**) "ANTH 331: Disaster Archaeology" [www.academia.edu/2178788/Archaeology_of_Disaster].

BGS (British Geological Survey) (**2021**) "How to classify a landslide." *Discovering geology – landslides* [www.bgs.ac.uk/discovering-geology/earth-hazards/landslides/how-to-classify-a-landslide/].

BULLARD, Fred M (**1979**) "Volcanoes and their activity", pp. 9-48 in *Volcanic activity and human ecology*, ed. by PD SHEETS & DK GRAYSON. New York: Academic Press.

Cabinet Office (**2015**) *Disaster management in Japan.* Tokyo: Government of Japan (in Japanese and English).

CIT Tectonics Observatory (**n.d.**) "What happened during the 2004 Sumatra earthquake?" [http://www.tectonics.caltech.edu/outreach/highlights/sumatra/what.html].

CLANCEY, Gregory (**2006**) *Earthquake nation: the cultural politics of Japanese seismicity, 1868-1930.* Berkeley, CA: University of California Press.

COOPER, Jago (**2012**) "Weathering climate change. The value of social memory and ecological knowledge." *Anthropological Dialogues* 19.1: 46-51 [DOI: 10.1017/S1380203812000062].

COOPER, Jago & Payson D SHEETS (eds) (**2012**) *Surviving sudden environmental change, answers from archaeology.* Boulder, CO: University Press of Colorado.

CULSHAW, Martin (**2010**) "Geological hazards: how safe is Britain?" Shell Lecture, 8 Dec 2010, at the Geological Society of London.

Diamond, Jared M (**2011**) *Collapse: how societies choose to fail or succeed*. New York: Viking.
Dilek, Yildirim; Yujiro Ogawa & Yasukuni Okubo (**2021**) "Characterization of modern and historical seismic – tsunamic events and their global–societal impacts", pp. 1-22 in *Characterization of modern and historical seismic-tsunamic events, and their global-societal impacts*, ed. by Y Dilek *et al.* GSL Special Publications 501. London: Geological Society of London [DOI: 10.1144/SP501-2021-17].
DPD (Disaster Prevention Division, Bureau of General Affairs, Tokyo Metropolitan Government) (**2020**) *Tokyo Metropolitan Government disaster prevention guide book*. Tokyo Metropolitan Government [https://www.bousai.metro.tokyo.lg.jp].
Egan, Rachel & Payson Sheets (**2018**) "Disaster archaeology", 4 pp. in *The encyclopedia of archaeological sciences*, ed. by SL López Varela [DOI: 10.1002/9781119188230.saseas0192].
Etienne, S (**2012**) "Marine inundations hazards in French Polynesia: geomorphic impacts of tropical cyclone Oli in February 2010", pp. 21-40 in Terry & Goff (eds) (2012).
Fei, Jie & Jie Zhou (**2016**) "The drought and locust plague of 942e944 AD in the Yellow River Basin, China." *Quaternary International* 394: 115-122 [DOI: 10.1016/j.quaint.2014.11.053].
Feng, Chen-Chieh & James P Terry (**2012**) "Exploratory spatial analysis of typhoon characteristics in the North Pacific Basin", pp. 187-194 in Terry & Goff (eds) (2012).
Fitzhugh, Ben (**2012**) "Hazards, impacts, and resilience among hunter-gatherers of the Kuril Islands", pp. 19-42 in Cooper & Sheets (eds) (2012*a*).
Force, Eric R (**2015**) *Impact of tectonic activity on ancient civilizations: recurrent shakeups, tenacity, resilience, and change*. Lanham, MD: Lexington Books.
Fujii, Toshitsugu & Hitoshi Yamasato (**2015**) "Integrated monitoring of Japanese volcanoes", pp. 445-459 in *Volcanic hazards, risks and disasters*, ed. by JF Shroder & P Papale. Amsterdam: Elsevier.
Fujimoto, Yu; Akihiro Kaneda & Yasuhisa Kondo *et al.* (**2011**) "Organizing diverse and dispersed information on the endangered cultural properties by a voluntary initiative: consortium for the earthquake-damaged cultural heritage (CEDACH)." Originally presented as a poster session at the 39th Annual Conference on Computer Applications and Quantitative Methods in Archaeology (CAA), Beijing, April 12–16, 2011 [www.researchgate.net/publication/267824574].
Fukushima, Daisuke & Kazuhiro Ishihara (**2005**) "Volcanic disaster prevention and community development: how to convert the volcano into a museum." *Annuals of Disaster Prevention Research Institute, Kyoto University* 48C: unpg. [www.dpri.kyoto-u.ac.jp/dat/nenpo/no48/48c0 /a48c0p17.pdf] (dead link).
Furuichi, Takahisa; Nobutomo Osanai & Shin-ichiro Hayashi *et al.* (**2018**) "Disastrous sediment discharge due to typhoon-induced heavy rainfall over fossil periglacial catchments in western Tokachi, Hokkaido, northern Japan." *Landslides* 15: 1645-1655 [DOI: 10.1007/s10346-018-1005-1].
Furumura, T; K Imai & T Maeda (**2011**) "A revised tsunami source model for the 1707 Hoei earthquake and simulation of tsunami inundation of Ryujin Lake, Kyushu, Japan." *Journal of Geophysical Research* 116: B02308.
Gerrard, Christopher; Paolo Forlin & Melanie Froude *et al.* (**2021**) "The archaeology of a landslide: unravelling the Azores earthquake disaster of 1522 and its consequences." *European Journal of Archaeology* First View, 24pp. [DOI: 10.1017/eaa.2021.4].
Goff, James & Walter Dudley (2021) *Tsunami: the world's greatest waves*. Oxford: Oxford University Press.
Goff, James; Jasper Knight & Daisuke Sugawara *et al.* (**2016**) "Anthropogenic disruption to the seismic driving of beach ridge formation: the Sendai coast, Japan." *Science of the Total Environment* 544: 18-23 [DOI: 10.1016/j.scitotenv.2015.11.106].
Goff, JR & BG McFadgen (**2002**) "Seismic driving of nationwide changes in geomorphology and prehistoric settlement—a 15th Century New Zealand example." *Quaternary Science Reviews* 21: 2229-2236.
Goff, James & Daisuke Sugawara (**2014**) "Seismic-driving of sand beach ridge formation in northern Honshu, Japan?" *Marine Geology* 358: 138-149.
Goff, J & JP Terry (**2012**) "Living with natural hazards in the Asia–Pacific region", pp. 1-2 in Terry & Goff (eds) (2012).
Goff, J; R Witter & J Terry *et al.* (**2020**) "Palaeotsunamis in the Sino-Pacific region." *Earth-Science Reviews* 210: #103352 [DOI: 10.1016/j.earscirev.2020.103352].

GOTO, Kazuhisa; Shigehiro FUJINO & Daisuke SUGAWARA *et al.* (**2014**) "The current situation of tsunami geology under new policies for disaster countermeasures in Japan." *Episodes* 37.4: 258-264 [DOI: 10.18814/epiiugs/2014/v37i4/005].

GOULD, Richard (**2004**) "Disaster archaeology at the West Warwick, Rhode Island nightclub fire scene." *The SAA Archaeological Record* 4.1: 6-11.

---- (**2007**) *Disaster archaeology*. Salt Lake City, UT: University of Utah Press.

---- (**2011**) "From living archaeology to disaster archaeology." *Ethnoarchaeology* 3.2: 187-202.

GRATTAN, John & Robin TORRENCE (eds) (**2007**) *Living under the shadow: cultural impacts of volcanic eruptions*. Walnut Creek, CA: Left Coast.

HAVENITH, Hans-Balder; Almaz TORGOEV & Anika BRAUN *et al.* (**2016**) "A new classification of earthquake-induced landslide event sizes based on seismotectonic, topographic, climatic and geologic factors." *Geoenvironmental Disasters* 3(#6): 24pp. [DOI: 10.1186/s40677-016-0041-1].

HEIDARZADEH, Mohammad; David R TAPPIN & Takeo ISHIBE (**2019**) "Modeling the large runup along a narrow segment of the Kaikoura coast, New Zealand following the November 2016 tsunami from a potential landslide." *Ocean Engineering* 175: 113-121 [DOI: 10.1016/j.oceaneng.2019.02.024].

HEKI, Kosuke (**2003**) "Snow load and seasonal variation of earthquake occurrence in Japan." *Earth and Planetary Science Letters* 207: 159-164.

HIGHLAND, Lynn & Margo JOHNSON (**2004**) "Landslide types and processes." USGS science for a changing world. Fact Sheet FS2004-3072 [DOI: 10.3133/fs20043072].

HUNTER, J & M COX (**2005**) *Forensic archaeology. Advances in theory and practice*. London: Routledge.

IGARASHI, Yosuke (**2007**) "Abstract: Verification of Japanese tsunami data in the Integrated Tsunami Database (ITDB)." *Tsunami Commission Workshop 2007* [https://silo.tips/download/verification-of-japanese-tsunami-data-in-the-integrated-tsunami-database-itdb].

ITO, Yoshihiko; Shusaku YAMAZAKI & Toshiyuki KURAHASHI (**2020**) "Geological features of landslides caused by the 2018 Hokkaido Eastern Iburi Earthquake in Japan", pp. 171-183 in Y DILEK *et al.* GSL Special Publications 501. London: Geological Society of London [DOI: 10.1144/SP501-2019-122].

IWABUCHI, Yoko; Hideharu SUGINO & Fumihiko IMAMURA *et al.* (**2012**) "Development of tsunami trace database with reliability evaluation on Japan coasts." *Doboku Gakkai Ronbunshū* B2 (Kaigan Kōgaku) 68.2: 工 1326-工 1330 (in Japanese with English title and abstract).

JAFFE, Bruce; Bruce RICHMOND & Helen GIBBONS (**2011**) "International team studies tsunami deposits in Japan to improve understanding and mitigation of tsunami hazards." *Sound Waves* June.

JMA (Japan Meteorological Association) (**n.d.*a***) "NWPTAC of Japan Meteorological Agency: Northwest Pacific Tsunami Advisory Center" [www.data.jma.go.jp/svd/eqev/data/nwptac/index.html].

---- (**n.d.*b***) "Volcanic warnings and volcanic alert levels" [www.data.jma.go.jp/svd/vois/data/tokyo/STOCK/kaisetsu/English/level.html].

---- (**2020*a***) "Weather, climate & earthquake information: earthquakes and volcanoes." [www.jma.go.jp/jma/en/menu.html].

---- (**2020*b***) "Real-time risk map." [www.jma.go.jp/bosai/en_risk/#lat:34.705493/lon:130.253906/ zoom:4/colordepth:normal/elements:land].

JOSEPH, Antony (**2011**) *Tsunamis: detection, monitoring, and early-warning technologies*. Amsterdam: Elsevier/Academic Press.

KAGAN, Yan Y & David D JACKSON (**2013**) "Tohoku earthquake: a surprise?" *Bulletin of the Seismological Society of America* 103.2B: 1181-1194 [DOI: 10.1785/0120120110].

Kamakura-shi (**2022**) "Kamakura city hazard map" [www2.wagmap.jp/kamakura-hazard/Portal].

KAMEDA, Jun; Hiro KAMIYA & Hirokazu MASUMOTO *et al.* (**2019**) *Nature Scientific Reports* 99(#13119): 7pp. [DOI: 10.1038/s41598-019-48820-y].

KANER, Simon; Junko HABU & Akira MATSUI (**2011**) "Rescuing archaeology affected by the Japanese earthquake and tsunami." *Antiquity* 85.335: 258-269.

KARLSRUD, Kjell (**2006**) *Tsunami risk mitigation measures with focus on land use and rehabilitation*. Oslo: Norwegian Geotechnical Institute.

KAWABATA, Yasunari (**1957**) *Snow Country* (translated by Edward G SEIDENSTICKER). Tokyo: Tuttle.

KEEFER, David K (**1984**) "Landslides caused by earthquakes." *Geological Society of America Bulletin* 95: 406-421.

KEYS, David (**1999**) *Catastrophe: an investigation into the origins of modern civilization*. London: Century Books.

KING, Simon (**2019**) "Hurricanes, typhoons & cyclones" [www.bbc.co.uk/newsround/24879162].

KINTIGH, Keith W; Jeffrey H ATSCHUL & Mary C BEAUDRY *et al.* (**2014**) "Grand challenges for archaeology." *American Antiquity* 79.1: 5-24.

KOBAYASHI, Masaru (**2019**) "Akita's three lakes: the legend of Hachirō Tarō and the dragon", pp. 112 in Barnes & Soda (eds) (2019).

KONDO, Y; T UOZU & Y SEINO *et al.* (**2013**) "Voluntary activities and online education for digital heritage inventory development after the Great East Japan Earthquake", pp. 391-396 in *International Archives of the Photogrammetry, Remote Sensing and Spatial Information Sciences*, XL-5/W2.XXIV International CIPA Symposium, 2 – 6 September 2013, Strasbourg, France.

KONDO, Yasuhisa; Akihiro KANEDA & Yu FUJIMOTO *et al.* (**2011**) "The CEDACH DMT: a volunteer-based data management team for the documentation of the earthquake-damaged cultural heritage in Japan." CAA2011 – Revive the Past: Proceedings of the 39th Conference in Computer Applications and Quantitative Methods in Archaeology, Beijing, China, 12-16 April 2011.

KOSHIMURA, Shunichi & Nobuo SHUTO (**2015**) "Response to the 2011 Great East Japan Earthquake and Tsunami disaster." *Philosophical Transactions of the Royal Society of London A: Mathematical, Physical and Engineering Sciences* 373(#2053) [DOI: 10.1098/rsta.2014.0373].

Kyodo (**2021**) "A new map shows Mount Fuji eruption could affect larger areas." *Japan Times* 26 March 2021 [www.japantimes.co.jp/news/2021/03/26/national/mount-fuji-eruption-map/].

LANCASTER, George (**2020**) "These 5 countries have advanced Tsunami warning systems." *Preventionweb* [https://www.undrr.org/news/these-5-countries-have-advanced-tsunami-warning-systems].

LIN, Aiming; Tadashi MARUYAMA & Takao MIYATA (**1998**) "Paleoseismic events and the 1596 Keicho-Fushimi large earthquake produced by a slip on the Gosukebashi fault at the eastern Rokko Mountains, Japan." *Island Arc* 7.4: 621-636 [DOI: 10.1046/j.1440-1738.1998.00214.x].

LIRITZIS, Ioannis; Alexander WESTRA & Changhong MIAO (**2019**) "Disaster geoarchaeology and natural cataclysms in world cultural evolution: an overview." *Journal of Coastal Research* 35.6: 1307-1330 [DOI: 10.2112/JCOASTRES-D-19-00035.1].

MAENO, F & F IMAMURA (**2007**) "Numerical investigations of tsunamis generated by pyroclastic flows from the Kikai caldera, Japan." *Geophysical Research Letters* 34: L23303.

MAENO, Fukashi; Fumihiko IMAMURA & Hiromitsu TANIGUCHI (**2006**) "Numerical simulation of tsunamis generated by caldera collapse during the 7.3 ka Kikai eruption, Kyushu, Japan." *Earth Planets Space* 58: 1013-1024.

MASANO, Atsuko (**2010**) "The immense cost of Japanese dams and dam-related landslides and earthquakes." *The Asia-Pacific Journal* 8.1, no. 2(#3280): 11pp. (in English, translated from Japanese).

MCGOVERN, Thomas H (**2012**) "Foreword", pp. vii-xii in COOPER & SHEETS (eds) (2012).

MEI, Estuning Tyas Wulan & Franck LAVIGNE (**2012**) "Influence of the institutional and socio-economic context for responding to disasters: case study of the 1994 and 2006 eruptions of the Merapi Volcano, Indonesia", pp. 171-186 in TERRY & GOFF (eds) (2012).

MINOURA, K; F IMAMURA & D SUGAWARA *et al.* (**2001**) "The 869 Jogan tsunami deposit and recurrence interval of large-scale tsunami on the Pacific coast of northeast Japan." *Journal of Natural Disaster Science* 23: 83-88.

MIYAMOTO, Kuniaki (**2010**) "Numerical simulation of landslide movement and Unzen-Mayuyama disaster in 1792, Japan." *Journal of Disaster Research* 5.3: 280-287 [DOI: 10.20965/jdr.2010.p0280].

MLITTJ (**n.d.**) (Ministry of Land, Infrastructure, Transport and Tourism of Japan) "Hazard Map Portal Site – investigate hazard risks around you" [https://disaportal.gsi.go.jp/index.html] (in Japanese).

MORI, S & Y OGAWA (**2010**) "Geohazards in coastal areas near the northernmost Sagami Trough, central Japan: review of neotectonic activity in onshore and offshore areas of the Izu island arc collision–subduction zone", pp. 87-110 in *Characterization of modern and historical seismic-tsunamic events and their global-societal impacts*, ed. by Y DILEK *et al.* GSL Special Publications No. 501. London: The Geological Society of London.

MURAKAMI, Yoshinao (**2019**) "Katakai-Ienoshita site, Akita Prefecture, buried by the lahar accompanying the 10th-century eruption of Mt Towada", pp. 97-113 in Barnes & Soda (eds) (2019).

NAGATA, Hidehisa (**2018**) "Studies on nontectonic geologic structure in Japan: recent progress and future perspective." *Chishitsu Zasshi* 124.11: 899-912 [DOI: 10.5575/geosoc.2018.0040].

NAKAMURA, Yoichi; Kazuyoshi FUKUSHIMA & Xinghai JIN *et al.* (**2008**) "Mitigation systems by hazard maps, mitigation plans, and risk analyses regarding volcanic disasters in Japan." *Journal of Disaster Research* 3.4: 297-304.

NAKATA, Setsuya (**2005**) "The forefront of volcanic disaster prevention software technology", pp. 130-137 in *Kazan funka ni sonaete*, ed. by Civil Engineering Association. Tokyo: Doboku Gakkai.

NCEI (**n.d.**) (National Centers for Environmental Information) "World-wide tsunamis: 2000 B.C. – to the present." NOAA>NESDIS>NCEI (formerly NGDC)>Natural Hazards [https://www.ngdc.noaa. gov/hazard/tsunami-db-intro.html].

NIED (**2016**) (National Research Institute for Earth Science and Disaster) and Commission on Mitigation of Volcanic Disasters Prevention, the Volcanological Society of Japan "Database on volcanic hazard maps and reference material (1983–)" [https://vivaweb2.bosai.go.jp/v-hazard/index-e.html].

NISHIMURA, Takeshi (**2017**) "Triggering of volcanic eruptions by large earthquakes." *Geophysical Research Letters* 44: 7750-7756 [DOI: 10.1002/2017GL074579].

NOAA (**2020**) "Deep-ocean assessment and reporting of tsunamis (DART®) description." National Data Buoy Center [www.ndbc.noaa.gov/dart/dart.shtml].

NORMILE, Dennis (**2018**) "Slippery volcanic soils blamed for deadly landslides during Hokkaido earthquake." *Science* [www.sciencemag.org/news/2018/09/slippery-volcanic-soils-blamed-deadly-landslides-during-hokkaido-earthquake].

NUNN, Patrick D (**1994**) *Oceanic islands.* Oxford: Blackwell.

OCHA ROAP (United Nations Office for the Coordination of Humanitarian Affairs – Regional Office for Asia Pacific) (**2006**) "Holocene eruptions and selected volcanoes in Asia Pacific." Map Ref: OCHA_ROAP_Volcanic_Eruptions_v1_060803 [https://www.preventionweb.net/publication/ holocene-eruptions-and-selected-volcanoes-asia-pacific].

---- (**2007**) "Japan: natural hazard risks." Map ref: OCHA_JPN_HAZARD_V1_070309 [https://reliefweb.int/map/japan/japan-natural-hazard-risks-09-mar-2007].

OETELAAR, Gerald (**2019**) "Volcanic ash and landscape evolution: reconstruction of a 7000-year old landscape on the northwestern Great Plains of North America", pp. 47-75 in Barnes & Soda (eds) (2019).

OGUCHI, Takashi; Kyoji SAITO & Hiroshi KADOMURA *et al.* (**2001**) "Fluvial geomorphology and paleohydrology in Japan." *Geomorphology* 39: 3-19.

OKAMURA, Katsuyuki (**2015**) "Tsunami and heritage after the 2011 Great East Japan Earthquake", pp. 245-256 in *Water and heritage: material, conceptual and spiritual connections*, ed. by WJH WILLEMS & HPJ VAN SCHAIK. Leiden: Sidestone Press.

OKAMURA, Katsuyuki; Atsushi FUJISAWA & Yasuhisa KONDO *et al.* (**2013**) "The Great East Japan Earthquake and cultural heritage: towards an archaeology of disaster." *Antiquity* 87(335): 258-269.

OKAY, Nilgun (**2019**) "Disaster Archaeology." Abstract for TURQUA Quaternary Symposium, Istanbul May 2018 [www.researchgate.net/publication/324991456_DISASTER_ARCHAEOLOGY/link/5af05c02a6fdcc8508b9becc/download].

OKUBO, T (**2016**) "Traditional wisdom for disaster mitigation in history of Japanese architectures and historic cities." *Journal of Cultural Heritage* 20: 715-724.

OSU (Oregon State University) (**2021**) "Laki, Iceland – 1783." *Volcano World* [https://volcano.oregonstate.edu/laki-iceland-1783].

PAKOKSUNG, Kwanchai; Anawat SUPPASRI & Fumihiko IMAMURA *et al.* (**2019**) "Simulation of the submarine landslide tsunami on 28 September 2018 in Palu Bay, Sulawesi Island, Indonesia, using a two-layer model." *Pure and Applied Geophysics* 176: 3323-3350 [DOI: 10.1007/s00024-019-02235-y].

PARIS, Raphaël; Patrick WASSMER & Franck LAVIGNE *et al.* (**2014**) "Coupling eruption and tsunami records: the Krakatau 1883 case study, Indonesia." *Bulletin of Volcanology* 76: 23pp. [DOI: 10.1007/s00445-014-0814-x].

PERES, Tanya M & Aaron DETER-WOLF (**2018**) "Finding the positive in the negative: archaeology and data collection in the face of natural disasters." *Advances in Archaeology Practice* 4: 288-297.

PLATH, David W (**1966**) "Japan and the ethics of fatalism." *Anthropological Quarterly (Special Issue on Fatalism in Asia: old myths and new realities)* 39.3: 161-170.

PLATT, John P; Haoran XIA & William Lamborn SCHMIDT (**2018**) "Rheology and stress in subduction zones around the aseismic/seismic transition." *Progress in Earth and Planetary Science* 5(#24): 12 pp. [DOI: 10.1186/s40645-018-0183-8].

PNSN (**n.d.**) "Volcanic earthquakes." *Pacific Northwest Seismic Network* [https://pnsn.org/outreach/earthquakesources/volcanic].

PTDB (Palaeotsunami Database) (**2017**) "The New Zealand palaeotsunami database." [https://ptdb. niwa.co.nz].

REDMAN, Charles L (**2012**) "Global environmental change, resilience, and sustainable outcomes", pp. 237-244 in COOPER & SHEETS (eds) (2012).

RICO, Trinidad (**2014**) "The limits of a 'heritage at risk' framework: the construction of post-disaster cultural heritage in Banda Aceh, Indonesia." *Journal of Social Archaeology* 14.2: 157-176.

RIEDE, Felix (**2016**) "Volcanic activity and human society." *Quaternary International* 394: 1-5 [DOI: 10.1016/j.quaint.2015.08.090].

RITCHIE, Hannah (**2018**) "What were the world's deadliest earthquakes?" *Our World in Data* [https://ourworldindata.org/the-worlds-deadliest-earthquakes].

ROMAN, Diana C & Katherine V CASHMAN (**2006**) "The origin of volcano-tectonic earthquake swarms." *Geology* 34.6: 457-460 [DOI: 10.1130/G22269.1].

RYZEWSKI, Krysta & John F CHERRY (**2012**) "Communities and archeology under the Soufrière Hills Volcano on Montserrat, West Indies." *Journal of Field Archaeology* 37.4: 316-327.

SAINO, Hirohiko (**2012**) "Evidence of agricultural disasters on the Sendai Plains: from Yayoi to Pre-modern periods", pp. 186-215 in *Collected writings on Tohoku history*. Tokyo: Seibundo Shinkosha (in Japanese).

SATAKE, Kenji (**2007**) "Tsunamis", pp. 484-510 in *Treatise on geophysics, v. 4: Earthquake seismology*, ed. by H KANAMORI. Amsterdam: Elsevier.

SATO, Shinji (**2015**) "Characteristics of the 2011 Tohoku tsunami and introduction of two level tsunamis for tsunami disaster mitigation." *Proceedings of the Japan Academy Series B* 91: 262-272.

SCHAMBACHA, L; ST GRILLIA & DR TAPPIN *et al.* (**2020**) "New simulations and understanding of the 1908 Messina tsunami for a dual seismic and deep submarine mass failure source." *Marine Geology* 421(#106093): 20pp. [DOI: 10.1016/j.margeo.2019.106093].

SCHLANGER, Nathan (**2012**) "Preventive archaeology", pp. 661-665 in *The Oxford companion to archaeology* (2nd edition), ed. by NA SILBERMAN. Oxford University Press.

SCHLANGER, Nathan; Laurent NESPOULOUS & Jean-Paul DEMOULE (**2016**) "Year 5 at Fukushima: a 'disaster-led' archaeology of the contemporary future." *Antiquity* 90(350): 409-424.

SEREY, Alejandra; Laura PIÑERO-FELICIANGELI & Sergio A SEPÚLVEDA *et al.* (**2019**) "Landslides induced by the 2010 Chile megathrust earthquake: a comprehensive inventory and correlations with geological and seismic factors." *Landslides* 16: 1153-1165.

SEREY, Alejandra; Sergio A SEPÚLVEDA & William MURPHY *et al.* (**2020**) "Developing conceptual models for the recognition of coseismic landslides hazard for shallow crustal and megathrust earthquakes in different mountain environments – an example from the Chilean Andes." *Engineering Geology and Hydrology* 54.2(2021): 19pp. [DOI: 10.1144/qjegh2020-023].

SHEETS, Payson (**1980**) *Archaeological studies of disaster: their range and value*. Boulder, CO: Institute of Behavioral Science, University of Colorado.

SHEETS, Payson & Jago COOPER (**2012**) "Learning to live with the dangers of sudden environmental change", pp. 1-18 in COOPER & SHEETS (eds) (2012).

SHIKI, Tsunemasa; Yoshinobu TSUJI & Koji MINOURA *et al.* (eds) (**2008**) *Tsunamiites: features and implications*. Amsterdam: Elsevier.

SHIMOYAMA, Satoru (**1997**) "On the range of the disaster archaeology—for taking measure of effects from disasters." *Hominids* 1 (March): 83-103 [in Japanese with English title and summary].

---- (**2002*a***) "Volcanic disasters and archaeological sites in southern Kyushu, Japan", pp. 326-341 in Torrence & Grattan (eds) (2002).

---- (**2002*b***) "Basic characteristics of disasters", pp. 19-27 in Torrence & Grattan (eds) (2002).

SHOZAWA, Shin'ichirō (**2014**) "Japan's volcanic research in distress: Ontake eruption exposes the dire state of volcano surveillance" [www.nippon.com/en/currents/d00146/].

SHUTO, Nobuo & Koji FUJIMA (**2009**) "A short history of tsunami research and countermeasures in Japan." *Proceedings of the Japan Academy Series B* 85.8: 267-275.

SIMKIN, Tom & Richard S FISKE (**1983**) "Krakatau 1883: a classic geophysical event." *Eos* 64.34: 3pp.

SIMKIN, Tom; Lee SIEBERT & Russell BLONG (**2001**) "Volcano fatalities – lessons from the historical record." *Science* 291 (5502): 255.

Smithsonian Institution (**n.d.**) "Global Volcanism Program: Aira > Photo Gallery." [https://volcano.si.edu/volcano.cfm?vn=282080].

SODA, Tsutomu (**2008**) "Evidence of earthquakes and tsunamis uncovered at archeological sites in Japan." *Kōkogaku Jaanaru* 577: 21-28 (in Japanese with English title).

SYNOLAKIS, Costas E; Jean-Pierre BARDET & José C BORRERO *et al.* (**2002**) "The slump origin of the 1998 Papua New Guinea Tsunami." *Proceedings of The Royal Society A* [DOI: 10.1098/ rspa.2001.0915].

TAJIMA, Fumiko & Takumi HAYASHIDA (**2018**) "Earthquake early warning: what does 'seconds before a strong hit' mean?" *Progress in Earth and Planetary Science* 5: 63 [DOI: 10.1186/s40645-018-0221-6].

TAMURA, Y (**2002**) "Hot fingers in the mantle wedge: new insights into magma genesis in subduction zones." *Earth and Planetary Science Letters* 197: 105-106.

TANAKA, Hiroyuki KM; Kenji SUMIYA & László OLÁH (**2020**) "Muography as a new tool to study the historic earthquakes recorded in ancient burial mounds." *Geoscientific Instrumentation Methods and Data Systems*: 19 pp. [DOI: 10.5194/gi-2020-13].

TAPPIN, Dave (**2015**) "Tsunamis – learning from recent events." *Proceedings of the Open University Geological Society* 2015.1: 57-61.

TAPPIN, David R; Stephan T GRILLI & Jeffrey C HARRIS *et al.* (**2014**) "Did a submarine landslide contribute to the 2011 Tohoku tsunami?" *Marine Geology* 357: 344-361.

TERRY, JP & J GOFF (eds) (**2012**) *Natural hazards in the Asia-Pacific region: recent advances and emerging concepts.* GSL Special Publications 361. London: The Geological Society of London.

THORDARSON, Th & S SELF (**2003**) "Atmospheric and environmental effects of the 1783–1784 Laki eruption: a review and reassessment." *Journal of Geophysical Research* 1089(D1).4011 [DOI: 10.1029/2001JD002042, 2003].

THORDARSON, Th; S SELF & N ÓSKARSSON *et al.* (**1996**) "Sulfur, chlorine, and fluorine degassing and atmospheric loading by the 1783–1784 AD Laki (Skaftár Fires) eruption in Iceland." *Bulletin of Volcanology* 58: 205-225.

TODA, Shinji; Ross S STEIN & Stephen H KIRBY *et al.* (**2008**) "A slab fragment wedged under Tokyo and its tectonic and seismic implications." *Nature Geoscience* 2008.1: 771-776.

Tokyo VAAC (**2021**) "Tokyo Volcanic Ash Advisory Center, Japan Meteorological Agency" [https://ds.data.jma.go.jp/svd/vaac/data/index.html].

TORRENCE, Robin & Trudy DOELMAN (**2007**) "Chaos and selection in catastrophic environments: Willaumez Peninsula, Papua New Guinea", pp. 42-65 in Grattan & Torrence (eds) (2007).

TORRENCE, Robin & John GRATTAN (eds) (**2002**) *Natural disasters and cultural change.* London: Routledge.

TOYOTA, Hirofumi & Susumu TAKADA (**2015**) "Geotechnical distinction of landslides induced by near-field earthquakes in Niigata, Japan." *Geography Journal* 2015: #359047, 11pp.

UEDA, T & A KATO (**2019**) "Seasonal variations in crustal seismicity in San-in district, southwest Japan." *Geophysical Research Letters* 46: 3172-2179.

UKAWA, Motoo; Eisuke FUJITA & Eiji YAMAMOTO *et al.* (**2000**) "The 2000 Miyakejima eruption: crustal deformation and earthquakes observed." *Earth Planets Space* 52.8: 19-26.

UNESCO *et al.* (**2018**) *Tsunami sources 1610 B.C. to A.D. 2017 from earthquakes, volcanic eruptions, landslides, and other causes.* Rome: UNESCO/IOC-ITIC-NOAA-ICSU-WDS.

UNESCO/IOC-NOAA (**2021**) "ITIC (International Tsunami Information Center): Pacific" [http://itic.ioc-unesco.org/index.php?option=com_content&view=category&layout=blog&id=1153&Itemid=1153].

USGS (**n.d.**) "Alert level icons." *Volcano Alerts* [www.usgs.gov/volcano/alert-level-icons].

USGS Earthquake Hazards Program (**2013**) "Magnitude 9.0 – near the east coast of Honshu, Japan." *Earthquake Hazards Program* [sti2013_arehart_usgsearthquakehazardsprogram-tohokueq summary.pdf].

VAN BUREN, Mary (**2001**) "The archaeology of El Nìno events and other 'natural' disasters." *Journal of Archaeological Method and Theory* 8.2: 129-149.

VERVAECK, Armand & James DANIELL (**2011**) "Japan tsunami follow up: total death toll drops below 23,000." [http://earthquake-report.com/2011/06/20/japan-tsunami-following-up-the-aftermath-part-16-june/] (dead link).

VOILAND, Adam (**2018**) "Landslides in Hokkaido." NASA Earth Observatory [https://earthobservatory.nasa.gov/images/92832/landslides-in-hokkaido].

WARRICK, Richard A (**1979**) "Volcanoes as hazard: an overview", pp. 161-194 in *Volcanic activity and human ecology*, ed. by PD SHEETS & DK GRAYSON. New York: Academic Press.
WARTMAN, J; L DUNHAM & B TWARI *et al.* (**2013**) "Landslides in eastern Honshu induced by the 2011 off the Pacific Coast of Tohoku earthquake." *Bulletin of the Seismological Society of America* 103.2B: 1503-1521.
WHITE, Lloyd T; Nicholas RAWLINSON & Gordon S LISTER *et al.* (**2019**) "Earth's deepest earthquake swarms track fluid ascent beneath nascent arc volcanoes." *Earth & Planetary Science Letters* 521: 25-36 [DOI: 10.1016/ j.epsl.2019.05.048].
WOLMAN, David (**2006**) "Mount Fuji overdue for eruption, experts warn." *National Geographic News* 17 (July): unpg.
YAMAMOTO, Takahiro & Setsuya NAKADA (**2015**) "Extreme volcanic risks 2: Mount Fuji", pp. 355-376 in *Volcanic hazards, risks and disasters*, ed. by JF SHRODER & P PAPALE. Amsterdam: Elsevier.
YAMASATO, Hitoshi (**2005**) "Classification of prior volcanoes and levels of volcanic activity", pp. 138-145 in *Kazan funka ni sonaete*, ed. by Civil Engineering Association. Tokyo: Doboku Gakkai (in Japanese).
ZHANG, Shuai; Ran LI & Fawu WANG *et al.* (**2019**) "Characteristics of landslides triggered by the 2018 Hokkaido Eastern Iburi earthquake, northern Japan." *Landslides* 12 June: 18pp. [DOI: 10.1007/ s10346-019-01207-6].
ZHAO, Dapeng (**2015**) "The 2011 Tohoku earthquake (Mw 9.0) sequence and subduction dynamics in Western Pacific and East Asia." *Journal of Asian Earth Sciences* 98: 28-49.

Chapter 12

True Jades, False Friends

This chapter represents a foray into a very different aspect of Tectonic Archaeology. Previous descriptions dealt with sanukite and obsidian sources produced during volcanic eruptions in the Japanese Islands (BOX 4 and Chapter 6). These stones were mainly used by prehistoric peoples for utilitarian objects. Here, we will look at the more ornamental stones that are referred to under the catch-all term 'jade'. For millennia, jades have been valued in many cultures in both China and Japan. The favoured types and sources of jades have changed over time, as has our knowledge of the stones themselves. One of the greatest problems in dealing with archaeological jades is the correct identification of the stones in order to trace their source and thereby understand the social relations underlying their patterns of procurement, production, and consumption. In the last 15 years, the field has been upset by increasingly detailed chemical analyses of so-called jadeites. This impacts the gemological world more than geological, but what ramifications does it hold for archaeology?

Plate Tectonics is crucial for understanding the formation and distribution of the true jades: nephrite and jadeitite. Their tectonic contexts determine the minerals available for their composition. True jade mineralogy will be detailed below in relation to their host rocks, dolomitic limestone (dolostone) and serpentenite. Metamorphic processes (Chapter 2.5) are integral to the formation of both host rocks and jade products, while the processes of exhumation and obduction (Chapter 2.5.3, 2.7) determine where the true jades are found in the world today.

Introduction

This chapter is organized hierarchically from wider concerns to specific mineralologies involving solid solution chemistry and mineral crystallization. We begin with the various rocks and their constitutent minerals and elements that make up both true and false jades. From there we move to a Plate Tectonics discussion of the locations of true jade formation and their host rocks, detailing their metamorphic transformations (particularly by **metasomatism**). Important in the latter is explaining how the host rocks and their contents end up on land, allowing the nephrite and jadeitite rocks to be mined for archaeological tools and ornaments and for modern sculptures and jewelry. The second half of the chapter is devoted to true jade mineralogy to the extent needed to understand the products that occur in the archaeological record and, yes, in the modern marketplace.

The category jade/yu

The first problem encountered in discussing 'jade' is defining what we mean by this term; there is a chequered development from the Latin *lapis nephriticus* leading to 'nephrite' and from the Portuguese *pedra de ilharga* to 'jade' via French *pierre de la jade* (Galopim de Carvalho 2020). Now 'jade' is applied to both nephrite and jadeite – what are considered to be the 'true jades'. In East Asia, 'jade' is commonly used to translate the Chinese word *yu* (玉), traditionally designating 'beautiful stones' worthy of

This chapter partly overlaps the previously published paper Barnes (2018) "Understanding Chinese jade in a world context" *Journal of the British Academy* 6: 1-63 but has been expanded to cover jadeite/*feicui* mineralogy. See the other paper for more Chinese cultural information and nephrite specific to China.

fashioning into ritual objects and personal ornaments. Indisputably, the most highly prized stones among *yu* have been scientifically identified as 'true jades' (Keverne 1995; Middleton & Ambers 2005: 403) – either the *rock* called 'nephrite', most common in China, or the *mineral* 'jadeite', which was only available in areas surrounding China: Japan, Siberia, Kazakhstan, and Burma/Myanmar. Jadeite was known and used in Japan from the Jomon period onwards, but Japanese jadeite was never used in China. Instead, jadeite from Burma/Myanmar began to appear from the 13c and was commercially imported after 1874 (Hughes *et al.* 2000). It was known in Chinese as *feicui* (翡翠), the common name of the multi-coloured kingfisher, due to the mineral's bright green colour. Recent analyses of these *feicui* jades, however, have disrupted understandings of the traditional category of jadeite, as discussed below.

Returning to *yu*, a whole series of other kinds of rocks and minerals are included in the jade/*yu* categories; thus, these categories do not designate a specific kind of stone but they refer to valuable and beautiful stones. It is clear that jade/*yu* do *not* include shiny gemstones of the kind currently favoured in most of the world, though Stern *et al.* (2013) have named jadeite and ruby as "Plate Tectonic gemstones". It would be a mistake to dismiss the importance of the myriad types of rocks and minerals that are included in *yu* but designated as 'false jades', 'hemi-jades', 'pseudo-jades', and jade 'imitations' or 'simulants'; these all occur in archaeological contexts and have their own stories to tell. Many such ornamental stones occur together with true jades in their geological setting. Thus, all of these false and true jades need to be dealt with together to understand their procurement, processing, and distribution.

Rock vs mineral

What is the difference between a rock and a mineral? This is not such an idle question when dealing with jades, as misunderstandings begin here. Rocks are made of minerals, and mineral lattices are composed of elements, but elements can also be 'free-floating' in a rock, as a vein or flecks in a rock (thinking of gold here). In minerals, elements that are not represented in the mineral's ideal formula can be substitutive in the crystal lattice or hide within mineral lattice interstices (cf. Table 12.1). Mineral names and ideal formulae are approved and standardized by the International Mineralogical Association for use, with updates every 6 months or so (e.g., IMA 2022). The current IMA listing contains over 5700 approved valid minerals, but many more that are not 'approved' reside in the literature.

Table 12.1 Composition analysis of a 'nearly ideal tremolite' mineral

Tremolite = $Ca_2Mg_5Si_8O_{22}(OH)_2$

oxide/element	element	%wt
Major elements >1%	= 97.69%	
SiO_2	silicon	58.51
MgO	magnesium	23.74
CaO	calcium	13.29
H_2O	hydrogen	2.15
Minor elements 0.1%–1%	= 1.52%	
$Fe^{+2}O$	iron(II)	0.99
$Fe^{+3}{}_2O_3$	iron(III)	0.20
Na_2O	sodium	0.20
MnO	manganese	0.13
Trace elements <0.1%	= 0.22%	
F	fluorine	0.07
Al_2O_3	aluminium	0.05
K_2O	potassium	0.04
Cl	chlorine	0.03
TiO_2	titanium	0.02
TOTAL		99.43%

In brief, nephrite is a *rock* that is comprised of two *minerals*, tremolite and/or actinolite, plus any other minor accessory minerals that may be present. If an object is almost exclusively tremolite or actinolite, then it may be termed a monominerallic rock. Often, however, the mineral composition is not clearly identified and merely referred to as tremolite-actinolite. Jadeite is a mineral; it also often exists in monomineralic form. But most often, other minerals are involved; jadeite then occurs as a *rock* called jadeitite. Although Harlow & Sorensen (2005: 113) state that the best jades for lapidary work are monomineralic rocks, such a focus on such 'pure' jades in an archaeological context is misplaced. The various minerals and elements – as well as isotopic varieties and Rare Earth Elements (REE) – that are incorporated into nephrite and jadeitite are most useful in identifying sources and time periods.

Within any mineral, there are major, minor, and trace elements. As an example, a nearly 'ideal' form of the mineral tremolite from Italy (Table 12.1) was shown to include a number of different elements beyond those in the formal tremolite formula that the authors give.[1] In this table, the elements (given as oxides where appropriate) are demarcated by solid horizontal lines and ordered by declining abundance. The minor and trace elements indicate so-called 'impurities' which may be important for colouring or distinctive for sourcing. Given these uses, they could instead be considered 'bonuses'.

Despite the rock/mineral differences between the two names of nephrite (rock) and jadeite (mineral), their characteristics are given in Table 12.2.[2] Nephrite and jadeite are both silicates (containing SiO_2) but their chemical formulas are very different: nephrite contains substantial amounts of calcium–Ca and varying amounts of magnesium–Mg and iron–Fe, while jadeite contains aluminum–Al and sodium–Na. Nephrite is often referred to as amphibole jade while jadeite is called pyroxene jade. Both are "polycrystalline aggregates" with "complex interlocking structures" (Edward LIU, in Galopim de Carvalho 2020). See Zhang & Yuan (2017) for how these terms are presented in Chinese and Japanese.

Table 12.2 Comparisons of the two 'true jades'

Rock	Nephrite	Monominerallic Jadeitite
Essential minerals	tremolite, actinolite	jadeite
Mineral group	calcic amphiboles	sodic pyroxenes
Crystalline structure	double-chain inosilicate	single-chain inosilicate
Composition	$\square Ca_2(Mg_{5.0-4.5}Fe^{2+}_{0.0-0.5})Si_8O_{22}(OH)_2$	$NaAlSi_2O_6$
Water content	hydrous	anhydrous
Density	2.9–3.01 g/cm^3	3.3–3.6g/cm^3
Hardness (Mohs scale)	6–6.5	6–7
Crystal habit	microcrystalline, fibrous	micro- to macro-crystalline, non-fibrous
Texture	silky	blocky
Other properties	tougher than jadeitite	harder than nephrite

Both nephrite and jadeitite are metamorphic rocks, identifiable by fibrous or interlocking crystals that give them strength and toughness. In fact, nephrite cannot be termed nephrite unless it has a felted texture (fibrous), without which it would just be the minerals tremolite/actinolite. The mineral jadeite is harder than nephrite, indicated by its Mohs scale numbers,[3] but it usually has interlocking fan-shaped or blocky crystals rather than a fibrous structure. Nephrite has a silky feel, and its matted, fibrous (often called 'felted') nature – the interlocking of short fibres – makes it tough, i.e., difficult to break. Felting

[1] Compare this formula with the IMA-approved formula in Table 12.2.

[2] Mineral names are standardized by the Commission on New Minerals, Nomenclature and Classification of the IMA (International Mineralogical Association) (Nickel & Nichols 1991, 2009). The IMA designations are 'recommendations', not strictly enforced; moreover, mineral statuses constantly change and many are contentious. Tremolite itself was just redefined in 2008 by the IMA, so that earlier references might not use the approved name and formula.

[3] On the hardness scale developed by Frederick Mohs in 1822 (MINSOCAM 2016); see hardness diagram in Apx 14: Fig. D.

is understood to be produced in two stages: first is the solidification of the crystals, and then mechanical processes which twist or intertwine the crystals under relatively high temperatures (400–500°C) but low pressure (Campbell *et al.* 2008). The crystals may be oriented in random directions, radial or unidirectional patterns, producing a specific nephritic 'material'. Tremolite (as shown in Table 12.1) and/or actinolite *without felting* is *not* nephrite.

The greatest difference between true jades and other rocks and minerals, however, is not their appearance – as there are many beautiful stones – but the difficulty of working true jades. Not just because they are hard – quartz is even harder at Mohs 7 – but because they cannot easily be formed by chipping and require specific time-consuming and laborious fashioning by abrasion (Sax *et al.* 2004; Sax *et al.* 2008). This investment increases their value proportionally; throughout most of (pre-)history, this added value could only be afforded by the elite.

New revelations from feicui

Although nephrite-jade, especially mutton-fat nephrite, has been the favoured status symbol in China for centuries, *feicui* jades superseded nephrite symbolically following their import from Burma/Myanmar after 1874. Since then, *feicui* jade has consistently been treated as jadeite (see GAHK 2006). In 2006, however, Chinese chemical studies of *feicui* were reported at the First International Gem and Jewelry Conference in Bangkok (Franz *et al.* 2014). In Europe Adamo *et al.* also published analyses of Po Valley jade (2006). Both works revealed that many rocks thought to be jadeite/jadeitite were actually other minerals (omphacite) or a mixture (jadeite + omphacite, or jadeite + kosmochlor, or indeed, all three together). The Chinese work led to a revision of the Hong Kong Trade Descriptions Ordinance (Cap 362, secion 22, Section 2, L.N. 265 of 2008) (as presented in Galopim de Carvalho *et al.* 2020: unpg.), which reads (*italics added*):

> The expression "fei cui", when used in the course of any trade or business to describe an article, means the article is a granular to fibrous polycrystalline aggregate which is composed solely or principally of any of the following or *any combination* of the following – (a) jadeite; (b) omphacite; (c) kosmochlor.

Thus, what is known as *feicui* does not have to be jadeite; it could also be omphacite or kosmochlor, or a mixture. For individuals who bought *feicui* thinking they were buying true jade (jadeite), this must have been a blow. The GIA lab at Gemological Institute of America headquarters in Carlsbad, California began testing all 'jadeite' passing through its hands and discovered that a considerable number of items were omphacite (McClure 2012). McClure states that jadeite and omphacite are indistinguishable by normal gemological testing methods – by appearance, color, texture, refractive index (RI), specific gravity (s.g.), and absorption spectra; to tell them apart, Raman spectroscopy was needed. Thus, gemologists and jewelers cannot make identifications in their work place: pieces must be sent to a lab with the correct instruments for identification.

Since the Chinese government acknowledges the variable nature of *feicui* – and allows *feicui* to be sold as true jade – is there any point in distinguishing between items comprised primarily of jadeite versus *feicui* comprised of the other two minerals, either alone or in combination with jadeite? For clarity's sake, McClure (2012) proposes that jadeite-jade and omphacite-jade can be treated equally, but the latter should only apply to green omphacite-jade to the exclusion of other omphacite colours. Franz *et al.* (2014) go further and accept jadeite-jade, omphacite-jade, and kosmochlor-jade as identified by

Raman and infrared spectroscopy, apparently regardless of colour; but gemologists classify blue omphacite from Guatemala, for example, as omphacite, not jade (GemSelect 2019).

For archaeologists of China, this is a moot question, since Burma/Myanmar jades did not become widespread in East Asia until the 19c. However, Japanese jades contain substantial amounts of omphacite, and there are other areas of the world where omphacite could be mistaken for jadeite (see Adamo *et al.* 2006). Given that jadeite and omphacite share formation processes, it is not surprising that they should occur together. Moreover, since they cannot be distinguished without careful lab analysis, should omphacite-jade be considered a true jade?? Why would we value one pyroxene mineral over another if they both are equally rare and valuable and fulfil the same functions?

True Jades and False Friends

Archaeological perspectives

Worldwide, more than 40 different rocks and minerals have been found to act as simulants of true jades. Most of these can easily be detected (see Hobbs 1982). 'Simulants' is a term that is often preferred to 'pseudojade', 'false jade', and 'imitations', which are negative in nature and likely to invite dismissal of these other rocks and minerals; but in fact, these materials can be extremely illuminating as to source areas and mining technologies. Many were exquisitely worked and included as *yu*. For two reasons, they should be considered together with variations in quality of the true jades: (1) they will inform on sources of raw materials, and (2) their relative distributions can be used to monitor the development of hierarchical social relations and trade routes in the archaeological record.

Much of the work on jades in East Asia has been conducted by Chinese researchers. As early as 1989, WEN Guang, of the Geological Research Institute in Beijing, lamented that the traditional Chinese word for jade (*yu* 玉), originally confined to tremolite, had been broadened to the extent that it was meaningless (Wen 1989). We can also say the same for the English term 'jade', as it may refer to several ornamental stones. From the late 1980s to early 1990s, WEN Guang and JING Zhichun singly and together published a series of geoarchaeological articles on jade compositions. Wen & Jing (1992) emphasized that the identity of the stone cannot be made precisely without investigation of its chemical composition *and* microstructure. They analysed the compositions of 268 archaeological jades from Chinese Neolithic sites (see Barnes 2018: table 2), finding that stones included as jade (*yu*) in the Neolithic archaeological reports consisted of 21 different rocks and/or minerals in addition to nephrite. Wen & Jing concluded from these data that true jade was dominant in the Neolithic samples: 204 of 268 pieces (76% as calculated here). Due to this dominance, they proposed that Early Neolithic peoples in northeastern China (from Xinle and Chahai sites, now included in the Xinglongwa culture) were already able to distinguish true jade from other stones, as their ornaments are 100 per cent nephrite. Recently, however, other ornamental stones (chalcedony and steatite) have been reported from Xinlongwa culture sites in this area (Tang, C 2008: 265). However, the breakdown of some of the site contents showed that among the 149 nephrite samples, 128 (86%) were tremolite, 16 (11%) were actinolite, and 5 (3%) were tremolite–actinolite. In a second study of 500 nephrite artefacts (Wen & Jing 1996), percentages of tremolite alone varied upward from 65%, the remainder being actinolite.

Wen & Jing further distinguished two kinds of non-nephrite stones used as substitutes for true jades in later cultures: 'hemijades' and 'pseudojades':

- Pseudo-jades, they said, "may be found associated with nephrite in natural deposits, which might cause ancient people to treat these jadelike minerals as nephrite" (Wen & Jing 1992: 258). The five most common were antigorite, sericite, quartz, talc, and calcite (Wen & Jing 1996: 67); others named were albite, amazonite, dickite, diaspore, enstatite, fluorite, kaolinite, lizardite, magnesite, muscovite, and pyrophyllite (Wen & Jing 1992).

• Hemi-jade is a term they introduced "to refer to the mixture of nephrite and associated minerals". Such associated minerals in hemi-jades are yet again different from "auxiliary minerals (impurities) such as magnetite and chromite", which often give jades their varying colours (Wen & Jing 1992: 255). Hemijades included albite+tremolite, calcite+tremolite, talc+tremolite, antigorite+ tremolite, amazonite + tremolite, and tremolite+prehnite (Wen & Jing 1992, 1996: 64). The common theme here is 'tremolite', which is the dominant mineral in the sample of archaeological jades analysed by Wen and Jing. Perhaps some would now like to include some *feicui* as hemi-jades, in combinations of jadeite+omphacite, jadeite+kosmochlor, or even omphacite+kosmochlor?

Another term, '*semi-nephrite*', has occasionally been used in the past for nephrites that span the tremolite/actinolite border (Iizuka 2012); but that term now denotes "either massive amphibole, which lacks the felted cohesive texture, or nephritic amphibole enclosing coarser crystals" (Harlow & Sorensen 2005: 120, 125). In other words, 'semi-nephrite' has the same chemical composition as tremolite–actinolite but not the felted texture.

These early results can be contrasted with recent studies, reviewed by WANG Rong in 2011. Wang openly acknowledges that ancient jades in China include many 'types' of jade: amphibole jade (i.e., nephrite, including both tremolite and actinolite), serpentine jade, turquoise jade, agate jade, anorthite jade, zoisite jade, and others (Wang, R 2011: 674). His study covers a wider range of time periods and geographical areas, from the Neolithic to the Han period (206 BC–AD 220) across northern and eastern China. Of note are the artefacts made of serpentine and agate in collections from early Neolithic sites on the East Coast, with nephrite varying from 29% to 54%. Thus, it is clear that nephrite did not dominate in all Mainland Neolithic cultures. In the succeeding Liangzhu culture, nephrite rose to 93%, but in the later periods and peripheral regions, nephrite decreased to between 66% and 73%.[4] These figures were culled from the existing literature and so might change with rigorous sampling. Examining the ornamental stones considered by archaeologists to be jades (*yu*), Wen & Jing count 20 different rocks and minerals, whereas Wang counts over 30 (Wen & Jing 1996: 67; Wang, R 2011). This corresponds to other situations around the world.

Modern marketing concerns

New Zealand greenstones, treasured by the Maori, include nephrite (*pounamu*) and bowenite (a hard variety of antigorite) called *tangiwai* (Tennant *et al.* 2005). In China, Soochow (Suzhou) jade is bowenite (Desautels 1986: 9); it has been carved to make some of the most exquisite jade sculptures in historical and contemporary China. Two of the most common jade types today, Lantian jade and Xiuyan jade, are primarily the serpentine mineral antigorite,[5] while lizardite, another serpentine mineral, has recently been identified for the first time in Lantian jade (Wang, YY *et al.* 2012). Some modern Lantian jade is also made of serpentinized marble, and a similar deposit was discovered in 2013 at Pizhou City, Jiangsu Province. This is now being called 'Lantian jade' even though it is not from Lantian County. These non-nephrites are different from Lantian tremolite-jade, to be discussed below. Jingbai jade is actually white agate (Cheng *et al.* 2004: 31, table 1), and Dushan jade (Nanyang jade) is an aggregate of plagioclase,

[4] Numbers of artefacts were reported; the percentages are my calculations.

[5] WANG, YY *et al.* (2012) (but they note this might not apply to ancient Lantian jade); LIU, Z *et al.* (2009), Liang *et al.* (2012).

zoisite, and hornblende.[6] These identities in themselves do not totally devalue these materials: Cheng *et al.* state that stones other than nephrite "such as serpentine, Dushan jade, Jingbai jade, even quartz, were also artistically valued" (Cheng *et al.* 2004: 30), while Wang & Li state that "the five most important materials of Chinese ancient jade are amphibole, serpentine, turquoise, agate/chalcedony, and anorthite–zoisite (Dushan jade) (Wang, R & Li 2011: abstract)." Wang & Zhang list amphibole, serpentine, turquoise, and quartz varieties as the four main materials of ancient Chinese jade (Wang, R & Zhang 2011). ZHANG ZW *et al.* (2020) propose that among Longshan cultures, the use of nephrite may have been limited to higher ranking sites; their analysis of artefacts from a site in Shandong province only identified antigorite, actinolite, clinochlore, turquoise, and muscovite but no nephrite.

Many minerals and rocks are currently marketed as jade: both Amazon jade and Colorado jade are amazonite (microcline); others are Andes jade and Korean jade (serpentine, bowenite), Australian jade (chrysoprase), Californite (massive vesuvianite); Guatamalan *jade negro* (omphacite–taramite), Indian and Mixian jade (aventurine quartz), Lushan Mountain jade (marble), Malaysian jade (quartz), Mexican and Shaanxi Moyu jade (calcite), New Caledonian jade (anorthite), Ophite or New jade (serpentine), Oregon or Swiss jade (green chalcedony), Pounamu (New Zealand greenstones), Rainbow jade *jade lila* (jadeite–pumpellyite rock), Shetaicui jade (aventurine/dolomite/quartzite), Tangiwai (bowenite), Transvaal jade (green hydrogrossular garnet), Xiuyu jade (serpentine), prehnite, grossular garnet, agate, jasper, serpentinite, albitite, metabasite, quartz schist ... the list could go on.[7]

Then there are the newly defined *feicui* jades (kosmochlor and omphacite), as well as Maw Sit Sit. Since the first two form in concert with jadeite, they will be dealt with below. Maw Sit Sit was identified as a metamorphic rock separate from jadeitite only in 1963 by Gublein (1965). Gublein notes that it was being marketed abroad as chloromelanite, though many pieces discovered/acquired were thought to be jadeite. Although Gublein was unable to identify all the component minerals, Maw Sit Sit is now known to be a multimineralic rock comprised mainly of kosmochlor, clinochlore, chromian jadeite, and albite, with minor amounts of eckermannite, chromite, and other minerals (King 2014). Its appearance is bright green, from the chromian jadeite and clinochlore, with black swirls of kosmochlor.

Having reviewed the vast range of rock and mineral types that have been included under the umbrella terms 'jade' and '*yu*', we now turn our focus to the true jades, nephrite (rock) and jadeite (mineral). But the question still remains, should we treat omphacite-jade and kosmochlor-jade as true jades or as false friends? Wait and see...

Plate Tectonics and Formation of True Jades

If ever there is justification for studying Plate Tectonics by archaeologists, it is here in the study of true jades. Plate Tectonics forms the basis not only for understanding the formation processes of both nephrite and jadeitite but also accounting for where these jade sources are located today. The true jades have different host rocks in which they form: in general, serpentinite can be the host for jadeitite or nephrite formation, while dolomitic marble (dolostone) is only host to nephrite formation. Serpentinite and marble are both metamorphic rocks that have been subjected to single or multiple instances of metamorphism. Two processes leading to those host rocks, serpentinization and dolomitization, are

[6] Xiao, Q *et al.* (2009); WANG, R & Li (2011); Zhang, G *et al.* (1989); it is sometimes called the rock 'saussurite' (Wen & Jing 1996).

[7] For interesting compilations, see Desautels (1986: ch. 2); Walker (1991), Middleton & Freestone (1995); Dept of Geological Sciences (2009), the 1980–2010 index to the journal *Gems & Gemology*, and [https://madcatwoman-enterprises. tumblr.com/post/ 74776783593/jade-as-there-are-a-very-large-variety-of].

elucidated in Appendix 14. Why and how do these host rocks, whose initial rock bodies (**protoliths**) formed in the ocean or deep in the Earth, end up on dry land so that we can obtain their jades? The answer lies in the tectonic processes of **exhumation** and/or **obduction** in the **subduction zone** or **supra-subduction zone** (cf. Chapter 2).

Jadeite is known to form primarily in the **subduction channel** within serpentinizing bodies of peridotite, the rock of the Earth's mantle (Figure 12.1: green dots). The process of serpentinization (Apx 14) is basically hydration of the mantle rock, peridotite, by 'seawater-like' fluids (Harlow & Sorensen 2005: 138).[8] Much of this water is held in sediments from the ocean floor and trench fill; most such sediments

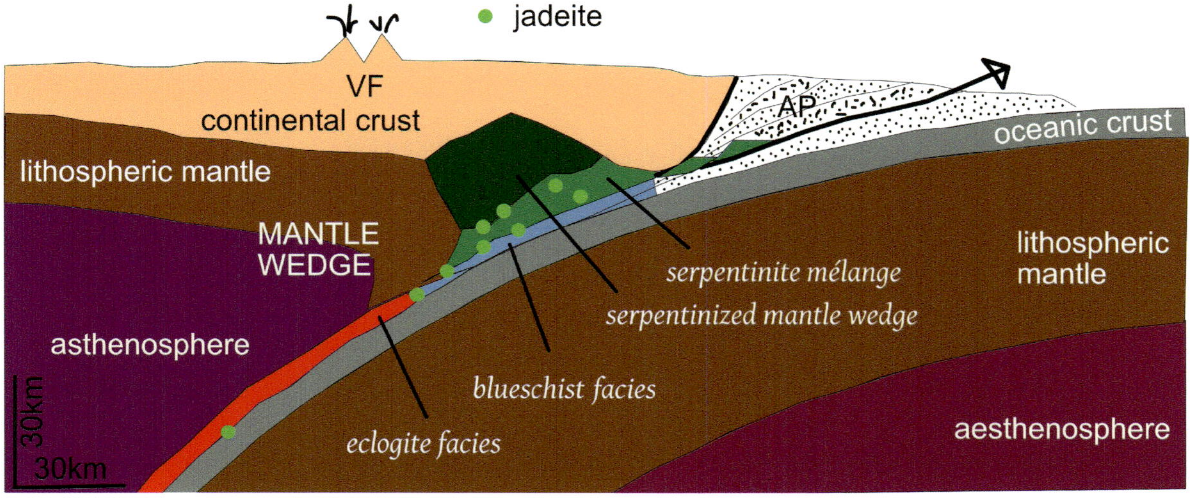

Figure 12.1 Loci of jadeite formation in the subduction channel
Volcanoes in Volcanic Front (VF) not to scale, AP = Accretionary Prism
Arrow shows direction and context of serpentinite mélange exhumation

are incorporated into the Accretionary Prism, but some are dragged down into the mantle with the downgoing oceanic slab. As they are dehydrated at depth within the blueschist **metamorphic facies** (see gradient III in Figure 2.8),[9] that water is exuded into the overlying mantle rock to form serpentinite (Figure 12.1: dark green). Note that serpentine is one form of false jade named above, and serpentine 'jade' should not be confused with 'jadeite from serpentinite'.

Serpentinite thus forms from peridotite at the surface of the mantle wedge facing the downgoing slab in the subduction channel; being a softer rock, much debris can be pressed into it, forming a **serpentinite mélange** (Figure 12.1: lighter green). This debris may include gabbro or basalt fragments from the oceanic crust or other ocean rocks, which are also metamorphosed to blueschist facies. Although Harlow & Sorensen (2001) identified highly faulted tectonic blocks of serpentinite or serpentinite mélange as hosts for jadeite formation, there are instances where jadeite forms within the blueschists

[8] Harlow *et al.* (2015: sec. 4) gives details of the kinds and sources of the water.

[9] Note that this dehydration event is separate from the dehydration of the oceanic slab in the eclogite facies; that dehydration causes the mantle rock to melt and rise through the crust to form volcanoes. However, slab dehydration (cf. Chapter 2.4.3) at blueschist-eclogite facies level probably contributes fluids for serpentinization as well (Harlow & Sorensen 2005: 138; Tsujimori *et al.* 2012; Harlow *et al.* 2015).

or metamorphosed mafic rocks (metabasites) that are incorporated into the serpentinite mélange (Harlow *et al.* 2015). These rocks are part of the ocean floor debris that gets pressed into the mantle wall serpentinite. In addition to jadeitites, omphacites based on the mineral omphacite can also form from reactions of fluids with basaltic blocks (Harlow & Sorensen 2001: 2).

Nephrite, on the other hand, can form from two different rocks, as described in the next section:

• from limestone that has first undergone dolomitization (see Apx 14) then metamorphosed to marble. Marble is formed in two contexts: in regional metamorphism by increasing pressure and temperature (P/T), as shown by trajectory II in Figure 2.8, or by contact metamorphism in trajectory I The latter situation occurs when an igneous intrusion of felsic magma encounters dolomitic limestone or marble, and the fluids exuded in this process contribute to nephrite formation, via metasomatism, based on the mineral dolomite.

• from serpentinite in contact with igneous rock or metamorphosed greywacke, a kind of fine-grained sandstone (Harlow & Sorensen 2005: 113). These occur in siuations where both serpentinite and graywacke have been obducted onto the continental mass.

Thus, we have the interesting situation where both of the host rocks (dolomitized marble, serpentinite) of true jades are metamorphic rocks, while nephrite and jadeitite are also metamorphic products. The hierarchy of successive changes are discussed further below. But before leaving this section, let us answer the question how these host rocks ended up on dry land.

Figure 12.1 further illustrates how serpentinite bodies from the subduction channel might be exhumed to the surface – along faults in the forearc of the volcanic arc. This can occur under either compressional or extensional conditions (cf. Chapter 2.6.1). To preserve jadeite, exhumation must be fast, so that the jadeite does not entirely convert to albite or analcime (Shigeno 2005) in retrograde metamorphism as it comes up from depth and reaches lower pressures towards the Earth's surface. Even so, recrystallizations from high pressure minerals in the blueschist/eclogite facies to lower pressure minerals in the greenschist and amphibole facies (Figure 2.8) are often found within jadeitites.

The dolomitic limestones that are parent rock to dolomarble, one of the host rocks of nephrite, form in the sea in a variety of contexts. As shallow-sea or shelf sediments, they are liable to be shoved into continental masses during the closure of oceans or the closure of back-arc basins such as the Japan Sea Basin (Figure 4.1). They are often found in association with ophiolites (Chapter 2.7) – slices of ocean floor (Figure 2.4) that have been obducted onto land (Figure 12.2).

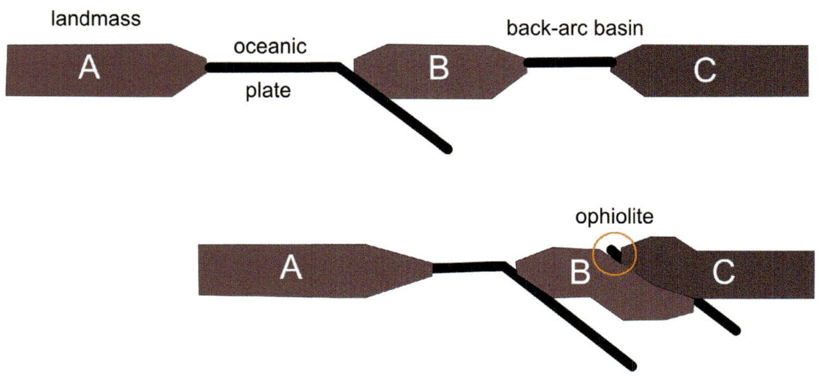

Figure 12.2 Back-arc basin closure in Tibet resulting in an ophiolite

Three landmasses are depicted in Figure 12.2: A is being drawn towards B by ocean plate subduction under B; B and C are separated by a back-arc basin in which an oceanic plate has formed. Pressure from A collides B into C, resulting in a fragment of the back-arc basin floor incorporated into the combined B+C landmass, forming an ophiolite, while the majority subducts under C. When A eventually collides with B, ocean floor fragments might also be obducted into B.

Two caveats are emphasized by Harlow *et al.* (2015: 109). First, ophiolites themselves, those slices of ocean floor that occur on land, are rarely hosts for jadeitite, even though their gabbro/basalt rocks (Apx 5: Table B) may be serpentinized. This is because ophiolite rocks are generally metamorphosed at lower pressures resulting in lizardite and chrysotile (Harlow *et al.* 2014: 33); they lack antigorite, the main serpentine mineral for jadeite formation, except at their base. Second, Harlow *et al.* also caution that tectonic blocks of serpentinite on land, exhumed from the subduction channel, should *not* be considered ophiolites, as they sometimes are. Unfortunately, subduction channel serpentinite and ophiolites often occur in pairs (Harlow *et al.* 2015: 210) because after all, they represent coordinated subduction zone processes even when currently embedded in continental contexts. However, the former is distinguished by being "either highly faulted or a melange and is always associated with a major fault which is usually a lateral fault on the wedge side of a fossilized convergent margin" (Harlow & Sorensen 2001: 2).

The worldwide distributions of jadeitite and nephrite have been mapped by Harlow and colleagues through successive publications that show their locational affinities with current and fossil subduction zones associated with mountain-building (orogeny): ophiolite-associated nephrite (Harlow & Sorensen 2005: fig. 1; Harlow *et al.* 2014: fig. 10-23), and blueschist-associated jadeitite (Harlow & Sorensen 2005: fig. 10; Harlow *et al.* 2014: fig. 10-1). The nephrite maps, however, do not do justice to the important and pervasive occurrences of nephrite on the China Mainland. Detailed maps of ophiolite and carbonate limestone distribution in China and the types of nephrites produced there are presented in Barnes (2018: figs. 8, 9). Tsujimori and colleagues take a different approach to jadeite location by the age of the orogens in which they occur (Tsujimori & Harlow 2012: fig. 1; Harlow *et al.* 2015: fig. 2). However, the known ages of the geological belts of Japan as discussed here in Chapter 3 are obscured by the scale of that map, whereas Japan has some of the oldest jadeite deposits in the world, dating to the Palaeozoic and paralleling those in the CAOB (Central Asian Orogenic Belt) (Tsujimori *et al.* 2012; Harlow *et al.* 2015: table 2).

Having discussed the generalities of different kinds of jades and the tectonic occurrences of the true jades, in the following sections, the mineralogy of the true jades will be examined, followed by a detailed look at their derivation from their host rocks. Although these sections are long and dense, and perhaps repetitive, even more information about the true jades is available in the many publications of Harlow, Tsujimori, Sorensen, and colleagues.

True Jade Minerals

To understand more about the formation of true jades and, in particular, jadeite's relationship to the *feicui* jades, let us begin by looking at nephrite and jadeite solid solution series. These utilize ternary diagrams to map out the relations between the nephrite minerals (tremolite and actinolite), and between jadeite and its pyroxene neighbours. The minerals in each ternary diagram are related to each other via similar crystal lattices but with differential distributions of elements that occupy the lattices.

Solid solution chemistry, however, does not account for all of jadeite occurrences; we will also look at mineral replacements, in which fluids still play a part, and solid state metamorphism of jadeite.

Tremolite, actinolite, and most jadeites are recognized to be products of metasomatism, a form of metamorphism involving fluids; a range of minerals can result depending on temperature/pressure conditions, the elements present in these fluids and their individual abundances, and the order in which the elements are integrated into mineral lattices. The easiest to comprehend is the nephrite minerals, tremolite and actinolite, which belong to a solid solution series named tremolite–ferro-actinolite. Jadeite, on the other hand, is related to several minerals, especially omphacite. It is interesting to note that in the tremolite–ferro-actinolite series, the variation is measured between reciprocal *elements* ($Mg/(Mg+Fe^{2+})$),[10] while in the jadeite-aegirine and jadeite-omphacite series, reciprocity is calculated by the co-existence of different *minerals*.

The Problem with Nephrite

Nephrite: rock and minerals

The term 'nephrite' is not considered by the IMA a 'valid mineral species', but it is often used as a nonspecies name (synonym) of the mineral tremolite – in the sense that mica is a synonym or common name for a muscovite mineral and rock salt is a synonym or common name for the mineral halite.[11] This accepted use of nephrite = tremolite has given rise to many misunderstandings about the nature of nephrite jade.

Researchers in the field describe nephrite as belonging to the tremolite–actinolite series (Wen & Jing 1992: 261, 1996; Wen 1994; Tsien 1996; Harlow & Sorensen 2005). However, this assignment is also a misnomer because a 'tremolite–actinolite *series*' implies that actinolite is an end-member of the series.[12] Instead, actinolite is actually the middle member of the series 'Tremolite–Actinolite–Ferro-actinolite'; this series has since been renamed as 'tremolite–ferro-actinolite', a new designation whereby only the end-members of the solid solution are given in binomial nomenclature.[13] Consequently, Harlow *et al.* revised their definition of nephrite as "a rock composed fundamentally of tremolite–ferro-actinolite" (2014: 340). This properly specifies the series, but it obscures the fact that there are currently no known prehistoric nephrite artefacts with ferro-actinolite as the essential mineral, though it is still a possibility.[14]

Felted ferro-actinolite is extremely rare (Bersani *et al.* 2014), but a source was discovered in Guangxi, China in 2011, which is now being used for carvings (Zhong *et al.* 2014). Analyzed samples were 95–99% ferro-actinolite, with lesser amounts accompanied by stilpnomelane and other minerals. Ancient ferro-

[10] This is sometimes turned around, e.g., Fe/(Mg+Fe) (cf. Jiang *et al.* 2020).

[11] See Nickel & Nichols 1991: appendix C "Synonymy of nonspecies names".

[12] Iizuka *et al.* 2007: 14, "actinolite, the iron-rich end-member of the calcium amphiboles." Furthermore, there are other calc-amphibole solid solution series (Wittke 2009), forming when calcium drops below 6%.

[13] George Harlow pers. comm. 26 November 2016, conforming to Hawthorne *et al.* 2012. Unfortunately Harlow *et al.* (2014: 36) state that "actinolite is no longer a valid mineral species" and that "tremolite [now] spans the compositions formerly termed actinolite" (my insertion); they refer to Hawthorne *et al.* 2013 on this issue, but the publication in question is Hawthorne *et al.* 2012, which does not contain these revisions, and Hawthorne himself says that the statement above is not true (Frank Hawthorne pers. comm. by email, 14 December 2016): actinolite is a valid mineral species as given in the July 2022 IMA list (IMA-CNMNC 2022).

[14] George HARLOW, Yoshiyuki IIZUKA, and Frank Hawthorne, pers. comm. by email, 26 November 2016 and 3 January 2017.

actinolite jade objects are either absent or not yet found. Using Energy Dispersive X-ray Fluorescence, Casidio *et al.* (2007) attributed a dark-coloured proto-historic *zhang* blade to ferro-actinolite, but according to Pat Daly at Gem-A London (pers. comm., 16 February 2017), this attribution is problematic because EDXRF must avoid preferred orientation of crystals, and no accommodation for this in the testing was mentioned by Casidio and colleagues; also no chemical analysis was undertaken to assess the blade composition. The authors Casadio and Douglas agree the results might be misleading (by email 27 February 2017 and 7 March 2017).

Tremolite-actinolite is sometimes referred to as the 'nephrite series' (Bersani *et al.* 2019), but this is again misleading because the minerals in the series are not always felted – and as stated above, felting is an immutable attribute of nephrite. Since the intermediate member of the series, actinolite, is still a valid mineral name, the series will be referred to as Tremolite–Actinolite–Ferro-actinolite (abbreviated here as TAF-a) (Table 12.3), with the understanding that nephrite is formed from only the first two, tremolite and/or actinolite, minerals, and that they must be felted to earn the name nephrite.

Table 12.3. Distinguishing tremolite, actinolite, and ferro-actinolite in the solid solution TAF-a = $Ca_2(Mg,Fe)_5Si_8O_{22}(OH,F,Cl)_2$

Mineral	$Mg/(Mg+Fe^{2+})$	composition range	colour*
Tremolite calcium magnesium silicate hydroxide	≧0.90	$\square Ca_2Mg_5Si_8O_{22}(OH)_2$ to $\square Ca_2Mg_{4.5}Fe^{2+}_{0.5}Si_8O_{22}(OH)_2$	colourless to grey
Actinolite calcium magnesium iron(II) silicate hydroxide	0.9–0.5	$\square Ca_2Mg_{<4.5}Fe^{2+}_{>0.5}Si_8O_{22}(OH)_2$ to $\square Ca_2Mg_{2.5}Fe^{2+}_{2.5}Si_8O_{22}(OH)_2$	pale green to dark green
Ferro-actinolite calcium iron(II) silicate hydroxide	<0.5	$\square Ca_2Mg_{<2.5}Fe^{2+}_{>2.5}Si_8O_{22}(OH)_2$ to $\square Ca_2Fe^{2+}_5Si_8O_{22}(OH)_2$	dark green to black

* colour designations are arbitrary and subjective.

How much of the rock needs to be nephrite when mixed with other elements/minerals? Harlow *et al.* (2007: 230) note that greater than 95% amphibole is 'typical' for nephrites. Obviously, the remaining percentages are allocated to accessory minerals and trace elements. Thus, it is of crucial importance for researchers to define their thresholds for determining mineral status and also to provide detailed chemical characterizations if possible to specify which nephrite mineral, tremolite or actinolite, is predominant and what other minerals are present.

Nephrite's solid solution minerals

The most important elements that determine whether a mineral is tremolite or actinolite are iron and magnesium. The way these operate can be seen when examining the TAF-a solid solution series. The TAF-a series belongs to calcic amphibole minerals as seen in Table 12.2, based on calcium Ca, hence the 'calcic', and their crystalline structure (C2/m). Minerals in the series have a silicate component Si_8O_{22}, and they contain a hydroxyl (-OH) group. The elements distinguishing the different minerals are the ratios of magnesium–Mg and ferrous iron-Fe^{2+} [also written as iron(II)]. The IMA-approved formulae for tremolite and ferro-actinolite only apply to their extreme ends (end-compositions) – while in reality most tremolite can include some iron. For the end-composition (at the diagram's borders), tremolite has zero iron–Fe, and end-composition ferro-actinolite has zero magnesium–Mg. Tremolite and actinolite within the left half of the series have varying amounts of both iron and magnesium.[15] There are other minor or trace elements (e.g., chromium–Cr, manganese–Mn, aluminium–Al, and sodium–Na) that might be incorporated into the chemical formulae, giving rise to a number of IMA unapproved mineral names such as fluoro-tremolite, soda-tremolite, chrome-tremolite, etc.

The truncated ternary diagram of the TAF-a solid solution series is shown in Figure 12.3 – truncated because minerals at the top of the diagram do not exist in reality. This series exhibits continuous chemical changes from end-member tremolite (on the left) through intermediate member actinolite (in the middle) to end-member ferro-actinolite (on the right). The changes involve reciprocal amounts of magnesium and iron, whose proportions are used to identify the minerals and their boundaries. Because the ions of Mg and Fe are of the same size and electrical charge, they can substitute for each other in the chemical formula, but they vary systematically and reciprocally across the series. Although the boundaries between these minerals are ultimately arbitrary – and varying definitions are possible – the IMA-CNMNC recommends thresholds in the magnesium and iron ratios of $Mg/(Mg+Fe^{2+})$ to distinguish them. Thus, the (arbitrary!) boundaries shown in Figure 12.3 are based the thresholds of $Mg/(Mg+Fe^{2+})$ = $\geqq 0.90$ for tremolite, and $Mg/(Mg+Fe^{2+})$ = <0.5 for ferro-actinolite, while actinolite has between 0.9–0.5

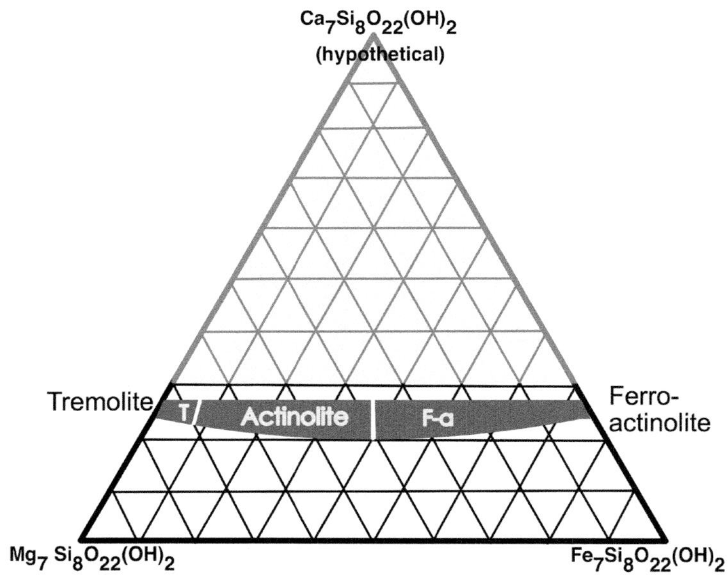

Figure 12.3 Ternary composition diagram for the tremolite–actinolite–ferro-actinolite (TAF-a) continuous solid solution series. The three major elements (calcium, magnesium, and iron) are at 100% in the triangle corners, comprising the minerals indicated by the formula; each parallel line moving away from the corner signals a 10% decrease in the specified element. The divisions between the TAF-a minerals, as calculated by the ratio $Mg/(Mg + Fe^{2+})$, are shown as white lines through the solid grey area. There are no known amphiboles that form at the top end of the triangle, hence its fadeout and the hypothetical composition.

$Mg/(Mg+Fe^{2+})$. Despite these divisions, researchers may refer to nephrite simply as tremolite-actinolite, even when the ratios are known. For example, three jades analyzed from a Liangzhu tomb in China proved to have $Mg/(Mg+Fe^{2+})$ values of 0.892, 0.871, and 0.869; these are all ostensibly in the actinolite range, but the researchers designated each object as tremolite-actinolite (Gan *et al.* 2010). However, the

[15] Siivola & Schmidt (2007) recommend that ferro-actinolite is properly abbreviated as Fe2-Act; however, the IMA-CNMNC approved mineral symbols list gives it as 'Fact' (Warr 2021).

newly discovered black ferro-actinolite nephrite samples from Guangxi have $Mg/(Mg+Fe^{2+})$ ratios as low as 0.343 and 0.419, clearly separating them from actinolite (Zhong *et al.* 2014: fig. 13).

Whole rock analyses such as XRF produce an average reading for the $Mg/(Mg+Fe^{2+})$ ratio that may mask some interweaving of distinct actinolite or tremolite crystals; in such cases, however, the crystals will generally not have very different ratio readings (BOX 13). It is seldom the case that one would find crystals in the white tremolite end-member range (0.99–1.00) interleaved with crystals of lower ratio readings unless the host rocks were, for example, a seamount (basalt) with fringing reefs (carbonate). It would then be conceivable for nephrites to have mixed tremolite (>0.96) crystals from carbonate together with actinolite crystals (IIZUKA Yoshiyuki, pers. comm. by emails, October 2021).

By these definitions, tremolite accounts for a very small portion of the TAF-a series only at the far left of the ternary diagram, but its composition may include up to about 10% iron in the Mg/Fe total ($Fe_{0.5}/Mg_{4.5}$). The complete lack of iron in the tremolite end-member makes it colourless; mutton-fat jade (white), the premier Chinese nephrite, would thus be close to this boundary. However, Cheng *et al.* (2004: 30) note that "all pure jadeite, nephrite, serpentine, Jingbai jade and calcite are white in color", so even the lack of colour is not diagnostic of material.

Nephrite colours

In general, nephrite colour is determined by the amount of iron-Fe, which also determines the species identity of the mineral, tremolite or actinolite. However, actinolite can have up to 50% Fe in the Mg/Fe total ($Fe_{2.5}/Mg_{2.5}$). The increasing amount of iron and decreasing magnesium produces hues ranging from light to dark green - and most nephrites are so coloured. Similar colours are produced by iron in serpentinite (Chen Q *et al.* 2014), so again, colour is not a clue to rock identity (nephrite vs serpentinite) nor to mineral identity (tremolite, actinolite, or serpentine).

Figure 12.4 shows the $Mg/(Mg+Fe^{2+})$ ratio divisions, chemical compositions, and colours that these ranges include. Nephrite colour is often used to assign a piece to one or the other mineral, but colour is not a good indicator of mineral content as mentioned. Within the colour range of Hetian nephrite jades, Liu *et al.* have documented colour differences across the tremolite-actinolite range: white tremolite-nephrite had ratios of $Mg/(Mg+Fe^{2+})$ of 0.98–1.00, while green tremolite measured 0.93–0.99; black nephrite was in the actinolite range with ratios of 0.63–0.90 (Liu *et al.* 2011*b*: tables 2, 3, 4; see also Shi *et al.* 2015). Note that tremolite can be green (from the iron content) or even black if the colour component is graphite as in the Qiemo–Ruoqiang region of southern Xinjiang (LIU Y *et al.* 2011*a*; see discussion of Chinese jade sources in Barnes 2018).

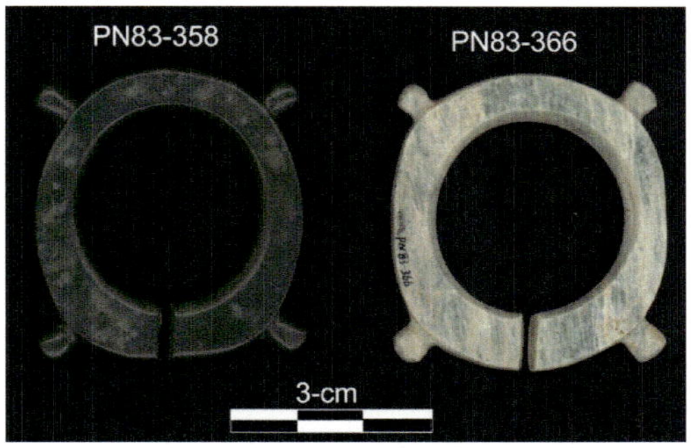

	Mg/(Mg+Fe)	FeO	Fe
PN83-358			
1	0.914	3.95	0.446
2	0.910	3.88	0.450
3	0.909	4.04	0.461
4	0.907	4.10	0.475
5	0.900	4.36	0.510
6	0.885	5.16	0.592
PN83-366			
1	0.918	3.76	0.422
2	0.915	3.85	0.434
3	0.910	3.93	0.464
4	0.886	5.14	0.587
5	0.886	5.14	0.590

Figure 12.4 LVSEM-EDS spot analyses on nephrite slit-rings

Note that there is almost complete overlap in Mg/(Mg+Fe) ratios between the two artefacts, and that there are relatively wide differences in iron content at different spots on a single artefact.

Spot analysis by non-invasive LVSEM-EDS of the two slit-rings from Peinan, Taiwan (Figure 12.4) give a range of $Mg/(Mg+Fe^{2+})$ ratios within each piece (Iizuka *et al.* 2005). These are highly correlated with the amount of iron-Fe at different spots on the artefact. Significantly, Fengtian nephrite also contains zincochromite (Zchr) as an accessory mineral; one artefact measured 6wt% zinc (Zn). While chromite (Chr) itself ($Fe^{2+}Cr_2O_4$ spinel) is often stated to be a major contributor to the bright green colour of *feicui* jade (but see below), zincochromite ($ZnCr_2O_4$ spinel) apparently does not contribute to the deeper green colour of Fengtian nephrite jade. Zincochromite is rare enough to be used to distinguish Fengtian nephrite from other Asian and Pacific nephrites (Iizuka *et al.* 2005).

Colour variations within an artefact do exist (Figure 12.4); but even when using point analyses such as EPMA or Raman Spectroscopy, it is difficult to identify the $Mg/(Mg+Fe^{2+})$ ratio of individual crystals. It might be said that the whiter ones are 'tremolitic' and the greener ones are 'actinolitic'. This, however, may be misleading, since tremolite can be consistently green, as demonstrated by the following study.

Wang & Shi (2021) have discovered that green Manas and Sayan nephrites are primarily tremolites. As reported in their tables 1 and 2, 21 Manas samples and 11 of 13 Sayan samples had $Mg/(Mg+Fe^{2+})$ ratios of $R^* \geqq 0.90$. Two Sayan samples were less than 0.90, indicating they were actinolite. However, the range of readings for these combined samples extended to 0.99, far greater than the Fengtian nephrites (BOX 13) which, although they also crossed the border between tremolite and actinolite, did not exceed 0.93. It was not stated which minor inclusions gave rise to the green colour of these nephrites, but both Manas and Sayan samples contained chromite and chlorite-group minerals. They were distinguished by Manas containing uvarovite and Sayan containing bornite.

Nephrite summary

Though 'nephrite' is not a valid *mineral* name, it is a very useful *rock* name that designates a specific felted fabric characteristic of the amphibole minerals, tremolite, actinolite and ferro-actinolite (Table 12.4). It is usually not possible to distinguish these minerals by eye, and colour is not a sure guide to identity. It takes chemical analysis to discover the exact ratio of $Mg/(Mg+Fe^{2+})$ to be able to identify an object's place on the definitional scale (as in Figure 12.3). The exact composition is an important way to source objects if the characteristics of the nephrite mine are known, as at Fengtian and Xiaomeiling in BOX 13. In addition to its *main* mineral constituent(s), nephrite may *contain* several *minor* minerals (cf. Table 12.1), separate from those that are associated with it (Table 12.4: right). The minor and trace minerals are best regarded as 'bonus ingredients' as they may be very informative. It is important to note that white tremolite-nephrite and green actinolite-nephrite are seldom mined from the same source; this is because they derive from different kinds of rocks (Iizuka 2021), as will be discussed below. Thus, although the chemistry in the solid solution series provides a continuum of Mg-Fe distribution, in reality the products are separated in time and space according to formation processes.

Table 12.4 Summary chart of nephrite definitions and list of some associated minerals

Nephrite rock = 95% amphibole (TAF-a series) composed variously of the following minerals:			
Mineral	**$Mg/(Mg+Fe^{2+})$**	**Texture**	**Attribution**
Tremolite	$\geqq$0.90	felted = nephrite not felted — not nephrite	
Actinolite	0.9–0.5	felted = nephrite not felted — not nephrite	
Ferro-actinolite	<0.5	felted = rare, nephrite not felted — not nephrite	

Tremolite-associated minerals in nephrite:
e.g., diopside, calcic garnet, carbonate, epidote, clinozolisite, apatite, allanite, quartz, pargasite, prehnite, andesine, titanite, zircon, calcite

Actinolite-associated minerals in nephrite:
e.g., grossular, spinel, chlorite, serpentine, chromite, magnetite, magnesite, datolite, monazite, graphite, rutile, pyrite

BOX 13 Taiwanese Nephrite Case Study

The chemical compositions of Taiwanese jade objects from the Fengtian nephrite source have Mg/(Mg+Fe^{2+}) ratios between 0.87 to 0.93 – clustering across the nominal border of 0.90 separating tremolite from actinolite cluster across the tremolite-actinolite border as shown here. The term 'semi-nephrite' has occasionally been used in the past for this combination (Iizuka 2012), but it now more commonly denotes tremolite/actinolite without the felted crystal habit (and therefore not nephrite jade). Iizuka notes that the Fengtian nephrite range is close to the Mg# readings for serpentinite and mantle-derived olivine (i.e., 0.88–0.92), the dominant mineral in mantle peridotite. This correlates with and substantiates the inference that the host rock for Fengtian nephrite formation is not a carbonate. In contrast, nephrite objects that have ratios of 0.99–1.00, particularly mutton fat jades, are always from carbonate sources (see Xiaomeiling [XML] distributions below).

The following figure compares white and green nephrite chemistries in eastern Asia, with all mineral data taken by EPMA (Iizuka & Hung 2005). The ratio Mg/(Mg+Fe) is plotted against the number of silicon atoms per formula unit; the variation in silica is caused by some substitutions of Al, Fe^{3+}, or Cr in the crystal lattice. Note that the Fengtian nephrites form a coherent body of material that crosses over the R*=0.90 demarcation between actinolite and tremolite.

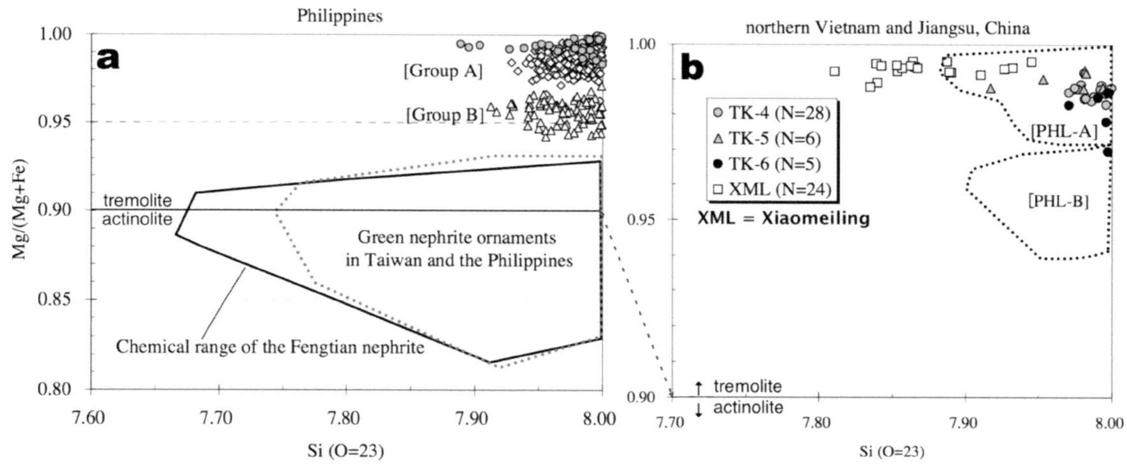

(a) illustrates the range of Fengtian nephrite as encompassing the Mg/(Mg + Fe) = 0.82 to 0.93; most green nephrite objects from Southeast Asia derive from this source. White (Group A) and white-green-brown (Group B) nephrites from the Philippines are clearly separated chemically.

(b) illustrates Groups A and B from the Philippiines (PHL) in comparison with white nephrites obtained from Trang Kenh (TK), nephrite workshop site in Hai Phong, northern Vietnam (circles, triangles), and Jiangsu (squares). The latter, raw jade sourced from the Xiaomeiling (XML) mines, clearly has a different composition than most of the former, believed to be from a local source in Vietnam.

As the graphs show, the chemical characterizations serve to identify the relative amounts of magnesium vs iron that correlate with colour (disregarding other causes), but more importantly, they can help identify sources of jade materials. The Mg/(Mg+Fe^{2+}) R* ratios are very different for Fengtian nephrite in Taiwan and the Philippines (upper figure: below 0.93) versus nephrite in northern Vietnam and Jiangsu Province, China (upper figure: above 0.93, and lower figure).

Jadeitite problems[16]

The predominant mineral in the rock jadeitite is the mineral jadeite,[17] though like the rock nephrite, the rock jadeitite can also include several accessory minerals and trace elements in addition to its main jadeite component. Jadeite has the crystalline structure of a clinopyroxene, and the presence of sodium-Na puts it in the sub-group of sodic clinopyroxenes. As a rule of thumb, jadeitite is a rock that "contains at least 90 vol% pyroxene, and the average pyroxene has at least 90 mol% jadeite" (Harlow & Sorensen 2005: 128). This leaves 10 vol% available for pyroxene minerals *other than* jadeite (such as aegerine or omphacite), and another 10 mol% available for minerals *other than* pyroxenes, including even tremolite and actinolite (see Table 12.6 for some associated minerals).

Jadeite-jade

Three types of the jadeite mineral are known depending on their mode of formation: first, by direct precipitation via solid solution with other pyroxenes (much like the TAF-a solid solution discussed above); second, by replacement of other minerals with jadeite; and third, jadeite+quartz P/T mutual transformations with albite. The first two are dependent on the circulation of hydrous fluids and are therefore a form of metasomatism. The third transformation is mainly due to changes in pressure and temperature (P/T). Tsujimori & Harlow (2012) developed the terms P-jadeite for precipitation and R-jadeite for replacement; the term P/T-jadeite is used here in contrast. Jadeite itself is a rare mineral and mostly forms as P-jadeites; R-jadeites are very rare. At the Earth's surface, jadeite+quartz is unstable, and this kind of jadeite does not survive well (Perkins *et al.* 2021), reverting to albite. Thus, the kind of jadeite/jadeitite most relevant to archaeologists and gemologists is P-jadeite.

P-jadeite

Tsujimori & Harlow (2012) note that "most jadeitites belong to the P-type and are considered to have precipitated directly from a Na-Al-Si-rich aqueous fluid in some cavity, crack, and fracture in serpentinized peridotite or a high-pressure metamorphic rock" (2012: 185). P-jadeite precipitates from fluids that are rich in sodium-Na, aluminum-Al, silica-Si, but lacking in quartz (Ibid. 2012: 376). The jadeite crystals are characterized by cryptic or rhythmic zoning – diagnostic of crystallization from an aqueous fluid (Harlow & Sorensen 2001). Figure 12.5 charts P-jadeite as an end-member of solid solutions that include as neighbours omphacite and aegirine. This is a far more complicated graph than that for the nephrite minerals; the three end-member points represent the QUAD minerals (Q) at the top, jadeite (Jd) in the lower left, and aegirine (Aeg) in the lower right. Of the 20 possible pyroxenes that are generated by solid solution chemistry and organized into six categories (I-VI) (Morimoto *et al.* 1988), only three minerals are of interest here: Jadeite-Jd and Kosmochlor-Ko – which together comprise half of the Na-pyroxenes based on sodium-Na (Group IV) – and Omphacite-Omp, one of only two Ca-Na pyroxenes based on calcium-Ca/sodium-Na (Group III) (Table 12.5).

For minerals located along the jadeite–aegirine axis, at the base of the diagram, the 50% mark is nominally used to differentiate Jd species and Aeg species. The ternary base represents no iron-Fe^{3+} in ideal Jd_{100} and no aluminium-Al in ideal Aeg_{100} (Nestola *et al.* 2006: table 1). Only the end-members are referred to by the singular jadeite or aegirine (Figure 12.5), while members along the axis are notated as e.g., $Jd_{100}Aeg_0 > Jd_{74}Aeg_{26} > Jd_{35}Aeg_{65} > Jd_0Aeg_{100}$, etc. As can be seen from these notations, the two minerals are reciprocal over the axis length, with different names in sectors 1–8 in Figure 12.5.

[16] First lamented by Yoder (1950) then Dobretsov (1984) and used by Harlow *et al.* (2012).

[17] These terms have been turned around in at least one archaeological publication (Hung *et al.* 2006: 213, ftn.1).

Morimoto *et al.* (1988: 71, with references to their own tables) advise that "An element specified as a modifier [such as ferrian or chromian] should be present as a general rule in a quantity larger than 0.1 or 0.01 atoms in the standard chemical formula of 60 or 4 metal atoms [their] Table 5 depending on the maximum content in [their] Table 4" (brackets inserted). Jade-rich rocks can also contain calcium from the QUAD axis, leading to the existence of calcian jadeite (1), calcian ferrian jadeite (2) and ferrian jadeite (6).

Table 12.5 Chemical formulae of minerals mentioned in jadeite section
End-members shown in bold

Name	**Abbrev**	**Ideal formula**	**Classification**
jadeite	Jd	NaAlSi2O6	Na-pyroxene
aegirine	Aeg	$NaFe^{3+}Si_2O_6$	Na-pyroxene
kosmochlor	Ko	$NaCr^{3+}Si_2O_6$	Na-pyroxene
omphacite	Omp	$(Ca,Na)(Mg,Fe,Al)_{Si2O6}$	Ca-Na pyroxene
aegirine-augite	Agt	$(Ca,Na)(Fe,Mg,Fe)_{Si2O6}$	Ca-Na pyroxene
diopside	Di	CaMgSi2O6	Ca pyroxene
hedenbergite	Hd	$CaFe^{2+}Si_2O_6$	Ca pyroxene
augite	Aug	$(Ca,Mg,Fe)_2Si_2O_6$	Ca pyroxene
enstatite	En	Mg2Si2O6	Mg-Fe pyroxene
ferrosilite	Fs	Fe2+2Si2O6	Mg-Fe pyroxene
wollastonite	Wo	$CaSiO_3$	pyroxenoid
albite	Ab	$Na(AlSi_3O_8)$	Plagioclase feldspar

Despite a seemingly continuous distribution of minerals throughout the ternary diagram, it has been noted since the 1980s that there are 'composi-tion gaps' between some of these minerals (Cortesogno *et al.* 2002; Tsujimori & Harlow 2012). These gaps shift and even disappear with increasing temperature and/or pressure conditions. Thus, under certain conditions, intervening values for mineral compositions may not be present so that when mineral samples are mapped onto the diagram, they may cluster with gaps inbetween. In particular, when mapped onto the ternary diagram, jadeite (5) and omphacite samples seem to be far separated, so there is no significant continuum between them. Nevertheless, they are often found together in rocks, including jadeitite.

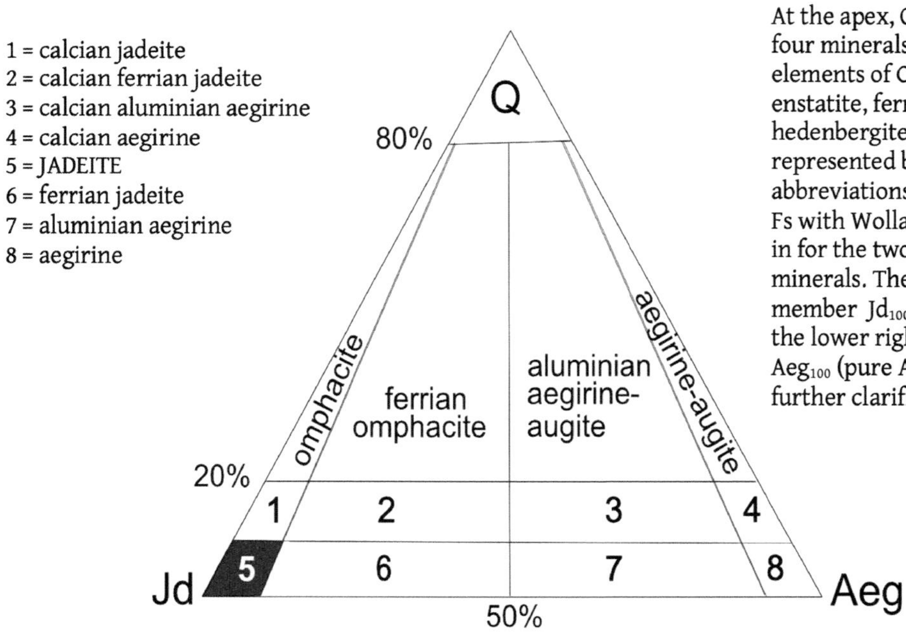

At the apex, Q (QUAD) stands for four minerals that have the major elements of Ca, Mg, and Fe: enstatite, ferrosilite, diopside, and hedenbergite; these are often represented by only mineral abbreviations En-Di-Wo or Wo-En-Fs with Wollastonite (Wo) standing in for the two other unnamed quad minerals. The lower left is the end-member Jd_{100} (pure Jadeite), and the lower right is the end-member Aeg_{100} (pure Aegirine). See text for further clarifications.

Figure 12.5 The ternary diagram for Q–Jd–Aeg

R-jadeite

R-jadeite is generated under high-pressure low-temperature changes in the minerals of an existing rock, which often "preserves textural, mineralogical, or geochemical evidence of a preexisting protolith" (Tsujimori & Harlow 2017: 185). As with P-jadeite, it is dependent on the circulation of fluids through host rocks. A tell-tale indication of replacement is the existence of relic "textural, mineralogical or geochemical evidence of a pre-existing protolith" (Tsujimori & Harlow 2012: 376). Fluids not only carry elements from elsewhere but they can leach more elements out of host rocks that can be used to replace those found in the rocks they encounter, causing new and/or different minerals to form. Harlow and colleagues (Harlow *et al.* 2014) predicted that transforming a protolith (original rock) such as plagiogranite into jadeitite would entail a complex 'mass balance' that would be difficult to achieve. However, Hertwig *et al.* (2021) have demonstrated that it could be accomplished by 'simple' de-silication of the igneous rock with the addition of minor amounts of Fe, Mg, Ca, and Na in the somatizing fluids. These researchers have identified a wide variety of jadeite-rich rocks resulting from replacement transformations. Such rocks, however, are typical metamorphic rocks with a variety of interlocking mineral crystals, jadeite included. R-jadeite is not only rare but relatively inconsequential for studying archaeological artefacts.

P/T-jadeite

The third type of jadeite is produced with minimal fluid input during changes in pressure and temperature; these may usefully be referred to as P/T-jadeites. This is the solid state chemical reaction between jadeite+quartz and albite, the sodium end-member of the plagioclase solid solution. The transformation happens on a sliding scale of P/T, with the formula $Na(AlSi_3O_8) \rightleftharpoons NaAlSi_2O_6 + SiO_2$; thus albite⇌jadeite+quartz. This change is reciprocal and occurs at specific combinations of pressure-temperature changes along a gradient (Figure 12.6). It is a P/T metamorphic transformation that occurs primarily in the blueschist facies, and it depends on the prior existence of one of the two participants. Despite this being mainly a P/T change, we should keep in mind that almost all metamorphic changes involve fluids to a greater or lesser degree (Newton 2014: 155). It can be seen that Jd+Q virtually disappears at pressures less than 5Kb; thus exhumation up to depths less than about 23 km will begin the transformation from Jd+Q to Ab – the only question is how long that will take for jadeite to completely disappear. An albitized jadeitite, known as Kaho no Hisui (Kaho jade) in Japan indicates that the transformation is not completely valueless (Tsujimori 2017: 220).

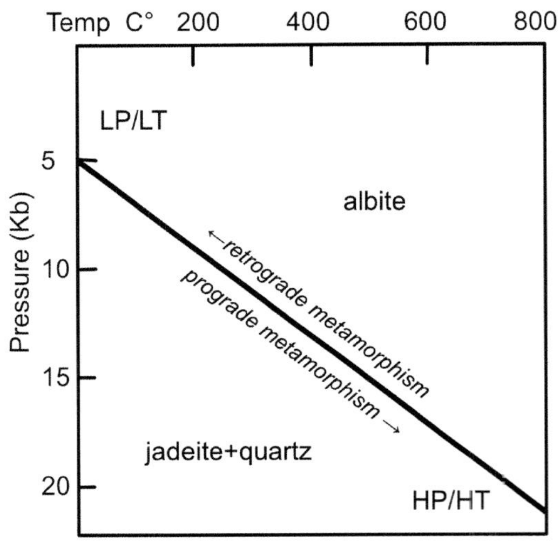

Figure 12.6 High (H) and Low (L) Pressure/Temperature conditions for albite vs jadeite+quartz formation

The other feicui *jades*

Kosmochlor-jade

Kosmochlor, previously known as ureyite (Harlow & Olds 1987), was initially identified in meteorites from which it derived its cosmic name change. Harlow & Olds identified it in several jadeites from worldwide locations, but it is in Burma/ Myanmar that it occurs massively to form kosmochlor-jade. Like P-jadeite, massive kosmochlor-jade forms from precipitating fluids; however, its relation to other pyroxenes is controversial. Morimoto (1988: 66) stated that it "show[s] little or no solid solution towards other end members", although Harlow & Olds stated that it has "complete... miscibility with jadeite and significant solid solution with omphacite" (1987: 126). Kosmochlor often forms around chromite, from which it derives chromium to yield its bright green colour; according to Lennartson (2014) chromium can also appear as a variety of colours depending on its various coordination compounds.

Omphacite-jade

Omphacite is a more complex mineral, containing more main elements than the Jd–Aeg series; for example, the formula for 'intermediate omphacite' is given as $(Na_{0.5}Ca_{0.5})(Al_{0.5}Mg_{0.5})Si_2O_6$ (Lin *et al.* 2020), indicating equal amounts of sodium, calcium, aluminium, and magnesium. Like actinolite in the TAF-a series, omphacite is not an end-member in Figure 12.5, with the 80% and 20% marks indicating decreasing amounts of the elements (Ca, Mg, Fe), which make up most of the QUAD minerals (Morimoto *et al.* 1988); these percentages mark the boundaries of omphacite definition.

Another mixture of 1:1:1 Aeg:Di:Jd, which falls in the omphacite range, has previously been termed 'chloromelanite'; however, this term is obsolete among mineralogists, while its use among gemologists has been replaced by the term Maw Sit Sit discussed above, a chromite-rich rock similar to jadeitite also occurring in Myanmar (Sieber 2021). Despite not occurring on the IMA approved mineral list, MINDAT states that chloromelanite is a variety of jadeite with the ideal formula

Table 12.6 Primary (P) and Secondary (S) minerals occurring in select worldwide **jadeitites**

M=Myanmar, G-1=N.Guatemala, G-2=S.Guatemala, Jn=Japan(Itoigawa), CA=California, Ru=Russia, Kz=Kazakhstan, BM=Borus mélange

	M	G-1	G-2	Jn	CA	Ru	Kz	BM
actinolite-Na				P				S
albite	S	P,S	S	P	P?,S	S	P,S	S
allanite		P	P					
amphiboles	P(sodic)							
analcime	S	P?,S		S	S	S	S	S
apatite		P	P	S(Sr)				
banalsite	S	S						
barroisite				P				
benitoite?							S?	
carbonate								S
celsian		S						
chlorite								S
chromite	P						P	
clinochlore	S	S		P				
copper					P?,S			
cymrite	S	S	S					
diopside			S					
eckermannite				P				
epidote								S
garnet								P
graphite	P	P					P	
"hornblende"								P
itoigawaite				S				
jadeite	P	P	P	P	P	P	P	P
kosmochlor	P							
K-feldspar		S						
lamprophyllite				S				
lawsonite			P					
magnetite							P	
matsubaraite				S				
natrolite				S	S	S	S	
nepheline		S						
omphacite	P	P	P	P	P	S	P	P,S
paragonite		P						
pectolite			P	S	S			
phengite		P	P				S?	
phillipsite						S		
phlogopite						S		P
preiswerkite		P						
pumpellyite			P				P?	
pyrite		P						
quartz			P,S	P?				
rengeite				S				
rutile		P	P	P			P	P
slawsonite				S				
taramite		S		P				
tausonite				S				
titanite		P,S	P	P			P	
tremolite			S			S	S	
vesuvianite			S					
xonotlite				S				

$Na(Al,Fe^{3+})Si_2O_6$, which puts it between jadeite and aegirine in the solid solution – to which other names have now been given (Figure 12.5). The confusion surrounding this unapproved mineral name is apparent in the discussion thread "Chloromelanite" on the Mindat.org website.

Jadeitite summary

Jadeite that is precipitated from metamorphic fluids containing sodium, aluminium, and silica is the most common type of jadeite mineral (P-jadeite), existing in rocks (jadeitite) that contain a variety of other mineral components (Table 12.6). The jade mineral and jadeitite rocks are primarily formed in serpentinized mantle rock within the subduction zone; they may also form within blocks of various oceanic floor rocks and granites that have beein incorporated into serpentinite mélanges. Jadeite forms in conjunction with other minerals in solid solution series, the most common association being between jadeite and omphacite. Another product of such solid solutions series is kosmochlor.

As we leave this section, the outstanding question is whether we should accept kosmochlor-jadeite and omphacite-jadeite as true jades or relegate them to the status of false friends despite their being solid solution neighbours to jadeite and included in *feicui*. The suggestions of Franz *et al.* (2014) for revising the jadeite nomenclature in parallel with jadeite jade are interesting, resulting in kosmochlor-jadeite jade and omphacite-jadeite jade, based on the mineral identified in 75% of spot tests (15 of 20) on a piece using Raman or FTIR analysis. Time will tell how these research results on Burma/Myanmar jades will impact the field. As yet, reference to either of these alternate forms of HP/LT 'jades' as 'jadeitite' rocks is not apparent; thus, we can assume that 'jadeitite' exclusively refers to rocks that "contains at least 90 vol% pyroxene, and the average pyroxene has at least 90 mol% jadeite" (Harlow & Sorensen 2005: 128).

True Jade Rocks and Their Host Rocks

It was established above that most jadeitites and nephrites are metasomatic rocks, resulting from the circulation of fluids either precipitating the minerals within or replacing existing minerals within pre-existing rock bodies. Those bodies are mainly serpentinite and dolomitic marble (dolomarble). Not only are nephrite and jadeitite metamorphic rocks, so too are serpentinite and dolomarble, their host rocks. Thus, this section begins by looking at the hierarchy of metamorphic processes acting on the parent rocks that produce the host rocks and then the jades.

Hierarchical metamorphic processes

Throughout the literature, one reads that jadeitite is formed in serpentinite while the nephrite minerals can be formed from either serpentinite or dolomitic rocks, as put forward early on by Wen & Jing (1992, 1994). These processes involve both the P/T metamorphic transformation of rock and metasomatism due to fluid interactions. Figure 12.7 is a simplification of such metamorphic relationships, and as usual for a simple diagram, more details await in the literature.

In Figure 12.7, metasomatism is represented by a solid line, and P/T metamorphism by a dashed line. Note that two kinds of P/T metamorphism are present: contact metamorphism due to igneous intrusion (cf. Figure 2.8, gradient I), and blueschist facies metamorphism taking place during subduction (cf. Figure 2.8, gradient III). The numbers in Figure 12.7 appear in brackets below:

- The main sets of protoliths (parent rocks) [1] in question are limestones (high calcium-Ca and high carbonate contents), and upper mantle (peridotite, dunite, lherzolite, harzbergite)/lower ocean floor rock (gabbro, basalt) of ultramafic and mafic compositions, respectively (cf. Apx 5: Table B). Other sedimentary and igneous rocks can be involved in jadeite formation (Harlow *et al.* 2015) and are represented here as 'ocean rocks'.

• These parent rocks can be subjected to three different processes to form three host rocks: *dolomitization* [2] of limestone to form dolomitic (magnesian) limestone aka dolostone; *serpentinization* [3] of dolostone/ peridotite/gabbro/basalt to form serpentinite; and *metamorphism* [4a] to the *blueschist-facies*.

• However, two other metamorphic processes can affect dolostone. First, entire tracts of dolostone can be converted to dolomarble during the P/T conditions of regional metamorphism (Figure 2.8, gradient II; e.g., the Dolomite Mountains in Italy). Second, contact metamorphism [4b] resulting from an igneous intrusion into dolostone can transform the dolostone via the application of heat into dolomitic marble (dolomarble) in addition to sending in fluids – such as those that make skarn minerals as well as nephrite.

• Serpentinite, on the other hand, can be subjected to increasing Pressure/Temperature, from the greenschist facies to the blueschist facies or even the eclogite facies (cf. Figure 2.8). A further solid state metamorphic possibility is jadeite formed from albite through prograde P/T metamorphism [4c], as seen at the lower right in Figure 12.7 and discussed above as P/T-jadeite.

• The results of these processes are three types of jadeites (P-, R-, P/T-jadeites) and two types of nephrite: D-nephrite formed from dolomarble and S-nephrites formed from serpentinite. S-nephrites, moreover, have three possible parent rocks that were serpentinized: ultramafic rocks, mafic rocks, and dolomarble. These nephrite types will be discussed in detail below.

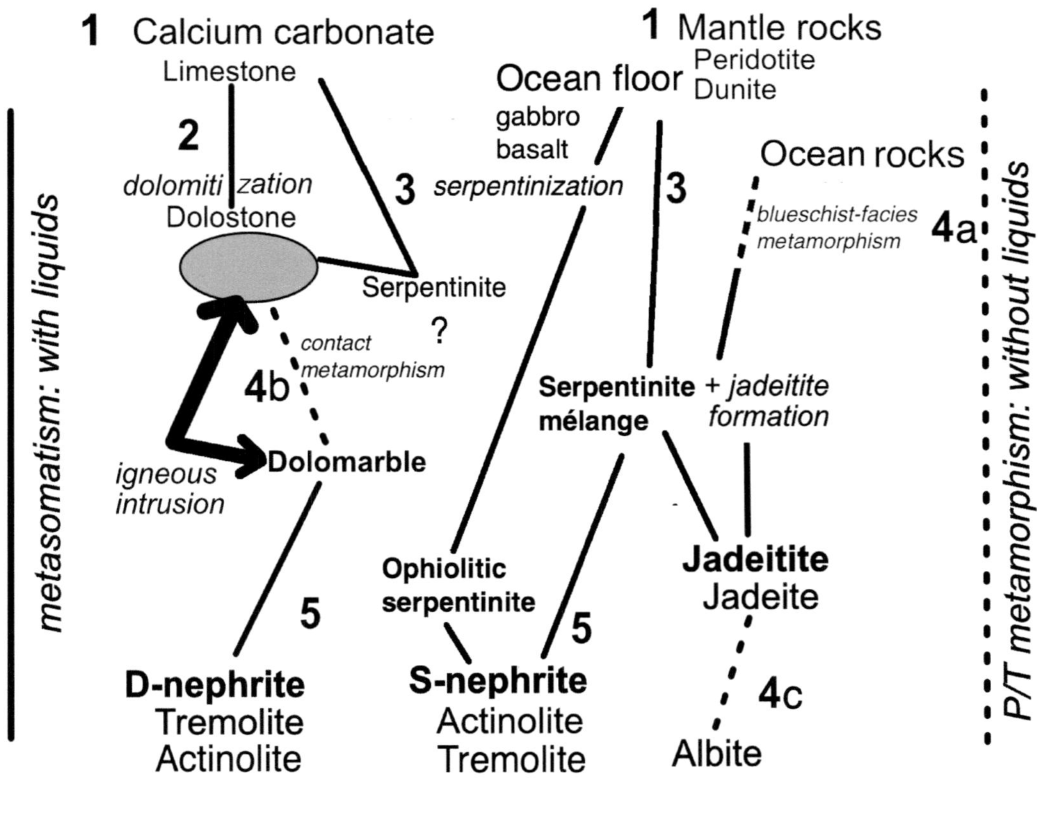

Figure 12.7 The hierarchical metamorphism of host rocks and jade mineral formation

The problem with metamorphism is that it can affect rocks many times throughout geological history and in many different ways, including reversal (retrograde metamorphism) or wanderings of the metamorphic trajectories. So, what we find today is a palimpsest and/or cumulative results of these previous processes.

Three derivations of nephrite

We have already seen that nephrite can be composed of two different minerals, tremolite and/or actinolite. Here we look at the three processes producing the two different types of nephrite, the rock, that can form from the host rocks of either dolomarble or serpentinite (stage [5] in Figure 12.7). For brevity's sake, these will be referred to as **D-nephrite** or **dN** for nephrite-from-dolomite and **S-nephrite** or **sN** for nephrite-from-serpentine.[18] See Appendix 14 for the processes of serpentinization or dolomitization of these host rocks. S-nephrite is formed by fluid interaction with serpentinite, while D-nephrite can be formed by fluid interaction directly between dolomarble and igneous rock *or* by replacement of an intervening mineral (LIU, Y *et al.* 2011*a*,*b*; Yui & Kwon 2002). In most cases, nephrite formation is judged to be a process of mineral replacement, with some retaining relics of host rock minerals (Harlow *et al.* 2014).

The difference between dN and sN production primarily has to do with the minerals inherent in their host rocks, and then minerals from the host rock environment.

- Dolostone, comprised mainly of dolomite $CaMg(CO_3)_2$, is replete with magnesium-Mg and some calcium-Ca but no silica (SiO_4). For D-nephrite formation, the silica must be supplied by contact and metasomatic exchange with an igneous rock (Frost & Frost 2014: 204). Dolomite has little iron, but some iron, as well as other elements, notated by the □ in the formula (Table 12.3), may also seep in from the igneous rock fluid intrusion. This results in D-nephrite having a composition generally with very little iron; thus, D-nephrites cluster at the tremolite end of the solid solution series (cf. Figure 12.3, Table 12.3) with *R>0.90.
- The serpentine minerals $□_3[Si_2O_5](OH)_4$ in serpentinite may have a variety of elements at position □ : magnesium-Mg, iron-Fe, nickel-Ni, manganese-Mn, aluminium-Al, and zinc-Zn; however, the most common serpentine minerals – lizardite, chrysotile, and antigorite – are Mg-dominant (Mindat.org: serpentine). Note that calcium-Ca is not on this list; however, it is present in the peridotite protolith, in the olivine minerals $□_2SiO_4$, with □ = Ca,Fe,Mn,Ni,Mg (Mindat.org: Olivine Group). However, both calcium and magnesium can be provided by circulating seawater, where Ca = 400 ppm and Mg = 1272 ppm (Stanford n.d.). Unlike dolomarble, serpentinite is rich in silica, perfect for S-nephrite formation.

D-nephrite from dolomite (dN)

For D-nephrite formation, silica must be obtained from elsewhere, which is where igneous intrusions are crucial. D-nephrite is formed via the invasion of quartz-rich aqueous fluids exuded from igneous rocks into dolomarble. Quartz (silicon-Si) is often an excess element in felsic magma (e.g., forming obsidian), and residual quartz may be transferred from an igneous intrusion into the country rock; this aqueous transfer usually results in skarn mineral formation – important for mining of metals. However, if the country rock is dolomarble that the igneous intrusion encounters, tremolite may be among the minerals newly formed. This is one of the outcomes of contact metamorphism combined with metasomatism. Although the processes causing felting into nephrite are seldom discussed, it is possible

[18] In parallel with Itype, Stype, specialized Stype, and Atype granites (Chappell & White 2001; Kumar & Singh 2014: 140-141).

to imagine that the pressure of the igneous intrusion against the country rock may be enough to cause the crystals to knit together during formation. Low-moderate pressure and temperatures in the greenschist to amphibolite facies conditions produce D-nephrite (Harlow *et al.* 2014: 42). Both tremolite and actinolite nephrite can result from metasomatism of dolomarble; tremolite is more common, but as the R* values for Hetian nephrites quoted above show, actinolite nephrite is also possible with greater iron infusion.

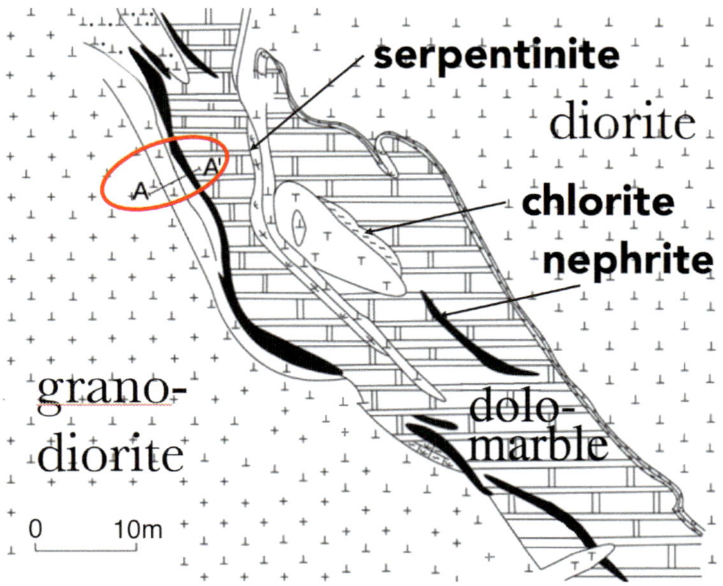

Figure 12.8a The Alamas ophiolite

Dolomarble metamorphosed by igneous intrusion of granodiorite on the left; transect A–A', circled above, is shown in 12.8b below.

In the Alamas ophiolite in north-western China (Figure 12.8a), nephrite is accompanied by other rocks (serpentinite) and minerals (chlorite) produced from the same contact metasomatism. A transect (A-A') exhibits the sequential changes in mineralogy from the igneous felsic rock granodiorite to the dolomarble (Figure 12.8b). It is understood among researchers that white tremolite-nephrite is exclusively D-nephrite, formed only in dolomarble from contact metasomatism. But this does not mean that other colours cannot be produced in this process. As Figure 12.8b shows, there is a gradation from green to white-green to white nephrite that can characterize D-nephrite. However, deeper green colours are usually a property of S-nephrite from serpentinite.

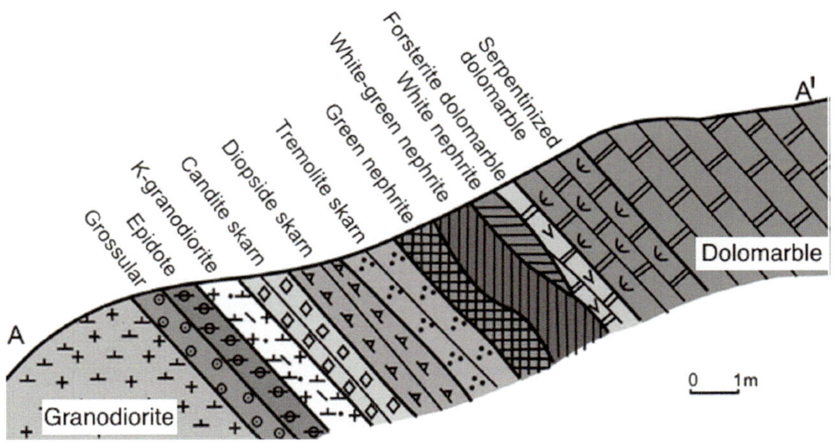

Figure 12.8b Zoned nephrite in the Alamas ophiolite

(b) successive zones of metamorphism from left to right after granodiorite: contact metamorphism resulting in new minerals; then metasomatic reactions producing various skarns; then three colours of nephrites; then altered marble adjacent to the original dolomarble.

D-nephrite via diopside

As with jadeite above, nephrite can also be formed by mineral replacement in an intervening step. In investigating the yellow Yinggelike nephrite of Xinjiang Province, China, researchers have identified three different stages of metasomatic activity (see equations in Jiang *et al.* 2020: 15-16). In Stage I, a new mineral formed from dolomite was diopside; in Stage II, dolomite was replaced by coarse tremolite, and that was in turn replaced by fine-grained tremolite in which the needles were randomly oriented and gave the nephrite its felting. Much of the diopside created in Stage I was also replaced by tremolite in either Stage I or II. In Stage III, calcite veins formed from hydrothermal circulation of fluids.

The significance of these findings is that nephrite minerals are not always precipitated directly from a fluid but clearly can form by replacing other metasomatic minerals. Jiang *et al.* state that "diopside exists in almost all D-type nephrites and is considered to be related to the nephrite formation" (2020: 14).

S-nephrite from serpentine (sN)

Serpentinite is a metasomatic rock comprised mainly of serpentine minerals (see Apx 12.2). As seen in Figure 12.7, two parent rocks can be metasomatized into serpentinite that produce S-nephrite: ultramafic mantle peridotite and derivatives, and mafic gabbro/basalts. Most S-nephrites are produced in serpentinite mélanges which incorporate a variety of silicic rock such as felsic igneous rocks, shale, and chert, etc. (Harlow *et al.* 2014: 44). The metasomatic process involves the replacement of serpentine via a calcium-Ca and silicon-Si enriched fluid. This occurs at greenschist facies P/T conditions (Ibid. 2014: 48). The mineral compositions of S-nephrite vary greatly, assisting in determining sources. For example, the main accessory mineral associated with Fengtian S-nephrite in Taiwan is chromite, while garnet (uvarovitic-grossular), chlorite, and serpentine occur occasionally (Iizuka *et al.* 2007); this mineral combination identifies nephrite derived from serpentinite, and occurrences conforming to this are Fengtian (Taiwan), New Zealand, Gansu in China, and British Columbia (Ibid: 13; Hung *et al.* 2007). Fengtian nephrite, however, is distinguished from these other locations by the higher amounts of zinc and manganese inclusions in its chromite crystals. Other locales of S-nephrite give insight to the variety of rocks involved in the formation of nephrite as well as other accompanying rocks in the 'metasomatic zone': e.g., diopsidite, epidotite, rodingite, chloritite, as well as the mineral talc (Harlow & Sorensen 2005: table 2).

Serpentinization of dolomitic limestone can also occur with the addition of silica and water, as indicated in Figure 12.7, resulting in serpentine minerals plus calcite (Rao & Murty 1980; Chakraborty *et al.* 2022). It is not clear whether nephrite can then form from the resulting serpentinite, but carvings made of serpentinite from dolomite are known in the Shu culture of ancient China (ZHANG, R 2022).

Distinguishing sN and dN

Since the turn of the millennium, Chinese scholars have used nondestructive PIXE (Particle Induced Xray Emission) analysis to distinguish S-nephrite and D-nephrite (CHEN, T-H *et al.* 2004). Major findings of the collective research by ZW ZHANG and colleagues (2010, 2011, 2012) are:

1) that the *general* dividing point among the sampled nephrites according to $Mg^{2+}/(Mg^{2+}$+Fe2+(3+)[19] [abbreviated as the R* ratio] is drawn at R* = 0.93, confirming earlier whole rock studies (Wen & Jing 1992; ZHANG, ZW *et al.* 2011); note that the empirical 0.93 is more than the arbitrary 0.90 used to divide tremolite and actinolite in the TAF-a series (cf. Figure 12.3).

2) that this measure is not entirely accurate in separating D-nephrites from S-nephrites, as was demonstrated early on (Wen & Jing 1992: fig. 6; ZHANG, ZW *et al.* 2011). One sample with very high Fe content (from presumed local geological circumstances) was grouped with serpentinite-derived nephrites, but its overall composition was closer to D-nephrites.

Consequently, factor analysis of element combinations was undertaken to provide better distinction, and the groups clearly separated into sN above and dN below, as shown in Figure 12.9. That study found that three trace elements (chromium–Cr, cobalt–Co, and nickel–Ni) together with the R* ratio were crucial in determining the separation. Two types of nephrite were thus recognized: Type I with chemical composition corresponding to tremolite [Dn], and Type II with chemical composition corresponding to actinolite [Sn]. Type II nephrites "always contain more transition metals such as Cr, Co, Ni and so on" (ZHANG, ZW *et al.* 2010: 367). In that study, it is notable that only nephrite from Manas County, Xinjiang was included in the S-nephrite group (Type II, actinolite) along with New Zealand and British Columbian nephrite; nephrites from Hetian, Yecheng, Xiuyan, Xiaomeiling, and Wenchuan were all D-nephrites (Type I, tremolite).

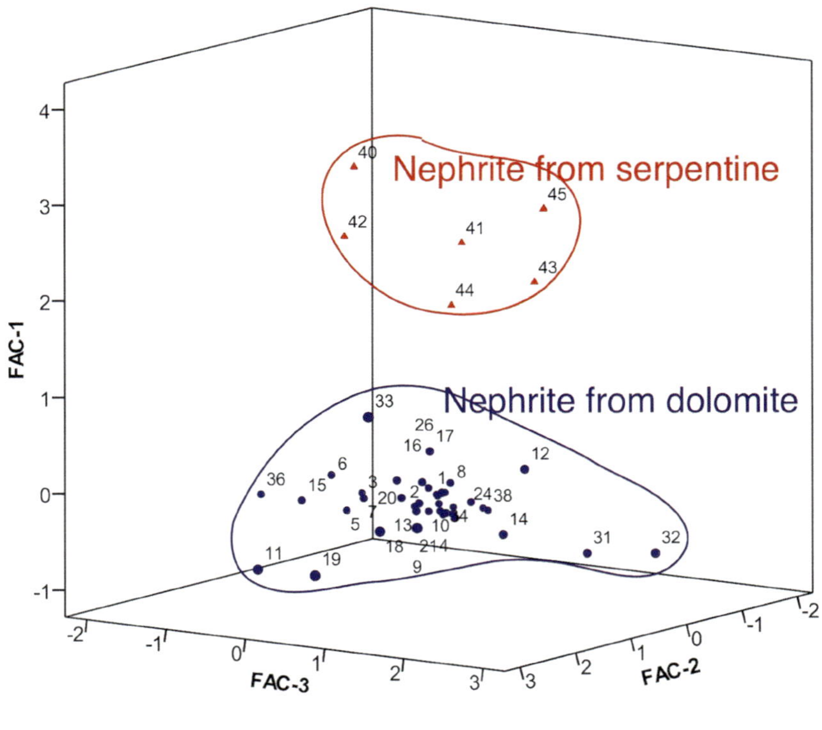

Figure 12.9 Distinguishing S-nephrite and D-nephrite by factor analysis
Numbered items are actual samples, with Manas nephrite (40, 41) the only S-nephrite from China (Xinjiang) among others worldwide. Wenchuan samples (31, 32) are slightly separated from the rest of the Chinese D-nephrite samples. The ratio of Mg/(Mg + Fe) and the three trace elements Ni, Cr, and Co were the important determinants of the factors.

[19] Note that this rendering is more specific than the standard Mg/(Mg + Fe) given above.

Two derivations of jadeitite

Jadeitite from serpentinizing peridotite

Harlow *et al.* (2015: 118-119) judge that most jadeite is formed by precipitation within peridotite that is being serpentinized into the serpentine mineral antigorite (cf. Apx 12.2). They present a model process dependent on brittle fracture of mantle peridotite during serpentinization, allowing fluids to flow into the cracks and fissures where they precipitate into jadeite and other minerals to make the rock jadeitite. The minerals can contain fluid inclusions, and successive fluid infiltrations can form what they call "multigenerational P-type jadeitite", often zoned in construction. Furthermore, this type of jadeitite is distinguished as P_S-type, or R_S-type in the case of replacement, indicating its serpentinite host – in contrast to jadeitite formation in other rocks (see below).

Typically, such veins of jadeitite can vary from 1.5 m to 50 m wide, and they are often enclosed by successive layers of other mineral aggregates (Figure 12.10). In this figure, the rock albitite is surrounded by the minerals chlorite and actinolite, while magnesite 'blobs' occur within the serpentinite. The mineral albite is an alteration of jadeite in retrograde metamorphism as the vein was exhumed to the Earth's surface (cf. Figure 12.6). The chlorite rim is said to form a 'blackwall', which is a chemical reaction boundary between the host or country rock and the new rock/mineral formations. Such blackwalls may be made of single or several minerals as here with chlorite, actinolite, clinochlore, and sometimes talc (Harlow & Sorensen 2005).

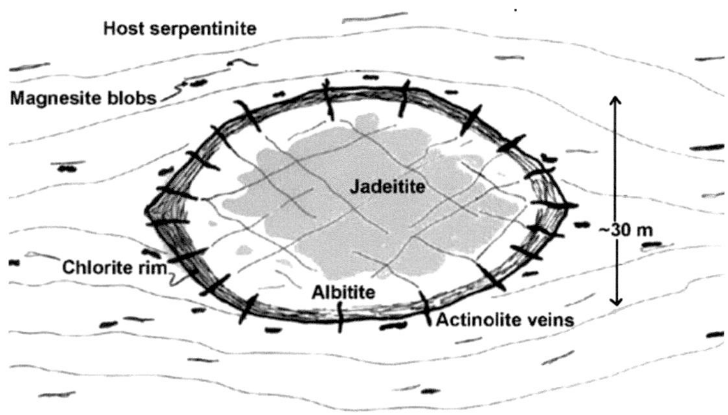

Figure 12.10 Cross-section of a jadeitite vein forming in serpentinite

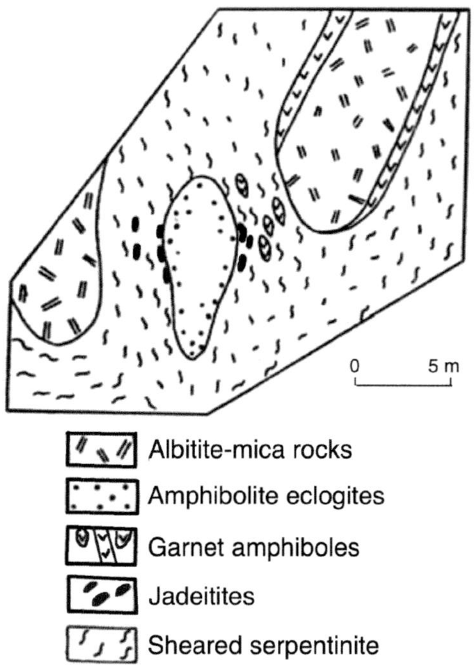

Figure 12.11 Jadeitite pod formation in serpentinite

Large blocks or small pods of jadeitite (Figure 12.11) can also occur. Jadeitites involve many different primary and secondary minerals that occur as crystals within the jadeitite body (Table 12.6), but some of these can also form independently with jadeitite. In this figure, jadeitite can be seen forming in association with eclogites, while albite-mica rocks formed a blackwall rim of garnet-amphibolites. The presence of eclogites indicates metamorphism within the eclogite facies at great depths and pressures. Whereas nephrites generally form under low pressure and medium to hot temperatures (emanating from the igneous intrusion), jadeitites form at high pressures and low temperatures (maximum $P < 2.5$ GPa and $T < 600$°C but more generally $P < 2.0$ GPa and $T < 500$°C) (Harlow *et al.* 2015: 118). These pressures correspond to between 20 and 85 km in depth and correlate with the blueschist and eclogite facies in the subduction trajectory (III in Figure 2.8), underwriting the statement of Shi *et al.* (2012: 364): "jadeitites are clearly the result of HP/LT conditions produced only in subduction zone environments."

Silica-Si can be extracted by circulating fluids from such mafic blocks of rock, contributing to the Na-, Al- Si-rich fluids that precipitate jadeite. This is not a one-time process, as many jadeite crystals exhibit zoning (growth under different chemical conditions); thus, the composition of the fluids may vary over time or by source. Further detailed discussion of rock-rock chemical interactions, the nature and derivation of the fluids that are necessary for jadeite formation, and the exact processes providing for these interactions and transformations can be found in Harlow & Sorenson (2001, 2005), Harlow *et al.* (2014), Tsujimori & Harlow (2018), and references therein.

Jadeitites within blueschist/eclogite rocks

Several other rocks have been named as hosts for jadeitite production; these are metabasites: "metamorphosed mafic rock that has lost all traces of its original texture and mineralogy owing to complete recrystallization" (Mindat.org: Metabasite). In the case of jadeitite formation, they have been metamorphosed to the blueschist or eclogite facies (Figure 2.8). Similar to serpentinite, cracks and fissures form in the rocks and are invaded by fluids, which can theoretically produce either precipitate or replacement jadeitite. These have been further distinguished in notation as P_B-type and R_B-type, the B indicating blueschist.

BOX 14 uses the Japanese jadeitites to illustrate some of the parameters of jadeite formation discussed above, as currently available in the literature. Descriptions of the geological belts listed can be found in Appendix 6. It is notable that the jadeitites found in the Renge Metamorphic Belt, associated with the Oeyama Ophiolite Belt, and in the Kurosegawa Belt are some of the oldest in the world, while those found in the later Sanbagawa and Kamuikotan Belts are much more recent.

Conclusions

Nephrite in general is formed through metasomatism of either dolomarble (D-nephrite) or serpentinite (S-nephrite). D-nephrite is a product of contact metasomatism involving high temperatures but low pressure, while S-nephrite forms at greenschist facies P/T.

D-nephrite from dolomite (dN) (white to light green) is found in areas where dolomitic limestone, (dolostone) a shallow-seas rock, has been obducted onto land and either marbleized to dolomitic marble (dolomarble) in regional metamorphism or by being heated through contact metamorphism. Nephrite, mainly monomninerallic tremolite, forms from fluids exuded by felsic igneous intrusion in contact with the dolomitic rock (the process of metasomatism), with tremolite replacing the dolomite. The tectonic activity of obduction usually involves the closing of a sea via oceanic plate subduction, often a back-arc basin sea, sandwiching both shallow limestone and slices of oceanic floor (ophiolites) into the accreted or collided landmasses. Thus, areas of ophiolite distribution in continental settings, particularly fossil subduction zones as in western China, are prime sources for finding tremolitic nephrite (dN).

S-nephrite, from serpentine (sN) (dark green), is most often produced in serpentinite mélanges which incorporate numerous other oceanic rocks. Contact between the serpentinite and rock inclusions by metasomatic fluids transfers calcium and silica to replace serpentine minerals with actinolite, the major mineral of S-nephrite. Such nephrite can also be produced by the replacement of the intrusive rock's

BOX 14 Tectonic Contexts of Japanese Jadeite

Itoigawa jade, Kotaki Workshop Matsumoto City [www.itoigawahisui.jp]

The most prominent source of jadeitite in Japan is the Itoigawa complex in Niigata Prefecture; Itoigawa jades have been used throughout prehistory for ornament production, particularly the curved bead seen here: *magatama*, one of the three "Imperial Treasures". The best source of jadeitites is on the Kotaki stream, a tributary of the Hime River that flows through Itoigawa City. Several colours of Itoigawa jadeitites are known: green from chromite or Fe^{2+}, and purple or blue from titanium or iron. But it is the white jadeite marbled with green (from Fe or Cr) or with dark green (from 2–8% omphacite, Abduriyim *et al.* 2017) that was most highly prized in prehistory.

Nine locations of jadeitite occurrence have been discovered in Japan:

Location	Geological Belt	Protolith	Host rock	Jadeite & accessory minerals		Dates
Itoigawa-Omi/ Happou (Niigata)	Oeyama/ Renge MB	mantle dunite, harzbergite. lherzolite	serpentinite (antigorite) mélange	P-type (no Q)	R-type relic horn-blende, gabbroic texture	zircon in jadeite U-Pb ca. 520 Ma, exhumed 340–320 Ma
Osa(yama) (Okayama)	Oeyama/Renge MB blueschists	mantle harzbergite, dunite	blueschists, in serpen-tinite (lizardite/ chrysotile) mélange	P-type (no Q) omphacite pectolite analcime titanite	R-type	U-Pb 521–451 Ma, exhumed 320 Ma
Oya/Sekinomiya Wakasa (Tottori)	Oeyama/ Renge MB blueschists	mantle spinel-harzbergite	serpentinite (antigorite) mélange	P-type (no Q)		?K-Ar 470–440 Ma exhumed at 290–280 Ma
Yatsushiro (Kumamoto)	Kurosegawa MB	gabbro	meta-gabbro			
Engyoji/ Inocho (Kochi)	Kurosegawa MB		serpentinite mélange			
Nishisonogi (Nagasaki)	Sanbagawa MB		serpentinite		R-type + Q, detrital Zrn	82 Ma, exhumed ca. 90–75Ma
Mikkabi (Shizuoka)	Sanbagawa MB		metagabbro			
Yorii/Chichibu (Saitama)	Sanbagawa MB		serpentine> actinolite			
Asahikawa (Hokkaido)	Kamuikotan MB		serpentine greenschist			

At least two series of highP/lowT metamorphic belts (MB) have yielded jadeite in Japan: Renge and Kurosegawa, dating early-to-middle Palaeozoic, and Sanbagawa and Kamuikotan, dating to the Cretaceous. Renge and Kurosegawa belts were described in Chapter 3 as belonging to the South Kitakami Terrane, which evidences the earliest subduction off the North China Block but is now broken up through western Japan. Except for Itoigawa-Omi, the content of other jadeite-rich sources generally contain between 50 and 80% jadeite (Abduriyim *et al.* 2017), with occasional values rising above 80%. Most do not qualify as jadeitite and none but Itoigawa-Omi are recognized sources for archaeological bead-working. However, the facts that Jd>80% can occur in other areas and that some Wakasa jadeitite approximates Itoigawa-Omi (see Tsujimori 2017: fig. 2) should keep us alert to the possibility of such hard stones being utilized as jade. Moreover, eclogite-blueschist facies rocks and serpentenite occur in more localities than above for future jadeite-rich rock discoveries. Since the South Kitakami Terrane is thought to continue into the Primorye region of Russia, there is good reason to think that jadeitite might be found there as well as in the Uralides and Altides as listed in Harlow *et al.* (2015: table 2).

minerals through metasomatism. S-nephrite is also distributed in relation to ophiolites, whether the serpentinite is derived from mantle rocks or from serpentinized ocean crust.

Jadeite in jadeitites is a product of high pressure / low temperature metasomatism in serpentinized mantle rock (i.e., serpentinite mélange) in the subduction channel. Jadeitites occur on land through exhumation processes and are accompanied by rocks that have been metamorphosed to the blueschist or eclogite facies. Jadeite can be a precipitate of fluids (P-jadeite) or results from mineral replacement (R-jadeite); when these form in serpentinite, they are annotated P_S-jadeite and R_S-jadeite. Jadeites formed in blueschist facies rock are indicated as P_B- and R_B- type jadeite. The distribution of blueschist-eclogite facies rocks in mountain chains around the world correlate with sources of the 19 currently known locales of jadeitite discovery.

The data presented in this Chapter are only a minor proportion of those available in the literature, where details of mineral associations and assemplages, as well as trace elements and isotopic variations, can be found. These are crucial to identifying and distinguishing sources of archaeological jades, but as yet, very few matches of artefacts and source have been made, giving great scope for future research on this topic. Knowledge of the mechanisms and location of formation of the true jades and the processes by which they are exposed on dry land aid in identifying sources where raw materials of nephrite and jadeitite might be sourced.

Acknowledgments

This Chapter draws heavily on George Harlow's work with his colleagues. He is a geologist with the American Museum of Natural History but is very much involved in archaeological issues. He stands as role model of an ideal relationship between geologist and archaeologist, and I appreciate the time he as given me in reading drafts and answering questions by email.

Figures & Tables Sources

Figure 12.1 redrawn from (Tsujimori & Harlow 2017: fig. 4)
Figure 12.2 redrawn from (Guynn *et al.* 2006: fig 5)
Figure 12.3 compiled from (Wittke 2009: appendix A; Nelson 2011)
Figure 12.4 courtesy of Iizuka Yoshiyuki (cf. Iizuka *et al.* 2005: plate 1)
Figure 12.5 after [www4.nau.edu/meteorite/Meteorite/Book-GlossaryP.html] (dead link)
Figure 12.6 redrawn from (Nelson 2011)
Figure 12.7 graph by the author
Figure 12.8 after (Zhang Q *et al.* 2016: fig. 2).
Figure 12.9 after (Zhang ZW *et al.* 2011: fig. 4)
Figure 12.10 modified from (Harlow & Sorensen 2005: fig. 12)
Figure 12.11 modified from (Harlow & Sorensen 2005: fig. 14)

BOX 13 illustration after (Hung *et al.* 2006: fig. 20.9)
BOX 14 photograph after [www.itoigawahisui.jp/page/20], with permission from the Itoigawa Jade Workshop (Itoigawa Hisui Kōbō)

Table 12.1 compiled from (Ballirano *et al.* 2008); definitions of major, minor, and trace elements taken from (Blake 2001: 87)
Table 12.2 compiled from (Deer *et al.* 1997*a,b*; Harlow *et al.* 2014; Desautels 1986; Jiang *et al.* 2020; Harlow & Sorensen 2001)
Table 12.3 compiled from the literature
Table 12.4 compiled from (Deer *et al.* 1997*b*: 136–138; Hawthorne *et al.* 2012: 2036)

Table 12.5 mineral abbreviations according to Siivola & Schmid (2007), formulae according to IMA-CNMNC (2022), classifications from (Morimoto *et al.* 1988: table 2)

Table 12.6 after (Harlow & Sorensen 2005: table 3)

BOX 14 table compiled from (Tsujimori & Harlow 2012: table 1, sec. 2.5-2.8; Abduriyim *et al.* 2017; Tsujimori 2017; Tsutsumi *et al.* 2010)

References

ABDURIYIM, Ahmadjan; Kazuko SARUWATERI & Yusuke KATSURADA (**2017**) "Japanese jadeite: history, characteristics, and comparison with other sources." *Gems & Gemology* 53.1: 48-67.

BALLIRANO, Paolo; Giovanni B ANDREOZZI & Girolamo BELARDI (**2008**) "Crystal chemical and structural characterization of fibrous tremolite from Susa Valley, Italy, with comments on potential harmful effects on human health." *American Mineralogist* 93: 1349-1355.

BARNES, Gina L (**2018**) "Understanding Chinese jade in a world context." *Journal of the British Academy* 6: 1-63 [DOI: 10.85871/jba/006.001].

BERSANI, D; A ANDÒ & L SCROCCO *et al.* (**2014**) "Study of the composition of amphiboles in the tremolite–ferro-actinolite series by micro-raman and sem-edxs." *11th International GeoRaman Conference* [www.hou.usra.edu/meetings/georaman2014/pdf/5063.pdf].

BERSANI, Danilo; Sergio ANDÒ & Laura SCROCCO *et al.* (**2019**) "Composition of amphiboles in the Tremolite–Ferro-Actinolite Series by Raman spectroscopy." *Minerals* 9.#491: 12pp. [DOI: 10.3390/min9080491].

BLAKE, Stephen (**2001**) *Melting the mantle*. Milton Keynes: The Open University.

CAMPBELL, Gareth P; Gordan M MISKELLY & James M CURRAN *et al.* (**2008**) *Elemental analysis of New Zealand nephrite jade (pounamu) for forensic purposes*. UdG Digital Repository, Universitat de Girona [dugi-doc.udg.edu].

CASADIO, Francesca; Janet DOUGLAS & Katherine T FABER (**2007**) "Noninvasive methods for the investigation of ancient Chinese jades: an integrated analytical approach." *Analytical and Bioanalytical Chemistry* 387: 791-801.

CHAKRABORTY, Kasturi; Praveer PANKA & Prabir Kumar MUKHOPADHYAY *et al.* (**2022**) "Lizardite – chrysotile veins from Vempalle Formation of Cuddapah Supergroup : a study in fluid-rock interaction." *Current Science* 122.7: 826-839.

CHAPPELL, BW & AJR WHITE (**2001**) "Two contrasting granite types: 25 years later." *Journal of Volcanology and Geothermal and Research* 48.4: 489-499.

CHEN, Qi; Yongya WANG & Fuxi GAN (**2014**) "The chemical compositions and origin characters of serpentine jade from different deposits." *Applied Mechanics and Materials* 624: 119-123.

CHEN, T-H; T CALLIGARO & S PAGÈS-CAMAGNA *et al.* (**2004**) "Investigation of Chinese archaic jade by PIXE and μRaman spectrometry." *Applied Physics A: Materials Science & Processing* 79: 177-189.

CHENG, HS; ZQ ZHANG & B ZHANG *et al.* (**2004**) "Non-destructive analysis and identification of jade by PIXE." *Nuclear Instruments and Methods in Physics Research B* 219-220: 30-34 [DOI: 10.1016/j.nimb. 2004.01.023].

CORTESOGNO, Luciano; Laura GAGGERO & Gabriella LUCCHETTI *et al.* (**2002**) "Compositions and miscibility gap in Na-Ca clinopyroxenes through high-pressure metamorphism." *Peridico di Mineralogia* 71.1: 1-25 [http://tetide.geo.uniroma1.it/riviste/permin/testi/V71/1.pdf].

DEER, William Alexander; Robert Andrew HOWIE & J. ZUSSMAN (**1997*a***) *Rock-forming minerals. Vol. 2A: Single-chain silicates [jadeite]*. London: Geological Society of London [brackets inserted].

---- (**1997*b***) *Rock-forming minerals. Vol. 2B: Double-chain silicates [nephrite]*. London: Geological Society of London [brackets inserted].

Dept of Geological Sciences, U Texas at Austin (**2009**) "Jade: jadite, nephrite" [www.geo.utexas.edu/courses/347k/redesign/gem_notes/jade/jade_main.htm].

DESAUTELS, Paul E (**1986**) *The jade kingdom*. New York: Van Nostrand Reinhold.

DOBRETSOV, Nikolai Leontyevich (**1984**) "Problem of the jadeite rocks, associating with ophiolites." *Mineralia slovaca* 16.1: 3-12.

FRANZ, Leander; Tay Thye SUN & Henry A HÄNNI *et al.* (**2014**) "A comparative study of jadeite, omphacite and kosmochlor jades from Myanmar, and suggestions for a practical nomenclature." *The Journal of Gemmology* 34.3: 210-229 [DOI: 10.15506/JoG.2014.34.3.210].

FROST, B Ronald & Carol D FROST (**2014**) *Essentials of igneous and metamorphic petrology*. Cambridge University Press.

GAHK (Gemmological Association of Hong Kong) (**2006**) *Standard methods for testing Fei Cui (Jadeite Jade)*. Hong Kong: The Gemmological Association of Hong Kong, Ltd.

GALOPIM DE CARVALHO, Rui; Edward JOHNSON & Edward LIU (**2020**) "Jadeite jade or 翡翠 feicui ?" *15th Home Gemmology Webinar* supported by CIBJO _ the World Jewellery Confederation [www.youtube.com/watch?v=eYB-ocSe3oA].

GemSelect (**2019**) "Omphacite jade gemstone information" [www.gemselect.com/gem-info/ omphacite-jade/omphacite-jade-info.php].

GUBELIN, Edward J (**1964-65**) "Maw-sit-sit, a new decorative gemstone from Burma." *Gems & Gemology* XI.8: 227-238, 255.

GUYNN, J; P KAPP & A PULLEN *et al.* (**2006**) "Tibetan basement rocks near Amdo reveal 'missing' Mesozoic tectonism along the Bangong suture, central Tibet." *Geology* 34.6: 505-508.

HARLOW, George E & E Peter OLDS (**1987**) "Observations on terrestrial ureyite and ureyitic pyroxene." *American Mineralogist* 72: 126-136.

HARLOW, GE & SS SORENSEN (**2001**) "Jade: occurrence and metasomatic origin." *Australian Gemmologist* 21.1: 7-10.

---- (**2005**) "Jade (nephrite and jadeitite) and serpentinite: metasomatic connections." *International Geology Review* 47.2: 113-146 [DOI: 10.2747/0020-6814.47.2.113].

HARLOW, GE; Sorena S SORENSEN & Virginia B SISSON (**2007**) "Jade", pp. 207-254 in *The geology of gem deposits*, ed. by LA GROAT. Quebec: Mineralogical Association of Canada.

HARLOW, George E; Sorena S SORENSEN & Virginia B SISSON *et al.* (**2014**) "Chapter 10: The geology of jade deposits", pp. 305-374 in *The geology of gem deposits*, ed. by LA GROAT. Québec: Mineralogical Association of Canada.

HARLOW, GE; T TSUJIMORI & SS SORENSEN (**2012**) "Introduction." *European Journal of Mineralogy* 24: 197-198.

---- (**2015**) "Jadeitites and plate tectonics." *Annual Review of Earth and Planetary Sciences* 43: 105-138 [DOI: 10.1146/annurev-earth-060614-105215].

HAWTHORNE, Frank C; Roberta OBERTI & George E HARLOW *et al.* (**2012**) "IMA Report: Nomenclature of the amphibole supergroup." *American Mineralogist* 97: 2031-2048.

HERTWIG, Andreas; Walter V MARESCH & Hans-Peter SCHERTL (**2021**) "Jadeitite and related rocks in serpentinite mélanges from the Rio San Juan Complex, Dominican Republic: evidence for both isochemical replacement and metasomatic desilication of igneous protoliths with fluid-assisted jadeite growth." *Russian Geology and Geophysics* 62.5: 496-524 [DOI: 10.2113/RGG20204265].

HOBBS, Jill M (**1982**) "The jade enigma." *Gems & Gemology* 1982.Spring: 3-19.

HUGHES, Richard W; Oliver GALIBERT & George BOSSHART *et al.* (**2000**) "Burmese jade: the inscrutable gem." *Gems & Gemology* 36.1: 2-26.

HUNG, Hsiao-chun; Yoshiyuki IIZUKA & Peter BELLWOOD (**2006**) "Taiwan jade in the context of Southeast Asian archaeology", pp. 203-215 in *Uncovering Southeast Asia's past: selected papers from the 10th International Conference of the European Association of Southeast Asian Archaeologists*, ed. by EA BACUS *et al.* Singapore: NUS Press.

HUNG, Hsiao-chun; Yoshiyuki IIZUKA & Peter BELLWOOD *et al.* (**2007**) "Ancient jades map 3,000 years of prehistoric exchange in Southeast Asia." *PNAS* 104.50: 19745-19750.

IIZUKA, Yoshiyuki (**2012**) "Archaeological mineralogy: analyses of jade by LV-SEM and their results." *Japan Electronics News* 44: 23-39 (in Japanese).

---- (**2021**) "Research on nephrite stone tools in Southeast Asia's prehistoric period." *Kokogaku Kenkyu* 68.1 (269): 106-109 +plates (in Japanese)

IIZUKA Yoshiyuki & Hsiao-Chun HUNG (**2005**) "Archaeomineralogy of Taiwan nephrite: sourcing study of nephritic artifacts from the Philippines. *Journal of Austronesian Studies* 1(1): 35-79 [https://mocfile.moc.gov.tw/files/201807/2.pdf]

IIZUKA, Yoshiyuki; Hsiao-Chun HUNG & Peter BELLWOOD (**2007**) "A noninvasive mineralogical study of nephrite artifacts from the Philippines and surroundings: the distribution of Taiwan nephrite and Implications for island Southeast Asian Archaeology", pp. 12-19 in *Scientific research on the sculptural arts of Asia*, ed. by J DOUGLAS *et al.* London: Archetype Publications.

IIZUKA, Yoshiyuki; Cheng-hwa TSANG & Kun-shiu LEE (**2005**) "Archaeomineralogy on nephritic jade artifacts from Peinan site, eastern Taiwan", pp. 38-57 in *Peinan jade artifacts in the collection of the National Museum of Prehistory*, ed. by CH TSANG & MC YEH. Taitung: National Museum of Prehistory.

IMA-CNMNC (International Mineral Association – Commission on New Minerals, Nomenclature and Classification) (**2022**) "The new IMA list of minerals – a work in progress – updated: July 2022" [http://cnmnc.main.jp/imalist.htm].

JIANG, Ying; Guanghai SHI & Liguo XU *et al.* (**2020**) "Mineralogy and geochemistry of nephrite jade from Yinggelike deposit, Altyn Tagh (Xinjiang, NW China)." *Minerals* 10.#418: 22 pp.

KEVERNE, Roger (ed.) (**1995**) *Jade*. London: Lorenz Books.

KING, Hobart M (**2014**) "Maw Sit Sit." *Geoscience News and Information* [https://geology.com/gemstones/maw-sit-sit/].

KUMAR, Santosh & Rishi Narain SINGH (eds) (**2014**) *Modelling of magmatic and allied processes*. Dordrecht: Springer.

LENNARTSON, Anders (**2014**) "The colours of chromium." *Nature Chemistry* 6: 942.

LIANG, Shuai; Yongchun WANG & Furong ZHAI (**2012**) "Yingkou YuLong jade ore forming conditions and mineralization analysis." *Advanced Materials Research* 356-360: 3020-3022.

LIN, Chenlu; Xuemei HE & Zhiyun LU (**2020**) "Phase composition and genesis of pyroxenic jadeite from Guatemala: insights from cathodoluminescence." *Royal Society of Chemistry* 10: 15937-15946.

LIU, Yan; Jun DENG & Guanghai SHI *et al.* (**2011*a***) "Geochemistry and petrology of nephrite from Alamas, Xinjiang, NW China." *Journal of Asian Earth Sciences* 42.3: 440-451.

---- (**2011*b***) "Geochemistry and petrogenesis of placer nephrite from Hetian, Xinjiang, northwest China." *Ore Geology Reviews* 41: 122-132.

LIU, ZhiYong; FuXi GAN & HuanSheng CHENG *et al.* (**2009**) "Mineral structure and nondestructive analysis of Xiuyan jade from Liaoning Province." *Acta Petrologica Sinica* 25.05: 1281-1287.

MCCLURE, Shane F (**2012**) "The jadeitie/omphacite nomenclature question." *GIA News from Research* [www.gia.edu/research-resources/news-from-research].

MIDDLETON, Andrew & Janet AMBERS (**2005**) "Case study: an analysis of nephrite jade using Raman microscopy and x-ray fluorescence spectroscopy", pp. 403-411 in *Raman spectroscopy in archaeology and art history*, ed. by HGM EDWARDS & JM CHALMERS. London: Royal Society of Chemistry.

MIDDLETON, Andrew P & Ian FREESTONE (**1995**) "The mineralogy and occurrence of jade", pp. 413-423 in *Chinese jade*, ed. by J RAWSON. London: British Museum Press.

MINDAT (Mineral Database) by Jolyon & Katya RALPH via Hudson Institute of Mineralogy [mindat.org].

MINSOCAM (Mineralogical Society of America) (**2016**) "Mohs' scale of hardness." [www.minsocam.org/MSA/collectors_corner/article/mohs.htm] (dead link).

MORIMOTO, N *et al.* (**1988**) "Nomenclature of pyroxenes." *Mineralogy and Petrology* 39: 55-56.

NELSON, Stephen A (**2011**) "Inosilicates (pyroxenes and amphiboles)." *Tulane University course material, EENS2110* [www.tulane.edu/~sanelson/eens211/inosilicates.htm].

NESTOLA, Fabrizio; Tiziana Boffa BALLARAN & Christian LIEBSKE *et al.* (**2006**) "High-pressure behaviour along the jadeite NaAlSi2O6–aegirine NaFeSi2O6 solid solution up to 10 GPa." *Physics Chemistry Minerals* 33: 417-425 [DOI: 10.1007/s00269-006-0089-7].

NEWTON, Robert C. (**2014**) "Book review: Metasomatism and the chemical transformation of rock." *Elements* April: 155.

NICKEL, Ernest H & Monte C NICHOLS (**1991**) *Mineral reference manual*. New York: Van Nostrand Reinhold.

---- (**2009**) *IMA/CNMNC list of mineral names*. Livermore, CA: Materials Data, Inc.

PERKINS, Dexter; Andrea KOZIOL & John BRADY (**2021**) "Thermodynamics." *Teaching Phase Equilibria* [https://serc.carleton.edu/research_education/equilibria/thermodynamics.html].

RAO, A.N. Vyasa & MS MURTY (**1980**) "A study of the serpentinisation in the Vempalle dolomitic limestones near Pulivendala, Cuddapah District." *Proceedings of the Indian Academy of Science (Earth Planet Sciences)* 89.1: 17-22.

SAX, Margaret; Nigel D MEEKS & Carol MICHAELSON *et al.* (**2004**) "The identification of carving techniques on Chinese jade." *Journal of Archaeological Science* 31.10: 1413-1428.

SAX, Margaret; Carol MICHAELSON & Nigel D MEEKS (**2008**) "A methodology to identify techniques of jade carving in China: the Neolithic period", pp. 279-295 in *New frontiers in global archaeology: defining China's ancient traditions*, ed. by T LAWTON. Beijing: AMS Foundation for the Arts, Sciences and Humanities.

SHI, Miao; Bingsong YU & Ying GUO (**2015**) "Structural and mineralogical characterization of green nephrite in Hetian, Xinjiang, China (abstract)." *Key Engineering Materials* 633: 159-164.

Shigeno, M; Y Mori & T Nishiyama (**2005**) "Reaction microtextures in jadeitites from the Nishisonogi metamorphic rocks, Kyushu, Japan." *Journal of Mineralogy and Petrology* 100: 237-246.

Sieber, K (**2021**) "Chloromelanite." *EPI Institut Für Edelstein Prüfung* [www.epigem.de/en-us/topics/chloromelanite.html].

Siivola, Jaakko & Rolf Schmid (**2007**) "12. List of Mineral Abbreviations." *British Geological Survey Papers* [www.bgs.ac.uk/scmr/docs/papers/paper_12.pdf].

Stanford (**n.d.**) "Mineral makeup of seawater."Stanford University, CA [https://web.stanford.edu/group/Urchin/mineral.html].

Stern, Robert J; Tatsuki Tsujimori & George Harlow *et al.* (**2013**) "Plate tectonic gemstones." *Geology* 41.7: 723-726.

Tang, Chung (**2008**) "Observations on the earliest slit rings in North China", pp. 265-275 in *Memorial to Professor SERIZAWA Chosuke: archaeology, ethnology, history papers*. Tokyo: Rokuichi Shobo.

Tennant, W Craighead; Rodney FC Claridge & Catherine A McCammon *et al.* (**2005**) "Structural studies of New Zealand pounamu using Mössbauer spectroscopy and electron paramagnetic resonance." *Journal of the Royal Society of New Zealand* 35.4: 385-398.

Tsien, Hsien Ho (**1996**) "Mineralogical studies of archaic jades." *Acta Geologica Taiwanica* 32: 5-8 (intro).

Tsujimori, Tatsuki & George E Harlow (2012) "Petrogenetic relationships between jadeitite and associated high-pressure and low-temperature metamorphic rocks in worldwide jadeitite localities: a review." *European Journal of Mineralogy* 24.March: 371-390 [DOI: 10.1127/0935-1221/2012/0024-2193].

---- (**2017**) "Jadeitite (jadeite jade) from Japan: history, characteristics, and perspectives." *Journal of Mineralogical and Petrological Sciences* 112: 184-196 [DOI: 10.2465/jmps.170804].

Tsutsumi, Yukiyasu; Kazumi Yokoyama & Ritsuro Miyawaki *et al.* (**2010**) "Ages of zircons in jadeitite and jadeite-bearing rocks of Japanese Islands." *Bulletin of the National Museum of Natural Sciences* Series C, 36: 19-30.

Walker, Jill (**1991**) "Jade: a special gemstone", pp. 19-41 in *Jade*, ed. by R Keverne. London: Anness Publishing 1991> Lorenz 1995 > Springer Science-Business Media, New York 1991.

Wang, R (**2011**) "Progress review of the scientific study of Chinese ancient jade." *Archaeometry* 53.4: 674-692.

Wang, Rong & Yuesheng Li (**2011**) "Multiexcitation Raman spectroscopy in identification of Chinese jade." *Spectroscopy letters: an international journal for rapid communication* 44.6 [DOI: 10.1080/ 00387010.2011.577885].

Wang, R & W-S Zhang (**2011**) "Application of Raman spectroscopy in the nondestructive analyses of ancient Chinese jades." *Journal of Raman Spectroscopy* 42.6: 1324-1329.

Wang, YY; FX Gan & HX Zhao (**2012**) "Nondestructive analysis of Lantian jade from Shaanxi Province, China." *Applied Clay Science* 70: 79-83.

Warr, LN (**2021**) "IMA-CNMNC approved mineral symbols." *Mineralogical Magazine* 85.3: 291-320 [DOI: 10.1180/mgm.2021.43].

Wen, Guang (**1989**) "Ancient jade of China." *China Non-metallic Mining Industry Herald* (March) [http://en.cnki.com.cn/Article_en/CJFDTOTAL-LGFK198903003.htm].

---- (**1994**) "Some features of nephrite and its prospecting significance (abstract)." *Institute of Geology, Chinese Academy of Geological Sciences Collected Works* 27:[http://cpfd.cnki.com.cn/Article/ CPFDTOTAL-ZGDJ199400003005.htm].

Wen, Guang & Zhichun Jing (**1992**) "Chinese Neolithic jade: a preliminary geoarchaeological study." *Geoarchaeology: an international journal* 7.3: 251-275.

---- (**1996**) "Mineralogical studies of Chinese archaic jade." *Acta Geologica Taiwanica 32: 55-84.*

Wittke, James (**2009**) *Meteorite Book*. College of Engineering, Forestry, and Natural Sciences, Northern Arizona University [http://nau.edu/search.aspx?q=Meteorite+book] (dead link).

Xiao, Qi-yun; Ke-qin Cai & Fu-jian Jiang (**2009**) "Tentative discussion on the mineral cataclasis jade-forming process of Dushan jade in Nanyang City, Henan Province." *Acta Geoscientica Sinica* 30.05: 607-615.

Yoder, Hatten S (**1950**) "The jadeite problem." *American Journal of Sciences* 248 (April, May): 225-248, 312-334 [DOI: 10.2475/ajs.248.4.225].

Yui, TF & ST Kwon (**2002**) "Origin of a dolomite-related jade deposit at Chuncheon, Korea." *Economic Geology* 97: 593-601.

Zhang, Baoshuai; Xiaotong Wu & Sun, Yufeng *et al.* (**2020**) "Complex raw materials and supply system: mineralogical and geochemical study for the jade artefacts of Longshan culture (2400~2000BC) from Sujiacun site in coastal Shandong, China." *Archaeometry* 63.3: 18 pp. [DOI: 10.1111/arcm.12634].

ZHANG, Guowei; Zaiping YU & Yong SUN *et al.* (**1989**) "The major suture zone of the Qinling orogenic belt." *Journal of Southeast Asian Earth Sciences* 3.1/4: 63-76.

ZHANG, Lili and YUAN, Xinqiang (2017) "Discussion on accurate Chinese translations of 'nephrite' and 'jadeite'." *International Journal of Information and Education Technology* 7.9: 707-711 [DOI: 10.18178/ijiet.2017.7.9.958].

ZHANG, Qichao; Yan LIU & He HUANG *et al.* (**2016**) "Petrogenesis and tectonic implications of the high-K Alamas calc-alkaline granitoids at the northwestern margin of the Tibetan Plateau: geochemical and Sr–Nd–Hf–O isotope constraints." *Journal of Asian Earth Sciences* 127: 137-151.

ZHANG, Rubai (**2022**) "A preliminary study on the jade materials of ancient Shu culture in the folk collection" *iNEWS* [https://inf.news/en/culture/06a23e3dde9ed2493c1febf498d4c2ab.html].

ZHANG, Zhu-wu; Fu-xi GAN & Huang-sheng CHENG (**2010**) "Chemical composition characteristic of nephrite formed by different metallogenic mechanisms and geological environments." *Acta Mineralogica Sinica* 2010.03 [http://en.cnki.com.cn/Article_en/CJFDTOTAL-KWXB201003016.htm].

---- (**2011**) "PIXE analhysis of nephrite minerals from different deposits." *Nuclear Instruments and Methods in Physics Research* B 269: 460-465.

ZHANG, ZW; YC XU & HS CHENG *et al.* (**2012**) "Comparison of trace elements analysis of nephrite samples from different deposits by PIXE and ICP-AES." *X-Ray Spectrometry* 41.6: 367-370 [DOI: 10.1002/xrs.2413].

ZHONG, Qian; Zongting LIAO & Ligian QI *et al.* (**2014**) "Black nephrite jade from Guangxi, southern China." *Gems & Gemology* 55.2: 198-215 [DOI: 10.5741/GemS.55.2.198].

Part II

Reflections

The chapters in Part II began as case studies of Japanese archaeology carried out under conditions of volcanic eruptions, earthquakes, and tsunami (Chapters 8, 9, 10). Far from remaining isolated sub-disciplines, however, they were illustrated in Chapter 11 to be interconnected by tectonic processes which can affect almost any site. This suggests that all archaeologists in Japan are tectonic archaeologists once their eyes have been opened to the possibilities and realities of tectonic effects. It is also the case that such archaeological studies have great relevance for modern Japan, as tectonic processes which affected life in the past are also affecting life in the present and future. Thus, both hazard research and mitigation and Disaster Archaeology (Chapter 11) illustrate the continuity between the past, present, and future.

By presenting these issues, it was not intended to establish firm disciplinary boundaries but to alert the greater archaeological community of the importance of tectonic processes in both the creation of our habitable space and the associated risks of inhabiting them. This volume has focussed on the subduction zone in which Japan sits, but many of the lessons herein are applicable to other areas – often thought to be stable or benign until some disaster is wrought upon them.

Chapter 12 represents a different aspect of Tectonic Archaeology that is relevant anywhere – the creation of stone resources used throughout prehistory and history. The true jades are the most extreme example, as they are directly a product of Plate Tectonics as carried out in the subduction channel itself (jadeitite) and in the supra-subduction zone (nephrite). Understanding not just where stone resources reside in the landscape but how they came to be there deepens our knowledge of the relationship between Earth processes and human utilization of their products. A full book could be written on the stone resources of Japan (or anywhere), their formation and current loci, and how these resources fit into the political economy of a certain time period. Such a picture needs to be built up from individual studies. But to integrate them into a society-wide economic system – together with other kinds of artefacts – is a task for the future. Here, a chapter on the true jades will suffice.

Part III

Nara Basin Studies

When I began my doctoral fieldwork in Nara, Japan in 1977, I had only Bill Farrand's University of Michigan course "Geology for Archaeologists" under my belt. Luckily in Nara, I was given copies of Nara Basin topographic sheets (1:25,000 scale) on which to locate the Yayoi-Kofun site survey remains available from the Prefectural survey maps of 1971-75 (Kashikoken 1971-75). Information for the site survey maps was inscribed onto 1:10,000 maps, then reduced to 1:15,000 for publication (34 map sheets in all). My topographic maps, pasted together for just the basin area, extended about 2 x 3 m, very unwieldly to work with.

As I worked through the survey data, I realized the surface scatters that were identified were related directly to canal and pond locations. Unlike surveys in arid areas, the areal locations and extents of surface scatters did not represent the area and extent of subsurface archaeological remains but were only indicative of where such might be found. Moreover, the canals and ponds had been laid out to conform to a 7–9c gridded *jōri* paddy field system, which not only masked the archaeological remains underneath but also totally altered the Basin's surface (Barnes 1986). In order to know more about the site finds, I applied to the Nara Kashihara Archaeological Institute (Kashikoken) to see the bags of sherds that were collected during the survey. I was taken to a large prefab building in which the bags were stored and was invited to look at anything I wanted. I asked for specific site remains, coded to the survey maps. Alas, the bags were not stored in any order...Thus, to make sense of earlier Yayoi and Kofun remains known purely from the survey maps, the contemporaneous topographic context of the finds needed to be explicated. There were hints in the current landscape to indicate the natural directions of river flow and the levees they produced, now occupied by villages and vegetable gardens. These would be the key to reconstructing at least the natural flow direction of the rivers before canalization in the *jōri* system from the 8c.

It was suggested by my archaeology colleagues at Tenri University, whose excavations I had participated in, that I get in touch with the historical geographers at Nara Women's University, Professor Takehisa Yoshihiko and Senda Minoru. Prof. Takehisa set me up with a stereoscopic reader and 1964-date aerial photographs from Kokudo Chiri-in, the Geospatial Information Authority of Japan [gsi.go.jp], whereupon I began tracing the evidence for land height variations onto my unwieldly map in order to identify landforms different from the ubiquitous paddy-fields. I checked these landforms by fieldwalking. This experience not only benefitted me greatly, but a few years later Takehisa drew his knowledge together and published a pamphlet for doing such research (Takehisa 1989). At a similar time, I was given an introduction to Ichikawa Koichiro, a geologist at Nara Educational University who researched the Median Tectonic Line. He opened doors to understanding what the landforms I had found in analyzing the aerial photos might mean. Indeed, Goff *et al.* (2021: 769) have stated that worldwide, "It is with aerial photography that landscape archaeology truly started to grow" – true for Japan as well.

Tutoring by and interactions with these scholars resulted in a map of landforms within the Nara Basin and an appendix on Nara geology in my submitted dissertation (Barnes 1983) but not in its published form (Barnes 1988). The map is presented here as Figures 13.2 and 14.1, while the appendix is updated and expanded as Chapter 13. Chapter 14 summarizes the tests of this landscape reconstruction. The first test was a coring project in the Asawa area of Tenri City in 1984 (JRG 1986, 1987), while the second was a coring and excavation project in Miwa, Sakurai City in 1988 and 1989 (Barnes & Okita 1993). The excavation was the first time that a foreigner (myself) at a foreign university (Cambridge, UK) was granted a licence to excavate in Japan. It is my ongoing regret that no-one has since followed in these footsteps. Chapter 15 compares soil samples from Nara with others to determine bone preservation as seen at the Shibu site.

References

BARNES, Gina L (**1983**) *Yayoi-Kofun settlement archaeology in the Nara Basin, Japan.* Ann Arbor, MI.: University Microfilms (ProQuest). Published as Barnes 1988.

---- (**1986**) "Paddy field archaeology." *Journal of Field Archaeology* 13.4: 371-379.

---- (**1988**) *Protohistoric Yamato: archaeology of the first Japanese state.* Anthropological Papers 78 and Modern Papers in Japanese Studies 17. Ann Arbor: Museum of Anthropology and the Center for Japanese Studies, University of Michigan.

BARNES, Gina L & Masaaki OKITA (eds) (**1993**) *The Miwa Project: survey, coring and excavation at the Miwa site, Nara, Japan.* BAR International Series 582. Oxford: Tempvs Reparatvm.

GOFF, James; Bruce MCFADGEN & Nick MARRINER (**2021**) "Landscape archaeology—the value of context to archaeological interpretation: a case study from Waitore, New Zealand." *Geoarchaeology an International Journal* 36: 768-779 [DOI: 10.1002/gea.21864].

JRG [Joint Research Group on the Geomorphological Recognition and Land Utilization of Pre- and Protohistoric Japanese Peoples] (**1986**) "Natural environments in the Nara Basin through the pre- and protohistoric ages I: geology and geomorphology." *Kobunkazai Kyoiku Kenkyu Hokoku* 15: 1-30 (in Japanese with English title).

---- (**1987**) "Natural environments in the Nara Basin through the pre- and protohistoric ages II: descriptions of core samples and analysis on biogenic materials." *Kobunkazai Kyoiku Kenkyu Hokoku* 16: 23-74 (in Japanese with English title).

Kashikoken (**1971-1975**) *Nara Prefecture archaeological site maps.* Nara: Kashihara Archaeological Institute.

TAKEHISA, Yoshihiko (**1989**) *Basic research for systematizing methods for analyzing historical landscapes through aerial photos.* Nara: Geography Department, Nara Women's University.

Chapter 13

Nara Basin Geology & Geomorphology

As background for the consideration of Nara Basin geoarchaeology, after an initial tour of the Nara Basin's topography, this chapter reviews the local geology and geomorphology. The major geological processes are referenced to Part I as they affected the development of western Japan, fitting the Basin into the larger picture. The borders between upland and lowland features in the Basin have been largely obscured by historic settlement and agriculture over the last 1500 years. Reconstruction of the natural lay of the land by aerial photograph was carried out in the late 1970s (Barnes 1982), allowing an explanation of the protohistoric settlement patterning (Barnes 1983, 1988). That reconstruction of Nara Basin uplands is consistent with the sedimentary units identified in 1993 and now illustrated in the 'seamless' online map available from the Geological Survey of Japan (GSJ 2020). The geology of the region is more complex than shown in the 'seamless' map, though the Nara Basin area is far less complex than some other regions of Japan comprised of Accretionary Complex basements (cf. Chapter 3). The faulted structure of the Basin explains the Basin's features and its vulnerability to further ground movement in both types of documented earthquakes.

Topographic Tour

The Nara Basin (Figure 13.1) is only a small portion of a much larger Nara Prefecture (cf. Apx 2: Fig. A). Relatively small, ca. 25~30 km NS and 10~15 km EW, it entails 300 km^2 and occupies only 12% of the Prefecture's area but contains 85% of its population as of 1996 (Ikeda & Ohashi 1996: 42) – but much more these days. The capital of Nara Prefecture, Nara City, sits at the northeast end of the Basin. As a basin, it is surrounded by mountains on three sides (W,S,E) and low hills on the north (ca. 100~200 m), separating it from Kyoto Basin. The Nara lowlands sit between 45 and 80 m and are connected to the Osaka Basin via the outlet of the Yamato River and some mountain passes; to the southwest, a narrow corridor leads over a low pass into the Kii River valley – the latter being the home of the Median Tectonic Line (MTL) of recent tectonic import (cf. Chapter 3). The mountains on the east are breached by the Hasé River in the southeast and, in the Kyoto Basin, by the Kizu River. Among these basins, Nara afforded the best extensive agricultural land in a very mountainous region, leading to its having become the home of the first Japanese state, Yamato (Barnes 1988). The Basin's mountains protected it from the main scourge of typhoons from the south, though the central Basin lowpoint was often "subject to violent floods during the typhoon period"; its location makes it free of the drought affecting the Inland Sea region in summer, though it has the lowest annual precipitation rate (40–60 inches) in the Prefecture (Hall 1932: 279, fig. 2).

As the current satellite photograph illustrates, modern urban sprawl has encroached everywhere up to the steepest slopes, but even now, there is more agricultural land in the Nara Basin than in the dense urban coverage of the Osaka Basin. In the late 1970s, Nara Basin had few cities, some large towns, and the rest of the Basin was dotted with farming hamlets 1-2 ha in size amongst a sea of paddy-fields. However, even then, the natural lay of the land was obscured by the agricultural infrastructure – a gridded field system (*jōri*) which had been instituted with the adoption of a Chinese administrative system in the 8c. Thus, to allow an understanding of the pre- and proto-historic settlement of the Basin, it was necessary to undertake a landscape reconstruction using available aerial photographs. These were mainly taken in the 1960s, but some post-war reconnaissance photos were also available from Kokudo Chiri-in, the Japanese Geospatial Information Authority. They were

This chapter supersedes Appendix I, "Nara Basin Geomorphology", in Barnes (1983).

made available to me through Nara Women's University, where I produced Figure 13.2 (Barnes 1982). That map conforms well to the geological units as later mapped by Itihara in 1993 (cf. his fig. 7.1) – an exercise that has been further refined in the 'seamless' map as presented below.

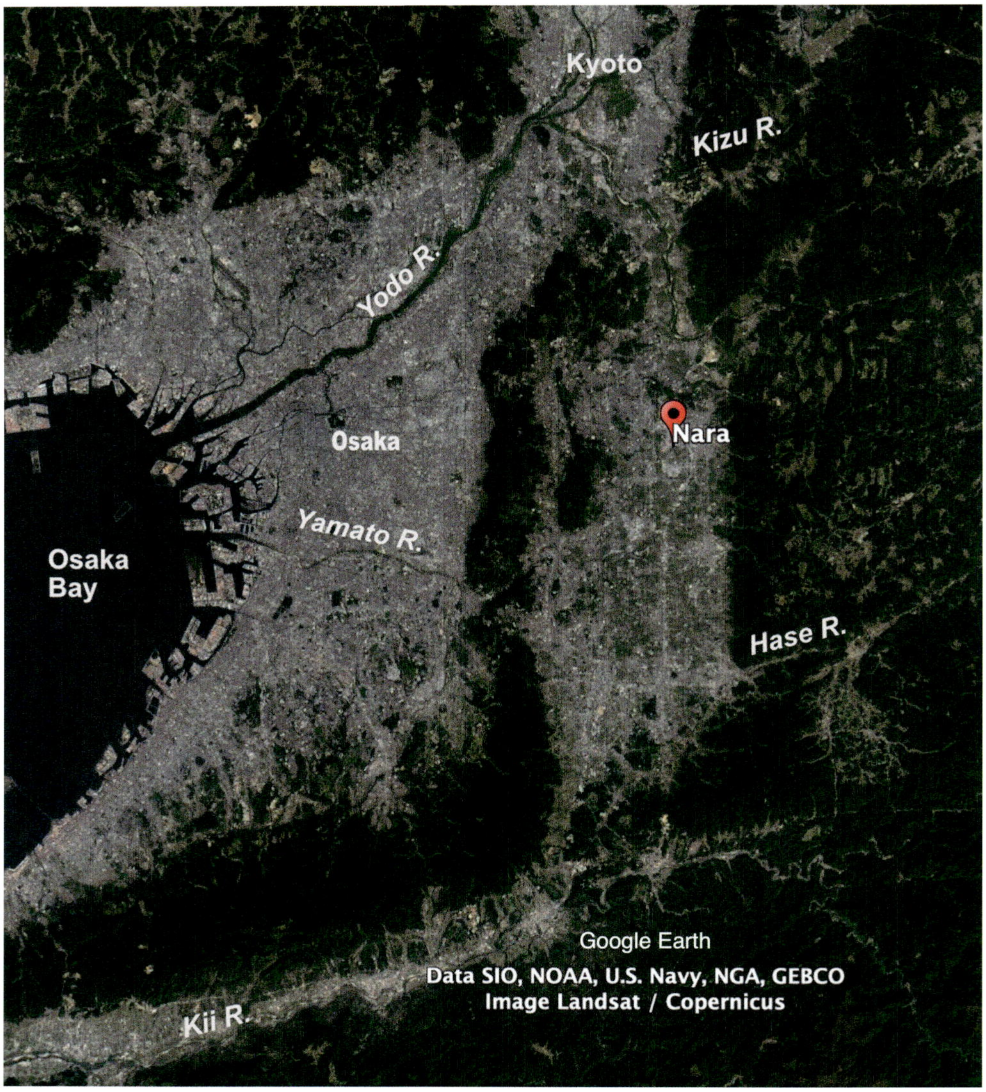

Figure 13.1 Modern view of Nara Basin and surroundings

In historical documents, the Nara Basin is known as the Central Land of Reed Plains, indicating the marshy centre where the main rivers converge to form the Yamato River (Figure 13.2) before draining into the Osaka Basin through the narrow Kamenosé corridor between the northern Ikoma Mountains (1) and the Miocene volcanic mountains represented by Mt Nijo (2). The eastern Basin is defined by sharply rising mountains beginning with Wakakusa-yama (7) (cf. Figure 7.12) and ending with Mt Miwa (6); these mountains, forming a green belt down the eastern side of the Basin, are now designated as the Yamato Aogaki (Green Wall) National Park,[1] referred to below as the Aogaki Range (headed by the Kasagi Range). This range levels out to the east to form the Yamato Plateau (ca. 400~600 m). The southern Basin is bordered by the three famous mountains confining the early

[1] Thanks to KITADA Hiroko for the placenames for Kamenosé Keikoku and Yamato Aogaki Kokutei Kōen.

capital on the Asuka River – Mts Unebi (3), Miminashi (4), and Kaguyama (5); they front a region of deep mountains (ca. 500~800 m) to the south, epitomized by Yoshino-yama. Comparison of the reconstruction with the satellite photo shows how the Umami and other hills, hill fingers, terraces, and alluvial fans and levees have been completely obscured by settlement.

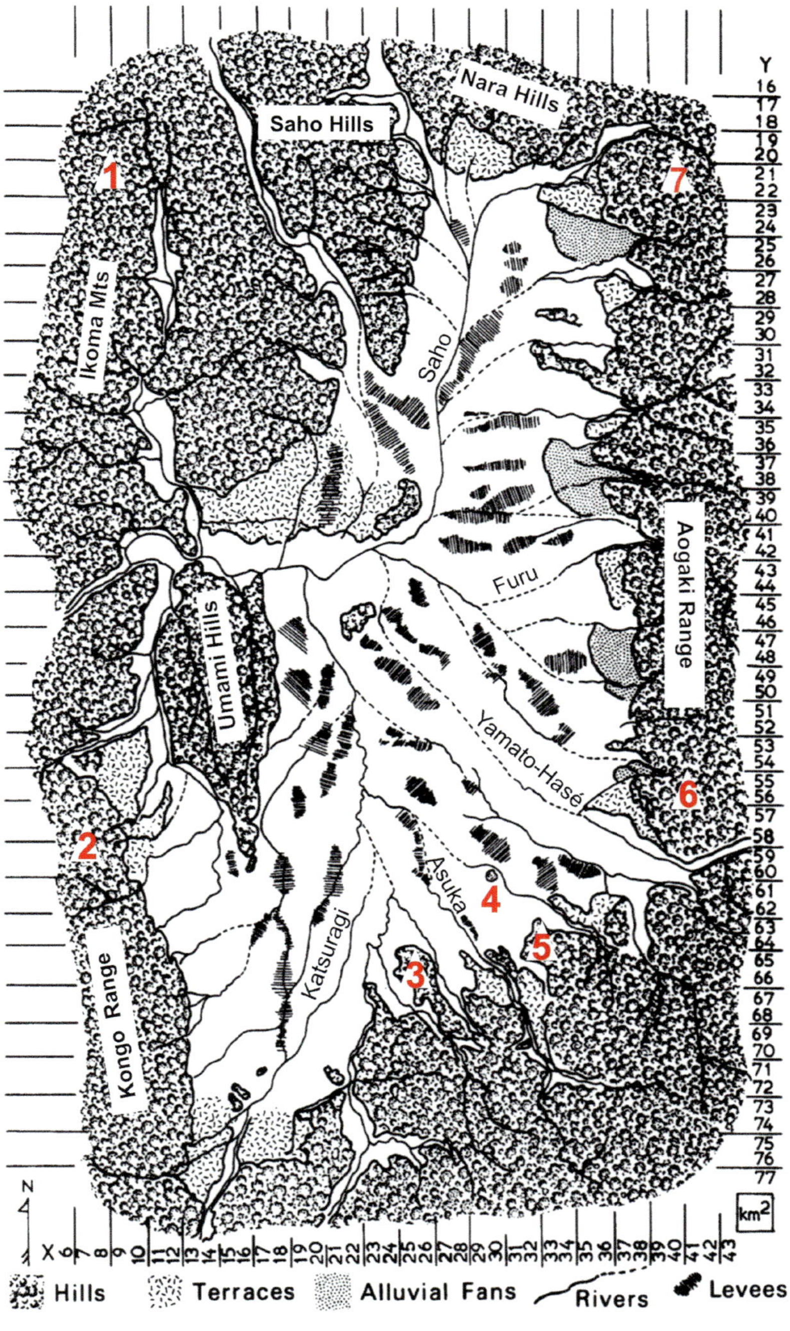

Figure 13.2 Nara Basin landscape reconstruction
0.5 km grid numbers, 1.0 km grid lines
Select Nara Mts: 1 = Ikoma, 2 = Nijo, 3 = Unebi, 4 = Miminashi, 5 = Kagu, 6= Miwa, 7 = Mikasa

Local Geology in Brief

Basement rocks

According to the discussion of the development of the bedrock belts of Japan (Chapter 3; Apx 6) and as shown in Figure 13.3, the Nara Basin (NB) sits primarily in the Ryoke Belt (Ry, light green), bordered on the north by the Mino-Tanba Belt (darker green) and on the south by the Sanbagawa Metamorphic Belt (MB) (Sb+Shm: pink). The whole of Nara Prefecture crosses the Ryoke and Sanbagawa belts into the Shimanto Belt, which forms the lower Kii Peninsula. The Ryoke and Mino-Tanba Belts were originally one and the same Accretionary Complex (AC) formed in the Jurassic period. However, with the development of the Ryoke batholith granites in the Cretaceous period, part of the Mino-Tanba belt was transformed by contact metamorphism (cf. Chapter 2.5.2), thus becoming the Ryoke Belt which entails both the granite plutons of the batholith and AC sediments that underwent regional metamorphism; these are referred to together below as the 'Ryoke rocks'. The Ryoke batholith (cf. Figure 3.8) stretches thousands of kilometers from Hong Kong to the top of Sakhalin. It represents the Volcanic Front along the edge of the Eurasian continent in the Late Cretaceous period (cf. Figure 3.7: inset) – though the volcanic edifices have eroded away leaving only the granites comprising magma chambers that have been brought to the surface through exhumation (cf. Chapter 2.5.3).

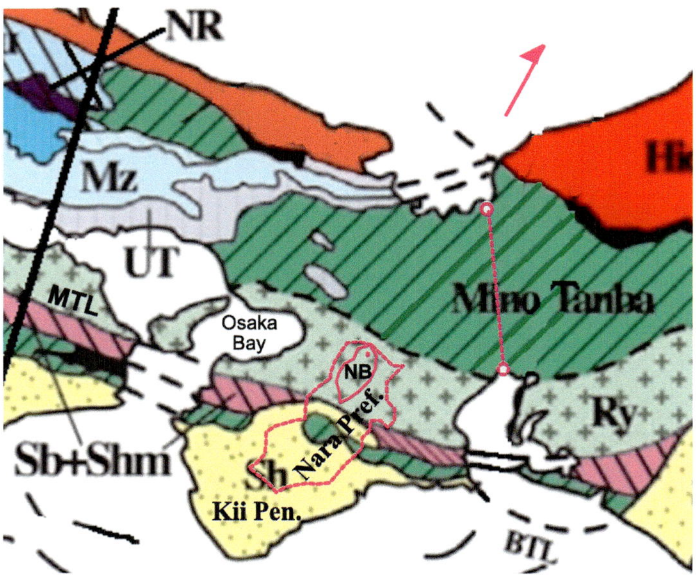

Figure 13.3 Nara Basin in the Ryoke Belt
NB = Nara Basin
• Nara City
MTL = Median Tectonic Line
Ry = Ryoke Belt
Sb = Sanbagawa Metamorphic Belt
Shm = Shimanto Metamorphic Belt
Sh = Shimanto AC
o——o = 100 km at Waist of Honshu

South of the Ryoke Belt, the Sanbagawa Belt was also an AC but was metamorphosed at a deep level on a regional scale during subduction, then exhumed to the surface. The Sanbagawa MB in Figure 13.3 contains both the Sanbagawa MB (Sb) *sensu stricto* and the recently recognized Shimanto MB (Shm) (cf. Chapter 3). The northern boundary of the Sanbagawa MB is the Median Tectonic Line (neo-MTL), a major recent right-lateral strike-slip fault which cuts a river valley through the Kii Peninsula into Shikoku Island. This valley can be reached most easily from the Nara Basin by following the Katsuragi River (cf. Figure 13.2) southwest out of the Nara Basin. Below the metamorphic belts is another strip of the Mino-Tanba Belt referred to as the Chichibu Belt, but the Kii Peninsula is primarily composed of the latest on-land AC, the Shimanto (Sh) Belt; Nara Prefecture extends deep into these terranes.

Rock types & sediments

How the rocks of these various geological belts are manifested around the current Nara Basin is very different from the broad brush of Figure 13.3, which simply places the Basin in the Ryoke Belt. A closer look in Figure 13.4, using the 1:200,000 'seamless' online geological map zoom facility (GSJ 2020), reveals that the alluvium of the central Basin (light blue, cf. Apx 15), topped with emergent river levees, is surrounded by 20 other units: 9 of these sedimentary, 10 igneous, and 1 metamorphic.

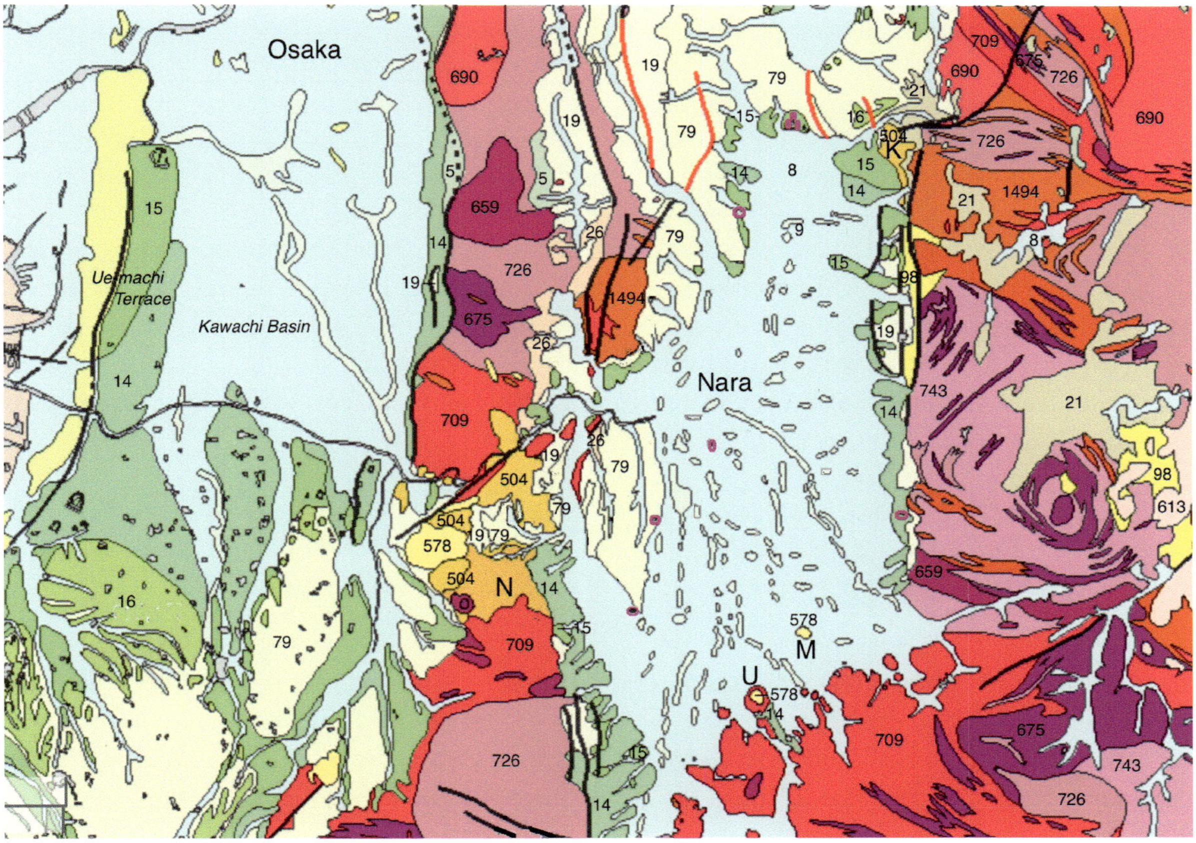

Figure 13.4 Geology of the Nara Basin region

Prominent keyhole-shaped tombs shown as pink circles
Unit numbers keyed to Appendix 15
Letters indicate volcanic rocks: N = Nijo+, U = Unebi, M = Miminashi, K = Kasuga

• Sediments: In general, the central alluvium of Holocene date (#8,9) bordered by Plio-Pleistocene hilly deposits (#79) belonging to the Osaka Group (Itihara 1993) of the #2 Setouchi Geologic Province (cf. BOX 6), and Late Pleistocene–Holocene terrace deposits (#5,14,15,16,19); however, there are also small areas exposed of Miocene marine and non-marine sediments (#21,26,98). This map does not illustrate the buried Miocene~Pleistocene sediments underlying the basins.

• Plutonic & metamorphic rocks: The surrounding mountains are composed of metamorphic rocks (#1494) of the Ryoke batholith metamorphic aureole, and a variety of plutonic rocks: one Early Cretaceous diorite/quartz diorite (#676) is notable in the mountains northeast of Nara City, but a majority of the igneous rocks belong to the Late Cretaceous (#659,675,690,708,709,726,743) – including gabbro, diorite/quartz diorite, granodiorite, and tonalite – with some exhibiting high-temperature metamorphism as gneiss or gneissose rock. The Cretaceous granitic plutons sort into Early and Late, belonging respectively to Older Ryoke and Younger Ryoke (cf. Apx 10).

• Volcanics: Lastly, representatives of Miocene volcanics (#504,578), associated with the establishment of the Philippine Plate subduction (cf. Chapter 4), are exposed in the northeast (K), south (M,U) and west (N) Basin. These four letters on Figure 13.4 designate famous Nara mountains composed of intermediate to felsic volcanic rocks (cf. Apx 5: Table B). The Mikasa andesite from Mt Kasuga is little known but has the same date of eruption as Mt Nijo, ca. 13 Ma (Tatsumi *et al.* 1980; Ozaki *et al.* 2000).

We have met Mt Nijo before and found that it was the source of an andesite (sanukite) that was used for prehistoric stoneworking (cf. Box 5). Unebi and Miminashi were also the source of a pink rhyolite (Sato et al. 2012) that erupted ca. 15 Ma and was used for polished stone reaping knives within the Basin during the Yayoi period.

Not shown here are the Muro volcanic rocks to the east, resulting from the eruptions of two or three Miocene calderas located on the eastern edge of the Kii Peninsula between 16 and 14 Ma (Sato *et al.* 2012: 64). Pyroclastic flow deposits alternating with volcanic ash fallout lithified as ignimbrite to form the southern half of the Yamato Plateau to the east of the Aogaki Range, while tuff resulting from both tephra fallout and redeposition by lahar and debris flow exists as far north as Nara City and east–west from Mie to Osaka Prefectures.

Sato et al. (2012) note that the tuff hosts famous rock-carved Buddhas near Mt Kasuga - after which it was named (Sekibutsu Tuff, 'rock Buddha' tuff). It is also exposed along the Yamato River in eastern Osaka where 7c rock-carved tombs were hollowed into it. The latter can be visited at Takaido Yokoana Park in Kashiwara City, Osaka.

One of the interesting features of the Nara region is that most of the Miocene and Plio-Pleistocene sediments rest directly and unconformably on Ryoke granites of Cretaceous date – which were originally emplaced on the continental edge. If continental rifting was initially caused by a magma plume, as discussed in Chapter 4, then this would have involved considerable uplift of the entire region *before* separation of the archipelago. Uplift on this scale would have resulted in widespread erosion, with Palaeogene (66–24 Ma) sediments that may have covered or accumulated on those granites eroding away. With rifting beginning in the Early Miocene, the exposed granite surface would have then been subject to lacustrine and marine sedimentation as the rifted island arc subsided underwater (cf. Figure 4.2). The unroofed granites were thus subject to sedimentation in the Late Miocene and Pliocene before the Rokko Movement began in the Pleistocene.

The next section reviews the depositional history of sediments in the Basin and the actual formation of the Basin itself in the Rokko Movement. It is important to realize that most of the sediments in the Basin pre-date the physical construction of the Basin in the Mid-Pleistocene. Thus, a long-term approach going back into the Miocene is needed to understand the presence and nature of these sediments, but the Rokko Movement accounts for their current distribution.

Basin Faulting and Sedimentation

Tectonic basin formation

If we backtrack to Chapters 4 and 5, we can see how the general processes described there affected the structures and deposition of sediments that are found in the Osaka and Nara Basin lowland hills and centres. E-W normal faulting has been discovered in the Ryoke granite bedrock of Nara dating to between 45 and 30 Ma (Mitamura *et al.* 2016); this was associated with the doming and rifting of the continent before separation of the archipelago. At the end of Early Miocene, NW–SE faulting set in (Nishioka *et al.* 2001). This is evidenced by modelling that suggests that a deep trough in the Nara Basin bedrock (Figure 13.5) angles NW to SE (Sekiguchi *et al.* 2018 fig. 1: right). Faulting in the opposite direction (NE–SW) occurred post-Pliocene (Nishioka *et al.* 2001). Altis (1999) is of the opinion that these 'transpressional' forces are the same that acted to separate the Japanese landmass from the continent, forming the Japan Sea. He proposes that right-lateral movement on the neo-MTL strike-slip fault caused clockwise rotation of up to 50° of blocks along the northern edge of the neo-MTL; this accounts for the NE-SW orientation of some of the major basins such as Osaka Basin (Altis 1999: fig. 3). The alternation of block edges in transposed positions accounts for the highs and lows between and in basins. Ozaki *et al.* (2001: fig. 5) pinpoint the formation of these basins just before 1.8 Ma due to this NE–SW reverse faulting, and several of the major river corridors in the Kinki region – such as the Hasé and Kizu Rivers, and the Kamenosé Corridor for the Yamato River follow this trajectory today.

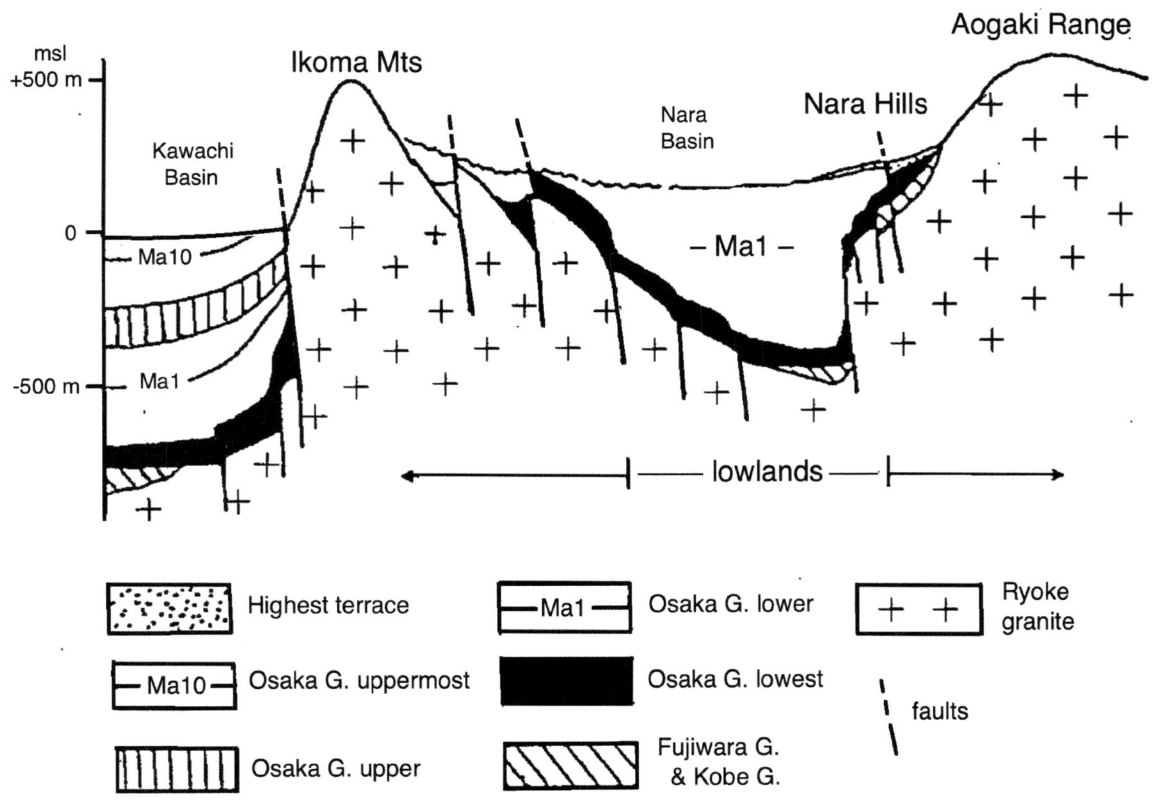

Figure 13.5 E-W section of Nara Basin at north end

From the early Pleistocene, E-W compression has produced N–S reverse faulting (Nishioka *et al.* 2001). This is the Rokko Movement that produced the current Nara Basin with the uplift of the Ikoma Mountains and Aogaki Range. Ikeda & Ohashi (1996) note that another change in direction of faulting from N–S to NW–SE occurred "after the middle of the Pleistocene" and as a result the bedrock is divided into small blocks and the Basin shape was "changed from a box to a diamond" (1996: 64). This produced what Ikeda & Ohashi call the "new" Nara Basin (1996: 53), bounded by the emergent Ikoma and Kongo Mountains and the Yamato Plateau – succeeding the "old" Nara Basin of earlier Plio-Pleistocene sediments. The palimpsest record of these fault activities can be seen in Altis (1999: fig. 3). According to Ikeda & Ohashi (1996), the current form of the Basin was not achieved until after the mid-Pleistocene, half a million years ago (1996: 42). Shiba (2021) has newly postulated that this consisted of uplift and sea level rise of perhaps >1000 m after 430 ka.

Sediment Groups & Formations

The sediment groupings discussed below derive from large-scale maps; regional investigations will have a much finer distinctions and datings. Groupings are named according to the system of formal 'lithostratigraphic' units with definitions given in Chapter 5 of the Stratigraphic Guide by the International Commission on Stratigraphy (ICS 2020):

- Group – two or more Formations
- Formation – primary unit of lithostratigraphy
- Member – named lithologic subdivision of a Formation
- Bed – named distinctive layer in a Member or Formation
- Flow – smallest distinctive layer in a volcanic sequence

These designations are generally but not strictly hierarchical, i.e., a Member may continue between Formations, and a correlative sedimentary body will have different names in different locations, while a whole series of other terms may be applicable (or inapplicable) as discussed in the Guide. This scheme has been adopted in Japan, and the 2nd edition of the Guide has been translated into Japanese (Salvador 1994); however, the stricture that "terms 'lower', 'middle', and 'upper' should not be used for formal subdivisions of lithostratigraphic units" (ICS 2020: Ch. 5-F.1) does not seem to have been followed in Japan.

One difficulty in discussing the sedimentary bodies occupying the basins is that more traditional scholarship is written within the Group–Formation–Member system, but descriptions for the 'seamless' map are not notated as such. For example, Osaka Group sediments (Mitamura 1993) are broadly exposed as hilly regions along the northern and western sides of the Nara Basin. These can be correlated with #19 and #79 (cf. Figure 13.4), which are described as "non-marine sediments" in the former case, and "brackish sediments or marine & non-marine sediments mixture" in the latter case (Apx 15). However, it is more difficult to correlate the Miocene Groups that occur around the Nara Basin edge: Nijo, Fujiwara, and Yamabe. Miocene sediments on the 'seamless' map are described as #21: "non-marine conglomerate"; #26: "non-marine sandstone, or alternation of sandstone & mudstone or sandstone & mudstone"; #98: "brackish, or marine & non-marine mixture sandstone, alternation of sandstone & mudstone" (cf. Apx 15). The only clear correlation here is #98 with the Fujiwara Group, which was mapped early on (Matsushita 1971: fig. 42). These are all discussed below.

Sediments with respect to changing land/seascapes

The majority of the sediments in the Osaka and Nara Basins were deposited either in the Early to Middle Miocene as Inland Sea #1, during the process of rifting the Japanese landmass from the continent, or in the Second Setouchi Geologic Province in Plio-Pleistocene times (Box 6).

• Inland Sea #1 deposits contain early freshwater sediments as rifting took place on land, and then finally marine sediments as the rifted archipelago was initially submerged (cf. Figure 4.2). Yoshida (1992) identified four stages (I–IV) in Inland Sea #1 development.

• The Second Setouchi Geologic Province was originally named Inland Sea #2, and Yoshida identified three stages (VII–IX). The sediments of Inland Sea #2 were originally considered to be the

Osaka Group, applying to the Osaka, Nara, and Kyoto Basins in three stages: Lowest, Lower, and Upper. The sediments of this series, however, were initially laid by freshwater (not a sea), while later freshwater sediments are intercalated with marine incursions.

Inland Sea #1

Yoshida (1992) documented a marine incursion around Ise Bay at 18 Ma, reaching halfway across the Kii Peninsula and up through western Honshu by 16 Ma. The volcanic activity in the Setouchi Volcanic Zone (Chapter 4), from ca. 16.5 to 12 Ma, provided the Nara Basin with its sanukite sources in the west (cf. Box 5), its famous rhyolitic mountains in the south (Unebi and Miminashi, Figure 13.2), and the vast pyroclastic flow of the Muro Formation in the southern Yamato Plateau into the eastern Basin (Nishioka *et al.* 2001: figs. 18, 25). With the cessation of rifting at 15 Ma, the tectonic regime shifted from N–S extension to N–S compression, leading to uplifting of the Japanese landmass as the Philippine Plate subducted northwards. The period between 15 and 3 Ma was characterized by stabilization (Takahashi 2006) and 'slow subduction' (Kimura *et al.* 2005; Pickering *et al.* 2013). A hiatus in sedimentation during this time led to the emergent surface being heavily eroded through planation (Altis 1999), with most Inland Sea #1 sediments lost – i.e., washed into the sea for later accumulation in the Nankai Prism (cf. Figure 2.7). Some Miocene sediments – the Fujiwara and Yamabe Groups (cf. Table 13.2) – lie at the base of the Nara Basin and are found along the eastern Nara Basin edge and in the Yamato Plateau. The Nijo Group is particularly notable in the Kamenosé Corridor, exposed at 150–160 m. In the Ikoma Mountains, at least six of these Miocene terrace surfaces are known, and eight have been identified in the Yamato Plateau (Ikeda & Ohashi 1996: 53); many more are found at elevations between 400 and 1000 m in the mountains of western Honshu (Itihara 2001).

The Seto Depression

Meanwhile, the new N–S compression of the western archipelago resulted in a long-wavelength (150 km) folding parallel to the Nankai subduction trench. By 5 Ma the folding had produced the Seto Depression bordered by mountains that now form the E–W granitic backbone ranges in Honshu and Shikoku (Yonekura *et al.* 2001: 303). The Seto Depression itself stretched from Ise Bay to Kyushu. Around 4 Ma, the rate of subduction of the Philippine Plate accelerated perhaps in conjunction with the formation of the Okinawa Trough west of the Ryukyu Archipelago (Kimura *et al.* 2005: 983), and at 3 Ma, the Plate began subducting to the WNW (Takahashi 2006). The neo-MTL responded to this pressure from the obliquely subducting Philippine Plate by reactivating as a right-lateral strike slip fault in the Late Pliocene~Early Pleistocene (Sangawa 1978; Toda 2016). This increased activity resulted in 'transpressive faulting', postulated by Altis to have broken up the Seto depression into individual basins oriented NE to SW, twisting clockwise in a pre-existing pattern with the right-lateral movement on the fault (Altis 1999: fig. 9). The Osaka Basin is one such NE–SW oriented basin, but its structure differs radically, as we shall see, from the Nara Basin.

Second Setouchi Geologic Province / Inland Sea #2

Shiba (2021) employs a new 4-stage classification of Pliocene through Holocene deposits in Osaka based on both tectonics and sedimentation processes. Using nearly 100 marker tephra in his figure 2, he correlates the Osaka stratigraphy with strata across central Japan, from the Kinki through Kanto to Hokuriku regions (cf. Apx 2: Fig. A). Marine incursions happened earliest in the Hokuriku region, next in Kanto, and belatedly in the Kinki region (Shiba 2021: fig. 2). Comparisons of his four stages with traditional understandings of the Osaka Group sediments, as shown for the Osaka Plains in his figure 2, are presented in Table 13.1. Since Osaka stratigraphy is the basis of Nara Basin stratigraphy (Table

13.2), descriptions here focus on the eponymous Osaka Group. Each stage is described in more detail below with additional data.

Table 13.1 Comparison of traditional and redefined Osaka Group stages, correlated with Marine Isotope Stages (MIS) and marine sediments (Ma)

Osaka Group (3.6 Ma – 430 ka)	Stage	**Shiba's four stages (3.6 Ma to 15 ka)**
mid-Pliocene~Middle Pleistocene	1	Osaka Low Sub-Group Pliocene~early Early Pleistocene 3.6 Ma – 1.3 Ma, >MIS 103~37, no marine sediments
Lowest 3.6 Ma – 1.8 Ma (MIS 63) MIS >63, no marine sediments		
Lower 1.8 Ma – 880 ka MIS ca. 63~22, Ma0~2	2	Osaka Middle Sub-Group late Early Pleistocene – early Middle Pleistocene 1.3 Ma – 680 ka, MIS ca. 37~17, Ma0~Ma5
Upper 880–430 ka MIS 21~12, Ma3~8		
	3	Osaka Upper Sub-Group late Middle Pleistocene 680–130 ka, MIS 16~6, Ma6-11
late Middle Pleistocene ~ present		
High Terraces 430–130 ka MIS 11~6, Ma9-11		
Middle Terraces 130–60 ka MIS 5~3.5, Ma12	4	Upper Pleistocene 130 – 15 ka MIS 5~2, Ma12
Low Terraces 60–15 ka MIS 3.5~1, no marine sediments		
Post-glacial 15 ka – present MIS 1, Ma13		**Post-glacial 15 ka – present** MIS 1, Ma13

Stage 1: Pliocene~early Early Pleistocene
From the Pliocene into the Early Pleistocene, the Osaka Basin filled with freshwater deposits (Table 13.1: Stage 1). Shiba (2021) describes this as a time of simultaneous archipelagic uplift and sea level rise. The Plio-Pleistocene sediments lie unconformably on bedrock and are composed of silt, sand, and sandy gravel – often forming off-shore deltaic sands in coastal areas (2021: 40, 47).

The earliest tephra date for Osaka Group sediments in the Nara Basin is the Unami Tephra (Table 13.2) with a fission-track date of 2.9±0.2 Ma (Kawamura 1993: fig. 12). Boring for hot springs near Hokkeji Temple in Nara City indicated the Osaka Group sediments were only ca. 600 m thick at the deepest point, lying on 46 m of Miocene sediments which in turn rest unconformably (the wiggly line in Table 13.2) on bedrock (i.e., Ryoke granites) (Mitamura 1993: 100). Since then, modelling of the Basin floor (Sekiguchi *et al.* 2018: fig. 1) suggests that the lowest portion of bedrock at 800 m deep lies in the northeastern Basin between Nara and Tenri Cities, with the lowest point – possibly greater than 1000 m deep – offset to very near the Basin's eastern flank (cf. Figure 13.5).

Stage 2: late Early Pleistocene – early Middle Pleistocene
The Marine clay "Utahime" in Nara correlates with the Ma1 clay layer elsewhere, which is overlain by the Pink Tephra redated to ca. 1 Ma by oxygen isotope stratigraphy (Satoguchi & Nagahashi 2017: 158). Sakamoto *et al.* (2001) warns that ash layers represent redeposition as turbidites by turbidity currents (usually caused by underwater landslides).

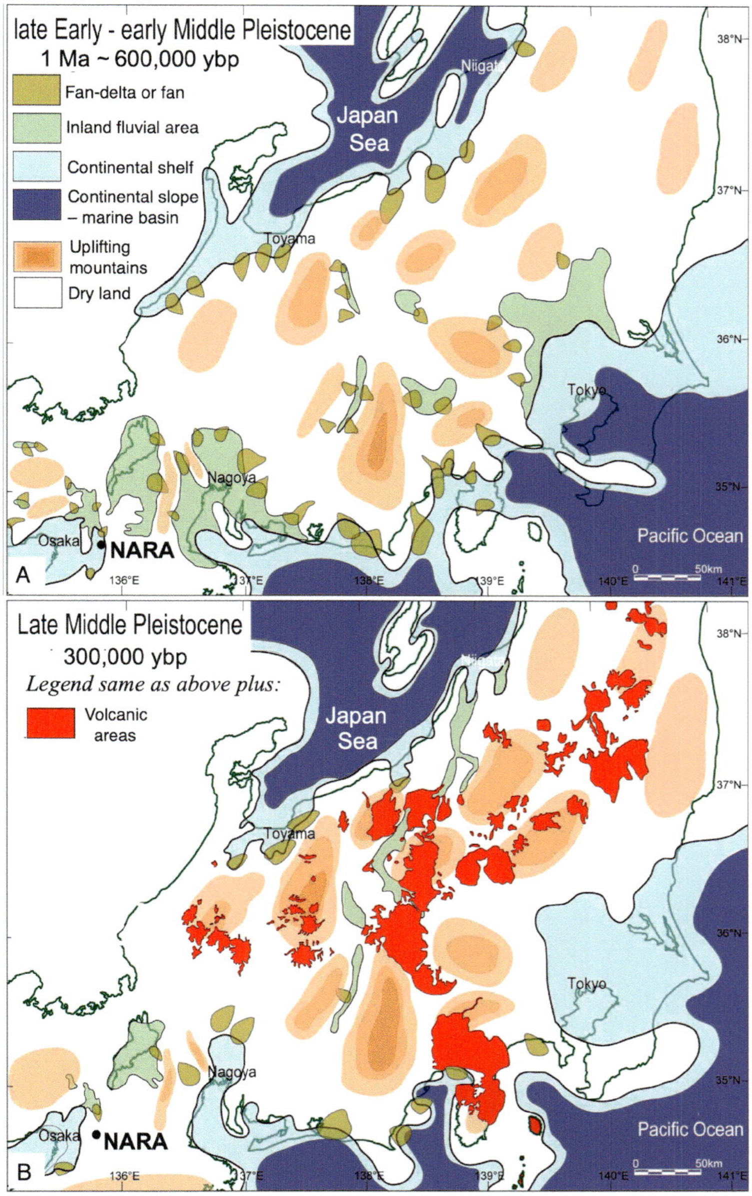

Figure 13.6 Landform changes in central Japan from Early to Middle Pleistocene

Between a million and 800,000 years ago, seawater had crept into the Nara Basin (Figure 13.6-A). Shallow coring in the central Basin (Mitamura 1993: 100) revealed the depths of Ma1, Ma2, and Ma3 (Table 13.2); Ma3 sandwiches the Azuki Tephra dated to 850 ka,[2] which marks the border between the Lower and Upper Osaka Group.

Stage 3: late Middle Pleistocene
Around 400,000 BP, differential basin subsidence and mountain uplift accelerated (cf. Figure 5.2), taking their modern shapes. By 300,000 BP (Figure 13.6-B), the Nara Basin had become dry land. The Quaternary volcanic cycle also began in this stage, while central Japan mountains rose even higher. The uplifting caused the formation of the High Terraces in the region between the marine incursions Ma9 and Ma11 (430–130 ka).[3]

One of the last major landforms to be created (not included in Shiga's Stage 4) was the Uemachi Terrace (cf. Figure 13.4), uplifted on a reverse fault as a N–S block in the Osaka Basin, forming a 'peninsula' that separated Osaka Bay from Kawachi Bay. The uplift of the Terrace occurred after the laying of the Ma12 marine clay, dating to MIS 5 between 524 and 478 ka; this can clearly be seen by the upbowing of the Ma12 clay layer in the Uemachi deposits (Takemura *et al.* 2017: figs. 3, 6).

[2] This tephra was dated to 0.87 Ma by fission track (Kinki/Tokai 1973: table 1), which Satoguchi & Nagahashi report as 0.85 Ma (2012: 159).

[3] This is contrary to Itihara's assessment (1993: 5) that the Highest and Middle terraces correlated with Ma12 and Ma13, respectively.

Table 13.2 Sedimentary divisions in the Nara Basin

Formation names apply to different parts of the Basin; their presentation here is neither strictly hierarchical nor comprehensive with approximate placement. Wiggly line = unconformity

A. Plio-Pleistocene ~ Holocene

Group 層群 (*sōgun*)	**Ma clay dates**	**Date** (tephra date)	**Formation** 累層 (*ruisō*)	**Sediments**	**Location**
	Ma13 (MIS 1)	Holocene (K-ah 7300 BP)	Ikaruga	landfill; sand, silt Akahoya Tephra	central Basin surface
			LGM peat		-7 m deep
		(A-Tn 30 ka)	Yamanobe	peat, silt, sand of chert, quartz Aira Tephra	central Basin
		L.Pleistocerne		gravel & silt fill	Lower Terrace (tl)
	Ma12 (MIS 5)	(Aso-3 133 ka)		Aso Tephra gravel fill	Middle Terrace (tm)
Mid-Plst event	Ma9–11 (MIS 7)	late Middle Pleistocene		gravel fill	Higher Terrace (th)
Osaka G. 0.425 Ma					
Upper 0.520 Ma 0.86 Ma	Ma7 Ma3@ -25~28.3m (MIS 21)	(SS-Az 850 ka)	Kokuzo Seika	sand/silt/gravel Sakura Tephra Azuki Tephra	NW hills Eastern Basin edge
Lower 1.8 Ma	Ma2@ -42.3~45.2m (MIS 25) Ma1@ -75~80.5m (MIS 31) Ma0	(Ss-PNK ca. 1 Ma) (1.1 Ma)	Shirakawa-ike Tanabe	Pink Tephra sand/sandy silt gravel; Yellow tephra	Eastern Basin edge NW hills
Lowest 3 Ma		2.9±0.2	Tomigaoka	sandy silt/gravel Unami Tephra	NW hills
Ryoke rocks		Cretaceous		both granites & metamorphics	

B. Miocene

Nijo G.	-600~646m msl in Basin	Miocene			N-S of Kamenosé at +150-160 m msl
		13 Ma	Tsukeno	Muro tephra	Sakurai region
Fujiwara G.		17.9±1.1 Ma	Toyoda Kokuzo Iwai	freshwater lake & marine deposits	exposed along foot of Aogaki Range
Yamabe G.		17.7±1.6 Ma	Hayama Iu, Sogo Oyamado	freshwater lake & marine deposits	SW Yamato Plateau
Ryoke rocks		Cretaceous		both granites & metamorphics	

Stage 4: Upper Pleistocene
This Stage began with MIS 5e and the marine incursion Ma 12 (130–120 ka), with the cutting of the riverine Middle Terraces as the sea level fell thereafter. From ca. 58 ka the Lower Terraces were formed, especially as sea level fell during the LGM. Then, during the Pleistocene–Holocene transition, the landscape underwent a "marked increase in erosive force" (Oguchi *et al.* 2001: 4). Corings by Matsuoka *et al.* (1984) recovered no sediments dating between 20,000 and 9,000 BP (cf. Table 14.3).

Landforms for human occupation
The Uemachi Terrace allowed the Kawachi Bay to form between it and the Ikoma Mountains. The Bay was flooded in the early Holocene, laying Ma13 sediments during the Jomon Transgression. The maximum was achieved at 5300 cal. BP; then by 3500 cal. BP, deltaic sands from the Yodo River in the north and the Yamato River in the south were filling in the Bay (Masuda *et al.* 2016: fig. 5). Kawachi Bay became a freshwater lake by 800 AD; in 1744 the Yamato River was rerouted in 1744 to cut directly west across the Uemachi Terrace, after which the lake was filled for urban development (Suito Osaka 2022).

Unlike other some other mountain basins in Japan, there has been relatively little sedimentation in the central Nara Basin during the period of human occupation. Corings reveal clustered sedimentation events, separated by a hiatus between 24 and 11.5 ka (Itoh *et al.* 2017). The first was prior to the LGM – the Yamanobe Formation (different from the Miocene Yamabe Group) dated 34,000 and 21,500 ^{14}C BP uncalibrated and including the AT tephra at cal. 29,000 BP; that Formation was surfaced with peat before the LGM (Kokawa & Yoshida 1964; Matsuoka & Nishida 1980; Ooi 1992). The second was in the Holocene (the Ikaruga Formation). Itoh *et al.* calculate that deposits since the LGM are no more than 5 m thick, and they interpret the Ikaruga formation (Table 13.2) – consisting of silt containing plant macroremains – to be "flood basin mud" (2017: 106). They also measured the depth of archaeological sites in the Basin, with all archaeological material occurring within 2.2 m of the present ground surface, and Yayoi remains mainly at 0.8m.

Historical geographer FUJIOKA Kenjirō early on drew attention to the relationships between prehistoric settlements and landforms (Fujioka 1941). He noted that in the Kawachi and Nara Basins, Jomon sites were located on alluvial fans below the eastern boundary fault lines, while Yayoi sites extended into the basins on deltaic sands. The former was demonstrated by the grid-mapping of Jomon ceramics in the Nara Basin (Barnes 1983: fig. 4.6), while Yayoi ceramics were found more densely scattered along the fans but also into the Basin center (Barnes 1983: fig. 4.7) – not to mention many archaeological site discoveries and excavations since then.

Thus, the four traditional units of the Nara Basin topography consist of Miocene deposits in upland areas, Osaka Group sediments forming rolling hills of the Basin flanks, later terraces and fans edging the Osaka Group, and the central Basin alluvium. Their distributions are shown in Figure 13.7, although these are grossly portrayed here and refinements can be found in the literature.

In Figure 13.7, Miocene sediments (purple) are seen west of Mt Nijo and along the northeast Basin edge, where they present as the Fujiwara Group, named after Fujiwara-cho in Nara City. These sediments date to 16.8–19 Ma by fission track and represent the freshwater lake occupying the rift in the continental edge before the opening of the Japan Sea Basin (Ozaki 2001). They sit unconformably on Ryoke granite, dipping to the west at 30–50°. A contemporaneous Yamabe Group occupies the southern edge of the Yamato Plateau. Both of these Groups consist of freshwater and marine-laid sediments (cf. Table 13.2).

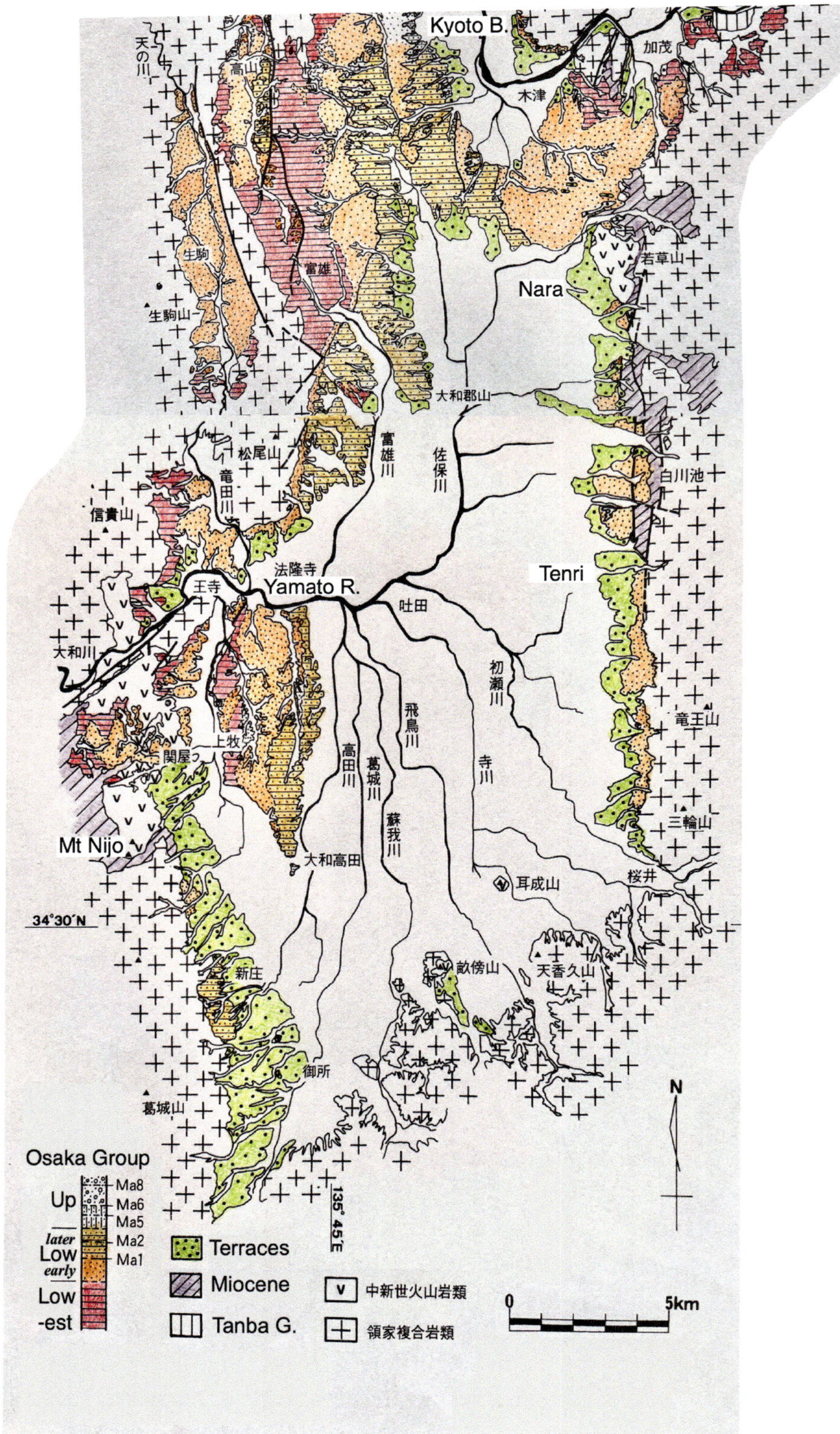

Figure 13.7 Important sedimentary groups on the flanks of the Nara Basin

Lowest Osaka Group sediments (red) are distributed primarily in the northwest Basin hills and around the Yamato River outlet. The Lower Osaka Group is divided into two by the occurrence of the first marine clay Ma1 – (orange) below that marker and (gold) above it. It is notable that the eastern Basin edge hosts mainly early Lower Osaka sediments, and the later Lower Osaka sediments occur mainly along the western Basin edge. The Upper Osaka sediments are sparse in the Nara Basin, though the Sakura Tephra dating to 520 ka has been recognized (Anon. n.d.). The central alluvium (Ikaruga Member) lies on an eroded Osaka Group surface of Pleistocene date; it incorporates the marker tephra K-Ah (Nishida 1992). The (yellow-green) portions of Figure 13.7 are termed 'terraces' in the original. This is problematic, as will be discussed below.

'Terraces'

The word for 'terrace' in Japanese is *dankyū*, meaning 'stepped hill'; Hataya (2017) discusses two terrace classification schemes, by origin (e.g., marine, riverine) or by location (along rivers, along shores), and cautions that these can be confused. *Dankyū* are included in *daichi* indicating 'upland' as opposed to both 'lowland' and hills/mountains per se. Despite *dankyū* being defined in terms of the standard types of coastal, riverine, and lacusterine terraces with flat surfaces and a frontal toe or scarp (Chigakudan 1981: 783b), maps often include *dankyū* and alluvial fans (*senjōchi*) together in a single legend (e.g., Yoshikawa *et al.* 1981: 6.8). So it is with Figure 13.7, where the first tier of uplands on the Basin's flanks is classed as 'terrace' (yellow-green). This is despite the case that Nara was identified as having substantial alluvial fans steeper than 3% and >5 km in length and width (Yoshikawa *et al.* 1981: 145-6, fig. 6.12).

Mather *et al.* (2017) lament the tendency for terraces (*sensu stricto*) to be analyzed separately from alluvial fans – arguing that these should be considered together in landscape evolution, especially because there is interaction between them. Here, the problem is the opposite: untangling the two forms to appreciate the different forces acting on them. Furthermore, this dichotomy hides the more numerous landforms that might be identified among fan-like forms: megafans, alluvial fans, tributary junction fluvial fans, debris cones, and debris flows (Mather *et al.* 2017: fig. 1). This is not the place to delve into the processes that produce these forms (though we have met debris flows in Chapter 6), but the generalities as can be gleaned from the standard presentations in Japanese are presented below.

Many such terraces began life as alluvial fans (*senjōchi-sei dankyū men*) (Nishioka *et al.* 2001: 3, 7) but have been subject to river incision and are perhaps cut across their toes by later river or fault action to form frontal terrace scarps. In the Nara Basin, they usually sit below Lower Osaka Group sediments, but some have formed directly against the Ryoke granite mountains, especially in the southwestern Basin. In that area, the terraces are covered with alluvial fan material, but these cannot be discussed here because that portion of the Basin is covered in the 1:50,000 map for Yoshinoyama, which is no longer available (AIST 2020). In the southern Basin, Ryoke granites outcrop widely; terraces are only noted around Unebiyama. The eastern Basin edge has the most continuous series (Figure 13.8), while the Saho River catchment in the northwestern Basin is also well populated.

Ikeda & Ohashi (1996: illust. 1) compared Nara Basin classifications of landform surfaces by seven different authors between 1973 and 1996; Wakita *et al.* do similar comparisons for the SE Kyoto Basin (2013: table 7.1) That tremendous variation settled into a standard for the 1:50,000 map series for the Nara Basin (Ozaki *et al.* 2000; Nishioka *et al.* 2001). The following scheme is based on coring analyses of terraces along the eastern edge of the Nara Basin (Figure 13.8). Three terrace levels, with sub-levels, are recognized, containing mostly sand and gravel with some clay layers (Sangawa 2001). These three groupings supersede the five named terrace groups previously proposed by Sangawa (1985). His highest 'Kokozo' terrace is no longer included, as the terrace base fill has been eroded away and only underlying hillslopes remain (Sangawa 2001: 78). The other named terrace surfaces have been incorporated into the current numbering system:

- Higher Terrace deposits (th), incorporating the Ma11 marine clays of the late Middle Pleistocene; these occur in highly restricted small patches along the eastern Basin rim at ca. 130–135 m msl (Figure 13.8: solid). These are equivalent to Sangawa's original Narazaka surface.
- Middle Terrace deposits (tm), incorporating the Ma12 marine clays; Middle to early Late Pleistocene; these show a patchy distribution and have been divided into two groups: tm1, tm2. Middle terrace deposits generally sit between 10 and 14 m above the central alluvium; they include Sangawa's original Shikanoen surfaces and a majority of the Wani surfaces.
- Lower Terrace deposits (tl); Late Pleistocene; these occur extensively down the eastern Basin rim, discontinuous in the northeast but almost continuous in the southeast; three groupings are recognized: tl, tl1, tl2. Lower terraces (tl) are located between 4 and 6 m above the central alluvium at 60–80 m msl. These include some Wani and the Ichinomoto surfaces.

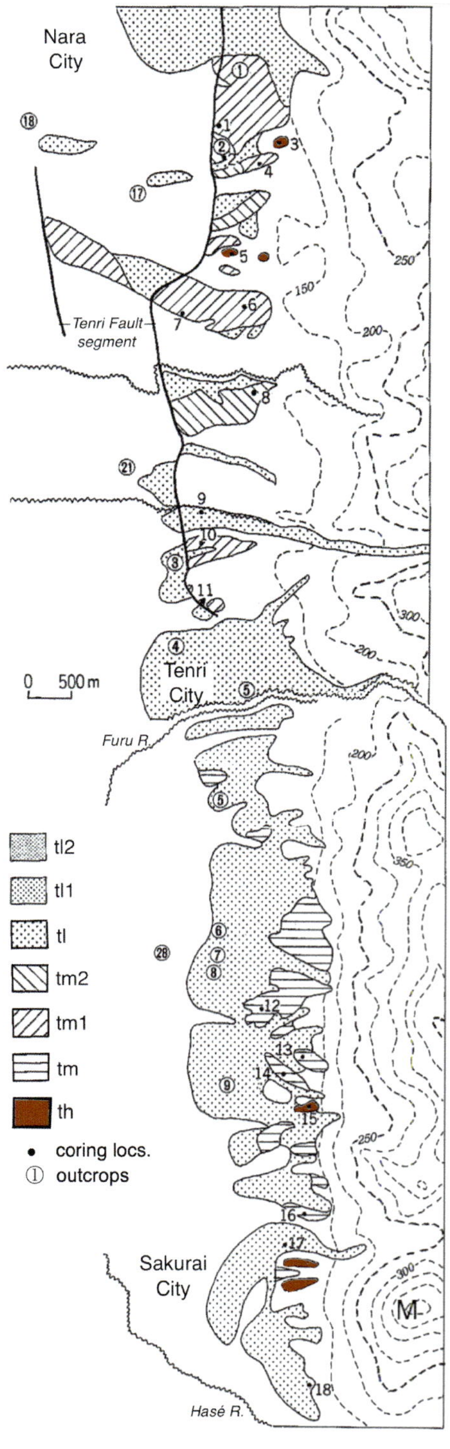

Figure 13.8 "Terrace" classifications along the eastern flank of the Nara Basin

The classes H, M, and L for terraces are used throughout Japan with equivalent datings to the above.

In contrast to normal usage, the "terraces" at the Nara Basin's edges (Figure 13.8) were formed as alluvial fans by small streams descending from the mountains (Sangawa 2001), and those in Figure 13.8 tend to dip to the west below the East Nara (Tōen) Fault system (Nishioka *et al.* 2001: 8). Many Lower and even some Middle Terrace deposits rest directly on Osaka Group sediments, but several Middle Terraces in the southeastern Basin rest directly on Ryoke granites.

The terraces have generally been correlated with Pleistocene glaciations. Sangawa (2001: 78) relates the Middle Terraces to ca. 130,000–800,000 BP, corresponding to the the Riss Glacial and following interglacial (MIS 6-5e). Lower Terrace fill is attributed to the Late Pleistocene, consisting primarily of coarse granite gravel eroded from the uplands to the east. Sangawa (2001: 78) links Lower Terrace 1 (tl1) in Nara to the surface of the Musashino Hills (Late Pleistocene) in the Kanto, and Lower Terrace 2 (tl2) to the surface of the Tachikawa Loam surface (Latest Pleistocene) in the Kanto region (cf. Figure 6.15). These terraces have little weathering or dissection (gullying).

Very few absolute dates are available to test these dating assumptions, but two recent coring projects contribute to our understanding. The most recent by Nihon Gennen (2019) was conducted on the Tsugaru Peninsula in the far north of Honshu in Aomori Prefecture (Apx 2: Fig. A). Datings of Middle Terrace fill indicate formation during MIS 5b–e, with Toya Tephra (115–112 ka) near the base.

The Lower Terrace deposits in the Nara Basin are linked to a layer underlying the Basin's centre. Recent coring investigations (SGSK 2014) identified a buried unit compatible with the Lower terrace deposits. It lies between 4.5-9.5 m deep under the basin alluvium and rests on top of the Osaka Group sediments; the unit pinches out to the east, affected by uplift of the Aogaki Range. It is a coarsening up unit (from clay/silt to gravel/sand) in Core 10, which is ca. 130 m due west of the lowest terrace. Core 9, ca. 60–70 m west of the lowest terrace, lacked gravel but incorporated a thicker layer of organic matter than Core 10, dated to 28,060±160 (cal. 30,526–29,447 BC) (SGSK 2014: figs. 28, 31), while Core 8, closer yet to the terrace, yielded a ^{14}C date of 24,360±130 (cal. 26,770–26,105 BC) (SGSK 2014: table 6). These dates are comparable with the uncalibrated dates obtained from peat layers in the coring project of 1977–1979 (cf. Figure 14.2). The organic matter indicates backmarsh accumulation before the LGM; during the LGM, there was heavy erosion which laid down coarse materials over the peat layers. This suggests that the Lower Terraces on the Basin fringe were uplifted and separated from the Basin fill after the LGM.

Holocene alluvium

The Basin floor surface consists of 6–10 m of finer sediments with landform differentiation into river levees (cf. Figures 13.2, 13.4) and backmarsh areas. The landscape reconstruction in Figure 13.2 was based on height differentials, and there are some very clear Holocene features that do not appear in either Figure 13.4 or 13.7. Particularly apparent are the terrace scarps bordering the Yamato River to the north as it exits the Basin in the central west (Figure 13.2: X13–21, Y40–41) and along the northern edge of the Hasé River as it exits the mountains in the southeast (X35–39, Y56–59). Field walking these scarps determined that the height differential was as little as 50–70 cm. Also not indicated at all in Figures 13.4 and 13.7 are alluvial fan features that had clear rounded toes in the aerial photographs. The end of sedimentary supply to the fans is correlated with the incision of the rivers. The lowering of the Basin's base level in the Holocene through subsidence or compaction of the central sediments (including the underlying Osaka Group) is possibly one factor to consider in addition to tectonic uplift along the eastern Basin edge.

River incision

There is some difference of opinion concerning river incision. Ikeda & Ohashi (1996) estimated that during the LGM, with a ~120 m drop in sea level, river incision evidenced by buried valleys reached 20 m below current sea level, resulting in the cutting of the lower terrace series. With the rise of the sea level into the Holocene, the current alluvium and river levees in the central Basin were formed (Ibid.: fig. 6). They estimate that these processes had stopped by about 3500 BP when peat layers formed in the Basin during the Late Jomon cold period. The central Basin, however, remained subject to flooding into the historic period with flood water deposition below ca. 45 m msl.

In contrast, Itoh *et al.* (2017) note that the basal sands of the Paleo-Yamato River incision in Osaka were determined to be -25 m from the current sea level at the point where it would have met the Yodo River emerging from the Kyoto Basin. But they reject the notion that fluvial incision would have extended back into the Nara Basin – also shown as incorrect in other parts of the world (Leeder & Stewart 1996); instead they note that "weak flood deposition may also have favored the widespread distribution of the peat" between the river courses (interfluves), especially in the cool climate leading up to and during the LGM (2017: 107-8).

In addition to disagreements about the dating of the peat layers (pre-LGM vs Jomon), these arguments focus solely on sea level change and do not take into consideration the rapid and continuing rise of the surrounding mountains. The linkage of tectonic uplift and river incision is well documented though not inevitable (Burbank & Anderson 2011), and the sporadic but rapid uplift events affecting central Japan are attested in Shiba (2021). Even today, the Ikoma Mountains have an Active Fault rate of 50 cm/ka and the Aogaki Range 20 cm/ka (Table 13.3). Thus, consistent if not smooth uplifting of the Basin edges in conjunction with sea level change over time has led to incision after the major terrace and fan depositions. The current rivers[4] cut through all of the sediments named above, from the Miocene at highest elevations through the Lowest and Lower Osaka Group and the various terraces fringing the Basin.

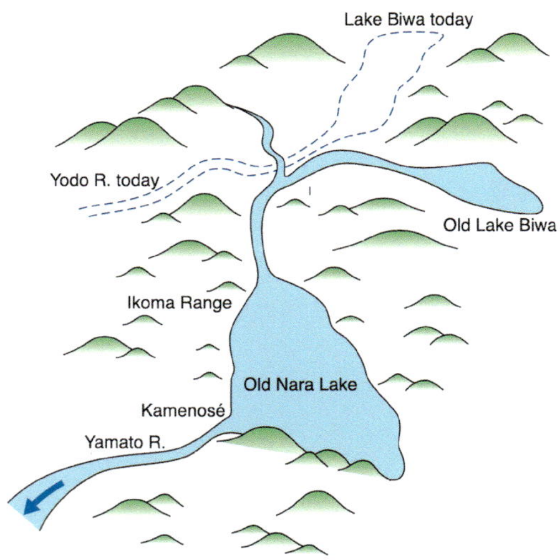

Figure 13.9 The mistaken concept of an 'old lake' in the Nara Basin

An old lake in Nara?

Before departing this section, I would like to bring up the problem of Ko-Nara Lake or Yamato Lake. These names have recently been attributed to standing water in the Nara Basin at 3 million years ago (Anon. 2016) and during the Jomon period (Yabuuchi 2021) – in works published by government institutions. It is also a concept promoted for the early historic period (Ando 2020; Yabuuchi 2021).

The illustration in Anon. (2016) (Figure 13.9) draws on Yoshida's sketch (1992: fig. 4) of Inland Sea #2 between 3 and 1 Ma, which depicts an inland area from Nagoya through Old Lake Biwa, to Nara/Kyoto Basins into Osaka Basin that is filled with 'freshwater' (*tansui*) deposits. Shiba (2021: fig. 6-2) adopts this for 2 Ma and labels them 'fluvial' (*kasen*) deposits. Yasuda's illustration is also presented in Anon (n.d.) as lakes but attributed to 2 Ma. As we know from the discussion above, however, there were many more times in the past that these basins contained standing water, and Kyoto Basin had a lake until recent historic times. But Nara?

Not only have several marine layers been discovered in Nara Basin's stratigraphy: the cores reaching minus 250 m from today's surface in the central Basin (Ozaki & Miyaji 2001: 66, figs. 36-37) indicate

[4] The rivers in Figure 13.7 show their canalized channels; their natural flow directions are shown in Figure 13.2.

that the Osaka Group is comprised of mainly sand and gravel in the east, sand in the west, and sand and 'mud' (*deisō*) in the centre. Given that the Basin is bowl-shaped with the eastern edge rising faster than the west, this distribution of sediments is understandable. The core diagrams (Ibid.: fig. 37: 1–5) show a preponderance of sand and gravel, with periodic intercalations of *deisō* (mud) and silt. However, unlike the marine layers, none of these *deisō* occur contemporaneously and extensively throughout the Basin lowlands, suggesting that these were ephemeral pondings in different areas and there was no single Old Nara Lake. The fact that lignite has been identified in Core H6 (Nishioka & Ozaki 2001: fig. 37: 1) at -160 m on the 60 m msl contour line indicates marsh formation – as lignite forms from peat; but this was only one core of 69 cores illustrated and only carbonaceous mud, not peat, has been found dating to the Jomon period.

Mention of marshes takes us to the later prehistoric and historic problem. The idea of a lake from Jomon into early historic times is apparently drawn from a poem in the 9c *Manyoshu* anthology and discussed by Ando (2020) and Yabuuchi (2021). This poem is attributed to Emperor Jomei (r. 629–641) standing on Kaguyama, one of Nara's three famous mountains in the southern Basin (Figure 13.2: 5) (NGS 1965: 3):

大倭（やまと）には　群山あれど、
とりよろふ　天の香具山。
登り立ち　国見をすれば、
国原は　煙立ち立つ。
海原は　鴎立ち立つ。
うまし国ぞ。あきづしま大倭の国は。

Countless are the mountains in Yamato
But perfect is the heavenly hill of Kagu;
When I climb it and survey my realm,
Over the wide plain the smoke-wreaths rise and rise
Over the wide lake the gulls are on the wing;
A beautiful land it is, the Land of Yamato!

The translators identified the lake as "Haniyasu at the northern base of Kaguyama. Though there remain only traces of it now, the lake was apparently very large" (NGS 1965: 3 ftn. 2). It is interesting that the character in the poem for 'lake' is *umi* 'sea' rather than *ike* or *-ko* meaning 'pond' or 'lake' respectively. Other poems in the *Manyoshu* speak of the dike or embankment (*tei*) of the Haniyasu-ike and of a 'hidden' marsh (*numa*) beside it. The location of these features between Kaguyama and Miminashi (cf. Figure 13.2: 4) suggest that a backmarsh formed between these two mountains where a reservoir was built later, as done throughout the Basin in the 5–10c.

Such backmarsh areas were demonstrated by the 1984 coring project (Joint Research Group 1987) even where a mountain did not stand in the way – at Asawa in the eastern central Basin just below the eastern terraces. This area hosted a wartime airstrip, which was built precisely because there were no villages there. The name Asawa could be a conflation of *asa(i)* 'shallow' and *sawa* 'marsh', but *sawa* also can mean 'stream'. Coring revealed two clear layers of carbonaceous mud dating to ca. 5000–3500 BC and 1000–0 BC respectively.[5] The phytolith remains were substantially *Phragmites* and bamboo. Although the species of the former could not be determined in this study, the 'common reed' *Phragmites australis* is "is a constructive species in the riparian areas in Japan" (Cao & Natuhara 2020). Non-Japanese members of the Asawa coring project hypothesized that reed beds were widely spread across the Basin, in accordance with the old epithet for Japan and particularly Yamato as "the fertile central land of reed plains" *toyo ashihara naka-tsu kuni* (CRD 2011). However, the Japanese members of the project believed that marsh areas were small but frequent across the lowland landscape. This latter interpretation would be valid in upstream areas that were characterized by river levees, but in the central Basin below 50 m msl where the Yamato River receives water from 156 tributaries

[5] The report attributes these dates to Early Jomon and Final Jomon respectively, though the datings of these periods have now changed (cf. Apx 1.1: Table A).

(Taniguchi 2014), flooding was a continuing historical problem – to the extent that residents had a saying 'One year sunshine, one year flooding'[6] (Yoshikoshi 1995: 111). Broad reed beds may have been extant for long periods of time in the central Basin before the rivers were canalized and the area was successfully drained and planted in rice between the 7th and 10th centuries, but flooding continues to occur due to embankment breaks and overflows (Ushiyama 2014: 196-197). It might finally be noted that gulls do frequent inland areas that do not necessarily hold marine waters.

Active Fault systems

Fault types and locations

Nishioka *et al.* (2001: 8) classify the faults in the Nara Basin as NW–SE trending normal faults, NE–SW trending reverse faults, and N–S trending reverse faults. The major Active Faults in Nara and Osaka are N–S trending reverse faults (Table 13.3). The two major systems involved in the formation of the Nara Basin are the Ikoma Fault system, running down the western side of the Ikoma Mountains in the north (max 642 m), and the East Nara (Tōen) Fault system fronting the Aogaki Range (max 497 m) along the eastern side of the Nara Basin. These fault systems form two of the numerous secondary fault systems in the Kinki Triangle, bordered on the north by the ATTL (Arima-Takatsuki Tectonic Line) and on the south by the MTL (Median Tectonic Line) (cf. Figure 5.11). The operating factor in emergence of the mountains was compression, which began in the Pliocene and continues into the Holocene.

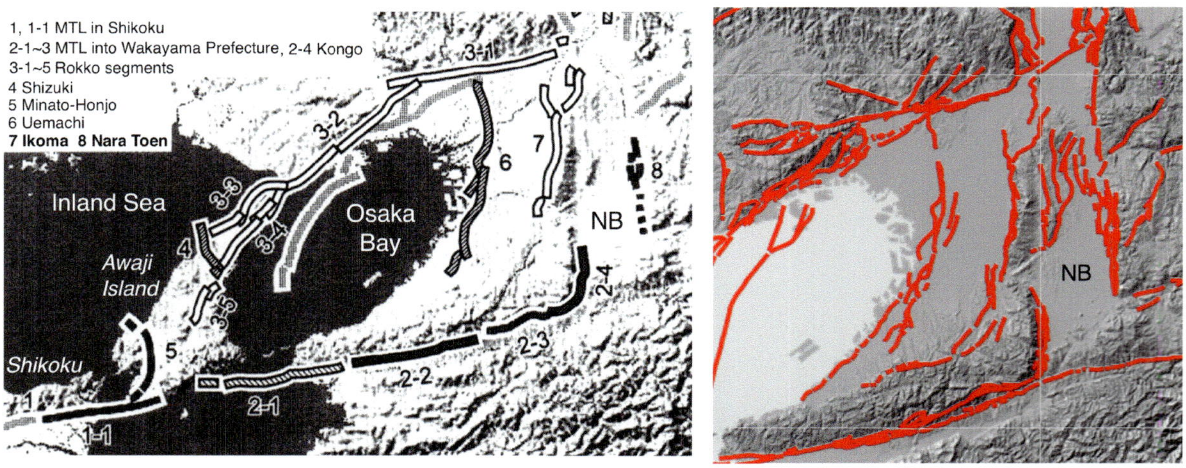

Figure 13.10 Kinki Triangle faults as of 2000 (left) and as documented in 2020 (right)
NB = Nara Basin, White grid lines at zoom setting 9 on right indicate 40 x 40 km squares

Monitoring Active Faults is increasingly important for hazard planning, and new ways of discovering Active Faults are continuously being developed (Kondo 2008; Suzuki 2013). Various maps illustrating fault locations have been produced throughout the past decades that vary widely in the detailed siting and naming of the faults in these systems,[7] and information is continually being updated. In 2000 for example, Active Fault zones (Figure 13.10: left) were identified for the Kinki Triangle (AFRG 2000), based on definitions in Matsuda (1990). This same region is shown in Figure 13.10: right as of 2012.

[6] 一年日照りで、一年洪水

[7] The use of the terms 'zone' and 'system' when discussing these are my additions in English; the Japanese terminology simply refers to these as 'faults' (*dansō*, which literally means 'cut layers').

Many fault lines have been identified, mapped, and grouped into 'behavioral segments' as termed by Matsuda (1990).[8] His criteria for designating several coordinated fault lines as these 'behavioural segments'[9] are (Figure 13.11):

① Isolated active faults over 10 km long with no other active faults within 5 km

② Multiple faults running in almost the same direction and aligned with a distribution gap no larger than 5 km in the strike direction

③ Fault group having mutual spacing of up to 5 km, running in parallel within a width of 5 km

④ Ancillary faults or branch faults with different strikes whose midpoint of the fault line is more than 5 km away from the main fault

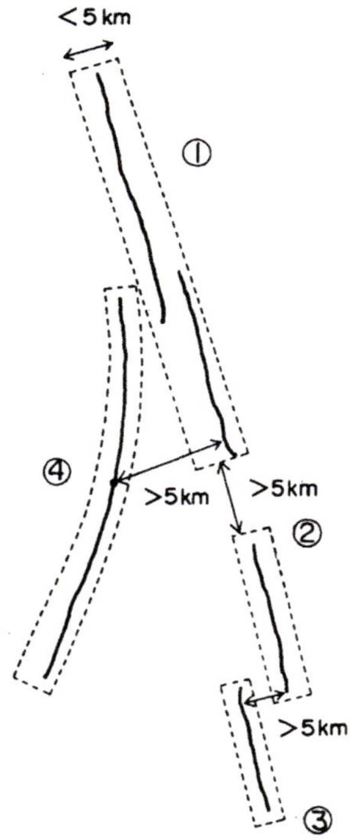

Figure 13.11 Fault segment definitions

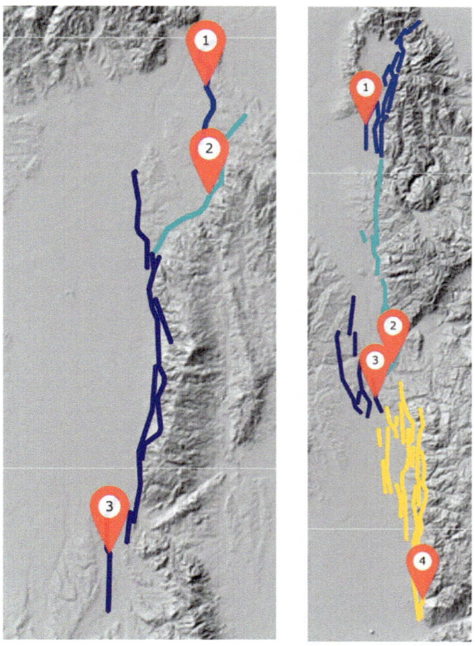

Figure 13.12 Faults along eastern Osaka, Kyoto, and Nara Basins

Left: Ikoma Fault zone segments: 1) Otokoyama; 2) Katano; 3) Ikoma

Right: Nara Toen (East Nara) Fault zone; segments: 1) Obaku; 2) Ide; 3) Sahoda; 4) Tenri; East Nara Tōen Fault zone extends from southern Kyoto Basin (segments #1, #2) into the Nara Basin, with the Sahoda segment (#3) affecting the Saho Hills, and the Tenri Segment (#4) affecting the entire east side of the Basin

Thus, the segments of Active Faults do not necessarily all move together. The Ikoma Fault zone, which runs down the west side of the Ikoma Mountains in the Osaka Basin (Figure 13.12: left) entails three segments, while the East Nara (Tōen) Fault zone, which extends along the east side of the Nara Basin into the southern Kyoto Basin to the north, is illustrated as having four active segments (Figure 13.12: right).

Each of these 'segments' is not a single fault line but has several sub-segments. Since the 'segment' is the unit that is seismogenic, an earthquake on one segment will energize movement on all its sub-segments but may not affect other segments. These sub-segments represent distribution of deformation during earthquakes in what has been termed a 'cascade', and both the size and amount of slip of an earthquake is determined by the length of the active part of the fault (Yoshioka 2009). Many of the sub-segments have historic names that occur on earlier maps, e.g., Obitoke as part of the Tenri Fault segment (Nishioka *et al.* 2001: fig. 38) or Gosukebashi as part of the Rokko Fault zone (Lin 1999). However, these sub-segment names are not incorporated into the formal Active Fault Database,

[8] The concept is widely cited as coming from McCalpin (1996), e.g., in Yoshioka (2009).

[9] Longitudinal or extended latitudinal bundles of fault lines are referred to as 'segments', using the English word, rather than 'sections' as in USGS (n.d.).

which gives names and numbers only to the segments, e.g., Rokkoyama segment 176-07 (AIST 2012). This makes it somewhat difficult to correlate fault names in the older literature with the way they are presented in the Active Fault Database. The hierarchy of zones and segments for Osaka and Nara Basins are listed in Table 13.2.

In addition to the segments in Nara shown in Figure 13.12, two more fault zones run down the western side of the Basin: West Nara Fault Zone and the MTL Fault Zone. The segments of these can generally be seen in Figure 13.10: right, and the segments are listed in Table 13.3 below.

Table 13.3 Details of some Active Faults in the Nara and Osaka Basins

Lg = length, m = metres, ka = per thousand years, betw. = between, Prob. = probable; timings of last movements are given for some time within calendar year ranges, – indicates no data. The [number] in the Segment column indicates the number of discrete fault lines (sub-segments) that can be counted on the map.

Segment	Code #	Trend	Dip	Lg km	Fault sense	risen side	Slip rate m/ka	Slip/ per event	Inter -val	Last move betw.	Prob. rup-ture
NARA BASIN (counterclockwise)											
East Nara (Tōen) Fault zone (174)											
Tenri [17]	174-04	N–S 0°	45° E	21	rev.	E	0.2	2.4 m	11 ka	895~ 1194 AD	0.04 –0.3%
Sahoda [6]	174-03	N–S 0°	45° E	11	rev.	E	0.0	–	–	–	–
West Nara Basin Fault zone (256)											
Matsuno -yama [4]	256-02	NNE-SSW 20°	60° W	23	rev.	W	0.0	–	–	–	–
Yata [6]	256-01	N–S 0°	60° E	21	rev.	E	0.0	–	–	–	–
OSAKA BASIN (clockwise)											
Ikoma Fault zone (177)											
Ikoma [8?]	177-03	N–S 0°	45° E	33	rev.	E	0.5	3.8 m	7.7 ka	450~934 AD	0– 0.4%
Habikono Fault zone (182)											
Tonda-bayashi[3]	182-01	NNE-SSW 20°	45° W	18	rev.	W	0.1	2.1 m	21 ka	–	ca. 0.1%
Kawachi-Nagano [3]	182-02	NE–SW 40°	45° W	10	rev.	W	0.0	–	–	–	–
MTL (Median Tectonic Line) Fault zone (183)											
Kongo [9]	183-01	N–S 0°	45° W	17	rev.	W	0.8	2 m	2.3 ka	69~294 AD	1–2%
Gojo-dani	183-02	ENE–WSW	45° N	28	rt-lat.	N	2.7	3.3	1.2 ka	1320~ 950 AD	2%– 16%
Shobu-dani	183-03	ENE–WSW	45° N	29	rev.	N	0.0	—	—	—	—
Uemachi Fault zone (178)											
Uemachi [18?]	178-01	N–S 20°	60° E	36	rev.	E	0.6	4.8 m	8 ka	580~ 1868 AD	–
Sennan [8]	178-02	NE-SW 40°	60° E	27	rev.	E	0.3	3.1	10 ka	287~186 8	0– 0.3%
Uchihata Fault zone (262)											
Uchihata [9]	262-01	ENE-WSW 70°	60° S	17	rev.	S	0.0	–	–	–	–

In the northern Nara Basin, the Sahoda segment crosses the Nara Hills into the Kyoto Basin, while the Matsuno-yama segment crosses that to run down the east side of the Ikoma Mountains and cuts the Kamenosé corridor for the Yamato River to drain the Basin. Within the Ikoma Mountains is the Yata Fault segment, cutting along the eastern edge of the Ikoma Mountains and down through their centre to form the Tomio River valley. The Kongo segment belonging to the MTL Fault Zone cuts from the Kii River valley through the passage into the Nara Basin to Mt Nijo in the Basin centre. Details of each of these segments can be accessed by clicking on each fault strip in the online map (AIST 2012), opening the label box wherein there is an option to look at "more details", or inputting the name or number into the search box.

It will be noted that all the segments in Table 13.3 are reverse faults; among them, the Ikoma Fault system is thought to have been the main mechanism for shortening the Kinki district and uplifting the Ikoma Range (Ishiyama 2003). Ishiyama interpreted the Ikoma Fault to have lifted the Ryoke granites up over Pliocene sediments in the Osaka Basin. Most of the faults in these basins dip to the east with the upthrown side (hanging wall, cf. Figure 2.10) on the east; this is consistent with compression from the east. However, some of the faults dip to the west; in particular, the secondary mountain range paralleling Mt Ikoma (entailing Kitayama and Yatayama) is bordered by the Yata Fault segment on the west and the Matsuno-yama on the east, with the former dipping eastwards and the latter dipping westwards. This suggests that this secondary range is a pop-up block that lies between the Tomio River valley and the Nara Basin proper. This secondary range continues in the southwestern Basin as the Umami Hills, but there are no known Active Faults described which continue into this area to the east of the Hills.

Other important mountains around the Basin that have been raised in the Rokko Movement by these reverse faults are (Ozaki *et al.* 2000): Mt Ikoma (642 m msl) towering over the northwest Basin, and the Aogaki Range extending almost uniformly from Mt Kasuga-oku (480 m)[10] in the northeast down to Mt Miwa (467 m) which anchors the southeast Basin (cf. Figure 13.2: 6, 7); the Aogaki Range segues into the Yamato Plateau (400–500 m) occupying the northeastern part of Nara Prefecture.

The Tenri Fault segment, at the eastern edge of the Nara Basin and part of the East Nara Fault system, was originally tracked only halfway down the Basin's eastern side. However, it is now known to extend all the way to the Hasé River, where the Miwa Project coring and excavation project took place (discussed in Chapter 14). A fault scarp that ran through the project area was confirmed through fieldwalking during the project (Soma 1993), leading to the important distinction between this natural fault scarp and two artificially leveled terrace surfaces on its west and east. Heretofore the scarp had been interpreted as an artificial adjunct to terracing operations. Calculations have been provided for the amount of uplift across the fault line (Nishioka *et al.* 2001: 105). In one instance, the eastern side is estimated to have uplifted 3 m during an M7.4 earthquake (SGSK 2014: 2).

The far southwestern edge of the Nara Basin is defined by a branch fault of the Median Tectonic Line, the Kongo Fault segment. Its relatively short cycle of earthquake repetition (2300 years) and relatively

[10] Three peaks are historically important in the northeast Nara Basin. Mt Wakakusa (342 m, formerly known as Mt Mikasa) has long been burned annually in January in a great celebration, and it remains bald (its name means 'young grass') the rest of the year. To its southeast is the highest mountain in the northeast, Mt Kasuga (480 m), which has a hillock (260 m) that backs the Kasuga Shrine and hosts the Mt Kasuga Primeval Forest, a UNESCO World Heritage site. All these belong to the Kasagi Mountain Range, which forms the northwestern portion of the Yamato Plateau and northern end of the Aogaki Range.

high slip rate per thousand years (0.8 m/ka) probably accounts for the elevation of Mt Kongo (at 1125 m) being the highest in the Basin, despite the slip distance per event (at 2 m) being lower than on the Ikoma Fault segment (3.8 m) or the Tenri Fault segment (2.4 m). Toda (2016) notes that although the Kongo Fault segment is oriented N-S, it is particularly active because it is accommodating right-lateral slip of the east–west MTL, that might otherwise have extended further east. The Kongo Mountains have a steep eastern side but slope gently to the west (dip 45°W) compared to the Ikoma Mountains in the northwestern Basin have steep western sides and slope gently to the east (dip 45°E).

In concluding this section, it is noteworthy that the recurrence rates for the immediate fault movements lining the Nara Basin are minimal. These are listed in the last column of Table 13.2. They differ slightly from those given by Yoshioka (2009), but most important is his determination that some of the faults have not moved since the Pleistocene; these are generally marked with a dash in Table 13.2. Sangawa *et al.* (1985) further note that there was a trade-off in activity between the western and eastern Nara Basin, with the western faults becoming inactive and the eastern faults becoming active sometime in the Middle Pleistocene. It is thus the eastern edge of the Basin that is rising, albeit at a slow rate (0.4m/ka), while both the recurrence rate and slip rate (0.5/ka) of the Ikoma Mountains are slightly higher.

The creation of a fault is generally accompanied by an earthquake, and just because projected recurrence rates on the faults surrounding the Basin are minimal for the near future, this does not mean that the Kinki region has not been subjected to earthquakes near and far in the past. The following section reviews both major earthquake occurrence and damage in Nara caused by forces outside the Basin.

Earthquake record

As discussed in Chapter 2, earthquakes in Japan have two different sources: those caused by movement on Active Faults and those resulting from the subduction of oceanic plates underneath the archipelago. It is often difficult to identify the source of prehistorical and historical earthquakes, and even when known, the source is often not identified in the literature. Nevertheless, earthquakes have affected Nara, as the list in Ushiyama (2014: 196-197) demonstrates.

Active Fault earthquakes

Among the faults surrounding the Nara Basin, the Tenri, Ikoma, Kongo, and Gojo-dani Fault segments have historically noted earthquake movement. Among these, the Gojo-dani Fault segment has the highest regional probability of rupture (between 2 and 16% probability) within the next 30 years (cf. Table 13.2). Both the Tenri and Ikoma Fault segments have lower probabilities (0~0.4%) of activity within the next 30 years. Kongo is on a shorter repeat cycle (every 2300 years) than the others (Tenri at 11,000 years, and Ikoma at 7700 years), accounting for its higher probability of rupture in the near future; given that its last recorded movement was between 69 and 294 AD, it is probably overdue.

The most recent and only significant earthquake in the region since 1923 (the Great Kanto Earthquake) is recorded as occurring in northern Osaka Prefecture on 18 June 2018; it registered M_{JMA}6.1 with an intensity of 6-weak (see Table 2.2) and with an epicenter less than 13 km deep where the Ikoma Fault Zone meets the Arima-Takatsuki Tectonic Line (GSJ 2018; Hallo *et al.* 2019). Unfortunately, the exact fault line on which the earthquake occurred could not be determined, as "small earthquakes have occurred across the entire Tamba region [of Hyogo Prefecture] and are not always confined to known faults" (Kato & Ueda 2019: 2, brackets added). This is a poignant lesson that, despite the ongoing research into Active Faults, there is still much unknown, and complacency concerning the Nara Basin should be avoided. Iwahashi (2010: 36) states that "In Japan, inland active faults often lie in the mountain footslopes of urban neighborhoods" as has been seen above for the Nara Basin and is characteristic the world over (the change in slope between plains and mountains is diagnostic of a fault line). Movement on the Kongo Fault segment will affect Katsuragi and Gosé cities along the southwestern Basin, while Nara, Tenri, and Sakurai cities on the eastern side would be subject to any movement in the East Nara Fault Zone.

Subduction zone earthquakes

It must be kept in mind that the above are all Active Fault earthquakes, which are different from subduction earthquakes caused by plate subduction and seismicity in the Wadati-Benioff zone (cf. Chapter 2.4.2 and Figure 2.5). Earthquakes in the Nankai Trough subduction zone west of the Izu Peninsula (cf. Figure 2.1: red) dramatically affect inland areas, and Nara has been subject to intense ground shaking and multiple aftershocks since it is only 50–60 km from the seismogenic Philippine Plate surface (Satake 2015; Sekiguchi 2018). According to Satake, the Nankai Trough has been divided into five earthquake zones west to east: A-E (cf. Figure 9.6); the Kinki Triangle basins (Osaka, Nara and Kyoto) fall into zone C. Each of these zones is divided north–south into shallow source earthquakes (<10 km), strong plate coupling earthquakes (1–25 km) and deep transition zone (>25 km) "where low frequency tremors and slow earthquakes occur" (Satake 2015: 10). Historic documents record earthquakes sourced from this zone beginning 2500 years ago; most damage is seen in coastal changes, tsunami deposits in lakes and lagoons, and destruction seen on coasts and recovered in archaeological sites (Satake 2015: fig. 5).

Earthquakes affecting Nara

Drawing on the sources for Table 13.4, a listing of damages can be made. It is interesting that the early records focus on temples without recording damage to person or property. It is only with the 1854 Ansei Earthquake that death tolls for Nara are available.

The earliest earthquake recorded in Japan was in 416 AD (Table 13.4), but the first for which damages were recorded occurred in 599. In 684 AD, a wall was toppled that had just been completed in 656. The construction was recorded in the 720 *Nihon Shoki* and recovered archaeologically in the early 1990s (Sangawa 2001: 115). The earthquake of 1096 toppled the bell of Todaiji Temple. In 1099, the Kofukuji Temple gate and corridor were damaged. In 1177 at Todaiji, the hair whorls of the Great Buddha fell off and the bell fell again. In 1361, the Yakushiji and Toshodaiji Temples were damaged. In 1494, the Todaiji, Kofukuji, and Yakushiji were damaged, and aftershocks continued for several years. In 1498, the Kofukuji Temple sustained damage. In 1586, an M7.9 earthquake in the Hida region of Gifu Prefecture in the Kanto knocked down a wall of the Kofukuji.

The 1596 Keicho-Fushimi Earthquake ruptured in the Niigata–Kobe seismic zone on the Arima-Takatsuki Tectonic Line and the Rokko Fault System; luckily the Nara Basin was beyond the area of greatest damage which had an intensity of over 10 MMI (JMA Intensity 6-strong, cf. Table 2.2) (Kamai & Sangawa 2011). Nevertheless, the Kofukuji, Toshodaiji, Hokkeji, Kairyuoji temples sustained much damage. The 1707 (Hoei 4) Iga-Ueno Earthquake left 63 dead and much damage to the Hokkeji Temple, including the destruction of the pagoda – despite its essential earthquake-proof construction (cf. Chapter 9). This was one of the most destructive earthquakes in Japanese history, though damage in Nara was limited.

The 1854 (Ansei 1) earthquakes caused shaking intensity of JMA 6 (cf. Apx 7) – enough for liquefaction and considerable damage to property, leaving 450–520 dead and hundreds of houses collapsed. The first (M8.4) occurred in the wee hours, breaking the dike of a reservoir and causing a debris flow; the second (M7.2) occurred the evening of the same day on the Hananoki fault east of Nara. Sangawa (2001: 110-111) tells of the destruction in Nara and Yamato Koriyama Cities. These two subduction earthquakes ruptured sequentially in the Nankai and Tokai regions within 23 hours of each other. The Nobi Earthquake of 1891 was one of the largest in Japan, with 7,273 fatalities and 140,000 houses destroyed. In Nara the damages were less with 1 person dead, 2 injured, and 16 houses demolished.

Earthquakes that have affected Nara since 1936 are in shaded rows in Table 13.4 (Nara Chihō Kishōdai n.d.). The 1936 earthquake (M6.4) left 9 dead, with 1,200 houses destroyed. In 1944 the M7.9 combined Tokai/Nankai Earthquake left 3 dead. The Nankai Earthquake of 1946 injured 13 and destroyed 37 houses. The Yoshino Earthquake of 1952 is said to have occurred in the upper reaches of the Yoshino River (Ushiyama 2014); but the only Active Fault in the area is Senmata, which is recorded not to have moved in the Holocene (AIST 2012). Regardless, it left 3 dead and 6 injured. The 1995 Kobe Earthquake (Hanshin/Awaji) registered a shaking intensity of 4 in Nara, injuring 12 and damaging 15 houses. In 2004, Nara was shaken by two successive subduction earthquakes off the Kii Peninsula; the first resulted in rock falls and landslides in the southern Prefecture, while the second injured 6 people.

Table 13.4 Instances of earthquakes felt in Nara

Date	Reign or period	Name	Type	Magnitude /Shaking Index	Depth km /Source
Northeast					
2004		Tokaido-oki	S	M7.4	
Kinki Region					
416	Asuka P.				
599	Asuka P.			M7.0	
684	Asuka P.		S	M8.4	Nankai
887	Heian P.		S	M8.6	Nankai
1096	Heian P.		S-tsu	M8.4	Nankai/Tokai
1099	Heian P.		S	M6.4/M8.2	Nankai
1361	Heian P.		S-tsu	M8.4	Nankai
1449	Hotoku			M6.1	
1498	Meio 7		S-tsu	M8.6	Tokai
1586	Tensho 14			M7.9	Chubu
1596	Keicho 1	Fushimi	A	M7.0	
1605	Keicho 9		S-tsu	M7.9	Nankai/Tokai
1707	Hoei 4	Iga-Ueno	A	M8.6	
1707.4.10	Hoei 4		S-tsu	>8.4	Nankai
1854.7.9	Ansei 1	Iga-Ueno	A	M7.2 M6.8	Kizu R. Hananoki
1854.12.24	Ansei 1		S-tsu	>M8.4	Nankai/Tokai
1891	Meiji 24	Nobi		M8.0	
1944			S	M7.9 / 5	Tonankai
1946		Nankai	S	M8.0 / 5	Nankai
1952		Yoshino		M6.7	MTL?
1995		Hanshin/Awaji		M7.3/4	Nojima Fault
2000		Mie-ken		M5.7	
2004		Kii Pen-oki	S	M6.9/4	Nankai
2004		Kii Pen-oki	S	M7.4/4-5weak	Nankai [cont'd]

Nara Basin				
1091	Nara Hills		M6.4	
1177	Nara Hills		M6.3	
1494	Oji		M6	
1936	Kawachi-Yamato	A	M6.4	
1962	Kii River	A?	M6/4	
2000	Kongo	A	M4/2	8
2007	Kongo	A	M4.1	12
Nara Prefecture				
1899	Kii-Yamato	S	M7	40–50/Nankai
1952	Yoshino		M6.7	60/Nankai

Interestingly, Nambu *et al.* (2016) discuss the vulnerabilty of traditional wooden houses to seismic disturbance by the next Nankai earthquake. The houses are located in Nara's southwestern-most city within the small Gojo Basin – situated south of Mt Kongo and crossing over the Yoshino River. The authors are seemingly unaware that two fault segments of the MTL (the Shobu-dani N-dipping reverse fault and the Gojo-dani E–W strike-slip fault) pass through the Basin. Although Shobu-dani Fault has not moved in the Holocene, the Gojo-dani fault has a 2–16% probability that it will move again in the next 30 years. Active Fault earthquakes have recurrence cycles ten times longer than subduction earthquakes, but they cannot be ignored – especially when earthquakes have occurred on the Kongo Fault segment to the north in 1936, 2000, and 2007.

In looking forward, investigations undertaken by Nara Prefecture (Nara-ken 2005) estimated the Nara Basin to be affected at shaking intensities 5 or greater on the Ikoma and Kongo faults, while ruptures on the East Nara Tōen fault zone may cause shaking exceeding Intensity 6 with over 5000 expected fatalities. Subduction earthquakes from the Nankai Trough are expected to cause shaking Intensity 5-strong, with some areas reaching 6-strong. The Probability of Liquefaction (PL number) was most severe in areas of the the Toen, Ikoma, Kongo, and Yamato River fault regions (exceeding PL15), while a smaller chance of liquefaction was expected from subduction zone earthquakes at PL15, affecting only 0.6% of the land area. Given that the central Basin is filled with unconsolidated material with a high water level, it is not surprising that liquefaction is the most likely outcome. The current Nara City hazard map (Nara-shi 2008) clearly identifies the central alluvium at most risk of an M7.4 earthquake, on the Nara Toen Fault; the effects of movement on the Ikoma Fault, Kongo Fault, as well as subduction earthquakes are also calculated. Luckily, no tsunami are expected in the Nara Basin.

Summary

The Nara Basin is a small intermontane faulted basin of relatively recent creation by N–S reverse faulting in the Mid-Pleistocene. Sitting in the Ryoke Belt of the Cretaceous granite batholith, the surrounding mountains consist mostly of granite and some metamorphosed sediments originally belonging to the Accretionary Complex of the Mino-Tanba Belt. Erosional products from these highlands fill the Basin floor up to 600 m deep and are exposed around the Basin flanks as low hills up to 300 m deep and fringing terraces and alluvial fans. Stratigraphically, the lowest sediments are Miocene erosional products, while only a few patches of Miocene deposits remain in the uplands. Most of the sediments in the Basin fill belong to the Osaka Group, a regional part of the Second Setouchi Supergroup, that began around 3 Ma as Pliocene fluvial deposits from freshwater drainages into the Seto Depression, which extended from the Nagoya region through to Kyushu. From 1 Ma, marine incursions into the Seto Depression periodically laid marine sediments in layers numbered from -1 to 13; the Nara Basin contains evidence of Ma0–3 and Ma7, but from ca. 500 ka the Basin became dry land. The Higher, Middle, and Lower Terrace fills, which often began as alluvial fans, are coordinated with the timings of Ma9–11, Ma12, and Ma13 (the Jomon Transgression), respectively; but these marine incursions did not reach into the Basin. A peat layer in the Basin fill at ca. 7 m depth formed just prior to the LGM, and above it, Holocene fill is relatively thin at less than 6 m thick. It is composed of sand and mud deposited in riverine and backmarsh locations. Prominent alluvial levees are keys to understanding the natural flow patterns of the Basin's rivers, which were canalized from the 8c AD.

The unconsolidated sediments in the central Basin represent the greatest earthquake risk, from liquefaction. As the Basin is bounded on east and west by Active Faults and as it sits in the Nankai Earthquake Zone, it has been subject to several large earthquakes in the past and one is expected in the Nankai Trough within 30 years. Among the Active Faults, the western Ikoma Fault zone and the East Nara (Tōen) Fault zone have similar probabilities of erupting, between 0 and 0.4% probability in 30 years time, while the Kongo (1–2%) and Gojo-dani (2–16%) in the far southwest leading into the Median Tectonic Line have much higher probabilities.

The Nara Basin is reasonably well protected from typhoons by its intermontane setting, and it is not subject to tsunami. Landslides from high precipation are feared around the edges of the Basin but are most common in the Kamenosé corridor in the west and the Hasé River corridor leading from the southeast Basin to the east. The marshy plains were excellent areas for the beginnings of rice cultivation in the Yayoi period, and the Basin became the homeland of the first state-level society in Japan under Yamato authority.

Figure & Table Sources

Figure 13.1 Google Earth with data from SIO, NOAA, U.S. Navy, NGA, GEBCO, and the landsat image from Copernicus; modified with placenames except Nara (in original)
Figure 13.2 created by GLB and first published in (Barnes 1982)
Figure 13.3 extracted from Figure 2.1a
Figure 13.4 extracted and modified from (GSJ 2020) under the CC-BY-4.0 license
Figure 13.5 after (Ikeda & Ohashi 1996: fig. 5) modified by GLB
Figure 13.6 courtesy of M. Shiba, modified by GLB
Figure 13.7 after (Mitamura 1993: 7.1a,b) modified by GLB
Figure 13.8 compiled from (Sangawa 2001: figs. 38, 41)
Figure 13.9 after (Anon. 2016)
Figure 13.10 left: extracted from (RGAFJ 2000), modified by GLB
right: extracted from AIST (2012) and cited as per instructions at [https://gbank.gsj.jp/ activefault/index], accessed February 2022
Figure 13.11 after (Matsuda 1990: fig. 4)
Figure 13.12 extracted from AIST (2012) Active Fault Database of Japan, February 28, 2012 version. Research Information Database DB095, National Institute of Advanced Industrial Science and Technology [https://gbank.gsj.jp/activefault/index_e_gmap.html]) accessed November 2020

Table 13.1 compiled from Table 13.2 herein and (Shiba 2021: fig. 2)
Table 13.2 compiled from (Nishida 1992; Mitamura 1993; Ozaki & Miyaji 2001; Ishida 2002; Sugimori *et al.* 2003; Satoguchi & Nagahashi 2012; Sekiguchi *et al.* 2018; Shiba 2021; Anon. n.d.)
Table 13.3 extracted from AIST (2012) Active Fault Database of Japan, February 28, 2012 version. Research Information Database DB095, National Institute of Advanced Industrial Science and Technology. [https://gbank.gsj.jp/activefault/index_e_gmap.html]
Table 13.4 compiled from (Nambu *et al.* 2016; Hallo *et al.* 2019; JMA 2007; Usami 1996; Nishioka *et al.* 2001; Ushiyama 2014; Nara Chihō Kishōdai n.d.)

References

AFRG (Active Fault Research Group of the Geological Survey of Japan) (**2000**) "Active Fault investigation in the Kinki Triangle: paleoseismicity and earthquake potential of major Active Faults." *Daiyonki Kenkyu* 39.4: 289-301 (in Japanese with English abstract and title).

AIST (National Institute of Advanced Industrial Science and Technology) (**2012**) "Active Fault Database of Japan, February 28, 2012 version, 2021 update." Research Information Database DB095, National Institute of Advanced Industrial Science and Technology [https://gbank.gsj.jp/activefault/index] (in Japanese and English).

---- (**2020**) *Catalogue of geological maps, data.* Tsukuba City: Geological Survey of Japan.

ALTIS, Sungat (**1999**) "Interpretations of a Middle Miocene and late Quaternary steady dextral transpression in SW Japan and the opening tectonics for the Japan Sea." *Tectonophysics* 302: 257-285.

ANDO, Sei (**2020**) "Was ancient Nara Prefecture submerged at lake bottom? Investigating the Manyoshu mystery." *Warakuweb Culture* [https://intojapanwaraku.com/culture/133249/] (in Japanese).

Anon. (**n.d.**) "3-3. Landforms • geology outline" [www.kkr.mlit.go.jp/nara/pi/press/pdf/iinkai1/03.pdf] (in Japanese).

Anon. (**2016**) "2 Touring the region of the Yamato River." *Ministry of Land, Infrastructure, Transport and Tourism Kinki Regional Development Bureau website* [https://www.kkr.mlit.go.jp/yamato/about/press/2016/pdf/2syou.pdf] (in Japanese).

BARNES, Gina L. (**1982**) "Prehistoric landscape reconstruction and spatial analysis of artifact discovery locations." *Kokogaku to Shizen Kagaku* 15: 113-131 (in Japanese with English abstract).

---- (**1983**) *Yayoi-Kofun settlement archaeology in the Nara Basin, Japan.* Ph.D. diss. University of Michigan. Ann Arbor, MI.: University Microfilms (ProQuest). [Published as Barnes 1988].

---- (**1988**) *Protohistoric Yamato: archaeology of the first Japanese state.* Anthropological Papers 78 and Modern Papers in Japanese Studies 17. Ann Arbor: University of Michigan.

BURBANK, Douglas W & Robert S ANDERSON (**2011**) *Tectonic geomorphology.* Chichester, UK: John Wiley & Sons.

CAO, Y & Y NATUHARA (**2020**) "Composition and distribution of common reed (*Phragmites Australis*) along an urbanized river: a case from central Japan." *Applied Ecology and Environmental Research* 18.3: 3935-3950 [DOI: 10.15666/aeer/1803_39353950].

Chigakudan (ed.) (**1981**) *Dictionary of Earth Sciences.* Tokyo: Heibonsha (rev. ed.) (in Japanese).

CRD (Collaborative Reference Database) (**2011**) "Toyo ashihara no mizuho no kuni" [Rice grain heads of the fertile reed plains]. *Collaborative Reference Database* [https://crd.ndl.go.jp/reference/modules/d3ndlcrdentry/index.php?page=ref_view&id=1000085724] (in Japanese).

FUJIOKA, Kenjiro (**1941**) "Kawachi Plains and Yamato Basin: a statement towards archaeo-geography history]." *Ritsumeikan Daigaku Ronso 2, Rekishi chirihen* 1: 97-131 (in Japanese).

GSJ (Geological Survey of Japan) (**2018**) "Announcements concerning the earthquake in northern Osaka Prefecture on June 6, 2018 (Heisei 30)." [https://www.gsj.jp/hazards/earthquake/osakafuhokubu2018/index.html] (in Japanese).

---- (**2020**) "Seamless digital geological map of Japan V2 (1:200,000)." [https://gbank.gsj.jp/seamless/v2/viewer/] (in Japanese and English).

HALL, Robert B (**1932**) "The Yamato Basin, Japan." *Annals of the Association of American Geographers* 22.4: 243-292.

HALLO, Miroslav; Ivo OPRSAL & Kimiyuki ASANO *et al.* (**2019**) "Seismotectonics of the 2018 northern Osaka M6.1 earthquake and its aftershocks: joint movements on strike-slip and reverse faults in inland Japan." *Earth, Planets and Space* 71: 21pp. [DOI: 10.1186/s40623-019-1016-8].

HATAYA, Yuta (**2017**) "Investigating long-term land movement: uplift and subsidence, sedimentation and erosion." *Horonobe Forum 2017* [www.jaea.go.jp/04/horonobe/topics/29/pdf/29forum_ tokubetsu2.pdf] (in Japanese).

ICS (International Commission on Stratigraphy) (**2020**) "Stratigraphic guide." [https://stratigraphy.org/guide/litho].

IKEDA, Hiroshi & Tsuyosi OHASHI (**1996**) "Geomorphological study of Nara Basin." *Nara Daigaku Kiyō* 25: 41-64 (in Japanese with English title and summary).

ISHIDA, Shiro (**2002**) "Cenozoic: the Osaka Group in Kyoto Prefecture." *Kyoto Prefecture Red Data Book 2002* [https://www.pref.kyoto.jp/kankyo/rdb2002/geo/soil_g.html] (in Japanese).

ISHIYAMA, Tatsuya (**2003**) "Geometry and kinematics of the Uemachi and Ikoma fault zones beneath metropolitan Osaka, central Japan." *Kojishin Kenkyu Hokoku* 3: 145-155 (in Japanese with English title and abstract).

ITIHARA [Ichihara], Minoru (ed.) (**1993**) *The Osaka Group.* Osaka: Sōgensha (in Japanese with English Title, Table of Contents & Conclusion).

---- (**2001**) "Osaka Formation: Palaeo-Inland Sea river and lake systems." *Urban Kubota* 39: 2-13.

ITOH, Nobuaki; Kazuaki HORI & Masashi TAKADA (**2017**) "Latest pleistocene to holocene alluvial basin construction: an example from the Nara Basin, central Japan." *Quaternary International* 455: 102-112 [DOI: 10.1016/j.quaint.2017.06.016].

IWAHASHI, Junko (**2010**) "1:25,000-scale Active Fault Map in Urban Areas published by GSI." *Bulletin of Geospatial Information Authority of Japan* 58: 29-31 (in English).

JMA (Japan Meteorological Association) (**2007**) "Earthquake near the border between Osaka and Nara Prefectures (in Nara Prefecture) on November 6." [https://www.jishin.go.jp/main/chousa/07dec/p14-e.htm] (in Japanese with English title).

JRG [Joint Research Group on the Geomorphological Recognition and Land Utilization of Pre- and Protohistoric Japanese Peoples] (co-author) (**1987**) "Natural environments in the Nara Basin through the pre- and protohistoric ages II: descriptions of core samples and analysis on biogenic materials." *Kobunkazai Kyoiku Kenkyu Hokoku* 16.(March): 23-74 (in Japanese).

KAMAI, Toshitaka & Akira SANGAWA (**2011**) "Landslides on ancient embankments in the Kinki district (Japan): Strong motion seismoscope of the 1596 Keichou-Fushimi earthquake." *Quaternary International* 242.90-105.

KATO, Aitaro & Taku UEDA (**2019**) "Source fault model of the 2018 Mw 5.6 northern Osaka earthquake, Japan, inferred from the aftershock sequence." *Earth, Planets and Space* 71: 9 pp. [DOI: 10.1186/s40623-019-0995-9].

KAWAMURA, Yoshinari (**1993**) "Plio-Pleistocene Osaka Group in the Nara Hills, central Japan." *Chishitsugaku Zasshi* 99.7: 503-523 (in Japanese).

KIMURA, Jun-Ichi; Robert J STERN & Takeyoshi YOSHIDA (**2005**) "Reinitiation of subduction and magmatic responses in SW Japan during Neogene time." *Geological Society of America Bulletin* 117.7/8: 969-986.

Kinki/Tokai Chihō Shinseidai Kenkyū Group (**1973**) "Plio-Pleistocene tectonic movements in the Kinki and Tokai regions." *Chikyū Kagaku* 27.2/3: 66-80 (in Japanese with English title and summary).

KOKAWA, S & J Yoshida (**1964**) "Absolute age of the herbaceous peat including Menyanthes seeds in Nara Basin: radiocarbon age of the Quaternary deposits in Japan XII." *Chikyu Kagaku* 73: 42-42 (in Japanese).

KONDO, H; S TODA & K OKUMURA *et al.* (**2008**) "A fault scarp in an urban area identified by LiDAR survey: a case study on the Itoigawa–Shizuoka Tectonic Line, central Japan." *Geomorphology* 101: 731-739 (in English).

LEEDER, MR & MD STEWART (**1996**) "Fluvial incision and sequence stratigraphy: alluvial responses to relative sea-level fall and their detection in the geological record", pp. 25-39 in *Sequence stratigraphy in British geology*, ed. by SP HESSELBO & DN PARKINSON. Special Publication, No. 103. London: Geological Society of London.

LIN, Aiming; Tadashi MARUYAMA & Takao MIYATA (**1999**) "Late Holocene activity of the Gosukebashi fault revealed by trench study at the Nishitakigatani valley in the eastern Rokko Mountains, Japan." *Katsudanso Kenkyū* 18: 31-36 (in English).

MASUDA, Fujio; Natsumi ITOMOTO & Kazuhiro OGAWA *et al.* (**2016**) "Geological characteristics of the Latest Pleistocene to Holocene "chuseki-so" in the Kawachi Plain, Osaka Basin, Japan." *The Harris Science Review of Doshisha University* 57.2: 32-42 (in Japanese with English title and abstract).

MATHER, AE; M STOKES & E WHITFIELD (**2017**) "River terraces and alluvial fans: the case for an integrated Quaternary fluvial archive." *Quaternary Science Reviews* 166: 74-90.

MATSUDA, Tokihiko (**1990**) "Seismic zoning map of Japanese Islands, with maximum magnitudes derived from Active Fault data." *Jishin Kenkyosho Iho* 65: 415-422 (in Japanese with English title).

MATSUOKA, K & S Nishida (**1980**) "The latest Pleistocene to Holocene sediments in the Nara Basin." *Bulletin of the Faculty of Liberal Arts, Nagasaki University (Natural Science)* 21: 35-47 (in Japanese with English abstract).

MATSUOKA, K; S NISHIDA & M KANEHARA (**1984**) "Upper Pleistocene strata of the Nara basin and the palaeoenvironment of Nara surrounding the Furu site." *Maizo Bunkazai Tenri-kyo Chosadan Kokogaku Chosa Kenkyu Chukan Hokoku* 10: 5–24 (in Japanese).

MATSUSHITA, Susumu (**1971**) *Kinki region*. Nihon Chihō Chishitsushi. Tokyo: Asakura (in Japanese).

MCCALPIN, James P (**1996**) "Application of paleoseismic data to seismic hazard assessment and neo-tectonic research", pp. 439-493 in *Paleoseismology*, ed. by JP MCCALPIN. San Diego, CA: Academic Press.

MITAMURA, Keisuke; Takamoto OKUDAIRA & Muneki MITAMURA (**2016**) "Meso-scale brittle deformation structures in and around the Ikoma fault zone." *Chishitsu Zasshi* 122.2: 61-74 (in Japanese with English title, summary, captions).

MITAMURA, Muneki (**1993**) "Nara Basin", pp. 87-100 in *Osaka sōgun*, ed. by M ITIHARA. Osaka: Sōgensha.

NAMBU, Yasuhiro; Mina SUGINO & Sunao KOBAYASHI *et al.* (**2016**) "Structural investigation and seismic performance evaluation of traditional wooden houses in Gojoshinmachi in Nara Prefecture in Japan." *Journal of Structural Engineering* 62: 251-258 (in English).

Nara Chihō Kishōdai (**n.d.**) "Nara Prefecture earthquake damage" [http://www.jma-net.go.jp/nara/ jisin/saigai _jisin.html] (dead link).

Nara-ken (**2005**) "III Results of estimating earthquake damage" [https://www.pref.nara.jp/secure/ 146611/06-1_shizen.pdf] (dead link).

Nara-shi (**2008**) "Nara City earthquake hazard map: ground acceleration map." [https://www.city. nara.lg.jp/ site/bousai-saigai/2028.html] (in Japanese with English title).

NGS (Nippon Gakujutsu Shinkōkai) (**1965**) *The 'Manyōshū' - the Nippon Gakujutsu Shinkōkai translation of 'One Thousand Poems' with the texts in romaji.* New York: Columbia University Press.

Nihon Gennen (**2019**) "Geological investigation results in relation to the classification of terraces." [https://www2.nsr.go.jp/data/000266253.pdf] (in Japanese).

NISHIDA, Shiro (**1992**) "Late Cenozoic tephrostratigraphy and chronostratigraphy in the Nara Basin and its surrounding area." *Bulletin of the Nara University of Education* 41.2: 5-22 (in Japanese).

NISHIOKA, Y; M OZAKI & A SANGAWA *et al.* (**2001**) "Geology of Sakurai district, with Geological Sheet Map at 1:50,000." 地域地質研究報告 京都 (I l) 第 64 号 NI-53-15-1, 141pp. [GSJ_MAP_G050_11064_ 12001_D.pdf] (in Japanese with English abstract).

OGUCHI, Takashi; Kyoji SAITO & Hiroshi Kadomura *et al.* (**2001**) "Fluvial geomorphology and paleo-hydrology in Japan." *Geomorphology* 39: 3-19.

OOI, Nobuo (**1992**) "Pollen spectra around 20,000 years ago during the last glacial from the Nara Basin, Japan." *The Quaternary Research* 31.4: 203-212 (in English).

OZAKI, M; A SANGAWA & K MIYAZAKI *et al.* (**2000**) "Geology of the Nara district, with Geological Sheet Map at 1:50,000." 地域地質研究報告 京都 (I l) 第 64 号 NI-53-15-1: 162pp. [GSJ_MAP_G050_ 11052_12000_D.pdf] (in Japanese with English abstract).

OZAKI, Masaki & Yoshinori MIYAJI (**2001**) "Osaka Group", pp. 58-71 in *Geology of Sakurai district, with Geological Sheet Map at 1:50,000*, ed. by Y NISHIOKA *et al.* [GSJ_MAP_G050_11064_2001_D.pdf] (in Japanese).

PICKERING, Kevin T; Michael B Underwood & Sanny SAITO *et al.* (2013) "Depositional architecture, provenance, and tectonic/eustatic modulation of Miocene submarine fans in the Shikoku Basin: results from Nankai Trough Seismogenic Zone Experiment." *Geochemistry Geophysics Geosystems* 14.6: 1722-1739.

RGAFJ (Research Group for Active Faults in Japan) [Katsudansō Kenkyūkai] (**2000**) "Active Fault investigation in the Kinki triangle: paleoseismicity and earthquake potential of major active faults." *Daiyonki Kenkyu* 39.4: 289-301 (in Japanese with English title and abstract).

SAKAMOTO, Takahiko; Kyoko KATAOKA & Yoshihiro MORIYAMA (**2001**) "The volcanic ash layer resedimented by turbidity currents in the Osaka Group, Osaka, Japan." *Chikyu Kagaku* 55.3: 173-181 (in Japanese with English title and abstract).

SALVADOR, Amos (ed.) (**1994**) *Kokusai sōjo gaido [International Stratigraphic Guide: a guide to stratigraphic classification, terminology, and procedure].* Translated into Japanese by the Geological Society of Japan. Trondheim, Norway & Boulder, CO / Tokyo: The International Union of Geological Sciences & The Geological Society of America / Kyōritsu (in Japanese).

SANGAWA, Akira (**1978**) "Geomorphic development of Izumi and Sanuki Ranges and relating crustal movement." *The Science Reports of the Tohoku University, 7th Series (Geography)* 28: 313-338 (In English).

---- (**2001**) "Terraces and terrace deposits (th, tm, tl, tl1, tl2)", pp. 77-86 in *Geology of Sakurai district, with Geological Sheet Map at 1:50,000*, ed. by Y NISHIOIKA *et al.* [GSJ_MAP_G050_ 11064_2001_D.pdf] (in Japanese).

SANGAWA, Akira; Yoshihiro KINUGASA & Koji OKUMURA (**1985**) "Neotectonics along the eastern rim of the Nara Basin, central Japan." *Daiyonki Kenkyu* 24.2: 707-722 (in Japanese with English abstract).

SATAKE, Kenji (**2015**) "Geological and historical evidence of irregular recurrent earthquakes in Japan." *Philosophical Transactions of the Royal Society A* 373.#2140375: 15 pp. (in English).

SATO, Takaharu; Takeshi NAKAJO & Yutaka WADA *et al.* (**2012**) "The Miocene Muro ignimbrite." *Chishitsugaku Zasshi* 118.Suppl: 53-69 (in Japanese with English title and figure captions).

SATOGUCHI, Yasufumi & Yoshitaka NAGAHASHI (**2012**) "Tephrostratigraphy of the Pliocene to Middle Pleistocene Series in Honshu and Kyushu Islands, Japan." *The Island Arc* 21.3: 149-169.

SEKIGUCHI, Haruko; Kimiyuki ASANO & Tomotaka IWATA (**2018**) "3D velocity structure model of Nara Basin." Disaster Prevention Research Institute, Kyoto University Annual Meeting, Poster session P19 [www.dpri.kyoto-u.ac.jp] (in Japanese with English title and abstract).

SGSK (Sangyō Gijutsu Sōgo Kenkūjo) (2014) "Heisei 25 'Active Fault supplementary survey' report of results" [www.jishin.go.jp/main/chousakenkyuu/tsuika_hokan/h25_nara.pdf] (in Japanese).

SHIBA, Masahiro (**2021**) "Characteristics of crustal uplift since the Pliocene in central Honshu, Japan, and sea level rise." *Chikyu Kagaku* 75: 37-55 (in Japanese with English and abstract).

SOMA, H (**1993**) "Landscape reconstruction around the Shikishima Church", pp. 27-28 in *The Miwa Project: survey, coring and excavation at the Miwa site, Nara, Japan*, ed. by GL BARNES & M OKITA. BAR International Series 582. Oxford: Tempvs Reparatvm.

SUGIMORI, Shinji; Kazuo MATSUI & Naoki SUGIYAMA (**2003**) "Example investigation of Active Fault hidden in a sedimentary basin – the case of Ujigawa Fault." *Zenchiren 'Gijutsu e-Forum 2003' Saitama* [www.zenchiren.or.jp/e-Forum/2003/039.PDF] (in Japanese).

Suito Osaka Consortium (**2022**) "Transformations of ancient Osaka." [www.suito-osaka.jp/special/history/history_2.html] (in Japanese).

SUZUKI, Yasuhiro (**2013**) "Tectonic geomorphological Active Fault studies in Japan after 1980." *Geographical Review of Japan, Series B* 86.1: 6-13 (in English).

TAKAHASHI, Masaki (2006) "Tectonic development of the Japanese Islands controlled by Philippine Sea Plate motion." *Chishitsu Zasshi* 115.1: 116-123 (in Japanese with English title and abstract).

TAKEMURA, Keiji; Naoko KITADA & Hiroko ITO *et al.* (2017) "Quaternary science and geoinformatics." *Dai Yonki Kenkyu* 56.5: 207-215 (in Japanese with English title and abstract).

TATSUMI, Yoshiyuki; Takuo YOKOYAMA & Masayuki TORII *et al.* (**1980**) "K-Ar ages of Setouchi volcanic rocks from Osaka and east Yamaguchi area: age determination for Setouchi volcanic rocks, no. 4." *The Journal of the Japanese Association of Mineralogists, Petriologists and Economic Geologists* 75: 102-104 (in Japanese with English title).

TODA, S (**2016**) "Crustal earthquakes", pp. 371-408 in *The geology of Japan*, ed. by T Moreno *et al.* London: Geological Society of London.

USAMI, Tatsuo (**1966**) "Chart of the major damaging earthquakes in Japan." *Jishin Kenkyūsho Ihō* 44: 1571-1622 (in Japanese with English title and abstract).

USGS (**n.d.**) "U.S. Quaternary faults." [https://usgs.maps.arcgis.com/apps/webappviewer/index.html?id=5a6038b3a1684561a9b0aadf88412fcf].

USHIYAMA, Motoyuki (**2014**) "Learning from history: Nara's disaster history." *Nara no Saigaishi 12*. Nara Prefecture: Governor's Office [www.pref.nara.jp/secure/118509/naranosaigaisi12.pdf] (in Japanese).

WAKITA, K; K TAKEUCHI & K MIZUNO *et al.* (**2013**) "The geology of southwestern Kyoto (1:50,000 map) explanation." Tsukuba City: AIST, Geological Survey of Japan (in Japanese).

YABUUCHI, Satoshi (**2021**) "Number 26 Nara-Lake" [www.pref.nara.jp/59787.htm] (in Japanese).

YONEKURA, Nobuyuki; Sohei KAIZUKA & Michio NOGAMI *et al.* (eds) (**2001**) *Regional geomorphology of the Japanese Islands, Vol. 1: Introduction to Japanese geomorphology.* Tokyo: University of Tokyo Press.

YOSHIDA, Shiro (**1992**) "Geologic development of the Setouchi Geologic Province since early Miocene, with special reference to the First and Second Setouchi Inland Sea times." *Bulletin of the Geological Survey of Japan* 43.1/2: 43-67 (in Japanese with English title and abstract).

YOSHIKAWA, Shusaku (**2012**) "Quaternary stratigraphy of the Osaka sedimentary basin, central Japan." *Dai Yonki Kenkyu* 51: 1-19 (in Japanese with English title and abstract).

YOSHIKAWA, Torao; Sohei KAIZUKA & Yoko OTA (**1981**) *The landforms of Japan.* University of Tokyo Press (in English).

---- (**1981**) "Tectonic landforms", pp. 39-72 in *Landforms of Japan*, ed. by T YOSHIKAWA *et al.* Tokyo: University of Tokyo Press.

YOSHIKOSHI, Akihisa (**1995**) "Water hazards of the Nara Basin." *Nara Daigaku Kiyō* 23: 111-122 (in Japanese with English title and summary).

YOSHIOKA, Toshikazu (**2009**) "Evaluation of earthquake occurrence from active faults - valuation of rupture probabilities of active faults using the Cascade Earthquake Model based on behavioral segmentation." *Synthesiology (English edition)* 2.3: 194-200.

CHAPTER 14

Geoarchaeological Studies in Nara, Japan: the Integrated Findings

Gina L. BARNES, NISHIDA Shiro, OKITA Masaaki

In this Chapter, the authors summarize 20 years of scholarship and two field projects on geoarchaeology in Nara, Japan, carried out by researchers from 12 different institutions in 4 countries. The research goal was to test an aerial photographic reconstruction of surface landforms in the Nara Basin with subsurface data. Project A was conducted at Asawa; it tested, through geological coring, whether a suspected swampy backmarsh in the eastern Basin existed and whether it would yield data on the transition to wet rice agriculture in the mid-1st millennium BC. Project B was conducted at Miwa; it tested, through geological coring and subsequent excavation, the nature of upland agricultural terrace formation in the southeastern Basin and whether the suspected existence of a 4c palace site could be confirmed.

Two layers of carbonaceous clay at Asawa were dated to the Early (5000–3500 BC) and Final (1000–300 BC) phases of the Jomon period. Pollen data revealed the establishment of an evergreen oak forest from 5,000 years ago and anthropogenic changes in forest cover from 2,000 years ago. Phytoliths from rice, millets, reeds, and bamboo were recovered in layers postdating the Final Jomon carbonaceous clay. A fault scarp with anthropogenic modification of the terraces was identified at Miwa. It was discovered that an incised stream valley had been infilled in the Medieval period at the same time surface layers were razed; the front of the terrace was extended in the premodern period. Remains were recovered from the Middle Yayoi (100 BC–AD 100) and the Medieval (1185–1603 AD) periods. However, as the terracing involved razing the early historic levels, no data were recovered on the alleged 4c palace site.

The significance of these findings lies in the identification of (a) a swampy backmarsh at Asawa, where initial agricultural efforts in growing wet rice in the Basin may have occurred, confirming the aerial photographic reconstruction; and (b) hillside terracing activities at Miwa, from the Medieval period onwards, which have radically changed the configuration of the natural topography.

Introduction

In the 1980s, two geoarchaeological field projects were undertaken to test a landscape reconstruction of the Nara Basin, Japan, based on aerial photograph data (Figure 14.1). The aerial photographic reconstruction, with a general discussion of its methodology, has been described in several publications (Barnes 1982, 1983, 1986, 1988) and is summarized below. The 1984 project tested this reconstruction in a low-land area of the east central Basin at Asawa, Tenri City (JRG 1986, 1987; Barnes *et al.* 1986). The 1988/1989 project tested the reconstruction in an upland area of the southeast Basin at Miwa, Sakurai City (Barnes 1990*a*; Barnes & Okita 1993). In addition, both of these projects assessed prehistoric land use and land morphology. In this Chapter, we describe these projects, and interpret the most important

This chapter was originally published in *Geoarchaeology: An International Journal* 20.8: 837-860 (2005), [DOI:10.1002/gea.20085], reproduced with permission. The article has been reformatted for presentation here, with some amendments to illustrations and text updates together with added references.

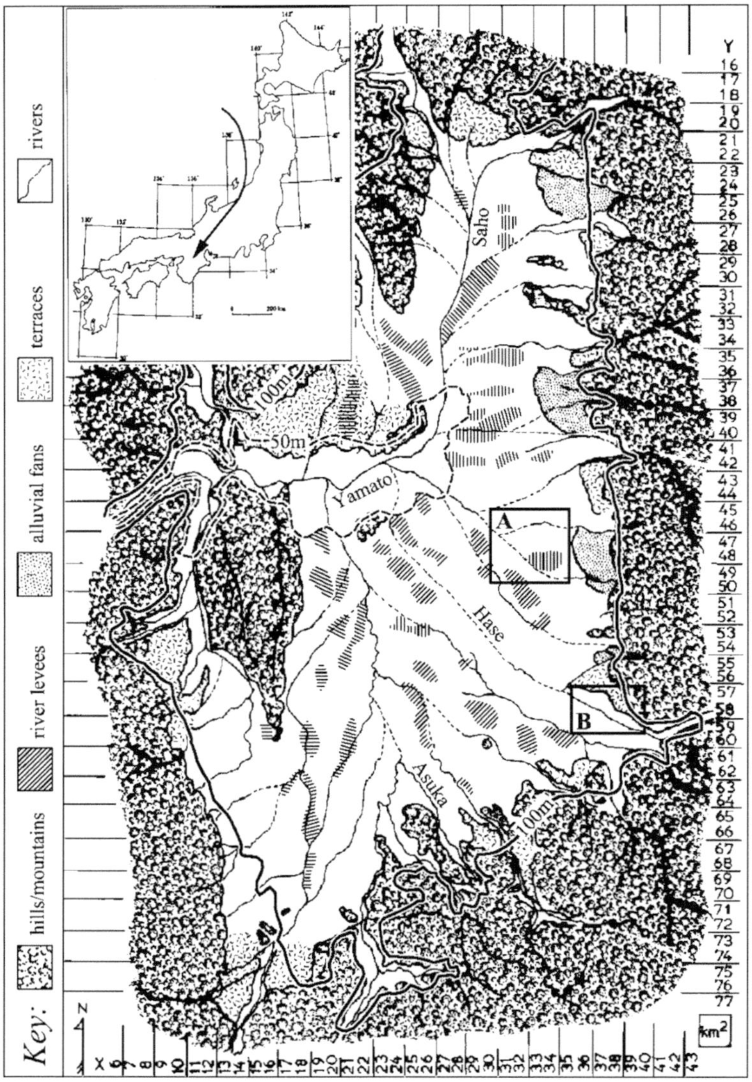

Figure 14.1 Aerial photographic reconstruction of natural topography in the Nara Basin

Most archaeologists map sites in the basin demarcated by 50 m and 100 m contour lines, which cross-cut the natural terrain as shown here and do not give a clear image of the relationships of sites to topography; grid numbers on X–Y axes mark off 1 km squares. A = Asawa project area (cf. Figure 14.3). B = Miwa project area (cf. Figures 14.8, 14.9).

findings as a coherent whole. Included are ^{14}C dates which have become available since previous publications, changes in the interpretation of the landforms resulting from research on Active Faults in the 1990s, and summaries of data from stratigraphic cores for both projects (Table 14.1).

Prehistoric Occupation & Landscape Reconstruction

The pattern of occupation in the Nara Basin follows the general scheme of Japanese prehistory. Though little is known of the Palaeolithic subsistence system, andesite (sanukite) resources from Mt Nijo were important for toolmaking in the terminal Paleolithic after 25,000 BP (see BOX 5). Postglacial occupation is characterized by ceramics and sanukite tools, produced during the long Jomon period (12,000–300 BC), with a subsistence system based on hunting and gathering. Wet rice agriculture was introduced from the continent into Kyushu Island in the first millennium BC, followed by other technical innovations in ceramics, tool-making, weaving, and metallurgical crafts. Consolidation of these new trends led to the establishment and spread of the Yayoi culture (~800 BC[1]–AD 250) through western Honshu Island. Agricultural society quickly developed hierarchical leadership, resulting in the construction of monumental mounded tombs for elite rulers from the early 3c in the Nara Basin. The relations of tombs and topography can be seen in Barnes (1988). The Kofun (old tomb) period (AD 250–710) witnessed the formation of the first state, Yamato, and its transformation in the 7c to a bureaucratic state, patterned on a Chinese administrative model.

[1] The Yayoi Period has been dramatically extended beyond 300 BC to 800/1000 BC (see Apx 2: Table B).

Table 14.1 Core summaries for Asawa (1984) and Miwa (1988) coring projects

1984 core #	length (m)	**1988** core #	length (m)
84-0	2.75	88-1	1.86
84-1	3.59	88-2	1.86
84-2	3.06	88-3	1.46
84-3	1.73	88-4	1.8
84-a	2.94	88-5	1.66
84-4	3.53	88-6	1.52
84-5	2.99	88-7	1.88
84-6	2.74	88-8	3.0
84-7	3.70	88-9	2.1
84-8	2.85	88-10	2.23
84-9	2.85	88-12	1.88
84-10	2.88	88-13	1.86
84-11	2.73	88-15	1.9
84-12	3.13	88-16	2.9
84-13	1.70	88-17	2.1
84-14	1.55	88-18	2.5
84-15	2.84	88-19	2.64
84-16	2.70	88-20	2.2
84-17	2.70	88-21	2.5
84-18	1.20	88-22	2.4
84-19	3.42	88-23	2.2
84-20	2.95	88-24	2.6
84-21a	4.40	88-25	2.52
84-21b	1.60	avg.	2.16 m
84-21c	0.81		
84-22	2.60		
84.23	3.50		
84-24	2.74		
84-25	3.00		
84-26	2.32		
84-27	3.20		
84-28	3.00		
84-29	2.87		
84-30	3.40		
avg.	2.76 m		

Prehistoric settlement of any area, Nara in particular, cannot be understood against the modern landscape background. It must be related to its contemporaneous landscape to assess the pattern of prehistoric activities across the then land surface. In areas of historically intensive agriculture, the landscape has often been transformed beyond recognition from its prehistoric state. Moreover, the very nature of historic agriculture affects the kind, amount, and location of archaeologically recoverable prehistoric remains (cf. Barnes 1986). Deep ploughing in the West and paddy-field construction in the East are two major agents of landscape transformation in the historic and modern periods.

In Nara, the Basin surface was transformed from the 8c through the institution of a gridded *jōri* system of field regularization. This involved large-scale land survey, new paddy-field construction oriented to the north–south grid, and canalization of the major rivers. Paddy-field construction effectively sealed the pre-8c landscape, and canalization terminated natural geomorphological processes of sedimentation and drainage within the riverine catchments. Agricultural land divisions in the present-day Nara Basin still follow the fossilized 8c grid system. Thus, to understand pre-8c activities, a reconstruction of the natural surface topography and riverine system was necessary.

Topographical reconstruction was accomplished through stereoscopic interpretation of aerial photographs of the Basin. These were mainly 1:10,000-scale aerial photographs from the 1960s, but some post-war 1:40,000-scale photographs were analyzed for large-scale patterns. Both show the Basin in its historic agricultural context, before urban sprawl obscured much of the topography. Landforms were identified by height differentials seen in stereoscopic view; as little as 50 to 70 cm could mark the boundaries of landforms. Such differentials often correlated with changes of orientation, size, or type of field divisions (e.g., paddy vs vegetable gardens). Topographic forms – such as hills, river terraces, and alluvial levees – were identified by height and/or landuse differences and plotted by the first author (GLB) in the late 1970s on corresponding 1:10,000 scale ordinance survey maps that used contours of 2.5 m intervals. Many of the fine-grained height differentials (<1 m) were field-checked at that time.

The resulting map (Figure 14.1)[2] shows alluvial fans, terrace scarps, river levees, and the direction of natural river courses before canalization, as well as the boundaries and morphologies of hilly areas along the Basin flanks according to landform rather than altitudinal contour lines. This reconstructed landscape is presumed to indicate the natural drainage system of the Basin before canalization of the rivers was carried out in conjunction with the 8c *jōri* field system. However, it is only a guide to the lay of the land before the 8c; rivers are always changing their exact courses and levees can be eroded away or covered by new floodplains. Nevertheless, this reconstruction is thought to provide a more accurate basis on which to project hypotheses about early landscape utilization and settlement patterns than the current gridded fields and canalized river system.

Lowland Geomorphology: the Asawa Project

One of the major questions this research addressed was the extent of the marshes in the lowland Basin prior to the *jōri* field system. The protohistoric epithet of the Basin – the "Central Land of the Reed Plains" used in the 8c chronicles – implies substantial wetlands in the Basin center. Natural marshes would not only have provided the reeds for thatching houses, but would also have served as the obvious context for early attempts at wet-rice agriculture. *Phragmites* and rice have similar environmental requirements: reeds can survive in water up to 40 cm deep, while rice must have a constant supply of clean, flowing water for at least 3 months during its summer growing season (cf. Barnes 1990*b*). Rice cannot tolerate stagnant water; therefore, a major challenge for early rice-growers was providing proper drainage to keep water circulating and eventually to dry out the beds, allowing the crop to ripen. Given these limitations, we viewed the central bottomland as less than ideal for wet-rice agriculture in the Nara Basin, as it was subject to excessive variations in water supply and to flood damage – from ponding

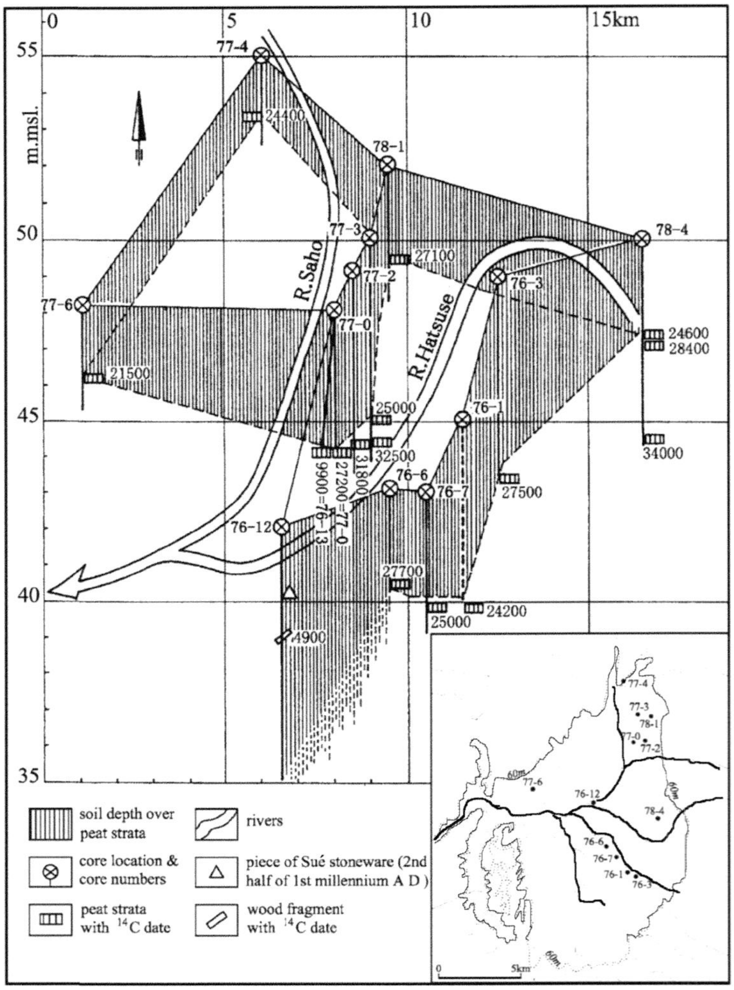

Figure 14.2 A three-dimensional fence diagram of sediments overlying peat layers in late 1970s' corings by the second author (NS). Note that Core #76-12 did not reach peat, hence the rendering of uncertain depth. The coring locations are shown relative to the Basin's major rivers (Saho and Hatsuse/Hase), and both core locations and rivers are identified geographically in the inset using an 80 m contour line.

[2] The original caption implied there were contour lines in the figure, but there were not; an example of the commonly used contour lines for site distribution maps appears in Figure 14.2: inset, an 80 m contour line.

and stagnation at the lowest point in the Basin, to rapid alluviation through flood deposition, or severe degradation through flood scouring.

Geological coring in the late 1970s revealed peat layers at 1.5 to 5.5 m below the present surface (Figure 14.2)[3] extending onto the alluvial apron (Nishida & Matsuoka 1977; Nishida *et al.* 1978, 1979). The majority of those layers were radiocarbon-dated to between 32,000 and 24,000 years ago, when the forests were changing from cold-temperate *Quercus-Alnus-Betula* to sub-boreal *Pinus* (haploxylon)-*Betula-Quercus*. These layers were too early to show evidence of the agricultural transition in the Basin, but peat discoveries did hint that open marshland extended well beyond the central bottomlands.

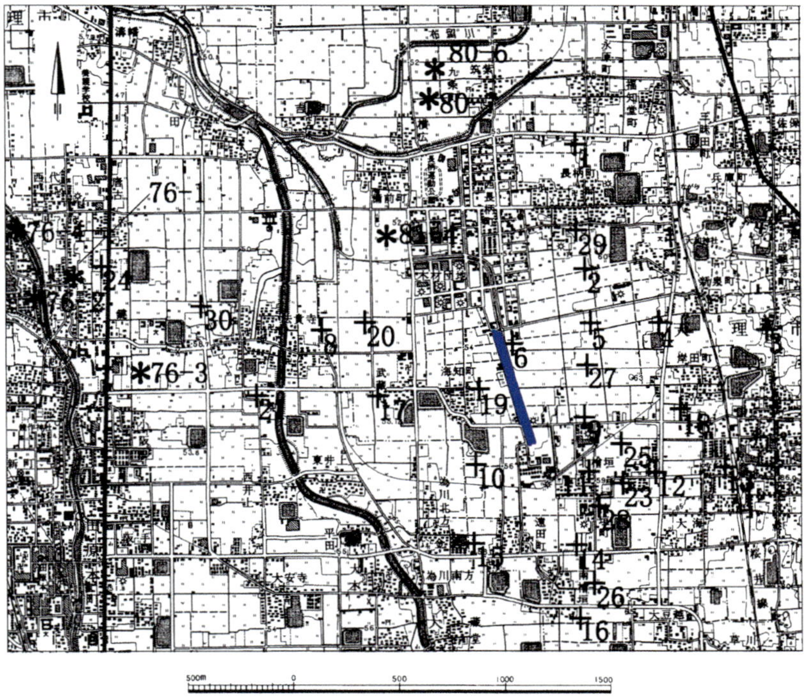

Figure 14.3 Locations of cores (numbered crosses) taken in the 1984 Asawa Project; stars mark previous coring locations. The wartime Asawa airstrip is shown in blue; the fact that it could have been built here indicates an area traditionally devoid of settlement, consistent with the presence of unsuitable ground water conditions.

Core, sediment, and water

In contrast to the 1970s deep coring projects noted above, the Asawa project concentrated on the shallow, uppermost strata in a limited geographical area. Thirty-four cores (Table 14.1: left) were taken with Dutch hand-augers within an approximately 3.1 x 3.8 km area across the old Asawa airfield[4] and into the surrounding villages (Figure 14.3). Core diagrams and sediment descriptions are given in JRG (1987: 53-74).

Coring of the upper sediments of the Basin in this area revealed two major layers of black carbonaceous clay but no peat. Their high carbon content allowed ^{14}C dating of the sediments themselves (JRG 1987). The dates clustered into two groups: 5250–3500 cal. BC, generally equivalent to the Early Jomon period, and 1250–200 cal. BC, coinciding with the Final Jomon period (Figure 14.4). These dates were augmented by the identification of known volcanic ash in several cores (JRG 1986):

1. Akahoya ash (7300 BP) in Core 84-4, occurring between strata ^{14}C-dated to 2340±90 (N-4882, uncalibrated) and 9050±205 (N-4883, uncalibrated); the K-Ah ash has been dated to 7300 BP by lake varve and 7330 BP by dendrochronological calibration (Hayakawa n.d.).

[3] Hatsuse River in Figure 14.2 = Hase River in Figure 14.1

[4] Stated in the original caption to be 'black' but was actually white; here it is presented in blue for clarification.

2. Sakate ash in Core 84-12, occurring under a stratum ^{14}C-dated to 2400±125 (N-4889, uncalibrated), and in Core 84-28, occurring under a stratum ^{14}C-dated to 7350±155 (N-4895, uncalibrated). Considerable controversy surrounds the age of the Sakate ash. Machida *et al.* (1984) put it at ~17,000 BP based on its similarity with Daisen Upper Hoki tephra; this dating was used in our interim publication (JRG 1987). In contrast, Oi (1992) adopts a date of ~10,000 BP based on a ^{14}C date from above a Sakate ash layer dating to 9550±210 BP. Two ^{14}C dates obtained for Sakate ash in 1997 were 15,800±200 and 15,900±200 (Takemura *et al.* 2000). However, the Kyoto Fission Track Laboratory has recently identified it as ash from Mt Sanbe in Shimane Prefecture (cf. Figure 6.10), obtaining a ^{14}C date of ~17,000 BP from samples in Kobe (Toru DANHARA, pers. comm. 9 October 2000). This date is compatible with that first proposed by Machida *et al.* (1984); thus, it does not challenge our interim interpretations.

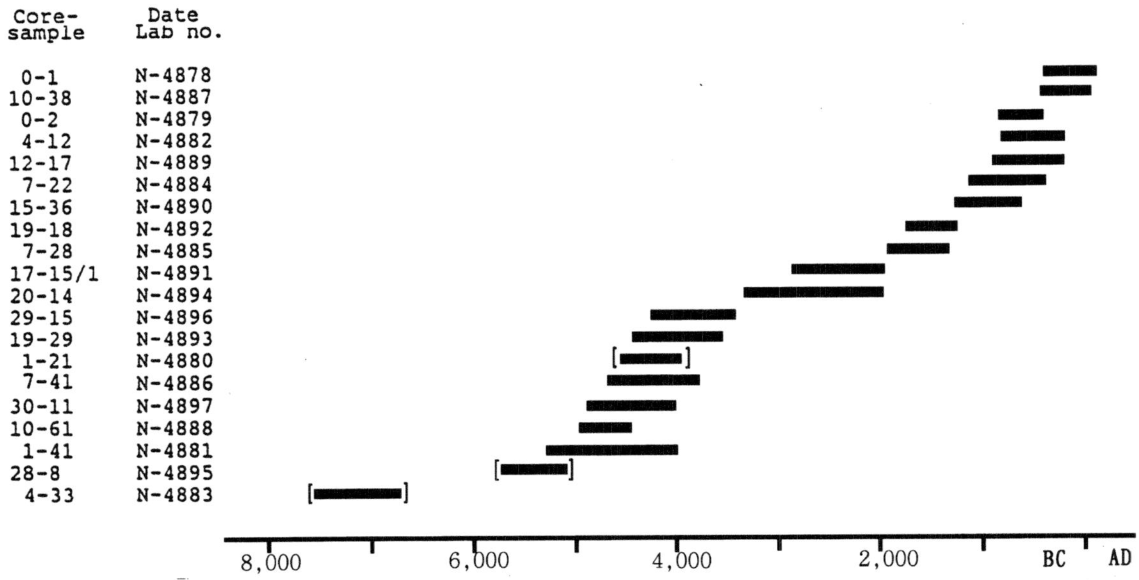

Figure 14.4 Radiocarbon dates of black carbonaceous clay deposits from the Asawa cores

Table 14.2 Depths of early-dated radiocarbon samples from lowest carbonaceous clay layers, indicating formation of clay layers during Climatic Optimum (4000–2000 BC)

core#	^{14}C lab#	date, calib. BC [uncalib.]	^{14}C depth (cm)	lowest black clay (cm)
84-19	N-4893	4410-3550	185-195	201-270
84-1	N-4881	5240-3950	265-270	203-320
84-30	N-4897	4850-3990	196-210	326-328
84-7	N-4885	4630-3730	210-215	206-350
84-28	N-4895	5710-5090	175-180	267-290
84-4	N-4883	[7510-6690]	307-314	316-332

The Early Jomon (~5000–3500 BC) and Final Jomon (~1000–500 BC) periods corresponded respectively with the postglacial Climatic Optimum, and the subsequent deterioration in climate. The climatic conditions in these two periods are thought to have been quite opposite (Fuji 1966). Warm temperatures and a rising sea level (the Jomon Transgression) characterize the first, and a cooler wet climate and a

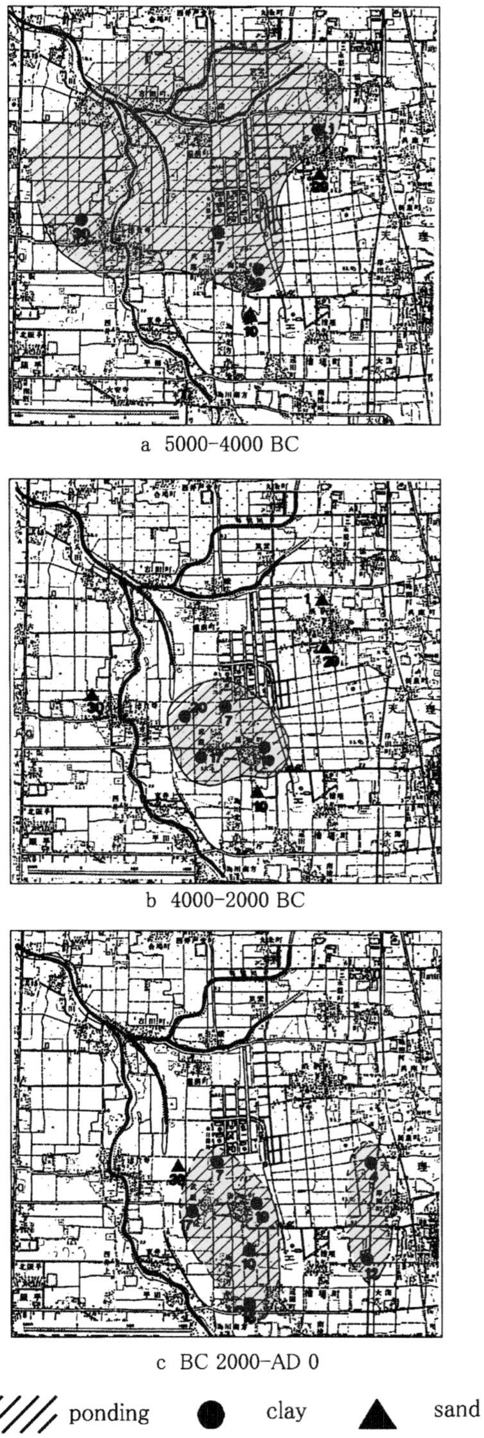

Figure 14.5 Extent of ponding through time at Asawa. Numbers indicate 1984 coring locations. See 'north' arrow in Figure 14.3 for directional orientation.

sea level of 2 to 5 m lower than today characterize the second (Barnes 1999).[5] Although it is not yet known how these different climatic conditions led to the formation of similar carbonaceous clay layers, it is clear that these periods witnessed low-energy deposition of fine-grained sediments in water in the Asawa region. Thus, the existence of standing water in the period of agricultural transition (Final Jomon) was documented in the Basin.

The gross depositional sequence indicates the nature of climatic change. From the Early Jomon, sediments underwent fining-up from fine sand or silt to clay, indicating ponding (Figure 14.5a), but some of the clay deposits terminated rather abruptly with the sudden deposition of coarse sand. Thus, in the higher-energy environment of the Middle Jomon, the extent of ponding was reduced in the east and west by levee, fan, and/or delta formation (Figure 14.5b). The scope of ponding then extended slightly to the south and southeast in the Final Jomon period (Figure 14.5c). Carbonaceous clay deposition was then abruptly terminated at some point after 0 BC/AD. This scenario may be overgeneralized due to the relatively small number of ^{14}C dates (Table 14.2) versus the multitudinous shifts in local depositional environments indicated by the complicated core stratigraphies.

In retrospect, the aerial photographic reconstruction identifying Asawa as a potential backmarsh area was correct. At different times in the post-glacial period, this part of the Basin held varying amounts of standing water, with the greatest extent of ponding coinciding with the latter half of the Climatic Optimum. It is known from a vast array of archaeological data that wet-rice agriculture was introduced to Japan from the Korean Peninsula in the early 1st millennium BC. To investigate whether such innovations were detectable in the Nara Basin record, biological samples from the cores were taken for further analyses.

Diatoms

Diatom analyses on three of the cores (84-5, 84-28, and 84-29) provided additional information on the

[5] Now rewritten as Barnes (2015).

stratigraphic formation of the Asawa area (JRG 1987). The majority of the 11 diatom samples contained a mixture of diatoms reflecting different environments including pond surfaces, pond bottoms, marshes, peat bogs, and mountain streams. Their combined presence suggests both temporal changes in upstream environmental conditions and different environmental sources of downwash. The lower sands of Core 84-29 yielded evidence of a warm water streambed (sample 29-14) located directly over a stratum, ^{14}C-dated to 4225–3410 cal. BC (N-4896), consistent with post-glacial climatic warming. In contrast, the upper clays of Cores 84-5 and 84-29 yielded cold water pond and bog varieties, showing the decline of temperatures after the Climatic Optimum. Both sand and clay layers yielded standing-water diatoms, some indicating stagnant conditions inimical to wet-rice cultivation. However, the predominance of marshy or lacustrine conditions with freshwater circulation indicates this area was suitable for wet-rice experimentation in the 1st millennium BC.

Pollen

The Basin's major floral phases (Table 14.3) had been established in a 1970s coring project (Matsuoka *et al.* 1984). A major hiatus in Basin stratigraphy, caused by erosional processes coinciding with the lowering of sea level during the Last Glacial Maximum, was dated to 20,000–9000 BP between Phases II and III. Sedimentation resumed after 9000 BP due to climatic warming, sea-level rise, and formation of coastal and Basin lowlands.

Table 14.3 Forest succession in the Nara Basin from the Late Pleistocene (I) through the Holocene to present (VI)

Phase VI	*Pinus* (diploxylon)-*Cyclobalanopsis* forest (ca. 2000 BP onwards) Anthropogenic secondary forest
Phase V	*Cyclobalanopsis* forest (ca. 5000–2000 BP) Warm temperate evergreen oak (*Laurelignosa*) forest
Phase IV	*Quercus-Cyclobalanopsis* forest (ca. 7500–5000 BP) Mid-temperate forest
Phase III	*Abies-Ulmus/Zelkova-Celtis/Aphananthe* forest (ca. 9000–8000 BP) Cold temperate broad-leaved deciduous forest without beech (*Fagus*)
[*no data for 20,000–9,000 BP*]	
Phase II	*Pinus* (haploxylon)-*Betula-Quercus* forest (ca. 30,000–20,000 BP) Sub-boreal needle-leaved forest, with some period of needle-leaved/ broad-leaved mixed forest
Phase I	*Quercus-Alnus-Betula* forest (ca. 35,000–30,000 BP) Cold temperate broad-leaved deciduous forest

The Asawa project's results contrasted in two ways from this earlier work. The recovery, identification, and dating of the Sakate ash provided a time reference showing that the erosional unconformity following Phase II began forming later in the Asawa area than in the central Basin. Moreover, the hiatus seems to have lasted longer in the Asawa area, because pollen from Phases III and IV was not recovered there. These differences conform to sequences of inland erosion and aggradation corresponding to sea-level change. During sea-level regression, erosion and incision work their way upstream, while during sea-level transgression aggradation works its way upstream.

Clear environmental shifts were documented at Asawa between Phases II, V, and VI. The successions in individual cores in Table 14.4 illustrate the absence of data from Phases III–IV in the Basin. The change from Phase II to V is apparent in Core 84-12 (JRG 1987). For Phase II, pollen samples from the clay layer immediately underlying the Sakate ash deposit (~17,000 BP) yielded *Abies*, *Pinus* (haploxylon), *Tsuga*, *Alnus*, *Betula*, *Quercus*, and *Celtis-Aphananthe*; the ratios of [arboreal pollen(AP):non-arboreal pollen (NAP):spore pollen(SP)] were 50–70% spore-dominant (Figure 14.6), indicating considerable bog formation. This botanic assemblage is compatible with the cold-climate Phase II forest (~30,000–20,000 BP), with *Alnus* as its major arboreal component. The Phase V evergreen oak forest is represented by pollen samples from the carbonaceous clay layer upwards. This typical primary forest composition of Holocene western Japan is dominated by *Cyclobalanopsis*, and, in general, has a much reduced spore count (mostly below ~12%). A ^{14}C date of 890–200 cal. BC (N-4889) was obtained from sediments low in the evergreen oak core sequence.

Table 14.4 Stratigraphic relationships of forest types in individual cores (in boxes); forest types are described in Table 14.3; ^{14}C dates are calibrated

Core #	Forest type:	Stratigraphic layering:
84-12	Phase V	forest sandwiching ^{14}C date of 890–200 BP, overlying:
	Phase II	forest pre-dating Sakate Ash (17,000 BP)
	Phases III-IV absent	
84-7	Phase VI	diploxylon pine and *Poaceae* (undated), overlying:
	Phase V	forest sandwiching ^{14}C dates of 1130–400 BP; 1935–1320 BP
84-8	Phase VI	diploxylon pine and *Poaceae* (undated), overlying:
	Phase V	forest (undated)
84-19	Phase VI	diploxylon pine and *Poaceae* (undated), overlying:
	Phase V	forest sandwiching ^{14}C dates of 1730–1225 BP; 4410–3550 BP
84-29	Phase VI	diploxylon pine and *Poaceae* (undated), overlying:
	Phase V	forest sandwiching ^{14}C dates of 4225–3410 BP

Several cores (84-6, 84-10, 84-19, 84-29) show a shift from Phase V to VI, with the disappearance of the primary forest cover and the beginning of anthropogenic disturbance resulting in the growth of secondary red pine forests. Decreased *Cyclobalanopsis* and spore counts are matched by dramatically increased diploxylon pine and *Poaceae*.[6] The Phase VI decrease in arboreal pollen, coupled with the higher average pollen yield of red pine, makes it clear that the pine forest cover was far less extensive than the earlier evergreen oak forests – a conclusion consistent with the opening of uplands for agriculture and settlement. Phase V dates of evergreen oak forest generally range from 2000 BC to 1000+ AD, covering the time period of the initiation of rice agriculture in the Nara Basin, as the Early Yayoi (Ongagawa phase) culture expanded from north Kyushu (cf. Barnes 1993*a*), and the expansion of early state society and its need of timber for architecture (cf. Totman 1989) and for fuel for the ceramic and iron industries.[7]

All three of the forest types below include *Poaceae* as a substantial component of their non-arboreal pollen. *Poaceae* includes both reeds (*Phragmites*) and rice (*Oryza*), but reed and rice pollen grains are

[6] The *Poaceae* references replace '*Gramineae*' in the original publication.

[7] The interpretation of the dates here differs from that in the original article.

unfortunately indistinguishable.[8] In investigating the presence of reed marshes and their use in the initiation of rice agriculture, phytolith identification of both genera, as developed by Fujiwara (1993), is the necessary tool.

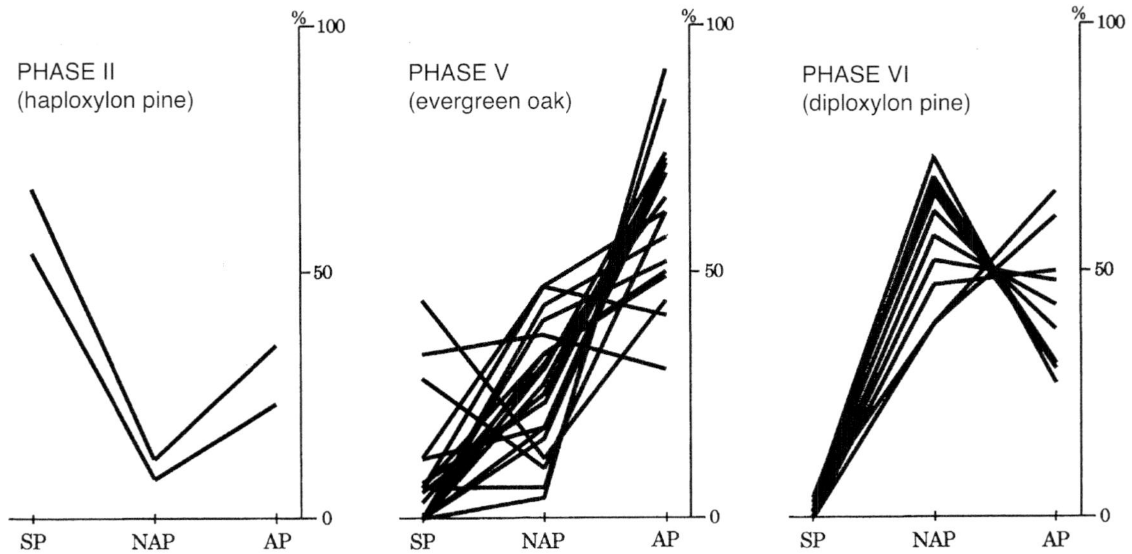

Figure 14.6 Changing proportions of SP, NAP, AP (pollen) in the 1984 Asawa cores for the three major forest types typified by Phases II, V, VI. Each line represents one sample composed of three percentage values for each of the three types of pollen (SP, NAP, AP); each line totals 100%. The Pleistocene *Alnus* forest Phase II ratio is dominated by spore-bearing plants at the expense of non-arboreal plants. The typical profile of the primary evergreen oak Phase V ratio comprised less that 12% spores and characteristically more arboreal pollen than non-arboreal pollen. In contrast, the red (diploxylon) pine secondary forest Phase VI typically has less than 5% spores and usually more non-arboreal than arboreal pollen.

Phytoliths

Phytolith analysis of selected sediments in three cores at Asawa (Figure 14.7) revealed the presence of the cultivated species of rice (*Oryza sativa*), *kibi* (*Panicum miliaceum*), and *awa* (*Setaria italica*), and of non-domesticated reeds (*Phragmites*), bamboo (*Bambusoidae*), and *ushikusa* (*Andropogon* sp.) (JRG 1987). Rice and reeds are water-based plants, while bamboo and *ushikusa* are dry land plants. None of the sampled cores (84-l6, 84-14, 84-27) were radiocarbon dated; therefore, direct temporal control is absent. However, some of the sediments can be related to other dated cores.

Selections of black clay layers in the three cores were tested for phytoliths, revealing bamboo and reeds present in varying proportions. The carbonaceous clay layer of Core 84-24 contained large numbers of reed phytoliths, to the virtual exclusion of bamboo. This core is located closest to the centre of the Basin, and the dearth of bamboo phytoliths places it some distance from dry land or from inflowing fresh water carrying upland plant remains into the marsh. The black clays in the other cores (samples 84-16-

[8] Although Nakamura (1974) had published differences between *Oryza* and *Phragmites*, other pollen grain-size based methods for *Poaceae* (formerly *Gramineae*) have been developed since then (e.g., Schüler 2011); see Inada (n.d.) for rice pollen photos following Nakamura's descriptions.

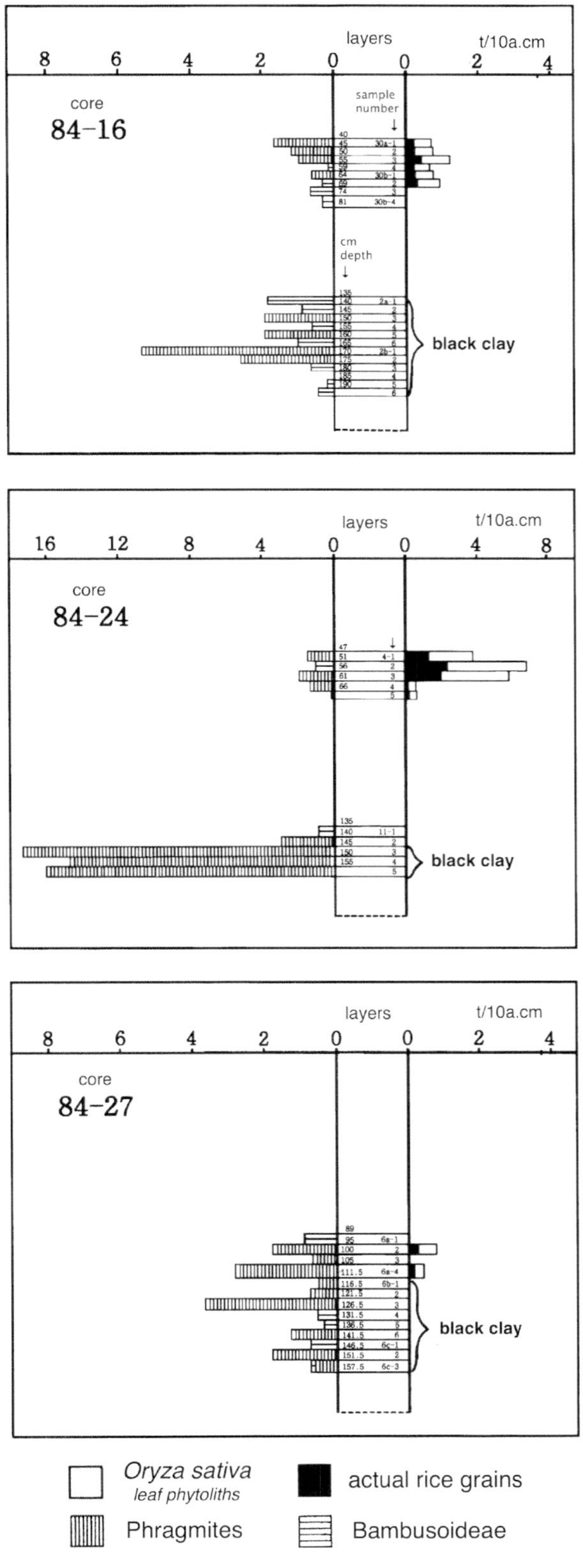

Figure 14.7 Relative frequencies of *Poaceae* phytoliths in three cores at Asawa

B2 and 84-27-6) show more of a mixture of reeds and bamboo, possibly consistent with their positioning closer to the Basin flank. Black clays in all three cores reveal increasing bamboo content towards their upper reaches, as more sand was deposited and the marsh disappeared.

Core 84-27 had two clearly separated thick black clay layers, but only the upper layer (sample 84-27-6) was selected for phytolith analysis, due to the probability that it corresponds to the Final Jomon period (~1000–500 BC), as described above. Directly over this lies clay containing fine micaceous sand. Phytolith testing shows a sudden introduction of cultivated rice at the point where the micaceous sand becomes prominent in the stratigraphy. Nevertheless, reeds continue to dominate the phytolith count. This should not be surprising: reeds and other weeds are known to have grown together with rice well into the Medieval period, occupying up to half the paddy-field surface area (Terasawa 2000).[9]

Other layers with rice phytoliths (samples 84-1-30, 84-24-4) consist mainly of sand. These sands were high in the core stratigraphies and well isolated from the black clays, and the relative proportions of phytoliths from cultivated and non-cultivated plants in them are quite different from Core 84-27. In Core 84-16 these proportions are nearly even, while in Core 84-24 rice is clearly dominant. It is likely that these upper samples represent Medieval or later rice production.

The phytolith evidence suggests that rice production in this part of the Basin directly followed the amelioration of the climate responsible for formation of the upper black clay. Rice phytoliths were recovered from the fine micaceous sands overlying the upper black clay; this change in sedimentary succession indicates a marginally higher-

[9] First noted by Fujiwara (1979).

energy environment bringing in coarser sediments from a different source. This suggests that water circulation improved, making the marshy areas more conducive to agriculture once the imported technology of wet-rice production became available in this area.

Upland Geomorphology: the Miwa Project

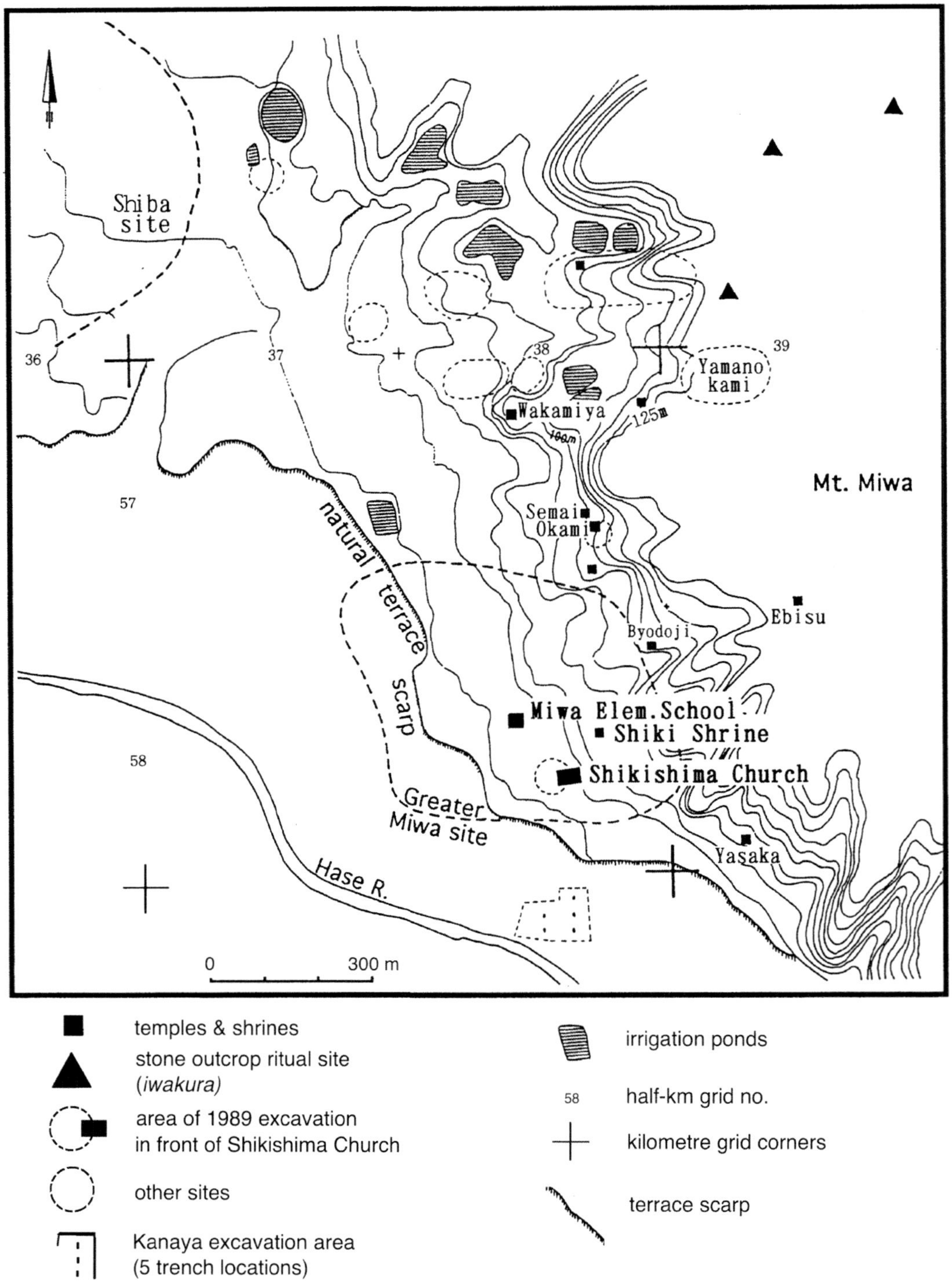

Figure 14.8 The Miwa site area and location of the Shikishima Tenri-kyo Church

The Miwa site location

The second project focused on the terrace structures at the base of Mt. Miwa in the southeastern Nara Basin – particularly the terraces on the grounds of the Shikishima Tenri-kyo Church in Sakurai City (Figure 14.8). The Miwa area was chosen for investigation because it is named in the 8c *Nihon Shoki* chronicles as the location of a 4c palace belonging to the sovereign Sujin, allegedly the first historical 'emperor' of Japan.

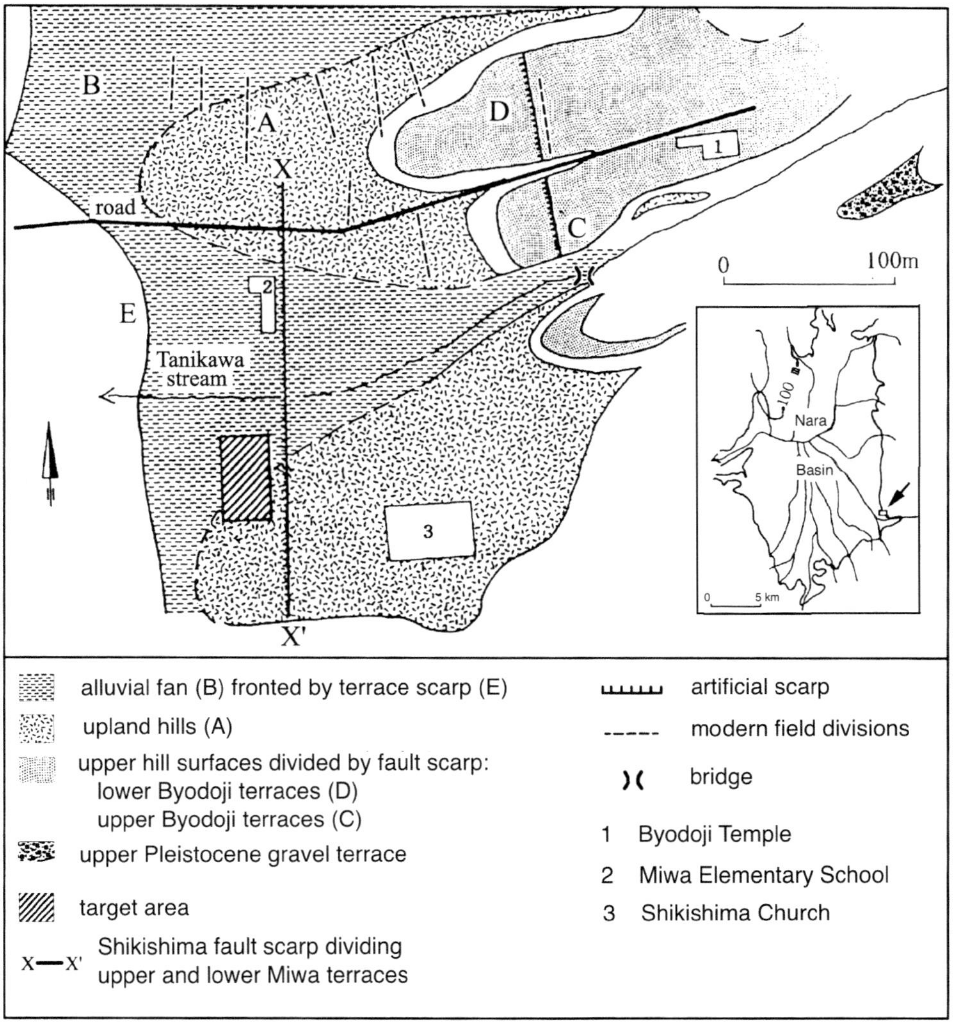

Figure 14.9 Reconstruction of landforms in the Miwa area
'Target area' indicates the site grid area for the Miwa Project on the lower terrace

The fans and terraces along the eastern Basin edge are known archaeologically to have supported two main phases of pre- and proto-historic occupation, between Early and Late Jomon (~8000–300 BC) and during the Kofun (250–710 AD) period. Much of this area also supports a considerable modern population in villages, towns, and even cities. In fact, settlement has completely obscured the boundary between the uplands and the Basin flank in the Miwa area, and aerial photography could only inform on the grosser aspects. Thus, testing of the aerial photographic reconstruction of the Basin's southeastern topography in this area focused on a target area (Figure 14.9) near the Shikishima Church to ascertain the nature of the Miwa terraces (natural or anthropogenic), and the extent of landform modification through urbanization and the survival of occupational strata from earlier periods.

Landform reconstruction

Field inspection suggested that the Shikishima Church occupies the promontory of a hill slope (area A in Figure 14.9) that has been almost entirely obscured by historic cut and fill activities (Soma 1993). Excavation at the site revealed a Medieval layer lying unconformably on a Yayoi layer, with no early historic layers in-between. Thus, it was assumed that the original land surface was first terraced in the Medieval period, resulting in the razing of the early historic layers. In our original article, Soma suggested that these Medieval activities produced a very long, continuous terrace scarp (x—x' in Figure 14.9) across the hill slope and continuing north across the stream valley to the opposing hill. It was surmised that a Medieval temple known as Byodoji, high up on the northern hill slope, might have conducted the initial terracing modifications to provide flat land for building space or fields. This was not a conclusion but a hypothesis suggested by the Miwa investigation.

Since then, due to raised consciousness about "Earthquake Archaeology" (Sangawa 1992) and detailed investigation of Active Fault systems (RGAFJ 1992), it is now hypothesized that the terrace scarps might be the product of movement along an Active Fault zone of the East Nara (Tōen) Fault System (HERP 2005). The "Active Fault Map in Urban Area: Sakurai" (Kokudo Chiri-in 1999) shows fault scarps following the same trajectories as the terrace scarps shown in Figure 14.9: x—x' *and* C, D. Thus, an anthropogenic origin for the terrace scarp x—x' itself as part of Medieval hillside levelling is now questioned, even though it is clear that the land above and below the scarp was razed during the Medieval period to provide level surfaces.

Prior to the 1989 excavation of the trenches at Miwa, cores were taken in 1988 to test the stratigraphy (Barnes 1993*b*). At the top of these cores were several layers of sand and gravel fill of modern deposition. Underneath were layers of black clayey sand which contained artefacts but were much too thick (up to 40 cm) for normal living-floor deposits or even normal paddy-field soils (usually 15~20-cm thick). These clayey sand layers extended from between 40–60 cm to 1.2–1.4 m below the present surface; beneath them, coarse sands and gravels of 600–2000-mm grain size were observed. The nature of the black strata was not understood at the time of coring, but excavation revealed consistent association with artefacts; they were thus considered 'artefact-bearing strata' (ABS) (*hōgansō*).[10] (This issue is revisited in Part III: Reflections.)

Both the corings and a subsequent electromagnetic resistivity survey evidenced contours interpreted as a buried stream valley cutting across the northern half of the target area, as suggested by the landforms in Figure 14.9. The area to the north and west in Figure 14.10 (to the left of core locations 88-5, 88-9, 88-18, and 88-20) had very low resistance compared to the ground in the east and south (to the right of those core locations). In two of the north–south boring transects, the depths of the coarse sands and gravel (as measured from an averaged uniform altitude of the middle terrace surface) dip to very low points in cores 88-2, 88-1, and 88-8, 88-19 (Figure 14.11). These correspond to locations where the northern portion of the middle terrace was of much lower resistivity than the southern part (Nishimura & Shell 1993). The lowest points of electric resistance could indicate damp sediments and the low position of coarse sands and gravels could indicate streambed deposits cutting into the regolith substrate; thus, the configuration is interpreted as a buried valley.

[10] The original article offered this explanation: "excavation at Miwa suggested derivation from varied human activities involving the addition of considerable organic material: rubbish deposition in the Yayoi period and Medieval dry-field manuring". Other aspects of this paragraph have also been revised to be more specific.

The buried valley was not excavated; therefore, its exact stratigraphy is unknown. However, if the Medieval stratum as recovered in excavation can be equated with the thick deposits recovered in the cores across this northern buried valley, then the valley would have had to be infilled during or by the

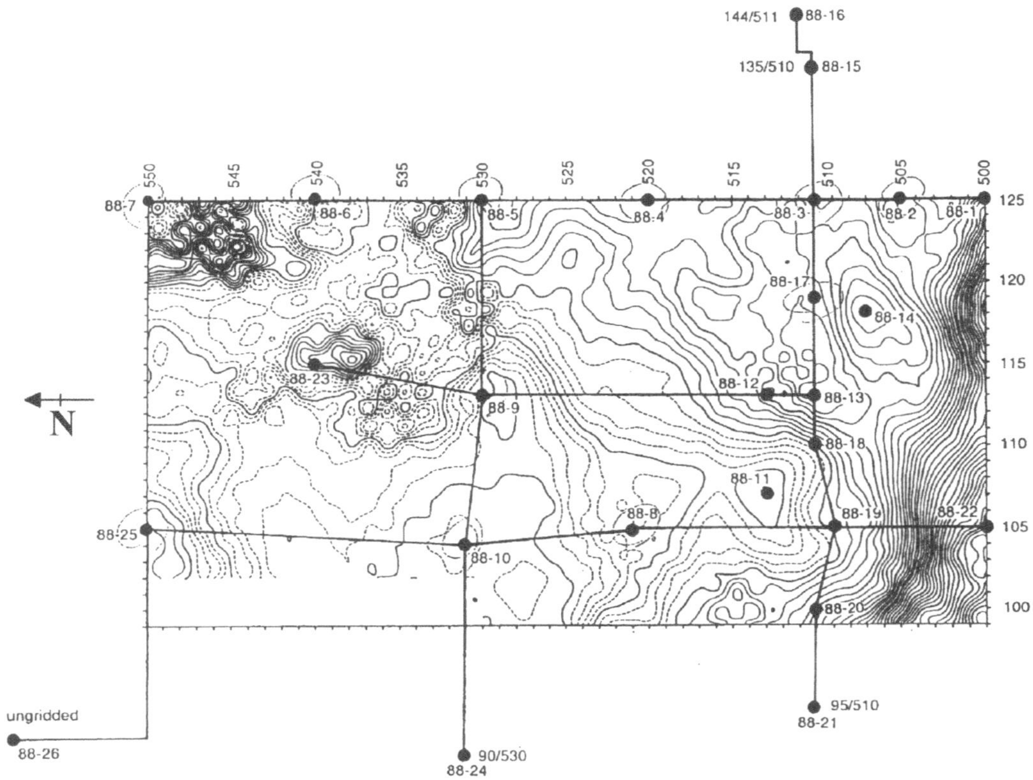

Figure 14.10 The target area showing the 1988 grid corings displayed on the 1989 resistivity survey

Medieval period to allow for overlying stratigraphic build-up. However, the front edge of the terrace was probably infilled much later as suggested by radiocarbon dates on stakes recovered from the bottom of Trench 3 (see below).

Terracing & re-terracing

The Miwa excavation was undertaken partly to investigate the transformation of the terraced area around the Church as it affected the survival of the prehistoric remains known from earlier accidental and construction finds. Excavations in Trenches 1–2 and 4–6 revealed the almost total absence of early historic material on site, suggesting the razing of the early historic layers in the Medieval period. The nature of the fill in Trench 3 revealed the premodern, western edge of that first terrace. The current surface of the Church terrace projects some 35 m beyond Trench 3 to the west, indicating that dump and fill activities have continued off the edge of the original terrace surface throughout the historic periods.

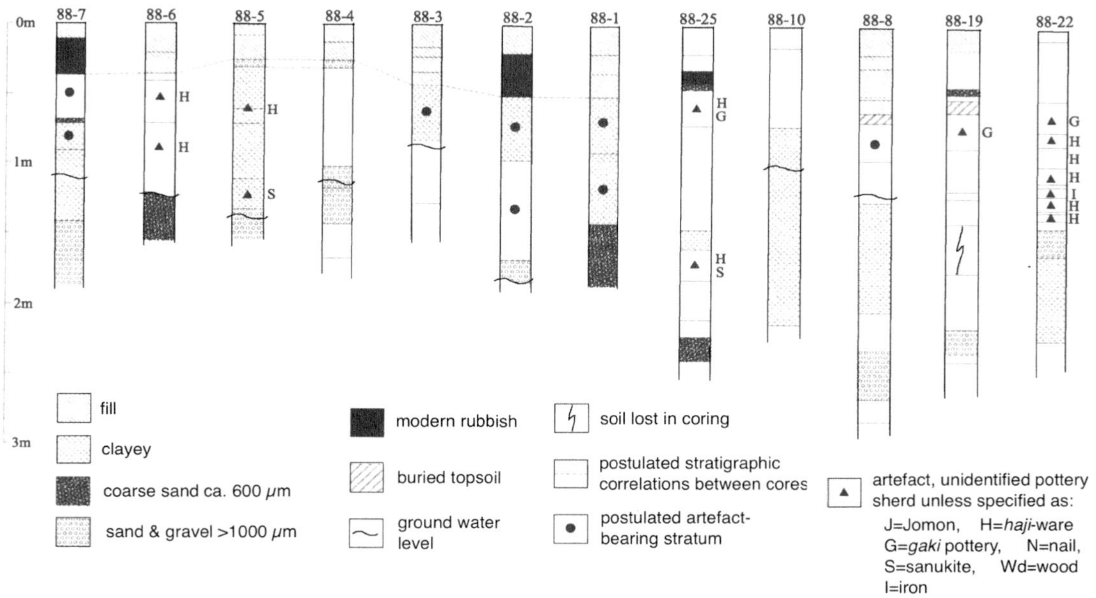

Figure 14.11 Core transects from the north (left) to south (right) demonstrating the positioning of basal sand overlying the buried stream valley on the Miwa terrace

The fill layers in the western part of Trench 3 (a 2 x 2 m excavation square) rested on a stratum dating to Early and Middle Yayoi (300 BC–AD 100).[11] This layer was cut north to south by a narrow, shallow ditch (F-31), on top of which lay a Chinese porcelain sherd of Ming dynasty date (1368–1644 AD) (Omi 1993). Underneath this stratum lay the basal trench layer into which three small Yayoi-period pits had been cut; these yielded Early and Middle Yayoi pottery (Kuwabara l993) plus millet and rice grains.

Eleven stakes (10 wood and 1 bamboo) were found in the eastern half of Trench 3, beginning from a depth of ~1.15 m and exposed for ~36 cm until excavation of the trench was abandoned at 1.58 m. Most of the stakes were within 20–30 cm of the east wall of the trench, and four abutted the wall so as to appear in the section drawing. Four of the stakes were removed for radiocarbon dating.

It was hypothesized that the wooden stakes recovered from Trench 3 would date to the Yayoi period in association with the pits in the bottom of the trench. However, the ^{14}C dates indicate their age as no earlier than 1660 AD (Table 14.5), near to the date of the Ming porcelain sherd found near the stakes. Thus, it is tempting to view the stakes as part of early premodern site usage – perhaps as a shoring of the Medieval terrace front. Nevertheless, it is clear that the land area in the western half of Trench 3 at the time of stake driving was not part of the Medieval terrace or the earlier hill promontory: it was situated below the hill/terrace on its western front. Thus, the extension of the Medieval terrace westwards through intermittent landfill is a recent, post-1660 phenomenon, possibly burying substantial Yayoi remains. The Church grounds seen today, therefore, are a much-expanded area concealing original landform diversity within its perimeter.

[11] The dating of this phase was later redefined as ca. 500 BC–AD 50 (cf. Barnes 2015: table 1.7)

Table 14.5 Radiocarbon dates on wooden stakes from Trench 3 at Miwa

Lab #	Sample	^{14}C date	calibrated at 2σ
GrN-17672	Miwa 1	150±30 bp	1670–1955 AD
GrN-17671	Miwa 2	180±30 bp	1660–1955 AD
	Pooled near:	165±23 bp	AD 1670–1955, T" = 0.44, X2 = 3.85
	Possible nears:		AD 1680, 1750, 1810, 1935, 1955
	1σ = 1678–1943		
	2σ = 1670–1955		

Local environment & site use

The immediate site environment was found to be essentially the same for both the Yayoi and Medieval layers. Pollen analysis revealed an abundance of various grasses (*Poaceae*); two shrubby bamboos (*Bambosoideae*) dominated the Yayoi and Medieval layers, and sedges (*Cyperaceae*) were also abundant in the Medieval period (Kanehara 1993). These open-land plants indicate that the immediate vicinity of the site had been cleared for human activity in both periods. Furthermore, based on the presence of the Goosefoot and Amaranth families (*Chenopodiaceae* and *Amaranthaceae*, respectively), the surface environment was naturally dry.

As usual for upland sites, the pollen from the Yayoi layers was so scarce and degraded that it was impossible to reconstruct the full environment of the time, but the prominence of open-area shrubs on the hill slope supports the general Basin flora as reconstructed above for Phase VI. It was noted that the extensive forest cover characterizing the evergreen oak Phase V retracted greatly in the anthropogenic red pine Phase VI. Therefore, the abundance of shrubby bamboo on the hill slope in the Yayoi period must have been related to intentional forest clearance.

The information from phytolith analysis contrasts with the evidence from pollen grains in suggesting some wet surface areas (Sugiyama 1993). All three phytolith samples (two from the Medieval period and one from Yayoi) yielded very high densities of rice phytoliths. The values (from ~11,400 to 21,400 grains/gram) greatly exceeded the threshold value of 5,000 grains/gram used for identifying former rice paddy-fields (Sugiyama 1993). Sugiyama is, therefore, of the opinion that rice was grown in Yayoi times (probably on the hillslope) and in Medieval times (probably in the terraced fields).

Other phytoliths recovered included bamboo (mainly *Pleioblastus* Sect. *nezawa*) and Andropogoneae (mainly *Miscanthus*), both dryland plants, and *Phragmites*, a wetland plant. Sugiyama resolves these contradictory data by postulating double-cropping, with rice plus a few reeds comprising the summer crop, and dry-field gardens replacing these in winter. Although isolated grains of both foxtail millet and bread wheat were recovered, no phytoliths of these or of other *Poaceae* crop plants (such as *Panicum*, *Hordeum*, *Coix*, and *Zea*) were in evidence (D'Andrea 1993). Such winter gardens might have been devoted to vegetable and root crops rather than alternative grains. Even so, it seems that such a scheme of proposed double-cropping can only be applied to the Medieval terraces, whose very construction would have necessitated the rerouting of stream water; this water could have been fed into terraced fields to support rice.

Such a scheme does not seem feasible for the Early-Middle Yayoi periods, the latter providing most of the ceramics at the site.[12] It is not known when the X–X' fault scarp was formed, but even if it existed in those Yayoi phases, the hillside would still have been sloping and unsuitable for wet-rice production. It was not until the Late Yayoi that higher ground in Japan began to be exploited through irrigation

[12] The original stated that "when the hill slope was presumably unterraced". This paragraph has been amended.

agriculture, with the increased availability of iron tools to dig ditches and construct level paddy-fields. Consistent with the above expectation, no special field structures or irrigation facilities were found at Miwa dating to the Yayoi period – or even the Medieval period, for that matter.

Though rice was indeed present on the hillside, a greater proportion of the phytoliths recovered from the Yayoi than the Medieval layers derived from grain husks, hairs, and beards. This could indicate the disposal of harvest or threshing waste, rather than the actual growing of rice. Virtually no research has been done in Japanese archaeology comparable to the pioneering work in Greece on identifying the loci of different stages of crop production and processing (Charles *et al.* 1997). It is quite possible that high phytolith densities in the Yayoi strata could be characteristic of deposits of threshing or even thatching waste as alternatives to paddy-field soils. Refined phytolithic research, therefore, might be extremely useful in reconstructing grain-processing stages that have previously only been investigated using seed and plant macroremains.

In sum, these considerations give rise to an interpretation of the Yayoi hillside being used as a trash dump – an interpretation consistent with the absence of any structural features in the stratigraphy and supported by an analysis of the sanukite flakes in the stratigraphy. Reynolds & Matsuzawa (1993) characterize the lithic debris as fresh chippings from stone tool-making that suggests their collection and deposition in an out-of-the-way area without further trammeling underfoot. If the debris had been left in an occupied area, one would have expected to see much damage on the chips from subsequent crunching underfoot.

A Final Valuation

The results of the Asawa and Miwa projects can be seen as having both confirmed and refined the original aerial photographic landscape reconstruction of the Nara Basin, adding environmental data to geomorphological interpretations.

Coring at Asawa confirmed previous research results that Late Pleistocene (20,000–9000 BP) deposits are generally missing from the Basin floor – as also discussed in Chapter 13 – but discovery of the Sakate ash layer (~17,000 BP) gives a more precise date to the beginnings of erosion along the Basin flanks during the post-glacial period. The Sakate ash was documented to have been unconformably succeeded by Holocene deposits belonging to the evergreen oak horizon. The existence of standing water in the Basin during this forest horizon was documented for two periods, 5250–3500 BC and 1250–200 BC, though in smaller areas of backmarsh rather than in a central Basin marsh. The stratigraphic position of early rice growing in these backmarsh areas was pinpointed above the second period (Final Jomon) of swampy carbonaceous clay deposits. This dating and positioning of rice in the micaceous sands correlates well with both the known spread of Yayoi culture through western Japan, and the known ecological preferences of wet-rice for growing in non-stagnant waters. Finally, the destruction of the evergreen oak forests and replacement by secondary red pine forests were documented for the late first millennium BC, as fields and settlement areas were cleared by the Yayoi agriculturalists.

At Miwa, coring and excavation led to the discovery of Medieval razing of the hillside landscape, resulting in the absence of early historic layers. Medieval layers were deposited directly onto Yayoi strata, and virtually no remains from the Kofun period were recovered in this area. The terrace scarp dividing the upper and lower Miwa terraces was re-interpreted as a fault scarp of unknown date rather than a feature accompanying the Medieval terracing operations. The existence of a buried valley cutting across the lower terrace was postulated from electromagnetic resistivity data, but as it remained unexcavated, its date is also unknown. The terrace front itself was documented to have been built out some 35 m to the west, indicating incremental historic landfill operations.

Botanical analyses demonstrated that the upland area was dry, open land during both the Yayoi and Medieval periods. Discovery of concentrated rice phytoliths in the Yayoi strata, however, has caused some interpretive controversy: whether the hillside was used for rice growing – wet or dry rice - or whether the phytoliths accumulated from crop processing activities. No field or irrigation facilities were recovered and the terrain was ill-suited to paddy-field construction in the Yayoi period; hence, we elect the latter interpretation. The results, however, point to a greater need for analyzing the spatial aspects of prehistoric crop processing in Japanese archaeology and greater differentiation between the occurrences of wet rice and dry rice.

Figure & Table Sources

Figure 14.1 by GLB
Figure 14.2 by GLB based on (Nishida & Matsuoka 1977; Nishida *et al.* 1978, 1979); inset enhanced and river names removed
Figure 14.3 based on (JRG 1986: fig. 2)
Figure 14.4 based on (JRG 1987: fig. 7)
Figure 14.5 based on (JRG 1987: fig. 8)
Figure 14.6 based on data in (JRG 1987)
Figure 14.7 based on (JRG 1987: fig. 6)
Figure 14.8 based on (Barnes & Okita 1993: fig. 1.3).
Figure 14.9 based on (Soma 1993: fig. 3.1)
Figure 14.10 based on (Nishimura & Shell 1993: fig. 4.1)
Figure 14.11 based on (Barnes 1993*b*: fig. 5.3)

Table 14.1 based on (JRG 1986; Barnes & Okita 1993)
Table 14.2 based on (JRG 1986: table 3); datings by Nihon Isotope Kyokai
Table 14.3 based on (JRG 1987: 7)
Table 14.4 based on (JRG 1987, 1987)
Table 14.5 analyzed at the Centrum voor Isotopen Onderzoek, University of Groningen, The Netherlands

References

Barnes, GL (**1982**) "Prehistoric landscape reconstruction and spatial analysis of artifact discovery locations." *Kokogaku to Shizen Kagaku* 15: 113–131 (In Japanese with English title and abstract).

---- (1983) *Yayoi-Kofun settlement archaeology in the Nara Basin Japan.* Ph.D. dissertation, Department of Anthropology, University of Michigan. Ann Arbor, MI: ProQuest.

---- (**1986**) "Paddy field archaeology." *Journal of Field Archaeology* 13: 371–379.

---- (1988) *Protohistoric Yamato: archaeology of the first Japanese state.* Ann Arbor, MI: Museum of Anthropology and the Center for Japanese Studies, University of Michigan.

---- (**1990*a***) *The Miwa site: preliminary excavation report.* Nara, Japan: The Miwa Project, Tenri Sankokan Museum.

---- (**1990*b***) "Paddy field soils." *World Archaeology* 22: 1–17.

---- (**1993*a***) "Miwa occupation in wider perspective", pp. 181-192 in *The Miwa project report: survey, coring and excavation at the Miwa site, Nara, Japan*, ed. by G Barnes & M Okita. Oxford: Tempvs Reparatvm (in English and Japanese)

---- (**1993*b***) "Geological coring at Miwa", pp. 39–48 in Barnes & Okita (eds) (1993).

---- (**1999**) *The rise of civilization in East Asia: the archaeology of China, Korea and Japan.* London: Thames and Hudson (paperback edition).

---- (**2015**) *Archaeology of East Asia: the rise of civilization in China, Korea and Japan.* Oxford: Oxbow Books.

Barnes, G; R Brandt & S Kaner *et al.* (**1986**) "Phosphate behavior in Japanese soils: a test case." *Kokogaku to Shizen Kagaku* 19: 59–68 (in Japanese with English title and summary).

BARNES, GL & M OKITA (eds) (**1993**) *The Miwa project report: survey, coring and excavation at the Miwa site, Nara, Japan.* BAR International Series 582. Oxford: Tempvs Reparatvm (in English and Japanese).

CHARLES, M; G JONES & JG HODGSON (**1997**) "FIBS in archaeobotany: functional interpretation of weed floras in relation to husbandry practices." *Journal of Archaeological Science* 24: 1151–1161.

D'ANDREA, C (**1993**) "Charred plant remains at Miwa", pp. 75–84 in Barnes & Okita (eds) (1993).

FUJI, N (**1966**) "Climatic changes of postglacial age in Japan." *Daiyonki Kenkyu* 5: 149–156 (in Japanese with English summary).

FUJIWARA, Hiroshi (**1979**) "Fundamental studies of plant opal analyses 3: estimation of the yield of rice in ancient paddy fields through quantitative analysis of plant opal." *Kōkogaku to Shizen Kagaku* 12: 29-42 (in Japanese with English title and summary).

---- (**1993**) "Research into the history of rice cultivation using plant opal analysis", pp. 147–158 in *Current research in phytolith analysis: applications in archaeology and palaeoecology*, ed. by DM PEARSALL & D PIPERNO. Philadelphia, PA: Museum Applied Centre for Archaeology (MASCA), University of Pennsylvania.

GSJ (Geological Survey of Japan) (**2020**) "Seamless digital geological map of Japan V2 (1:200,000)" [https://gbank.gsj.jp/seamless/v2/viewer/]

HAYAKAWA, Y (**n.d.**) *Hayakawa's one-million year tephra catalog.* Retrieved March 4, 2005, from [www.edu.gunma-u.ac.jp/~hayakawa/catalog/mastertephra/akahoya.html] (In Japanese).

HERP (Headquarters for Earthquake Research Promotion) (**2005**) "Evaluation of the Kyoto Basin – Nara Basin fault zone southern end (Nara Basin Toen Fault system)." *Jishin Chosa Iinkai* [www.jishin.go.jp/main/chousa/01jul_keina/].

INADA, Akira (**n.d.**) "Jomon rice cultivation and vegetation destruction 3: Gramineae pollen identification and detection of rice cultivation." *Sayama Natural History* [www7b.biglobe.ne.jp/~sayama_naturalhistory/vegitationalhistory/jomon-noukou/inekafun/inekafun.html] [sic] (dead link).

JRG (Joint Research Group on the Geomorphological Recognition and Land Utilization of Pre- and Protohistoric Japanese Peoples) (**1986**) "Natural environments in the Nara Basin through the pre- and protohistoric ages I: geology and geomorphology." *Kobunkazai Kyoiku Kenkyu Hokoku* 15: 1–30. (In Japanese with English title).

---- (**1987**) "Natural environments in the Nara Basin through the pre- and protohistoric ages II: descriptions of core samples and analysis on biogenic materials." *Kobunkazai Kyoiku Kenkyu Hokoku* 16: 23–74 (in Japanese with English title).

KANEHARA, M (**1993**) "Miwa site pollen analysis", pp. 75–84 in Barnes & Okita (eds) (1993).

Kokudo Chiriin (1999) *Active fault map in urban area: Sakurai (1:25,000).* Tokyo: Geospatial Information Authority of Japan.

KUWABARA, H (**1993**) "Diagnostic Yayoi shards excavated at Miwa", pp. 133–142 in Barnes & Okita (eds) (1993).

MACHIDA, H; F ARAI & S SUGIHARA *et al.* (**1984**) "Tephra and Japanese archaeology", pp. 865–928 in *Conservation science relating to antiquities and anthropogenic natural science: a synthetic report*, ed. by Nihon Bunkazai & Henshu Iinkai. Kyoto: Dohosha (in Japanese).

MATSUOKA, K; S NISHIDA & M KANEHARA (**1984**) "Upper Pleistocene strata of the Nara basin and the palaeoenvironment of Nara surrounding the Furu site." *Maizo Bunkazai Tenri-kyo Chosadan Kokogaku Chosa Kenkyu Chukan Hokoku* 10: 5–24 (in Japanese).

NAKAMURA, Jun (**1974**) "Some palynological notes on grass pollen, with special reference to *Oryza sativa*." *Daiyonki Kenkyu* 13.4: 187-199 (in Japanese with English title and abstract) [DOI: 10.4116_jaqua.13.187].

NISHIDA, S & K MATSUOKA (**1977**) "Holocene natural history in the Nara Basin, I." *Kobunkazai Kyoiku Kenkyu Hokoku* 6: 65–81 (in Japanese with English title and abstract).

NISHIDA, S; K MATSUOKA & Y NOGUCHI *et al.* (**1978**) "Holocene natural history in the Nara Basin, II." *Kobunkazai Kyoiku Kenkyu Hokoku* 7: 69–89 (in Japanese with English title and abstract).

---- (**1979**) "Holocene natural history in the Nara Basin, III." *Kobunkazai Kyoiku Kenkyu Hokoku* 8: 31–44 (in Japanese with English title and abstract).

NISHIMURA, Y & C SHELL (**1993**) "Geophysical surveys at Miwa", pp. 29–3 in Barnes & Okita (eds) (1993).

OI, N (**1992**) "Pollen spectra around 20,000 years ago during the last glacial from the Nara Basin, Japan." *Daiyonki Kenkyu* 31: 203–212 (in Japanese).

OMI, T (**1993**) "Medieval pottery in Miwa", pp. 153–180 in Barnes & Okita (eds) (1993).

RGAFJ (Research Group for Active Faults in Japan) [Katsudanso Kenkyukai] (**1992**) *Maps of active faults in Japan with an explanatory text*. Tokyo, Japan: University of Tokyo Press (in Japanese and English).
Reynolds, TEG & T Matsuzawa (**1993**) "Flaked lithics at Miwa", pp. 123–132 in Barnes & Okita (eds) (1993).
Sangawa, A (**1992**) *Earthquake archaeology: sites tell of earthquake history*. Tokyo: Chuko Shinsho 1096 (in Japanese).
Schüler, Lisa & Hermann Behling (**2011**) "Poaceae pollen grain size as a tool to distinguish past grasslands in South America: a new methodological approach." *Vegetation History and Archaeobotany* 20: 83-96 [DOI: 10.1007/s00334-010-0265-z].
Soma, H (**1993**) "Landscape reconstruction around the Shikishima Church", pp. 27–28 in Barnes & Okita (eds) (1993).
Sugiyama, S (**1993**) Miwa site phytolith analysis", pp. 65–74 in Barnes & Okita (eds) (1993).
Takemura, K; A Hayashida & M Okamura *et al.* (**2000**) "Stratigraphy of multiple piston-core sediments for the last 30,000 years from Lake Biwa, Japan." *Journal of Paleolimnology* 23: 185–199.
Terasawa, Kaoru (**2000**) *The birth of kingly authority*. Tokyo: Kodansha (in Japanese).
Totman, Conrad (**1989**) *The green archipelago: forestry in preindustrial Japan*. Berkeley, CA: University of California Press.

CHAPTER 15

Acid Soils and Acid Rocks: Misunderstood Implications for Bone Preservation in Japan

This Chapter addresses the problem of the general misunderstanding that few skeletal remains (fossil or more recent) are known from Japan *because of the acidity* of its *volcanic soils*. Fieldwork was undertaken to test the acidity of soils in three ecozones and the results are compared with general descriptions of Japanese soils. It was found that volcanic soils are actually less acidic than soils of other types. The problem in interpretation seems to have arisen from an unwarranted conflation of soil acidity with what used to be called 'acid' volcanic rocks. The natures of both of these are discussed and clarified below.

The Problem

Publication after publication on Japanese archaeology states that *the reason* there are so few human remains from any period in Japan, particularly the Palaeolithic, *is due to* the destructive nature of the country's 'acid volcanic soils'. Such publications range...

from the public media:

> "The acidic soil conditions common in Japan due to volcanic eruptions have made the paddy field the most productive farming method" [https://en.wikipedia.org/wiki/Paddy_field] (accessed 20 May 2022).

to the specialist literature:

> "Although the extensive distribution of volcanic ash is favourable for research focused on stratigraphy and relative chronology, it does not favour research into the palaeoenvironment as human and animal bones are not preserved in the acidic soil" (Imamura 1996: 28).
>
> "soils indicating strong acidity because of the volcanic islands..." (Gakuhari *et al.* 2020: pp. 2 of 10).

Even when admitting the role of precipitation, volcanic ash was thought intrinsically at fault:

> "In general, the soils of Japan are acidic, reflecting their origin from volcanic ash and the relatively high amount of rainfall (1500-2000 mm/year at least). Consequently, almost the whole territory is unfavorable to the preservation of bony substance" (Nakahashi & Nagai 1986-1987: 15).

I would like to take issue with this reasoning on three grounds:

- pH acidity is entirely unrelated to the typological category of 'acid igneous rock' – which is, in any case, an outdated term;
- not all soils in Japan are volcanic; and
- almost all soils in Japan are acidic in pH, but volcanic and limestone soils are *less acidic* than most others.

This Chapter begins with analyses of raw sediments from three different geographical locations and geological contexts in Japan, compares those results with the standard characteristics of Japanese soil types, and then describes the chemical differences between so-called acid and alkaline igneous rocks and pH acidity. Examples of bone preservation are examined for the conditions leading to their survival. Coming to

This Chapter was originally presented as a paper at the 2004 Society for American Archaeology (SAA) conference in Montreal, Canada, and was published as a working paper in the panel proceedings (JAA 2004).

grips with these issues should lead to greater understanding of the preservation biases in the Japanese archaeological record and their true causes.

Acidity: the pH measure

In order to begin this investigation, it is necessary to have a full understanding of the pH scale (Figure 15.1), how it is structured, and exactly what it measures. Box 15 describes the fundamentals of pH chemistry. It should be noted that pH is a measure of a very specific state: the hydrogen ion (H^+, the hydronium molecule) content, and the labels 'Basic' and 'Acid' are relative to the hydronium content of pure water, which is regarded as having a 'Neutral' pH (see Box 15). It is more difficult to measure the pH of solid materials that cannot be dissolved in a solvent, be it water, alcohol, amine, or acid. In this discussion, only solutions using water are referred to.

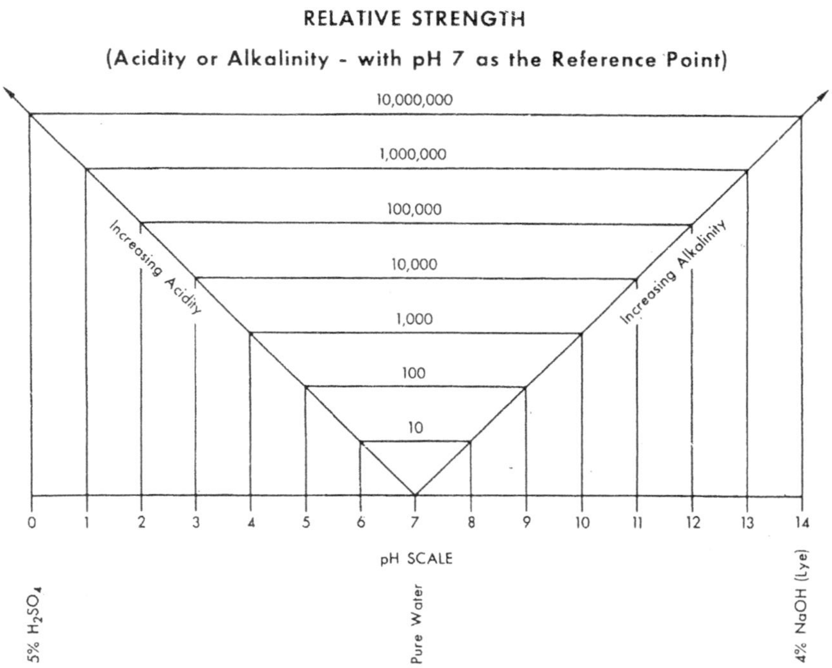

Figure 15.1 The pH logarithmic scale

In measuring the acidity of Earth materials, is worth remembering the distinction between soils and sediments. The technical definition of a soil is a vertical succession of A-B-C horizons, where A is surface material supporting plant growth, C is parent material, and B is an in-between horizon showing the effect of both the weathering of C and influence of plant-induced changes from A (cf. Figure 7.9). A sediment, on the other hand, does not have a vertical profile either because it has not developed into a soil, or it is a sample from a particular horizon. Weathered sediment between the solid bedrock and an upper soil horizon is called the 'regolith', but the term can also include sediments that have been transported and accumulated elsewhere (Taylor & Eggleton 2001: table 1.1). Regolith is often included in discussions of soil types (cf. Figure 7.2). This section discusses the pH of both soils and sediments.

Sediment acidity

The fieldwork conducted for this investigation utilized sediment samples from three different geographical and ecological zones: the Nara Basin, the Niigata coastal region, and upland Gunma (Apx 16). The geological contexts of the Nara samples were Holocene alluvial and Pleistocene lacustrine strata (i.e., Osaka Group, cf. Chapter 13); those of the Niigata coast came from Holocene sand dune formation, while the upland Gunma samples consisted of volcanic ash. The samples were analyzed within two weeks after collection for pH level, using standard lab procedures and the equipment in the Faculty of Agriculture, Niigata University. The samples analyzed consisted of 4 tephra samples from Gunma, 4 from Niigata (2 sand and 2 topsoil), and 9 samples from Nara (3 sand, 3 topsoil, 1 paddy soil, 1 silt, and 1 clay).

Box 15 The Principles of pH

The 'p' in pH stands for 'power' while the H stands for 'hydrogen'. Put simply, pH is the measure of the power or concentration of hydrogen in a material; more accurately, it is a measure of hydrogen cation (H^+) activity (a) on a negative logarithmic scale: $pH = -\log_{10} a_{H^+}$. Further refined, it is a measure of relative presence of the hydrogen cation (H^+) and the hydroxyl anion (OH^-). The scale is a measure of relative acidity or alkalinity over 14 points; pure water is used as the neutral central referent (7) because the hydrogen cation (H^+) and the hydroxyl anion (OH^-) concentrations are equal (at 25°C); that is, water (H_2O) can be dissociated into equal amounts of H^+ and OH^- ($H_2O \rightarrow H^+ + OH^-$). However, it is unlikely that hydrogen cations (H^+) exist unattached in a solution; in an aqueous solution they usually combine with the water molecule to form the hydronium ion (H_3O^+), also called the oxonium ion. The use of the hydrogen cation (H^+) as a cover term for its combination in solution (not only in water as H_3O^+ but also in alcohol [ROH_2^+], amine [RNH_3^+], or acid [$RCOOH_2^+$] solvents) is a standard semantic gloss in pH analyses. Thus, the equation for pH can also be written as: $pH = -\log[H^+(aq)]$, indicating hydrogen in an aqueous solution as hydronium.

As pure water is used for the standard for Neutral (7), a 5% H_2SO_4 solution is used as the standard of extreme acidity (0), and a 4% NaOH solution (lye) is used for the standard of extreme alkalinity (14). The H^+ and OH^- balances for these points on the scale are expressed as negative powers of 10, with 10^0 the value of 1M (molar mass), and all other powers as correspondingly less than 1M. Acids of pH 0 therefore have 1M H^+ and Basics of pH 14 have 1M OH^-:

Acid pH 0	Neutral pH 7	Basic pH 14
$H^+ = 10^0$	$H^+ = 10^{-7}$; $OH^- = 10^{-7}$	$H^+ = 10^{-14}$

Accordingly, materials having an excess of H^+ over OH^- are defined as Acid, those with an excess of OH^- over H^+ are defined as Basic, and the deviation from concentrations of 10^{-7} is represented by the 14 points on the scale. It is not necessary, however, for both hydronium and hydroxyls to be present in complementary amounts in every sample. The pH scale is essentially a measure of hydronium activity and the greater its presence, the more Acid the sample – from a concentration of 1×10^0M of H_3O^+ (pH 0) to 1×10^{-14}M of H_3O^+ (pH 14). There are acids and bases that have acidity and alkalinity values beyond the 14-point scale, but they are infrequently measured. It is important to note that the extreme ends of both the Acid and Basic readings on the pH scale can be very destructive; this is not a property of acids alone, as lye – the standard for extreme alkalinity – is also very caustic.

The logarithmic nature of the scale means there is a factor of 10 between each pH point, so that there are 10,000,000 units difference between the ends of the scale (Figure 15.1). The scale is standardized for measurements made at 25°C; at other temperatures, the pH error per °C has been calculated, ranging 0.001 error between 0°C (0.0037) and 100°C (0.0027). The value of Neutral varies with temperature (pH 7.5 at 0°C, and 6.5 at 60°C). The descriptive scale for soil pH (Table 15.2) therefore has a range of values ascribed to Neutral, as modified by temperature and other conditions. Temperature also affects Basic pH readings, but it has little effect on Acid pH values.

The results of the pH analyses (Table 15.1) indicate that the pH range for the seventeen samples was between 4.15 and 6.62 – from 'extremely acid' to 'neutral', with the volcanic samples having three of the four highest (least acidic) values. The pH values for the volcanic samples ranged from 5.82 'moderately acid', to 6.19 and 6.23 'slightly acid', to 6.62 'neutral' (qualitative values as listed in Table 15.2). The numbers of samples from each source type, each geological context, and each geographical location above are acknowledged to be small; but the results are fairly indicative in showing a certain trend that can be compared with standard measures of soil pH in Japan (Apx 17), as published in the *Ecological Atlas of Soils in Japan* (Nagatsuka 1997), to be discussed below.[1] The data above constitute sediments and cannot be compared directly with measures generalized for various soil types, especially because the *Atlas* compiles pH readings for soil horizons up to 1 m depth. Although among the above samples, Heijo (Loc. A) and Kogenji-Higashi I (Loc. L) were represented by two or more samples for the stratigraphic sequence, measurements were not taken at any location continuously to a metre's depth. Nevertheless, both datasets have certain trends in common.

Table 15.1 pH distributions of samples by sediment/soil type

	topsoil	sand	silt	clay	tephra
Nara					
A-1		5.66			
A-2	5.24				
B-1			5.80		
B-3		6.06			
C-1	5.5				
D-1				4.15	
E-1	5.0				
F-1	5.89				
Niigata					
G-1		5.37 (wet)			
H-1	5.67				
J-1		6.52 (dry)			
K-1		5.17			
Gunma					
L-1					6.25
L-2					6.62
L-3					5.82
L-4					6.19

Soil acidity

The *Ecological Atlas of Soils in Japan* (Nagatsuka 1997) systematically described 101 soil types found in Japan. Not all have pH recorded for them, but most of the descriptions are accompanied by a chart showing the levels of pH through a stratigraphic section, to the usual sampling depth of 100 cm. A visual comparison of the various pH tables for soil types (represented by roman numerals) presented in these sections illustrated in the *Atlas* reveals that pH readings for soils, regardless of the absolute values, characteristically vary over the stratigraphic section:

1. Some increase from lower readings at the surface to attain a steady higher reading or slightly increasing reading through the lower stratigraphy.
2. A few, however, start low at the surface, drop suddenly and then increase to higher readings.
3. Others remain fairly steady throughout the section
4. While some start low, bulge and then fall back to levels as high or higher than the starting point.
5. Some dip lower in the middle of the section but are fairly steady otherwise.
6. And finally, others start higher and drop lower.

It is important to note that similar patterns occur in different soil types; secondly, any particular soil stratigraphic section, as opposed to a sediment sample, cannot be characterized by a single pH reading; and thirdly, that the position of our samples in their original stratigraphic context will have influenced the pH

[1] The *Atlas* tables show pH calculated by two different methods: water (H_2O) and hydrochloric acid (HCl); the method used for the samples reported on here were analyzed by the water method (H_2O), so only that method as reported in the *Atlas* will be used for comparison.

readings obtained for them. To judge their representativeness, further comparisons of absolute pH values and the variables which determine them must be undertaken.

Table 15.2 gives the classification of pH values, ranging from 0-14 with their qualitative descriptions on the left, for the different soil types represented by Roman numerals; anything below Neutral pH values (6.6-7.3) are Acid. The value measurements of the stratigraphic sections of the 101 soils in the *Atlas* are arrayed against these measures. The quantitative ranges of pH values and descriptors of different soil samples in the types indicated here in Roman numerals are given in Appendix 17.

Table 15.2 Ranges of pH values of soil types I to XXII (> XXII uncharted here)

I Podzols; II Brown Earths; III-IV Fersiallitic; V-VI Andosols; VII-VIII Stagnant-water soils; IX-X Lowland; XI Gley; XII-XIII Paddy; XIV Peat; XV-XIX Calcmagnesian; XX-XXI Underdeveloped; XXII-XXV Raw; XXVI-XXVII Undifferentiated; XXIX-XXXII Anthropogenic.

		I	II	III	IV	V	VI	VII	VIII	IX	X	XI
0-4.4	Extremely acid	x	x	x	x	x	x		x			x
4.5-5.0	Very strongly acid	x	x	x	x	x	x	x	x		x	x
5.1-5.5	Strongly acid	x	x	x	x	x	x	x	x	x	x	x
5.6-6.0	Moderately acid		x	x	x	x		x		x	x	x
6.1-6.5	Slightly acid			x	x	x				x	x	x
6.6-7.3	**Neutral**			x		x				x	x	x
7.4-7.8	Slightly alkaline										x	x
7.9-8.4	Moderately alkaline										x	
8.5-9.0	Strongly alkaline											
9.1-14.0	Very strongly alkaline											

		XII	XIII	XIV	XV~XIX	XX	XXI	XXII
0-4.4	Extremely acid			x			x	
4.5-5.0	Very strongly acid	x		x			x	
5.1-5.5	Strongly acid	x	x	x	x	x	x	x
5.6-6.0	Moderately acid	x	x		x	x	x	x
6.1-6.5	Slightly acid	x	x		x	x	x	x
6.6-7.3	**Neutral**		x		x		x	x
7.4-7.8	Slightly alkaline				x			
7.9-8.4	Moderately alkaline				x			
8.5-9.0	Strongly alkaline				x			
9.1-14.0	Very strongly alkaline							

There are several things to note concerning these pH ranges:

- The ranges cannot be averaged for a particular soil or class because the distributions of values shift with depth, and the dominant value may be quite different from the lowest or highest value.
- The ranges, however, can be compared between soil examples and between classes. Clearly, as expected, Calcmagnesian soils derived from limestones (Types XV-XIX) have generally higher pH ranges than other types, as do some Andosols (V) generally derived from volcanic parent materials. Lowland soils (IX-X) also have higher pH value ranges as do Gleys (XI). On the other hand, Podzols (I) and Brown-earth soils (II) and other Andosols (VI), some Stagnant-soils (VIII) and Peat (XIV) generally have very low pH value ranges.
- Soils derived from volcanic parent material occur in many of the soil classes – not just andosols – and have a wide variety of pH values. Overall, the pH of any volcanic-derived soil is consistent with the other soils in its class; thus, parent material cannot be named as the main factor influencing the soil pH except for the Calcmagnesian soils (XV-XIX) of limestone origin and for the kurobokudo Andosols (V) of generally volcanic origin. The reason the kurobokudo Andosols have higher pH values is not only due to their volcanic

nature – the values are highly dependent on the presence of non-crystalline colloids such as allophane (cf. Chapter 7), which serve to tie up aluminium ions rather than leave them as loose ions contributing to soil acidity. They also bind humus into the colloids and metal complexes, delaying decay and release of humic/fulvic acids (Apx 13: Tables A, B). This is an important factor because Andosols are a recent type of soil postdating Jomon anthropogenic changes in the upland landscape. According to Shoji *et al.* (1993: 54) the cutting of forests and firing of upland areas in the Jomon period encouraged the growth of secondary cover comprised mainly of the *Miscanthus sinensis*, Japanese pampas grass. This grass is considered an andisolizer agent, although a certain type of Andosols can also form under beech forest. Andosols can be divided into those containing amorphous colloids (allophonic andosols) and those that do not (non-allophonic andosols). The latter are clearly more acid than the former (cf. Table 7.4).

In sum, most soils in Japan are pH acidic in nature, with some reaching pH Neutral status and only a few types – notably volcanically derived allophanic Andosols, limestone-derived Calcmagnesian, and Lowland soils – attaining Basic status. The sediments analyzed for this Chapter follow this pattern with the volcanic samples having the higher pH readings than most other samples of different origin. Moreover, many soil types (I-VI, VIII, XI, XIV and XXI) do have 'extremely acid' pH (4.0-4.4) readings in their profiles; however, many do not (VII, IX, X, XII-XIII, XV-XIX and XX), and some are not even 'very strongly acid' (IX, XIII, XV~XX, and XXII). Thus, common characterization of Japanese soils as between pH 4.0-4.5, if at all true, must have to do either with the dominant pH of topsoil rather than the soil profile as a whole or with the relative areal extents of more acidic soil types – issues which would need further investigation. From this project, however, we can learn one important thing: almost all soils – not only volcanic soils – in Japan are on the Acid side of Neutral pH, so blaming bone disintegration on volcanic soils alone is totally misplaced. In fact, volcanic soils are one of the lesser culprits. Why is this reversal so?

Acid Igneous Rocks vs. pH Acidity

Igneous rock classifications

I believe that the attribution by archaeologists of pH acidity to (only) volcanic soils in Japan comes from a deep misunderstanding of igneous rock classification, in which Japan has been characterized as having acid volcanic rocks. The typology of basic - intermediate - acid for igneous rocks, however, was a classification of the amount of *silica* (SiO_2) they contain – it is NOT a measure of pH acidity. It is not known why the term 'acid' was applied to igneous rocks of high silica content when the typology was established (R. Thompson, pers. comm. July 2003), but one circulating legend has it that silica was thought to dissolve directly into silicic acid $Si(OH)_4$ – which is now known not to be true. Best emphasizes this for magmatic rocks: "Acid and basic have no reference whatsoever to hydrogen ion content, or pH, as used in chemistry. (Long ago it was erroneously believed that SiO_2 occurred as silicic acid and [that] metallic oxide components, such as CaO and FeO, [occurred] as bases in magma.)" (2003: 31).

The terminology for igneous rock classification now uses 'felsic' instead of 'acid', and 'mafic' instead of 'basic'. '**Fel**sic', referring to the light-coloured **fel**dspar minerals, is used for characterizing high-**si**li**c**a (>66%) igneous products; if rigorously followed, this would avoid the misunderstanding that the term 'acid' encourages. The term 'basic' is similarly replaced by '**maf**ic' (the dark-coloured **ma**gnesium and **f**erri**c** minerals), indicating igneous products having 45–52% silica and a lot of dark-coloured minerals; the term 'ultrabasic' (<45% silica content) has been generally replaced by 'ultramafic' (cf. Apx 5: Fig. F).

Silica, SiO_2 silicon dioxide, occurs naturally in two forms: 1) the hexagonal crystalline Alpha-form (quartz) which is one of the most durable earth products, as it is cycled and recycled again and again through erosion, subduction and extrusion, undergoing physical abrasion but not chemical change; and 2) the Beta-form which is not soluble in water but is subject to hydration (e.g., obsidian hydration), where water enters the crystalline lattice and the water molecules attach themselves to silica tetrahedra, often to form clays. The Beta-form of silica (glass) in volcanic ash is highly subject to weathering and degradation because of the small size of the glass shards and their accessible surface areas (Shoji *et al.* 1993: 102). The weathered products of volcanic ash are types of clay (cf. Figure 7.7), not silicic acid H_4SiO_4 – which in any case is unstable in water and tends to combine with other elements.

Japan actually has all three types of igneous rocks in its geologic make-up: *mafic* gabbros and basalts, *intermediate* diorites and andesites, and *felsic* granites and rhyolites (cf. Apx 5: Fig. F). These all formed at different times under different tectonic conditions over 450 million years, but the volcanics associated with rifting of the continental rim into the islands ca. 24–15 Ma (Chapter 3) and the Quaternary volcanics beginning ca. 750 ka (Chapter 6) are the two most important for soil formation. There is a sequencing of volcanic products in the Quaternary regime that shows increasing silica content through time, as shown by the Towada caldera eruptions in Tohoku (Table 15.3; cf. BOX 7). Quaternary volcanic activity has produced most of the current surface cover of the Japanese Islands, leaving hardly a square inch untouched by ash fallout. But pre-existing this regime are other important igneous surface rocks such as the granitic (felsic) backbone of western Honshu and the andesite (intermediate) products of Miocene volcanoes in the Inland Sea area. Combined, these various igneous rocks account for 60% of the surface area (under the topsoil) in Japan today – but that is not 100%, and there are other types of sedimentary and metamorphic rocks as well that cover the other 40%.

Table 15.3 Towada caldera Holocene tephras

10,000 bp	Ninokura tephra (Ni)	basaltic andesite
8600 bp	Nanbu tephra (Na)	andesite pumice
5000 bp	Chuseri tephra (Ch)	dacite pumice and ash
2000 bp	Towada-b tephra (Tb, To-b)	rhyolitic ash
1000 bp	Towada-a (Ta, To-a)	rhyolitic ash

In sum, the basic-to-acid classificatory scheme of igneous rocks is out-dated; these classes indicated the relative amount of *silica* that the rocks (and volcanic ash) contain, and these now are more reliably referred to as mafic-intermediate-felsic. Referral to the old scheme of basic-to-acid does *not* mean that they are acid or basic on the pH scale. It is actually quite difficult to determine the pH of newly erupted volcanics before weathering changes their chemical structures. Therefore, it is not easy to characterize igneous rock types by their pH, and indeed, texts on igneous petrology do not deal at all with pH. However, we can approach this topic by looking at another dimension of igneous rock classification based on inclusion of base metals rather than silica.

Alkaline igneous rocks vs. pH alkalinity

A different classificatory scheme for igneous rocks depends on their inclusion of base metals, particularly K, Na, Ca, Mg, and Li. According to the Periodic Table (Apx 5: Fig. A), potassium (K), sodium (Na) and lithium (Li) belong to the Alkali Metals, while calcium (Ca) and magnesium (Mg) belong to the S-2 group of Alkali(ne) Earth Metals. Members of these two groups are all metals, and as part of their properties, "in chemical reactions [they] react with acids, form basic oxides and form positively charged ions" (Lewis & Evans 2001: 17). In other words, they are considered alkaline (base) metals because they will neutralize acid solutions to form salts, and a salt is specifically defined as "a substance made by the neutralisation of an acid with a base" (Ibid.: 459). The issue of ion formation is important because it is this feature that makes the base metals soluble in water – to be discussed further below.

Volcanic rock classification according to base metal presence is presented in Appendix 5: Figure G; and as can be seen in Table 15.4, the most prominent minerals in igneous rocks contain considerable quantities of these base metals (emboldened). Alkaline products have "higher concentrations of alkalis than can be accommodated in feldspars alone, the excess appearing as feldspathoids, sodic pyroxenes, sodic amphiboles and other alkali-rich phases" (Fitton & Upton 1987: ix). Sub-Alkaline products, on the other hand, are not so saturated with alkalis and are divided into Calc-Alkaline and Tholeiitic categories (Apx 5: Fig. G).

Table 15.4 Minerals occurring in igneous rock types and their chemical formulae
++ predominant, + present, but not all minerals in each category will occur
in each rock sample; base metals = **emboldened**

MINERAL	FELSIC	INTERM.	MAFIC	CHEMICAL FORMULA
alkali feldspar	++	+		**(K,Na)**$AlSi_3O_8$
quartz	++	+		SiO_2
plagioclase feldspar	++	++	++	**(Na,Ca)**$\{(Al,Si)_4O_8\}$
muscovite	+			4[**(K,Na)**$Al_2(AlSi_3)O_{10}(OH,F)_2$
biotite	+	+		**K(Mg,**$Fe)_3(Si_3Al)O_{10}$
olivine			+	4**(Mg,**$Fe)_2\{SiO_4\}$
amphibole	+	+		**NaCa$_2$(Mg,**$Fe)_5(Si,Al)_8O_{22}(OH)_2$
pyroxene		+	++	$(W)_{1-p}(X,Y)_{1+p}A_2O_6$, where W=**Ca,Na**; X=**Mg**, Fe^{2+}, Mn, Ni, **Li**; Y=Al, Fe^{3+}, Cr, Ti; Z=Si, Al; p=0-1

Three different tectonic settings have been identified that create igneous products with differing quantities of base metals (Best 2003: 35; Fitton & Upton 1987). Rifting characteristically produces Alkaline volcanics; such rifting may occur on land, as initially accompanying the Great Rift Valley of Africa. In Japan, Zone 4 along the Japan Sea coast is alkaline (Figure 6.2) as a result of rifting from the continent. Alkaline products can also be formed in subduction zones when descending slabs are so dehydrated that they can no longer stimulate the formation of Calc-Alkaline magma (Fitton & Upton 1987: xi). Subduction zones characteristically produce Tholeiitic products in island arcs at the beginning of their activity, and as the arcs mature, the products become more Calc-Alkaline; this process parallels that of silica concentration – in that initially erupted materials are usually basaltic, becoming more andesitic then rhyolitic over time (cf. Table 15.3).

In Japan, rifting was the process that separated the Japanese landmass from the continent and opened up the Japan Sea (Chapter 4); thus, Alkaline volcanic products most likely accompanied these rifting activities between 24 and 20 Ma in the Miocene (cf. Apx 9). Most recent igneous activity of Japan since the mid-Quaternary is classified as belonging to the typical Calc-Alkaline volcanic regime characteristic of maturing island arc formation (Best 1982: 51); but the Izu Arc is a new island arc formed with mainly Tholeiitic products. By definition, Calc-Alkaline products are enriched in silica and alkalies (Na_2O, K_2O) (Best 2003: 35), but Japan is recognized as having relatively low amounts of K. It is known that potassium content varies with the thickness of the Earth's crust in particular regions such as the Andes (45-km thick crust, high K) and New Zealand (35-km crust, medium K) (Ibid.), whereas the crust of northeastern Honshu, which supports the major Quaternary volcanoes, is thinner (ca. 30 km thick, low K) (Hunahashi *et al.* 1979: fig. 1b-1).

If Japanese igneous rocks are reasonably high in base metal content, can this affect pH values? According to Lewis & Evans (2001: 194-198), atoms of Alkali and Alkaline Earth metals tend to lose one or two electrons from their outer shell and become positively charged (as cations), and compounds they form with other elements tend also to be electrically charged (ionic). The Alkali Metals "react readily with water", as exemplified by potassium (K) or sodium (Na) below; two atoms of solid (s) metal and two molecules of water (l=liquid) yield two atoms of the metal cation and two molecules of the hydroxl in aqueous solution (aq) plus the release of two atoms of hydrogen as gas (g) (Lewis & Evans 2001: 195):

$$K(s) + 2H_2O(l) \longrightarrow 2K^+(aq) + 2OH^-(aq) + H_2(g)$$

$$Na(s) + 2H_2O(l) \longrightarrow 2Na^+(aq) + 2OH^-(aq) + H_2(g)$$

The fact that the product in solution has two hydroxyl molecules (OH^-) and no hydronium (H_3O^+) (see Box 15) indicates that the resulting solution is Basic in pH. Hydroxides formed with Alkali Metals are also very soluble in water. The Alkaline Earth Metals (Mg and Ca) are also water soluble, with cations forming an oxide or hydroxide and hydrogen gas (Ibid.: 197):

$$Mg(s) + H_2O(l) \rightarrow MgO(s) + H_2(g)$$

$$Ca(s) + 2H_2O(l) \rightarrow Ca(OH)_2(aq) + H_2(g)$$

The factor that makes all these Alkali Metals and Alkaline Earth Metals have a pH in the basic range is that there are no hydronium molecules formed in these solutions (see Box 15); instead, hydrogen is liberated as a pure gas.

One might note, however, that the above formulae take pure elements as their starting point, whereas in igneous rocks, such elements are integrated into minerals. Soluble elements are soluble because they can be attracted out of mineral crystalline lattices in the weathering process by the polar charges on the water molecule (H_2O) – the oxygen atom having a negative charge (as an anion) and the two hydrogen atoms having a positive charge (as cations). Depending on the ionic charge of the metal ionically bonded into the mineral, it may be more attracted to one or the other end of the water molecule than it is to the other atoms in the mineral lattice.

Volcanic products of the Calc-Alkaline regime thus contain considerable amounts of Alkali and Alkaline Earth Metals, and their major minerals can produce alkaline solutions upon dissolution of the base minerals in their lattices. But igneous products are not exclusively made up of their primary magmatic minerals; there are often minerals of xenoliths (fragments of country rock that are incorporated into the magma) and numerous secondary and accessory minerals present. All these will contribute to the pH of a volcanic product if it is tested at the moment of extrusion. Some of the volcanic ash in the Green Tuff Zone of the Japan Sea coastal zone (cf. Figure 4.4), extruded in Miocene times, originally had high pH readings. Even after millions of years weathering of volcanic ash into clays, pH readings of 8.0 and above have been obtained on buried layers: pH 8.4 on grey clays at 25 m depth (Yoshimura *et al.* 1979: 44-45); pH 8.9 on greenish blue clays at 2.2 m depth, and from pH 8.9 to 9.5 between 3.5 and 9 m depth.

The problem with volcanic ash is that it is very fine grained, and as soon as it hits the atmosphere (or the hydrosphere in a submarine eruption), it encounters water which begins to dissolve the soluble base metals out of the rock and mineral mass. The longer volcanic products are exposed to the atmosphere or hydrosphere, the more their pH will decrease in value. Volcanic materials that have already been colonized by plant growth and turned into soils are, by definition, aged products. In Japan, they have been exposed to very high precipitation rates (from 1016 mm to 2540 mm) and temperatures (exceeding 35°C) in many areas. Both encourage the leaching of alkaline elements; although most Japanese volcanic soils were Calc-Alkaline in their original chemistry, they are quickly deprived of their alkaline status through the leaching out of base metals. Thus, <u>it is rainfall that is responsible for acid soil status in Japan</u>, not the chemistry of the volcanic products themselves.

Acidification by Climate and Plant Activity

A standard statement about acidic soils is given by Bickelhaupt (n.d.: unpg.):

> Soils tend to become acidic as a result of: (1) rainwater leaching away basic ions (calcium, magnesium, potassium and sodium); (2) carbon dioxide from decomposing organic matter and root respiration dissolving in soil water to form a weak organic acid; (3) formation of strong organic and inorganic acids, such as nitric and sulfuric acid, from decaying organic matter and oxidation of ammonium and sulfur fertilizers. Strongly acid soils are usually the result of the action of these strong organic and inorganic acids.

To this list, it can be added that the chemical change from Fe^{3+} to Fe^{2+} also tends to make soils more acidic (Taylor & Eggleton 2001: 76).

Around the world, soil scientists are aware that areas with high rates of precipitation produce acidity in all types of soils – not just volcanic – because the rain leaches out all the base elements that might have originally been there (Chesworth 1992: 26):

> The acid trend is the common one in weathering and is found in regions of humid climate on materials with a generally unimpeded drainage....The chief characteristic of acidification of the weathering environment is the progressive replacement of cations on exposed mineral surfaces by protons [H^+].

However, rainwater with dissolved CO_2 contributes only to mild soil acidity (pH 6 to 5), while lower pH values are due to the presence of acids, both organic and inorganic: "The main sources of acidity in soil are organic acids, especially aliphatic and aromatic, coming either from organic matter decomposition or more directly from secretions of living organisms" (Chesworth 1992: 73-4). It follows then that andosols, which bind humus into colloids like allophone or to aluminium (cf. Table 7.7), do not allow the decay of that humus into humic and fulvic acids (Apx 13), therefore maintaining a slightly higher pH than other acid soils. Thus, tephrogenic soils are shown by the above experiment and *Atlas* readings to sometimes be less acidic than other soils.

Bone Preservation in Japan

This is not the place to itemize the growing database of human skeletal material retrieved from Japanese archaeological sites – but contrary to expectations, there are thousands of surviving skeletons that allow for analysis of differential muscle development, pre-mortem and fatal wounds, DNA processing, and strontium and carbon isotope analysis, etc. Such skeletal remains are almost always found in sheltered environments, two of the most obvious being caves and shellmound sites – the latter such as those listed in the Jomon shellmound database (Oikawa 1980; Oikawa *et al.* 1980). Shigehara (1994) noted that earlier Jomon skeletal material was usually found in inland cave or rock shelter sites; but from Middle Jomon, most are recovered from shellmounds, which also yield faunal remains. Several other kinds of sites have also acted to preserve human bone; examples of some of the less ordinary preservation conditions are offered here. Because good preservation may be caused by one or more factors, the examples will not be grouped by cause below.

Japan has relatively few limestone outcrops that provide for cave sites. The Akiyoshidai karst (BOX 3), and its Shuho Cave, is an example of some of the shallow-sea limestones that have been obducted onto land; others occur with ophiolites or the accreted South Kitakami Terrane (cf. Chapter 3) (e.g., Iwate-ken Kyoi 2001). Serizawa (1979) listed 43 cave sites (without specifying their rock types) that yielded Palaeolithic, Jomon, and Yayoi materials, but very few of them had good bone preservation – as affirmed more recently by Abe (2012); however, caves in Kyushu have yielded material good enough for DNA analysis (Shinoda *et al.* 2017). Most uplifted shelf limestones occur in the far southwest: Okinawa and Kagoshima prefectures. These are characterized by Calcmagnesian soils (XV-XIX in Nagatsuka's listing), but Late Pleistocene human skeletal remains have been recovered from limestone fissures and caves themselves, not from burials in soil (e.g., Suzuki & Hanihara 1982; Nakagawa *et al.* 2010).

Limestone caves and shellmounds, however, are not the only facilitators of bone survival. The following cases cover several circumstances that are not generally thought of in assessing the potential for skeletal survival, such as burial caves that were maintained for a short period of time (Figure 15.3) (see RDH 2016).[2]

<u>The Oyaji Cave site</u> (#33 in Serizawa 1979: fig. 1) is an unusual rock shelter, weathered 30 m wide x 13 m deep x 12 m high into the Oya Tuff in Utsunomiya, Tochigi. As described in BOX 4, Oya Tuff is a member of the Green Tuff which was exuded in the alkaline zone during the rifting of the archipelago from the continent (Chapter 4). The cave walls host four Buddhist sculptures, and excavations of the cave floor below

[2] For prefectural and district names, see Apx 2: Fig. A.

revealed its use from Jomon into modern times. The earliest stratum yielded human bones, particularly a well-preserved flexed burial (Figure 15.2) (Hashimoto 2019). Originally dated to about 11,000 BP, the burial was redated in 1998 to 17,000 BP, the beginning of Incipient Jomon (Apx 2: Table A) (*Yomiuri Shimbun* 29 May 1998).

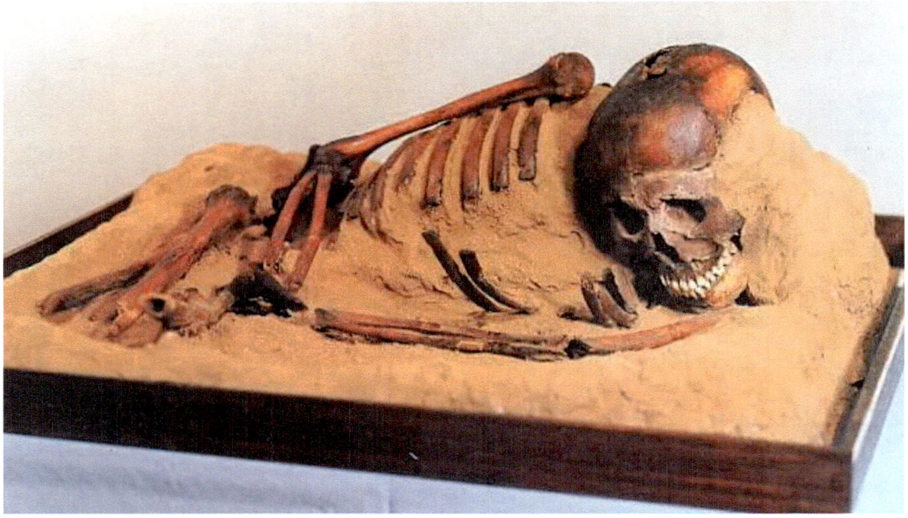

Figure 15.2 Incipient Jomon burial in Oya tuff, on display at Oya Temple, Tochigi

The Iwade Yokoana burial caves of late Kofun period date in Kanazawa City, Ishikawa, were dug into a hillside of Pleistocene marine deposits (GSJ 2020: Q1-21_so)[3]. One of the caves yielded human skeletal remains that have provided DNA evidence that the Japanese genome was consolidated in the Kofun period (Cooke *et al.* 2021). Since Kanazawa faces the Japan Sea, it is not surprising that marine layers line the coast; however, the Kitamura site considered below is far inland but benefits from a similar situation.

Figure 15.3 An example of Late Kofun burial caves at Yoshimi, Saitama

The Kanai Higashi-ura site in Shibukawa, Gunma, is another instance where tephra layers protected skeletal remains: six humans and three horses as well as an organic woven fence 3 m high and 20–30 cm thick were recovered from under a pyroclastic flow on the flank of Mt Haruna. Their preservation, after 1500 years since the eruption, was facilitated by the tephra acting as a 'capillary boundary', not allowing rain to penetrate the lower layers (Sugiyama 2019: 176).

The Doigahama site in Shimonoseki, Yamaguchi consists of a Early-to-Middle Yayoi-period burial site on a sand dune (JSG 2020: H_ssd) facing the Japan Sea (Doigahama 2014). Over 300 burials have been excavated, most directly in the sand but some enclosed in stone cists. Nakazawa (1997) classifies sand dunes as 'arenosols', and the silicious sand sample from Tottori registered pH 5.8-6.7 (Apx 17). Coral sands register even higher (pH 8.2–9.0), so it is logical that the more calcareous material (shells) is included in the sands,

[3] These numbers refer to the 'seamless' online map of Japanese geological distributions.

the higher the pH will be. If the sand dunes are covered with grass, their root mat somewhat protects the sand below (eg., Apx 16: J-1).

From skeletal morphology, the Doigahama people are known to have been continental immigrants, and the remains were sufficient for mtDNA testing which identified most of the sample to be related to Shandong inhabitants on the China Mainland 2500 years ago (Igawa *et al.* 2009).

The Kitamura site, situated on a riverine terrace in Akishina, Nagano, yielded more than 300 Mid-to-Late Jomon burials containing skeletal remains (Nagano Maibun 1993; Shigehara 1994). The reasons for bone preservation were listed as (Anon. 1993: 139):

- There was no strongly acidic loam
- The groundwater was slightly alkaline, and the bones were buried in soil with abundant water and little oxygen
- The site is located just below the Bessho Formation exposure in the mountains to the east; water passing through this Formation contains abundant calcium ions
- The dense sediments of Stratum III formed an impermeable layer, protecting the burials underneath from rainwater
- Strata III and IV did not allow ventilation and cut off oxygen supply

These conditions can be understood with reference to the geology of the area. The Itoigawa-Shizuoka Tectonic Line (cf. Chapter 3) runs through the centre of the Matsumoto Basin, slightly to the west of the site. This puts the Kitamura site to its east in the Fossa Magna (cf. Chapter 4), which was open to the sea during the Miocene. The Bessho Formation consists of deep-sea shales (NCEC 1988: 75-77; GSJ 2020: N1_som, N2_sosma), which absorbed much calcium from calcareous sea life. Spring water emanating from the Bessho Formation has a pH of 7.6 (Anon. 1993: table 10), and groundwater at different Kitamura site locations ranged from pH 6.8 to 7.3. The burials were excavated at 4 m depth, while the groundwater level at the site ranged from 0.6 m to 2 m below the surface. Thus, not only were the burials saturated with nearly neutral to slightly alkaline groundwater – they were also protected by densely packed sediments which formed a barrier to rainwater penetration and led to anoxic conditions, similar to the protective layers of dense sediments at the Kanai Higashi-ura site.

The Shibu site in Nara yielded rare finds of both human and animal bones: a double burial (Figure 15.4) dating to the terminal Middle Yayoi was preserved in a simple earthen pit covered with a wooden lid (Katayama & Sugihara 1998), while 422 animal bone fragments (mostly wild boar and shika deer) were water-preserved in wells and ditches (Tada 1998). Although skeletal material is often found in enclosed

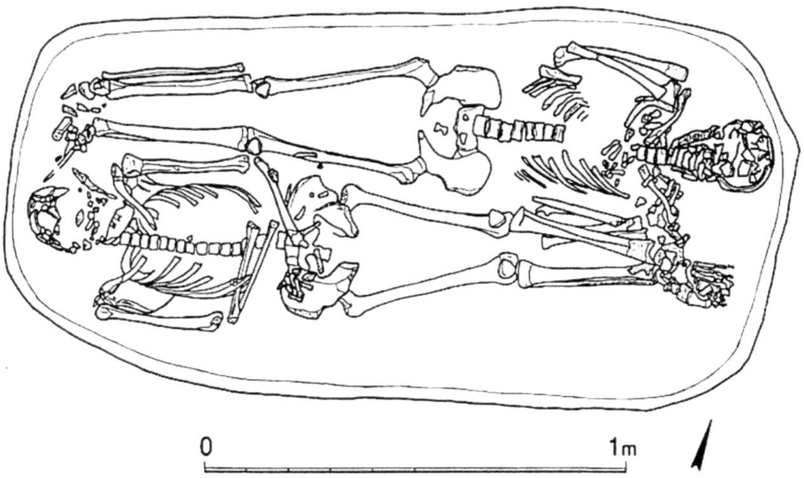

Figure 15.4 Double burial at Shibu site, Nara, in earthen pit covered with a wooden lid

burials such as jar burials and caskets, the Shibu burial is unusual because it was dug into Holocene alluvial sediments (GSJ 2020: H_sad). Nagatsuka (1997) gives a reading of pH 5.4–6.8 (from strongly acid to neutral) for lowland brown soils (Apx 17). Thus, depending on the local circumstances and nature of the burial facilities (wooden lid cap), the effects of precipitation in acidifying the soil may be deterred.

The Hirota site on Tanegashima Island, Kagoshima, was also a burial ground on sand dunes, dating to the Yayoi–Kofun transition. Of the 157 individuals excavated, at least 46 have been sexed (Yamochi 1999). The burial facilities consisted of slabs of rock, balanced on supporting rocks, covering bodies laid on the sand; many graves contained shell bracelets and other ornaments decorating the deceased. Thus, this site had three factors in favour of bone preservation: burial in sand, protected by a rock cover, and accompanied by more than 44,000 shell products. The site was designated as an Historic Site in 2008 (Kagoshima Kyōi n.d.).

Conclusions

This Chapter has been devoted to dismissing the widespread misconception that the soils of Japan are acidic *because* they are volcanic. On the contrary, it does not matter what type of soil it is: if the precipitation rates are high enough, any soil will be leached of its base elements. High temperatures, as found in Japan, increase this process of dissolution. Tephrogenic soils are fairly limited in distribution (cf. Figure 7.1, 7.3) and thus cannot account for a nationwide statement that volcanic deposits are the cause of soil acidity. In fact, almost all soils in Japan have acidic ranges in their profiles (Apx 17). Instead, the causes of acidic soils are a high level of precipitation, which can exceed 2000 mm/yr, and warm temperatures that increase the rate of leaching, and the release of humic and fulvic acids through plant decay.

The volcanic ash extruded in explosive eruptions in Japan are intermediate to rhyolitic in composition (Apx 5: Fig. F); they are classified as felsic – which used to be referred to as 'acid' igneous rocks because of their high *silica* content. This classification had nothing to do with acid pH values; in fact, tephra can have significant alkaline pH values when first extruded. However, the base metals are quickly leached from the tephra as soon as it is exposed to the moisture in the air and by high precipitation rates once deposited.

Tephra is as much a victim as the bone material buried within it, as rain waters seep through and remove calcium and other base elements. As volcanic ash layers are colonized by plants, their acidity is increased through the production of organic acids and the tephra weathers to clay; but it was demonstrated above that volcanic soils containing allophane are less acidic than others. Thus, originally alkaline volcanic ash becomes, with time, acid clayey soil, which cannot protect bones buried within unless under extraordinary circumstances – as in the site examples given above. These leaching and weathering processes should be clearly distinguished from the nature of volcanic products as originally extruded because weathering and leaching affect all soils in Japan, not just volcanic ones.

We must thus clearly distinguish between the meanings of 'acid' and 'basic' in the two different scientific disciplines of igneous petrology (as an out-dated measure of silica content) and solution chemistry (as a measure of hydronium concentration). It is my belief that the archaeological attribution of destructive powers to 'acid volcanic soils' in Japan has been an unfortunate confusion of the meaning of 'acid' in these two disciplines, and the acid-to-basic classification of igneous rocks should be thoroughly discarded in favour of the mafic-intermediate-felsic classification scheme. I hope that from now on, we can rightly attribute the general lack of human remains and fossils in Japan to the effects of high precipitation and high temperatures, leading to rapid leaching out of the base elements of such calc-alkaline materials in *both* volcanic parent material and bones, especially when bones are left unprotected by other structures such as caves or tombs or burial in calcium-rich deposits. All soils in Japan, not just tephrogenic, are subject to high precipitation and warm temperatures, resulting in them being on the acidic side. Among them, however, the tephrogenic andosols can be less acid than some other non-volcanic soils in Japan.

Figure & Table Sources

Figure 15.1 after (Wescott 1978: fig. 1.1)
Figure 15.2 after (Inoue n.d.)
Figure 15.3 by 京浜にけ - 投稿者撮影, CC -by- 3.0, [https://ja.wikipedia.org/w/index.php? curid=2228043]
Figure 15.4 after (Katayama & Sugihara 1998: fig. 6)

Table 15.1 compiled by author
Table 15.2 pH ranges of soil types according to Nagatsuka (1997); soil acidity gradations and descriptors according to Bickelhaupt (n.d.)
Table 15.3 compiled from (Shoji *et al.* 1993: 42-43)
Table 15.4 compiled from various standard sources

Box 1 compiled from (Bates 1973; Westcott 1978)

References

ABE, Yoshito (**2012**) "Researches on the Paleolithic human bone and the trace of their life at the cave site." *Kagaku Kenkyūhi Hojokin Kenkyū Seika Hōkokusho* 2011: 5pp. (in Japanese).

Anon. (**1993**) "Supplement – conditions preserving human bone", pp. 139-157 in *Nagano-ken Maizō Bunkazai Sentaa Hakkutsu Chōsa Hōkokusho 14, vol. 1 (honbun)*, ed. by Nagano Maibun. Nagano-ken: Nagano Archaeological Research Center (in Japanese).

BATES, Roger G (**1973**) *Determination of pH: theory and practice.* New York: John Wiley & Sons.

BEST, Myron G (**1982, 2003**) *Igneous and metamorphic petrology*. 2nd and 3rd editions. San Francisco: W.H. Freeman.

BICKELHAUPT, Donald (**n.d.**) "Soil pH: what it means." *Around Your World* [www.esf.edu/pubprog/brochure/soilph/soilph.htm].

CHESWORTH, W (**1992**) "Weathering systems", pp. 19-70 in IP MARTIN & W CHESWORTH (eds) *Weathering, soils and paleosols.* Amsterdam: Elsevier.

COOKE, Niall P; Valeria Mattiangeli & Lara M CASSIDY *et al.* (**2021**) "Ancient genomics reveals tripartite origins of Japanese populations." *Science Advances* 7: #eabh2419 + supplement.

Doigahama Anthropology Museum (ed.) (**2014**) *Doigahama site*. Shimonoseki-shi Bunkazai Chōsa Hōkoku 35. Yamaguchi: Shimonoseki City Board of Education (in Japanese).

FITTON, JG & GJ UPTON (**1987**) "Introduction", pp. ix-xiv in *Alkaline igneous rocks*, ed. by Fitton & Upton. London: Geological Society of London.

GAKUHARI, Takashi; Shigeki Nakagome & Simon Rasmussen *et al.* (**2020**) "Ancient Jomon genome sequence analysis sheds light on migration patterns of early East Asian populations." *Communications Biology* 3: 10pp. [DOI: 10.1038/s42003-020-01162-2].

GSJ (Geological Survey of Japan) (**2020**) "Seamless digital geological map of Japan V2 (1:200,000)." [https://gbank.gsj.jp/seamless/v2/viewer/] (in Japanese and English).

HASHIMOTO, Sumiro (**2019**) "Ooyaji cave site – footprints left by people in the Ooya soil." *Ooya-ishi bunka no reimei – ancient and medieval Ooya* [https://oya-official.jp/bunka/wp-content/uploads/sites/2/2019/ 05/1.大谷寺洞穴遺跡一大谷の地に足跡を残した人々ー.pdf] (in Japanese).

HUNAHASHI, Mitsuo; Jun WATANABE & Masao MINATO *et al.* (**1979**) "The basement complexes of northern Honshu", pp. 6-11 in M MINATO *et al.* (eds) *Variscan geohistory of northern Japan: the Abean orogeny*. Tokyo: Tokai University Press (in English).

IGAWA, Kazunari; Yoshitaka MANABE & Joichi OYAMADA *et al.* (**2009**) "Mitochondrial DNA analysis of Yayoi period human skeletal remains from the Doigahama site." *Journal of Human Genetics* 54.10: 581-588.

IMAMURA, Keiji (**1996**) *The prehistory of Japan and its position in East Asia*. UCL & University of Hawaii Press.

INOUE, S (**n.d.**) "Oyaji cave (rock shelter) site" [http://inoues.net/ruins2/ohyaiseki.html] (in Japanese).

Iwate-ken Kyōi (**2001**) *Cave sites of Iwate Prefecture*. Morioka: Iwate Prefectural Board of Education (in Japanese).

JAA (Japanese Archaeological Association) (**2004**) *Recent Palaeolithic studies in Japan: proceedings for tainted evidence and restoration of confidence in the Pleistocene archaeology of the Japanese archipelago*. Tokyo: JAA.

Kagoshima Kyōi (**n.d.**) "Hirota Site" [https://www.pref.kagoshima.jp/ba08/documents/5898_20120229135349-1.pdf] (in Japanese).

KATAYAMA, Kazumichi & Kiyoshi SUGIYAMA (**1998**) "Yayoi-period human remains excavated from a joint burial at Shibu site." *Nara Bunkazai Kenkyūsho Nenpō* 1998-II: 8-11 [http://repository.nabunken.go.jp/ dspace/handle/11177/5462] (in Japanese).

LEWIS, Rob & Wynne EVANS (**2001**) *Chemistry*. 2nd ed. Basingstoke, UK: Palgrave.

Nagano Maibun (**1993**) *Kitamura Site. Nagano-ken Maizō Bunkazai Sentaa Hakkutsu Chōsa Hōkokusho* 14 (vol. 1 *honbun*). Nagano City: Archaeological Research Center of Nagano Prefecture (in Japanese).

NAGATSUKA, Shizuo (**1997**) *Genshoku Nihon dojo seitai zukan* [Ecological atlas of soils in Japan]. Tokyo: Fuji Technosystem. (in Japanese with English title)

NAKAGAWA, Ryohei; Naomi DOI & Yuichiro NISHIOKA *et al.* (**2010**) "Pleistocene human remains from Shiraho-Saonetabaru Cave on Ishigaki Island, Okinawa, Japan, and their radiocarbon dating." *Anthropological Science* 118.3: 173-183 (in English).

NAKAHASHI, Takahiro & Masafumi NAGAI (**1986-1987**) "Preservation of human bone in prehistoric and historic sites in western Japan." *Asian Perspectives* 27.1: 15-27.

NCEC (Nihon no Chishitsu Editorial Committee) (**1988**) *Nihon no chishitsu 4: Chūbu chihō I.* Tokyo: Kyoritsu (in Japanese).

OIKAWA, Akifumi (1980) "Shellmound database." Archived at [www.jinbun-db.com/database/archives/517] by the Council for Databases in the Humanities, Japan (in Japanese).

OIKAWA, Akifumi; Sadaaki MIYAMOTO & Shuzo KOYAMA (**1980**) "A Jomon shellmound database: its compilation and application." *Bulletin of the National Museum of Ethnology* 5.2: 439-470 [https://minpaku.repo.nii.ac.jp/?action=repository_action_common_download&item_id=4533&item_no=1&attribute_id=18&file_no=1] (in Japanese with English title and abstract).

RDH (Rissho Daigaku Hakubutsukan) (**2016**) *Special exhibition #11: Yokoanabo.* Kumadani, Saitama: Oriental Bell Museum [www.ris.ac.jp/museum/profile/lvhgqo0000004877-att/yokoanabo_compressed.pdf] (in Japanese).

SERIZAWA, Chosuke (**1979**) "Cave sites in Japan." *World Archaeology* 10.3: 340-349.

SHIGEHARA, Nobuo (**1994**) "Human skeletal remains of the middle Jomon period excavated from the inland Kitamura site, Nagano prefecture." *Anthropological Science* 102.4: 321-344 (in English).

SHINODA, Ken-ichi; Hideaki KANZAWA-KIRIYAMA & Tsuneo KAKUDA *et al.* (**2017**) "Mitochondrial DNA analysis of human skeletal remains excavated from the Iwashita Cave and Shimo-motoyama rock shelter sites." *Anthropological Science* 125.1: 49-63 [DOI: 10.1537/asj.170] (in Japanese with English title and abstract).

SHOJI, Sadao; Masami NANZYO & Randy DAHLGREN (**1993**) *Volcanic ash soils: genesis, properties and utilization.* Amsterdam: Elsevier.

SUGIYAMA, Hidehiro (**2019**) "Investigations into the Kofun period disasters caused by Mt Haruna eruptions", pp. 168-182 in *TephroArchaeology in the North Pacific*, ed. by GL BARNES & T SODA. Oxford: Archaeopress.

SUZUKI, Hisashi & Kazuro HANIHARA (eds) (**1982**) *The Minatogawa Man: the Upper Pleistocene man from the island of Okinawa.* Tokyo: University of Tokyo Press.

TADA, Yumiko (**1998**) "Faunal remains excavated from the Shibu site." *Nara Bunkazai Kenkyūsho Nenpō* 1998-II: 8-11 [http://repository.nabunken.go.jp/dspace/handle/11177/5462] (in Japanese).

TAYLOR, G & RA EGGLETON (**2001**) *Regolith geology and geomorphology.* Chichester, UK: John Wiley & Sons.

WESTCOTT, C Clark (**1978**) *pH measurements.* New York: Academic Press.

YAMOCHI, Kumie (**1999?**) "The burials at Hirota site, Tanegashima: focusing on types of burial facilities." *Tenri Daigaku Kōkogaku Kenkyūshitsu Kiyō* 4: 35-51 [https://opac.tenri-u.ac.jp/opac/repository/metadata/search/030331004/] (in Japanese).

YOSHIMURA, Takahisa; Shigeyuki WAKABAYASHI & Junji AKAI (**1979**) "Kaolin-montmorillonite mixed layer minerals in the weathering zone of acid clay deposits from Itoigawa, central Japan. *Kobutsugaku Zasshi* 14 (special issue): 42-57 (in Japanese with English title and abstract).

Part III

Reflections

In this final section, I would like to discuss two findings of the Miwa Project (Barnes & Okita 1993) that are still unresolved – but may have been resolved at the time of the project *if I had already been aware* of the geology presented here in Part I and the new information in Chapter 13. The first finding to be discussed involves the source of pebbles found in the Miwa site stratigraphy, which reflects on the sedimentary history not only of the Basin but of the entire Second Setouchi Geological Province and faulted basin formation therein. The second finding involves the unusual thick black stratum in the Miwa site stratigraphy, which relates to the volcanic nature of Japanese soils as covered in Chapter 7.

Pebbles

According to excavation at the Miwa site in the southeastern Basin (cf. Figure 14.8), rounded pebbles of ca. 1 cm in length were incorporated in a pollen sample from the lower Yayoi layer in Trench 2. They were identified as to rock types and favourably compared to sediments of the Fujiwara Group (Kanehara & Okuda 1993), dating to ca. 18 Ma (cf. Table 13.2). Fujiwara Group sediments dating to the Miocene were presumed to have been the source of these pebbles, "re-deposited by fluvial action in recent times as gravels" (Kanehara & Okuda 1993: 111). But the authors could not choose which river might have been responsible – the Hasé River flowing below the Miwa terraces from the east (cf. Figure 14.8) or the Tanikawa stream that flowed through the Miwa site itself (cf. Figure 14.9).

The authors of this analysis did not mention the contemporaneous Yamabe Group, which is widespread in the mountain basins east-northeast of Mt Miwa. Like the Fujiwara Group, the Yamabe Group derives from sediments of the 1st Inland Sea (Yokoda 1978: table 2; CDK 1996: 1332) (see BOX 6) and are consistent with the types of rocks found in the pebble collection: chert, sandstone, and mudstone – all indicative of oceanic floor deposits. Furthermore, two pebbles were 'muddy schist', i.e., metamorphosed mudstone that were likely derived from the Ryoke metamorphic rocks. These rock identifications suggest that the pebbles derived from both the Yamabe Group *and* the Ryoke metamorphics of the Yamato Plateau. Given that the Basin's Lower terraces appear to have been uplifted and separated from the central alluvium after the LGM in the Pleistocene (Chapter 13) and the fact that the pebbles occurred *within* the Yayoi cultural layer dating around 0 AD, the pebbles must have been deposited by the Tanikawa stream, flowing along the north edge of the site (cf. Figure 14.9). The Miwa terraces are too high to have been flooded in the Yayoi period by the Hasé River flowing below them.

Knowledge of the timing of tectonic changes affecting the Nara Basin are important in determining erosional patterns and sediment sources. As documented in Chapter 13, terrace deposits derive from highland areas, and those deposits themselves can be redeposited again in the central Basin. The terraces and highlands have been uplifted by fault activity, while the central Basin alluvium is subject to compaction and subsidence as well as having undergone severe erosion during the LGM, e.g., Table 14.3 states that there is no information on forests in the Basin between 20,000 and 9,000 BC. Thus, all these factors must be taken into account when assessing the distribution of archaeological materials in and around the Nara Basin.

One element, however, that has become clear is that LGM sea level change has not affected the Basin. Riverine incision on the eastern Basin flanks was judged to be too far upstream from the nick point in the Kaménosé Corridor where the Yamato River flowed into the then Kawachi Bay; incision probably occurred during combined uplift of the Basin flanks and the lowering of the Basin floor during the

Holocene (Chapter 13). However, earlier Marine deposits have been identified in the Higher Terrace (th) and Middle Terrace (tm) deposits (cf. Table 13.2); the former contained Ma12 from the early Late Pleistocene and Ma11 from the late Middle Pleistocene. These areas thus must have been intermittently underwater before the Rokko Movement raised the surrounding Basin mountains through reverse faulting (Figure 13.5). This solves a question I had when initially pursuing this research, thinking the Basin terraces might have been marine terraces. However, by the time Palaeolithic peoples occupied Japan after 40 ka, the terraces had been formed and were high and dry for their occupation potential. Much of the highest terrace deposits, however, were eroded during the LGM, so if there were any artefactual remains up there, these now reside downstream in buried redeposited positions.

Black stratum

One of the most puzzling aspects of geological coring at the Miwa site was the discovery of a very thick (ca. 40 cm) layer of black sediments – much thicker than the usual ABS (Artefact-Bearing Stratum) or cultural layer usually found in archaeological sites. Upon further reflection, it should be considered whether this black layer was a buried andosol (cf. Chapter 7). Many tephra layers have been identified in Nara Basin stratigraphy (Nishida 1992), though none have been thick enough to account for *kurobokudo* formation. Nishida *et al.* (2015) report that *kurobokudo* soils comprise only 0.1% (37 ha) of Nara Prefecture, but this does not include buried soils. That *kurobokudo* exist in Nara at all is testimony to the possibility of thick tephra deposits. As a further qualification, both *Miscanthus* reeds (cf. Figure 7.8) and *nezasa* shrubby bamboo (cf. Figure 7.11) are characteristic of tephrogenic soils, as discussed in Chapter 7. That phytoliths of these plants dominated both the Yayoi and Medieval layers at Miwa (Kanehara 1993) supports the possibility that the black sediment was a *kurobokudo* soil.

Two candidates for the origin of such a thick tephra layer are gneissose granites underlying the site or pyroclastic sediments from a nearby volcanic eruption: Figure 7.1 indicates andosol distributions bordering on and into Nara Prefecture, while Cretaceous volcanics are spread from Mie Prefecture east of Nara to Osaka prefecture west of Nara (GSJ 2020) – and it must be remembered that the land surface was flatter then than it is now. The Rokko Movement (Chapter 5) interrupted this flatter distribution by faulted (tectonic) basin formation from the beginning of the Pleistocene; however, the terrace that Shikishima Church stands on is listed in the 'seamless' map (GSJ 2020) as belonging to K22_pan_a rock type (gneissose granite): the latest Cretaceous plutonics that belong to the Ryoke Belt granites (Apx 6). The gneissose granites and andosol distributions of Figure 7.1 match almost perfectly. Thus, if andosols (cf. Chapter 7) were derived from igneous processes east of Nara and are known up to the border of Nara Prefecture, then there is reason to believe that what the thick dark layers in Miwa site stratigraphy might represent are buried andosols (*kurobokudo*) on the flanks of Mt Miwa. A correction is in order here: Mt Miwa is not a volcano, as I misrepresented it in Barnes & Okita (1993), but it is part of the Ryoke batholith eroded in a cone-shaped form.

Another likely candidate for black soil formation is the Muro Pyroclastics, deposited in mid-Miocene (ca. 13–15 Ma, Nishida 1992: 20), probably from a caldera in the Chichibu Belt of the central Kii Peninsula (Harakawa 2013). Muro is located in the Hasé River drainage just 10 km east of Miwa; the pyroclastic layers are distributed from central Nara Prefecture to the Mie Prefecture border in an area NS 15 x EW 28 km, reaching 400 m depth maximum.

These findings provide further hypotheses for investigating the survival and distribution of buried *kurobokudo* soils on the flanks of the mountains surrounding Nara Basin. These soils may have been developed, as elsewhere in Japan, through the intentional firing of *Miscanthus*/shrubby bamboo stands (cf. Chapter 7) to provide better hunting grounds by prehistoric peoples. My 1983 study of site and artefact distributions revealed that Palaeolithic stone tools are found on the Late Pleistocene Lower terraces (tl) (cf. Figure 13.8), while Jomon stone tools and ceramics are typically found on the alluvial fans, particularly in the southeast and southwest. Sanukite lithic scatters are predominant around the andesite resources of Mt Nijo on the border between Osaka and Nara Prefectures. Thus, the upper regions of the Basin were exploited in these periods and may have entailed intentional firing of the flora, which contributed to the andolization of whatever volcanic deposits remained in these areas.

Conclusions

I was fortunate in being introduced to 'geology for archaeologists' early in my studies by Bill Farrand at the University of Michigan. This stimulated me to ask geological questions once I arrived in Nara for my dissertation fieldwork. Through introductions to geologists and historical geographers at Nara Educational University and Nara Women's University, I was able to create a geoarchaeological approach to Nara Basin finds that was unprecedented in research thusfar. Even so, a greater knowledge of the local geology (Ryoke Belt) (Ch3), the nature of tephrogenic soils (Ch7), the East Nara (Tōen) Fault System (Ch13), and micro-morphological analysis of soils, would have immensely benefitted the interpretations made by myself as Principal Investigator of the Asawa and Miwa Project results, even though specialists were involved in material identifications.

It is never too late to learn more about the region in which one works, and I thank all my professors and colleagues who assisted me in my Nara Basin research and continue to do so in my geological explorations.

References

BARNES, Gina L & Masaaki OKITA (eds) (**1993**) *The Miwa Project: survey, coring and excavation at the Miwa site, Nara, Japan.* BAR International Series 582. Oxford: Tempvs Reparatvm (in English and Japanese).

CDK (Chigaku Dantai Kenkyūkai) (ed.) (**1996**) *Cyclopedia of Earth Sciences.* Tokyo: Heibonsha (rev. ed.) (in Japanese with English title and index).

GSJ (Geological Survey of Japan) (**2020**) "Seamless digital geological map of Japan V2 (1:200,000)." [https://gbank.gsj.jp/seamless/v2/viewer/] (in Japanese and English).

HARAKAWA, Tomoya (**2013**) "Chemical analysis of Murō Pyroclastics and associated tuffs and consideration of their source (abstract)." *MA thesis, Kwansei Gakuin University* [https://kwansei.repo.nii.ac.jp/?action=repository_action_common_download&item_id=21291&item_no=1&attribute_id=20&file_no=1] (in Japanese).

KANEHARA, Masaaki & Hisashi OKUDA (**1993**) "Sourcings of pebbles in pollen sample #1709", pp. 111-112 in Barnes & Okita (eds) (1993).

NISHIDA, Ippei; Katsuya TAMADA & Isao TAKENAKA (**2015**) "Quantity of soil carbon storage in arable farmland in Nara Prefecture." *Bulletin of the Nara Agricultural Research Center* 46: 27-30 (in Japanese).

NISHIDA, Shiro (**1992**) "Late Cenozoic tephrostratigraphy and chronostratigraphy in the Nara Basin and its surrounding area." *Bulletin of the Nara University of Education* 41.2: 5-22 (in Japanese).

REYNOLDS, TEG & T MATSUZAWA (**1993**) "Flaked lithics at Miwa", pp. 123-132 in Barnes & Okita (eds) (1993).

YOKODA, Shuichiro; Kazumi MATSUOKA & Masuhiro YASHIKI (**1978**) "Miocene sediments of the Shigaraki and Yamato Plateau and their problems: Shigaraki and Yamato Plateau neotectonics research no. 1." *Chikyū Kagaku* 32.3: 133-150 (in Japanese).

VOLUME CONCLUSIONS

This volume has been aimed at geoarchaeologists who work in subduction zones, using Japan as a case study but appropriate to the Mediterranean, Caribbean, and New Zealand among other regions. Part I covered the formation of the Japanese landmass in a processual framework rather than as descriptive geology, while Part II reviewed the several approaches developed by Japanese archaeologists to the remains of hazards generated directly or indirectly by subduction zone processes: volcanic eruptions, earthquakes, and tsunami. Part III reviewed my personal geoarchaeological work carried out in conjunction with research into the formation of the first state in Japan in the Nara Basin.

Many of the lessons herein also apply to intra-continental areas that incorporate fossil subduction zones, such as northwestern North America, western China, or the Central Asian Orogenic Belt (CAOB). The resources are similar to current subduction zone resources, and the zones also represent areas where large faults separating geotectonic terranes can be reactivated, causing earthquakes. Compression involving Active Faults is a serious source of seismicity, as demonstrated in Chapter 9. Intra-plate seismicity is also caused by extensional rifting mechanisms, where a fault develops into a rift – exemplified by the Reelfoot Rift Zone giving rise to the New Madrid seismism that affects seven southern American states. Two other, less known, parallel inter-continental seismic zones are in the American west: the Rio Grande Rift, beginning close to the Wyoming border then down the western slopes of the Rockies into Mexico; on the west, the Intermountain Seismic Belt (ISB) running from Montana into Utah and Arizona. The three major seismic zones in Europe comprise the Anatolian Peninsula, the area west of the Black Sea, and surround the Adriatic Sea. Surprisingly, the Rhine valley is also at higher risk, together with Pyrenees. Earthquakes and seismic damage can be expected in these areas.

When rifts widen further, usually caused by underlying magmatic upwelling, the continental crust may split open to allow magma to form new oceanic plate material, as is happening in the Japan Sea Basin (Chapter 3) and Red Sea Basin. Before that final step, rift volcanics produce both low silica (basalt) and high silica (rhyolite) lava through volcanoes and other vent outpourings as in the Great African Rift System.

Such rifting magmatism is joined by intraplate hot-spot magmatism, accounting for the great igneous provinces of the Deccan and Siberian Traps, for example. Yellowstone Park is an active volcanic crater overdue for eruption as the latest addition to the string of volcanoes through the Snake River Plains emanating from the British Columbia Basalt Province. The San Francisco volcanic field, in Arizona at the southern edge of the Colorado Plateau, is also thought to be fed by hot-spot magma and is predicted to erupt again.

In conclusion, this volume argues for a far closer integration of geological processes into archaeological research beyond the usual sedimentary realm of most geoarchaeological work. Knowledge of tectonic forces and processes and how they condition the regional geology and geomorphology does far more than inform on local resources. Especially in subduction zone regions, the geology is far more complicated than that of mid-continent. While glaciation is not a factor in Japan, the effects of both direct and indirect Plate Tectonic processes in forming the Japanese landmass and the continuing capacity of those processes to affect it are huge determinants of the nature of archaeology there. Although Japanese archaeologists are generally trained as cultural historians (including myself), integrating geological coursework at an early stage would vastly increase understanding of the natural setting of archaeological sites and the potentialities of their being affected by tectonic processes; and it would allow a greater understanding of raw material resources available in their geological setting.

APPENDIX 1

Abbreviations

Certain abbreviations are used in this volume ideosyncratically. All century numbers are specified with a lower case 'c' after the number, e.g., 6c = sixth century. The Latin 'cf.', though it means 'compare' is used here to replace 'see' and designates figures and tables referred to out of numerical sequence.

For element abbreviations, see the Periodic Table (Apx 5: Fig. A).

For Japanese geological belt abbreviations, see Apx 6.

BC and AD are used here instead of BCE and CE, as they are more visually distinctive and since *both* are derived from the Christian calendar – which is irrelevant to East Asia and ancient Earth processes.

The list below consists of abbreviations occurring in the text but not necessarily in figure/table legends or bibliographies.

□ = in chemical formulae, the square stands for other minor elements that might be included
Å = Angstrom (one hundred-millionth of a cm)
AA = Atomic Absorption (spectroscopy)
AC = Accretionary Complex
AD = Anno Domani, the Christian era
AG = Arc Granite
AHRC = Arts and Humanities Research Council, England
AIST = National Institute of Advanced Industrial Science and Technology, Japan
aka = 'also known as'
AM = endomycorrhizal fungi
AMS = ^{14}C Accelerator Mass Spectrometry
AP = Arboreal Pollen
Apx = Appendix
As-B, As-C = Asama tephra
AT = Aira-Tanzawa tephra, also A-Tn
ATGL = Active Tectonics and Geomorphology Lab (Virginia Tech)
ATTL = Arima–Takatsuki Tectonic Line
ATWC = Alaska Tsunami Warning Center
AQF – Archaeoseismic Quality Factor
B&R = Basin-and-Range
B-SG = Beppu–Shimabara Graben
BABB = Back-Arc Basin Basalt
BAR = British Archaeological Reports
BC = Before the Christian Era
BFS = Brown Forest Soils
BGS = British Geological Survey
BP = Before Present (present = 1950)
B-SG = Beppu-Shimabara Graben
B-Tm = Baekdu (Paektu)–Tomakomai tephra
^{14}C = radiocarbon
c = century, e.g., 5c = 5th century
ca. = circa
cf. = refer to, used to flag out-of-sequence figure and table referents
cal. = calibrated (radiocarbon date)
Caltech = California Institute of Technology
CAOB = Central Asian Orogenic Belt
CAT = Computerized Axial Tomography
CCD = Carbonate Compensation Depth
CDMC = Central Disaster Management Council, Japan
CEDACH = Consortium for Earthquake-Damaged Cultural Heritage, Japan
ch., chs. = chapter(s); capitalized for this work (Ch), lowercase for others' works
Chigakudan = Chishitsugaku Dantai (Geological Commission), Japan
Chikyuken = Sōgo Chikyū Kankyō Kenkyūjo, same as RIHN
CIT = California Institute of Technology
cm = centimeters
cmolc = centimoles of charge: a unit for calculating exchange capacity for soils
CNMNC = Commission on New Minerals, Nomenclature and Classification (of the IMA)
COLS = Center for Obsidian and Lithic Studies, Japan
COSPEC = correlation spectrometer
CRD = Collaborative Reference Database, Diet Library, Japan
DART = Deep-Ocean Assessment and Reporting of Tsunamis
DAS = Durham Archaeological Service, UK
DCAN = Disaster Concerned Archaeologists' Network, Japan
DCRC = Disaster Control Research Center, Japan
DKP = Daisen Kurayoshi Pumice
dN = nephrite from dolomite, also D-nephrite
DOAS = differential optical absorption spectroscopy
DOI = Digital Object Identifier
DRE = Dense Rock Equivalent (of erupted tephra)
EAE = Earthquake Archaeological Effects
EDXRF = Energy Dispersive X- Ray Fluorescence
EEW = Earthquake Early Warning system, Japan
EM = ectomycorrhizal fungi; see also AM
EPMA = Electron Probe Micro-Analyzer
ESR = Electron Spin Resonance
FA = Fulvic Acids
FAO-UN = Food and Agriculture Organization, United Nations
FCS = Frequent Cool Small fires
FDMA = Fire Disaster Management Agency, Japan
FGDC = Federal Geographic Data Committee, USA
FIBS = Functional Interpretation of Botanical Surveys

FIL = Frequent Intense Large fires
FRI = Fire Return Interval
FTIR = Fourier-Transform InfraRed spectroscopy
Ga = (*giga annum*) = billion (thousand million) years *ago*; e.g., 4.6 Ga = 4600 mya (when the Earth formed)
GEBCO = General Bathymetric Chart of the Oceans
GGL = Great Gradient Line (separating east and west China)
GIA = Gemological Institute of America
GIS = Geographic Information System
GLB = the author, Gina L. Barnes
GNSS = Global Navigation Satellite System
GPa = gigapascal; pascal is a measure of force per unit area, defined as one newton per square metre
GPS = Global Positioning System
GSA = Geological Society of America
GSJ = Geological Survey of Japan
HA = Humic Acids
HERP = Headquarters for Earthquake Research Promotion, Japan
HI-net = High Sensitivity Seismograph Network, Japan
HMA = High-Magnesium Andesites
HP = High Pressure
hPa = hectopascal (measure of air pressure)
Hr = hour
Hr-AA = Haruna Arima Ash
Hr-FA = Haruna Futatsudake Ash
Hr-FP = Haruna Futatsudake Pumice
HS = *hakuhen-sentoki*
HVZ = Hohi Volcanic Zone
IAB = Island Arc Basalt
IBM = Izu-Bonin-Mariana (Trench)
ICS = Intermediate Cool Small fires
ICP = Inductively Coupled Plasma (spectroscopy)
ICU = International Christian University
IGCP = International Geoscience Program
Ik = Ikeda-ko tephra
IMA = International Mineralogical Association
INAA = Instrumental Neutron Activation Analysis
IOC = Intergovernmental Oceanographic Commission of UNESCO
I-STL or ITL = Itoigawa-Shizuoka Tectonic Line
ICG/ITSU = International Coordination Group for the Tsunami Warning System in the Pacific, now termed the ICG/PTWS
ICG/PTWS Intergovernmental Coordination Group for the Pacific Tsunami Warning and Mitigation System
IMA = International Mineral Association
IOC = Intergovernment Oceanic Commission of UNESCO
ICS = International Commission on Stratigraphy
ITIC = International Tsunami Information Center, USA
IUGS = International Union of Geological Sciences
IUGG-TC = International Union of Geodesy and Geophysics Tsunami Commission
IAVCEI = International Association of Volcanology and Chemistry of the Earth's Interior
JAS = Journal of Archaeological Science
JMA = Japan Meteorological Agency
JNES = Japan Nuclear Energy Safety Organization
K = Köppen index (also potassium)
K-CTL/KTL = Kashiwazaki–Choshi Tectonic Line
K-ah = Kikai Akahoya tephra
K-Ar = Potassium-Argon isotope series
KiK-net = Kiban Kyoshin Nework (of seismometers)
K-NET = Kyoshin Network (of seismometers)
ka = (*kilo annum*) = thousand years *ago*
Kashikoken = Kashiwara Institute for Archaeology, Nara
Kc-Sr = Kucharo Shoro tephra
Kc-Hb = Kucharo Haboro tephra
km = kilometres
Kr-M = Kirishima Goike tephra
Kr-UsA = Kirishima Ushinosune Ash
K-PR = Kyushu-Palau Ridge
LA-ICP-MS = Laser Ablation Inductively Coupled Plasma Mass Spectrometry
LGM = Last Glacial Maximum
LIDAR = Light Detection And Ranging, laser remote sensing method
LIP = Large Igneous Province
LP = Long-Period (earthquakes)
LVSEM-EDS = Low Voltage Scanning Electron Microscope-Energy Dispersive X-ray Spectroscopy
Kb-Ks = Kobayashi-Kasamori tephra (includes Sakura)
KLRG = Kanto Loam Research Group
m = metres
M = Magnitude (of earthquakes)
Ma = (*mega annum*) = million years *ago* (also mya)
MB = Metamorphic Belt
MEXT = Ministry of Education, Culture, Sports, Science and Technology, Japan
MIS = Marine Isotope Stage
MLITTJ = Ministry of Land, Infrastructure, Transport and Tourism of Japan
MOHO = Mohorovičić discontinuity
Mohs = the Mohs scale of mineral hardness, developed by Fredrich Mohs in 1773
MORB = Mid-Ocean Ridge Basalt
MMI = Modified Mercalli Intensity
msl = mean sea level
Mt = Mount(ain), equivalent to *-yama* in Japanese
MTL = Median Tectonic Line
MTZ = Mantle Transition Zone
mya = million years ago
Nabunken = Nara Research Institute for Cultural Properties, Japan
NAP = Non-Arboreal Pollen
NARO = National Agriculture and Food Research Organization, Japan
NASA = National Aeronautics and Space Administration, USA
NB = Nara Basin
NCB = North China Block

NCC = North China Craton
NCEI = National Centers for Environmental Information, USA
n.d. = no date (in published work).
NEIC = National Earthquake Information Center, USA
NGA = National Geospatial-Intelligence Agency, USA
NGDC = National Geophysical Data Center
NGDC/WDS = National Geophysical Data Center/World Data Service, USA
NIED = National Research Institute for Earth Science and Disaster Prevention, Japan
NIAES = National Institute for Agro-Environmental Sciences, Tsukuba (part of NARO)
Nichibunken = International Research Center for Japanese Studies
N-KTL = Niigata–Kobe Tectonic Line
NOAA = National Oceanic and Atmospheric Administration, USA
n.p. = no publisher (in published work)
NSF = National Science Foundation, USA
NTL = Novosibirsk Tsunami Laboratory, Russia
NWPTAC = Northwest Pacific Tsunami Advisory Center, Japan
OBE = Order of the British Empire
OCHA ROAP = Office for the Coordination of Humanitarian Affairs (United Nations) Regional Office for Asia and the Pacific
ODP = Ocean Drilling Project
OIB = Ocean Island Basalt
OLI = Operational Land Imager
On-Pm1 = Ontake #1 Pumice
OPS = Ocean Plate Stratigraphy
OSL = Optically Stimulated Luminescence
OxCal = radiocarbon calibration program of Oxford, UK
O-KTL = Oita–Kumamoto Tectonic Line
P/T = pressure/temperature designations of metamorphic conditions
P/T-jadeite = jadeite formed by pressure/temperature
P-jadeite = precipitate jadeite
PGPR = Plant Growth Promoting Rhizobacteria
pH = measure of acidity/alkalinity
PIC = Pacific Island Countries
PIXE = Particle Induced Xray Emission
PL = Probability of Liquefaction
PNSN = Pacific Northwest Seismic Network, USA
pp. = pages
ppm = parts per million
PTDB = Palaeotsunami Database, New Zealand
PTWC = Pacific Tsunami Warning Center, USA
QUAD = four minerals that have the major elements Ca, Mg, and Fe (cf. Figure 12.5)
*R = ratio of $Mg/(Mg+Fe^{2+})$ in nephrite
R-jadeite = replacement jadeite
RCS = Rare Cool Small fires
REE = Rare Earth Elements
RGAFJ = Research Group for Active Faults of Japan
RIHN = Research Institute for Humanity and Nature, Japan
RIL = Rare Intense Large fires
ROAP, see OCHA
SAR = Synthetic Aperture Radar
SCB = South China Block
SCC = South China Craton
SCSTAC = South China Sea Tsunami Advisory Center, China
sec. = section
s.g. = specific gravity
SIO = Scripps Institution of Oceanography
SK = Sakate tephra, Mt Sanbe
sN = nephrite from serpentine, also S-nephrite
SOM = Soil Organic Matter
SP = Spore Pollen
Spfa-1 = Shikotsu pyroclastic flow #1
SROC = Short Range Order Clays
ss = senso stricto
SSA = Seismological Society of America
SS-Az = Shishimuta Azuki tephra
SS-PNK = Shishimuta Pink tephra
Sz-11 = Sakurajima tephra #11
TAF-a = Tremolite–Actinolite–Ferro-Actinolite series
tf = tephric properties
TL = Tectonic Line
To-a = Towada-a tephra
TTL = Tanakura Tectonic Line
TTT = triple junction
TWC = Tsunami Warning Center
UCL = University College London
UNESCO = United Nations Educational, Scientific and Cultural Organization
unpg. = unpaginated (in published work).
U-Pb = uranium-lead radioactive decay series
URL = Uniform Resource Locator
U-series = uranium decay series
UTL = Ululun Tectonic Line (near the Korean Peninsula)
USDA = United States Department of Agriculture
USGS = United States Geological Survey
V = velocity
VAAC = Volcanic Ash Advisory Center
VEI = Volcanic Explosivity Index
VF = Volcanic Front
vi = vitric properties
VOIC = Volcanic Observations and Information Centers, Japan
VSJ = Volcanological Society of Japan
VT = volcano-tectonic
WAC = World Archaeology Congress
WI = Warmth Index
WinITDB = Windows platform for International Tsunami DataBase, Russia
XRF = X-ray Fluorescence

APPENDIX 2

Japanese Placenames & Periodizations

Table A Jomon periodization

Jomon phase	calibrated dates BP	calibrated dates BC
Incipient	15,000–12,500	13,350–10,550
Initial (Earliest)	12,500–7000	10,550–5050
Early	7000–5470	5050–3520
Middle	5470–4420	3520–2470
Late	4420–3200	2470–1250
Final (Latest)	3200–2920	1250–970 (SW Japan)
	3200–2350	1250–400 (NE Japan)
		600~ Satsumon (Hokkaido)
Epi- (Zoku-) Jomon		340 BC–700 AD (Hokkaido)

Dates of Jomon phases based on calibrated AMS ^{14}C dates: the phase dates below derive from more than 2000 AMS ^{14}C datings throughout the archipelago. The name of the period derives from the ceramics used: 'cord-marking' (*jōmon*) was a prominent motif. Note that Incipient Jomon began in the Pleistocene, but this scheme does not include any early finds of pottery before 15 ka (13,350 BC); these are assigned to the Palaeolithic.

Jomon subsistence was based on hunting, gathering, and fishing, and some horticulture. Northwestern and northeastern Japan are distinguished by the early arrival of wet-rice agriculture in the west, leading into the Yayoi period after 970 BC, and by the time lag of rice technology diffusion through Honshu by the end of Middle Yayoi. Jomon lifeways survived in Hokkaido through the Yayoi and Kofun periods but were sidelined by dry-field farmers using pottery who arrived from northern Honshu after ca. 600 AD to establish the Satsumon culture in Hokkaido.

Table B Yayoi–Kofun period divisions

Yayoi		Kofun	
Initial (Earliest)	1000–800 BC	Early	250–400 AD
Early	800–450 BC	Middle	400–500 AD
Middle	450 BC–50 AD	Late	500–710 AD
Late	50–250 AD	(Asuka)	552–710 AD

Note that the southwestern Yayoi were dependent on the newly imported wet-rice technology and used bronze tools and weapons. During the Early Yayoi, the northeast followed Zoku-Jomon lifeways, but agriculture was gradually introduced through the Middle Yayoi period. Dry-field crops were more important to the northeastern Yayoi, who also survived generally without bronze. The Kofun period saw the formation of the early state, Yamato. Scholars consider either from the 5c, after 552, or after 710 as the emergence of state society. There is currently a tendency to limit the Kofun (Old Tomb) period to 552, when Buddhism was established in Japan Islands and Chinese administrative structures began to be adopted – despite the tomb-building tradition continuing into the early 8c.

During the Kofun period, the northeastern farmer-hunters were known to the Yamato Court as *emishi*, and many became accomplished horse-riders who harried expansionist troops during the Nara and early Heian periods. Others migrated into Hokkaido after 600 AD and established the Satsumon culture.

Figure A Japanese Placenames & Periodizations

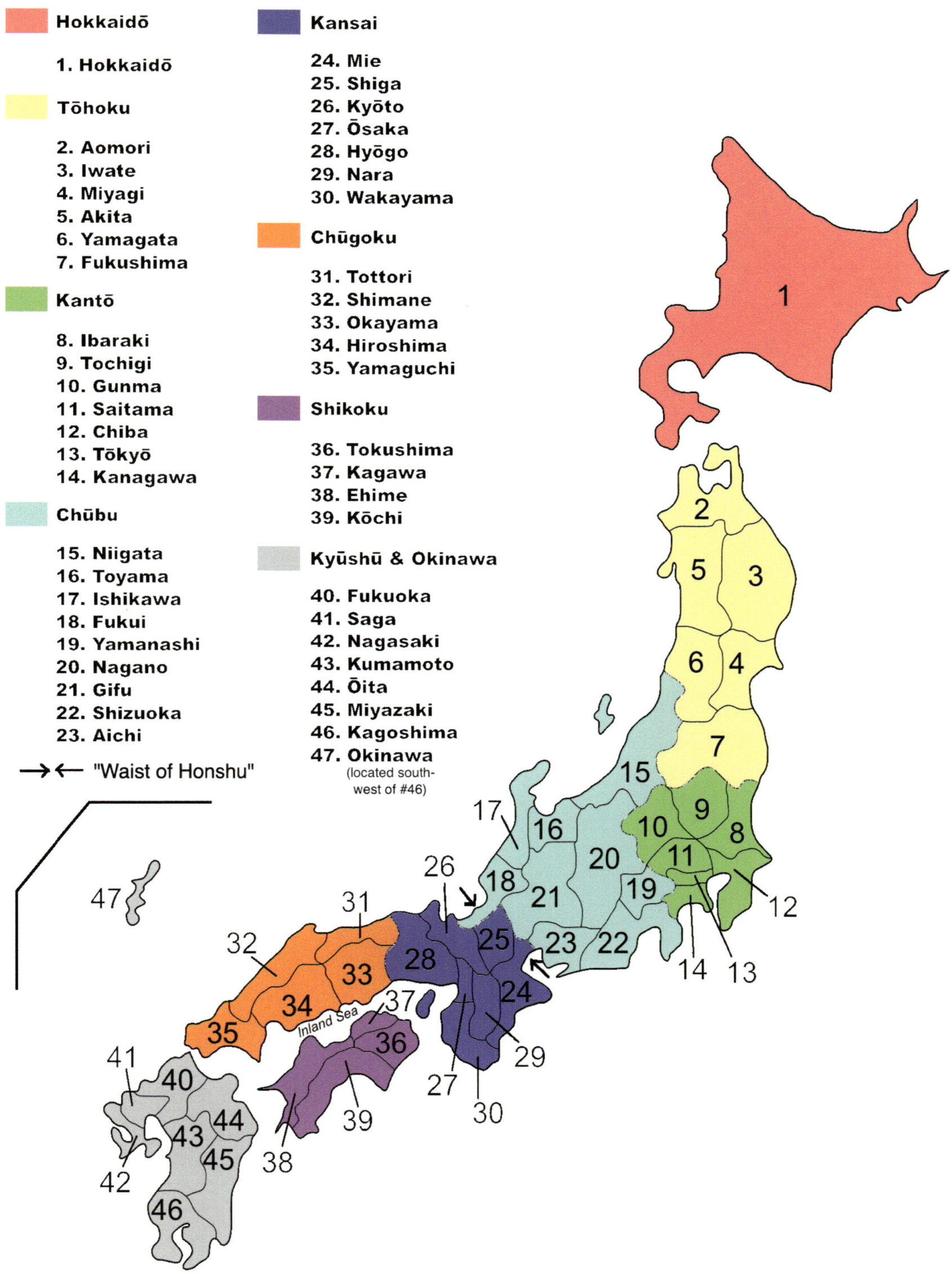

Source: by TheOtherJesse [Public domain], via Wikimedia Commons, modified by GLB [https://commons.wikimedia.org/wiki/File:Regions_and_Prefectures_of_Japan.svg]

APPENDIX 3

The Formal Timeline for Geological Chrono-stratigraphic Divisions

COHEN, KM; HARPER, DAT & GIBBARD, PL (2022). ICS International Chronostratigraphic Chart 2022/02. International Commission on Stratigraphy, IUGS. www.stratigraphy.org (visited: 25 MAY 2022)

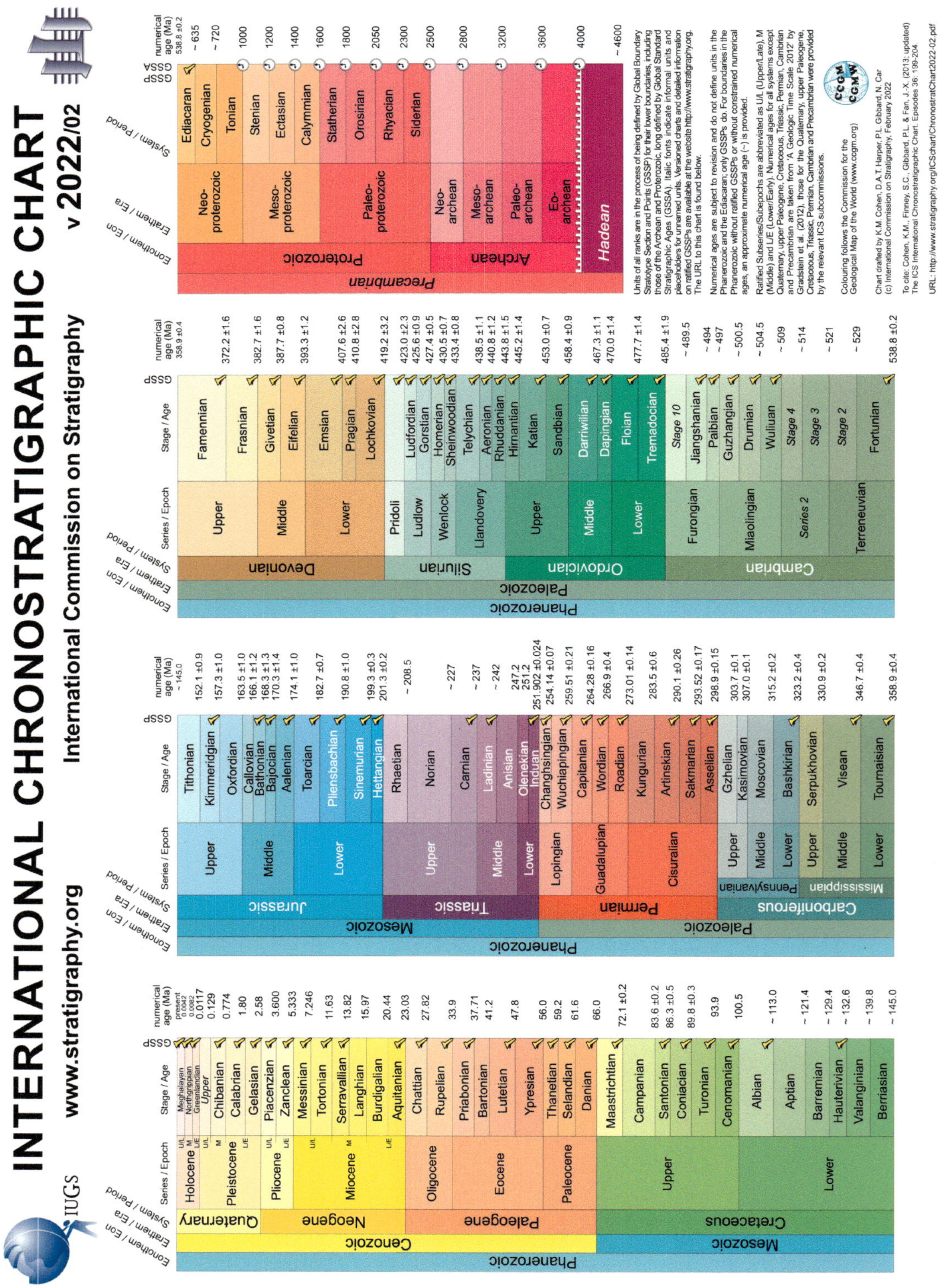

APPENDIX 4

Reading Geological Maps

Basic Considerations

Geological maps are different from topographic maps in that they attempt to illustrate bedrock and often tectonic movement affecting the area — such as fault slippage, direction of folding, etc. There are widespread conventions in illustrating these maps (Figure A), but national and regional variations can be seen. A chapter on "Geological Maps" in *Geology for Archaeologists* briefly describes how geological maps are made (Allen 2017).

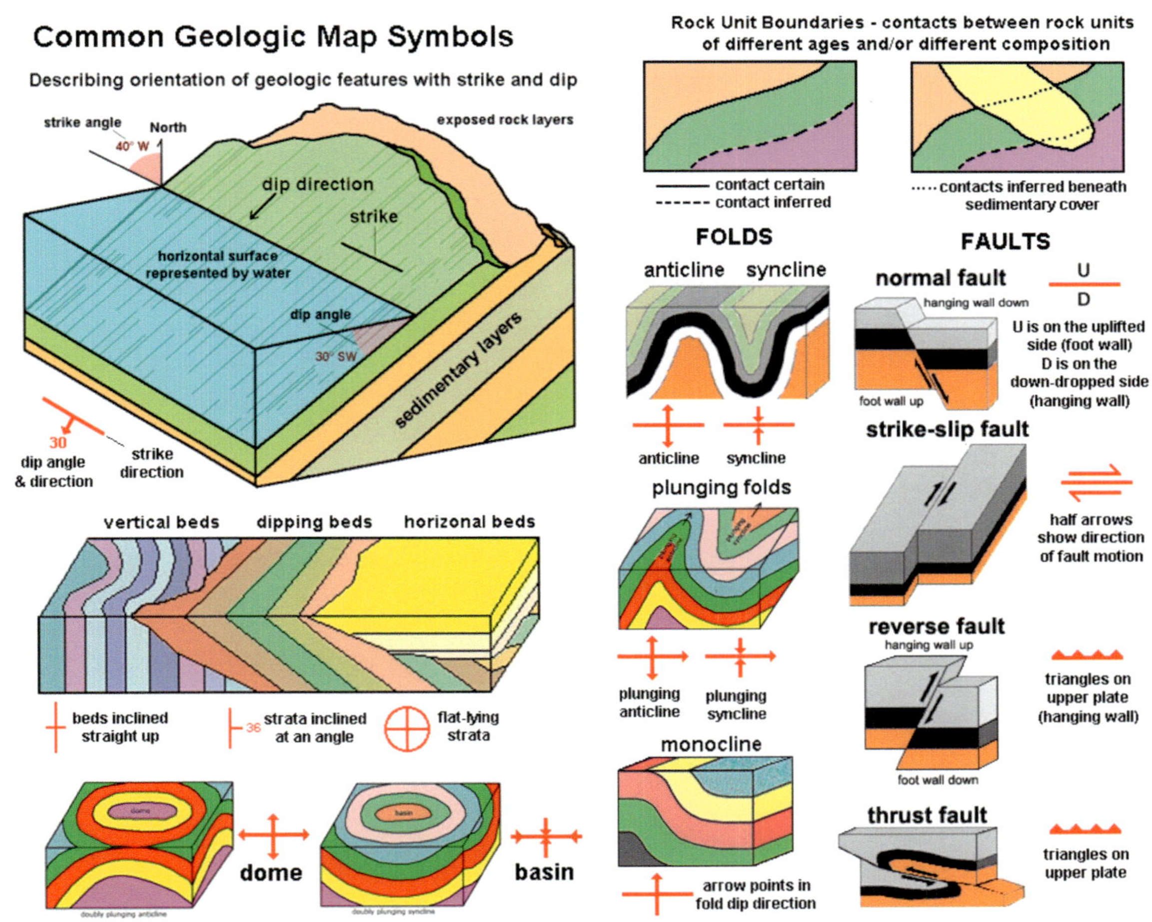

Figure A A simple introduction to geological map symbols

Gilmour & Widdowson recommend that bedrock currently exposed to the air be called an 'exposure', and buried bedrock to be considered an 'outcrop'. Even though a bedrock outcrop might be covered with surface sediments, it notionally could be viewed if all the "soil, vegetation and buildings [were] removed" (1999: 14). These *senso stricto* definitions, however, are often not followed; in archaeological usage, 'exposure' is termed an 'outcrop'; such outcrops are important sources of rocks and minerals for artefact production — obsidian being a prime example. Even within some geological writings, exposures of bedrock on the surface are termed outcrops.

It takes certain skills and knowledge to figure out how to read geological maps (Maltman 1998). This is due to the necessity of both understanding symbol usage that is different from topographical maps and having knowledge of the chronological processes behind formation and deformation of the geological strata. Since it is tectonic movement and its associated processes that largely creates and shifts around the rock units, an understanding of Plate Tectonics is extremely helpful in interpreting geological maps. The reason buried bedrocks are considered 'outcrops' *sensu stricto* is that they are often layered but tilted, so that only the edges of a certain stratum outcrops among the others, as shown in Figure A for vertical and dipping beds. One of the primary tricks in reading a geological map is to understand the underlying structure of such outcrops and particularly how outcrops of the same stratum are related to each other through unrelated topographic space. The illustrations of anticline & syncline folds, basins, and domes in Figure A are examples of the complex relationships that are not visible from the exposure surface.

National Resources

There is great variation in the amount of information that is presented on a geological map. Sometimes only the period of those units is given; others are designated by Group and Formation names without listing their contents. Rock units other than igneous are usually presented in terms of hierarchically constructed Supergroups > Groups > Formations > Members > Beds (from highest rank to lowest) (BGS 2020; GSJ 2001). Both sedimentary and metamorphic rocks fall into this 'lithostratigraphic' scheme but not igneous rocks. The Geological Survey of Japan has also published its standards for research and survey (GSJ 2001). Not all countries have agencies responsible for geological survey (Otto 1995); the maps of only three will be mentioned here.

In 1996, a map book of the Japanese island geology was published in full colour and accompanied by a CD (Editorial Group 1996). This has now been superseded by a 'seamless' online geological map of Japan at 1:200,000 scale [https://gbank.gsj.jp/seamless/v2/viewer/], based on existing data from several sources (previous maps, ongoing research); it is offered interactively online in English (GSJ 2020*a*) and was used here in Chapter 13, with the legend separately accessible (Apx 15). The geological information there is superimposed on a topographic map with the usual urban information; the opacity of the former can be adjusted to see where the geological units are positioned in relation to landmarks on the topographic maps. Toggle options show fault & boundaries, and legend symbols. Lat/longs and 'heights' are listed for the cursor location. In the 'simple' representation, dates are given to the Era level (cf. Apx 4), and general rock types are listed as igneous, sedimentary, metamorphic, and Accretionary Complex. If further detail is needed, one can switch to the 'original' in Japanese with fuller data on time period to the Epoch level and a description of the rock types. The legend box is not adjustable, so if the legend is not fully readable, descriptions of the icons and geological content can be found in the online document [https://gbank.gsj.jp/ seamless/v2/ legend.html] or in Appendix 15 here. However, the rock divisions visible in the online map are not classified into Geological Belts or Terranes as presented in Figure 3.1 herein, making it challenging to relate the rocks shown online to the simplified geological map Figure 3.1. An illustration of this 1:5 million map can be seen in Wakita (2013: fig. 1). Traditional geological paper maps, including a Tectonic Series, are available from GSJ (n.d.).

For the Quaternary alone, the Japan Association for Quaternary Research (JAQR 1991) has published a paper map series covering:

- Ia,b,c Landforms, geology and tectonics (1:1,000,000) three sheets, ed. by JAQR (1987)
- IIa Active faults and seismicity around the Japanese Islands (1:3,000,000) one sheet, ed. by RGAFJ (1991)
- IIb Prehistoric remains and palaeogeography [environment] (1:4,000,000) one sheet, ed. by JAQR (1987)

Parts I and IIb, which were previously published before being included in this collection, have their own explanatory text (JAQR 1987). Of course, working with these older maps, one must be aware of revised datings and interpretations since then, but the data are generally sound though archaeological information has greatly increased.

Finally, the Geological Survey of Japan has compiled a map of Active Faults (GSJ 2020*b*) (see Ch. 2.6.3) now online. As with the 'seamless' map described above of geological rock types, the Active Faults map also has an English version. However, the button to access this is currently off the screen to the right. There are a few tricks to using the map even in Japanese. The 'search' facility presents with an open menu on the left side; close it by clicking on the menu icon in the upper lefthand corner. Then there are two blue buttons on the righthand side; click the 'map' button and experiment with the options. The zoom facility operates from levels 1 to 13, but within a level, the screen can be enlarged with the fingers. At any level, click on a red fault line and a box will open with the fault code number, name, publications, and risk level; at the bottom of the box are blue letters for 'more details' that will take you to details about the fault, some of which, for example, are compiled here in Table 13.3.

The British Geological Survey (BGS) now has an interactive map at 1:650,000 scale that describes both superficial deposits and bedrock, individually or together for any chosen pinpoint (BGS ongoing). A new 3-D version is also available from that map site that has a zoom facility for larger scales. Map symbols and legends are published in Mawer (2002), available for download.

The United States Geological Survey (USGS) has integrated vast numbers of paper maps from different State sources into their viewable database (USGS ongoing). These have been constructed over time at different resolutions and detail. As a guide to reading them, the "FGDC digital cartographic standard for geologic map symbolization" prepared by the USGS for the Federal Geographic Data Committee is downloadable from (USGS & Geologic Data Subcommittee 2006). This series of maps is entirely different from the two interactive ones described for Japan and Britain. They have not been integrated into a single whole and have no interactive facility. The most one can do with the maps is view the original paper maps as they were published. In addition to the figure above, Dawes & Dawes (2011) give a simple explanation of geological maps from an American perspective.

Figure Sources

Figure A public domain, courtesy of Phil Stoffer [http://geologycafe.com/images/ map_symbols.jpg]

References

ALLEN, JRL (**2017**) *Geology for archaeologists: short introduction*. Oxford: Archaeopress.

BGS (British Geological Survey) (**2020**) "Lithostratigraphic hierarchy - explanatory note." [www.bgs.ac.uk/products/digitalmaps/lithostrat.html].

---- (**ongoing**) "Geology of Britain viewer." [http://mapapps.bgs.ac.uk/geologyofbritain/home.html].

DAWES, Ralph L & Cheryl D DAWES (**2011**) "Geology 101—Introduction to physical geology: basics table — geologic map symbols" [https://commons.wvc.edu/rdawes/G101OCL/Basics/BscsTables /geomapsymb.html].

Editorial Group (ed.) (**1996**) *Computer graphics, geology of Japanese Islands*. Tokyo: Maruzen (in Japanese).

GILMOUR, Iain & Mike WIDDOWSON (**1999**) *Maps and landscape*. Milton Keynes, UK: The Open University.

GSJ (Geological Society of Japan) (**n.d.**) "Catalogue of geological maps." [www.gsj.jp/Map/index_e.html].

---- (**2001**) International stratigraphic guide. Tokyo: Kyoritsu (in Japanese).

---- (**2020*a***) "Seamless digital geological map of Japan V2 (1:200,000)" [https://gbank.gsj.jp/seamless/v2/viewer/].

---- (**2020*b***) "Active Fault database" [https://gbank.gsj.jp/activefault/search].

JAQR (Japan Association for Quaternary Research) (**1987**) *Quaternary maps of Japan, explanatory text*. Tokyo: Tokyo University Press (in English)

---- (**1991**) *Quaternary maps of Japan*. Tokyo: Tokyo University Press (in Japanese).

MALTMAN, A (**1998**) *Geological maps: an introduction*. Chichester, UK: Wiley.

MAWER, CH (**2002**) *Cartographic standard geological symbol index, version 3*. British Geological Survey Research Report, RR/01/01. Keyworth, Nottingham: British Geological Survey.

OTTO, James M (**1995**) "National geological surveys: policies and practice." *Resources Policy* 21.1: 27-35.

RGAFJ [Katsudansō Kenkyūkai] Research Group for Active Faults in Japan (ed.) (**1991**) New Edition Active Faults in Japan - distributions and explanations. Tokyo University Press (in Japanese and English).

USGS (**ongoing**) "The national geologic map database." [https://ngmdb.usgs.gov/ngmdb/ngmdb_home.html].

USGS & Geologic Data Subcommittee, FGDC (**2006**) *FGDC digital cartographic standard for geologic map symbolization*. FGDC Document Number FGDC-STD-013-2006. Reston, VA: Federal Geographic Data Committee.

WAKITA, K (**2013**) "Geology and tectonics of Japanese Islands: a review - the key to understanding the geology of Asia." *Journal of Asian Earth Sciences* 72: 75-87 (in English)

APPENDIX 5

Elements, Minerals & Rocks

Elements

98.5% of the rocks of the Earth's crust are made up of only eight elements (Table A: column 1), which are the basic constituents of magma. Oxygen and Silicon are the most abundant, and when combined, they form silicon dioxide (SiO_2), commonly called silica, the basis of all silicate rocks formed from magma. See the Periodic Table (Figure A) for element abbreviations.

- O and Si are non-metal elements belonging to Groups 16 and 14 in the Periodic Table respectively, though Si is sometimes referred to as a 'metalloid'.
- Al is a metal of Group 13 and Fe is a transition metal (Group 8).
- Mg belongs to the alkaline earth metals (Group 2)
- Ca, Na, K are alkali metals (Group 1).

The elements are often distinguished based on their densities, with lighter elements (L) <5 g/cm^3 and heavier elements (H) >5 g/cm^3, though this distinction may vary according to author.

Minerals

The main elements of the Earth's crust combine to form minerals: the six most common are called the rock-forming minerals (Table A: column 6), characterizing igneous rocks (MEMPR 2017). Since there are approximately 5300 known mineral species (and thousands more varieties),[1] these six rock-forming minerals comprise only the first stages of rock formation from magma, with other minerals forming through other inorganic, metamorphic, or biological processes. Most of the rock-forming minerals are based on silica and therefore called 'silicates', derived from magma. The other major rock groups are sedimentary and metamorphic; they have their own common mineral types (see Davidson n.d.) such as the 'carbonates' ($CaCO_3$, $MgCO_3$), which form calcium or magnesium limestones. Metamorphic minerals form as existing rocks undergo unique pressure/temperature conditions or from fluid interaction with existing rocks (metasomatism) (cf. Ch. 2.5, Figure 2.8).

Of the six rock-forming minerals, quartz is a single mineral, silica (SiO_2) which may occur in different forms: as quartz crystals formed of silica tetrahedrons (see illustration in Table B) or non-crystalline (amorphous) glass. Quartz crystals grow in slow-cooling igneous rocks (granite) and in pegmatite formations. Volcanic glass is encountered in high-silica lava flows (forming obsidian) where the silica has not had the chance to crystalize. It is also the primary constituent of tephra, though other mineral crystals and country rock fragments with different chemical constitutions may also be present in tephra. Moreover, the glassy component itself usually includes elements other than silicon; these are helpful in characterizing the tephra and contribute to soil formation (Chapter 7 herein). Minerals, like elements, are also divided into light and heavy – the latter with a density (specific gravity s.g.) ≥2.8–2.9 g/cm^3. Heavy mineral crystals occurring in tephra are also useful for characterization of individual tephras.

The order in which minerals precipitate or crystallize from magma is from mafic to felsic minerals, as represented by the Bowen series (Bowen 1928) (Figure B). The specific minerals formed depend on the elemental composition of the magma as seen in Figure D. Often there is silica left over, so it may form amorphous quartz and be extruded as a volcanic glass (e.g., obsidian), or it may form large crystals as a pegmatite. The weathering of minerals also occurs in this order (Goldich 1938) (Figure C), with quartz being the most durable.

Quartz, in the form of diamond, is also the hardest mineral, as represented in the Mohs scale (Figure D). The true jades are approximately the hardness of quartz (Table 12.2), i.e., Mohs 6–7. The scale is logarithmic, with gypsum and talc soft enough to be scratched with a fingernail. Other objects can be used in the field to test the hardness of minerals, as indicated by the US Forest Service in Figure D.

[1] Figures vary in the literature, some of it quite old, from 3000 to 5300; new minerals are constantly being discovered. The International Mineralogical Association (IMA) recommends mineral names and approves validity; their list numbers 5809 minerals as of May 2022 (IMA-CMNNM 2022).

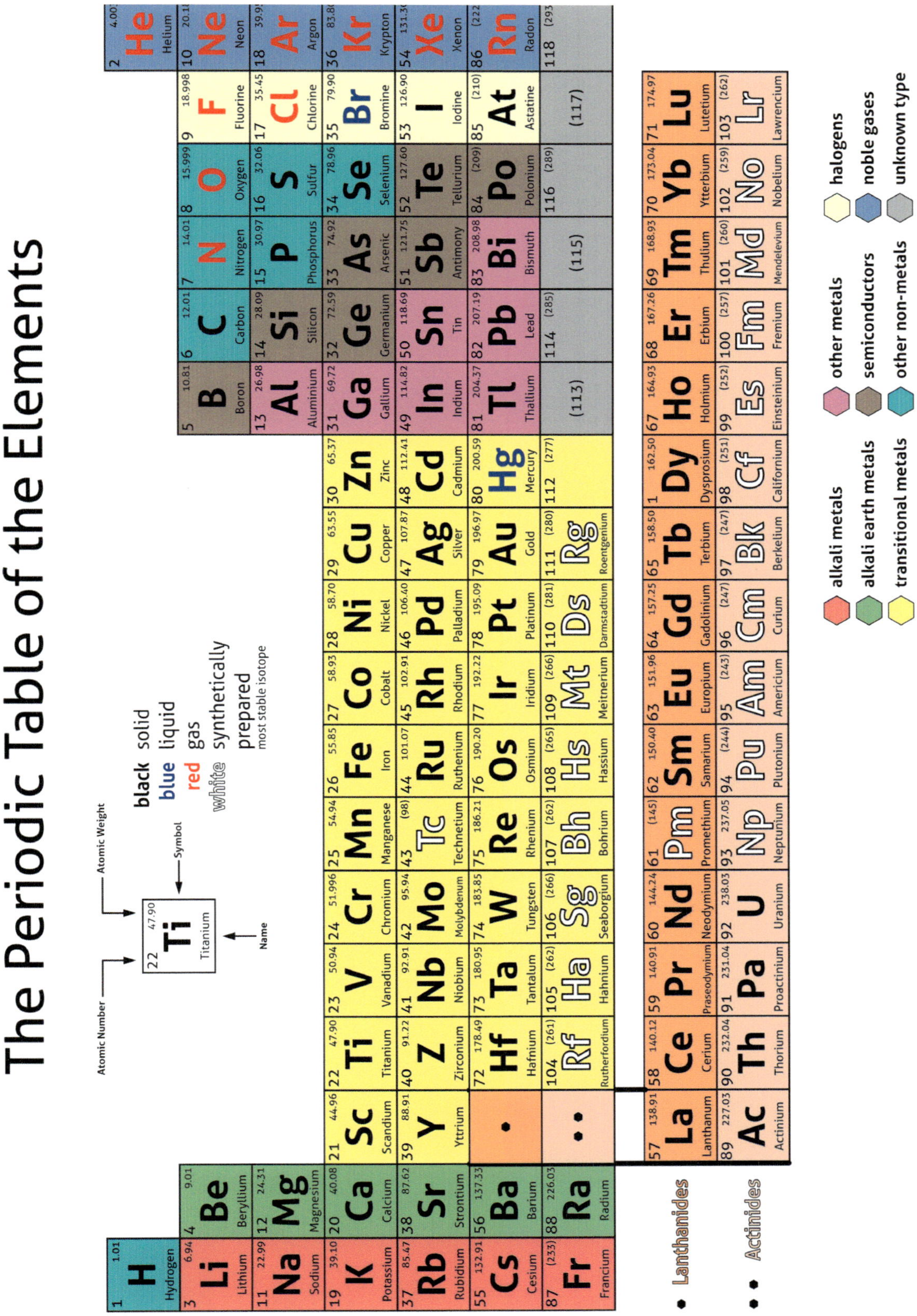

Figure A The Periodic Table of elements
(courtesy of Steve Smith - steve@destinysagent.com)

Table A Major elements and minerals discussed in the text

Key: L = light element, H = heavy element (the designations vary in the literature)
column 1 numbers: Group membership in the Periodic Table * Density given for O at 0° Celsius
column 2: (ai) = alkali metals, (ae) = alkaline earth metals, column 5: m = mafic, f = felsic

Earth's crust's eight major elements	Abbrev	Avg. % in magmas	Density g/cm3	**Rock-forming *minerals* /** MINERAL FAMILIES	Chemical composition specific gravity (s.g.)
oxygen (L, Periodic Table group 16)	O	46.6	1.429 g/l*	SILICATES	
silicon (L, 14)	Si	27.7	2.33	*quartz* (f)	SiO2 s.g. 2.6–2.7 (L)
aluminium (L, 13)	Al	8.1	2.7	FELDSPARS:	s.g. 2.6–2.7 (L)
iron (H, 8)	Fe	5.0	7.874	*orthoclase* (f) aka *K-feldspar*	$KAlSi_3O_8$ s.g. 2.6 (L)
calcium (L, 2)	Ca (ae)	3.6	1.55	*plagioclase* (f)	$NaAlSi_3O_8$–CaAl2Si$_2$O$_8$ s.g. 2.62–2.75 (L)
sodium (L, 1)	Na (ai)	2.8	0.968	PYROXENES: *augite* (m)	(Ca,Na)(Mg,Fe,Al,Ti)(Si,Al)2O$_6$ s.g. 3.2–3.6 (H)
potassium (L, 1)	K (ai)	2.6	0.856	*hypersthene* (m)	(Mg,Fe)SiO$_3$, s.g. 3.4–3.9
magnesium (L, 2)	Mg (ae)	2.1	1.738	AMPHIBOLES: *hornblende* (im)	Ca$_2$(Mg,Fe)4 Al[AlSi$_7$ O$_{22}$](OH)$_2$ s.g. 2.9–3.2 (H)
		tot 98.5		MICAS:	s.g. 2.8–3.0
For all the abbreviations of all elements, see the Periodic Table (Figure A)				*biotite* (m)	K(Mg,Fe)$_3$ AlSi$_3$O$_{10}$(OH)$_2$ s.g. 2.7–3.4 (~H)
				muscovite (f)	KAl$_2$ AlSi$_3$O$_{10}$(OH)$_2$ s.g. 2.8–2.9 (~H)
				OLIVINES (m):	(Mg,Fe)2SiO$_4$ s.p. 3.3–4.4 (H)
				APATITES	Ca$_5$(PO$_4$)$_3$(F,Cl,OH) s.g. 3.1–3.3 (H)

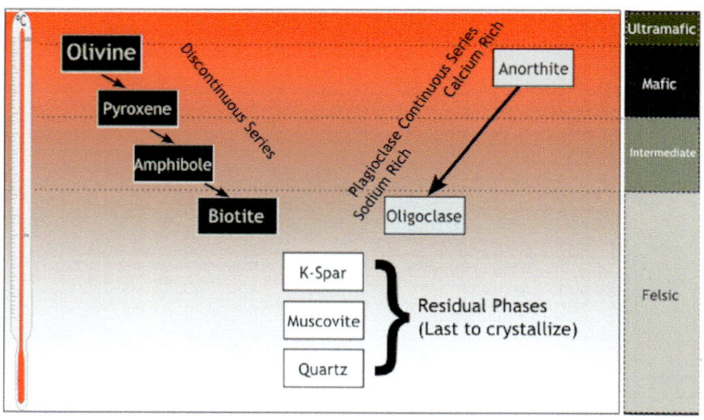

Figure B The Bowen Reaction Series of mineral crystallization

Olivine
Calcic plagioclase
Augite
Calci-alkalic plagioclase
Hornblende
Alkali-calcic plagioclase
Alkalic plagioclase
Biotite
Potash feldspar
Muscovite
Quartz

Figure C Mineral stability series in weathering

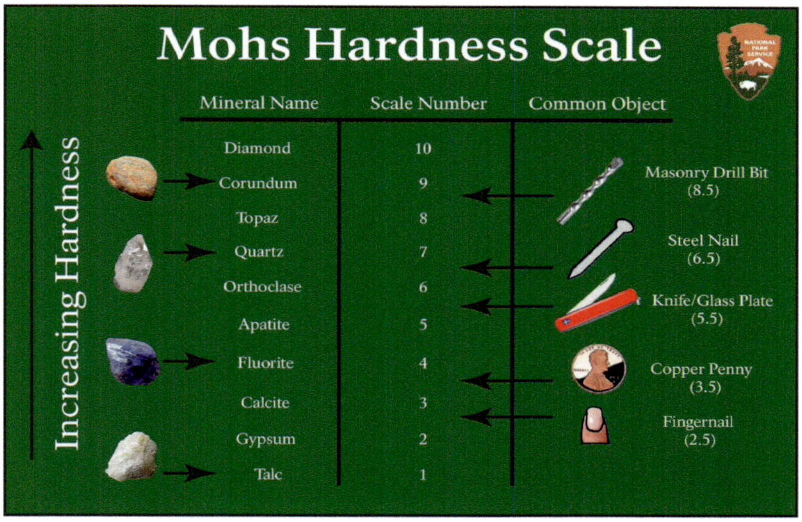

Figure D The Mohs scale of mineral hardness

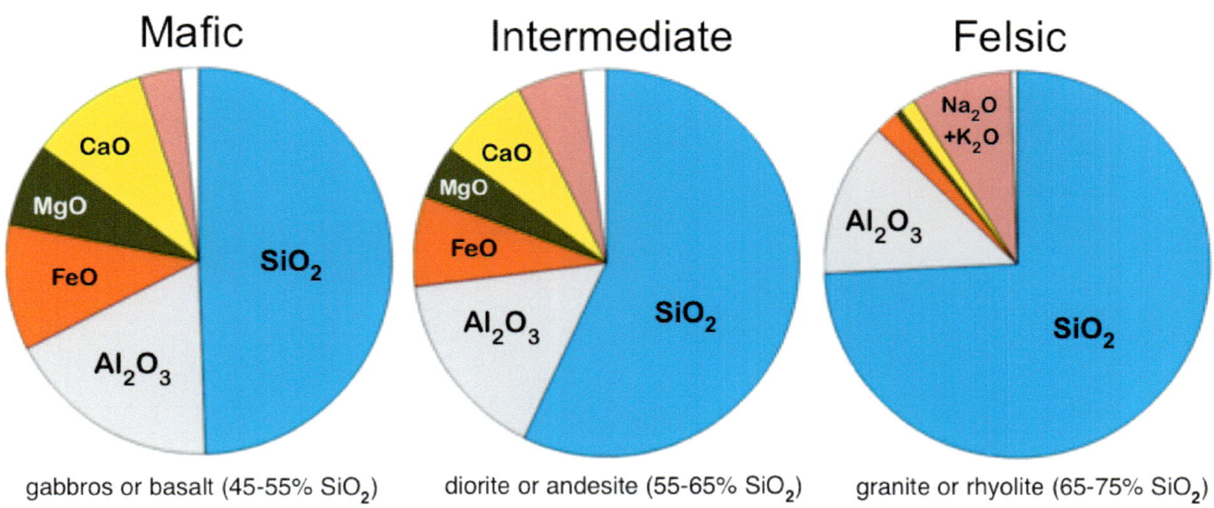

Figure E Magma composition according to silica content

Magma Types

Magma is basically molten rock from which minerals crystallize in order over time (Figure B). A magma body may also be mixed, with different sources supplying magmas of different chemical compositions. Table A: column 3 lists averaged percentages of the main elements comprising magma, but each of the main types have specific silica ranges (Figure E, Table B). The compositions in these pie charts are presented as oxides (combined with oxygen) in silica ranges from 45 to 75%; silica content below 45% characterizes ultramafic rock, and if the silica content is greater than 75% silica, the product is glass.

The silicate minerals form via linkages between silica tetrahedra (Table B: illustration); a silicon atom has a positive charge (4^+), while oxygen atoms have a negative charge (2^-). Thus, one silicon atom can link to four oxygen atoms and balance its electric charge; however, each oxygen atom is still negatively charged (1^-) and it is therefore wont to link to other ions including other oxygen atoms of other silicon tetrahedra. The latter thus form chains of different shapes; clays typically have a layered structure, as illustrated in Figure 7.4, incorporating the silicon tetrahedra (Table B below).

Igneous Rocks

Table B gives the silica percentages for classification of igneous rocks. The types of rocks formed in these categories (Figure F) can occur as either extrusive (extruded through volcanoes or fissures onto the Earth's surface) or intrusive (crystallize in magma chambers within the Earth's crust). The exception is the 'ultramafic' category, which refers to viscous mantle rocks composed mainly of peridotite and its derivatives (lherzolite, dunnite).

One classification scheme of *only* volcanic rocks is based on the relative proportions of alkali metals vs silica (SiO_2) (Figure G), dividing volcanic rocks into Alkaline and Sub-Alkaline. Alkaline rocks are particularly relevant to the kinds of volcanic activities that accompanied the rifting of the Japanese landmass in the Miocene (see Chapter 4; Apx 9), and they still characterize the western zone of the modern Volcanic Fronts (Figure 6.2).

Table B Classification of igneous rocks by silica (SiO_2) content and grain-size

Grain size	Rock type				Context
	felsic	intermediate	mafic	ultramafic	
Fine <0.25 mm	Rhyolite	Andesite	Basalt		Extrusive (volcanic)
Medium 0.25–2 mm	Microgranite	Microdiorite	Dolerite		Intrusive (plutonic)
Coarse > 2 mm	Granite	Diorite	Gabbro	/Peridotite	Intrusive (plutonic/mantle)
wt.% SiO_2	75%	65%	55%	45%	
Silica content	<--------------increasing-------------decreasing------------->				

silica tetrahedron construction:

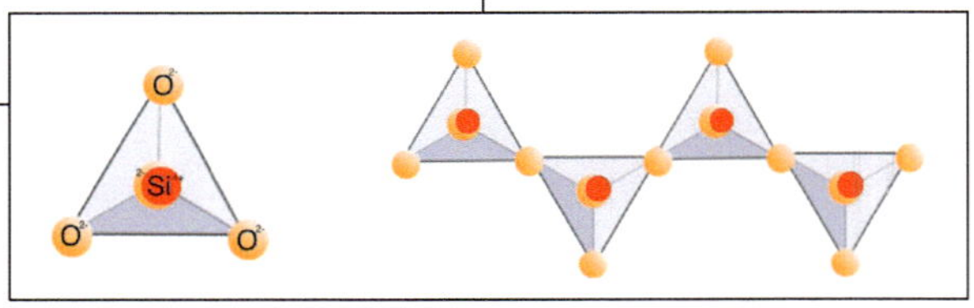

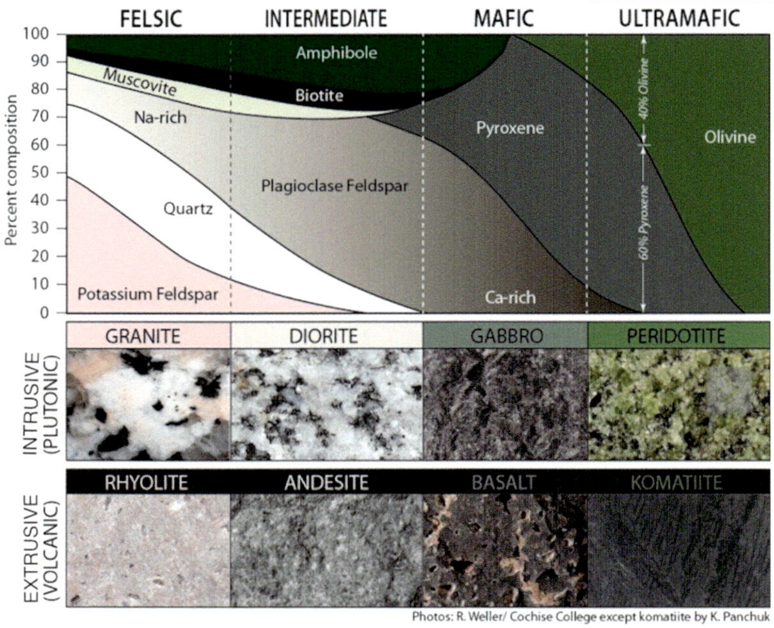

Figure F Igneous rock classification by silica content and their minerals

It should be noted that the two systems mafic–intermediate–felsic and basaltic–andesitic–rhyolitic overlap but are not equivalent. The former refers to the mineral complex in a particular silica range, whether the rock is plutonic or volcanic; while the latter refers to volcanic rocks of a certain chemical composition. Rock on the borderlines can be characterized as, e.g., 'basaltic andesite'. And the same adjectival forms exist for the plutonic rocks as well, e.g., 'gabbroic', etc. Granitoid is a catch-all term for different granites, though the term 'granitic' is used in this volume.

The Sub-Alkaline distribution is often broken down further into two sub-divisions: tholeiitic and Calc-Alkaline rocks. However, the category 'calc-alkaline' is difficult to define, having a convoluted history; Sheth *et al.* (2002: 686) accept Middlemost's statement (1985: 118) that:

> most petrologists would...agree that the volcanic rocks of this association consist of a comagmatic suite of subalkalic silica-oversaturated rocks, and that: (a) they tend to contain more Al_2O_3 than the normal rocks of the tholeiitic association; (b) their intermediate members [andesites] do not normally develop any significant enrichment in iron; and (c) orogenic andesite is the most characteristic member of the association.

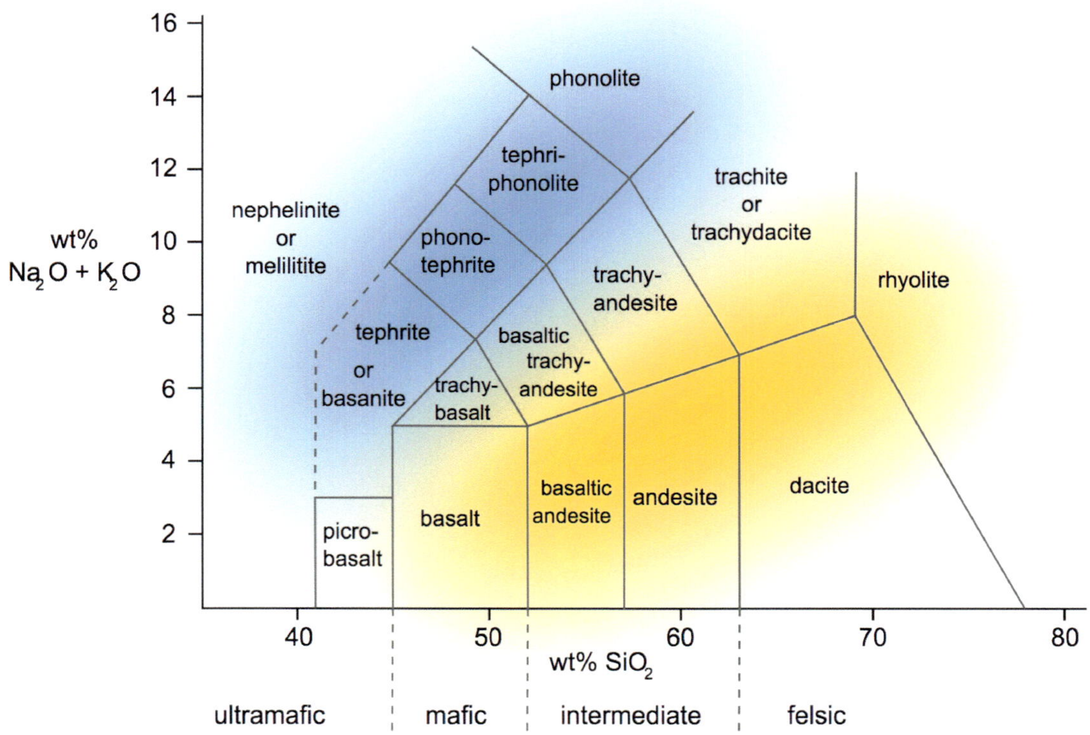

Figure G Modified IUGS classification of volcanic rocks

X-axis by silica content: ultramafic to felsic
Y-axis by sodium + potassium content

Blue = alkaline rocks
Yellow = sub-alkaline rocks

Experimental work has shown that the reciprocal ratios of Al and Fe to SiO_2 are influenced by the amount of H_2O in the magma: higher Al levels are correlated with higher H_2O levels in magma, and higher H_2O levels lead to repressed Fe levels (Tatsumi & Suzuki 2009: fig. 9). Although it has been general practice to identify calc-alkaline volcanic rocks as subduction zone products – in contrast to alkaline rocks which are generally regarded as extensional products (e.g., Fitton 1987) – Sheth and colleagues argue that calc-alkaline rocks can occur in extensional contexts as well; thus, they warn that "not all orogenic andesites are calc-alkaline, and not all calc-alkaline andesites are orogenic" (Sheth *et al.* 2002: 697).

The fact that calc-alkaline rocks, whatever their genesis, tend to have high Al and low Fe content is extremely important for ceramic manufacture (cf. Tite et al. 1992). The difference between earthenware and stoneware is that the high Al/high H_2O ratios allow clays derived from calc-alkaline deposits to be fired above 1000°C without slumping/melting, while iron in earthenware clays acts as a flux, causing the clay to 'melt' at 1000°C. Stoneware hardens through the new formation of mullite minerals based on the Al content: mullite formulae $3Al_2O_32SiO_2$ *or* $2Al_2O_3SiO_2$.

Finally, a word must be said about the difference between alkali and alkaline elements. Group I elements of the Periodic Table are known as the Alkali Metals (Li, Na, K, Rb, Cs, Fr) (Figure A above); individual members are known as 'alkalis'. Group II elements are the Alkali(ne) Earth Metals (Be, Mg, Ca, Sr, Ba, Ra). Despite the label alkali(ne) belonging to Group II, 'alkaline volcanics' usually refer to the presence of Group I elements Na and K (e.g., Figure E above); the additional presence of Ca from Group II thus requires the prefix 'calc'-alkaline.

Figure & Table Sources

Figure A courtesy of Steve Smith - steve@destinysagent.com

Figure B by Colivine (Own work) [CC0=Public Domain], via Wikimedia Commons [https://commons.wikimedia.org/wiki/File%3ABowen's_Reaction _Series.png]

Figure C after (Goldich 1938: table 18)

Figure D courtesy of the National Parks Service, USA, Public domain [https://www.nps.gov/articles/mohs-hardness-scale.htm]

Figure E & F from (Panchuk 2017: fig. 7.11), some font replaced for readability; distributed via CC-by-4.0 International License ("Read Panchuck 2017 for free" at [https://openpress.usask.ca/physicalgeology/] (required attribution for figure in Table B and Figures D and E, but link has since become inaccessible). [https://creativecommons.org/licenses/by-nd/ 4.0/deed.en _GB] (dead link); revised in 2021 as [https://paleolimbot.github.io/physical-geology/igneous-rocks.html] as fig. 7.10. Rock photos by special permission of Roger Weller (13jan'20)

Figure G by Woudloper – Own work, CC BY-SA 3.0, after Le Maitre 2002 [https://commons.wikimedia.org/w/index.php?curid=10311666]

Table A compiled by author

Table B compiled by author; illustration from (Panchuk 2017) with required attribution as above.

References

BOWEN, Norman Levi (**1928**) *The evolution of the igneous rocks*. Princeton, NJ: Princeton University Press.

DAVIDSON, Cameron (**n.d.**) "Common rock-forming minerals." [www.people.carleton.edu/~cdavidso/Geo110/CommonMinerals.pdf].

FITTON, JG & GJ UPTON (**1987**) "Introduction", pp. ix-xiv in *Alkaline igneous rocks*, ed. by FITTON & UPTON. London: Geological Society of London.

GOLDICH, Samuel S (**1938**) "A study in rock weathering." *Journal of Geology* 46.1: 17-58.

IMA – CNMNC (International Mineral Association – Commission on New Minerals, Nomenclature and Classification) (**2022**) "The new IMA list of minerals – a work in progress – updated: May 2022." pdf available at [http://cnmnc.main.jp/].

MEMPR (Ministry of Energy, Mines and Petroleum Resources) (**2017**) "Common rock-forming minerals." [www.empr.gov.bc.ca/Mining/Geoscience/PublicationsCatalogue/InformationCirculars/IC1987-05/Pages/contain.aspx].

MIDDLEMOST, Eric AK (**1985**) *Magmas and magmatic rocks: an introduction to igneous petrology*. Harlow, Essex: Longman Scientific & Technical.

PANCHUK, Karla (**2017**) *Physical geology* [https://openpress.usask.ca/physicalgeology/] (dead link).

SHETH, Hetu; Ignacio S TORRES-ALVARADO & Surendra P VERMA (**2002**) "What Is the 'Calc-alkaline Rock Series'?" *International Geology Review* 44: 686-701.

TATSUMI, Y & T SUZUKI (**2009**) "Tholeiitic vs Calc-alkalic Differentiation and Evolution of Arc Crust: Constraints from Melting Experiments on a Basalt from the Izu–Bonin–Mariana Arc." *Journal of Petrology* 50.8: 1575-1603 [DOI: 10.1093/petrology/egp044].

TITE, MS; GL BARNES & C DOHERTY (**1992**) "A technological study of earthenware and stoneware from southern Korea", pp. 64-69 in *Science and Technology of Ancient Ceramics: Proceedings of the International Symposium*, ed. by SRSSTAC. Shanghai: Shanghai Research Society of Science and Technology of Ancient Ceramics.

APPENDIX 6

Linked to Chapter 3

Major Geological Belts of Japan

Modified from Isozaki *et al.* (2010: fig. 1); data from Moreno *et al.* (2016) incorporated in sans serif font, other sources identified with icons linked to endnotes; for detailed descriptions of these geological belts/terranes, see Wallis *et al.* (2020).

# = the fifteen most important belts named in Isozaki (1996), as presented in Barnes (2003: fig. 9);
Ga=billion years; Ma=million years; '=' means 'equals' or 'is the same as'; AC=Accretionary Complex;
MB=Metamorphic Belt; OPS=Ocean Plate Stratigraphy; sed=sediments; *ss* = *sensu stricto*; meta=metamorphics;
HP=high pressure; S in Belt name=South; N in Belt name=North
{Dates} numerals = metamorphic dates in column 4; letters = period dates
{Dates} (see Apx 3 for period timespans): Paz=Palaeozoic, Mez=Mesozoic; ∈=Cambrian, O=Ordovician, S=Silurean, D=Devonian, C=Carboniferous, P=Permian, Tr= Triassic, J=Jurassic, K=Cretaceous, TT=Tertiary, Pg=Palaeogene, Ng=Neogene, Mc=Miocene, e=Early, m=Middle, l=Late

BELT (#)	ABBREVIATION/NOTES [BELT ABBREV][1] = 'same as'/'related to'	NATURE {DATES} = period	{DATES} meta-=metamorphosed date; for metamorphic 'facies' see Figure 2.8
ABUKUMA 阿武隈	Ab=Ry= Abukuma+Gs[2]	{J} AC + {eK} granites	meta-AC
AKIYOSHI #5 秋吉	Ak	AC, basalt, limestone, chert, tuff, *trench fill sed:**	*{P-T}† w/ {eC-mP} exotic limestone**meta-{lP} weak, greenschist
CHICHIBU #11+13 秩父	[Ch] outlier of M-T; +Kr Cs #11=Nchichibu, Cn #13=SChichibu	AC{lPaz-mMez}; 3 sub-belts: Cs, Cn, Ks (klippe) pelagic: deep-water chert>clastics; shallow-water limestone-dolomite, basalts; mélanges	
CHIZU 智頭	Cz; separated from Suo Belt by zircon dates	AC {T} MB silty sandy greenschist	meta-{180Ma} HP greenstone-glaucophane schist
GOSAISHO 御斉所	Gs=Ry	AC {J} granite batholith & metamorphic aureole	{K} meta-weak
HIDA #2 飛騨	Hd=Ok [HO]	continental fragment; *m-lPaz shallow marine/non-marine sed;* *granites {330-300, 270-250 Ma}‡* {C} protoliths, *but no {P*}*	{Pre-∈ (3.4Ga, 2.6Ga, 2.0-1.7Ga, 1.1Ga, 580Ma zircons); meta- {240–280Ma}
HIDA GAIEN 飛騨外縁	HG=Ng-Rn *part of SKitakami Terrane**	shelf facies, AC w/{∈–O} granite& metam rocks; eS tuff, limestone; {S-D} volcanics; {lD-eC} limestone	+Ōmi serpentinite mélange
HIDAKA 日高	Hdk=Sh [Hk]	2 ACs with opposite polarity metamorphic domain [Hk(m)]	{K}, partially meta- Hk(m)
HIGO 肥後	Hg includ. in Ryoke	Qinling-Dabieshan-Sulu suture fragments; {C-P} granites, limestone, {∈} volcanics	{3.7-1.9Ga} (detrital zircons) in {230-260 Ma} granites; medium meta-{230Ma}

BELT	ABBREVIATION/NOTES	NATURE {DATES}	METAMORPHIC {DATES}
HITACHI-TAKANUKI 日立-高貫	HT [Ht] (Ht-Tk) =Hg	northeastern fragment of Hg, {є} & {C} sed, igneous island arc rocks dating to {510–500 mya}; shelf facies possibly related to Sk	meta- medium
IDONAPPU イドンナップ	[Id] related to Kk	AC+mélange+Poroshiri ophiolite newly defined from [Hk]	strongly deformed
JOETSU 上越	[Jo=Ak+Mz]		{P}
KABATO-樺戸-REBUN-礼文	KR	newly defined; arc volcanics sub-belt in NK(Os)= M-T in Hokkaido	
KAMUIKOTAN 神居古潭	Kk=Sb [k]	AC >MB, sub-belt of SY, related to Id lower: {K} AC > HP/LT MB; upper: Horokanai ophiolite > meta- > forearc sed	meta-{K} HP/LT
KITAKAMI N – 北上, 北部 – OSHIMA 渡島	Nk-Os [O] =M-T, Ch	AC-m +[MM]Matsugataira-Motai metam + Miyamori-Hayachine ophiolite [Hy]	{J} Partially metam.
KITAKAMI S 北上, 南部	Sk	{Paz} thick shelf sediments, {440-410mya} arc granites *AC, part of SKitakami Terrane**, AC on {є–O} ophiolitic complex; {400mya} granite/ blueschist blocks in serpentinite/mudstone mélanges; {eS} tuff, limestone; {S-D} volcanics; {lD-eC} limestone	*{eO-lD *}*
KUROSEGAWA #12 黒瀬川	Kr=NR [Ks]	sandwiched in Ch; *part of SKitakami Terrane**; AC shelf facies on granite-metam basement;{ є} metam rocks, {O} granite rocks; {eS} tuff, limestone; {S-D} volcanics; {lD-eC} limestone; serpentinite/mudstone mélange	*{P*} {S-D} HP†*
MAIZURU #7 舞鶴	Mz closely related to UT; *island arc system°*	{P} collided intra-oceanic island arc, back-arc basin deposits, continental crust; Yakuno ophiolite; *trench fill sed.**	{P~T}, greenschist *{eO-lD*}*
MIHOSAN 三宝山	subset of Ch	AC + seamounts	{eK}
MINO 美濃–丹波 TANBA (ASHIO) #8	M-T=Ch [MT(As)]	{J}AC, fault-bounded chert–clastic stacks + mélange; {P} cap limestones on basalt; {P-J} OPS, {eJ} melangés; hotspot oceanic island setting for {T} basalts; tuff, trench-fill; *{P&T} exotic limestone**	{end T~J} *{eJ-eC*}*
MIYAMORI-HAYACHINE 宮守-早池峰	MH=NR [MM]?	Ophiolite + serpentinite mélange	{O} {480-470Ma} ophiolite; highP/ rocks in mélange
NAGASAKI 長崎	[Na]=Sb		
NAGATO 長門-蓮華 RENGE[3] #4	NR=Kr[Ks]	AC {O-D} > MB; Paz granite & blueschist blocks serpentinite/mudstone mélanges,includes Hida Gaien [HG] & Renge schist; {400mya} granitoid gneiss as tectonic blocks in west	{Paz} meta-{340-450Ma} lowT/highP; crystalline schists occur as tectonic blocks in east

BELT	ABBREVIATION/NOTES	NATURE {DATES}	METAMORPHIC {DATES}
NEDAMO 根田茂	Nd {P} part same as Ak	AC trench-fill sed;	{C-P}
NEMURO 根室	Nm	shelf sed; forearc sed with volcanics rotated 70-120° from continent	{lK-ePg}
OEYAMA #3 大江山	Oe	forearc ophiolite, oceanic crust	{580 mya}
OKI[2] #1 隠岐	Ok=Hd [HO]	continental fragment, silt, sand, *basalt, limestone‡* Jurassic granite intrusions	{Pre-Є} (3Ga, 2Ga, 1.7Ga) meta-{280-240Ma} medium
OSHIMA see Kitakami N			
RENGE 蓮華 #4	Rn=Re [orig. in Sangun Belt]	MB schists; regional HiPressure metamorphic belt associated with Oeyama Belt§	*{C-P}†*
RYOKE 領家 #9	Ry=Gs includ. older Hg; Younger Ryoke granites include San-yo granites	{J} M-T AC> lowP/T MB (20-30%)+ 80% granitoids: Older granite intrusions {>95-90mya} causing HighP gneisses; Younger granites intrusions {95-85/68 mya} causing weak contact metam	{102-98mya}highP + Hg medP +metam. aureole around {K} batholith
SANBAGAWA #10 三波川	Sb=[Kk] *s.s.;* previously incorporated Shm MB	AC>MB includ. *igneous plateau {lJ} fragments✣* [m] ophiolite, silty sandy greenschist covered w/eclogite schist *{P&T} exotic limestone**	AC={140-130Ma}+ meta-{mid-K, peak at 110-120Ma}+ lowT/highP greenschist~eclogite Mikabu greenstone
SANBOSAN 三宝山		protolith AC of Sanbagawa *ss* MB, now = SChichibu	
SANGUN #6 三郡	Sn [Sa] old name, now split into Renge (Re) schists and Suo (Su) schists		{220Ma} High P/T blueschist
SAN'YO 山陽		{K} igneous, north of MTL	
SHIMANTO 四万十	Shm, divided from Sb *s.s.* part of Sh-N that metamorphosed	{<80Ma} AC>MB silty sandy greenschist	meta-{K, 60Ma peak} lowT/highP greenschist~eclogite *includes Nomo ophiolite‡*
N SHIMANTO #14 北部四万十	Sh-N	*{lK}** +AC of unmetamorphosed pelagic-hemipelagic and trench fill sediments; mélange: sandstone mudstone (chert, greenstone)	+{100–70Ma AC} weak to strong metamorphism of part > Shimanto MB
S SHIMANTO #15 南部四万十	Sh-S (previously part of Sanbagawa MB)	unmetamorphosed {Pg} AC, southern end is {Mc} AC; sandstone, mudstone, (chert, greenstone); *{lC-eNg, no P}**	
SORACHI-YEZO 空知-蝦夷	[SY]=Sb, w/Kamuikotan Kk [k]sub-belt and Idonnappu [Id] sub-belt	{K} forearc zone sed over Poroshiri ophiolite sequence *{lJ} igneous plateau fragments & ophiolite✣*	*Sorachi greenstone✣*
SUO 周防	Su; includ. in Sangun *meta-Akiyoshi Belt°*	{P~T} AC *trench fill sed** silt, sand, AC>MB	*{eO-lD*}* meta-{240Ma} lowT/highP; greenschist ~glaucophane schist

BELT	ABBREVIATION/NOTES	NATURE {DATES}	METAMORPHIC {DATES}
TOKORO 常呂	Tk=Sb	MB	[K] lowT/highP
ULTRA-TANBA 超丹波	UT related to Mz	{m-lP} AC *{eO-lD} trench fill sed**	weak
UNAZUKI 宇奈月	*Hd eastern margin ‡*	*{300mya} ancient rifted continental shelf sandstone, mudstone, limestone, basalt‡*	med-meta

[1] Abbreviations in square brackets are used in Taira *et al.* (2016: fig. 1.6).

[2] Descriptions in sans serif font are derived from Moreno *et al.* (2016} and can be found via that book's index; apologies to authors for not citing titles and page nos.

ICONS: * Tazawa 2001; ° Wakita *et al.* 2018; † Domeier *et al.* 2016; ‡ Wakita 2013; Þ Watanabe *et al.* 2016; ✣Ichiyama *et al.* 2014; + Aoki *et al.* 2010; § Tsujimori & Itaya 1999 and Tsujimori 2017

References

AOKI, Kazumasa; Shigeru OTOH & Shuichi YANAI *et al.* (**2010**) "Existence of new independent regional metamorphic belt in the Sanbagawa Metamorphic Belt: orogenic evolution of Japan from Cretaceous to Tertiary." *Chigaku Zasshi* 119.2: 313-332 (in Japanese with English title and abstract).

BARNES, Gina L (**2003**) "Origins of the Japanese Islands: the new 'Big Picture'." *Japan Review* 15: 3-50.

DOMEIER, Mathew; Pavel V COUBROVINE & Trond H TORSVIK *et al.* (**2016**) "Global correlation of lower mantle structure and past subduction." *Geophysical Research Letters* 43.10: 4945-4953.

ICHIYAMA, Y; A ISHIWATARI & JI KIMURA *et al.* (**2014**) "Jurassic plume-origin ophiolites in Japan: accreted fragments of oceanic plateaus." *Contributions to Mineralogy and Petrology* 168.1: 1-24 (in Japanese with English title and abstract).

ISOZAKI, Y (**1996**) "Anatomy and genesis of a subduction-related orogen: a new view of geotectonic subdivision and evolution of the Japanese Islands." *The Island Arc* 5.3: 289-320.

ISOZAKI, Yukio; Shigenori MARUYAMA & Kazumasa AOKI *et al.* (**2010**) "Geotectonic subdivision of the Japanese Islands revisited: categorization and definition of elements and boundaries of Pacific-type (Miyashiro-type) orogen." *Chigaku Zasshi* 119.6: 999-1053 (in Japanese with English title and abstract).

MORENO, Teresa; Simon WALLIS & Tomoko KOJIMA *et al.* (eds) (**2016**) *The geology of Japan*. London: Geological Society of London.

TAIRA, Asahito; Y OHARA, Y & SR WALLIS *et al.* (**2016**) "Geological evolution of Japan: an overview", pp. 1-24 in *The geology of Japan*, ed. by T MORENO *et al.* London: Geological Society of London.

TAZAWA, Jun-ichi (**2001**) "Middle Permian brachiopod faunas of Japan and South Primorye, Far East Russia: their palaeobiogeographic and tectonic implications." *Geosciences Journal* 5.1: 19-26.

TSUJIMORI, Tatsuki (**2017**) "Early Paleozoic jadeitites in Japan: an overview." *Journal of Mineralogical and Petrological Sciences* 112: 217-226.

TSUJIMORI, Tatsuki & Tetsumaru ITAYA (**1999**) "Blueschist-facies metamorphism during Paleozoic orogeny in southwestern Japan: phengite K–Ar ages of blueschist-facies tectonic blocks in a serpentinite melange beneath early Paleozoic Oeyama ophiolite." *Island Arc* 8: 190-205.

WAKITA, K (**2013**) "Geology and tectonics of Japanese Islands: a review – the key to understanding the geology of Asia." *Journal of Asian Earth Sciences* 72: 75-87.

WAKITA, Koji; Takanori NAKAGAWA & Masahiro SAKATA *et al.* (**2018**) "Phanerozoic accretionary history of Japan and the western Pacific margin." *Geological Magazine* [DOI: 10.1017/S0016756818000742].

WATANABE, Yasushi; Tetsuichi TAKAGI & Nobuyuki KANEKO (**2016**) "Mineral and hydrocarbon resources", pp. 431-456 in *The Geology of Japan*, ed. by T MORENO *et al.* London: Geological Society of London.

APPENDIX 7

Linked to Chapter 2

APPENDIX 8

Geological Events Relating to Japan

Linked to Chapter 3

Ma = million years ago, Ka = thousand years ago, keyed to Chapters 2 and 3; period dates according to Apx 3; life forms according to Marshall, Michael (2009) "Timeline: the evolution of life." *New Scientist* 14 July 2009 [https://www.newscientist.com/article/dn17453-timeline-the-evolution-of-life/]

DIVISIONS	DATES	EVENTS (cf. Chapter 3)	LIFE FORMS
PRECAMBRIAN **4600-541 Ma**			
Archaean	4000–2500 Ma	• 3800 Ma oldest rock fragment in Japan	3.8 Ga beginning of life
			3.5 Ga single-cell organisms
			2.4 Ga Great Oxidation Event
			2.3 Ga Snowball Earth
			2.15 Ga photosynthesis
Proterozoic	2500–541 Ma	• 2000–1800 Ma age of Hida-Oki Belt detrital zircon datings, from NChina Block	?2 Ga eukaryotic cells
			?1.5 Ga plant, animal, fungi eukaryotes
		• 1000–900 Ma Yangtze and Cathaysia Blocks joined to form SChina Block [cf. Figure 3.3]	
		• 1000 Ma detrital zircon dates of SChina Block in Japan	900 Ma multicellular life
		• 900≥750 Ma Proterozoic supercontinent 'Rodinia'	800 Ma sponges
		• 875–720 Ma breakup of Rodinia; formation of proto-Pacific (Panthalassa Sea); liberation of Yangtze and Sino-Korean cratons	770 Ma Snowball Earth 680 Ma jellyfish
		• 600 Ma shift from passive rift to active subduction margin on Yangtze and Sino-Korean cratons	630 Ma bilateral symmetry
		• 600 Ma oldest continental granites in Japan (Hida Belt) [cf. Apx 10]	575 Ma Ediacarans
PALAEOZOIC **541–359 Ma**		• 520 Ma beginning of subduction with Nomo Plate 520–450 Ma [cf. Table 3.1]	570 Ma invert/vertebrate split
lower		• formation in Palaeozoic of Panthalassa Sea & Nedamo, Nagato-Renge, Kurosegawa & South Kitakami Belts [cf. Apx 6]	535 Ma Cambrian explosion 530 Ma true vertebrates
Cambrian	541–485 Ma	• first fossils from Japan	trilobites
		• >500 oldest ophiolite in Japan [cf. Table 3.2]	first land animals
Ordovician	485–444 Ma		489 Ma Great Ordovician Biodiversification Event
		• 480–310 accretion of island arcs to North China Block [cf. Figure 3.4]	465 Ma plants colonize land
		• 450–340 subduction of Renge Plate	460 Ma bony/cartilaginous fish
Silurian	444–419 Ma		
upper:			
Devonian	419–359 Ma		400 Ma insects & woody stems
Carboniferous	359–299 Ma	• 350 ridge subduction metamorphosing the Renge Belt	397 Ma 4-legged animals
		• 340–240 subduction of Farallon Plate	385 Ma oldest fossilized tree

		• 300 consolidated east China & Korean Peninsula [cf. Figure 3.5]	340 Ma amphibians
		• 320 Ma formation of Pangea	
		• oldest recognized unmetamorphosed AC, the Nedamo Belt [cf. Apx 6]	
		• 300 Ma ridge subduction and exhumation of Renge metamorphic belt; beginning of Farallon Plate subduction	
Permian	299–252 Ma	• formation of Akiyoshi, Maizuru, Ultra-Tanba Belts [cf. Figure 3.4; Apx 6]	250 Ma Permian mass extinction
		• accretion of Akiyoshi-Sawadani palaeo-seamount chain	ammonites & marine reptiles
MESOZOIC	252–66 Ma	• 240–120 Ma subduction of Izanagi Plate	210 Ma birds?
		• 240–225 Ma final suturing of NChina & SChina Blocks [cf. Figure 3.5]	
		• 200–180 Ma rifting and then break-up of Pangaea	200 Ma Triassic mass extinction
Triassic	252–201 Ma	• exhumation of Sangun metamorphic belt with passing of Farallon ridge	dinosaurs
			warm-bloodedness
Jurassic	201–145 Ma	• by 180 Ma break-up of Pangea	180 Ma egg-laying mammals
		• formation of Mino-Tanba, Ryoke, Chichibu Belts [cf. Figure 3.6; Apx 6]	168 Ma feathered, flightless dinosaur
		• 120–80 Ma Kula Plate subduction	150 Ma Archaeopterys, "first bird"
		• ca. 190 formation of Pacific Plate	140 Ma placental/marsupial split
		• accretion of Akasaka-Kuzū seamount cluster	130 Ma flowering plants
		• 146 Ma beginning of Japan's volcanic rock sequence	
Cretaceous	145–66 Ma	• formation of North Shimanto Belt [cf. Figure 3.7; Apx 6]	
		• accretion of Mikabu-Sorachi & Izanagi ocean plateaus	
		• 120–0 Ma tectonic quiescence with oblique plate subduction	100 Ma peak dinosaur time
		• 120–80 metamorphosing of the Sanbagawa Belt [cf. Apx 6)	93 Ma ocean anoxic event
		• 90–60 Ryōke igneous activity; exhumation of Sanbagawa Belt [cf. Figure 3.7 inset]	75 Ma primates and rodents split
		• 90 Ma Pacific Plate grows westward; forms present Pacific Ocean	70 Ma grasses appear
		• 60 Ma formation of Philippine Plate	65 Ma K/T extinction event
		• formation of Okhotsk Plate through trench jumping [Figure 3.7: inset]	Extinction of dinosaurs
CENOZOIC **65 Ma–present**			rise of mammals
Palaeogene	66–23 Ma		55 Ma Palaeocene/Eocene extinction
Palaeocene	66–56 Ma	• 56 Ma begin subduction of Pacific Plate	50 Ma start of whale evolution
Eocene	56–34 Ma	• 45 change of direction of Pacific Plate, NNW > WNW	40 Ma New World monkeys
		• 40–15 Ma migration north and rotation of Philippine Plate [cf. Figure 3.9]	
		• 40–45 Ma formation of Izu Arc at Philippine Plate edge [cf. Figure 3.9]	
Oligocene	34–23 Ma		25 Ma apes
Neogene	**23–2.68 Ma**	**(cf. Chapter 4)**	
Miocene	23–5 Ma	• early Miocene continental rifting	
		• subsidence of Tōhoku region	
		• 21–13 Green Tuff Movement (cf. Figure 4.4); bimodal submarine volcanism	
		• 19–15 Ma opening of Japan Sea Basin [cf. Figure 4.3], Japan & Kurile Arc formation	
		• 15 subduction of Philippine Plate and Izu Islands [cf. Figures 4.6, 4.7]	

		• Median Tectonic Line & Tanakura Tectonic Line formation	
		• 15 Ma back basin formation behind Izu Arc [Figure 4.1]	
		• 15 Ma assembly of Hokkaido parts; intrusion of Kurile Arc [cf. Figure 3.7]	
		• 13–5 Ma tectonic basin formation	
		• 13 Ma shift from extensional to compressional stress in NE Japan	
		• by 11 Ma Izu Arc collision	
		• 8 Ma emergence of Tōhoku as dry land	
		• uplift of Hida, Echigo & Hidaka ranges	7 Ma gorillas
		• 6–4 Ma extensional depression forming Palaeo-Inland Sea	6 Ma hominins
Pliocene	5–2.58 Ma	• Plio-Pleistocene volcanic eruptions	
Quaternary	**2.58 Ma–present**	**(cf. Chapters 5,6,7)**	
Pleistocene	2.58 Ma–11.7 ka		2 Ma *Homo erectus*
	2.58 Ma–700 ka (Early)	• formation of Fossa Magna & growth of Mt Fuji and Mt Hakone	
		• formation of Okinawa Trough	
		• 2/3s present-day relief uplifted in Quaternary	
		• 1 Ma northward tilting of SW Japan	
	700–13 ka (Middle)	• 700 ka beginning of Quaternary volcanics	200 ka *Homo sapiens*
		• 300–200 ka Rokkō movements [cf. Figure 4.10]	
	13 ka–present (Late)		extinction of megafauna
Holocene	11.7 ka–present		

Appendix 9 Chronology of Japan Sea Basin Rifting and Rift Volcanics[1]

Linked to Chapter 4

Red area = pre-rifting volcanics (55–25 mya), Green area = rifting phase (25–15 mya), Blue area = marine transgressions (53 mya, 16–15 mya). Row header numbers within chart give date (mya) for individual events; all datings are approximate, for guidance only, unless footnoted.

mya[2]	N. Kyushu-Chugoku-Shikoku	Kinai-Hokuriku-Kanto	Tohoku-W. Hokkaido
135–110		**Formation of Median Tectonic Line:** **beginning of left-lateral movement**	
120–65	Hiroshima Tectonics emplacement of the Hiroshima granite batholith and Ryoke granite in Ryoke Belt		
53	Cainozoic maximum high sea levels[2]		
50	55 Rifting begins subsidence of sedimentary basins Volcanic front: San'in to S. Korea deep magma/high alkali volcanics erupted on dry land = precursors of GreenTuff	Most active volcanics Second phase of MTL movement	
40–37 35	Effect of large-scale plate tectonics	Small-scale granite emplacement	Beginning of Pacific Plate subduction formation of two volcanic chains General uplift Volcanic extrusions on dry land
32–27	30 Philippine plate subduction opening Shikoku / Tsushima basins movement of SW Japan to south	32[3] Sado Tectonics: volcanic extrusions on dry land	granite emplacement in Hidaka Belt Okushiri Island basalts, andesites theoleiitic basalt north of Okushiri
25–20	Intrusion of Western Sea	28 Noto (andesite) 27 Nyu (rhyolite) 25 Kanazawa (rhyolite) Freshwater lake formation 23-21 Uplift, mafic volcanics	25 Half-graben faulting NNE-SSW Heavy erosion of uplifted land **Green Tuff Tectonics** Freshwater lakes Volcanic front moved 100 km E Plate subduction at higher angle and lower speed
	24 **Green Tuff Tectonics**	22 **Green Tuff Tectonics** 21 Formation of #1 Inland Sea and Kanto collapse basins	

20–18		Felsic volcanics erupt NW-SE Andesites in freshwater Rhyolite/dacite pyroclastic flows 18-14 Setouchi volcanics then uplift	Continued rifting **Marine transgression** E-W extension Massive subsidence except Kitakami / Abukuma mts Western deep rift basin fills w/ coarse gravels
18–17	Uplift and regression		
17–14	Rapid formation of sedimentary: basins and transgression Local volcanic activity : felsic volcanics fill basins basalt / dolerite intrusions Formation of Ryukyu Trough	Large sedimentary basins formed **Marine transgression** to form shallow sea Volcanic tuffs / pyroclastics 14 Disappearance of #1 Inland Sea	
16–14	Deepening and broadening of ocean floor beginning of Nishi-Kurosawa Transgression		
16–15	greatest extent of Neogene sea[3] deposition of marine layers		
15	**End of Japan Sea back-basin rifting**		
	NW Kyushu submerged Rapid plate subduction: felsic then mafic volcanics 15-14 andesitic volcanics 13 Sanuki andesites & alkali granites	Basalts at basin margin	Compression from ENE uplift of northern Hidaka Belt
		Folding / faulting of submarine sedimentary basins	alkali mafic igneous rocks assoc. with Chishima back-basin formation
12 10	12-11 high alkali basalts Uplift of Chugoku backbone regression of sea	Emplacement of granite plutons general uplift, esp. of Hida Range then cessation of uplift & erosion	colder, lowered sea levels
8		Uplift and erosion Kanto sedimentary centres move north	Present-day backbone volcanic chain formed & uplifted to dry land separating east/west Tohoku
7–5		Rhyolite, andesite & basalt lavas Formation from south of #2 Inland Sea in present trough Continued uplift on north and south	40 eruptions forming calderas
5	**End of Green Tuff Tectonics**		

[1] *Nihon no Chishitsu*, vol 1. (1990: 238-45); vol. 2 (1989: 244-52); vol. 4 (1988: 265-74); vol. 5 (1988: 238-42); vol. 6 (1987: 239-44); vol. 7 (1987: 223-9); vol. 8 (1991: 214-15); vol. 9 (1992: 315-17). Tokyo: Kyoritsu. [2,4] According to the composite marine $\delta^{18}O$ record for the Cainozoic (Wilson, RCL et al. 2000 *The great ice age: climate change and life*. London: Routledge & OU, fig. 4.15). [3] Okamura, Y. *et al.* 1995 "Rifting and basin inversion in the eastern margin of the Japan Sea" *The Island Arc* 4: 166-181.

APPENDIX 10

Select Granitic Belts, Plutons, and Batholiths Relating to Subduction Events

Linked to Chapter 6

Period	Event	Belt (location)	Granite name (rock type)	Date of granite Ma
____________540 Ma			Hida	600
Cambrian	*subduction begins*			520*
		(central Kyushu)	Hikawa (tonalite)	500*
		(north Kanto	Hitachi (meta-granitoid)	500*
		(Kitakami Mts)	Shoboji-Isawa-Horei (granitoids)	510–490*
____________485 Ma				
Ordovician		(WJapan Sea coast)	(arc granites)	450 Ma*
		Kurosegawa	Mitaki +5 locs (arc grantoids)	450 Ma*
		Abukuma	Hikami (arc grantoids)	450 Ma*
____________445 Ma				
Silurian				
____________420 Ma				437–
Devonian		Maizuru	Maizuru	–363 Ma
____________360 Ma				
Carboniferous		(east Kyushu)		ca. 300 Ma
		Yakuno Ophiolite	(plagiogranite)	285-280 Ma
____________250 Ma				
Triassic	*Farallon Plate subduction*			*250–180 Ma*
		S. Kitakami		250–180 Ma
		Hida, Oki	(plutono-metamorphic)	260–190 Ma
____________200 Ma	100 Ma			
Jurassic				
		Hida	Funatsu/Daebo	180 Ma
____________145 Ma				
Cretaceous	*Izanagi/Kula Plate subduction*			
		Ryoke	(plutono-metamorphic)	
			older (concordant)	L.Jur–E.Cret.
			younger (discordant)	L.Cret. 115–65 Ma
		(San'yo)	(volcano-plutonic)	110–70 Ma
		(San'in)	(volcano-plutonic)	110–30 Ma
		Abukuma	(plutono-metamorphic)	115–100 Ma
		Kitakami	(volcano-plutonic)	130–115 Ma
		(Kyushu)		115–100 Ma
	Hiroshima movement			100–70 Ma
			Hiroshima, Rokko, Suzuka	100–70 Ma
____________66 Ma				
Palaeogene		(San'in)		60–30 Ma
	Hokkaido assemblage			30-16 Ma
		Hidaka		30–16 Ma
			(see below for references)	

APPENDIX 11

Major Pre-Miocene Volcanic Rock Bodies

Linked to Chapter 6

Date	Name	Deposition (location)	Geochemistry	Rock type
Ordovician				
483, 465, 458–453* Ma				
	Hitoegane			tuff
>460 Ma*				
	Oeyama Ophiolite (Nagato-Renge Belt)			oceanic crust
>460 Ma*				
	Hayachine Ophiolite (South Kitakami Belt)			oceanic crust
460 Ma*		volcanics on top of above ophiolites		
470–460 Ma*		(WKyushu)		
Silurean-Devonian				
436, 360 Ma				
	Ofunato	volcanic ash	basaltic	schalstein (tuff)
		volcanic ash	felsic	tuff
		lava	basaltic	basalt
Carboniferous		(Kitakami Belt)		
333–290 Ma				
	Hikoroichi	lava, pyroclastic flows	basaltic	porphyrite tuff
	Arisu	lava, pyroclastics flows	basaltic	porphyrite tuff
	Ohdaira	lava, pyroclastics flows	basaltic	porphyrite tuff
	Nagaiwa	lava, pyroclastics flows	basaltic	porphyrite tuff
Early Cretaceous				
141–100 Ma				
	U. Kanmon	lacustrine		
		pyroclastic flow	andesitic-dacitic	tuffaceous sandstone
		pyroclastic flow	andesitic	ignimbrite
		lava	andesitic-dacitic	andesite, dacite
	Rikuchu	submarine		
		lava	andesitic, dacitic	andesite, dacite
		pyroclastic flow	andesitic, dacitic	
		lava	basaltic	basalt
Late Cretaceous Hiroshima Movement volcanics				
90–65 Ma				
	Nohi	pyroclastic flow	rhyolitic	welded tuff
		pyroclastic flow	dacitic	welded tuff
	Arima	lava	rhyolitic~rhyo-dacitic	rhyolite-dacite
		pyroclastic flow	rhyolitic~rhyo-dacitic	welded tuff, tuff, block tuff
	Abu	lava	rhyolitic~rhyo-dacitic	rhyolite-dacite
		pyroclastic flow	rhyolitic~rhyo-dacitic	welded tuff, tuff, block tuff
		pyroclastic flow	andesitic-dacitic	ignimbrite, tuff gravel
		lava	andesitic-dacitic	andesite, dacite

References

Appendix 10 compiled from:

ISOZAKI, Yukio (**1997*a***) "Contrasting two types of orogen in Permo-Triassic Japan: accretionary versus collisional." *The Island Arc* 6.1: 2-24 (in English).

---- (**1997*b***) "Jurassic accretion tectonics of Japan." *The Island Arc* 6.1: 25-51 (in English).

---- (**2019**) "A visage of early Paleozoic Japan: geotectonic and paleobiogeographical significance of Greater South China." *The Island Arc* 28.e12296: 17pp. (in English).

KIMURA, Toshio; Itaru HAYAMA & Shizuo YOSHIDA (**1991**) *Geology of Japan*. Tokyo University Press (in English).

MARUYAMA, Shigenori (**1997**) "Pacific-type orogeny revisited: Miyashiro-type orogeny proposed." *The Island Arc* 6.1: 91-120 (in English).

MIKI, Kozo & Masakazu FURUTANI (**1983**) *An illustrated guide to rocks and bedrock for the civil engineer*. Tokyo: Shikajima (in Japanese).

NAKAJIMA, Takashi; Masaki TAKAHASHI & Teruyoshi IMAOKA *et al.* (**2016**) "Granitic rocks", pp. 251-272 in *The geology of Japan*, ed. by T MORENO *et al.* London: Geological Society of London.

Appendix 11 compiled from:

ISOZAKI, Yukio (**2019**) "A visage of early Paleozoic Japan: geotectonic and paleobiogeographical significance of Greater South China." *The Island Arc* 28.e12296: 17pp. (in English).

KIMURA, Jun-ichi; James B GILL & Tomoyuki KUNIKIYO *et al.* (**2014**) "Diverse magmatic effects of subducting a hot slab in SW Japan: Results from forward modeling." *Geochemistry Geophysics Geosystems* 15: 691-739 [DOI: 10.1002/2013GC005132] (in English).

MIKI, Kōzō & Masakazu FURUTANI (**1983**) *An illustrated guide to rocks and bedrock for the civil engineer*. Tokyo: Shikajima (in Japanese).

MINATO, Masao; Hiroyuki TAKEDA & Toru HASHIMOTO *et al.* (**1962**) "On the volcanic rocks in the Japanese Paleozoic, first report – Gotlandian and Devonian." *International Geology Review* 4.4: 468-475 (in English).

MINATO, Masao; Hiroyuki TAKEDA & Makoto KATO (**1962**) "On the volcanic rocks in the Japanese Paleozoic, second report – Carboniferous." *International Geology Review* 4.5: 590-594 (in English).

NAKAJIMA, Takashi; Masaki TAKAHASHI & Teruyoshi IMAOKA *et al.* (**2016**) "Granitic rocks", pp. 251-272 in *The geology of Japan*, ed. by T MORENO *et al.* London: Geological Society of London (in English).

TAIRA, Asahito; Y Ohara, Y & SR Wallis *et al.* (**2016**) "Geological evolution of Japan: an overview", pp. 1-24 in *The geology of Japan*, ed. by T MORENO *et al.* London: Geological Society of London (in English).

APPENDIX 12

Clay Groups and Their Characteristics *

Linked to Chapter 7

CLAY GROUP Sub-group	Species	Struc Fig. 3	Shape	Chemical formula	Genesis	Alteration M=mont-morillonite
KAOLIN-SERPENTINE						
Kaolin	Kaolinite	1:1	platy	$H_2Al_2Si_2O_8$ H_2O or $Al_2Si_2O_5(OH)_4$ or $Al_2H_4O_9Si_2$	from volcanic glass, feldspars, phyllosilicates; warm, humid climate.	hydrates to form halloysite
Kaolin	Halloysite	1:1	platy, tubular, or spheroidal	$Al_2Si_2O_5(OH)_4$ -$2H_2O$ (interlayer water of varying quantities)	from volcanic glass, feldspars, phyllosilicates; warm, humid climate.	dehydrates to form kaolinite

SMECTITE			spongy, honeycomb		
Di-octrahedral smectite	Montmor-illonite**	2:1	flakes	$(Na,Ca)_{0.33}(Al,Mg)_2(Si_4O_{10})(OH)_2 \cdot nH_2O$ interlayer Na^+, Ca^{2+}, K+, Mg^{2+}	feldspar → sericite → illite→ M, plagioclase → sericite → M
Tri-octrahedral smectite	Saponite	2:1	ribbon-shaped +isometric particles	$Ca_{0.1}Na_{0.1}Mg_{2.25}Fe^{2+}_{0.75}Si_3AlO_{10}(OH)_2 \cdot 4(H_2O)$	
MICA					
Mica	Illite	2:1	lath or hexagonal	$(K,H_3O)(Al,Mg,Fe)_2(Si,Al)_4O_{10}[(OH)_2,(H_2O)]$	diagenic or hydrothermal
	Biotite	2:1	lamellae	$K(Mg,Fe)_3(AlSi_3O_{10})(F,OH)_2$	
AMORPHOUS					
	Allophane	gel	hollow nano spherules	$(Al_2O_3)(SiO_2)_{1.3-2}\,2.5\text{-}3H_2O$	
	Imogolite	gel	nano-tubes		

* The literature on clay types is convoluted and often contradictory; a simplified scheme is adopted here, but omitting vermiculite and chlorite

** also called bentonite (Slaughter, M & JW Early 1965. *Mineralogy and geological significance of the Mowry Bentonites, Wyoming.* GSA Special Paper 83. New York: Geological Society of America).

APPENDIX 13

Volcanic Soils Geochemistry[2]

Linked to Chapter 7

Physical Properties

In general, tephrogenic soils are valued for their friable texture due to their small grain size, low bulk density, good water retention, and good drainage. These latter two seem contradictory, but water retention is especially good in pumice-derived soils because the pumice grains are able to hold water film in their vesicles. Such pumice soils in Japan are called *miso-tsuchi* (coarse 'soybean-paste earth'), having both the colour and consistency of brown *miso*, a fermented soybean cooking ingredient. Being too porous can lead to rainwater leaching out soluble metals, while being too saturated can lead to water-logging and stickiness. One problem with water supply is posed by the presence of the non-crystalline clays allophane and imogolite. Their own chemical structures bind water (H_2O) or hydroxyl (OH^-) molecules, making the moisture unavailable to crops. One method of mitigation is to dehydrate the colloids by drying out the soil (e.g., by covering it over), which releases water from the colloids to the plants (Shoji *et al.* 1993: 236); colloids once dehydrated cannot be rehydrated (Neall 2006).

Friable texture depends on good water balance and the lack of large rock particles. When fine-grained tephrogenic soil is deep enough, long-rooted crops are well-suited for cropping, e.g., Japanese radish (*daikon*, ca. 30–35 cm in length) or edible burdock (*gobō*, 45–50 cm in length). This rootability depends on the absence of large stones or cemented layers; unweathered tephra glass shards, however, can prevent earthworm habitation (Neall 2006).

[2] This appendix was previously published in *TephroArchaeology in the North Pacific*, ed. by GL BARNES & T SODA (eds), pp. 294-299. Oxford: Archaeopress.

Humus & Humic and Fulvic Acids

Soil scientists make a distinction between soil organic matter (SOM) and humus (Pettit n.d.); SOM is the sum total of plant and animal matter added to soils naturally or artificially; this includes living micro-organisms and animals that live in the soil. Most soils contain less than 2% SOM, though to be fertile, a soil should have >2.8% SOM (Pettit n.d.). Humus is the decomposed organic matter that has lost all its cellular structure. It makes up about 65–75% of SOM, while the remainder is comprised of various carbohydrates, fats, waxes, alkanes, peptides, amino acids, proteins, lipids, and organic acids – things that Pettit helpfully describes as having chemical formulae, whereas there is no chemical formula for humus. A gardening rule of thumb is that humus should comprise between 5% and 10% of soils for good fertility (Bond 2017). Andosols can vary widely in their humus content: 3–11% for allophanic and 4–22% non-allophanic andosols. Both, nevertheless, are indicative of good fertility and are far above the average SOM contents of soils.

Humus can be further broken down into three components (Table A): insoluble humins, and the soluble acids humic acid (HA) and fulvic acid (FA). Humins in particular give andosols many of their unique characteristics, such as increased water capacity, soil stability, increased soil structure, and increased fertility.

The high humic content in andosols is due to humus stabilization, accomplished through the formation of Al-humus complexes or attachment to colloidal clays and ferrihydrite, a mineral ($5Fe_2O_3{\bullet}9H_2O$); in addition, humus can be physically protected within soil micro-aggregates, or chemically protected by the lack of P or absence of Al toxicity, both of which affect the micro-organisms that biodegrade organic matter (Ugolini & Dahlgren 2002). Under humid conditions and soil pH <5, humus is much more likely to form complexes with Al, making the aluminium unavailable for allophane formation (Shoji *et al.* 1993: 145). This is the anti-allophane effect referred to above.

Humic and fulvic acids comprise two of the three groups of humic substances that result from the decay of organic matter; these are soluble in water of different pH values, while the third humic group, humin, is insoluble. The acids are of interest here in soil formation.[3]

• Humic acids (Table A) are structurally diverse and complicated, consisting of carbon chains and rings. Their activities in soils include 1) binding with clay minerals to form stable organic clay 'complexes', 2) binding other organic chemicals in their molecular lattices, and 3) chelation: binding with inorganic elements in a manner utilizable by plants. Chelated elements are generally the micronutrients Fe^{2+}, Cu^{2+}, Zn^{2+}, Ca^{2+}, Mn^{2+} and Mg^{2+}. Note these are all metals with a valency of 2^+, while Al has a valency of 3^+, forming a more complicated chelate not only with humic acid but with other organic acids as well (see below under Al toxicity).

Four types of humic acid have been identified in Japanese tephra-derived soils (A, B, P, and RP) (see Shoji *et al.* 1993: fig. 6.10), with A-type humic acid being the most highly 'humified' and forming the dominant type in melanic andosols, while P-type are more common in fulvic andosols. A-type humic acid forms organic complexes with Al and Fe.

• Fulvic acids (Table B) are similarly structured of carbon chains and rings, but they are more chemically active than humic acids (because of their more numerous hydroxyl COH and carboxyl COOH groups).

Once the hydroxyl and carboxyl groups have expelled their hydrogen, then CO- and COO- are available to bind with free metal cations of 2+ valency, forming humate salts; thus HA and FA serve to preserve metals in the soil (Pettit n.d.) (FA not to be confused with Hr-FA tephra).

Andosol Nutrients

As we have seen above, andosol chemistry is first determined by type of magma (cf. Apx 5: Fig. E) deposited as tephra. Basaltic and andesitic tephras are noted for their productivity in Java and Japan. The general fertility of tephra-derived soils is most likely due to a continued high weathering rate of glassy materials which furnishes plant nutrients such as K and P (Shoji *et al.* 1993: 210). Plants need macronutrients (N, P, K), secondary nutrients (Ca, Mg, S) and micronutrients (B, Cu, Fe, Cl, Mn, Mo, Zn). Depending on their magma chemistry, volcanic products can contain all these except N, but they also contain significant amounts of Al, Na, and Io which can be toxic.

[3] Information mainly from Pettit n.d. unless otherwise stated.

Table A Humic substances, their relationships and characteristics

SOM soil organic matter *burning* → minerals, trace elements

decomposition

15–35%
NON-HUMIC SUBSTANCES *(rapid degradation)*:
= *main energy sources*
carbohydrates, fats, waxes, alkanes,
N in amino acids, nucleic acids and amino sugars
peptides, proteins, lipids, organic acids

65–85%
HUMUS = HUMIC SUBSTANCES (*slow degradation*):

insoluble:

Humins
- macro-organic
- most resistant to biodegration
- maintain soil stability
- improve: water-holding capacity, soil structure, fertility
- serve as cation exchange systems
- natural residence time in soil: 1140 years

soluble: :

Humic Acid (HA)
- weak organic acid
- soluble in pH>7 water
- binds to clay minerals
- forms plant useable salts with trace minerals
- serves as ion exchange & chelating systems
- dominant in neutral & alkaline soils
- natural residence time in soil: 1235 years

Fulvic Acid (FA)
- weak organic acid
- soluble at all pH
- more chemically reactive than HA
- high exchange capacity 2x that of HA
- small molecules so can enter plant systems
- dominant in leached / acid soils
- residence time: 870 years

Table B Comparison of humic and fulvic acids

	HUMIC ACIDS (HA)	FULVIC ACIDS (FA)
formula	C+H+O	$C_{135}H_{182}O_{95}N_5S_2$
pK1 value	4–8	
	charge density	
solubility	soluble in water of alkaline pH and precipitated at pH<2	soluble in water of all pH values
	fewer COOH and COH groups ∴ less active than fulvic acids	more COOH and COH groups ∴ more active than humic acids
		relatively low molecular mass
		more acidic than humic acid
		forms strong complexes with Fe^{3+}, Al^{3+}, and Cu^{2+}

Nitrogen-N

It was noted above that pioneer andolizers have the ability to fix nitrogen for their own use; however, once dead plant (and animal) matter is added back into the soil, it forms a pool of organic N. The organic compounds containing N are mostly complex proteins from rotting organic matter (non-humic substances in Table 4) and are not accessible to plant growth until this *organic* N has been converted into inorganic forms by soil microbes as part of the 'nitrogen cycle' (PASSEL n.d.*a*). Through a complicated set of chemical reactions, the proteins are broken down by micro-organisms to produce ammonium NH_4^+ in a process termed 'mineralization'. In a second multi-stage process, 'nitrification', micro-organisms convert organic N to nitrate NO_3^-. Both of these forms of *inorganic* N, nitrate and ammonium, are useful to plants. Inorganic N production processes occur best at pH levels around 7; greater or lesser pH causes a decline in inorganic N production, as does lack of oxygen or moisture for micro-organism metabolism (PASSEL n.d.*a*).

Nitrate moves freely in soil solutions, thus being immediately available to plants; but it is also therefore susceptible to leaching. Nitrate leaching is exacerbated in acid soils, and high precipitation rates will carry nitrates down into the subsoil where is it unavailable to plant roots. Japan has a humid temperate climate with precipitation of 725–2474 mm/yr, the same as the northwestern United States, though less than Indonesia (975–4974 mm/yr), but more than New Zealand (975–1474 mm/yr), and much more than the Mediterranean region (475–974 mm/yr). One reason for extensive leaching in andosols is that they have very little positive charge in their A horizons (cf. Figures 7.9, 7.10) for the nitrate anion to combine with; thus, nitrate can move downwards in the soil profile as fast as 10 cm per 100 mm of rainfall (Shoji *et al.* 1993: 177).

Ammonium, on the other hand, reacts with clays and organic matter in exchangeable or non-exchangeable forms. NH_4^+ molecules attach to clay mineral surfaces and to humic particles as exchangeable ions, and these can be released through ion exchange for plant use (Nielsen 1972). However, the ions can also be adsorbed into pores in the clay mineral crystal lattices, becoming fixed and non-exchangeable. Ammonium competes with K for these from aqueous solutions (ECIFM n.d.; Alshameri *et al.* 2018; Bot & Benites 2005: ch. 2). The efficiency of moving ammonium from the soil solution into clay relates to clay type (cf. Apx 12) and soil pH: experiments on six clay types showed an increase in removal between 2 to 6 pH, peaking between 6–7 pH and declining thereafter; however, the maximum removal figures ranged from ca. 18% to 90% between halloysite and vermiculite (Alshameri *et al.* 2018: fig. 3). Another study documented a 91% uptake of ammonium by montmorillonite, non-crystalline clays 45–54%, and only 10% by halloysite (Bajwa 2008). Thus, andosols with significant amounts of non-crystalline clays will adsorb (fix) significant amounts of ammonium, making it inaccessible for plant use.

Phosphorus-P

In contrast to nitrogen, "phosphorus is not the limiting nutrient element in revegetation of tephra-covered lands" (Shoji *et al.* 1993: 218). On volcanic soils, inorganic phosphate (P_i) is provided mainly by the phosphate-containing apatite minerals (calcium phosphates), more common in felsic than mafic tephras; apatites $Ca_5(PO_4)_3(F,Cl,OH)$ dissolve relatively rapidly in acid solutions, becoming immediately available for plant growth (Ibid.: 217). In some tephras, P_i supply is constant: the regosolic soils (cf. Figure 7.2) of Mt Usu revealeld that P levels after 25 years were "almost comparable with that of the fresh ash" (Shoji & Takahashi 2002: 114). Plants also supply phosphorus in an organic form (P_{org}) produced by the decay of biological matter; this form of P_{org} is held in organic compounds and is only useable after its conversion (mineralization) to P_i by micro-organisms.

P_i circulates freely in the soil solution where it is taken up by plants as $H_2PO_4^-$ and $HPO_4^=$ (PASSEL n.d.*b*); Under strongly acidic conditions, non-ionic phosphoric acid H_3PO^4 dominates; as the conditions move towards basic conditions, hydrogen atoms are dissociated to form H_2PO^{4-} then H_2PO^{4-} and finally the phosphate anion PO_4^{3-} in strongly basic conditions. The uptake of P is best between soil pH of 6.5–7; below pH 6.5, it tends to be fixed by Fe, Al, and Mn; above pH 7, it fixes mainly with Ca (PASSEL n.d.*b*). P that is removed from the soil by fixation or plant uptake needs resupply by transferring P_i from the 'Active Pool' in the soil if albitite sources are unavailable. The Active Pool holds organic and inorganic P in solid compounds which must be broken down into P_i. P_{org} is bound with humic and fulvic acids, and it can be stabilized in Fe/Al-humus complexes in acid soils or with Ca and Mg in neutral or alkaline soils (Marakov & Malysheva 2006). P_i also associates with these humus acids using Al and Fe as metal bridges (Bedrock *et al.* 1997); Tan (2008: 432) explains that 'aluminum bridging' helps the storage of phosphates and preserves humic molecules in Indonesian andosols. P_i can react directly with iron and aluminium; some of these reactions are reversible, but P_i is only released under certain biological or chemical conditions, so they are not a good source of P nutrient. Irreversible insoluble chelation reactions of P with Fe and Al are considered part of the 'Fixed Pool' where P is virtually unavailable for plant use. This pool also contains P_{org} that is resistant to mineralization (Pagliari *et al.* 2017). A separate problem is phosphate fixing due to coatings of hydrous Fe or Al oxide developing on phosphates adhering to mineral surfaces (Anon. 2017).

The high content of Fe and Al complexes in andosols is responsible for the sorption of phosphate (Table C); as tephra weathering increases, these active elements combine with phosphate to form insoluble compounds and lead to P deficiency for growing crops (Shoji *et al.* 1993:171). One of the criteria for andic soils as seen in Table 7.1 is a high P retention of ≥85%; however, this retention rate depends Al+½Fe % content. Acid, non-allophanic soils bind twice as much P with 5 times tighter bonds. P-sorption increases with temperature rise, so it is a greater problem in summer and sub/tropical climates (see Shoji *et al.* 1993: ch. 6.5.3 for details).

Table C Comparison of P-sorbtion between andosols

	Non-allophanic andosols	Allophanic andosols
P retention % at Al+½Fe %	P=48% at Al+½Fe=0.4%	P=38% at Al+½Fe=0.5%
max P=85% at Al+½Fe %	at Al+½Fe=1%	at Al+½Fe=2%

Potassium-K

Potassium is supplied naturally in tephras from the weathering of volcanic glass (Shoji *et al.* 1993: 222-224); rhyolite tephras contain higher levels than basaltic tephra, and K content also varies between calc-alkaline and alkali tephras. K variability is one of the difficult parameters to address in andosol fertility. In non-allophanic andosols of humid climates, the weathering of non-coloured glass provides K for the formation of 2:1 layer clays (see Apx 12); the allophane in allophanic andosols does not preferentially retain K. Both non-allophanic and allophanic soils sequester K in humic matter.

Soil pH

Soil acidicity or alkalinity is simply measured by pH on a scale from 0 to 14, reflecting the amount of hydrogen ions H+ in the soil solution (BOX 15). The neutral value is pH 7.0, and most vegetable crops grow in the pH range from 4.5 to 8. Volcanic ash deposits may start out with high pH due to the presence of the several alkali and alkaline base metals in felsic magma products, but these can be leached out quickly by exposure to moisture, and the weathering of volcanic glass as discussed above releases base metals through time.

Changes in soil pH may occur over time due to several processes: leaching out of nitrates as well as base metals, dissolution of aluminium compounds, and the neutralization of humic and fulvic acids. For example, when Al^{3+} combines with water, it produces aluminium hydroxide and a H+ as a by-product: $Al^{3+} + H_2O \rightleftarrows AlOH^{2+}(aq) + H+$; while the dissociation of hydrogen from the carboxyl (COOH) and hydroxyl (OH) groups on the surfaces of humic and fulvic acids occurs in the reactions COOH ⟶ COO− + H+ and COH ⟶ CO− + H+ (Pettit n.d.).

Non-allophanic andosols are more acidic than allophanic andosols (Table 7.2). In general andosols have a pH less than 7.0 (neutral) (Shoji *et al.* 1993: 165). But this pH does not reflect the original alkalinity of most tephra in Japan when new (see Chapter 15).

Al toxicity & tolerance

Most studies mention free Al^{3+} as the specific agent of aluminium toxicity in soils, since free Al^{3+} reacts with the cell walls and membranes of plant roots, restricting their growth (CCMA 2013). Aluminium becomes a problem when soil acidifies. As pH drops below 6.2 into acidic soil conditions, Al becomes increasingly soluble (Yokel 2002), dissolving from its bound forms and entering into the soil solution as free Al^{3+} – for example, as in the reversible reaction we saw above: $AlOH^{2+}_{(aq)} + H^+ \rightleftarrows Al^{3+} + H_2O$.

Al toxicity of soils can be tested by extracting it from soil solutions using potassium chloride (KCl). KCl extractable Al in amounts of 5–11 cmolc/kg can be toxic to plant roots (Shoji & Takahashi 2002);[4] Dadamouny (2015) gives the amounts of 0.5 ppm (parts per million) as troublesome and 1.0 ppm as toxic. Since non-allophanic andosols are generally more acidic than allophanic andosols, the former are more at risk of Al toxicity (Shoji *et al.* 1993: 232-233). Methods to mitigate Al toxicity are:

1) adding carbonates: lime (calcium carbonate, $CaCO_3$) so that it forms aluminium carbonate $Al_2(CO_3)_3$ (dawsonite), or gypsum (calcium sulfate, $CaSO_4$) so that the dissolved Ca pushes Al deeper into the subsoil (Davidson 2014); or

2) adding organic matter to the topsoil to form the salts Al fulvate and Al humate (Blamey 1999).[5] Either liming or green manuring inactivates aluminium by removing it from the soil solution through chelation so that it cannot further attack plant roots.

[4] cmolc = centimoles of charge: a unit for calculating exchange capacity for soils.

[5] Pettit (n.d.) uses 'humate' to include salts of both humic and fulvic acids; salts are merely electrically neutral compounds of a base and an acid.

References

ALSHAMERI, Aref; Hongping HE & Jianxi ZHU *et al.* (**2018**) "Adsorption of ammonium by different natural clay minerals: characterization, kinetics and adsorption isotherms." *Applied Clay Science* 159: 83-93 [DOI: 10.1016/j.clay.2017.11.007].

Anon. (**2017**) "Phosphorus Fixation." *Numerator Technologies* [www.numeratortech.com/tech_info/Phosphorus_Fixation1.pdf].

BAJWA, MI (**2008**) "Clay mineralogical and ammonium fixation characteristics of some tropical upland rice soils." *Communications in Soil Science and Plant Analysis* 13.11: 945-955 [DOI: 10.1080/0010362820 9367323].

BEDROCK, CN; MV CHESHIRE & CA SHAND (**1997**) "The involvement of iron and aluminum in the bonding of phosphorus to soil humic acid." *Communications in Soil Science and Plant Analysis* 28.11-12: 961-971.

BLAMEY, P (**1999**) "16.5.2 Aluminium and manganese toxicities", n.p. in *Plants in action adaptation in nature, performance in cultivation*, ed. by BJ ATWELL *et al.* Melbourne: Macmillan Education [http://plantsinaction.science.uq.edu.au/edition1/?q=content/16-5-2-aluminium-and-manganese-toxicities].

BOND, Chris (**2017**) "Do you need to add more humus to your garden beds?" *Maximum Yield* [www.maximumyield.com/do-you-need-to-add-more-humus-to-your-garden-beds/2/3445].

BOT, Alexandra & José BENITES (**2005**) *The importance of soil organic matter: key to drought-resistant soil and sustained food production.* Rome: Food and Agriculture Organization of the United Nations [www.fao.org/docrep/009/a0100e/a0100e05.htm].

CCMA (Corangamite Catchment Management Authority) (**2013**) "Aluminium." *Brown Book* [www.ccmaknowledgebase.vic.gov.au/brown_book/21_Aluminium.htm].

DADAMOUNY, MA (**2015**) "What are reliable criteria for identification of Al toxicity in an acid soil? Answer" [www.researchgate.net/post/What_are_reliable_criteria_for_identification_of_Al_ toxicity_in_an_acid_soil].

DAVIDSON, DJ (**2014**) "Gypsum and how it corrects aluminium toxicity in the subsoil." *EcoGEM* [www.eco-gem.com/gypsum-corrects-aluminum-toxicity-subsoil/].

ECIFM (**n.d.**) "Nitrogen & Phosphate." Reading University [www.ecifm.rdg.ac.uk/nutrient.htm].

MAKAROV, MI & TI MALYSHEVA (**2006**) "Phosphorus in humus acids." *Eurasian Soil Science* 38.11: 1208-1216.

NEALL, VE (**2006**) "Volcanic soils", pp. 23-45 in *Land use, land cover and soil sciences, Vol. VII in Encylopedia of Life Support Systems*, ed. by W VERHEYE. Paris: UNESCO & Eolss Publishers [www.eolss.net].

NIELSEN, J Dissing (**1972**) "Fixation and release of potassium and ammonium ions in Danish soils." *Plant and Soil* 36.71-88.

PAGLIARI, Paulo H; Daniel E KAISER & J ROSEN *et al.* (**2017**) "The nature of phosphorus in soils." *The University of Minnesota Extension: Nutrient Management* FO-6795-C [www.extension.umn.edu/ agriculture/nutrient-management/phosphorus/the-nature-of-phosphorus/].

PASSEL (**n.d.*a***) "Soils – part 5: nitrogen as a nutrient." *Plant & Soil Sciences Elibrary* [http://passel.unl.edu/pages/informationmodule.php?idinformationmodule=1130447042&topicorder=2&maxto=8].

---- (**n.d.*b***) "Soils – Part 6: phosphorus and potassium in the soil." *Plant & soil sciences elibrary* [http://passel.unl.edu/pages/informationmodule.php?idinformationmodule=1130447043&topicorder=3&maxto=15].

PETTIT, Robert E (**n.d.**) "Organic matter, humus, humate, humic acid, fulvic acid and humin: their importance in soil fertility and plant health." *Mesa Verde Humates* [https://humates.com/pdf/ORGANICMATTERPettit.pdf].

SHOJI, Sadao & Tadashi TAKAHASHI (**2002**) "Environmental and agricultural significance of volcanic ash soils." *Global Environmental Research* 6: 113-135.

SHOJI, Sadao; Masami NANZYO & Randy DAHLGREN (**1993**) *Volcanic ash soils: genesis, properties and utilization.* Amsterdam: Elsevier.

TAN, Kim H (**2008**) *Soils in the humid tropics and monsoon region of Indonesia.* Boca Raton, FL: CRC Press.

UGOLINI, FC & RA DAHLGREN (**2002**) "Soil development in volcanic ash." *Global Environmental Research* 6.2: 69-81.

YOKEL, RA (**2002**) "Alumium chelation principles and recent advances." *Coordination Chemistry Reviews* 228.2: 97-113.

APPENDIX 14

Metasomatic Processes

Linked to Chapter 12

Dolomitization

Limestone, generally the rock of our shallow ocean reefs, is formed primarily of fragmentary shells and skeletal fragments of marine life, including algae in a matrix of carbonate mud.[6] Limestones and carbonate muds in shallow seas, lagoons, and coastal flats may be subject to dolomitization by magnesium-rich seawater during the sedimentation process affected by sea level changes (Machel 2004: 11, 46). Once deposition is completed, limestones and carbonaceous muds may become dolomitized thousands or millions of years later through reaction with circulating fluids. Dolomitization of lagoon or basin carbonate muds is effective on a large areal scale (Moore 1989: 159-160), and "virtually all volumetrically large, replacive dolostone bodies are postdeposition and formed during some degree of burial" (Machel 2004: 7). Dolomitic limestone is often called 'dolomite' or 'dolostone'; here, 'dolomite' will be reserved for the mineral while 'dolostone' will be adopted in parallel with 'limestone'.

The major minerals of limestone are calcite and aragonite, which both have the chemical formula of calcium carbonate $CaCO_3$, but different crystalline structures. Where limestone or lime mud is exposed to magnesium-rich water, some calcium may be replaced by magnesium; magnesite $MgCO_3$, another carbonate mineral (which does not contain calcium) can also form. If more than one quarter of the calcium is replaced by magnesium, usually by circulating seawater, the rock becomes dolostone (dolomitic limestone), comprised mostly of the mineral dolomite: calcium magnesium carbonate $CaMg(CO_3)_2$.

Once dolostone is formed, it can be metamorphosed by heat and pressure, resulting in dolomitic marble (dolomarble). Impurities in the original dolostone can form new minerals when the dolostone recrystallizes under regional metamorphic conditions to form dolomarble (cf. Figure 2.8, gradient II). Tremolite marble forms in the lower to mid-amphibolite facies (ICL 2013: Glossary: Dolomarble).

Serpentinization

The processes and results of serpentinization are much more complicated and chemically diverse than with dolomitic rocks. Two types of rocks usually undergo serpentinization: ultramafic peridotite (and its relations) and basaltic gabbro/basalt. Peridotite is the main rock type of the Earth's upper mantle; it forms the underlay for the oceanic floor rocks (cf. Figure 2.4), and it supports the continental plates. A source of the olivine birthstone, peridot, peridotite has several varieties depending on the relative amounts of its three major constituent minerals:

- the olivine solid solution series $(Mg^{2+}, Fe^{2+})_2SiO_4$
- orthopyroxene minerals $(Ca,Mg,Fe)_2Si_2O_6$
- silica SiO_4

Peridotites are serpentinized with the addition of seawater, which reacts with the silicates to produce new minerals, mainly the serpentines, and expands the rock's volume. In essence, serpentine is simply hydrated olivine. Olivines are converted to serpentines by the reaction (formulae by Strekeisen 2020):

$$2Mg_2SiO_4\ \textbf{(olivine)} + 3H_2O = Mg_3Si_2O_5(OH)_2\ \textbf{(serpentine)} + Mg(OH)_2\ \textbf{(brucite)}$$

If carbon dioxide is present, the serpentine can be further converted:

$$2Mg_3Si_2O_5(OH)_4\ \textbf{(serpentine)} + 3CO_2 = Mg_3Si_4O_{10}(OH)_2\ \textbf{(talc)} + 3MgCO_3\ \textbf{(magnesite)} + 3H_2O$$

These simple formulae mask very complex and variable interactions to form serpentinite. The conversion of serpentine to magnesite in particular is a very important 'carbon sink' and is currently being investigated for carbon capture.

[6] Here, 'mud' is a grainsize designation and does not imply a significant amount of silt; the grains are primarily carbonates CaO_3 (both calcite and aragonite), and sometimes dolomite $CaMg(CO_3)_2$

Gabbro, derived from peridotite, is coarsely crystallized basaltic magma, generally from a magma chamber; it usually contains Ca, Mg, and Fe as well as plagioclase feldspar, which contains Na, Ca, Al, and silica. Seawater, reacting with gabbro to make serpentine, carries the Na, Ca, and some silica out of the igneous rock to enrich the circulating fluids. Few studies have been made on gabbro serpentinization, but what has been discovered is complicated stepwise transformations producing magnetite ($Fe^{2+}Fe^{3+}{}_2O_4$) instead of magnesite and lacking the brucite associated with peridotite serpentinization (Nozaka *et al.* 2017).

Minerals of the serpentine sub-group have the general chemical formula $D_3[Si_2O_5](OH)_4$ with D = Mg, Fe, Ni, Mn, Al, Zn (MINDAT 2022); these form during serpentinization as the serpentine polymorphs antigorite, lizardite, and/or chrysotile (the last one widely known as the major asbestos mineral) (Malpas 1992: 8).[7] Serpentine minerals are softer than the jade minerals, ranging from 2.5 to 3.5 on the Mohs scale of hardness (MINSOCAM 2016) (see Apx 5: Fig. D). A rock formed primarily of any of these three serpentine minerals is termed serpentinite; this rock can yield 'ornamental serpentine', an easily carved material and often called 'serpentine jade' (e.g., Chen *et al.* 2014). This is not to be confused with nephrite formed in serpentinite (Snephrite). Bowenite is a different, particularly hard form of serpentine that is also suitable for carving. Antigorite is used for ornaments in China and termed 'new jade' (*xinyu*) or 'new mountain jade' (*xinshanyu*).

Serpentinite is thus composed of a variety of minerals: predominantly the serpentine minerals, minor magnetite, and variable minor brucite, magnesite, calcite, dolomite, and talc. Minor chromite, clinopyroxene, orthopyroxene, and olivine may remain as relics of the parent rock. Serpentinite itself and the serpentine minerals are highly variable in colour from green to brown to black-green; the green is caused by iron and black spots are usually caused by magnetite. Many of the trace elements and minerals of serpentinite are inherited from its parent rock (usually peridotite) and can be passed on to its jades.

References

CHEN, Qi; Yongya WANG & Fuxi GAN (**2014**) "The chemical compositions and origin characters of serpentine jade from different deposits." *Applied Mechanics and Materials* 624: 119-123.

ICL (Imperial College London) (**2013**) "Imperial College rock library: the glossary." [www.imperial.ac.uk/earthscienceandengineering/rocklibrary/glossary.php].

IMA-CNMNC (**2022**) (International Mineral Association - Commission on New Minerals, Nomenclature and Classification) "The new IMA list of minerals - a work in progress - updated: November 2022." [http://cnmnc.main.jp/imalist.htm].

MACHEL, HG (**2004**) "Concepts and models of dolomitization: a critical reappraisal", pp. 7-64 in *The geometry and petrogenesis of dolomite hydrocarbon reservoirs*, ed. by CJR BRAITHWAITE *et al.* London: Geological Society of London.

MALPAS, J (**1992**) "Serpentine and the geology of serpentinized rocks", pp. 7-30 in *The ecology of areas with serpentinized rocks: a world view*, ed. by BA ROBERTS & J PROCTOR. Dordrecht: Kluwer Academic Publishers.

MINDAT (**2022**) "About serpentine sub-group." [www.mindat.org/min-11135.html].

MINSOCAM (Mineralogical Society of America) (**2016**) "Mohs' scale of hardness." [www.minsocam.org/ MSA/ collectors_corner/article/mohs.htm].

MOORE, CH (**1989**) *Carbonate diagenesis and porosity*. Amsterdam: Elsevier.

NOZAKA, Toshio; Robert P WINTSCH, and Romain MEYER (**2017**) "Serpentinization of olivine in troctolites and olivine gabbros from the Hess Deep Rift." *Lithos* 282/283: 201-214.

STREKEISEN, Alex (**2020**) "Alex Strekeisen: I vetrini della mia fantasia." [www.alexstrekeisen.it/english/ index.php].

YANG, Hsiao-Yu; Ruei-Hao SHIE & Pau-Chung CHEN (**2013**) "Carving of non-asbestiform tremolite and the risk of lung cancer: a follow-up mortality study in a historical nephrite processing cohort." *Occupational Environmental Medicine* 70: 852-857 [DOI: 10.1136/oemed-2013-101404.]

[7] Chrysotile is considered a growth 'habit' of lizardite (George Harlow, pers. comm., 2 March 2017); however, it has been redefined (Rd) as an approved mineral in the IMA mineral list (IMA-CNMNC 2022).

APPENDIX 15

Legend for Nara Basin 'Seamless' Geological Map

keyed to Figure 13.4

Most information was extracted from [https://gbank.gsj.jp/seamless/v2/legendSub.html?group=0&lang=en])[1]

COLOUR # unit	UNIT CODE[5]	DESCRIPTION	DATE
Central Basin — Sedimentary rocks			
8[2] Light Blue[3] rgb=230,255,255[4]	H_sad	valley floor, intermountain basin, river & coastal plain deposits	Holocene
9 Light Seagreen rgb=245,255,230	H_svd	natural levee deposits	Holocene
Basin margins — Sedimentary rocks (from top clockwise)			
5 Grey Green rgb=220,240,200	Q33-H_sfd	fan or talus deposits	late Late Pleistocene–Holocene
14 Bluish Seagreen rgb=200,255,170	Q32-33_std	terrace deposits	middle Late Pleistocene–late Late Pleistocene
15 Light Green rgb=200,255,122	Q-31_std	terrace deposits	early Late Pleistocene
16 Light Leafgreen rgb=220,255,100	Q-22_std	terrace deposits	late Middle Pleistocene
19 Beige rgb=255,255,220	Q1-21_sn	non-marine sediments	Pleistocene Gelasian–early Middle Pleistocene
21 Light Taupe rgb=230,226,172	N2-snc	non-marine conglomerate	M. Miocene, late Serravallian–Tortonian
26 Tan rgb=255,230,190	N3_sns	non-marine sandstone, or alternation of sandstone & mudstone or sandstone & mudstone	L. Miocene Messinian–Pliocene
79 Light Tan rgb=255,255,210	Q1-21_sb	brackish sediments or marine & non-marine sediments mixture	Pleistocene Gelasian–early Middle Pleistocene
98 Dark Yellow rgb=255,250,83	N1_sbs (prev. N12_sbs)	brackish, or marine & non-marine mixture sandstone, alternation of sandstone & mudstone	E. Miocene, late Burdigalian–early Langhian
Basin uplands — Igneous rocks (from top clockwise)			
504 Gold (N,U,M,K)[6] rgb=243,206,8	N2_vis_al (prev. N22_vis_al)	andesite & basaltic andesite lava & pyroclastic rocks	Miocene, late Serravallian–Tortonian
578 Gold rgb=255,239,119	N2_vas_al (prev. N22_vas_al)	dacite & rhyolite lava & pyroclastic rocks	Miocene, late Serravallian–Tortonian
613 Pink rgb=255,225,221	N2_vas_ap (prev. N12_vas_ap)	volcanic: dacite & rhyolite pyroclastic rocks	Miocene late Burdigalian - early Langhian
659 Mulberry rgb=166,19,112	K21_pbg_a	gabbro, island arc & continental[7]	Late Cretaceous Cenomanian–Santonian
675 Dark Violet rgb=159,27,155	K21_pbd_a	diorite & quartz diorite, island arc & continental	Late Cretaceous Cenomanian–Santonian
676 Dark Violet rgb=154,23,151	K12_pbd_a	diorite & quartz diorite, island arc & continental	Early Cretaceous Aptian–Albian

690 Lighter Red rgb=244,63,101	K22_pim_a	massive granodiorite & tonalite island arc & continental	Late Cretaceous Campanian–Maastrichtian
708 Darker Red rgb=243,33,82	K22_pin_a	gneissose granodiorite & tonalite, island arc & continental	Late Cretaceous Campanian–Maastrichtian
709 Blood Red rgb=241,29,77	K21_pin_a	gneissose granodiorite & tonalite, island arc & continental	Late Cretaceous Cenomanian–Santonian
726 Mauve rgb=221,155,173	K22_pam_a	massive granite, island arc & continental	Late Cretaceous Campanian–Maastrichtian
743 Violet rgb=223,155,193	K22_pan_a	gneissose granite, island arc & continental	Late Cretaceous Campanian–Maastrichtian
Metamorphic Rocks			
1494 Burnt Sienna rgb=208,105,3	K2-Pg12_mgmpe_lks (Prev. mgmpe_lkc)	pelitic gneiss & migmatite, in the sillimanite-Kfeldspar zone of low P/high T regional metamorphic rocks	Late Cretaceous–Paleocene Selandian

[1] As of October 2020, the "seamless" map is undergoing revision; both V1 and V2 are available online. The legend documents have not yet been fully updated to match the V2 map, thus the notation 'prev.' (previously) above. The codes that appear in the online V2 seamless map (GSJ 2020) are correct. The V2 map is accessible at [https://gbank.gsj.jp/seamless/v2/viewer/]. Reproduced here under 'fair use' policy.
[2] The unit numbers in the colour column are not consecutive, as they are drawn from the larger list of codes (totalling 2398) for all the rock units found in the whole of the Japanese Island, listed in "legendCsv.txt", available from the [soudan.gsj-ml@aist.go.jp].
[3] The colour names given here are subjective descriptions not linked to official hex colour names.
[4] The rgb numbers are given in "legendCsv.txt".
[5] Unit Codes are compiled by:
Period: **H**=Holocene, **Q**=Quaternary, **Pg**=Palaeogene, **N**=Neogene, **K**=Cretaceous
Rock type:
s=sedimentary: **sad**=alluvial deposits, **svd**=levee deposits, **sfd**=fan deposits, **std**=terrace deposits, **sn**=non-marine, **snc**=non-marine conglomerate, **sns**=non-marine sandstone, **sb**=brackish (mix of marine/non-marine), **sbs**=brackish marine/non-marine sandstone
v=volcanic: **vis_al**=andesite/basaltic andesite, **vas_al**=dacite/rhyolite
p=plutonic: **pbg_a**=gabbro, **pbd_a**=diorite/quartz diorite, **pim_a**=massive granodiorite/tonalite, **pin_a**=gneissose granodiorite/tonalite, **pam_a**=massive granite, **pan_a**=gneissose granite
m=metamorphic: **mgmpe_lks**=pelitic gneiss & migmatite, sillimenite=Kfeldspar zone
[6] The letters stand for the main volcanic edifices in and around Nara Basin from 9 o'clock counterclockwise: N=Mt Nijo, U=Unebiyama, M=Miminashi, K=Mt Kasuga.
[7] 'island arc' refers to the types of igneous rocks generated in forming island arcs but not hot spots; 'continental' in this case refers to igneous rocks generated within continental crust, not igneous rocks deriving from oceanic plates.

Important points for geologists:
• There are only three types of metamorphic rocks naturally found in the Basin: pelitic gneiss of sedimentary derivation, migmatites, and gneissose granite; however, it will be noted that the gneissose granite is not classified here under 'm' for metamorphic but under 'p' for plutonic.

There are several 'take-away' points concerning this geological map for archaeologists:
• The earliest rocks in and around the Nara Basin are relatively recent: some early Cretaceous but mostly Late Cretaceous and later. This means that any rocks of older derivation that have been used for cultural materials in the Basin must have either come from outside the Basin or been incorporated into the original Accretionary Complex forming part of the Ryoke Belt; in the latter case, however, their metamorphic date will be later than both their date of genesis and their date of accretion.
• The volcanic rocks surrounding the Basin are intermediate to felsic in composition (see Apx 5: Fig. F). These are characteristic of volcanic arcs whether at the continental edge or in island arc formations. Basalts are

particularly lacking in this region, though one prominent igneous rock unit (#659) is stated to include gabbro, the plutonic rock from a magma chamber that would have produced basalt without fractionation.
• The presence of pyroclastic rocks indicates there are sources of tuff that could be used for carving structural materials, e.g., stone scarcophagi and small stone 'pagodas'.
• The occurrence of rhyolite lava might indicate there were also obsidian flows, but Nara is not known as a source of archaeological obsidian.

Reference

GSJ (Geological Survey of Japan) (2020) "Seamless digital geological map of Japan V2 (1:200,000)" [https://gbank.gsj.jp/seamless/v2/viewer/].

APPENDIX 16

Analyses of sediment pH from Nara, Gunma, and Niigata

Linked to Chapter 15

The samples were processed according to standard procedures using the pH Meter TOA DKK HM=25G equipment in the laboratory of the Faculty of Agriculture, Niigata University in late September 2003, courtesy of Prof. Mutsuyasu ITO. The equipment was prepared by the lab assistant, but the analyses were done by G. Barnes. The samples were first dried and pulverized. 10 grams of each sample were then mixed with 25ml distilled water and left for one hour; each was stirred and let settle for 30 seconds before being measured with the pH Meter at approximately 26°C room temperature. Locations of sample-taking were recorded by sketch map and photographs (not included here).

Samples for pH analyses from Nara, Japan collected September 2003

Nine samples were taken during one day of car travel by G. Barnes with geologist Dr. Shiro NISHIDA (retired from the Earth Sciences Department, Nara Educational University), who chose the sites. The objective was to sample strata that were currently *not* subject to agricultural activities and which were as old and 'natural' as possible.

Loc. A Heijo (2 samples)
Samples taken in a grove of trees from the stream bank flowing N-S just south of the old open excavation exhibition building; bank was chest-height from stream bed.

A-1 **pH 5.66** from east bank, coarse sand at -40 cm; already quite hard. Nishida says it is latest alluviation on top of Osaka Formation. At its depth, it is not thought to have been farmed but is derived from stream deposits which may, however, have been subjected to leaching of agricultural fertilizers out of the stream water.

A-2 **pH 5.24** topsoil under trees on west bank, at ca. -20 cm; soil medium brown, quite friable with lots of roots. This topsoil could theoretically have been farmed before the purchase of the Heijo Palace site in the 1960s, but it could also be quite newly formed humus from leaf mold in the grove of trees.

Loc. B Ring Road (3 samples)
Samples taken from exposure on east side of north-south track leading into abandoned housing development on west, entered to the south from the ring road (*soto kanjosen*) jogged opposite from Enos service station on northeast corner.

B-1 **pH 5.80** from silt under iron deposit
B-2 untested thin iron deposit
B-3 **pH 6.06** from fine sand over iron deposit

Loc. C Karako (1 sample)
Sample taken from topsoil under an old tree growing at the northwest corner of Karako Pond beside a small roadside shrine.

C-1 **pH 5.5** from topsoil under tree; light brown and friable; very similar to A-2.

Loc. D Nukatabe (1 sample)
Sample taken from a roadside cut under a bamboo grove,at a level +15 cm from the road surface; near house 352-2 (formerly 397) in Nukatabe Kitamachi, south of sowa Elementary School behind Hirahata station.

D-1 pH 4.15 from yellow clay in local high spot; Nishida says it may be Osaka Formation

Loc. E Kashiwara (1 sample)
Sample taken from the forest floor to the left of the entrance to Unebe-ryo (Emp. Jinmu's burial mound) facing west. The forest is Evergreen Oak type.

E-1 **pH 5.00** dark topsoil, quite humic

Loc. F Tomimoto (1 sample)
Sample taken from wet disused paddy field growing *hie*; had notice that the field received a government subsidy for non-production; located south of where electric pylons cross Asuka River in Tomimoto.

F-1 **pH 5.89** from wet soil in field

4 samples for pH analyses from Niigata, Japan collected September 2003

These samples were taken during one day of bicycle travel west of Niigata University by G. Barnes. The sites chosen were non-agricultural in a variety of vegetational settings but all on dune sand.

Loc. G Sagata Lagoon (1 sample)
This lagoon is maintained as a nature park and is lined with reed beds; very sandy with no organic matter; top 3 cm of deposit at water's edge were heavily oxidized (red) but unoxidized grey sand underneath.

G-1 **pH 5.37** gray sand @ -3 cm from shoreline

Loc. H Uchino Temple (1 sample)
Sample taken from boundary of temple grounds under a large tree of Evergreen Oak forest type; gardens were cultivated at a short distance from the temple in the adjoining property.

H-1 **pH 5.67** topsoil from under tree

Loc. J Ikarashi shoreline (1 sample)
Sample taken from dune on coast in Ikarashi, Niigata, just under grass growth line on exposed, eroded side of dune.

J-1 **pH 6.52** dune sand under grass roots

Loc. K Niigata University (1 sample)
Sample taken in pine forest next to International House; pine needle mat ca. 25 cm thick, sandy underneath.

K-1 **pH 5.17** from exposed sand bank

4 samples for pH analyses from Gunma, Japan collected September 2003

Samples were taken during one day on an archaeological site in Shibukawa, Gunma, with Dr. Tsutomu SODA of the Palaeoenvironment Research Laboratory, Co. Ltd., Maebashi, courtesy of the site director, Mr. Masayuki OE. All strata were volcanic ash layers of differing colours; they are presented here in the stratigraphic succession from top to bottom.

Loc. L Kogenchi-Higashi I site

L-1 **pH 6.25** white tephra (FP), mid-6c date

L-2	**pH 6.62** brown tephra (FA), early 6c date
L-3	**pH 5.82** black tephra
L-4	**pH 6.19** yellow tephra

APPENDIX 17

Value Ranges of pH for Soil Profiles of Japanese Soil Types

Linked to Chapter 15

Nagatsuka (1997) identifies 32 different soil types in Japan within 14 soil classes, far more than illustrated in Figure 7.2. The data compiled here from his book are:

Type	Soil type number
pH	Values as shown in soil profile charts (examples without pH are not listed)
Parent	Material of C horizon
N...A	N = Natural growth, A = Agriculture; A-F = forestry, A-R = ranching, A-G = garden or dry field crops, A-P = paddy fields
Location	Prefectural names (cf. Apx 2: Fig. A)
Soil	Name according to Japanese scheme (FAO-Unesco and USA equivalents Are also given in Nagatsuka 1997.)
Range	*Qualitative evaluations* are used in Table 15.2

Type	**pH**	**Parent material**	**N...A**	**Location**	**Soil name**
PODZOLS CLASS:					
I-1	3.8-4.8	sandstone	N	Saitama	Typic Podzolic
I-2	4.3-5.3	sand-gravel	N	Hokkaido	Typic Podzolic
I-3	**3.9-5.3**	**volcanic**	N	Iwate	Peaty Podzolic
I-4	**3.8-5.2**	**volcanic**	N	Iwate	Wet Humus Podzolic
I-5	**3.9-4.8**	**volcanic**	N	Akita	Wet Iron Podzolic
I-6	**3.8-5.4**	**volcanic**	N	Nagano	Alpine Podzol
Range	3.8-5.4 = *from extremely to strongly acid*				
BROWN EARTHS CLASS:					
II-1	4.8-5.1	sandstone/slate	A-F	Miyazaki	Acid Brown Forest
II-2	3.8-5.2	slate	N	Saitama	Acid Brown Forest
II-3	4.9-5.8	slate/sandstone	N	Yamanashi	Acid Brown Forest
II-4	**4.8-5.7**	**volcanic**	N	Saitama	Acid Brown Forest
II-5	4.4-5.1	sandstone	N	Saitama	Acid Brown Forest
II-6	4.4-5.1	sandstone	N	Saitama	Acid Brown Forest
II-7	4.2-4.5	sandstone/shale	N	Hyogo	Acid Brown Forest
II-8	5.0-5.4	sand/siltstone	A-F	Hokkaido	Humic Brown Forest
II-9	**4.2-5.0**	**volcanic**	N	Akita	Surface-gleyed Brn. Forest
II-10	3.5-4.8	slate	N	Saitama	Podzolized Brown Forest
II-11	4.9-5.2	alluvium	N	Hokkaido	Pseudogleyed Brn. Forest
II-12	**4.7-5.4**	**volcanic**	N	Hokkaido	Reddish Brown Forest
Range	3.5-5.8 = *from extremely to moderately acid*				
FERSIALLITIC SOILS CLASS:					
Yellow-Brown Forest Soils					
III-1	3.9-4.9	granite	N	Mie	Typic Yellow-Brn. Forest

III-2	4.9-5.6	chert gravel	N	Shizuoka	Typic Yellow-Brn. Forest
III-3	4.6-6.9	gabbro	N	Shizuoka	Typic Yellow-Brn. Forest
III-4	5.6-5.8	mafic rock gravels	N	Shizuoka	Typic Yellow-Brn. Forest
III-5	4.8-6.6	muddy sand	N	Aichi	Pesudogleyed Yell-Brn For.
III-6	3.8-4.2	terrace gravel	N	Nagoya	Surface-gleyed Yell-Brn For.
range	3.8-6.9 = *from extremely acid to neutral*				
Red-Yellow Soils					
IV-1	4.6-4.8	terrace	N	Okinawa	Typic Red-Yellow
IV-2	4.6-5.3	arkose gravel	N	Aichi	Typic Red-Yellow
IV-3	4.9-5.1	mafic rock gravels	N	Shizuoka	Typic Red-Yellow
IV-4	4.2-5.7	chert gravel	N	Kobe	Pseudogleyed Red-Yellow
IV-5	5.0-6.2	sandstone	N	Okinawa	Surface-gleyed Red-Yellow
IV-6	4.1-5.1	sand-gravel	N	Okinawa	Bleached Red-Yellow
IV-7	4.6-5.0	terrace	N	Shizuoka	Yellow
Range	4.1-6.2 = *from extremely to slightly acid*				
ANDOSOL CLASS:					
Kuroboku Soils					
V-1	**5.4-6.1**	**volcanic**	N	Ibaraki	Ordinary Kuroboku
V-2	**6.3-6.8**	**volcanic**	N	Shizuoka	High-humic Kb w/duripan
V-3	**5.8-6.0**	**volcanic**	N	Kumamoto	Thick Kuroboku
V-4	**5.1-5.9**	**volcanic**	N	Tochigi	Thick high-humic Kuroboku
V-5	**5.7-6.0**	**volcanic**	A	Nagano	Paddy Kuroboku
V-6	**4.1-5.8**	**volcanic**	N	Hokkaido	High-humic Wet Kuroboku
V-7	**4.5-5.9**	**volcanic**	N	Ibaraki	Light-coloured Kuroboku
V-8	**5.5-7.0**	**volcanic**	N	Oita	Cumulic Kuroboku
Range	4.1-7.0 = *mainly from strongly acid to neutral*				
Para-Kuroboku Soils					
VI-1	4.3-5.1	terrace	N	Shizuoka	Ordinary Para-Kuroboku
VI-2	**4.2-5.2**	**volcanic**	N	Miyagi	thick high-humic Para-Kb
VI-3	**5.2-5.5**	**volcanic**	N	Tsukuba	Ordinary Wet Para-Kb
Range	4.2-5.5 = *from extremely to strongly acid*				
STAGNANT-WATER SOILS CLASS:					
Pseudogleys					
VII-1	**5.1-5.8**	**volcanic**	N	Hokkaido	Typic Pseudogley
VII-2	**5.0-5.8**	**volcanic**	A-R	Hokkaido	Humic Pseudogley
Range	5.0-5.8 = *from very strongly to moderately acid*				
Stagnogleys					
VIII-1	4.9-5.4	diluvial clay	N	Hokkaido	Typic Stagnogley
VIII-2	**4.1-5.1**	**volcanic**	N	Hokkaido	Peaty Stagnogley
Range	4.1-5.4 = *from extremely to strongly acid*				
LOWLAND SOILS CLASS:					
Brown lowland soils					
IX-1	5.4-6.8	alluvium	A-G	Ibaraki	Typic Brown Lowland
Range	5.4-6.8 = *from strongly acid to neutral*				

Gray lowland soils

X-1	5.1-5.6	fluvial deposit	A-R	Hokkaido	Typic Gray Lowland
X-2	7.0-8.0	fluvial/marine	A-G	Saga	Gleyic Gray Lowland
X-3	4.7-5.9	lacustrine mud	A-P	Hokkaido	Peaty Gray Lowland
Range	4.7-8.0 = *from very strongly acid to moderately alkaline*				

GLEY SOILS CLASS:

XI-1	6.4-7.5	marine	A-P	Okayama	Typic Gley
XI-2	5.8-7.8	marine	A-P	Okayama	Unripened Gley
XI-3	**5.5-6.4**	**volcanic**	A-P	Chiba	Andic Gley
XI-4	2.6-6.7	marine	N	Shimane	Thionic Gley
XI-5	5.0-6.5	alluvial	N	Okinawa	Potentially Thionic Gley
XI-6	7.4-7.8	alluvial	N	Okinawa	Calcareous Gley
Range	2.6-7.8 = *from extremely acid to slightly alkaline*				

PADDY SOILS CLASS:

Illuvial paddy soils

XII-1	5.2-6.3	alluvial	A-P	Ibaraki	Typic Illuvial Paddy
XII-2	**5.4-6.3**	**granite**	A-P	Nagano	Bleached Illuvial Paddy
Range	5.2-6.3 = *from strongly to slightly acid*				

Grayed paddy soils

XIII-1	5.5-6.6	fluvial deposit	A-P	Saitama	Typic Grayed Paddy
XIII-2	5.9-6.3	fluvial deposit	A-P	Kagawa	Bleached Grayed Paddy
XIII-3	5.8-7.1	fluvial/marine	A-P	Saga	Aquic Grayed Paddy
XIII-4	6.8-7.3	fluvial	A-P	Kagawa	Sub-aeric Grayed Paddy
Range	5.5-7.3 = *from strongly acid to neutral*				

PEAT SOILS CLASS:

XIV-1	4.0-5.1	peat	N	Hokkaido	High-moor Peat
XIV-2	4.0-4.2	peat	N	Hokkaido	Transitional-moor Peat
XIV-3	4.8-5.2	peat	A-P	Hokkaido	Low-moor Peat
Range	4.0-5.2 = *from extremely to strongly acid*				

CALCIMAGNESIAN SOILS CLASS:

XV-1	7.9-8.0	limestone	N	Kagoshima	Typic Rendzina-like
XV-2	7.5	limestone	N	Kagoshima	Brown Rendzina-like
XVI-1	7.0-7.6	limestone	N	Okinawa	Typic Terra Fusca-like
XVI-2	6.8-8.2	limestone	N	Okinawa	Pseudogleyed TF-like
XVII-1	5.3-7.8	limestone	A-G	Okinawa	Terra Rossa-like
XVIII-1	6.2-6.8	serpentine	N	Shizuoka	Chocolate Brown
XIX-1	8.4-8.5	mudstone	A-G	Okinawa	Typic Grumusol-like
XIX-2	7.6-8.2	mudstone	N	Okinawa	pseudogleyed Grumusol-like
Range	5.3-8.5 = *from strongly acid to strongly alkaline*				

UNDERDEVELOPED SOILS CLASS:

Rankers

XX-1	5.5-6.4	serpentinized gabbro	N	Shizuoka	Ranker (on mafic rock)
Range = *from strongly to slightly acid*					

Regosolic Kuroboku soils

XXI-1	**5.2-6.8**	**volcanic**	N	Hokkaido	Typic Regosolic Kb (pumice)

XXI-2	**3.5-5.8**	**volcanic**	N	Gunma	Typic Regosolic Kb (scoria)
Range	3.5-6.8 = *from extremely acid to neutral*				
RAW SOILS CLASS:					
Volcanogenous Regosols					
XXII-1	**5.5-6.7**	**volcanic**	N	Hokkaido	Typic Vol. Regosol (pumice)
XXII-2	**6.1-6.9**	**volcanic**	A-G	Tokyo	Typic Vo. Regosol (scoria)
Range	5.5-6.9 = *from strongly acid to neutral*				
Lithosols					
XXIII-1	4.5-4.8	chert	N	Shizuoka	Lithosol (chert)
XXIII-2	6.0-6.5	sandstone	N	Okinawa	Lithosol (sandstone)
Range	4.5-6.5 = *from very strongly to slightly acid*				
Regosols					
XXIV-1	4.3-4.7	granite	N	Aichi	Regosol (granite)
XXIV-2	6.0-6.9	sand/gravel	N	Shizuoka	Regosol (sand and gravel)
Range	4.3-6.9 = *from extremely acid to neutral*				
Arenosols					
XXV-1	5.8-6.7	sand dune	A-F	Tottori	Arenosol (silicious sand)
XXV-2	8.2-9.0	coral sand	N	Okinawa	Arenosol (coral sand)
Range	5.8-9.0 = *from moderately acid to strongly alkaline*				
UNDIFFERENTIATED SOILS:					
Alpine grassland soils					
XXVI-1	4.4-5.1	-	N	Nagano/Gifu	Alpine Grassland
Range = *from extremely to strongly acid*					
Volcanogeneous dark red soils					
XXVII-1	4.5-4.7	volcanic	A-F	Shizuoka	Volcanogeneous Dark Red
Range = *very strongly acid*					
ANTHROPOGENIC SOILS:					
XXIX-1	5.0-5.9	terrace gravels	A-G	Shizuoka	Hortisol
XXXI-4	**6.2-6.5**	**volcanic**	A-G	Hokkaido	Cut-and-fill
XXXII-1	6.7-8.0	marine	A-F	Chiba	Dredged soil
XXXII-2	**7.0-8.5**	**volcanic**	A-F	Chiba	Dredged soil w/ andosol
XXXII-3	**6.0-6.5**	**volcanic**	A-F	Chiba	Dredged w/ andosol, sand
Range	5.0-8.5 = *from very strongly acid to strongly alkaline*				

Reference

NAKATSUZUKA, Shizuo (**1997**) *Ecological atlas of soils in Japan*. Sendai: Fuji Technosystem (in Japanese with English title).

CLASSIFIED INDEX

Proper names and some entries (pages, chapters, figures) are given in ***bold*** – the latter where discussion is most relevant; details in SMALL CAPS. A = Appendix; B = Box; Ch = Chapter; F, F- = Figure; ftn = footnote; Pt = Part; T , T- = Table; ~ = repeats main entry word(s).

[1] The four main Japanese Islands (in ALL CAPS) are only referenced to their figures and tables, which guide the reader to discussion in the text.

GLOSSARY OF EAST ASIAN WORDS

This glossary gives the characters for Japanese (unmarked) and Chinese (C.) terms appearing in the text, together with simple translation, if literal then in double quotes. The locations of where they occur are given as chapter references, e.g. Ch5; Box; Apx = Appendix

akahoya 赤ホヤ seasquirt, sea pineapple; Ch6

akatsuchi 赤土 yellow-red sediments underlying the top humus layer of soil; Ch7

aki-no-enokoro-gusa 秋の狗尾草 Japanese bristle-grass, *Setaria faberi*; Ch7

aki no nanakusa 秋の七草 the "seven grasses of autumn"; Ch7

ando 暗土 "dark earth", same as *kurotsuchi*; Ch7

aome 青め greenish; Box 4

awa 粟 foxtail millet (*Setaria italica*); Ch7

buna ぶな beech, *Fagus crenata, Fagus japonica*; Ch7

bunkasō 文化層 a "cultural layer" in which artefacts are found; Ch1

daichi 台地 terrace, plateau, tableland; Chs5, 13

daikon 大根 large long white radish (mooli); Ch7, Apx 13

dankyū 段丘 terrace, see also *kashi dankyū*; Chs5, 13

dankyū-men 段丘面 terrace surface; Ch13

dansō 断層 "cut strata" = fault; Ch13

deisō 泥層 mud layer; Ch13

doronoki 泥の木 Japanese poplar (*Populus maximowiczii*); Ch7

emishi 蝦夷 a term used to indicate northern peoples outside the jurisdiction of the Nara state; Ch8, Apx 2

feicui (C.) 翡翠 rocks treated as jade in China, containing jadeite, omphacite, kosmochlor; Ch12

gobō 牛蒡 edible burdock (*Arctium lappa*); Ch7

hagi 萩 a flowering shrub (*Lespedeza*); Ch7

haikaki-yama 灰掻山 "mountains of raked tephra": a method of removing tephra from fields into mounds; Ch7

hakuhen-sentoki 剥片尖頭器 lithic tool that is made by re-touching both laterals of the proximal end of a thick, pointed blade or elongated flake with a sharp edge and tip; Ch8

hanrangen 氾濫原 floodplain; Ch5

harugasumi 春霞 springtime haze; Ch6

hatake 畑 dry fields; Ch7

heiya 平野 plain; Ch5

hie 稗 barnyard grass (*Echinochloa crus-galli → E. utilis/esculenta*); Ch7

hi-ire 火入れ setting control fires to manage plant communities; Box 3, Ch7

hisui 翡翠 jadeitite; Ch12

hōgansō 包含層 "(artefact)-bearing stratum" (ABS); Ch14

ike 池 pond; Chs. 4, 8, 13

jishin kōkogaku 地震考古学 earthquake archaeology; Chs. 1, 9, 13, 14

jōri 条理 the Nara-period land divisioning system based on the Chinese model resulting in a 100-m grid, each grid square divided into house plots in urban areas or six or eight paddy-fields; Ch13

kabanoki 樺の木 Japanese white birch, *Betula platyphylla* var. japonica; Ch7
kasen 河川 fluvial; Ch13
kashi dankyū 河岸段丘 riverine terrace; Ch13
kawa 川 river, suffixed to many placenames, e.g., Yamatogawa, Yamato River
kazanbai kōkogaku 火山灰考古学 tephroarchaeology (lit. "volcanic ash archaeology"); Chs1, 8
kibi きび millet (*Panicum miliaceum*); Ch14
konseki 痕跡 trace; Ch11
-ko 湖 lake, as a suffix, i.e., Biwako = Lake Biwa
kōya 荒野 "rough meadows" used for hunting; Ch7
kura 蔵 traditional storehouse; Box 4
kurobokudo 黒ボク土 "black fluffy earth", i.e., an andosol; Ch7, PtIII Reflections
kurobokudo-dosō 黒ボク土層 black "fluffy soil" stratum; Ch7
kuroko 黒鉱 black ores found at seafloor hydrothermal vents; Ch4
kurotsuchi 黒土 the black humus layer of soil; Ch7
kyūryō 丘陵 hills; Ch5
magaibutsu 磨崖仏 carved-cliff Buddhas; Box 4
masado 真砂土 regolith; Ch10
medake 女竹 Simon bamboo (*Pleioblastus simonii*); Ch.7
mehi-shiba 目日芝 crabgrass (*Digitaria adscendens*); Ch7
miso 味噌 fermented soybeans; Box 4, Apx 13
miso-tsuchi 味噌土 tephrogenic soil with the colour and texture of *miso*; Apx 13
nami 波 wave; Ch10
nāpai Okinawan for 'rituals'; Ch10
nezasa 根笹 a dwarf bamboo (*Plioblastus variegatus* Makino); Ch7, PtIII Reflections
numa 沼 marsh, Ch 13; suffix for place names, Ch10
onoe yanagi 尾上柳 Sakhalin willow (*Salix sachalinensis*); Ch7
ryūsui kyakudo 流水客土 "flowing-water guest soil": a method of introducing new soil into paddy-fields; Ch8
saigai kōkogaku 災害考古学 "disaster archaeology"; Ch11
sankakusu 三角州 delta; Ch5
sasa ささ bamboo grasses (*Sasa* spp.); Ch7
satoyama 里山 "village mountain": hillslopes behind villages providing resources; Ch7
satsuma-imo さつま芋 sweet potato, named after the Satsuma Peninsula, Kyushu; Ch7
sawa 沢 usually means a small stream valley but can apply to a marsh; Ch13
senjōchi 扇状地 alluvial fan; Chs5, 13
senjōchi-sei 扇状性 alluvial fan-created characteristics; Ch13
shirome 白目 whitish; Box 4
shiso 紫蘇 beefsteak herb (Perilla frutescens var. crispa); Ch7
-sō 層 a stratum, see also '*hōgansō*'
soba 蕎麦 buckwheat (*Fagopyrum esculentum*); Ch7
suiden 水田 wet-rice paddy-fields; Ch7
suname 砂目 sandy; Box 4
susuki 須々吉 eulalia or Chinese silvergrass (*Miscanthus sinensis*); Ch7
tai 帯 geotectonic "belt", terrane; Ch3
tansui 炭水 freshwater; Ch13
tei 堤 embankment; Ch13
tenchi-gaeshi 天地返し "turn the earth over": a method of burying tephra to make soil usable for planting; Ch8
toyo ashihara naka-tsu kuni 豊葦原中津国 "fertile central land of reed plains" = the Nara Basin; Ch13
tsu 津 harbour; Ch10
tsunami-ishi 津波石 boulders deposited by tsunami; Ch10
tsunami kōkogaku 津波考古学 tsunami archaeology; Preface, Chs. 1, 8, 9, 11, 13
umi 海 sea; Ch13
une 畝 a mounded ridge used for planting garden crops; Ch8
unema 畝間 the space between *une*; Ch8
ushikusa 牛草 *Schizachyrium brevifolium (Sw.) Nees*; Ch14
yabukōji 薮こうじ coralberry, *Ardisia japonica*; Ch7
yakihata 焼畑 "burned-field" agriculture, i.e., slash-and-burn or swidden; Ch7
yama 山 mountain, affixed to many mountain names, e.g., Fujiyama = Mt Fuji; sometimes read as -san, e.g., Fuji-san
yasechi 痩せ地 "infertile soils"; Ch7
yu 玉 (C.) beautiful ornamental rocks and minerals including jadeite; Ch12
zhang 璋 (C.) a stone blade, often produced in nephrite; Ch12

GEOLOGICAL TERMS: GLOSSARY & INDEX

Glossary terms are flagged by sans serif *font only for the first instance in each chapter; these assume cumulative knowledge through the chapters, so that only relatively new or obscure terms are flagged in later chapters. For elements, and major rock and mineral names, see A5 or Mindat.org or Strekeisen (2020). Where the term is given greater context or explanation, the occurrences are marked in bold as e.g.,* **Ch2.4**, *i.e., Section 2.4 in Chapter 2. Proper names are also given in* **bold.** *Terms defined elsewhere in the Glossary are given in 'single quotes' except within quotations. Icons in parentheses (*; Ø; §; ‡; ¶; †) provide citations, listed at the end of the Glossary. Font types are eccentric according to entry needs.*

~ indicates main vocabulary item, A = Appendix, B = Box, Ch = Chapter, F = Figure, T = Table
e.g., A5: F-A, A5: T-A = Appendix 5: Figure A, Appendix 5: Table A

accretion – *a term used in various ways in geology, basically indicating something adhering to something else*: 28, 30, 117, **Ch2.4.4.**

accretion tectonics – *the growth of a 'continental plate' through accretion to its edge of 'trench' and 'ocean floor' sediments and oceanic terranes such as 'ophiolites', 'island arcs' and 'ocean plateaus'; often related to mid-oceanic ridge subduction*: 210-13, 215, **Ch2.4.4**, **47;** EPISODIC 47, 49, 110; RATE 48. See also 'island arcs, accreted'.

Accretionary Complex (AC) – *a 'terrane' that derived from the bulldozing of 'oceanic plate' fragments, 'trench fill' and some 'metamorphic rocks' from the 'subduction channel'; when forming, called an 'accretionary prism'*: 4, 30, 74, **77-8**, 81, 96-7, 106, 140, 210, 377, 456, **A6**, **B1**, B2, **F2.5**ff, F3.1, T3.3; METAMORPHISM OF 78. See locations of geological belts in F3.1 and details in **A6**.

PLACENAMED: **Abukuma** A6; **Akiyoshi** 86, 88, 92, 212, A6, F3.4; **Chichibu** 89-90, 117, A6, F3.6, F5.9; **Chizu** A6, **Gosaisho** A6, F3.6; **Hida Gaien** F13.3 (black); **Joetsu** F3.4; **Maizuru** 86, 92, 117, 212, F3.4, F5.9, F13.3; **Mino-Tanba** 89, 92, 380, 403, F3.6, F5.9, F13.3; **Nagato-Renge** F3.4, F13.3; **North Shimanto** 213; **North Kitakami-Oshima** 90, F3.6; **Nedamo** 85, F3.4; **Ryoke** 89, F3.6, F13.3, F13.3; **Sanbosan** 92-**93**; **Sangun** 212, F3.4; **Shimanto** 94, F5.9, F13.3; **South Kitakami** 211; **Ultra-Tanba** 86, 92, F3.4, F13.3.

~ **cycle** – *uplift→ erosion→ accretion*: 92.

~ **orogen** – *"A period of deformation and continental growth involving suturing of exotic terrane to a continental margin during collision and convergence"* (¶).

~ **prism** – *the wedge of accreted material bulldozed into the continental edge from the ocean 'trench' and 'ocean floor'; a fossilized one on land is an 'Accretionary Complex'*: 47, 63, 77, 86, 210, 212, 345, **F2.5**, F3.4 (inset), F5.5, F12.1. PLACENAMED: **Idonappu** 94; **Kamuikotan** 94; **Nankai prism** 16, 48, 81, 385, **F2.8**, **F3.1**.

~ **wedge** (= 'accretionary prism'): **Ch2.4.4**, **F2.5**ff.

Active Faults – *in Japan, faults that have caused earthquakes within the Quaternary*: 119, 216, 219, 310, 410, 422, T13.3; MAPS 456-7. See also 'faults'.

Active Fault earthquakes – *in Japan, shallow earthquakes in the Earth's crust that have been active within the Quaternary period*: 216, 248, 255, 258, 268, 289, 318-19, 395-400, 403-4, **Ch2.4.2.1**, **Ch2.6.3.2**, **T2.1**. Compare 'subduction zone earthquakes'. See also 'earthquake' for lists of both types.

active margin (= convergent plate boundary) – *"A boundary between two plates that are moving toward each other"* (*), *resulting in 'subduction', 'accretion', or 'collision'*: 3, 33, 41, 83, 109, **Ch2.3.1**.

Active Volcanoes – *in Japan, volcanoes that have erupted within the Holocene*: 5, **145-7**, 157, 222-3, 305; MONITORING 305-7; **Special Measurement Law for Active Volcanoes** 145.

aerosols, volcanic – *"solid particles or liquid droplets"* (¶) *in gases erupted from 'volcanoes'*: 14.

alkali elements – *Group 1 'Alkali Metals' of the Periodic Table, e.g., sodium-Na, cessium-Cs*: 106, 189, 317, 436-8, 458, 464, **A5: F-A**, A5: T-A, **Ch2.4.3**.

alkaline elements – *Group 2 'Alkaline Earth Metals' of the Periodic Table, e.g., magnesium-Mg, calcium-Ca, strontium-Sr, boron-Bo*: 189, 317, 436-8, 458, 464, **A5: F-A**, A5: T-A, **Ch2.4.3**, T15.2. See also 'pH'.

alkaline volcanics - *a product of rifting; in Japan, Zone 4 along the Japan Sea coast*: 110, 114, 145, 155-6, 430, A5: F-F~G, A9, F6.2, T4.1, T6.2.

ALKALINE – *magma containing alkaline elements (K, Na)*: 436-7, 462-3, **A5: F-G.**

CALC-ALKALINE – *magma containing alkaline elements plus Ca*: 45, 109, 145, 436-7, 442, 463, T4.1, T6.2.

SUB-ALKALINE – *not so saturated with alkalies* 45, 437, 463, **F6.2**, **A5: F-G**.

THEOLEIITIC – *alkali-poor magma* 106, 436-7, 463.

allophane – *an amorphous clay mineral gel or colloid developing from volcanic ash, found in some tephrogenic soils*: 15, 163, 177, 181-2, 184-5, 198-200, 216, 435, 439, 442, A12, A13, **Ch7**, **F7.3**, F7.7, T7.1, **T7.2**; ALLOPHANIC SOILS 177-8, 188, 190, 197, 435, A13: T-C, **F7.9**; NON-ALLOPHANIC SOILS 177-8, 190, 197, 435, A13: T-C.

alluvial fan, see 'landscape: alluvial fans'.

andic – *soils that have phosphate retention of ≥85%, low bulk density of ≤0.90 kg dm*$^{-3}$ *and* Al_{ox}½Fe_{ox}^{2} *value of ≥2%*: 15, 176, 178, 180-1, 184, 193, **T7.1, T73**; ALU-

~ **plates** – *the tectonic plates of the Earth's surface made of 'continental crust'*: 14, 33, **Ch2.3.1**. Compare 'oceanic plates'. See list under 'tectonic plates'.

~ **rifting** – *the splitting of a continent, usually through magma upwelling from the 'mantle', causing doming that stretches the 'continental crust' until it breaks open, allowing 'oceanic crust' to form from the magma in the rift basin floor*: 105, 382-3, 389, 436-7, 439, 462, A8, A9, **Ch2.8**.

~ **shelf** – F13.6.

~ **slope** – *the slope at the end of the continental shelf, sloping downward between 1.5° and 6° (¶) towards the ocean floor:* F13.6. See also 'shelf sediments', 'turbidites'.

convergent margin (= 'active margin'): 347, B1. Includes 'subduction zone' , 'collision zone'.

convergent plate boundary (= 'subduction zone').

convoluted bedding – *a layer of wavy or folded sedimentary laminae, often produced as a form of 'soft sediment deformation' from earthquake shaking.*

core – *the heart of the Earth, it "consists of two parts, an outer liquid section and an inner solid section, both chiefly of iron and nickel with about 10 percent lighter elements. It is surrounded by the mantle" (Ø)*: 214 (figure), B1, **F2.2**.

core-mantle boundary – *transition from Earth's 'core' to 'mantle' at ca. 2,890 km depth*: 214-15, B1.

co-seismic – *an event that occurs simultaneously with and caused by an earthquake, e.g., landslide*: 268, 279, 289, 318, 321. See also 'syn-'.

country rock – *the rocks surrounding a magmatic intrusion into the 'continental crust'*: 53, 106, 361, 458, **Ch2.4.3**.

crater – *usually, the open top of a stratovolcano or a 'caldera' that may hold hot lava or a lake*: 157; CRATER LAKE 146, 309. PLACENAMED: **Ikeda** 227, 236; **Kusatsu-Shirane** F6.7; **Nakanoumi** B7; **Shikotsu** F6.16; **Showa** 157; **Sunset** 220; **Towada** B7.

craton – *"The stable portions of the continents that have escaped orogenic activity for the last 2 billion years" (Ø), often associated with 'mobile belts' or granite-greenstone belts*: 3, 6, 8, **Ch2.3.2.1**. See also 'block', 'shield'. PLACENAMED: **North China** 37, 46, 96, A8, T3.3; **Siberian** 37; **Sino-Korean** A8; **South China** 37, 46, A8, T3.3; **Wyoming** 34; **Yangtze** 37, A8.

crust – *the upper solid surface of the Earth whose base is defined by the 'Moho discontinuity' below which is the 'lithosphere'*: 28, 255, 310, 437, B1, **Ch2.3.1**, **2.3.2**.

crustal growth – *adding area to continents either through terrane accretion or collision, or by the formation of 'accretionary prisms' that become 'accretionary complexes'*: 45-6.

cryptotephra – *microscopic 'tephra' grains:* 15, 17.

crystal – *standardized structural arrangements of atoms in 'minerals'*. See 'tetrahedra'.

~ **habit** – *the exterior shape of mineral crystal growth (a mineral can have more than one growth habit)*: **T6.4**.

crystallite – *"the initial stage of crystallization of a magma or of a glass" (¶) before its shape can be discerned*: T6.4.

crystallization series – *formation and weathering sequences*: A5: F-B~C, A5: T-A.

microcrystalline – *crystals only visible under a microscope*: T6.4.

phenocrysts – *conspicuously large crystals in a rock, set in a fine-grained matrix*: T6.4.

Curie point – *"The temperature above which thermal agitation prevents spontaneous magnetic ordering" of 'ions' in a material" (¶)*: T6.4.

deep-sea trench, see 'subduction trench'.

debris cones – *deposits of sediments that have a slope >10°*: 391.

debris flows – *mixture of rocks, soil, and mud; coarser than a mudflow 'lahar'; also called "debris avalanches"*: 160, 185, 225, 227, 229, 234, 242, 280, 303, 310, 317, 321, 382, 391, 402, F6.4, F8.10.

delamination – *the loss of crust below a continent*: 108.

Dense Rock Equivalent (DRE) – *the amount of rock (km³) estimated to produce a certain amount of tephra*: 147, 158, 160.

detrital zircon – *zircon mineral crystals that have survived erosion to be deposited in sediments dating later than the zircons*: 46, Ch2.4.1, **Ch2.4.4**; F3.5; DATING 49, 82-3, 93, 465, A8.

diagenesis - *the initial compressive process of forming sedimentary rock involving chemical, physical, and biological changes*: **Ch2.5.3**, F2.8, 55. See also 'lithification'.

divergent margins - *two tectonic plates drawing away from each other by 'sea-floor spreading'; same as 'passive margins' - though current research indicates they are more active than thought*: 34.

doline – *a limestone sinkhole or cenote*: B3.

dolomarble – *metamorphosed 'dolostone', i.e. 'magnesian limestone', based on the mineral dolomite*: 346, 358-61, 365, F12.7, F12.8.

dolomitization - *the conversion of calcareous limestone into magnesian limestone, the latter having a prepondernace of the mineral dolomite instead of calcite*: 344, 346, 359, **A14**, F12.9.

dolostone – *magnesian limestone/dolomitic limestone where Mg replaces Ca in normal calcareous 'limestone'*: 359, 365, F12.7.

downward younging polarity – *a term used to describe the counterintuitive assessment of age of accretionary prisms and complexes which become younger the lower the stratigraphic position*: 47, 77, 212, **Ch2.4.4**.

downwelling – *the downward circulation of magma in the upper 'asthenosphere'*: 74, 214.

dyke – *a sheet of magma emplaced vertically, usually in a fracture*: 37, 39, 253, F2.4. Compare 'sill'.

and Antarctica. The name means the 'Land of the Gonds', a people in India: 81, T3.1.

granite types – *"Light colored, coarse grained, intrusive igneous rock characterized by the minerals orthoclase and quartz with lesser amounts of plagioclase feldspar and iron-magnesium minerals. Underlies large sections of the continents"* (Ø): **141**, 214, **A10**; VOLCANO PLUTONIC (arc-type) 141; PLUTONO-METAMORPHIC 141, 446; RADIOACTIVE 214-215. **Ryoke** 140-141, 380, 382-383, 389, 391, F13.5, older 382, younger 382; **San'in** 140-141; **San'yo** 140-141. See A10 for more names. See also 'igneous rocks'.

graphitization – *the natural formation of the mineral graphite from organic matter through a series of chemical transformations*: 51.

graben – *a linear depression of regional proportions bounded by normal faults that dip towards each other*: 59-60, 63, 126, 157, A9, B6, F2.13, F3.1. Compare 'horst'. PLACENAMED: **Beppu-Shimabara** 136, 156-7, 159, 311, F5.11, **F6.10**; **Kagoshima** 136, 156, 159, F5.11, **F6.10**.

gravity – in relation to isostacy 60, Ch2.6.6, **F2.11**; in relation to delamination 108. See also 'specific gravity'.

greenstone – *dark-green metamorphosed 'mafic' rock, e.g., ocean floor, metasomatized into 'serpentinite'*. See also 'metamorphic rocks'.

green tuff – *submarinely extruded, metasomatized, volcanic tuff*: 439, A9, **F4.4**, **F4.5**; ARTEFACTS F4.5; LANDSCAPE F4.5; OYA-STONE 439, B4.

Green Tuff Movement – *submarine volcanics associated with the rifting of the Japan Sea Basin*: **110-12**, 140, A8.

Green Tuff Zone – *area of distribution of green tuff in Japan*: 112-14, 130, 141, 438, **F4.4**.

Group, lithostratigraphic – *the second highest order of lithostratigraphic classifications, below 'Supergroup' but above 'Formation' and 'Member'*: 74, 384, 456, T13.2. PLACENAMED: **Fujiwara** 384-5, 389, 445, F13.5, T13.2; **Kazusa** 130, F5.7; **Kobe** F13.5; **Kobiwako** B6; **Nijo** 384-5, T13.2; **Osaka** 381, 384-6, 389-391, 394-5, 403, B6, B10, F13.5, F13.7, T13.1, T13.2; **Tanba** F13.7 **Tokai** B6; **Yamabe** 384-5, 389, T13.2.

Group, periodic – *divisions of the Periodic Table*: 458, A5: F-A, **A5: T-A**.

hanging wall – *the side of the rock that moves in a fault*: 58, 133, **F2.10**.

hazard/geohazard – 215-16, 219, 317, 322-3, 325, 448, **Ch11**; FREQUENCY 304; HUMAN CONTRIBUTIONS TO 323; MAPS 18, 303, 308-9, 316, 321, 396, 403, F11.2; PLANNING 323-4, 327, 329, 396; RISK 303-4, 309-310, 312, 323, F11.1.

heat – *measurement of how close magma is to the Earth's surface*: **F2.11**, 63.

hemipelagic – *pertaining to the shallow sea or continental shelf regions*: 38, A6. Compare 'pelagic'.

high-silica, see 'felsic'.

horst – *an emergent, usually linear, block of land bounded by normal faults that dip away from each other*: 59, F3.1, F5.8a. Compare 'graben'.

hot springs – 220, 386.

hot spot – *the upwelling of a hot 'magma plume' through the 'mantle' to the surface outside of a 'subduction zone', often causing 'rifting', and forming 'volcanoes' or ' large igneous provinces*: 8, 77, 108+ftn1, 155, 448, **F2.2**. See also 'seamount'.

humic acid – *a weak organic acid*: 190, 199, 435, 439, 479, T7.4, A13: T-A&B. Compare 'fulvic acid'.

humus – *organic component of the top layer of soil*: 177-8, 180, 184-5, 189-90, 199, 439, T7.2, T7.4, T7.7, A13.

hydro – *having to do with water or hydrogen.*

hydration – *the absorption of water, as in obsidian hydration*: 16, 147, 183, 435.

hydrolysis – *the decomposition of silicate rocks/ minerals by water via the absorption of H^+ or OH^-*: 183.

hydronium – *the chemical compound H_3O^+*: 437-8, 442.

hydroxide – *a chemical compound entailing hydroxyls (-OH)*: 437.

hydroxyl – *the unit (OH^-)*: 349, 437, 478-9, 482.

hypocentre, see 'earthquake hypocentre'.

igneous processes – *the creation, movement, and solidification of 'magma' and its structures*: **Ch2.4.3**.

igneous rocks – *rocks crystallized from 'magma'*: 278 (figure), 77, 141, 167, 185, 210, 359, 381-2, 456, 458, **462-4, A5: T-B, A5: F-F**, A6, A9, A10, A15, B1, **Ch2.4.3**, F12.9; ACIDIC VS BASIC 435; GRAIN SIZE A5: T-B. See also 'crystals'.

INTRUSIVE (= plutonic) – *rocks crystallized from magma intrusions into the Earth's crust*: 38, 96, 381, 446, A10, **F1.7**; **diorite** 16, 38, 381-2, 436, F12.8b; **dolerite** 62; **dunite** 37, 358; **gabbro** 37, 49, 54, 62, 141, 345, 347, 359, 362, 382, 436, F2.4; **granite** 12, 16, 38, 132, **141**, 211, 288, 436, F5.2a, F5.9; **granodiorite** 382, F12.8; **harzbergite** 359, 462; **lherzolite** 37, 358, 462; **peridotite** (mantle rock) 37, 55, 62, 345, 359-60, 362, 364, 462, F2.4, F12.1; **tonalite** 382. See also 'batholith', 'granite types'.

EXTRUSIVE (≈ volcanic) – *rocks crystallized from magma that has been extruded onto the Earth's surface*: **F1.7**, **A5: F- G, A11**; **adakite** 155; **andesite** 16, 38, 107, 128, 153, 156, 141, 382, 436, 463, B4, T4.1, T6.2; EMA 114; SANUKITE 114-15, 155, 338, 382, 385, 410, B5; **basalt** 17, 45, 49, 54-5, 62, 63, 79, 88, 90, 93-4, 105-7, 109, 114, 155-6, 211, 215, 345, 347, 351, 358-9, 362, 436, 448, B3, F2.4, T4.1, T6.2, T7.2; **dacite** T4.1, T6.2, T7.2; **lava** 16, 111, 147, 448; **obsidian** 12, 20, 147, 149, 153, 338, 360, 455, B5, F6.5.; **rhyolite** 38, 382, 436, 448, F5.8b, T6.2, T7.2. See also 'felsic', 'sedimentary rocks: pumice, tuff'. PLACENAMED: **Koshidake** 156; **Noto** A9; **Shirataki** 149, F6.5 (inset); **Wada Toge** F6.9; **Kanazawa** A9; **Muro** 382, 385, 446; **Nohi** 142, A11; **Nyu** A9.

~ **floor sediments** – *silicious or carbonaceous sediments deposited on the ocean floor by sea animals*: 47, B6, **F2.5**; VOLCANIC MOUNDS 110, **F2.14**.
~ **floor spreading** – *the creation of new oceanic crust by volcanism in the 'rift' in an 'oceanic ridge'*. See 'mid-ocean ridge', 'spreading centre'.
~ **floor slice**, see 'ophiolite'.
~ **plateau** – *a flat-topped submarine feature formed of 'flood basalt' by volcanic activity*. See 'large igneous province'.
~ **crust** – *the 'basaltic' crust of 'oceanic plates' usually 7 km in thickness with a standard stratigraphy*: 14, 211, 345, Ch2.3.1, **Ch2.3.2.2**, 37, **F2.2**, **F2.4**, **F12.1**; IN BACK-ARC BASIN 105. Contrast *'continental crust'*. See also 'mid-ocean ridge basalt'.
~ **plates** – *one of two kinds of 'tectonic plates', the other being 'continental plates'*: 4, 30, 400, **Ch2.3.1**, **F2.1**, **F2.5**. See list under 'tectonic plates'.
~ **plate stratigraphy** (OPS) – *the standard sequence of rock types that make up the ocean floor underneath any sedimentary cover*: A6, **F2.4**.
~ **ridge** – 211, 310. PLACENAMED: **Daito** F6.11; **Kyushu-Palau** 49, **95**, 117, 136, 153-4, 159, F3.1, **F3.9**, F5.1, **F5.11**, F6.11; **Oki-Daito** F6.11 See also 'mid-oceanic ridge'.
~ **ridge subduction** – *subduction of an 'oceanic ridge', often causing increased uplift, erosion, and accretion of 'trench fill' into an 'accretionary prism'*: 30, **48-9**, 78, 85, 88, 93, 105, 156, 212; CYCLE 211.
~ **slab** – *'oceanic crust' being drawn down into the 'subduction channel'*: 155, 215, 222, 256, **F2.5**.

ophiolite – *segments of 'oceanic crust' or fragments of 'ocean plateaus' or seamounts, and their associated 'sedimentary rocks', that have been 'obducted' onto land and trapped within an 'accretionary complex'*: 4, 55, 77-8, 86, 140, 210-11, 347, 365, 367, 439, 467, A8, A11, **Ch2.8, F12.2**, F12.7; OPHIOLITE SUITE 37, 62. PLACENAMED: **Alamas** 361, F12.78; **Horokanai** 91, 465; **Isua** F2.12; **Miyamori-Hayachine** 85, 466, A11, F3.4, T3.2; **Nomo** 62, 467, F3.4, T3.2; **Oeyama** 83, 85, 106, 365, 467, A11, F3.4, T3.2; **Poroshiri** 466-7; **Troodos** 14, 62; **Yakuno** 86, F3.4.

orogen/orogenic – *"Linear to arcuate in plan, intensely deformed crustal belt associated with mountain building"* (Ø): 347. Compare 'craton', 'shield'.
~ **belt** – *a longitudinal landmass usually created by volcanic activity over a subduction zone*. See also 'orogeny', 'fossil subduction zone'. PLACENAMED: **Altides** 37; **Nipponides** 4, 37.
~ **event** – *the time period in which an instance of orogeny takes place*.
~ **zones** (= 'orogenic belt'): 8, Ch2.3.1, **Ch2.3.2.1**; **Central** (China) 37; **Jiangnan** 37.

orogeny (orogenic, orogenesis) – *often limited to mean the process of "mountain-building," but it "is a complex concept that refers not only to mountain building but also to the formation of 'continental crust', the recrystallization of previously-formed 'orogens', the formation of major orogenic structures, and sedimentation related to mountain building"* (‡:92): 30, 211, 347; COLLISION-TYPE 212; CORDILLERAN-TYPE 212; PACIFIC-TYPE 93, 212, **B1**. PLACENAMED: **Abean** 211; **Akiyoshi** 211; **Sakawa** 211.

overthrust, see 'fault type: thrust'.

oxide – *the chemical combination of an element with oxygen, e.g., silica SiO_2 = silicon + oxygen= silicon dioxide*: 340, 461, **A5: F-E**, T12.1.

Pacific-type orogeny – *mountain-building in a subduction zone rather than resulting from collision*: 93, **B1**. See also 'orogeny'.

paired metamorphic belts – *high P/T and low P/T metamorphic belts together characteristic of 'Pacific-type orogeny'*: **B1**.

palaeomagnetism – *the Earth's past magnetic values as recorded in rocks*: 31, 109, 116-17, 289, F4.3.

palaeosols – *soil profiles that have subsequently been buried by other sediments such as 'tephra'*: 163, 185, 194, F8.10.

passive margin (= divergent plate boundary) – *"A continental boundary formed by rifting and continental rupture"* (*), *as on both sides of the Atlantic Ocean*: 83, **Ch2.3.1**.

pegmatite – *igneous plutonic rock, usually 'felsic' with crystals larger than 1 cm, sometimes up to a metre in length*: 458.

pelagic – *pertaining to the deep sea*: 38, A6. See also 'hemipelagic'.

Periodic Table – 436. See A5: F-A & T-A.

pH – *a measure of acidity vs alkalinity, with water being neutral*: 177-8, 185, 189, 482, **A16**, **B15**, **Ch15**, F15.1, T7.2, T15.2.

phenocryst – *unusually large mineral crystals in a finer-grained matrix, e.g., pink K-felspar crystals in granite; if >1 cm, they comprise the rock 'pegmatite'*: T6.4.

phyllosilicates – *clays organized in sheets*: 181-2, Ch7, F7.4. Compare the amorphous clays 'allophane', 'imogolite'.

planation – *'erosion' resulting in a planar surface, usually regional in scope*: 385.

plate, tectonic – *"A rigid segment of the Earth's 'lithosphere' that moves horizontally and adjoins other plates along zones of 'seismic' activity. Plates may include portions of both continents and ocean basins"* (Ø) : **1**, **3**, **2.3**. See also 'continental plates', 'oceanic plates'.
~ **margins** – *the edges of tectonic plates*: 33, Ch2.3.1. See also 'active margins', 'passive margins'.
~ **boundaries** – *plate edges defined by active 'subduction' or 'collision' zones or 'mid-oceanic ridges', or in the case of the San Andreas fault, a 'transform fault'*: **Ch2.3.1**, **F2.1**, **F2.3**, **F2.5**.

Plate Tectonics – *"A theory of global tectonics according*

edge(s), and closure of the ocean through 'oceanic plate' consumption via subduction (§:12): 36, 50.
xenolith – *a piece of 'country rock' that breaks off into a 'magma' chamber and becomes incorporated in a 'plutonic' or 'volcanic' rock*: 438.

Citation icons:

* = (Jackson 1997)
Ø = (Richardson 2006-2017)
§ = (Condie 1997)
‡ = (Maruyama *et al.* 1997*b*)
¶ = (Neuendorf *et al.* 2005)
† = (USGS n.d.)

References

Condie, Kent C (**1997**) *Plate tectonics and crustal evolution*. Oxford: Butterworth-Heinemann.
Jackson, Julia (**1997**) *Glossary of geology*. 4th ed. Alexandria, VA: American Geological Institute.
Maruyama, Shigenori; Yukio Isozaki & Gaku Kimura *et al.* (**1997*b***) "Paleogeographic maps of the Japanese Islands: plate tectonic synthesis from 750 Ma to the present." *The Island Arc* 6.1: 121-142 (in English).
Neuendorf, Klaus KE; James P Mehl Jr & Julia A Jackson (eds) (**2005**) *Glossary of geology*. Alexandria, VA: American Geosciences Institute [https://glossary.americangeosciences.org/].
Richardson, Stephen M (**2006-2017**) "Illustrated glossary of geologic terms" [http://www.ge-at.iastate.edu/glossary.shtml] (dead link).
Strekeisen, Alex (**2020**) "Alex Strekeisen: I vetrini della mia fantasia" [www.alexstrekeisen.it/english/index.php].
USGS (United States Geological Survey) (**n.d.**) "Earthquake Glossary." *Earthquake Hazards Program* [https://earthquake.usgs.gov/learn/glossary/].

TECTONIC ARCHAEOLOGY

Subduction Zone Geology in Japan and its Archaeological Implications

Gina L. Barnes

Access Archaeology

Archaeopress Access Archaeology

ARCHAEOPRESS PUBLISHING LTD
Summertown Pavilion
18-24 Middle Way
Summertown
Oxford OX2 7LG
www.archaeopress.com

ISBN 978-1-80327-399-0
ISBN 978-1-80327-400-3 (e-Pdf)

Cover: Geology of Asia 225Ma, by Fama Clamosa - Own work, CC BY-SA 4.0 [https://creativecommons.org/licenses/by-sa/4.0/legalcode] [https://commons.wikimedia.org/w/index.php?curid=85632155], modified by GLB with words and circle; the Japanese landmass did not look like this at that time, but its positioning is approximate. See Figure 3.5 instead.

This book is available direct from Archaeopress or from our website www.archaeopress.com

Dedicated to the memory of
William R. Farrand
27 April 1931 — 22 March 2011

"William R. Farrand, Ph.D., professor of geological sciences, director of the Exhibit Museum, and curator, Museum of Anthropology, College of Literature, Science, and the Arts, will retire from active faculty status on June 30, 2000.

"Professor Farrand received his B.S. and M.S. degrees in 1955 and 1956, respectively, from The Ohio State University and his Ph.D. degree in 1960 from the University of Michigan. From 1960-64, he was on the faculty of Columbia University and in 1964-65 he was a visiting professor at the University of Strasbourg in France. He joined the faculty of the University of Michigan in 1965 as an assistant professor and was promoted to associate professor in 1967 and professor in 1974. He was named curator in the Museum of Anthropology in 1975 and director of the Exhibit Museum of Natural History in 1993.

"Professor Farrand's early work centered on the glacial history of Michigan and the American/Canadian Midwest. He studied landforms and their evolution; the crustal rebound that occurs after the ice sheet load is removed from the earth's surface, and the history of the glacial advances and retreats. He was among the first to apply the techniques of radiocarbon dating to elucidate the timing of some of these events. Much of Professor Farrand's scholarship lay at the interface between geology and archaeology; in fact, his career helped to define the field of geoarchaeology. He was particularly interested in the sedimentology, stratigraphy, and paleoclimatology of archaeological and early man sites around the Mediterranean and has spent extensive periods of time working on sites in Syria, Jordan, Lebanon, Israel, Turkey, Greece, and elsewhere in Africa and the Far East.

"Professor Farrand taught courses at all levels and assisted with directing and teaching in the department's summer field program at Camp Davis, Wyoming. He published more than 120 papers, has served on the editorial boards of 4 journals, and has been president and secretary of the American Quaternary Association. Professor Farrand is a fellow of the American Association for the Advancement of Science and has been honored with the Archaeological Geology Award of the Geological Society of America.

"The Regents now salute this faculty member by naming William R. Farrand professor emeritus of geological sciences and curator emeritus."

University of Michigan
Regents' Proceedings 344